Progress in Botany/Fortschritte der Botanik 49

49 Progress in Botany

Structural Botany Physiology
Genetics Taxonomy Geobotany

Fortschritte der Botanik
Struktur Physiologie Genetik
Systematik Geobotanik

Editors
H.-Dietmar Behnke, Heidelberg
Karl Esser, Bochum
Klaus Kubitzki, Hamburg
Michael Runge, Göttingen
Hubert Ziegler, München

Springer-Verlag
Berlin Heidelberg New York
London Paris Tokyo

With 26 Figures

ISBN-13:978-3-642-73025-2 e-ISBN-13:978-3-642-73023-8
DOI: 10.1007/978-3-642-73023-8

2131/3130-543210

Contents

E. GEOBOTANY

List of Editors

Section A	Professor Dr. H.-D. BEHNKE, Zellenlehre, Universität Heidelberg, Im Neuenheimer Feld 230, D 6900 Heidelberg
Section B	Professor Dr. H. ZIEGLER, Institut für Botanik und Mikrobiologie, Technische Universität München, Arcisstr. 21, D 8000 München 2
Section C	Professor Dr. Dr. h.c. K. ESSER, Lehrstuhl für Allgemeine Botanik, Ruhr-Universität, Postfach 10 21 48, D 4630 Bochum 1
Section D	Professor Dr. K. KUBITZKI, Institut für Allgemeine Botanik und Botanischer Garten, Universität Hamburg, Ohnhorststr. 18, D 2000 Hamburg 52
Section E	Professor Dr. M. RUNGE, Lehrstuhl für Geobotanik, Systematisch-Geobotanisches Institut der Universität, Untere Karspüle 2, D 3400 Göttingen
Section F	Professor Dr. K. KUBITZKI, Institut für Allgemeine Botanik und Botanischer Garten, Universität Hamburg, Ohnhorststr. 18, D 2000 Hamburg 52

A. Structural Botany

I. General and Molecular Cytology: The Cytoskeleton: Microfilaments

By UDO KRISTEN

1. Introduction

The first part of this survey on the recent progress in the research of cytoskeletal structures was presented 1 year ago (KRISTEN: Progr. Bot. 48, 1), and dealt exclusively with microtubules. Other components of the cytoskeleton are actin filaments, intermediate or 10-nm filaments, myosin filaments and superthin filaments. All of them are microfilaments and can be distinguished by their thickness.

The intermediate filaments (for review see LAZARIDES 1982a; TRAUB 1985; SCHLIWA 1986) have a thickness in the range of 8 to 12 nm and are therefore intermediate in size between actin filaments (6 nm) and myosin filaments of muscles (16 nm). With regard to molecular composition, the class of intermediate filaments can be subdivided into five tissue-specific filaments: the vimentin, desmin, cytoceratin, glia- and neurofilaments. None of these filaments have yet been found in plant cells. They appear to occur exclusively in vertebrate cells, thus they will not be dealt with in this chapter.

Nevertheless, the diagnostic importance of the intermediate filaments in human cancer disease (ALTMANNBERGER et al. 1984) should be mentioned here. Classification of these filaments according to conventional and monoclonal antibodies can be used to distinguish the major classes of human tumors (OSBORN and WEBER 1983).

Myosin is the most abundant protein in striated muscle cells. There it forms myosin filaments which interact with actin filaments in the contraction of sarcomeres via sliding (SHETERLINE 1983). In smooth muscles and non-muscle cells of vertebrates it is present, but relatively less in relation to actin. However, myosin is rarely to be found in organisms which do not belong to the animal kingdom. It occurs in small amounts in the plasmodia and amoebae of slime moulds (PELTZ et al. 1985) and in yeast cells (WATTS et al. 1985), whereas plant cells have not yet been unequivocally shown to contain myosin.

Superthin filaments have an obviously smaller diameter (2 - 3 nm) than actin, intermediate or myosin filaments. They co-exist, as a novel class of cytoskeletal elements, with actin and myosin filaments within the sarcomeres of muscle cells (WANG 1984; WANG et al. 1984) and have recently been shown to also occur in the slime mould *Physarum polycephalum* (GASSNER et al. 1983, 1985).

Actin is very common in cells of all eukaryotes. It represents 5 - 10% of the total cellular protein and forms actin filaments which play an important role in the intracellular movements, even of higher plants.

In order to avoid confusion, the terms "microfilaments" and "actin filaments" shall henceforth not be used synonymously (as is so frequently done, especially in the botanical literature).

Progress in Botany, Vol. 49

2. Molecular Organization of Actin Filaments

Nearly all data concerning the substructure and molecular composition of actin filaments have been obtained from isolates of non-plant tissue (for review see POLLARD and COOPER 1986). Very little information in this field is available from the literature dealing with plant objects.

The actin filament (F-actin) consists of identical globular protein monomers (G-actin) each of which is composed of a single polypeptide chain with 374 or 375 amino acid residues. The molecular weight of the actin monomer was calculated at 41.8 kD from the amino acid composition. The shape and structure of the actin molecule has been studied by X-ray diffraction and by electron microscopy of crystalline actin sheets induced by gadolinium (AEBI et al. 1981; SMITH et al. 1983). The actin molecule was shown to be a distinctly asymmetrically wedge-shaped body with dimensions of 5.6 × 3.3 × 4.0 nm and with a slight groove dividing the molecule into two lobes. It has one binding site for ATP and one high affinity site for calcium or magnesium.

In the electron microscope, the actin filaments appear as long, curved strands with a width of approximately 6 - 8 nm. However, X-ray diffraction of actin filaments presented evidence for a native filament diameter of 9 - 10 nm (EGELMAN and PADRON 1984). This finding supports the suggestion that the long axis of the wedge-shaped actin monomer is nearly perpendicular to the filament axis (EGELMAN et al. 1983). Another model proposed for the arrangement of actin monomers within the filament is based on a longitudinal orientation of the monomer axis, resulting in filaments 7 - 8 nm in diameter (FOWLER and AEBI 1983). The question, which of the two models is correct, is not yet fully answered although the existence of the intermolecularly cross-linking amino acid residues of the actin monomers has recently been indicated (ELZINGA and PHELAN 1984; SUTOH 1984).

Electron micrographs from negatively stained actin filaments revealed a 2-start, right-handed, double-stranded helix with an axial repetition of 36 nm. However, recent observations revealed some disorder in the arrangement of actin monomers within the filament (EGELMAN et al. 1982, 1983). This probably results from small deviations of the normal angle (167°) between two adjacent actin monomers (EGELMAN et al. 1982). A change of the angle from 167° to 166° would cause a change from 38.2 to 35.5 nm in the distance between crossover points along the filament. A certain degree of structural flexibility, probably due to such changes, seems to be important for the insertion of actin filaments into filament bundles and into other high order structures (DE ROSIER and TILNEY 1984).

There are several taxa- and tissue-specific types of G-actin proteins (α-, β-, γ-actin) slightly differing from each other by only a few amino acid substitutions (e.g. SCHEDL and DOVE 1982). Multiple forms of actin are not confined to higher eukaryotes. In the acellular slime mould *Physarum polycephalum*, actin occurs in three major forms which were distinguished by their isoelectric points (PAHLIC 1985). The distribution of these forms among the cytoplasm and the nuclei of this organism appears to be unequal. In differentiating *Dictyostelium* aggregates, the unequal actin distribution was suggested to be due to locally different levels of actin mRNA (BARKLIS and LODISH 1985).

The different actin types appear to be functionally similar and interchangeable, as has been shown by the studies of actin genes and their expression of different vertebrate tissues and species (FYRBERG et al. 1981; GUNNING et al. 1984). The number and distribution of actin genes in the genome varies from species to species. Only one actin gene has been found in yeast but 6 in *Drosophila* (FYRBERG et al. 1981), more than 6 in the slime moulds *Dictyostelium* and *Physarum*

(McKEOWN and FIRTEL 1981; SCHEDL and DOVE 1982), 20 in sea urchins (JOHNSON et al. 1983) and up to 30 in mice and men (SORIANO et al. 1982; MINTY et al. 1983). In some species these genes are highly dispersed in the genome, in others they are closely linked (SORIANO et al. 1982; McKEOWN and FIRTEL 1981; LEE et al. 1984).

Numerous actin-associated proteins have been isolated from non-plant tissue and slime moulds (for literature see review by POLLARD and COOPER 1986). Many of them have been chemically characterized and put into different categories according to their binding sites and functions. From their functions, at least four classes can be distinguished among the actin-binding proteins (WEEDS 1982; POLLARD and COOPER 1986; SCHLIWA 1986; WITKE et al. 1986). However, they will not be discussed here since none of them has yet been unequivocally shown to occur in plant cells.

3. Assembly of Actin Filaments

In vitro polymerization of G-actin to form filaments involves three steps: monomer activation, nucleation and filament elongation.

Addition of magnesium or potassium chloride to a solution of G-actin in a low ionic strength buffer causes a slight conformational change of the monomers which seems to be necessary for their polymerization (PARDEE and SPUDICH 1982; COOPER et al. 1983). At least three of the thus activated monomers must then assemble to form a short F-actin fragment which probably serves as a nucleus for the elongation process (FRIEDEN 1983; TOBACMAN and KORN 1983; COOPER et al. 1983; LAL et al. 1984). Elongation proceeds by addition of actin monomers to the ends of the F-actin fragment. This process is fast in comparison to the nucleation if excess ATP is present. Under this condition, every actin monomer added uses hydrolysis of one ATP molecule to ADP.

Actin monomer addition is more rapid at one end of the elongating filament than at the other. The elongation rate of the actin filament depends on the G-actin concentration of the solution. In 0.5 mg ml^{-1} actin, this rate was determined to be 0.24 µm s^{-1} after an addition of about 70 monomers at the fast-growing and about 20 at the slow-growing end (POLLARD and MOOSEKER 1981). The release of monomers from each end was negligible in this actin concentration.

The growth preference of one end is based on the structural polarity of the actin filament. This polarity can be demonstrated by the decoration of F-actin with heavy meromyosin (JACKSON 1982). Polarized filament elongation in vivo has been demonstrated using actin bundles in *Chara* (WILLIAMSON et al. 1984; WILLIAMSON and HURLEY 1986). The bundles were locally damaged by strong irradiation, and their regrowth was observed after immunofluorescence labelling. The elongation occurred from the "barbed" but not from the "pointed" end of the filaments, according to their previously established polarity by decoration with heavy meromyosin.

In the steady state phase of filament assembly the rate of monomer addition is equal to the rate of monomer loss at each end of the actin polymer. In other words, the association events at the fast-growing end are counterbalanced by dissociation events at the slow-growing end. The filament is then said to "treadmill" (for a discussion of the treadmilling hypothesis of actin assembly/disassembly, see HILL and KIRSCHNER 1982). However, the phenomenon of treadmilling may be only a special case in the dynamic equilibrium of actin assembly/disassembly because there are discrepancies in the rate constants for G-actin addition and release and also for the rate constants reported from the attending ATP hydrolysis (BRENNER and KORN 1983, 1984). It has been shown that during F-actin assembly, ATP hydrolysis significantly lags behind filament elongation (CARLIER et al. 1984). As a consequence, ATP actin caps are formed at the ends of the filament. Moreover, it has been suggested that the rate of actin polymerization depends on the type of nucleotides (ATP or ADP) bound to the terminal actin monomers that form the elongation site of the filament (PANTALONI et al. 1985). The role of ATP-actin capping in the process of actin polymerization and filament stability was recently discussed by SCHLIWA (1986).

4. Visualization of Actin Filaments

Over the past 2 decades, several methods (in addition to conventional electron microscopy) have been developed to detect actin filaments in cells from animals, humans and slime moulds (NBD-phallotoxin fluorescence: BARAK et al. 1981; SCHROEDER and CHRISTEN 1982; immunofluorescence: LAZARIDES 1982b; WANG et al. 1982; electron microscopy of negatively stained actin filaments: SMALL 1981; electron microscopy of HMM-decorated actin filaments: JACKSON 1982; electron microscopy of HMM-decorated actin filaments in detergent-extracted cells: HIROKAWA and TILNEY 1982; high voltage electron microscopy: SCHLIWA 1982; LOFTUS et al. 1984). Among these methods, the techniques of phallotoxin fluorescence and immunofluorescence microscopy provided powerful tools for the intracellular localization of actin in plants.

Phallacidin and phalloidin, two phallotoxins of the fungus *Amanita phalloides*, bind specifically to F-actin. When labelled with the fluorophore, 4-chloro-7-nitrobenz-2-oxa-1,3-diazole (NBD), both these actin probes are well suited for the visualization of actin filaments by light microscopy (BARAK and YOCUM 1981). Thus, F-actin bundles have been detected in algal cells (BARAK et al. 1980; NOTHNAGEL et al. 1981) and in parenchymatous cells of the vascular tissue in conifer roots (PESACRETA et al. 1982). In the latter the F-actin distribution reflected the pattern of cytoplasmic streaming.

Rhodamine-labelled phalloidin has been used to visualize three-dimensional networks of actin filaments in various vascular plant cells. These included epidermal and fruit cells, root hairs, staminal hairs, pollen grains and pollen tubes (PARTHASARATHY 1985; PARTHASARATHY et al. 1985; WITZTUM and PARTHASARATHY 1985; HESLOP-HARRISON et al. 1986; PIERSON et al. 1986) as well as differentiating, meristematic and dividing cells (CLAYTON and LLOYD 1985; DERKSEN et al. 1986; SEAGULL et al. 1987). A rhodamine-labelled derivative of phalloidin, phalloin, has been used to localize F-actin in the fungus *Uromyces phaseoli* (HOCH and STAPLES 1983). A fluorescent analogue technique (WANG et al. 1982), employing rhodamine-labelled phalloidin and low-light-level microscopy, was used to investigate the morphology and dynamics of actin filament-membrane associations in microplasmodia of *Physarum polycephalum* (KUKULIES et al. 1985).

Fluorophore-labelled actin antibodies have frequently been used for the light microscope visualization of F-actin in cells of plants, fungi and slime moulds. In this way F-actin was shown to be present in the statocytes of cress roots (HENSEL 1986a). Immunofluorescence of the actin filaments was demonstrated in the cortical cytoplasm of *Chara* (WILLIAMSON and HURLEY 1986; WILLIAMSON et al. 1986) and at different cytoplasmic sites of *Bryopsis* (MENZEL and SCHLIWA 1986a) and *Acetabularia* (MENZEL 1986). In the latter F-actin seems to be involved in various morphogenetic processes. Both techniques, NBD-phallacidin and immunofluorescence microscopy were used to investigate the cytoskeleton of the homobasidiomycete *Schizophyllum commune* (RUNEBERG and RAUDASKOSKI 1986) and of the slime mould *Physarum polycephalum* (UYEDA and FURUYA 1985).

A combination of immunolabelling and electron microscope techniques, including negative staining and critical point- and freeze-drying, was shown to be suited for the visualization of microtubules and microfilaments in a slime mould (CLAVIEZ et al. 1986). High voltage electron microscopy of the cytoskeleton in whole plant cells revealed three-dimensional filamentous networks (COX et al. 1986). However, neither microtubules nor actin filaments could be unequivocally detected within these networks.

Another possibility for visualization of actin filaments in plant cells, the decoration with heavy meromyosin (HMM), was frequently used some years ago (see literature cited by JACKSON 1982).

5. Actin Filaments and Intracellular Movements

a) Involvement of Actin Filaments in Organelle Movement

Long-distance transport of organelles within plant cells is generally considered to proceed along microtubules or bundles of actin filaments (for the role of microtubules in organelle movement, see KRISTEN, Progr. Bot. 48, 1). In the siphonous green alga *Bryopsis*, chloroplast movement has recently been demonstrated to require organized arrays of both of these cytoskeletal elements (MENZEL and SCHLIWA 1986b). It has been suggested that microtubules control the directionality of the chloroplast movement in this alga, whereas the actin bundles are responsible for chloroplast motility. In this regard, the structural feature of the actin filament-microtubule association in *Bryopsis* resembles that of *Reticulomyxa*, an amoeba which also exhibits colinear alignment of actin filaments and microtubules (KOONCE 1985; KOONCE and SCHLIWA 1986). The probability of interaction between both these structures during mitotic sister chromosome separation was focused in the work of SEAGULL et al. (1987). They additionally reported on drastic changes of the actin filament arrays in higher plant cells during the cell cycle.

A cooperation of actin filaments and microtubules has also been suggested to occur in statocytes of cress roots. In these cells, responsible for the positive orthogravitropic growth of the root, microtubules probably serve as anchoring sites for actin filaments which mediate the translocation of the endoplasmic reticulum (HENSEL 1985, 1986b).

The involvement of actin filaments in the distribution and anchoring of endoplasmic reticulum in anion bulb scale cells was circumstantially evidenced by QUADER et al. (1987). They demonstrated that the ER concomitantly redistributed with F-actin reorganization after centrifugational dislocation. KACHAR (1985) demonstrated the unidirectional movement of membrane-bound organelles along actin bundles isolated from *Chara* internodal cells. Their velocity was determined by slow motion or frame-by-frame playbacks of recorded video images to be in the range of 11 to 62 $\mu m\ s^{-1}$ when a calcium-free medium with ATP was used.

The favoured mechanism of organelle movement is based on a force-generating ATPase attached to the organelle membrane and moving along a track provided by either actin filaments or microtubules (WILLIAMSON 1986). In the case of actin filaments, the force-generating ATPase is thought to be myosin. It is generally assumed that myosin couples tightly onto the filament when ATP is absent and detaches again in the presence of ATP. Moreover, myosin appears to account for the movement of organelles along the filaments towards their fast-growing (or plus) ends.

Strong evidence for this idea has been provided from experiments using small myosin-coated plastic beads which were brought into contact with characean actin bundles. These beads were shown to move unidirectionally along the bundles in a manner similar to that observed with organelles (SHEETZ and SPUDICH 1983; SHIMMEN and YANO 1984, 1985; KOHAMA and SHIMMEN 1985). The velocity of the movement was reduced with increasing calcium concentration (VALE et al. 1984; KOHAMA and SHIMMEN 1985). This shows that the force-generating mechanism is calcium-sensitive, but in a way opposite to that in muscle cells where calcium is known to stimulate actin-myosin filament contraction (HIGHSMITH and COOKE 1983). The site of Ca^{2+} action in the myosin-mediated sliding of beads along *Chara* actin bundles is not yet known. Recent observations, however, indicated that this action may be linked to myosin (SHIMMEN and YANO 1986).

All the experiments with myosin-coated beads demonstrate the feasibility of the theory suggesting myosin-coated organelles. However, they do not show that organelles actually have myosin at their surface. Therefore, localization of myosin or myosinlike ATPase in plant cells requires further experimentation.

b) Involvement of Actin Filaments in Cytoplasmic Streaming

The importance of actin filaments in cytoplasmic streaming is widely accepted, although direct evidence is still lacking. However, some results strongly supporting the idea of actin filament-driven movements of cytoplasmic compounds have been added since ROBARDS (Progr. Bot. 45, 1) presented a review in this field in 1983.

In various plant cells the pattern of F-actin distribution was shown to coincide with the site of cytoplasmic streaming, just as the filament orientation was shown to coincide with the direction of cytoplasmic streaming (WILLIAMSON et al. 1984, 1986; PARTHASARATHY 1985; PERDUE and PARTHASARATHY 1985).

Characean internodal cells are still the most commonly used subjects for the study of cytoplasmic streaming. In these cells, actin filament bundles are anchored on the stationary chloroplast files. Organelles smaller than chloroplasts, and probably equipped with myosinlike protein, slide along the filament bundles. The unidirectional organelle movement is thought to cause rotational cytoplasmic streaming (see literature cited by TAKAGI and NAGAI 1983).

This idea is supported by the observations that increasing calcium concentration reduces the velocity of both cytoplasmic streaming (TOMINAGA et al. 1983) and organelle movement along actin cables (KOHAMA and SHIMMEN 1985). Actin-based, probable myosin-mediated, cytoplasmic contraction was recently reported to occur in siphonous green algae after local wounding (MENZEL 1987). The contraction could be blocked by the anti-F-actin agents cytochalasin D or N-ethyl-maleimide, but not by microtubule inhibitors. It was shown by immunofluorescence that actin filament bundles were quickly formed around the affected area after wounding.

Actin-myosin interaction, similar to that present in non-muscle cells of animals and in slime moulds, has frequently been suggested to be also involved in intracellular movements of plants (JACKSON 1982). Support for the participation of myosin or myosinlike protein arose from recent investigations of the velocity distributions of characean cytoplasmic streaming (PIDDINGTON and ROSS 1985). The jerky advance of organelles, detected by laser light scattering and analyses of video film, probably resulted from the statistical pulsatile action of myosin-type cross-bridges attached to the organelles.

c) Involvement of Actin Filaments in Vesicle-Mediated Secretion

The intracellular transport of non-cellulosic polysaccharides and proteins to be secreted occurs by vesicles derived from the ER or from the Golgi apparatus. These vesicles usually follow a certain route called the classical endomembrane flow pathway (for literature see review by ROBINSON and KRISTEN 1982). Treatment with cytochalasins, a group of anti-F-actin agents (see Sect. 7), was shown to inhibit the transfer of Golgi vesicles within this pathway. This causes accumulations of vesicles around the dictyosomes (PICTON and STEER 1981; VOLKMANN and CZAJA 1981; KRISTEN and LOCKHAUSEN 1983). These observations support the theory that actin filaments are a prerequisite to the mechanism of Golgi vesicle movement from the dictyosomes to the plasma membrane and are therefore involved in vesicle-mediated secretion. However, direct evidence for the involvement of F-actin in vesicle transfer is still lacking.

6. Actin Filaments and Cell Motility

For some time, differently prepared portions of slime mould plasmodia have been used as models for the study of actin-myosin interaction in morphodynamic changes and locomotion of cells (YOSHIMOTO et al. 1981; YOSHIMOTO and KAMIYA 1984; OGIHARA 1982; WOHLFARTH-BOTTERMANN et al. 1983; ACHENBACH 1985; ACHENBACH and WOHLFARTH-BOTTERMANN 1986; PIES 1985; PIES and WOHLFARTH-BOTTERMANN 1984, 1985, 1986). Recent investigations of plasmodial actomyosin predominantly deal with the role of calcium in the mechanism of motive force generation. The results obtained are controversial. Some supported the theory of calcium-promoted contraction (e.g. OGIHARA 1982, WOHLFARTH-BOTTERMANN et al. 1983), others suggested an inhibitory effect of calcium (e.g. PIES 1985; ACHENBACH and WOHLFARTH-BOTTERMANN 1986). Recent studies of cryosections of *Physarum polycephalum* plasmodia clearly indicated an inhibitory effect of low Ca^{2+} concentrations on the cytoplasmic actomyosin contraction (PIES and WOHLFARTH-BOTTERMANN 1986). This finding is in good agreement with the model of KOHAMA and SHIMMEN (1985). They suggest an inhibition of the actomyosin-based contraction activity in *Physarum* by the coupling of Ca^{2+} to light chain myosin.

One of the crucial points in the control of motile activities of non-muscle cells seems to be the regulation of the intracellular actin polymerization. The distribution of actin filaments changes in relation to amoebid locomotion, as has been shown for moving *Dictyostelium* amoebae (RUBINO et al. 1983; RUBINO and SMALL 1987; ROOS et al. 1986).

It has been reported that the chemotactic stimulation of this cellular slime mould was followed by alterations of the quantitative ratio between G- and F-actin (McROBBIE and NEWELL 1984). These changes of actin organization and distribution were partially related to the locomotion of *Dictyostelium* amoebae (McROBBIE and NEWELL 1985). However, the dynamic control of the spatial and temporal shifts in the G-actin/F-actin ratio and their relation to amoebid movements is still far from being fully understood.

7. Anti-F-Actin Agents

Certain natural compounds inhibit either polymerization or depolymerization of actin and thus interfere with a broad range of cellular mechanisms.

At least five cytochalasins (a-e), heterocyclic fungal metabolites, are known to bind to F-actin at a ratio close to one molecule per actin filament (FLANAGAN and LIN 1980; McLEAN-FLETCHER and POLLARD 1980). The fast-growing (or plus) end of the actin filament appears to be the binding site because filament elongation is blocked by low concentrations of cytochalasins at that end when F-actin fragments, decorated with heavy meromyosin, are used as seeds in vitro and in vivo (BRENNER and KORN 1980; WILKINS and LIN 1981). This retardation of filament elongation in vitro and in vivo is the main effect of cytochalasins (POLLARD and MOOSEKER 1981; FOX and PHILLIPS 1981; BROWN and SPUDICH 1981; BONDER and MOOSEKER 1986). A second effect is the acceleration of overall actin polymerization. This is probably due to an increase in the concentration of cytochalasin-induced F-actin nucleation seeds (BRENNER and KORN 1981; FUSSMANN and DANCKER 1986). A third result of cytochalasin action, the reduction of the average actin filament length, is probably based on filament breakage (MARUYAMA et al. 1980; SCHLIWA 1982). In any case, cytochalasins usually cause a complete disruption of the cellular F-actin system with the result that actin filament-based processes are inhibited or, at least, deviated.

Because of this property, cytochalasins have been frequently used as tools to test whether actin filaments are involved in certain cellular mechanisms investigated and to differentiate between F-actin- and microtubule-mediated actions (PALEVITZ 1980; NOTHNAGEL et al. 1981; BRAWLEY and ROBINSON 1985; HENSEL 1985; WITZTUM and PARTHASARATHY 1985; MENZEL and SCHLIWA 1986a,b; WILLIAMSON and HURLEY 1986; QUADER et al. 1987; MENZEL 1987).

Phallotoxins, cyclic peptides isolated from the fungus *Amanita phalloides*, show effects opposite to those of cytochalasins. They bind to F-actin (BARAK and YOCUM 1981) and stabilize actin filaments against depolymerizing stimuli possibly by reducing the actin dissociation rate constants at both filament ends (COLUCCIO and TILNEY 1984). Phallotoxins are highly specific tools for the visualization of actin filaments when they are used in a form covalently labelled with fluorescent dyes (see Sect. 4).

Two toxins, Latrunculin A and B (isolated from the Red Sea sponge *Latrunculia magnifica*), have been reported to disorganize actin filaments and filament bundles in cultured mouse cells (SPECTOR et al. 1983). Normal filament organization was shown to recover when the toxins were removed.

Another toxin, Botulinum C 2, produced by certain strains of *Clostridium botulinum*, has recently been shown to cause ADP-ribosylation of G-actin but not of F-actin in vitro and in vivo (AKTORIES et al. 1986; REUNER et al. 1987). Moreover, experiments with chicken embryo cells revealed disintegration of the actin filament network after treatment with this toxin. From these results it has been suggested that Botulinum C 2 toxin may stabilize the G-form of actin and thereby decrease the availability of the monomer for its polymerization into actin filaments (REUNER et al. 1987).

Both these toxin types, Latrunculins and Botulinum toxins, may provide novel, promising tools for the study of actin filament assembly and action even in plant cells.

References

ACHENBACH, F.: Eur. J. Cell Biol. (Suppl. 7) 36, 3 (1985). - ACHENBACH, F., WOHLFARTH-BOTTERMANN, K.E.: Eur. J. Cell Biol. 40, 135-138 (1986). - AEBI, U., FOWLER, W.E., ISENBERG, G., POLLARD, T.D., SMITH, P.R.: J. Cell Biol. 91, 340-351 (1981). - AKTORIES, K., BÄRMANN, M., OHISHI, I., TSUYAMA, S., JAKOBS, H.K., HABERMANN, E.: Nature 322, 390-392 (1986). - ALTMANNBERGER, M., OSBORN, M., DROESE, M., WEBER, K., SCHAUER, A.: Klin. Wochenschr. 62, 114-123 (1984).

BARAK, L.S., YOCUM, R.R.: Anal. Biochem. 110, 31-38 (1981). - BARAK, L.S., YOCUM, R.R., NOTHNAGEL, E.A., WEBB, W.W.: Proc. Natl. Acad. Sci. USA 77, 980-984 (1980). - BARAK, L.S., YOCUM, R.R., WEBB, W.W.: J. Cell Biol. 89, 368-372 (1981). - BARKLIS, E., LODISH, H.F.: Exp. Cell Res. 159, 479-486 (1985). - BONDER, E.M., MOOSEKER, M.S.: J. Cell Biol. 102, 282-288 (1986). - BRAWLEY, S.H., ROBINSON, K.R.: J. Cell Biol. 100, 1173-1184 (1985). - BRENNER, S.L., KORN, E.D.: J. Biol. Chem. 255, 841-844 (1980); - ibid. 256, 8663-8670 (1981); - ibid. 258, 5013-5020 (1983); - ibid. 259, 1441-1446 (1984). - BROWN, S.S., SPUDICH, J.A.: J. Cell Biol. 88, 487-491 (1981).

CARLIER, M.-F., PANTALONI, D., KORN, E.D.: J. Biol. Chem. 259, 9983-9986 (1984). - CLAVIEZ, M., BRINK, M., GERISCH, G.: J. Cell Sci. 86, 69-82 (1986). - CLAYTON, L., LLOYD, C.W.: Exp. Cell Res. 156, 231-238 (1985). - COLUCCIO, L.M., TILNEY, L.G.: J. Cell Biol. 99, 529-535 (1984). - COOPER, J.A., BUHLE, E.L., WALKER, S.B., TSONG, T.Y., POLLARD, T.D.: Biochemistry 22, 2193-2202 (1983). - COX, G., VESK, M., JUNIPER, B.: Nord. J. Bot. 6, 641-649 (1986).

DERKSEN, J., TRAAS, J.A., OOSTENDORP, T.: Plant Sci. 43, 77-81 (1986). - DE ROSIER, D.J., TILNEY, L.G.: J. Mol. Biol. 175, 57-73 (1984).

EGELMAN, E.H., PADRON, R.: Nature 307, 56-58 (1984). - EGELMAN, E.H., FRANCIS, N., DE ROSIER, D.J.: Nature 298, 131-135 (1982); - J. Mol. Biol. 166, 605-629 (1983). - ELZINGA, M., PHELAN, J.J.: Proc. Natl. Acad. Sci. USA 81, 6599-6602 (1984).

FLANAGAN, M.D., LIN, S.: J. Biol. Chem. 255, 835-838 (1980). - FOWLER, W.E., AEBI, U.: J. Cell Biol. 97, 264-269 (1983). - FOX, J.E.B., PHILLIPS, D.R.: Nature 292, 650-652 (1981). - FRIEDEN, C.: Proc. Natl. Acad. Sci. USA 80, 6513-6517 (1983). - FUSSMANN, B., DANCKER, P.: Z. Naturforsch. 41c, 781-786 (1986). - FYRBERG, E.A., BOND, B.J., HERSHEY, N.D., MIXTER, K.S., DAVIDSON, N.: Cell 24, 107-116 (1981).

GASSNER, D., SHRAIDEH, Z., WOHLFARTH-BOTTERMANN, K.E.: Cell Biol. Int. Rept. 7, 905-913 (1983); - Eur. J. Cell Biol. 37, 44-62 (1985). - GUNNING, P., PONTE, P., KEDES, L., HICKEY, R.J., SKOULTCHI, A.I.: Cell 37, 709-715 (1984).

HENSEL, W.: Protoplasma 129, 178-187 (1985); - Naturwissenschaften 73, 510-511 (1986a); - Planta 169, 293-303 (1986b). - HESLOP-HARRISON, J., HESLOP-HARRISON, Y., CRESTI, M., TIEZZI, A., CIAMPOLINI, F.: J. Cell Sci. 86, 1-8 (1986). - HIGHSMITH, S., COOKE, R.: Evidence for actomyosin conformational changes involved in tension generation, 207-238. In: Cell and Muscle Motility, Vol. IV, eds. R.M. DOWBEN, J.W. SHAY. New York: Plenum Press 1983. - HILL, T.L., KIRSCHNER, M.W.: Int. Rev. Cytol. 78, 1-125 (1982). - HIROKAWA, N., TILNEY, L.G.: J. Cell Biol. 95, 249-261 (1982). - HOCH, H.C., STAPLES, R.C.: Eur. J. Cell Biol. 32, 52-58 (1983).

JACKSON, W.T.: Actomyosin, 3-29. In: The Cytoskeleton in Plant Growth and Development, ed. C.W. LLOYD. London: Academic Press 1982. - JOHNSON, P.J., FORAN, D.R., MOORE, G.P.: Mol. Cell Biol. 3, 1824-1833 (1983).

KACHAR, B.: Science 227, 1355-1357 (1985). - KOHAMA, K., SHIMMEN, T.: Protoplasma 129, 88-91 (1985). - KOONCE, M.: Proc. Electron Microsc. Soc. Am. 43, 488-489 (1985). - KOONCE, M.P., SCHLIWA, M.: J. Cell Biol. 103, 605-612 (1986). - KRISTEN, U., LOCKHAUSEN, J.: Eur. J. Cell Biol. 29, 262-267 (1983). - KUKULIES, J., BRIX, K., STOCKEM, W.: Eur. J. Cell Biol. 39, 62-69 (1985).

LAL, A.A., KORN, E.D., BRENNER, S.L.: J. Biol. Chem. 259, 8794-8800 (1984). - LAZARIDES, E.: Annu. Rev. Biochem. 51, 219-250 (1982a); - Methods Cell Biol. 24, 313-332 (1982b). - LEE, J.J.L., SHOTT, R.J., ROSE, S.J., THOMAS, T.L., BRITTEN, R.J., DAVIDSON, E.H.: J. Mol. Biol. 172, 149-176 (1984). - LOFTUS, J.C., CHOATE, J., ALBRECHT, R.M.: J. Cell Biol. 98, 2019-2025 (1984).

MARUYAMA, K., HARTWIG, J.H., STOSSEL, T.P.: Biochim. Biophys. Acta 626, 494-500 (1980). - McKEOWN, M., FIRTEL, R.A.: J. Mol. Biol. 151, 593-606 (1981). - McLEAN-FLETCHER, S.D., POLLARD, T.D.: Cell 20, 329-341 (1980). - McROBBIE, S.J., NEWELL, P.C.: J. Cell Sci. 68, 139-144 (1984); - Exp. Cell Res. 160, 275-286 (1985). - MENZEL, D.: Protoplasma 134, 30-42 (1986); - Eur. J. Cell Biol. (Suppl. 17) 43, 37 (1987). - MENZEL, D., SCHLIWA, M.: Eur. J. Cell Biol. 40, 275-285 (1986a); - ibid. 40, 286-295 (1986b). - MINTY, A.J., ALONSO, S., GUENET, J.-L., BUCKINGHAM, M.E.: J. Mol. Biol. 167, 77-101 (1983).

NOTHNAGEL, E.A., BARAK, L.S., SANGER, J.W., WEBB, W.W.: J. Cell Biol. 88, 364-372 (1981).

OGIHARA, S.: Exp. Cell Res. 138, 377-384 (1982). - OSBORN, M., WEBER, K.: Lab. Invest. 48, 372-394 (1983).

PAHLIC, M.: Eur. J. Cell Biol. 36, 169-175 (1985). - PALEVITZ, B.A.: Can. J. Bot. 58, 773-785 (1980). - PANTALONI, D., HILL, T.L., CARLIER, M.-F., KORN, E.D.: Proc. Natl. Acad. Sci. USA 82, 7207-7211 (1985). - PARDEE, J.D., SPUDICH, J.A.: J. Cell Biol. 93, 648-654 (1982). - PARTHASARATHY, M.V.: Eur. J. Cell Biol. 39, 1-12 (1985). - PARTHASARATHY, M.V., PERDUE, T.D., WITZTUM, A., ALVERNAZ, J.: Am. J. Bot. 72, 1318-1323 (1985). - PELTZ, G., SPUDICH, J.A., PARHAM, P.: J. Cell Biol. 100, 1016-1023 (1985). - PERDUE, T.D., PARTHASARATHY, M.V.: Eur. J. Cell Biol. 39, 13-20 (1985). - PESACRETA, T.C., CARLEY, W.W., WEBB, W.W., PARTHASARATHY, M.V.: Proc. Natl. Acad. Sci. USA 79, 2898-2901 (1982). - PICTON, J.M., STEER, M.W.: J. Cell Sci. 49, 261-272 (1981). - PIDDINGTON, R.W., ROSS, D.A.: Aust. J. Plant Physiol. 12, 27-37 (1985). - PIERSON, E.S., DERKSEN, J., TRAAS, J.A.: Eur. J. Cell Biol. 41, 14-18 (1986). - PIES, N.J.: Eur. J. Cell Biol. (Suppl. 7) 36, 50 (1985). - PIES, N.J., WOHLFARTH-BOTTERMANN, K.E.: Cell Biol. Int. Rep. 8, 1065-1068 (1984); - Cell Tissue Res. 239, 365-374 (1985); - Eur. J. Cell Biol. 40, 139-149 (1986). - POLLARD, T.D., COOPER, J.A.: Annu. Rev.

Biochem. 55, 987-1035 (1986). - POLLARD, T.D., MOOSEKER, M.S.: J. Cell Biol. 88, 654-659 (1981).

QUADER, H., HOFMANN, A., SCHNEPF, E.: Eur. J. Cell Biol. 44, 17-26 (1987).

REUNER, K.H., PRESEK, P., BOSCHEK, C.B., AKTORIES, K.: Eur. J. Cell Biol. 43, 134-140 (1987). - ROBINSON, D.G., KRISTEN, U.: Int. Rev. Cytol. 77, 89-127 (1982). - ROOS, U.P., DE BRABANDER, M., NUYDENS, R.: Cell Motil. 6, 176-185 (1986). - RUBINO, S., SMALL, J.V.: Protoplasma 136, 63-69 (1987). - RUBINO, S., FIGHETTI, M., UNGER, E., CAPPUCCINELLI, P.: J. Cell Biol. 98, 382-390 (1983). - RUNEBERG, P., RAUDASKOSKI, M.: Eur. J. Cell Biol. 41, 25-32 (1986).

SCHEDL, T., DOVE, W.F.: J. Mol. Biol. 160, 41-57 (1982). - SCHLIWA, M.: J. Cell Biol. 92, 79-91 (1982); - Cell Biol. Monogr. 13, 1-326 (1986). - SCHROEDER, T.E., CHRISTEN, R.: Exp. Cell Res. 140, 363-371 (1982). - SEAGULL, R.W., FALCONER, M.M., WEERDENBURG, C.A.: J. Cell Biol. 104, 995-1004 (1987). - SHEETZ, M.P., SPUDICH, J.A.: Nature 303, 31-35 (1983). - SHETERLINE, P.: Mechanisms of Cell Motility. Molecular Aspects of Contractibility. London: Academic Press 1983. - SHIMMEN, T., YANO, M.: Protoplasma 121, 132-137 (1984); - Proc. Jpn. Acad. Ser. B. Phys. Biol. Sci. 61, 86-89 (1985); - Protoplasma 132, 129-136 (1986). - SMALL, J.V.: J. Cell Biol. 91, 695-705 (1981). - SMITH, P.R., FOWLER, W.E., POLLARD, T.D., AEBI, U.: J. Mol. Biol. 167, 641-660 (1983). - SORIANO, P., SZABO, P., BERNARDI, G.: EMBO J. 1, 579-583 (1982). - SPECTOR, I., SHOCHET, N.R., KASHMAN, Y., GROWEISS, A.: Science 219, 493-495 (1983). - SUTOH, K.: Biochemistry 23, 1942-1946 (1984).

TAKAGI, S., NAGAI, R.: J. Cell Sci. 62, 385-405 (1983). - TOBACMAN, L.S., KORN, E.D.: J. Biol. Chem. 258, 3207-3214 (1983). - TOMINAGA, Y., SHIMMEN, T., TAZAWA, M.: Protoplasma 116, 75-77 (1983). - TRAUB, P.: Intermediate Filaments: A Review. Berlin, Heidelberg, New York: Springer 1985.

UYEDA, T.Q.P., FURUYA, M.: Protoplasma 126, 221-232 (1985).

VALE, R.D., SZENT-GYORGI, A.G., SHEETZ, M.P.: Proc. Natl. Acad. Sci. USA 81, 6775-6778 (1984). - VOLKMANN, D., CZAJA, A.W.P.: Exp. Cell Res. 135, 229-236 (1981).

WANG, K.: Cytoskeletal matrix in striated muscle. The role of titin, nebulin and intermediate filaments, 258-306. In: Contractile Mechanisms in Muscle, eds. G.H. POLLACK, H. SUGI. New York: Plenum Press 1984. - WANG, K., RAMIREZ-MITCHELL, R., PALTER, D.: Proc. Natl. Acad. Sci. USA 81, 3685-3689 (1984). - WANG, Y.L., HEIPLE, J.M., TAYLOR, D.L.: Fluorescent analog cytochemistry of contractile proteins, 1-11. In: Methods in Cell Biology, Vol. 25, ed. L. WILSON. New York: Academic Press 1982. - WATTS, F.Z., MILLER, D.M., ORR, E.: Nature 316, 83-85 (1985). - WEEDS, A.: Nature 296, 811-816 (1982). - WILKINS, J.A., LIN, S.: Biochem. Biophys. Acta 642, 55-66 (1981). - WILLIAMSON, R.E.: Plant Physiol. 82, 631-634 (1986). - WILLIAMSON, R.E., HURLEY, U.A.: J. Cell Sci. 85, 21-32 (1986). - WILLIAMSON, R.E., HURLEY, U.A., PERKIN, J.L.: Eur. J. Cell Biol. 34, 221-228 (1984). - WILLIAMSON, R.E., PERKIN, J.L., McCURDY, D.W.: Eur. J. Cell Biol. 41, 1-8 (1986). - WITKE, W., SCHLEICHER, M., LOTTSPEICH, F., NOEGEL, A.: J. Cell Biol. 103, 969-975 (1986). - WITZTUM, A., PARTHASARATHY, M.V.: Eur. J. Cell Biol. 39, 21-26 (1985). - WOHLFARTH-BOTTERMANN, K.E., SHRAIDEH, Z., BARANOWSKI, Z.: Cell Struct. Funct. 8, 255-265 (1983).

YOSHIMOTO, Y., KAMIYA, N.: Cell Struct. Funct. 9, 135-141 (1984). - YOSHIMOTO, Y., MATSUMURA, F., KAMIYA, N.: Cell Motil. 1, 433-443 (1981).

Prof. Dr. UDO KRISTEN
Institut für Allgemeine Botanik
Universität Hamburg
Ohnhorststraße 18
D-2000 Hamburg 52, FRG

II. Cytology and Morphogenesis of Fungal Cells

By HANS R. HOHL

This report covers the period of 1984 through 1986. Some citations date back to 1983 and occasionally to earlier years provided they have not been discussed before in this series. Emphasis has been put on the following topics: (1) cytoplasmic components, in particular the cytoskeleton, (2) hyphal tip growth and synthesis of cell wall components in the growing, vegetative hypha, and (3) nuclear events, predominantly mitosis and meiosis. Morphogenetic events related to spore structure, formation and germination have been summarized in the last report (JELKE, Progr. Bot. 46, 13-24 1984).

1. Books and Reviews

The book "Fungal differentiation", edited by SMITH (1983), provides an excellent synthesis of fungal structure and differentiation. Similar topics are addressed in several chapters of "Microbial cell wall synthesis and autolysis", edited by NOMBELA (1984), in "Fungal nutrition and physiology" by GARRAWAY and EVANS (1984), in "The ecology and physiology of the fungal mycelium", edited by JENNINGS and RAYNER (1984), and to a lesser extent in "Ecology of saprotrophic fungi" by COOKE and RAYNER (1984). MOORE et al. (1985) edited the proceedings of a fine symposium on "Developmental biology of higher fungi". A wealth of information is contained in the symposium volume on "Yeast cell biology" edited by HICKS (1986). Morphological and cytological aspects of symbiosis involving fungal partners are treated in "VA Mycorrhizae", edited by POWELL and BAGYARAJ (1984), in "The lichen-forming fungi" by HAWKSWORTH and HILL (1984), in "Lichen physiology and cell biology", edited by BROWN (1985), and in "The cereal rusts" by BUSHNELL and ROELFS (1984). "Fungal virology" is covered by BUCK (1986). There are monographs on "The biology and technology of the cultivated mushroom" (FLEGG et al. 1985), on "*Allomyces* - a different fungus" (OLSON 1984), and on "*Phytophthora*: its biology, taxonomy, ecology and pathology" (ERWIN et al., eds. 1983), the latter containing excellent chapters on cytology and morphogenesis. "Fungal protoplasts" is edited by PEBERDY and FERENCZY (1985). "Ultrastructure techniques for microorganisms", edited by ALDRICH and TODD (1986), contains several chapters dealing with fungi. "Fungal dimorphism: with emphasis on fungi pathogenic to humans" has been edited by SZANISZLO (1985) and covers biological and medical aspects of this increasingly attractive field. Another volume from this field is "Filamentous microorganisms, biomedical aspects", edited by ARAI (1985). COLE (1986) reviews models of cell differentiation in conidial fungi and DEMOULIN (1985) discusses the progress made over the last 10 years regarding the red algal - higher fungi phylogenetic link.

Progress in Botany, Vol. 49

2. Techniques

Increasing use is being made of newer cell biological methods. Detailed technical information on freeze substitution for TEM is given by HOCH (1986), on results obtained with this method from *Saprolegnia* by HEATH et al. (1985), from *Basidiobolus haptosporus* by MCKERRACHER and HEATH (1985b), from mitosis in *Microsporum canis* by KANBE and TANAKA (1985), from haustoria of *Erysiphe graminis* f.sp. *hordei* by HIPPE (1985), from *Ustilago avenae* by HIPPE and HERMANNS (1986), from haustoria of *Erysiphe graminis* and subcuticular stroma of *Venturia inequalis* by DAHMEN and HOBOT (1986). Enzyme histochemical methods for fungi are treated by POWELL (1986), and the low-temperature scanning electron microscopy of fully hydrated, rapidly frozen specimens by BECKETT and READ (1986). Papers by e.g. ALDRICH et al. (1984) on elemental distributions in gamma particles of *Allomyces* and *Blastocladiella* or by GARRISON and FISKIN (1984) on elemental analysis of microbody populations in *Basidiobolus haptosporus* employ dispersive X-ray analysis. Most promising are techniques based on hapten binding including gold- or FITC-labelled lectins (PENDLAND and BOUCIAS 1984; HORISBERGER et al. 1985; BENHAMOU and OUELLETTE 1986a; HOHL and BALSIGER 1986), and gold- or FITC-labelled enzymes to localize their respective substrates (pectins with gold-labelled pectinases in *Ascocalyx abietina* by BENHAMOU and OUELLETTE 1986b, chitin with a chitinase-gold complex in *Fusarium* f.sp. *lycopersici* by CHAMBERLAND et al. 1985, and in *Wangiella dermatitidis* by HARRIS and SZANISZLO 1986), or immunological (mostly indirect immunocytochemistry) methods employing polyclonal or monoclonal antibodies (RUNEBERG et al. 1986 with *Schizophyllum commune*; HORISBERGER et al. 1985 with *Saccharomyces rouxii*, HORISBERGER and ROUVET-VAUTHEY 1985 with *Schizosaccharomyces pombe*, DAY et al. 1986 with fimbriae of *Ustilago heufleri* and *Peronospora hyoscyami* f.sp. *tabacina*, SVIRCEV et al. 1986 with fimbriae of *Ustilago violacea*). Quantitative methods for estimating lectin binding on zoospores and cysts of *Phytophthora* (WILLIAMS et al. 1984; BACIC et al. 1985) have also been developed.

3. Cytoplasmic Structures

a) Cytoskeleton

Profiting from studies with mammalian systems, fungal cytologists are increasingly turning their attention to the fungal cytoskeleton. This field has experienced some of the most exciting developments during the period covered by the present report. Of the three classes of cytoskeletal subcomponents, microtubules and actin filaments (microfilaments) have been beautifully visualized in fungal cells while the presence of intermediate filaments is uncertain (SCHRANTZ 1982) and largely unstudied. The fungal cytoskeleton has mostly been examined in rusts (*Uromyces*) by HOCH and STAPLES (1983b, 1985) and HOCH and BOURETT (1986), in yeasts (*Candida*, *Saccharomyces*, *Schizosaccharomyces*) by GREER and SCHEKMAN (1982), MARKS and HYAMS (1985), KILMARTIN and ADAMS (1984), ADAMS and PRINGLE (1984), ANDERSON and SOLL (1986), and NOVICK and BOTSTEIN (1985), in *Schizophyllum commune* by RUNEBERG et al. (1986), in *Neurospora crassa* by SIKORA and MARZLUF (1982) and ORBACH et al. (1986), and in *Aspergillus nidulans* by WEATHERBEE et al. (1985). Interestingly, the presence of myosin as a necessary complement for actin to form contractile elements has received little attention in fungi. Evidence for a myosinlike protein based on the Ca^{2+} ATPase activity in the absence of Mg^{2+} has been obtained recently in growing mycelia of *Neurospora crassa* (TUINEN et al. 1986).

Microfilaments. Further evidence accumulates that in the fungi microfilaments play a role in cytoplasmic movement and contraction, in the induction of nuclear division (STAPLES and HOCH 1982), septum formation (microfilamentous septal belt, see also Runeberg et al. 1986), budding (SOLL and MITCHELL 1983), hyphal tip growth or wall synthesis.

Following the chemical identification of actin from fungal cytoplasm, e.g. from *Neurospora* (SIKORA and MARZLUF 1982) or from *Saccharomyces* (GREER and SCHEKMAN 1982), several investigators have visualized actin in fungal cytoplasm. Using rhodamine-conjugated phalloin together with appropriate controls HOCH and STAPLES (1983b) have localized F-actin in *Uromyces phaseoli*. The structures observed in the fluorescence microscope were filaments likely identical with bundles of microfilaments identified by TEM, peripheral plaques and intranuclear inclusions. The plaques have been tentatively equated with filasomes (microvesicles surrounded by a filamentous matrix). In a similar study RUNEBERG et al. (1986) demonstrated the presence of actin at the apices of the basidiomycete *Schizophyllum commune*. Since actin filaments were also found in places where septa formed it was suggested that actin-containing structures are necessary in places of new cell wall deposition, as suggested by earlier TEM studies (SOLL and MITCHELL 1983). Further support for this view has been obtained from work with actin-filament inhibitors such as cytochalasin A (e.g. ALLEN et al. 1980; GROVE and SWEIGARD 1980; DARGENT et al. 1986; HAROLD and HAROLD 1986), from actin mutants (NOVICK and BOTSTEIN 1985) and from some studies in yeast. In *Saccharomyces uvarum* and *S. cerevisiae* actin granules cluster at the site of evagination and in the ensuing buds, while actin fibres aligh themselves in the mother cell – bud axis. At the end of the cycle, actin appears to cluster or form a band at the mother cell – bud junction (KILMARTIN and ADAMS 1984; ADAMS and PRINGLE 1984).

ANDERSON and SOLL (1986) found differences in actin localization during bud and hypha formation in *Candida albicans*. Using rhodamine-conjugated phalloidin they observed that actin granules clustering at the site of evagination spread through the cytoplasmic cortex during bud formation. In marked contrast to this, with continued hyphal growth, the majority of actin granules were restricted to the hyphal apex. Thus, differences in zones of expansions in the two cases are correlated with actin granule distribution patterns.

Microtubuli. In fungal hyphae microtubules are generally oriented parallel to the main axis and may also extend into the hyphal apex (HOCH and STAPLES 1985; RUNEBERG et al. 1986), as on occasion has been observed previously. They have been and continue to be implicated in nuclear migration (MCKERRACHER and HEATH 1985b), induction of nuclear division (STAPLES and HOCH 1982), organelle positioning (HERR and HEATH 1982) and movement (O'DONNELL and MCLAUGHLIN 1984b), septum formation, conidial development (WEATHERBEE et al. 1985) and localized cell wall formation (KILMARTIN and ADAMS 1984).

HOCH and STAPLES (1985) studied possible nucleation sites for microtubule polymerization. Repolymerization of microtubules in germlings of *Uromyces phaseoli* subjected to depolymerizing agents, such as cold, demecolcine, griseofulvin and nocodazole, occurred first in the hyphal apices and not near the nuclei or spindle pole bodies (SPBs). It was concluded that the microtubule nucleating region in such fungal cells is located in the apical regions and that in these undifferentiating, undividing cells the SPBs do not function as microtubule organizing centres (MTOCs). Cell differentiation in *Uromyces appendiculatus* is inhibited with D_2O and taxol, implicating the microtubular system in these processes (HOCH and BOURETT 1986). These authors also obtained evidence that chemically (by K^+) and thigmotropically (by sharp stomatal edges) induced cell differentiation are different processes.

b) Plasma Membrane and Endomembrane System

In his book on "Plant membranes" ROBINSON (1985) and in a recent review HARRIS (1986) embody a wealth of information on the structure, function and development of the ER, nuclear envelope, microbodies, Golgi apparatus, tonoplast and vacuoles, plasma membrane and coated vesicles of the plant cell, which is of importance for the understanding of the corresponding but less well-studied fungal structure.

Zoospores bound only ConA, cysts bound SBA and ConA. This indicates that accessible α-D-glucosyl/α-D-mannosyl-containing glycoconjugates predominate at the zoospore surface, while the cyst surface has additional galactosyl and/or N-acetyl-D-galactosaminosyl residues. The outer surface of the plasma membrane of *Phytophthora* zoospores carries predominantly accessible α-D-mannosyl- or α-D-glucosyl-containing glycoconjugates since of several lectins tested only ConA bound substantially (BACIC et al. 1985).

FUJIKAWA and MIURA (1986) found distinct intramembrane particle aggregation in slowly freezing hyphae of the basidiomycete *Lyophyllum ulmarium*. These aggregations were restricted to points where there was direct contact between ice crystals and the plasma membrane and appear to be a direct consequence of this mechanical stress. HIPPE and LUETH (1986) observed the formation of large hexagonal clusters of intramembrane particles in the plasmalemma of *Ustilago avenae* as a result of fungicide treatment. They ascribe the effect to changes in parameters of the lipid bilayer or to a fluid-gel phase transition of the lipid matrix.

Recent interest in the H^+-ATPases of the fungal plasma membrane, isolated mostly from *Neurospora*, *Schizophyllum* and *Saccharomyces* has been reviewed by BOWMAN and BOWMAN (1986). All three H^+-ATPases, from plasma membrane, vacuole and mitochondria appear to be proton pumps in fungi. SMITH and SCARBOROUGH (1984) describe the large-scale isolation of the *Neurospora* plasma membrane H^+-ATPase.

Endocytosis in *Saccharomyces cerevisiae* may be visualized by soluble or particulate markers (MAKAROW 1985, 1986; RIZEMAN et al. 1986). The transport of a fluorescent macromolecule into the vacuole has been observed to occur via endosomes comparable to mammalian systems (MAKAROW and NEVALAINEN 1987). Coated vesicles likely to be involved in cell transport have been detected in yeast cells by MUELLER and BRANTON (1984).

Improved preservation of the Golgi apparatus and of wall vesicles by freeze-substitution techniques have by now been demonstrated also for an oomycete, *Saprolegnia* (HEATH et al. 1985). The study revealed both differentiation of a specific, single cisterna in each Golgi body and greater differences in membrane thickness throughout the endomembrane system. Localization of carbohydrates, inosine diphosphatase and thiamine pyrophosphatase has been demonstrated in typical dictyosomes of *Achlya bisexualis* and in single Golgi cisternae of *Hypomyces chlorinus* (DARGENT et al. 1982), adding to our still fragmentary knowledge of the functional complexity of this organelle in fungi.

A peculiar, tubular, smooth endoplasmic reticulum has been described in three aquatic fungi, *Blastocladiella emersonii*, *B. britannica* and *Catenaria anguillae* (BARSTOW et al. 1985). According to the authors it possibly represents a reservoir of membrane material that is formed during the growth phase and is converted into other endomembranes during sporogenesis. Multilaminate bodies have been proposed to represent membrane reserves in trapping fungi of *Dactylaria brochopaga* (DOWSETT and REID 1983).

The ionophore, monensin, is known to disrupt endomembrane functions. In *Allomyces macrogynus* it strongly disturbs zoosporogenesis by interfering with the cleavage process (SEWALL et al. 1986). Swollen cisternae are observed in treated cells possibly representing dictyosome equivalents which, when undisturbed, possibly play a critical role during zoospore formation.

c) Microbodies, Woronin Bodies

The question of microbodylike organelles as taxonomic markers among oomycetes has been again reviewed by POWELL et al. (1985). The presence in zoospores of an alga of an organelle resembling K-bodies of oomycetes supports the phylogenetic association between algae and oomycetes, as does the structure of the transition region of the axoneme of *Olpidiopsis saprolegniae* var. *saprolegniae* which resembles the transitional helix of chromophytous algae (BORTNICK et al. 1985).

In *Neurospora crassa* (slime mutant) at least some microbodies are pleomorphic and may wrap around vacuoles (WANNER and THEIMER 1982). Not uncommon microbodies may contain inclusions of amorphous, rod- or plate-shaped, semi-crystalline or crystalline nature. Several types of microbodies or microbodylike organelles may exist in the same organism at the same developmental stage, such as in zoospores of *Phytophthora palmivora* (POWELL and BRACKER 1986) or *Olpidiopsis* (BORTNICK et al. 1985). Large inclusions may cause deformations of the microbodies (WANNER and THEIMER 1982). Conspicuous inclusions are commonly associated with microbodies of some ascomycetes and in some cases were similar in appearance to the Woronin bodies of these species.

FISKIN et al. (1986) have localized alkaline phosphatase activity in membranes of microbodies of *Neurospora crassa* and *Aspergillus nidulans*. Hexagonal inclusions found in the cytoplasm of *N. crassa* and believed to arise from microbody precursors were without cytochemical staining for microbody marker oxidases. The additional demonstration of putative alkaline phosphatase activity with microbodies of the zygomycetous *Basidiobolus* (GARRISON and FISKIN 1984, 1986a) implies that this association may be common to a wider variety of filamentous fungi. There is no information on yeasts as yet.

The presumed major function of Woronin bodies to rapidly seal septal pores has been experimentally tested by COLLINGE and MARKHAM (1985) in *Penicillium chrysogenum*. Woronin bodies were found to have plugged 38.6% of the septal pores in the damaged region of the colony within 3.6 s and approximately 94% within 20 s of cutting. Sealing was found to extend along the hyphae to at least those septa removed third from the severed compartment. The plugging mechanism was found to be capable of sealing septal pores only 180 nm from a cut.

d) Storage Products, Miscellaneous Structures

Metachromatic (volutin) granules are vacuolar structures, which respond metachromatically with toluidine blue, and which are rich in (poly)-phosphates. Their molecular composition in yeast has been investigated by JACOBSON et al. (1983). They are a common feature of the fungal hyphae of such symbiotic systems as ectomycorrhizas, ectendomycorrhizas, endomycorrhizas and lichens but are also found in all major free-living groups of fungi with the exception of the oomycetes (CHILVERS et al. 1985). Their formation following phosphate uptake by hyphae of a mycorrhizal fungus has recently been studied by LAPEYRIE et al. (1984). They formed within 3 h after feeding of orthophosphate, and several cations

(Ca^{2+}, Mg^{2+}, K^{+}, and Na^{+}) had no influence on their formation. Hyphae deprived of phosphorus during growth lacked these granules (CHILVERS et al. 1985).

Concentric bodies (CBs) are found preferentially in lichenized mycobionts but also occur in a few groups of free-living ascomycetes. According to PONS et al. (1984) positive records have so far been restricted to the Dothideales, Ostropales, Helotiales, Lecanorales, Verrucariales, Caliciales, Teloschistales, and Peltigerales. The function of CBs still remains elusive. A differential distribution of these spherical structures is reported for different parts of the thallus of *Pseudopeziza trifolii*, whereas they occurred in all types of cells investigated in *Sphaceloma rosarum* (BEILHARZ 1985). PEVELING et al. (1985) propose a modified model on the ultrastructure of CBs based on older versions, with the outer shell formed from rays composed of particles. Concentric granules of somewhat different morphology are found in *Allomyces*, *Microallomyces*, and *Blastocladia* (MEYER 1987). Based on EDAX analysis they contain phosphorus and perhaps sulphur. In *Allomyces* they function in septal pore plugging comparable to Woronin bodies of ascomycetes.

Inclusions of viruslike particles are known for a variety of fungi. They are now reported from hypovirulent strains of *Endothia parasitica* (NEWHOUSE et al. 1983), and recently also from a convert strain (ELLZEY et al. 1986). They have been detected in additional species. Two types of particles were found in *Basidiobolus haptosporus* and *B. ranarum* (GARRISON and FISKIN 1986b), probably the first report of viruslike particles with fungi of the Entomophthoraceae. Their occurrence in an oomycete (*Phytophthora drechsleri*) is described by ROOS and SHAW (1985).

e) Additional Studies

Some papers not further discussed in the text may indicate the range of ultrastructural studies of the past few years. Phycomycetes: the water moulds *Sommerstoffia spinosa* (SAIKAWA and HOSHINO 1986) and *Olpidiopsis varians* (MARTIN and MILLER 1986b), the oomycetes *Phytophthora palmivora* (HEMMES and LERMA 1985), *Pythium ultimum* (HEMMES and STASZ 1984), and *Petersenia palmariae* (PUESCHEL and VAN DER MEER 1985), the latter a parasite of a red alga which spends almost all of its asexual cycle intracellularly and is wall-less during its vegetative phase; the Zygomycetes *Sporodiniella* by GBAJA and YOUNG (1985) and *Ellisomyces anomalus* (both Mucorales) by BEAKES and CAMPOS-TAKAKI (1984); and *Endogone pisiformis* by GIBSON et al. (1986a). Ascomycetes: *Monascus* (WONG and CHIEN 1986a,b). Basidiomycetes: HALE and EATON (1985) on soft rot fungi, HOCH and STAPLES (1983a) on the rust fungus *Uromyces phaseoli* var. *typica*.

4. Nucleus, Mitosis and Meiosis

Advances in this field now require refined methods such as serial sectioning and 3D reconstructions. Topics mostly deal with complete descriptions of nuclear cycles including the complex structural metamorphosis of the nucleus-associated organelle (NAO) or spindle pole body (SPB), the fate and changes of the nuclear envelope or the quantification of spindle microtubule numbers and positions.

Plasmodiophoraceae. Mitotic divisions during sporangiogenous plasmodial cleavage in *Woronia* are similar to the cruciform nuclear divisions of other plasmodiophoraceous taxa (DYLEWSKI and MILLER 1984). The mitotic divisions during formation of sporangia are centric, closed with polar fenestrae and characterized by the formation of intranuclear membranous vesicles. Interestingly, these membrane fragments derived from the original nuclear envelope are incorporated into the new daughter nuclear envelopes forming within the confines of the original nuclear envelope.

Oomycetes. Freeze substitution of mitotic spindles in *Saprolegnia* (HEATH et al. 1984) revealed an intranuclear centriole-associated lamina whose behaviour suggests a role in centriole migration. During early anaphase the non-kinetochore microtubules form typical divergent interdigitating half-spindles with ca. 10% pole-to-pole microtubules and only a low and variable number of close interactions between opposite, interdigitating microtubules. The authors suggest that during the dispersed state these microtubules interact with the nuclear matrix to spread tension forces acting on the spindle poles, whereas during the fasciated (bundled) stage they are linked by short bridges whose cycle rate limits spindle elongation.

In *Saprolegnia*, kinetochore microtubules persist throughout mitosis but disappear in meiosis coincidently with the formation of synaptonemal complexes at pachytene and reform at metaphase I (TANAKA and HEATH 1984). This could provide a clue to the observation that in some fungi chromosomal segregation is random in meiosis and non-random in mitosis. In late metaphase all kinetochores are attached to a single microtubule during both meiosis and mitosis.

Mitosis in *Olpidiopsis varians* (MARTIN and MILLER 1986a) is similar to other oomycetes, with an intact nuclear envelope characterized by polar fenestrae, persistent nucleolar material and an intranuclear spindle. Based on the presence of a persistent interzonal bridge and a fenestrated polar region, *O. varians* represents a species with intermediate characteristics between the oomycetes and the chytridiomycetes.

Zygomycetes. There is considerable structural heterogeneity in NAOs among members of the zygomycetes which may have taxonomic significance (e.g., MURRIN et al. 1984). Highly intriguing is the NAO of *Basidiobolus*. According to earlier studies, it is an apparent exception from the general rule that only flagellated organisms have centrioles as NAOs. The situation has been restudied by MCKERRACHER and HEATH (1985a). They found that the NAO of *Basidiobolus* is in fact composed of a cylinder of 11 - 12 singlet microtubules embedded in an osmophilic matrix. However, this structure is clearly different from centrioles which posses triplets of microtubules and show a nine-fold symmetry.

The closed, intranuclear spindle of the aphid-pathogenic fungus *Erynia neoaphidis* (Entomophthorales) has chromosome permanently attached by kinetochore microtubules to the poles during mitosis (BUTT and BECKETT 1984a). The spindle develops as the SPBs migrate and separate. At metaphase the eccentric spindle apparently contains only kinetochore microtubules and is located in a relatively chromatin-free zone. Paired sister kinetochores are arranged in a broad metaphase plate. During anaphase kinetochore microtubules shorten, astral and non-chromosomal microtubules develop and elongate and the interpolar distance increases. The SPB has an unusual "pan and handle" structure and consists of extra- and intranuclear components (BUTT and BECKETT 1984b).

Basidiomycetes. During mitosis in *Tilletia foetida* (HOFFMANN and GOATES 1985) chromatin surrounds a linear spindle that persists after nuclear separation. The SPB substructure has a crystallinelike lattice of longitudinally and transversely oriented electron-dense striations not observed in SPBs of other organisms. SPB division is similar to that reported in *P. malvacearum* and perhaps also in yeast. Regeneration of the interphase SPB structure starts from a small, spheroid body which develops into a bar and, by apposition of new SPB material to each end of the bar, into the double disc of the interphase nucleus. The situation is similar to the few other cases closely studied so far and appears to be a common feature in basidiomycetes (see also O'DONNELL and MCLAUGHLIN 1984a, TAYLOR 1985). Several morphological features of nuclei in *T. foetida* are similar to those found in the Uredinales which suggests a close relationship.

Phylogenetic implications of mitosis or meiosis (and septum formation) are also discussed for *Filobasidiella neoformans* by MOCHIZUKI et al. (1986), for *Helicobasidium mompa* (Auriculariales) by BOURETT and MCLAUGHLIN (1986) and for *Tulasnella araneosa* by TAYLOR (1985). In *H. mompa* the extranuclear interphase SPB contains two, three-layered discs. During prometaphase, the SPB discs move into the plane of the nuclear envelope where they occupy gaps. The spindle pole is enclosed by a cap of endoplasmic reticulum. The overall data are in agreement with the idea that the parasitic Auriculariales are closely related to the rusts and ancestral to many advanced basidiomycetes. In *T. araneosa* mitosis is unusual in that the SPB develops asymmetrically; the SPB middle piece is large and transversely curved, and the nuclear envelopes of adjacent late anaphase nuclei fuse. Available evidence indicates that the SPB of *T. araneosa* is more different from that of Uredinales than it is from representatives of the other four orders of Basidiomycotina that have been thoroughly studied.

Continuing their series of detailed investigations on nuclear cycles, O'DONNELL and MCLAUGHLIN (1984a,b) provide a 3D reconstruction of the SPB during meiosis in *Ustilago maydis*. The nuclear and SPB cycles of *U. maydis* support the inclusion of the Ustilaginales in the basidiomycetes, and these features are similar to those reported for the heterobasidiomycetous yeasts. In addition, the authors provide a summary of the nuclear and SPB features which, with the exception of the Uredinales, are shared by all basidiomycetes that have been studied in detail. (1) The SPB consists of two globular, subglobular, or bar-shaped elements connected by an electron-dense middle piece. (2) The spindle pole body is extranuclear but associated with the other membrane of a differentiated region of the nuclear envelope. (3) A two-layered intranuclear element is sandwiched between a chromatinlike hemisphere and the differentiated region of the nuclear envelope. In *U. maydis* the intranuclear element is discernible at mitotic prophase but not at meiotic prophase. (4) A ribosome-free zone immediately surrounds the cytoplasmic face of the SPB. (5) During early stages of nuclear migration, the longitudinal axis of the SPB is situated near the apex of a teardrop-shaped nucleus. (6) Cytoplasmic microtubules are most frequently oriented parallel to the direction of nuclear migration. (7) At the onset of meiosis and mitosis, nuclear volume is reduced several-fold. (8) Where information is available, the globular elements approximately double in size and the middle piece becomes elongate and is lost during late prophase I to metaphase I. During this transition the globular elements enter either a transiently disrupted or permanently fenestrated nuclear envelope.

5. Wall Structure and Formation

Since the review by GOODAY (1983) a series of advances have been made regarding the enzymes involved, products formed and mechanisms of growth. Additional fine structural analyses of hyphal apices are given by e.g. CLEARY et al. (1986) for *Allomyces macrogynus*, HEATH et al. (1985)

for *Saprolegnia* and HOCH and STAPLES (1985) for *Uromyces phaseoli*. Models for the generation of a transcellular proton current capable of accounting for polarized transport of wall vesicles are studied by GOW et al. (1984), HAROLD et al. (1985) and TURIAN et al. (1985).

a) Hyphal Tip Growth

Mechanism. The long-standing theory (BARTNICKI-GARCIA 1973) that hyphal tip growth in fungi is governed by a delicate balance between wall-synthesizing and wall-lysing enzymes has recently been challenged. In particular, the essential role of wall-lysing enzymes has been questioned in view of more recent experimental evidence. According to WESSELS and co-workers (WESSELS et al. 1983; WESSELS 1984; SIETSMA et al. 1985; SONNENBERG et al. 1985) hyphal tip growth in *Schizophyllum commune* may be controlled by successive chemical modifications of the major cell wall components from the tip to the subapical region. This occurs by (1) an increase in crystallization of chitin which during its formation is transiently non-crystalline (VERMEULEN and WESSELS 1986), (2) an increase in β-1,6-cross-linkages subapically from 10 - 60% of the chitin – β-glucan complex and (3) by additional subapical synthesis of wall glucans. The arising complex is thought to endow the delicate, highly plastic wall at the hyphal tip (VERMEULEN and WESSELS 1984) with increasing rigidity as it moves laterally, which can explain the presumed gradient in rigidity required in the extension zone for hyphal morphogenesis.

Enzymes, Chitosomes. Glucan synthases have been studied by GIRARD and FÈVRE (1984) in *Saprolegnia* and by QUIGLEY and SELITRENNIKOFF (1984) in *Neurospora crassa*. They are membrane-bound and, in the case of *Saprolegnia*, associated with the plasma membrane and with at least part of it on the outside surface. Furthermore, apical and subapical parts of hyphae may differ in the kind of wall polymers they synthesize.

Chitin synthase has recently been purified by KANG et al. (1984) in *Saccharomyces cerevisiae*, by BARTNICKI-GARCIA et al. (1984) in *Neurospora crassa*, by HAENSELER et al. (1983a) in *Agaricus bisporus* and by MONTGOMEREY et al. (1984) in *Coprinus cinereus*. Chitin polymerization can be activated in vivo using calcofluor white and Congo red (RONCERO and DURAN 1985), or by monensin (POLI et al. 1986). The enzyme from yeastlike cells in *Ceratocystis multiannulata* was found to be more sensitive to proteolytic activation than the mycelial form (DEWERCHIN and VAN LAERE 1985). In *Mucor rouxii* the synthesis of chitin and chitosan is coordinated (DAVIS and BARTNICKI-GARCIA 1984). The *Agaricus* enzyme is similar to that of *Mucor* (RUIZ-HERRERA et al. 1984), as is the case for other chitin synthases, and its activity can be rather specifically stimulated by digitonin. This may be causally related to its antimycotic activity. The topic of chitin synthases as targets for fungicides is explored by FURTER and RAST (1985).

Using molecular cloning techniques BULAWA et al. (1986) found that the *Saccharomyces cerevisiae* structural gene for chitin synthase is not required for chitin synthesis in vivo. This is evidence for a second type of chitin synthase. According to the authors this chitin synthase II has properties differing from those of chitin synthase I, e.g., it is undetectable under standard assay conditions for chitin synthase I, and it seems to be involved in septum formation.

The characterization of chitosomes as miniorganelles of 40 - 70 nm diameter transporting chitin synthases to the plasma membrane has been advanced by several studies. According to RUIZ-HERRERA et al. (1984), in yeast cells of *Mucor rouxii* about 80 - 85% of the chitin synthase activity is recovered in the chitosomal peak, similar to the findings in *Neurospora crassa* (BARTNICKI-GARCIA et al. 1984). Dissociation of chitosomes into 16 S subparticles and their reconstitution into activity chitin-synthesizing chitosomes has been achieved by HAENSELER et al. (1983b). The fact that chitosomes lack Concanavalin-A-binding sites distinguishes the chitosome membrane from that of the plasma membrane and of other organelles (BARTNICKI-GARCIA et al. 1985).

Interestingly, melanin synthesis is also compartmentalized (HEGNAUER et al. 1985). It takes place within and upon membrane-bound particles of 200-nm diameter which contain several smaller (60-nm diameter) granules. The 200-nm particles may be analogous to the animal "melanosome". A possible relationship between the small granules and chitosomes is an intriguing question to be pursued further.

An investigation of wall-lysing enzymes by HUMPHREYS and GOODAY (1984) revealed two forms of chitinase in *Mucor mucedo*, a soluble (probably in vacuoles) and a membrane-bound form. The chitinase occurs in the form of a zymogen which can be activated by trypsin. Inactivation takes place after treatment with phospholipases, suggesting that microsomal chitinase requires phospholipid-enzyme interactions for activity. Both characters are shared with the chitin synthase, indicating that the membrane-bound form might be instrumental in the regulation of hyphal growth and branching. Cellulases and β-glucanases in *Phytophthora infestans* were studied by BODENMANN et al. (1985).

b) Wall Regeneration

The increased use of protoplasts as tools in molecular genetics and biotechnology has led to additional investigations of wall regeneration (YANAGI et al. 1985 in *Coprinus macrorhizus*, GIRARD et al. 1984 in *Saprolegnia monoica*, CABO et al. 1986 in *Mucor rouxii*, GAUGY and FÈVRE 1985 in different species of *Penicillium*, BROWN et al. 1986 in *Penicillium pinophilum*, KOPECKA and KREGER 1986 in baker's yeast, JACOBSEN et al. 1985 in *Geotrichum candidum*, and DOUGLAS et al. 1984 in *Aspergillus awamori*).

The polysaccharides produced during wall regeneration in *Aureobasidium pullulans* (QUIGLEY and SIEHR 1984) formed a loose, thick coating on the cells. The glycans that were formed were similar to those in normal walls; however, they lacked the branching that is normally found. This finding is consistent with those of others on reversion of yeast protoplasts. According to TANAKA et al. (1985), the formation of chitin clearly precedes that of glucan during regeneration of protoplasts from *Pyricularia oryzae*.

c) Septum

Septa appear trilamellate in TEM sections. They have been viewed as single chitin-rich structures, with the central portion representing densely packed purer chitin, and the staining layers on either side of this being chitin mixed with other wall material such as glucans. GOODAY and GOW (1983) propose an alternate model for the hyphal septum of *Candida albicans*. They view the trilaminate appearance as two plates of densely staining, chitin-rich material separated by a lightly staining area: Chitin fibrils are arranged predominantly tangentially (GOW and GOODAY 1983), and the septa lack some components of the lateral walls. This model may be applicable to septa of other fungi. Polyoxin D, a competitive inhibitor of chitin synthase, inhibited formation of septa and disturbed growth of lateral and yeast cell walls (HILENSKI et al. 1986).

In the basidiomycete *Coriolus versicolor*, septal sealing in damaged mycelia is a two-stage process (AYLMORE et al. 1984). The first is the instantaneous plugging of the pore channel with electron-dense material originating in situ, unlike the preformed Woronin bodies of ascomycetes (e.g. KIMBROUGH and CURRY 1985, 1986). The second, beginning several minutes later, involves the detachment of the septal apparatus present in the ruptured compartment and remodelling of the septal swelling on the other side of the wall to give a permanent seal. The process is finished within 20 min, and parenthesomes play no part in the plugging response.

Septal pore ultrastructure continues to be applied to taxonomic problems (e.g. KHAN and KIMBROUGH 1982; O'DONNELL and MCLAUGHLIN 1984b; OBERWINKLER 1985 and BOURETT and MCLAUGHLIN 1986 for basidiomycetes; KIMBROUGH and CURRY 1985 for Ascobolaceae; KIMBROUGH and CURRY 1986 for two tribes of the Pyronemataceae; SMITH and BATENBURG - VAN DER VEGTE 1985 for *Blastobotrys* and *Sporothrix*, confirming the heterogeneity of the latter genus; GIBSON et al. 1986b for *Glaziella aurantiaca*, which is being transferred from the zygo- to the ascomycetes). However, as has been pointed out before, septal structures and pluggings from different parts of the thallus may differ considerably (KHAN and KIMBROUGH 1982 for some basidiomycetes; PATRIGNANI et al. 1984 for *Tremella mesenterica*; KIMBROUGH and CURRY 1986 for some ascomycetes).

d) Dimorphism

260 papers on fungal dimorphism published up to about 1979 were reviewed by BEMMANN (1981) with emphasis given to *Aureobasidium pullulans*, *Candida albicans*, *Blastomyces brasiliensis*, *B. dermatitidis*, *Histoplasma capsulatum*, *H. farciminosum* as well as to species of *Mucor* and *Mycotypha*. The field has been summarized more recently by STEWART and ROGERS (1983). More recent investigations also include *Coccidioides immittis* (COLE and SUN 1985), a peculiar diphasic fungus characterized by distinct saprobic and parasitic cycles. A drastic increase in chitin content of the mycelial form over that of the yeast form has been found in *Wangiella dermatitidis* using histochemical methods (HARRIS and SZANISZLO 1986), confirming the general rule that hyphal forms have higher chitin contents than yeast forms.

In *Ceratocystis ulmi* calcium ions regulate dimorphism which is influenced by Ca^{2+}-calmodulin. Calcium is needed for hyphal growth. Without it the organism reverts to the yeast phase (MUTHUKUMAR and NICKERSON 1984, 1985).

Candida albicans has received most of the attention in recent years (ODDS 1985; SOLL 1985,1986). Dimorphism in this species can be regulated by a pH regime (BUFFO et al. 1984) or rather by the glucose present in the medium (POLLACK and HASHIMOTO 1987). The programs of protein synthesis accompanying bud and hypha formation are strikingly similar, and regulation of the dimorphic behaviour appears to occur via subtle spatial, temporal and quantitative differences in the zones of cell wall expansion (STAEBELL and SOLL 1985; SOLL et al. 1985).

References

ADAMS, A.E.M., PRINGLE, J.R.: J. Cell Biol. 98, 934-945 (1984). - ALDRICH, H.C., TODD, W.J., eds.: Ultrastructure Techniques for Microorganisms, 548 pp. New York, London: Plenum Press 1986. - ALDRICH, H.C.,

BARSTOW, W.E, LINGLE, W.L.: Arch Microbiol. 139, 102-104 (1984). - ALLEN, E.D., AIUTO, R., SUSSMAN, A.S.: Protoplasma 102, 63-75 (1980). - ANDERSON, J.M., SOLL, D.R.: J. Gen. Microbiol. 132, 2035-2047 (1986). - ARAI, T., ed.: Filamentous Microorganisms, Biomedical Aspects, 460 pp. Tokyo: Japan Sci. Soc. Press 1985. - AYLMORE, R.C., WAKLEY, G.E., TODD, N.K.: J. Gen. Microbiol. 130, 2975-2982 (1984).

BACIC, A., WILLIAMS, M.L., CLARKE, A.E.: J. Histochem. Cytochem. 33, 384-388 (1985). - BARSTOW, W.E., LINGLE, W.L., LOVETT, J.S.: J. Ultrastruct. Res. 93, 168-178 (1985). - BARTNICKI-GARCIA, S.: Fundamental aspects of hyphal morphogenesis, 245-267. In: Microbial Differentiation, eds. J.M. ASHWORTH, J.E. SMITH, 450 pp., 23rd Symp. Soc. Gen. Microbiol. Cambridge University Press 1973. - BARTNICKI-GARCIA, S., BRACKER, C.E., LIPPMAN, E., RUIZ-HERRERA, J.: Arch. Microbiol. 139, 105-112 (1984). - BARTNICKI-GARCIA, S., BARTNICKI, I., SENTANDREU, R.: Current Microbiol. 12, 141-146 (1985). - BEAKES, W.L., CAMPOS-TAKAKI, G.M.: Trans. Br. Mycol. Soc. 83, 607-614 (1984). - BECKETT, A., READ, N.D.: Low-temperature scanning electron microscopy, 45-86. In: Ultrastructure Techniques for Microorganisms, eds. H.C. ALDRICH, W.J. TODD, 548 pp. New York, London: Plenum Press 1986. - BEILHARZ, V.C.: Trans. Br. Mycol. Soc. 84, 79-85 (1985). - BEMMANN, W.: Zentralbl. Bakteriol. II. Abt. 136, 369-416 (1981). - BENHAMOU, N., OUELLETTE, G.B.: J. Histochem. Cytochem. 34, 855-867 (1986a); - Histochem. J. 18, 95-104 (1986). - BODENMANN, J., HEINIGER. U., HOHL, H.R.: Can. J. Microbiol. 31, 75-82 (1985). - BORTNICK, R.N., POWELL, M.J., BANGERT, T.N.: Mycologia 77, 861-879 (1985). - BOURETT, T.M., MCLAUGHLIN, D.J.: Can. J. Bot. 64, 130-145 (1986). - BOWMAN, B.J., BOWMAN, E.J.: J. Membr. Biol. 94, 83-97 (1986). - BROWN, D.H., ed.: Lichen Physiology and Cell Biology, 362 pp. New York, London: Plenum Press 1985. - BROWN, J.A., OGAWA, K., WOOD, T.M.: Enzyme Microb. Technol. 9, 527-532 (1986). - BUCK, K.W.: Fungal Virology, 336 pp. Boca Raton, FL: CRC Press 1986. - BUFFO, J., HERMAN, M.A., SOLL, D.R.: Mycopathologia 85, 21-30 (1984). - BULAWA, C.E., SLATER, M., CABIB, E., AU-YOUNG, J., SBURLATI, A., ADAIR, W.L., ROBBINS, P.W.: Cell 46, 213-225 (1986). - BUSHNELL, W.R., ROELFS, A.P.: The Cereal Rusts, Vol. I, 560 pp. New York, London: Academic Press 1984. - BUTT, T.M., BECKETT, A.: Protoplasma 120, 72-83 (1984a); - ibid. 120, 61-71 (1984b).

CABO, A.P., REYES, E., NOVAES-LEDIEU, M., MENDOZA, C.G.: Trans. Br. Mycol. Soc. 86, 545-549 (1986). - CHAMBERLAND, H., CHAREST, P.M., OUELLETTE, G.B., PAUZE, F.J.: Histochem. J. 17, 313-321 (1983). - CHILVERS, G.A., LAPEYRIE, F.F., DOUGLASS, P.A.: New Phytol. 99, 203-210 (1985). - CLEARY, A., YOUATT, J., O'BRIEN, T.P.: Aust. J. Biol. Sci. 39, 241-254 (1986). - COLE, G.T.: Microbiol. Rev. 50, 95-132 (1986). - COLE, G.T., SUN, S.H.: Arthroconidium-spherule-endospore transformation in Coccidioides immitis, 281-333. In: Fungal Dimorphism: With Emphasis on Fungi Pathogenic for Humans, ed. P.J. SZANISZLO, 395 pp. New York, London: Plenum Press 1985. - COLLINGE, A.J., MARKHAM, P.: Exp. Mycol. 9, 80-85 (1985). - COOKE, R.C., RAYNER, A.D.M.: Ecology of Saprotrophic Fungi, 415 pp. London, New York: Longman 1984.

DAHMEN, H., HOBOT, J.A.: Protoplasma 131, 123-143 (1986). - DARGENT, R., TOUZÉ-SOULET, J.-M., RAMI, J., MONTANT, C.: Exp. Mycol. 6, 101-114 (1982). - DARGENT, R., EL MOUGITH, A., TOUZÉ-SOULET, J.-M.: C.R. Acad. Sci. 302, III; 277-282 (1986). - DAVIS, L.L., BARTNICKI-GARCIA, S.: J. Gen. Microbiol. 130, 2095-2102 (1984). - DAY, A.W., GARDINER, R.B., SMITH R., SVIRCEV, A.M., MCKEEN, W.E.: Can. J. Microbiol. 32, 577-584 (1986). - DEMOULIN, V.: BioSystems 18, 347-356 (1985). - DEWERCHIN, M.A., VAN LAERE, A.J.: Biochem. Physiol. Pflanzen 180, 507-513 (1985). - DOUGLAS, C.M., SYNAN, T.R., BOBBITT, T.F., NORDIN, J.H.: Exp. Mycol. 8, 146-160 (1984). - DOWSETT, J.A., REID, J.: Mycologia 75, 1094-1099 (1983). - DYLEWSKI, D.P., MILLER, C.E.: Protoplasma 121, 42-53 (1984).

ELLZEY, J.T., HAMMONS, T.L., COOPER, M.O.: Mycologia 78, 313-316 (1986). - ERWIN, D.C., BARTNICKI-GARCIA, S., TSAO, P.H., eds.: *Phytophthora*: Its Biology, Taxonomy, Ecology, and Pathology, 392 pp. St. Paul, MN: Am. Phytopathol. Soc. 1983.

FISKIN, A.M., GARRISON, R.G., FIGUEROA, M.: Cytobios 48, 185-194 (1986). - FLEGG, P.B., SPENCER, D.M., WOOD, D.A.: The Biology and Technology of the Cultivated Mushroom, 347 pp. Chichester, New York: Wiley 1985. - FUJIKAWA, S., MIURA, K.: Cryobiology 23, 371-382 (1986). - FURTER, R., RAST, D.M.: FEMS Microbiol. Lett. 28, 205-211 (1985). -

GARRAWAY, M.O., EVANS, R.C.: Fungal Nutrition and Physiology, 401 pp. New York, Chichester: Wiley 1984. - GARRISON, R.G., FISKIN, A.M.: Ann. Microbiol. 135B, 227-250 (1984); - Ann. Inst. Pasteur 137A, 213-222 (1986a); - Mycopathologia 95, 139-144 (1986b). - GAUGY, D., FÈVRE, M.: Microbios 44, 285-293 (1985). - GBAJA, I.S., YOUNG, T.W.K.: Microbios 42, 263-272 (1985). - GIBSON, J.L., KIMBROUGH, J.W., BENNY, G.L.: Mycologia 78, 543-553 (1986a); - ibid. 78, 941-954 (1986b). - GIRARD, V., FÈVRE, M.: Planta 160, 400-406 (1984). GIRARD, V., CAPELLANO, A., FÈVRE, M.: J. Gen. Microbiol. 130, 2367-2374 (1984). - GOODAY, G.W.: The hyphal tip, 315-356. In: Fungal Differentiation, ed. J.E. SMITH. Mycological Series, Vol. 4, 624 pp. New York, Basel: Dekker 1983. - GOODAY, G.W., GOW, N.A.R.: Exp. Mycol. 7, 370-373 (1983). - GOW, N.A.R., GOODAY, G.W.: Protoplasma 115, 52-58 (1983). - GOW, N.A.R., KROPF, D.L., HAROLD, F.M.: J. Gen. Microbiol. 130, 2967-2974 (1984). - GREER, C., SCHEKMAN, R.: Mol. Cell Biol. 2, 1270-1278 (1982). - GROVE, S.N., SWEIGARD, J.A.: Exp. Mycol. 4, 239-250 (1980).

HALE, M.D., EATON, R.A.: Mycologia 77, 447-463 (1985). - HAENSELER, E., NYHLÉN, L.E., RAST, D.M.: Exp. Mycol. 7, 17-30 (1983a); - Biochim. Biophys. Acta 745, 121-133 (1983b). - HAROLD, R.L., HAROLD, F.M.: J. Gen. Microbiol. 132, 213-219 (1986). - HAROLD, F.M., KROPF, D.L., CALDWELL, J.H.: Exp. Mycol. 9, 183-186 (1985). - HARRIS, N.: Annu. Rev. Plant Physiol. 37, 73-92 (1986). - HARRIS, J.L., SZANISZLO, P.J.: Mycologia 78, 853-857 (1986). - HAWKSWORTH, D.L., HILL, D.J.: The Lichen-Forming Fungi, 166 pp. Glasgow: Blackie 1984. - HEATH, B.I.: RETHORET, K., MOENS, P.B.: Eur. J. Cell Biol. 35, 284-295 (1984). - HEATH, I.B., RETHORET, K., ARSENAULT, A.L., OTTENSMEYER, F.P.: Protoplasma 128, 81-93 (1985). - HEGNAUER, H., NYHLÉN, L.E., RAST, D.M.: Exp. Mycol. 9, 221-229 (1985). - HEMMES, D.E., LERMA, A.P., Jr.: Mycologia 77, 743-755 (1985). - HEMMES, D.E., STASZ, T.E.: Mycologia 76, 924-935 (1984). - HERR, F.B., HEATH, M.C.: Exp. Mycol. 6, 15-24 (1982). - HICKS, J., ed.: Yeast Cell Biology. UCLA Symposia on molecular and cellular Biology, N.S. Vol. 33, 671 pp. New York: Liss 1986. - HILENSKI, L.L., NAIDER, F., BECKER, J.M.: J. Gen. Microbiol. 132, 1441-1451 (1986). - HIPPE, S.: Protoplasma 129, 52-61 (1985). - HIPPE, S., HERMANNS, M.: Protoplasma 135, 19-30 (1986). - HIPPE, S., LUETH, H.: J. Theor. Biol. 121, 351-366 (1986). - HOCH, H.C.: Freeze-substitution of fungi, 183-212. In: Ultrastructure Techniques for Microorganisms, eds. H.C. ALDRICH, W.J. TODD, 548 pp. New York, London: Plenum Press 1986. - HOCH, H.C., BOURETT, T.M: Eur. J. Cell Biol. 41, 290-297 (1986). - HOCH, H.C., STAPLES, R.C.: Mycologia 75, 795-824 (1983a); - Eur. J. Cell Biol. 32, 52-58 (1983b); - Protoplasma 124, 112-122 (1985). - HOFFMANN, J.A., GOATES, B.J.: Can. J. Bot. 63, 86-96 (1985). - HOHL, H.R., BALSIGER, S.: Bot. Helv. 96, 289-297 (1986). - HORISBERGER, M., ROUVET-VAUTHEY, M.: Experientia 41, 748-750 (1985). - HORISBERGER, M., ROUVET-VAUTHEY, M., RICHLI, U., FARR, D.-R.: Eur. J. Cell Biol. 37, 70-77 (1985). - HUMPHREYS, A.M., GOODAY, G.W.: J. Gen. Microbiol. 130, 1359-1366 (1984).

JACOBSON, L., HALMAN, J., YARIV, J.: Biochem. J. 201, 473-479 (1983). - JACOBSEN, T., JENSEN, B., OLSEN, J., ALLERMANN, K.: Can. J. Microbiol. 31, 93-96 (1985). - JENNINGS, D.H., RAYNER, A.D.M., eds.: The Ecology and Physiology of the Fungal Mycelium. Symp. Br. Mycol. Soc., 564 pp. Cambridge University Press 1984.

KANG, M.S., ELANGO, N., MATTIA, E., AU-YOUNG, J., ROBBINS, P.W., CABIB, E.: J. Biol. Chem. 259, 14966-14972 (1984). - KHAN, S.R., KIMBROUGH, J.W.: Mycotaxon 15, 103-120 (1982). - KANBE, T., TANAKA, K.: Protoplasma 129, 198-213 (1985). - KILMARTIN, J.V., ADAMS, A.E.M.: J. Cell Biol. 98, 922-933 (1984). - KIMBROUGH, J.W., CURRY, K.J.: Mycologia 77, 219-229 (1985); - ibid. 78, 735-743 (1986). - KOPECKÁ, M., KREGER, D.R.: Arch. Microbiol. 143, 387-395 (1986).

LAPEYRIE, F.F., CHILVERS, G.A., DOUGLASS, P.A.: New Phytol. 98, 345-360 (1984).

MAKAROW, M.: EMBO J. 4, 1861-1866 (1985); - Endocytosis in Saccharomyces cerevisiae. Internalization of soluble and particulate markers into cells and spheroplasts, 451-459. In: Yeast Cell Biology, ed. J. HICKS. UCLA Symposium on Molecular and Cellular Biology, N.S. Vol. 33, 671 pp. New York: Liss 1986. - MAKAROW, M., NEVALAINEN, L.T.: J. Cell Biol. 104, 67-75 (1987). - MARKS, J., HYAMS, J.S.: Eur. J. Cell Biol. 39, 27-32 (1985). - MARTIN, R.W., MILLER, C.E.: Mycologia 78, 11-21 (1986a); - ibid. 78, 230-241 (1986b). - MCKERRACHER, L.J., HEATH, I.B.: Mycologia 77, 412-417 (1985a); - Protoplasma 125, 162-172 (1985b). - MEYER, R.J.: Mycologia 79, 44-54 (1987). - MOCHIZUKI, T., TANAKA, S., WATANABE, S.: Mycopathologia 94, 31-38 (1986). - MONTGOMERY, G.W.G., ADAMS, D.J., GOODAY, G.W.: J. Gen. Microbiol. 130, 291-297 (1984). - MOORE, D., CASSELTON, L.A., WOOD, D.A., FRANKLAND, J.C., eds.: Developmental Biology of Higher Fungi. Br. Mycol. Soc. Symp., Vol. 10, 614 pp. New York: Cambridge University Press 1985. - MUELLER, S.C., BRANTON, D.: J. Cell Biol. 98, 341-346 (1984). - MURRIN, F., HEATH, I.B., NEWCOMBE, W.: Protoplasma 120, 84-90 (1984). - MUTHUKUMAR, G., NICKERSON, K.W.: J. Bacteriol. 158, 390-392 (1984); - FEMS Microbiol. Lett. 27, 199-202 (1985).

NEWHOUSE, J.R., HOCH, H.C., MACDONALD, W.L.: Can. J. Bot. 61, 389-399 (1983). - NOMBELA, C., ed.: Microbial Cell Wall Synthesis and Autolysis. Proc. FEMS Symp., Vol. 27, 328 pp. Amsterdam, New York: Elsevier 1984. - NOVICK, P., BOTSTEIN, D.: Cell 40, 405-416 (1985).

OBERWINKLER, F.: Bot. Jahrb. Syst. 107, 541-580 (1985). - ODDS, F.C.: CRC Crit. Rev. Microbiol. 12, 45-94 (1985). - O'DONNELL, K.L., MCLAUGHLIN, D.J.: Mycologia 76, 468-485 (1984a); - ibid. 76, 486-502 (1984b). - OLSON, L.W.: *Allomyces* - A Different Fungus, Vol. 73, 96 pp. Copenhagen: Council for Nordic Publ. in Bot., Opera Bot. 1984. - ORBACH, M.J., PORRO, E.B., YANOFSKY, C.: Mol. Cell Biol. 6, 2452-2461 (1986).

PATRIGNANI, G., PELLEGRINI, S., GEROLA, F.M.: Caryologia 37, 77-86 (1984). - PEBERDY, J.F., FERENCZY, L., eds.: Fungal Protoplasts: Applications in Biochemistry and Genetics, Mycol. Series, Vol. 6, 364 pp. New York: Dekker 1985. - PENDLAND, J.C., BOUCIAS, D.G.: Mycopathologia 87, 141-148 (1984). - PEVELING, E., ROBENEK, H., BERNS, B.: The architecture of the concentric bodies in the mycobiont of *Peltigera praetextata*, 275-285. In: Lichen Physiology and Cell Biology, ed. D. BROWN, 362 pp. New York, London: Plenum Press 1985. - POLI, F., PANCALDI, S., VANNINI, G.L.: Eur. J. Cell Biol. 42, 79-84 (1986). - POLLACK, J.H., HASHIMOTO, T.: J. Gen. Microbiol. 133, 415-424 (1987). - PONS, N., GAY, J.L., SUTTON, B.C.: Trans. Br. Mycol. Soc. 83, 181-183 (1984). - POWELL, C.I., BAGYARAJ, D.G.: VA Mycorrhiza, 288 pp. Boca Raton, FL: CRC Press 1984. -

POWELL, M.J.: Cytochemical techniques for the subcellular localization of enzymes in microorganisms, 421-445. In: Ultrastructure Techniques for Microorganisms, eds. H.C. ALDRICH, W.J. TODD, 548 pp. New York, London: Plenum Press 1986. - POWELL, M.J., BRACKER, C.E.: Mycologia 78, 892-900 (1986). - POWELL, M.J., LEHNEN, L.P., Jr., BORTNICK, R.N.: BioSystems 18, 321-334 (1985). - PUESCHEL, C.M., VAN DER MEER, J.P.: Can. J. Bot. 63, 409-418 (1985).

QUIGLEY, D.R., SELITRENNIKOFF, C.P.: Exp. Mycol. 8, 202-214 (1984). - QUIGLEY, D.R., SIEHR, D.J.: Appl. Environm. Microbiol. 47, 882-883 (1984).

RIEZMAN, H., CHVATCHKO, Y., DULIĆ, V.: TIBS 11, 325-328 (1986). - ROBINSON, D.G.: Plant Membranes. Endo- and Plasma Membranes of Plant Cells, 331 pp. New York, Chichester: Wiley 1985. - RONCERO, C., DURAN, A.: J. Bacteriol. 163, 1180-1185 (1985). - ROOS, U.-P., SHAW, D.S.: Trans. Br. Mycol. Soc. 84, 340-344 (1985). - RUIZ-HERRERA, J., BRACKER, Ch.E., BARTNICKI-GARCIA, S.: Protoplasma 122, 178-190 (1984). - RUNEBERG, P., RAUDASKOSKI, M., VIRTANEN, I.: Eur. J. Cell Biol. 41, 25-32 (1986).

SAIKAWA, M., HOSHINO, J.: Mycologia 78, 554-561 (1986). - SCHRANTZ, J.-P.: Biol. Cell 46, 269-276 (1982). - SEWALL, T., OLSON, L., LANGE, L., POMMERVILLE, J.: Protoplasma 133, 129-139 (1986). - SIETSMA, J.H., SONNENBERG, A.M.S., WESSELS, J.G.H.: J. Gen. Microbiol. 131, 1331-1337 (1985). - SIKORA, L., MARZLUF, G.A.: J. Gen. Microbiol. 128, 439-445 (1982). - SMITH, J.E., ed.: Fungal Differentiation. A Contemporary Synthesis, 624 pp. Basel: Dekker 1983. - SMITH, M.Th., BATENBURG-VAN DER VEGTE, W.H.: Antonie Leeuwenhoek J. Microbiol. 51, 121-128 (1985). - SMITH, R., SCARBOROUGH, G.A.: Anal. Biochem. 138, 156-163 (1984). - SOLL, D.R.: Candida albicans, pp. 167-195. In: Fungal Dimorphism: With Emphasis on Fungi Pathogenic for Humans, ed. P. SZANISZLO, 395 pp. New York: Plenum Press 1985; - BioEssays 5, 5-11 (1986). - SOLL, D.R., MITCHELL, L.H.: J. Cell Biol. 96, 486-493 (1983). - SOLL, D.R., HERMAN, M.A., STAEBELL, M.A.: J. Gen. Microbiol. 131, 2367-2375 (1985). - SONNENBERG, A.S.M., SIETSMA, J.H., WESSELS, J.G.H.: Exp. Mycol. 9, 141-148 (1985). - STAEBELL, M.A., SOLL, D.R.: J. Gen. Microbiol. 131, 1467-1480 (1985). - STAPLES, R.C., HOCH, H.C.: Exp. Mycol. 6, 293-302 (1982). - STEWART, P.R., ROGERS, P.J.: Fungal dimorphism, 267-313. In: Fungal Differentiation: A Contemporary Synthesis, ed. J.E. SMITH. Mycology Series, Vol. 4, 624 pp. New York, Basel: Dekker 1983. - SVIRCEV, A.M., SMITH, R., GARDINER, R.B., RACKI, J.M., DAY, A.W.: Exp. Mycol. 10, 19-27 (1986). - SZANISZLO, P., ed.: Fungal Dimorphism: With Emphasis on Fungi Pathogenic for Humans, 395 pp. New York, London: Plenum Press 1985.

TANAKA, H., KAMIMIYA, S., MIKI, T., KOMINATO, M.: Agric. Biol. Chem. 49, 1393-1396 (1985). - TANAKA, K., HEATH, I.B.: Protoplasma 120, 36-42 (1984). - TAYLOR, W.J.: Protoplasma 126, 1-18 (1985). - TUINEN, D., ORTEGA PEREZ, R., TURIAN, G.: Bot. Helv. 96, 299-302 (1986). - TURIAN, G., TON-THAT, T.C., ORTEGA PEREZ, R.: Bot. Helv. 95, 311-322 (1985).

VERMEULEN, C.A., WESSELS, J.G.H.: Protoplasma 120, 123-131 (1984); - Eur. J. Biochem. 158, 411-415 (1986).

WANNER, G., THEIMER, R.R.: Two types of microbodies in *Neurospora crassa*, 269-282. In: Peroxisomes and Glyoxysomes, eds. H. KINDL, P.B. LAZAROW. Ann. N.Y. Acad. Sci., Vol. 386, 550 pp. New York: N.Y. Acad. Sci. 1982. - WEATHERBEE, J.A., MAY, G.S., GAMBINO, J., MORRIS, N.R.: J. Cell Biol. 101, 706-711 (1985). - WESSELS, J.G.H.: Apical hyphal wall extension. Do lytic enzymes play a role?, 31-42. In: Microbial Cell Wall Synthesis and Autolysis, ed. C. NOMBELA. Proc. FEMS Symp. 27, 328 pp. Amsterdam: Elsevier 1984. - WESSELS, J.G.H., SIETSMA, J.H., SONNENBERG, A.S.M.: J. Gen. Microbiol. 129, 1607-1616 (1983). - WILLIAMS, M.L., BACIC, A., CLARKE, A.E.: Exp. Mycol. 8, 238-244 (1984). - WONG, H.-C., CHIEN, C.-Y.: Mycologia 78, 593-599 (1986a); - ibid. 78, 713-721 (1986b).

YANAGI, S.O., MONMA, M., KAWASUMI, T., HINO, A., KITO, M., TAKEBE, I.: Agric. Biol. Chem. 49, 171-179 (1985).

Prof. Dr. HANS R. HOHL
Institut für Pflanzenbiologie
der Universität
Zollikerstr. 107
CH-8008 Zürich
Schweiz

III. Vergleichende Morphologie des Vegetationskörpers

Von Wolfgang Hagemann

1. Allgemeines

An kritischen Auseinandersetzungen mit der Pflanzenmorphologie ist kein Mangel (SATTLER 1986; RUTISHAUSER u. SATTLER 1986; MEEUSE 1986). MEEUSE (1986) sieht zwar in seiner "Anatomy of Morphology" das Konzept der Urpflanzen im Sinne von TROLL als "an abhorrence in a science supposed to be more or less exact", doch sei eine moderne Form des Bauplan-(Typus-) Prinzips sehr viel rationaler als SATTLERS Formenkontinuum. SATTLER (1986) streitet in seiner "Biophilosophy" der "klassischen Cormuslehre" den Erfolg nicht ab, doch plädiert er für komplementäre Betrachtungsweisen, wie z.B. seiner "Continuum-Lehre" oder auch der von RUTISHAUSER (1986) ebenfalls für fruchtbar gehaltenen "Phytonlehre". In Fallstudien wie z.B. *Acacia longipedunculata*, galioiden *Rubiaceen*, *Hydrothrix gardeneri* oder den *Utricularia*-Arten u.a. wird gezeigt, daß man mit dem Blattbegriff in erhebliche Schwierigkeiten kommt (SATTLER 1986: 101 ff; SATTLER u. RUTISHAUSER 1986; vgl. aber auch KAPLAN 1984). Dies ist richtig, doch sind die angeführten Fallstudien nicht wirklich klar. Die "Cormuslehre" wird von SATTLER (1986: 102) als die klassische morphologische Theorie ausgegeben. Die sogenannten Grundorgane können bei der Mehrzahl aller höheren Pflanzen beschrieben werden. Sie bilden unabhängig von ihrer Form ein Gefügesystem (Typus), in dem sie durch ihre Lagebeziehungen zueinander definiert, d.h. homolog gesetzt werden können. Weder die Grundorgane, noch der Typus und die Homologie sind also Theorien, allenfalls Konzepte, sondern erst die Antwort auf die Frage, warum man diesen Typus so häufig beobachten kann. Wenn man nun den Wert auf die Kontinuität der Formen bei den Pflanzenorganen legt, diese beschreibt und nach der Begründung sucht, dann wird man eine andere Antwort und somit andere Theorie erhalten, bei welcher der Begriff der Homöosis eine zentrale Rolle spielen könnte (COONEY-SOVETTS u. SATTLER 1986). In dieser Arbeit wurde unter dem Gesichtspunkt der Homöosis die Entwicklung der Phyllocladien bei den Asparagaceen sehr detailliert dargestellt. Der Homologie-Begriff ist dabei unwesentlich. Immer aber muß die Theorie eine Phylogenetische sein und insofern brauchen sich beide nicht auszuschließen. Sie sollten komplementär im Sinne von RUTISHAUSER und SATTLER sein. Das Verhältnis zwischen Beschreibung und Theorie ist, wie die Beispiele zeigen, sehr eng, was bestätigt, daß theorieneutrale Beschreibung auch dann nicht möglich ist, wenn die Beschreibung der Theorie vorausgeht. Die Art der Beschreibung grenzt die möglichen Theorien ein. Sehr starker Einfluß auf die Vorstellung über die Organisation und Abstammung der Pflanzen geht seit jeher von der Palaeobotanik aus. Wie unsicher indessen die Grundlagen der Palaeobotanik sind, erhellt aus der überraschenden Neubeschreibung von *Rhynia major* als *Aglaophyton major* durch EDWARDS (1986), die ihrer Organisation nach nicht mehr als Pteridophyt gelten kann. Somit bleibt *Rhynia gwynne vaughani* als einzige Art dieser Gattung übrig, deren oft bezweifelte Sporophytennatur EDWARDS erst 1980 bestätigt hatte. Viele anregende Gedanken wurden von RAVEN (1986) zusammengetragen, in denen vor allem die herkömmlichen Vorstellungen einer kritischen Bewertung unter phy-

Progress in Botany, Vol. 49

sikalischen Gesichtspunkten unterzogen werden. Weniger glücklich verfährt der Autor mit morphologischen Argumenten. Begriffe wie "parenchymatisch", "pseudoparenchymatisch" sind wenig brauchbar, seit die Feinstrukturanalyse der Zellteilungsmechanismen bei Grünalgen als wesentliche Unterscheidung etwa "Siphonoblasteme", "Phycoblasteme" und "Phragmoblasteme" zu unterscheiden gestatten, und wenn schon nur noch Charophyceen als Vorfahren der Archegoniaten in Frage kommen, wie es auch RAVEN hervorhebt, können Algen wie *Fritschiella*, die ein Phycoblastem besitzen, nicht mehr als Modelle für höhere Landpflanzen in Frage kommen. Auch können diese ursprünglich keine terminalen Angien besessen haben, weil bei der offenen Konstruktion ursprünglich thallöser Landpflanzen nach dem Muster von *Coleochaete* bzw. *Parka*, die selbst keine terminalen Angien besaßen, das Meristem ein undeterminiertes war (vgl. Progr. Bot. 45: 32; 42: 40-41).

Weitere neue anregende Argumente über die Abstammung der höheren Landpflanzen haben REMY u. HASS (1986) ausgehend von den devonischen Gametophyten *Sciadophyton* und *Lyonophyton* in die Diskussion gebracht, wobei sie leider nicht an die moderne Grünalgenforschung als vielmehr an die kaum noch aktuellen Vorstellungen FRITSCHs anknüpfen. Auch NIKLAS' (1984) Bemühungen durch Vermessung früher Landpflanzen Gesetzmäßigkeiten der Stelenevolution herauszufinden, hängt von der Richtigkeit der hypothetischen Basis ab, *Rhyniophytina*, *Trimerophytina* und Progymnospermen als eine Abstammungsreihe zu sehen.

Wenn man sich mit den leidenschaftlichen Diskussionen über die Phylogenie der Gymnospermen der letzten Zeit beschäftigt, entsteht nicht selten der Eindruck, als wären dies Fragen, die allein von Palaeobotanikern entschieden werden könnten (vgl. z.B. MEYEN 1984; BECK 1985; MILLER 1985; ROTHWELL 1985; WALKER u. WALKER 1984; HUGHES 1984 u.a.). Der Palaeontologe FONDI (1986: 228) z.B. spricht aus, daß über den Abstand zwischen zwei verwandten Formen nur die Palaeontologie urteilen könne. Die Tatsache, daß nur ein kleiner Anteil der Pflanzenarten, die gelebt haben, fossile Reste hinterlassen haben, und die Wahrscheinlichkeit, daß für die Evolution der Pflanzen wesentliche Differenzierungsprozesse sich unter Bedingungen abgespielt haben könnten, die keine Fossilisierung erlaubten, wird verdrängt. Wenn man sieht, wie jeder noch so unscheinbare fossile Rest einer Pflanze mit beträchtlichem Aufwand an Phantasie in phylogenetische Spekulationen eingebaut wird (vgl. z.B. MEYEN 1984), so ist mein Eindruck, daß die Grenze zur Fiktion längst überschritten ist. Das sehr anregende Buch von PROBST (1986) über die Biologie der Moos- und Farnpflanzen verstärkt diesen Eindruck. Die gegensätzlichen Hypothesen zur Entstehung der Landpflanzen werden hier in drastisch vereinfachten Formenreihen gegenüber gestellt, die letzlich nur unser Unwissen demonstrieren. Die reichlich vorhandenen praktischen Hinweise zur Beobachtung der Pflanzen sollte man zu allererst beherzigen, denn unsere Vorstellung von der Konstruktion und der Ontogenese des Pflanzenkörpers bedürfen einer Revision von der Bedeutung eines Paradigmawechsels. "Verwachsung" als sog. Elementarprozeß bedeutet nach PROBST nicht Verwachsung, sondern eine genetisch bedingte morphologische Veränderung im Lauf der Stammgeschichte, die dazu führt, daß aus einem stark aufgeteilten Vegetationskörper ein kompakter, oder flächiger Vegetationskörper entsteht. Dies verraten uns nicht fossile, sondern rezente Pflanzen. Die Konstruktion des Phragmoblastems ermöglicht den Übergang von fädiger Gestalt nach den von *Coleochaete*-Arten vorgeführten Muster zu flächiger Gestalt. Hierbei handelt es sich nicht um eine Verwachsung von Fäden, sondern um eine Folge der Beweglichkeit des Phragmoplasten, der die Zellplatte so anordnen kann, daß anstelle der Fäden der am Rande, d.h. terminal wachsende Flächenthallus von *Coleochaete scutata* resultiert. Telome entstehen durch Verzweigung (Meristemfraktionierung). Eine Wiedervereinigung von Telomen zwecks Erzeugung von Blattflächen ist vollkommen überflüssig. Die nun schon seit

Jahren geführte Diskussion gerade über diese Frage ist von KLOTZ (1985) in seinen Bemerkungen zur Typologie des Blattes vollkommen übersehen worden. *Coleochaete* als Modell für die Ableitung von Landpflanzen wird mit einer zwingenden Fülle von Argumenten aus der Grünalgenforschung weiter gestützt (GRAHAM 1985; GRAHAM and TAYLOR III 1986; MATTOX and STEWART 1984; MELKONIAN 1984 u.a., vgl. Prog. Bot. 47: 31). Geradezu deprimierend sind für Morphologen viele entwicklungsphysiologische Arbeiten zu lesen. Ein besonders negatives Beispiel gibt der Symposiumsband "Control of Leaf Growth" (BAKER et al. 1985) mit der Einleitung "Although cell division lays down the foundations of leaf architecture, the thousands of different cell types, air spaces and recognizable external form have their bases in the magnitude and direction of cell wall expansion at the level of the individual cell". Mit unserer aus dem letzten Jahrhundert stammenden Sicht eines aus Elementarorganismen zusammengesetzten Pflanzenkörpers können wir die Konstruktion der Höheren Pflanzen nicht begreifen, weshalb auch die Überprüfung moderner biologischer Theorien an Pflanzen immer weniger möglich wird, wie sie beispielsweise in dem von BROOKS u. WILEY (1986) vorgelegten Buch "Evolution as Entropy" in leicht verständlicher Form dargelegt worden ist. Wenn diese Autoren aufgrund physikalischer Überlegungen zu dem Schluß kommen, "Evolution is survival of the adequate and not of the most fit", dann wird klar, daß Selektion überbewertet und den Prozessen der Diversifikation zu wenig Aufmerksamkeit zugewendet wird (HAGEMANN 1978).

2. Meristeme

a) Sproßscheitelmeristeme

Die Arbeiten der letzten zwei Jahre auf diesem Gebiet haben das Verständnis der Konstruktion und Funktion kaum mehr wesentlich vertieft. Arbeiten, die sich mit Beschreibung und Zonierungsmustern befassen, sind seltener geworden. Sie finden sich in Arbeiten mit spezieller Thematik. So ist die Beschreibung der Scheitelmeristeme bei einigen *Similax*-Arten durch MARTIN und TUCKER (1985) nur die Vorarbeit zur Klärung der Blattentwicklung bei diesen Pflanzen. KUMAR et al. (1984) haben eingehend den Scheitel von *Agave wightii* beschrieben. CHRETIEN u. VIETH (1985) haben Knospen von *Picea pungens* in vitro kultiviert. Aus histologisch noch wenig differenzierten Blattanlagen und Meristemgewebe bildet sich kallöses Gewebe, in dem neue Scheitelmeristeme entstehen. Aus diesen entstehen zuerst neue Blattanlagen, worauf das Zonierungsmuster folgt. Insbesondere bei der vergleichenden histochemischen Untersuchung spielen die Zonierungsmuster eine wichtige Rolle. So haben GOYAL et al. (1985) eine experimentelle Studie über die Scheitelmeristeme von *Matthiola incana* mit dem nicht neuen Ergebnis vorgelegt, daß bei vegetativen Scheiteln und Infloreszenzscheiteln die großzelligen Zentralmutterzellen vom kleinzelligen Flankenmeristem umgeben sind, welche sich durch hohen Protein, RNA- und DNA-Gehalt auszeichnen, wobei der Blütenstengel insgesamt dem Flankenmeristem entspricht. Die ersten beiden sind offene, undeterminierte Scheitel, wogegen Blütenscheitel geschlossen, bzw. determiniert sind.

Einen größeren Raum nehmen Untersuchungen über Enzymaktivitäten in den Zonierungsmustern der Scheitelmeristeme ein, welche diese bestätigen. So haben ORR (1984, 1985) und ORR u. SCHONEMANN (1986) Enzymaktivitäten in den Scheitelmeristemen von *Brassica campestris* gemessen und in einer Serie von Publikationen mitgeteilt, daß sie bei Cytochromoxydase, Glucose-6-Phosphat-Dehydrogenase und 6-Phosphatgluconat-Dehydrogenase im Flankenmeristem und beim Übergang zur Inflorenzbildung im Meristemmantel stark zunehmen. Glucose-6-Phosphatase-Aktivität fanden sie bei vegetativen Scheiteln in allen Zonen gleichermaßen, im Infloreszenzscheitel dagegen nur noch im Meristemmantel und dort verstärkt. NOUGAREDE et al. (1985) haben die Rolle von Adenylat-Cyclase und 5-Nucleotidase bei der Aktivierung der ruhenden Cotyledonarknospen bei Erbsen untersucht und gefunden, daß die Enzyme durchweg an den Zelloberflächen einschließlich der Plasmodesmen ruhender und aktiver Scheitel nachweisbar sind und somit die Aktivierung der Knospen nicht primär vom Auftreten dieser Enzyme ausgelöst werden kann. Bei demselben Objekt fand die Arbeitsgruppe SCHATT et al. (1985) im ruhenden Samen im Wurzelscheitel Zellen in G_1 und G_2, im Sproßscheitel dagegen nur Zellen in G_1. Entsprechend traten bei der Keimung die ersten Zellteilungen bei Wurzelspitzen schon nach 36 Stunden ein, während sie im Sproßscheitel erst nach 48 Stunden zu beobachten waren. Daraus schliessen die Autoren auf verschiedene Hemmsysteme in Sproß- und Wurzelspitzen desselben Embryos. Beim Sproßscheitelmeristem fanden REMBUR u. NOUGAREDE (1987) im Fall von *Chrysanthemum* in den großkernigen Zellen der Initialzone ("axial zone") ein Maximum an Dekondensation der DNA in der G_1-Phase, womit möglicherweise eine verstärkte Transkription bei verminderter Replikation angezeigt wird. Initialzone und organogenetische Zone des vegetativen Scheitelmeristems besitzen Populationen nicht zyklierender und zyklierender Zellen, wobei die durchschnittliche Länge des Zellzyklus in der Initialzone größer ist (GONTHIER et al. 1985). Beim Übergang zur Infloreszenzbildung nimmt die Länge des Zellzyklus durchschnittlich in beiden Zonen ab. Die Zonierung bleibt jedoch ebenso erhalten, wie die Tatsache, daß in beiden Zonen Populationen nicht zyklierender Zellen zu beobachten sind (GONTHIER et al. 1987).

Auch Untersuchungen über die Ultrastruktur der Scheitelmeristeme sind selten. SAWHNEY u. SEKHAR (1985) und SEKHAR und SAWHNEY (1985) fanden bei *Lycopersicon esculentum* über den ganzen Scheitel hinweg eine durch elektronendichtes Material auffallende äußere Zellage mit großen Vakuolen und sich differenzierenden Plastiden. Die Zellen dieser Lage zeigten auch keineswegs selten auftretende Periklinalteilungen. Die inneren Zellen des Meristems wiesen keine bemerkenswerte Zonierung auf. Auch BERGGREN (1984, 1985) fand bei *Salix*-Knospen in der äußeren Zellschicht einige Periklinalteilungen. Im übrigen besaßen aktive Knospen eine gut ausgebildete Zonierung, keine Reservestoffe und viel rauhes ER, die ruhenden Winterknospen hingegen nur schwach ausgebildete Zonierung und in den Zellen Reservestoffe, Gerbstoffe, Phytoferritin, Ca-Oxalat, autophage Vakuolen und viel glattes ER.

PULAWSKA (1986) fand auch bei *Actinidia arguta* in der äußersten Tunicalage Periklinalteilungen, niemals jedoch in den vier Initialen, welche diese Schicht erzeugen. Sie beschreibt Ursprung und Wachstum dieser Schicht im Hinblick auf eine mögliche Chimärenbildung. Eine knapp gehaltene und klar gefaßte Übersicht über den derzeitigen Stand der Chimärenforschung gibt ein Buch von TILNEY-BASSET (1986). CUNNINGHAME u. LYNDON (1986) zeigten, daß bei *Pisum sativum* Periklinalteilungen im Zusammenhang mit der Blattbildung etwa 1/2 Plastochron vor der entsprechenden Formveränderungen des Scheitels bei *Silene coeli*-rosa sogar 2 Plastochrone vorher zu beobachten seien. Sie stehen damit Vermutungen GREEN's (vgl. Progr. Bot. 47: 33) kritisch gegenüber. In ihren Diagrammen sind allerdings auch die in Restmeristemen gehäuften Periklinalteilungen einbezogen, die in erster Linie mit der Leitgewebedifferenzierung zu tun haben. Schwer verständlich erscheinen die Schlüsse, die CHRISTIANSON (1986) aus der Entwicklungsgeschichte "semigametischer" chimärischer Embryonen von *Gossypium* gezogen hat. Hier ist zu fragen, wie weit die Definition des "Proembryos" (= ontogenetischer Abschnitt von der Zygote bis zur Ausbildung Protoderms) Sinn hat, und wieweit Differenzierung von Blättern und Scheitelmeristemen nebeneinander aus Zellen des Proembryos erfolgt, wo doch Blattbildung eine Funktion des Sproßscheitelmeristems ist (vgl. Prog. Bot. 47: 40). 1984 hatten

KLEKOWSKI u. KAZARINOVA-FUKSHANSKY (Prog. Bot. 47: 33) errechnet, daß bei geschichteten Scheitelmeristemen, wie sie bei der Mehrzahl der Angiospermen und einigen Gymnospermen vorliegen, die Möglichkeit zur Selektion mutierter Zellen gegeben ist. KLEKOWSKI et al. (1985) setzten die Diskussion über derartige Selektionsmechanismen anhand weiterer Modellrechnungen fort. RUTH et al. (1985) fanden am Beispiel von *Juniperus*-Chimären, daß weißzellige Sektoren einen Selektionsvorteil gegenüber grünen besitzen. Hat das noch Sinn, wo doch in natürlichen Populationen Chimären nicht einmal adäquat im Sinne von BROOKS u. WILEY (1986) sind? So großartig das Mittel der Chimärenforschung ist, Klarheit über den Verschiebebahnhof für die Meristemzellen im Sproßscheitel zu erhalten, so kommen doch bei ihrer Perspektive die Prozesse der primären Morphogenese im "Phragmoblastem" gänzlich aus dem Blick.

Mit dem Scheitelmeristem der Kurztriebe von *Pinus longaeva* haben sich EWERS u. SCHMID (1985) befaßt. Sie fanden an der Langtriebbasis nur abgestorbene, verkorkte Meristeme, weiter aufwärts lebende und am oberen Ende schließlich solche, die nach der Bildung der Nadelblätter noch Schuppenblätter ausgebildet hatten und zu den Langtrieben an der Spitze des Triebabschnittes überleiten. Die rasterelektronenmikroskopische Untersuchung der Sproßscheitel von *Vitis vinifera* durch POSLUSZNY u. GERRATH (1986) hat gegenüber früheren Untersuchungen zu keinen neuen Gesichtspunkten im Streit über monopodialen oder sympodialen Bau der Sproß- und Rankensysteme geführt.

Die Frage, ob die *Marattiales* Sproßscheitelzellen besitzen, ist umstritten. IMAICHI (1986) bejaht die Frage bei *Angiopteris lygodiifolia*, deren junge Sprosse klar eine dreischneidige Scheitelzelle mit NOMARSKI-Optik erkennen lassen. Die abgebildeten älteren Exemplare lassen Zweifel an der Interpretation von IMAICHI zu. 1984 und 1985 hat sie die Blattbildung und Verzweigung bei *Dennstaedtia scabra* und *Histiopteris incisa* behandelt. Von Interesse ist die Beobachtung, daß die Blattspur von *Histiopteris* (bei bolivianischen Herkünften im Gegensatz zu japanischen) ohne jegliche Blattlücke als geschlossene Röhre aus dem Bündelrohr des Rhizoms auszweigt. IMAICHI stößt im übrigen auf erhebliche Schwierigkeiten bei der Abgrenzung von Dichotomie und phyllogener Verzweigung, was bei der prinzipiellen Unabgrenzbarkeit von Blatt und Rhizom zu erwarten ist. Man kann ihr nur beipflichten, daß viel zu wenig Farne morphologisch wirklich bekannt sind. Es liegt im Zuge der Zeit, daß beispielsweise in dem umfangreichen Symposiumsband "Biology of Pteridophytes" (DYER u. PAGE 1985) auf vergleichend morphologische Arbeiten völlig verzichtet wird. SEILHEAN u. MICHAUX-FERRIERE (1985) haben die klar zonierten Scheitelmeristeme der *Nephrolepis*-Stolonen untersucht und gezeigt, daß der Zellzyklus in der großzelligen Initialzone schneller abläuft, als im kleinzelligen Flankenmeristem. Auch die Scheitelzelle, die histologische Merkmale starker Differenzierung besitzt, hat einen kurzen Zellzyklus.

Beim Vergleich von vegetativen und reproduktiven Scheitelmeristemen der heterophyadischen *Equisetum*-Arten sieht HAUKE (1985) parallele Entwicklungen mit den Angiospermen. Der Übergang erfolgt bei maximaler Erstarkung. Während die Blattwülste durch schmale Ringwülste angelegt werden, werden die Sporangiophore durch einzelne domförmig gestaltete Meristeme angelegt, wobei der Meristemmantel des breiten Kegels ähnlich mancher Infloreszenzbildung bei den Angiospermen aufgebraucht wird.

b) Blattrandmeristeme

Daß die Blattentwicklung wirklich von allem Anfang an von einem mehr oder weniger weit das Sproßscheitelmeristem umgreifenden Randmeristem ausgeht, wird durch eine Reihe sehr gründlicher Untersuchungen bestätigt. Besonders eindrucksvoll sind die Arbeiten von POETHIG u. SUSSEX (1985a,b) über die Blattentwicklung beim Tabak. Ihre "klonalen Analysen" (vgl. Progr. Bot. 47: 34) haben gezeigt, daß die Blattanlagen aus mehr als 100 Zellen bestehen, die halbmondförmig den Scheitel umgeben und deren Züge im späteren Blattrand enden. Damit werden ältere Darstellungen von DULIEU (1968) im wesentlichen bestätigt. Wenn hier vom Randmeristem nicht die Rede ist, so deshalb, weil die Analyse auf die Entwicklung des ganzen Blattes abgestellt ist und dabei die Region nicht besonders hervortritt. Dennoch ist gerade in den frühesten Stadien allein durch Anordnung der Ausgangszellen ein "Randmeristem" gegeben. Wenn man damit die Form des Blattmeristemgewebes beschreiben will.

MERRILL (1986a-c) hat sich in einer Serie von Publikationen eingehend mit der Entwicklung der Heterophyllie bei Sämlingen von *Fraxinus pennsylvanica* auseinandergesetzt. Die noch voll meristematischen Blattanlagen bilden schon früh ein umlaufendes kleinzelliges Blattrandmeristem, indem eine erhöhte Mitoserate nachgewiesen wird. Form und Größenunterschiede in vergleichbaren Stadien in der Anlagefolge bringen nach Auffassung MERRILL's eine heteroblastische Entwicklung zum Ausdruck, die die Ableitung der ungefiederten Primärblätter von gefiederten ausschließen. Die Argumentation überzeugt jedoch nicht, weil die beobachteten Unterschiede nicht qualitativer Natur sind. Kritisch ist auch anzumerken, daß Ganzrandigkeit der Primärblätter etwas anderes ist, als Ganzrandigkeit bei Blättern erstarkter Pflanzen, was typologisch arbeitenden Autoren wie GOEBEL, TROLL u.a. durchaus klar war. Die Kritik MERRILL's an der typologischen Methode sticht daher nicht, denn es ist ein Unterschied, ob eine Blattanlage mangels Masse beim Sämling das zur Fiederbildung notwendige Ausmaß nicht erreichen kann, oder ob die Fiederbildung bei Blattanlagen voll erstarkter Pflanzen unterdrückt wird. Eine Hemmung des Randmeristems zwischen den Fiederanlagen konnte MERRILL übrigens nicht nachweisen. Nach PERIASAMY u. MURUGANATHAN (1985) soll das erste Fiederpaar des basipetalen digitaten Blattes von *Tabebuia pentaphylla* bereits gebildet werden, bevor die Blattanlage ein Randmeristem besitzt.

Das Problem der Unifacilität bei Blättern ist in verschiedenen Arbeiten diskutiert worden, wobei sich die Auffassung mehr und mehr durchsetzt, daß es echte Unifacialität bei Blättern nicht gibt. In einer sehr ausführlichen vergleichenden Studie über die Entwicklung flacher und runder *Allium*-Blätter hat KRAEHENBUEHL schon 1983 gezeigt, daß die jungen Blattanlagen durchgehende Randmeristeme besitzen. So müssen auch die schlauchförmigen Spreiten von *Allium cepa* als äquifacial aufgefaßt werden und können nicht mit unifacialen Vorläuferspitzen homolog gesetzt werden. Anders sieht KRAEHENBUEHL jedoch in den Cotyledonen der *Allium*-Arten unifaciale Organe, was er mit ihrer terminalen Stellung im Rahmen der Continuum-Lehre SATTLER's (1986) in Zusammenhang bringt. M. E. wäre aber Unifacialität der Keimblätter gegenüber Laubblättern eine neue Qualität und daher mit der Continuum-Lehre nicht vereinbar.

In Verbindung mit der Bildung sog. gehörnter Schuppenblätter bei Bracteen, Hochblättern und Perianthgliedern, die zwar schon in die reproduktive Region des Pflanzenkörpers gehören, aber im Zusammenhang mit dem Blattrandmeristem hier Erwähnung finden sollen, hat KUNZE 1984 beobachtet, daß sie allesamt aus einfachen, mit Randmeristemen umgebenen Blattanlagen hervorgehen. Dabei können je nach Objekt Querzonen am oberen Ende des Unterblattes, am Spreitengrund oder am Grund der Vorläuferspitze gebildet werden. Bei *Pilea* entsprechen die Schuppen den Medianstipeln der Laubblätter, ihr Anhängsel somit dem gesamten Oberblatt. Bei den Übergangsformen zwischen Schuppen- und Blütenorganen wird die ganze ungegliederte Blattanlage zur Schuppe. Bei *Passiflora*, *Impatiens* und *Begonia* entspricht das Anhängsel der Vorläuferspitze, die

Schuppe mithin dem gesamten übrigen Blatt. Auch VISHNU et al. (1984) haben im Fall der sog. unifacialen Vorläuferspitzen bei Agaven keine echte Unifacialität konstatiert, sondern die Rundung auf die Tätigkeit eines frühzeitig arbeitenden Ventralmeristems zurückgeführt. Ein weiterer Beitrag zur Organisation von Blattanlagen stammt von HOPPE (1985), welcher die Entwicklung der den interfoliaren Stipeln der Involucralblätter homologen Drüsen der Euphorbiaceencyathien untersucht hat. Die Drüsen entstehen aus dem Spitzenbereich der Stipel, und zwar dem noch urmeristematischen Rand, an dem sich der selbst urmeristematische und ringförmige Drüsenrand ventral und dorsal ausbildet. Die Drüsenanhänge verschiedener Form verdanken ihren Ursprung einem neugebildeten dorsalen, einem Blattrandmeristem ähnlichen Meristem, das aber durch Reembryonalisierung angelegt wird. Dieses wird mit einer Querzone verglichen und die Drüsenanhänge deshalb als "hypopeltate Strukturen" bezeichnet. In der sehr anregenden Diskussion scheint mir indessen der Begriff des Blattrandmeristems überstrapaziert. Zu bedenken ist zunächst, daß es sich nicht um Blätter, sondern die Ränder von Stipeln handelt, die allerdings aus dem primären Blattanlagenrand direkt hervorgehen. Drüsen sind histologische Strukturen, die gleichwohl massiv angelegt, während der Entwicklung an ihren Rändern wie ein Blattrandmeristem aussehen können. Wesentlich für eine kritische Morphologie ist gerade, daß man nicht alles, was optisch gleich ist, in den selben Topf steckt. Das neugebildete System der Drüsenanhänge könnte nur dann als hypopeltater Auswuchs gelten, wenn es durch Inkorporation vom echten Randmeristem her entstehen würde. Daß den Blattrandmeristemen ähnliche Meristeme möglicherweise auch deren Fähigkeiten übernehmen könnten, wäre ein gutes Beispiel für Homöosis im Sinne SATTLER's s.o. Ähnliches gilt auch für die vom üblichen abweichenden Fälle von Phyllodien bei *Acacia baueri*, *A. verticillata*, *A. longipedunculata* und *A. conferta*, die RUTISHAUSER (1986) (vgl. auch RUTISHAUSER u. SATTLER 1986; KAPLAN 1984) behandelt hat. Auch hier kann der Begriff der Homöosis mit Gewinn eingesetzt werden, womit zugleich geklärt wird, daß es sich um Neubildungen handelt, die sich vom Typus nicht nach dem Prinzip von den variablen Proportionen ableiten lassen, sondern eine Typogenese bedeuten, in deren Verlauf zusätzliche Bauelemente entstehen (vgl. auch KUNZE 1985).

JEUNE (1986) (vgl. Progr. Bot. 47: 35) entwickelte für die von ihm schon früher bearbeiteten basipetalen Fiederblätter ein mathematisches Entwicklungsmodell. Es läßt Wachstumskomponenten, die in der Phylogenese in den Entwicklungstypus eingegangen sind, z.B. ternate Fiederbildung, interkalare Wachstumszonen und deren Verschiebung außer Betracht, deren Resultante der heute konstatierte Entwicklungstypus ist. Das Modell ist eine Beschreibung des beobachteten Entwicklungsablaufs. Sinnvoller wäre, ein Modell nicht von den registrierten Wachstumsvorgängen des Objektes aus, sondern aus einer phylogenetischen Theorie heraus zu entwickeln, wobei die Theorie durch den Vergleich mit der Natur falsifiziert oder verifiziert werden könnte.

Die "offenen Blätter" der *Guarea*-Arten *(Meliaceae)* sind bei den Angiospermen im Gegensatz zu den Farnen eine seltene Ausnahme und alles andere als ursprünglich. STEINGRAEBER u. FISHER (1986), sowie FISHER (1986a) haben gezeigt, daß es sich um akropetale Fiederblätter handelt, bei denen die Blattspitze meristematisch bleibt und schubweise über Jahre hinweg neue Fiederpaare erzeugt. Die Fiedern reifen indessen schneller aus, als neue Fiedern gebildet werden, so daß sich die Spitze schließlich erschöpft und wie z.B. bei *Nephrolepis* abstirbt, ohne eine Endfieder zu bilden. In der Rhachis tritt sekundäres Dickenwachstum auf. Damit entwickelt sich dieses Blatt analog zu den begrenzt wachsenden Kurztrieben mancher *Phyllanthus*-Arten, über die NOZERAN et al. (1984) eine Übersicht gegeben haben.

Ein origineller Versuch von TURLIER (1984) ist im Fall von *Cabomba aquatica* ein fünfseitiges Blattprimordium anzunehmen. Die Fiedern des Wasserblattes sollen aus dem Spitzenbereich und den beiden Flanken, die peltate Spreite des Luftblattes aus den Flanken, Dorsal- und Ventralseite entstehen.

GOULD u. CUTTER (1986) haben die Studie von YOUNG (vgl. Progr. Bot. 47: 35) an Mutanten von *Pisum*-Blättern fortgesetzt und einige abweichende Formen an in vitro kultivierten Blättern beschrieben. Die schon geäußerte Kritik am beschreibenden Begriffsystem gilt auch hier.

Präzise Studien über die Entstehung von Blattprimordien und ihre Randmeristeme sind selten. Zu nennen lohnt sich nur die Arbeit von IMAICHI u. NISHIDA (1986) über die Entwicklung des fertilen 3-dimensionalen Blattes von *Botrychium ternatum*. Sie finden Ähnlichkeiten mit manchen hoch abgeleiteten Angiospermenblättern, etwa bei der Scheidenbildung, fehlende Einrollung an der Spitze, massive Randmeristeme, Entwicklung von Scheitelzellen an den Fiederspitzen. Die Fiederanlagen haben die Tendenz, langzylindrisch zu bleiben. Die fertile Fieder wird auf der Ventralseite des Blattprimordiums angelegt. Sie hat keine Beziehung zum ursprünglichen Blattrandmeristem, weshalb sie die Blattkonstruktion mit dem "Phyllophor" coenopteroider Farne in Zusammenhang bringen. Die Isolierung und Kultivierung jüngster Blattanlagen von *Matteuccia struthiopteris* aus dem Scheitelmeristem hat gezeigt, daß die Determination zu Blättern schon sehr früh erfolgt (VON ADERKAS u. HICKS 1985). Bereits P_1 entwickelt sich in der Regel anders als bei *Dryopteris* als Blatt weiter. Des öfteren dagegen wurde Sproß- und Wurzelbildung an der Basis explantierter Blattanlagen beobachtet.

c) Wurzelscheitelmeristeme

Unter der Herausgeberschaft JACKSON's (1986) entstand ein wohlgelungenes Buch, das in kompakter Form über einen Großteil der Adventivwurzelbildung und ihre exogenen wie endogenen Ursachen informiert. Insbesondere werden darin die Aspekte der Wurzelmeristemanlegung von PETERSON u. PETERSON für die Bildung von Seitenwurzeln in ihren Mutterwurzeln dargestellt, während LOVELL u. WHITE in einer kurzen und einprägsamen Übersicht über die Anlageorte, ihre histologisch frühesten Veränderungen und die ersten Stadien der sproßbürtigen Wurzeln berichten.

ROBERTS et al. (1985) haben Wurzelspitzen von *Halophila ovalis* untersucht und sich dabei besonders mit der Bildung des Lakunensystems befaßt. Der Vergleich der Scheitelmeristeme von Seitenwurzeln bei *Pistia*, *Hydrocharis* und *Eichhornia* zeigt, daß sie aus Endodermis und Perizykel der Mutterwurzel entstehen (CLOWES 1985). Die Rhizodermis geht bei *Pistia* und *Hydrocharis* aus der Endodermis, bzw. dem Perizykel hervor und bildet eine geschlossene Meristemschicht, während sie bei *Eichhornia* zusammen mit der Rinde, ein dem Perizykel entstammendes Meristem bildet. In Gefäßen gezogene Sämlinge von *Picea glauca* hatten ein weniger differenziertes Wurzelsystem als unter natürlichen Bedingungen wachsende Pflanzen (JOHNSON-FLANAGAN u. OWENS 1985). Wurzelhaare wurden nur von kleineren Seitenwurzeln gebildet. Bei den rhythmisch wachsenden Langwurzeln geht das Meristem nach Bildung einer metacutinisierten Zellschicht, die proximal brückenartig in die Endodermis übergeht, vorübergehend in Ruhe (braune Wurzeln) und kehrt später, das Abschlußgewebe durchbrechend, wieder zum Wachstum zurück (weiße Wurzeln). Hohe AS-D-Esterase Aktivität in Zellen, die das Ruhezentrum in *Pisum sativum* Wurzelspitzen proximal ummanteln, betrachten GAHAN u. RANA (1985) als Indikator für die Determination zu Stelengewebe. Die Frage, ob die Determination mit einer Mitose verbunden ist, wird verneint.

Simulationsmodelle zum Wachstum von Zellreihen in Wurzelmeristemen wurden von BERTAUD et al. (1986), BERTAUD u. GANDAR (1986) sowie von

HARTE u. LINDEMAYER (1986) publiziert. Bei diesen wird besonders auf die mitotische Aktivität und die Nachbarschaftseinflüsse auf den Zellzyklus geachtet.

Die zur Zeit rege geführte Diskussion über das Orientierungsvermögen der Wurzeln im Boden hat auch zur Untersuchung von Wurzelspitzen auf die damit in Zusammenhang gebrachten Strukturen geführt. In der Hypothese von AUDUS (1979) und neuerdings PICKARD (1985) sind dies im Rahmen der alten Statolithen-Theorie die Amyloplasten, die Druck auf die vielfach beobachteten ER-Geflechte in den zentralen Columellazellen der Wurzelhaube ausüben sollen. FJELL (1985) fand diese Strukturen in der Calyptra der Seitenwurzeln von *Salix viminalis* und stellte fest, daß nur die Amyloplasten der Columella rasch sedimentieren und dabei nur dann, wenn die Wurzel senkrecht steht, auf dem ER-Geflecht liegen, welches sich polar über der distalen Zellwand befindet. In anderen Calyptra-Portionen sedimentieren die dort ebenfalls vorhandenen Amyloplasten nicht. Nach PICKARD's (1985) Vorstellung, soll durch den Druck der Amyloplasten auf das ER Calcium in das Plasma gelangen. Dieses könnte Phytohormone, wie etwa den Quertransport von IAA beeinflussen (MULKEY u. VAUGHAN 1985). BARLOW u. RATHFELDER (1985) haben Wurzelspitzen und Krümmungsvorgänge in ihrer Nähe vermessen. Wie schon KUTSCHERA (1983) vermutet hatte, fanden jedoch MOORE und seine Mitarbeiter in zahlreichen Feinstrukturuntersuchungen keine Hinweise für die Statolithen-Theorie. In keinem Fall konnte MOORE (1984, 1985a-d, 1986a,b) und MOORE u. McLELEN (1985a) in Untersuchungen an *Zea mays*, *Allium cepa*, *Rhizinus communis*, *Helianthus annuus* und *Phaseolus vulgaris* eine Berührung der Amyloplasten mit dem ER entdecken. Auch andere Beobachtungen oder Versuche führten nicht zum Ziel (MOORE 1985e,f; MOORE u. McLELEN 1985b), doch erwiesen sich Größe der Wurzelspitzen, ihre Schleimproduktion, bei *Allium cepa* auch die Beobachtung, daß Wurzeln im Wasser nicht, in Luft jedoch auf Schwerkraft reagieren, als wichtig (MOORE 1985e, MOORE u. FONDREN 1986). Schließlich entwickelte MOORE (1985g, MOORE u. EVANS 1986) aufgrund der Beobachtung, daß bei der Verschleimung der Calyptrazellen erhebliche Mengen an Calcium in den Haubenschleim gegangen und hier unter dem Einfluß der Schwerkraft an der Calyptra-Unterseite angereichert werden, seine neue Hypothese, nach welcher der Ca-Gradient im Wurzelhaubenschleim einen steuernden Einfluß auf den IAA-Transport nehmen soll. Amyloplasten und ER in den Columella-Zellen hätten danach keine sensorischen Funktionen, sondern wären lediglich mit der Schleimbildung der Calyptra in Zusammenhang zu bringen.

Die Wurzelspitzen von *Regnellidium diphyllum*, insbesondere deren Calyptra wurden von EASTMAN u. PETERSON (1985) einer sorgfältigen histologischen und histochemischen Untersuchung unterworfen. KURTH u. GIFFORD (1985) untersuchten den DNA-Gehalt der Scheitelzellen bei *Azolla filiculoides* und fanden ihn bei jungen Wurzelanlagen am höchsten, während er später mit der Abnahme der Teilungsrate ebenfalls abnimmt. KARRFALT konnte 1984 Jungpflanzen von *Isoetes andicola (= Stylites andicola)* auf ihr Wurzelbildungsmeristem hin untersuchen und die Vermutungen von RAUH u. FALK (Fortschr. Bot. 22: 13-15) bestätigen, nach denen es bei *Isoetes* und *Stylites* übereinstimmt. Im Gegensatz zu diesen rechtfertigt er jedoch mit diesem Befund den Einzug der Gattung *Stylites* durch GOMEZ (1980).

d) Anlegung des Leitgewebesystems

Wie schon in früheren Beiträgen, sollen hier nur Arbeiten referiert werden, welche sich mit den Beziehungen zwischen der Morphogenese des Vegetationskörpers und der Anlegung des Leitgewebesystems befassen. Weitere Arbeiten von LARSON (1985, 1986a,b) beschreiben mit der gewohnten Präzision die Leitgewebesysteme der Araliacee *Polyscias quilfoylei* und der Fiederblätter von *Fraxinus pennsylvanica*. Die multilakunäre *Polyscias* enthält im Stamm zwischen den sich entwickelnden Blattspuren von 5 Fiederblättern weitere Stränge, zu denen noch keine Blattanlagen nachweisbar sind. Sie werden funktionell wie Hilfsbündel ("subsidiary bundles") eingeschätzt. Auch die Entwicklung der Leitbündelsysteme der

Fiederblätter werden entwicklungsgeschichtlich bearbeitet, doch fehlt hierbei ein präziser Bezug zur primären Morphogenese, wie sie z.B. in den schon erwähnten Arbeiten MERRILL's (1986a-c) dargestellt worden ist. Für SUGIYAMA (1984) ist die Blattform bei Lardizabalaceen, die pinnat, palmat oder ternat sein kann, ein unwesentliches Merkmal, weil die Leitbündelsysteme bei allen Formen im Prinzip gleich seien. Den Zusammenhang zwischen Blattbildung und Leitgewebedifferenzierung hat besonders HECKER (1985) herausgestellt, welche sich mit der Entwicklung des Leitbündelsystems in den Keimlingen von *Helianthus*, *Ricinus* und *Tilia* beschäftigt hat. Die Unterschiede zwischen diesen Keimlingen sind nur quantitativer Natur. Alle unterscheidbaren Stränge im Achsenkörper bilden zusammen ein monozentrisches Sproßbündel, in welchem sich Teilstränge unter dem Einfluß der Blätter differenzieren. Dem gegenüber ist das Wurzelbündel polyzentrisch. Die polare Stele des Embryos gibt die Verschiedenheiten von Sproß und Wurzel erst in genügendem Abstand von der Übergangszone klar zu erkennen. In einem Überblick über die Musterbildung im Leitgewebe von Achsenkörpern hat SACHS (1985) auf den Zusammenhang mit dem polaren Auxintransport verwiesen. Dieser geht im Gewebe den ersten sichtbaren Anzeichen der Leitgewebedifferenzierung voraus. SACHS beobachtet allerdings die Xylemdifferenzierung, die der Phloemdifferenzierung erst folgt. Die Vorstellungen über die Entstehung eines Auxintransportsystems sind indessen spekulativ. SACHS hat an Kanalisierungseffekte gedacht, die in der Umgebung der Kanäle gleiche Differenzierung hemmen. So resultiert ein Muster von Elementarsträngen, das an die Hilfsbündelmuster im Sinne LARSON's erinnert. Hier mag eine Studie von MAGENDANS (1985) neue Gesichtspunkte beisteuern, der die Differenzierung der Nervenenden im Blatt von *Hedera helix* untersucht hat. Während bei der primären Morphogenese in massiven Procambiumsträngen unter dem Einfluß des aktiven Baustofftransportes Phloem zuerst gebildet wird, erscheint in den letzten Auszweigungen der Blattnerven zuerst Xylem, so daß möglicherweise ein Zusammenhang mit der Transpiration besteht. Dabei wird auch für die frühzeitige Anlegung von Interzellularen und ersten Stomata hingewiesen, welche für die Musterbildung entscheidend sein können.

Die Einheit des sich differenzierenden Leitbündelsystems in der wachsenden Sproßspitze wird in einer sehr anregenden Arbeit von MEICENHEIMER (1986) hervorgehoben, in der er die Entwicklung des Bündelsystems von *Linum usitatissimum* mit neuen Mitteln beschreibt. In der Sproßspitze würde das Leitgewebe als zylindrisch angeordnetes Restmeristem/ Procambium (als zwei kontinuierlich verbundene Entwicklungsphasen seien beide nicht zu trennen) gegenüber dem Grundgewebe von Rinde und Mark "determiniert". In diesem Zylinder werden in Zusammenhang mit der Blattbildung die Stränge durch die Differenzierung der Blattlücken in Verbindung mit dem interfasciculären Parenchym ("primary rays") abgegrenzt. Somit müssen auch die Blattstellungsregeln auf die Bündelanordnung zutreffen. Leider wird der Fortgang der Entwicklung nicht mehr berücksichtigt, weshalb die Zweckmäßigkeit der entwickelten Terminologie nicht an der Gesamtentwicklung getestet werden kann. So bleibt es bei Inkonsequenzen im Gebrauch der Begriffe Determination und Differenzierung. Schon ein polar wachsender Sproß ist differenziert. Das Terminalmeristem ist ein undeterminiertes, auf Wachstum differenziertes Gewebe im Gegensatz zum determinierten, auf verschiedene Dauerfunktionen differenzierten Dauergewebe. Bei der Histogenese des Sprosses wird Mark und Rinde zunächst determiniert, während das Restmeristem/ Prokambium zwar differenziert, aber nicht determiniert wird. Bei der weiteren Entwicklung wird wiederum das Blattlückenparenchym determiniert. Das Prokambium differenziert sich weiter zu Kambium, welches seinerseits undeterminiert bleibt. Seine Differenzierung in Fusiform- und Strahleninitialen könnte dem undeterminierten Kambium vom determinierten Gewebe her aufgezwungen werden. Determiniert werden außen das Muster der Phloemstränge wie innen das Xylem ebenso und gleichzeitig mit dem interfasziculären Parenchym, dessen kambiale Partien jedoch undeterminiert bleiben. Interessant ist auch die Beobachtung, daß die Musterbildung möglicherweise mit der Determination von Parenchym zu tun hat, wobei die in der Zielsetzung und von ihren Objekten her so verschiedenen Arbeiten von MAGENDANS u. MEICENHEIMER zu vergleichbaren Ergebnissen geführt haben, die im Gegensatz zu SACH's Auffassung stehen.

Den Zusammenhang zwischen Blattstellung und Leitbündelanordnung im Achsenkörper hat KIRCHOFF (1984) in einer vergleichenden Studie hervorgehoben, wobei er nachweist, daß von wenigen Ausnahmen abgesehen, die Anzahl der Blattspursympodien im Achsenkörper mit dem Nenner des phyllotaktischen Bruches übereinstimmt. JAIN (1984) hat die Differenzierung des Bündelsystems bei *Adenocalymna* und MIALOUNDAMA u. PAULET (1986) haben dasselbe bei *Gnetum africanum* beschrieben. Das anomale sekundäre Dickenwachstum von *Amaranthus* und *Achyranthes* soll sich nach BHAMBIE u. SHARMA (1985) vom Perizykel der Wurzel ausgehend basipetal und akropetal entwickeln, wobei der gesamte Perizykel in ein zusätzliches Kambium verwandelt wird. Nach dem ersten Perizykelkambium werden nach und nach gleichartige Kambien in der innersten Rinde angelegt. Dieses würde gegen die übliche Darstellung sprechen, nach der ein einseitig nach innen arbeitendes Kambium alle weiteren erzeugt. Nicht wirkliche Klarheit schafft die Studie von STIEBER u. BERINGER (1984) über die Organisation des sekundären Dickenwachstums in der Zuckerrübe. Die Bildung der akzessorischen Kambien wird mit der Blattbildung in Zusammenhang gebracht. Gewisse Parallelen zu Monokotylen sind unverkennbar, doch mangelt es weiterhin an einer wirklich klaren entwicklungsgeschichtlichen Studie. Bei den Monokotylen mit sekundärem Dickenwachstum ist das durch die Arbeiten DEMASON's und ihren Mitarbeitern (Progr. Bot. 47: 37) geschehen, denen DEMASON u. WILSON (1985) eine weitere Studie über *Cordyline terminalis* mit übereinstimmenden Ergebnissen haben folgen lassen. Im primären Meristemmantel angelegte Blattspurbündel, bilden abwärts mit dem amphivasalen sekundären Bündeln eine entwicklungsgeschichtliche Einheit. Prinzipiell scheinen mit diesen Befunden auch die zu sekundärer Verdikkung befähigten Iridaceen übereinzustimmen (RUDALL 1984). Nur mehr verhältnismäßig oberflächlich wird neben der Rhizomorphologie auch das Bündelsystem von *Smilacina racemosa* dargestellt (LA FRANKIE 1985a), indem sich ein aus kräftigen Bündeln zusammengesetztes zentrales und ein zylindrisches, aus dünnen Bündeln bestehendes corticales System unterscheiden lassen. Die Sympodialglieder sind auf breiter Basis mit den Bündelsystemen der jeweiligen Muttersprosse verbunden. Eine Vorstellung von der Entwicklung wird jedoch nicht gegeben. Eine vorläufige Übersicht über die Bündelsysteme der Gattung *Philodendron* haben FRENCH u. TOMLINSON (1984) in Fortsetzung ihrer vergleichenden Arbeiten über die Araceenbündelsysteme gegeben.

Vernachlässigt wurden bisher beim Studium der Leitbündelsysteme die Anschlußverhältnisse der Wurzeln an die Sproßstele. CHAU (1985) hat diese Kenntnislücke angesprochen und versucht, in einer vergleichenden Studie diese Verhältnisse für die Ophioglossaceen und einige wenige leptosporangiate Farne zu klären. Er fand dabei zwei Anschlußtypen, den "graduellen" und den "abrupten". Die Wahl einer sehr isoliert stehenden Pflanzengruppe, der Mangel eines morphologischen Konzeptes und das Fehlen entwicklungsgeschichtlicher Untersuchungen lassen jedoch Zweifel an der Zweckmäßigkeit solcher Untersuchungen zu.

e) Embryoentwicklung

Eindrucksvoll vermittelt RAGHAVAN's (1986) Buch über den heutigen Stand der Forschung auf dem Gebiet der "Embryogenesis", in das Morphologie und Entwicklungsphysiologie, besonders auch bei experimentell erzeugten somatischen Embryonen, einbezogen worden sind. Wie weit seit früheren Berichten (vgl. Progr. Bot. 40: 40, 47: 39) auf diesem Gebiet Fortschritte erzielt worden sind, bleibt jedoch fraglich. Solange sich nicht die Einsicht durchsetzt, daß schon eine Zygote, genauso wie eine somatische Zelle, die einen Embryo bildet, bereits ein polarer Sproß in seiner kleinsten Form ist, bleibt Zelldifferenzierung im wachsenden Embryo unverständlich und solange wird auch "Embryogenesis" eine wachsende Datensammlung bleiben, die sich aus blinden Experimenten speist. WILLIAMS u. MAHESWARAN (1986) kommen zu dem Schluß, daß es gleichgültig sei, ob somatische Embryonen aus einer primär oder sekundär (an einem nicht weiterentwickelten Embryo) aus einer Zelle oder einer Gruppe von Zellen entstehen. Dabei sei noch nicht klar, ob die in jedem Fall zu beobachtende Isolierung des Embryos durch Unterbrechnung der Plasmodesmen zu den Nachbarzellen und Ausbildung einer Cuticula erster Schritt oder ein Folgeschritt der Embryobildung ist. Isolierung und Polarisierung bilden jedenfalls zusammen ein entscheidendes Geschehen, welches eine Zelle oder eine Zellgruppe zum Embryo (= Sproß?) macht. Diese Einheit sprechen auch BRUCK u. WALKER (1985a,b) an, wenn sie bei ihren Schnittexperimenten festgestellt haben, daß die Epidermisbildung ein Einmalereignis sei, und sie nach Beschädigung nicht erneut gebildet werden kann. Auch der Embryo repariert wie der Sproß eine beschädigte Epidermis durch ein Periderm. Schon die Zygote bildet eine Cuticula, die später differenzierte Epidermis übernimmt sie, nicht aber die endogen gebildete Wurzel (vgl. Progr. Bot. 45: 40). Aufgrund von Messungen bei somatischen Embryonen von *Daucus carota* haben SCHIAVONE u. COOKE (1985) vorgeschlagen, in die übliche Entwicklungsreihe Kugelstadium, Herzstadium und Torpedostadium ein Stadium des "verlängerten Embryos" zwischen Kugel- und Herzstadium einzufügen, weil die Hypokotverlängerung deutlich vor der Kotyledonenausgliederung einsetzt.

Die Korrelationen zwischen Sproß und Wurzelpol hat KONDRATIEVA-MELVILLE (1986) an verschiedenen Dikotylen-Beispielen behandelt. Sie sind während der Embryoentwicklung sehr eng, werden aber später entkoppelt. Eine experimentelle Studie zur Bildung von Grenzwurzeln (dieser auf TROLL u. WEBER, vgl. Fortschritte Bot. 12: 33, zurückgehende Ausdruck wird nicht verwendet) im Kragen von *Impatiens balsamina* haben BARKER et al. (1986) publiziert. Sie zeigten, daß der Kragen nach Entfernung regeneriert und weitere Wurzelanlagen am Hypokotyl angelegt werden. Die Embryogenese von *Najas marina* haben VIJAYARAGHAVAN u. KAPOOR (1985) unter morphologischen und histochemischen Gesichtspunkten dargestellt. Der Embryo keimt ohne Ruheperiode bei der Samenreife aus. Histochemische Untersuchungen bei der Embryoentwicklung von *Ottelia*, *Vallisneria* und *Limnophyton*, die zeigten, daß die Köpfchenzellen vor der Cotyledoanlegung keinerlei Unterschiede aufwiesen, ließen SHAH (1983) das Köpfchen als Sproßscheitel deuten, der einen seitlichen Cotyledo bildet, wobei ein wegen seiner Kleinheit scheinbar seitlich liegendes Restmeristem übrig bleibt.

Zwei erwähnenswerte Studien stellen mit REM-Techniken den Grasembryo dar, nämlich VAN LAMMEREN (1986) *Zea mays* und NEGBI (1986) *Avena fatua*. Bei *Zea mays* wird überzeugend die klassische Auffassung bestätigt (vgl. Progr. Bot. 47: 40), nach der Scutellum und Coleoptile eine Einheit bilden. Die Coleoptile wird eindeutig als Wulst vom Scutellum angelegt. Aus verständlichen Gründen läßt diese Methode bei der Wurzelbildung keine eindeutige Klärung zu. NEGBI hat vor allem die Weiterentwicklung des Scutellums bei der Keimung des Windhafers verfolgt. Weil dabei kein Wachstum mehr stattfindet, die Oberflächenzellen aber durch papillenartiges Auswachsen dem zurückweichenden Endosperm folgen, nimmt er die außerordentliche Scutellumlänge von *Avena fatua* als Merkmal, daß dieser Pflanze als Unkraut unter Kultursorten einen Vorteil bringen könnte.

Nur wenige Arbeiten über Farngametophyten, keine wichtigen zur Sporenphytenentwicklung wurden im Berichtszeitraum publiziert. VON ADERKAS u. RAGHAVAN (1985) fanden, daß sich *Schizaea pusilla* nach dem *Vittaria*-Typ nach NAYAR u. KAUR entwickelt. NESTER (1985) hat die Sporenkeimung von *Anemia mexicana* mit dem REM dargestellt. Die erste Zellteilung, die noch bei geschlossener Spore stattfindet, führt zur Rhizoidbildung. CHENOU et al. (1986) erzeugten durch Kältebehandlung abweichende Spindelorientierung und unvollständige Wandbildung in den Zygoten von *Marsilea vestita*. Ein Überblick über die Gametophyten der Ophioglossaceen haben PANT et al. (1984) gegeben.

3. Blattstellung und Blattmorphologie

Abies balsamea ist in ihrer Blattstellung äußerst variabel. Nicht weniger als 12 verschiedene Muster und Übergänge zwischen diesen fanden ZAGORSKA-MAREK (1985). Ein Computermodell für die Erzeugung von Blattstellungsmustern entwickelten SCHWABE u. CLEWER (1984) für den Fall, daß ein polar transportierter Hemmstoff beteiligt wäre. Auch wurde nach mathematischen Methoden gesucht, die Entwicklung mannigfaltiger Blattformen zu erfassen. Durch Vergleich von ca. 120 abgrenzbaren, dreieckigen Elementarfeldern während der Entwicklungszeit und die Bestimmung von Vektoren der Wachstumsgeschwindigkeiten, die über eine wachsende Blattfläche verteilt sind, läßt sich die Entstehung von Formenverschiedenheiten ausgehend von den untereinander noch gleich aussehenden 2 cm langen Spreiten (also lange nach dem Abschluß der primären Morphogenese) einer normalen und einer anormalen Blattform von *Vitis vinifera* beschreiben, wobei beide Methoden zu übereinstimmenden Ergebnissen führen (WOLF et al. 1986). Vorschläge zur Vermessung von Blattflächen unter Berücksichtigung ihrer Form zum Zweck, systematisch relevante Formen zu finden, haben auch DICKINSON et al. (1987) gemacht. Derartige mathematische Beschreibungen tragen indessen nur wenig zum Verständnis der Blattformen bei.

Wieder aufgegriffen hat ELLENBERG (1985) das Problem der "Träufelspitzen" bei Blättern und kommt aufgrund von Geländebeobachtungen wie Laborexperimenten zu dem Schluß, daß der Begriff wenig sinnvoll sei. Er schlägt die Bezeichnung "Vorspitzen" oder "Apiculae" vor und bringt sie mit der schnellen Entwicklung großflächiger Blätter unter feuchtwarmen Bedingungen in Zusammenhang, bei denen besonders Knospenschutzvorrichtungen weitgehend fehlen.

GUEDÈS (1984) hat sich mit dem Problem der Rhachis-Unifacialität bei Umbelliferen-Blättern auseinandergesetzt und fand bei *Conium maculatum* und einigen weiteren Umbelliferen in den Fiederjochen einen Randmeristemverlauf, der den Vorstellungen TROLL's entspricht. Unifacialität sah er entgegen neueren Darstellungen (s.o.) als ein primäres Merkmal an, ohne sich freilich durch das Studium selbst zu informieren.

Der Blattdimorphismus bei *Eichhornia crassipes* wird vermutlich durch das Phytochrom-System gesteuert (RICHARDS u. LEE 1986). Während in dünnen Populationen die Knospen der Pflanzen vom vollen Licht getroffen werden, sinkt das Verhältnis R:FR wegen der Beschattung durch eigene Blätter in dichten Populationen auf 0,28 ab, was dünnere und verlängerte Blattstiele ohne die sonst für die Pflanzen typische Anschwellung im Stiel zur Folge hat. Ihre Untersuchungen über den Blattdimorphismus von *Callitriche heterophylla* haben DESCHAMP u. COOKE (1985, vgl. Progr. Bot. 47: 42) fortgesetzt und gefunden, daß er ebenfalls außengesteuert ist und über den Turgor reguliert wird. Bei gleichen Blattanlagen hält das Randwachstum bei Landblättern länger an.

Eine Übersicht über die ultraviolett reflektierenden Zonen, die bei den meisten Fallenblättern der Insektivoren vorhanden sind, haben JOEL et al. (1985) gegeben. CHIARETTO (1984) hat die Blattorgane der Iridaceen-Gattung *Trimezia* vergleichend morphologisch und anatomisch untersucht. Die drehrunden Spreiten werden entsprechend der Interpretation durch ARBER als Blattstielhomologa und Phyllodien gedeutet. Da die Studie nicht entwicklungsgeschichtlich ist und ein Spreitenrudiment nicht existiert, bleibt eine solche Deutung fragwürdig. In diesem Zusammenhang mag erwähnt werden, daß WALTERS u. BARTHOLOMEW (1984) in einer Untersuchung bei Acacien-Blättern gezeigt haben, daß Phyllodien und Fiederblätter von ihrer Leistung zu urteilen nahezu übereinstimmen. Phyllodien enthalten auf die Oberfläche bezogen mehr Chlorophyll als Fiederblätter, die letzten dagegen mehr Protein und Ribulosediphosphat.

In einer demographischen Arbeit über das Verhalten der Blattanlagen von *Matteuccia struthiopteris* von VON ADERKAS u. GREEN (1986) wurden Entwicklungsdauer, Differenzierung der Blätter und Laubblätter, Sporophylle und Kataphylle untersucht. Die Differenzierung der vieljährigen Blattanlagen erfolgt erst mit ihrem Austrieb, weil nach Entblätterung Organe, die sonst Sporophylle oder Kataphylle gebildet hätten, ersatzweise Laubblätter bilden. Der Ersatz geernteter Blätter erfolgt sofort aus dem Vorrat von Blattanlagen der Knospe auf Kosten der nächsten Jahrgänge.

4. Verzweigung und Wuchsformen

a) Allgemeines und dikotyle Pflanzen

Seit dem Erscheinen des Buches von HALLÉ et al. (Progr. Bot. 40: 47, 45: 46-47) hat die Beschäftigung mit Verzweigungssystemen einen bemerkenswerten Auftrieb erhalten. Eine Übersicht über die Verzweigung von Bäumen hat FISHER (1986b) gegeben. TOMLINSON (1987) hat darauf hingewiesen, daß zur Verzweigung wenig Grundsätzliches bekannt sei und in gängigen Lehrbüchern einschlägige Kapitel fehlten. Provisorische Begriffe seien unvermeidbar, doch müsse ihre Notwendigkeit sorgfältig geprüft werden. In dieser Hinsicht gebe der von STERN u. PRIDGEON (1985) und STERN et al. (1985) vorgeschlagene Begriff des Ramicauls für hochdifferenzierte Rhizomsysteme der Orchideen ein negatives Beispiel, das auch bei RASMUSSEN (1985) auf kritische Ablehnung stieß. TOMLINSON's Klage aber dürfte bei Kennern der umfangreichen Literatur, z.B. TROLL, RAUH, MEUSEL u.a., auf wenig Verständnis stoßen. Die Wissenslücke bei den Verzweigungen der tropischen Pflanzen ist zwar verständlich, aber kein Grund, das Geleistete zu ignorieren. TROLL ist die Feststellung zu danken, daß vegetative und reproduktive Verzweigung nur im Zusammenhang verständlich gemacht werden können. Geht man von Zuwachseinheiten ("Moduln" sensu HALLÉ et al., vgl. hierzu WALLER u. STEINGRAEBER 1985), wie es MUELLER (1985) am Beispiel von *Alstonia* tat, so sind diese Einheiten, die während der vegetativen Phase erstarken und in der Regel mit einer Blüte oder einem Blütenstand enden, geschlossene Einheiten. Die Erstarkung bewirkt bei den Seitenknospen eine Akrotonie, wobei beim Übergang in die reproduktive Entwicklungsphase durch Eintritt sylleptischen Knospenaustriebs ein Blütenstand angelegt wird. Die basalen, nur schwachen Seitenknospen bleiben als gehemmte "Augen" zurück, eine Erscheinung, die mit MÜLLER-DOBLIES u. WEBERLING (vgl. Progr. Bot. 47: 43) als "Opsigonie" bezeichnet werden kann. Der Ausdruck "Prolepsis", der hier von HALLÉ et al. verwendet wurde, ist von der Wortbedeutung her völlig sinnverdreht. Im Fall vieler tropischer Pflanzen sind sylleptische vegetative Fortsetzungstriebe zu beobachten, bevor der Hauptsproß in einem Blütenstand endet. Unter den Bedingungen des Jahreszeitenklimas dienen jedoch im akrotonen System die stärksten gehemmten Knospen der Innovation im folgenden Jahr. Dies ist "Katalepsis". Auch in diesem Fall bleibt Opsigonie wirksam, denn i.d.R. finden sich auch bei kataleptischen Triebsystemen Reserveknospen an den Sproßbasen. Prolepsis, d.h. vorzeitiger Austrieb, ist auf Katalepsis zu beziehen, womit im klassischen Fall ein Verhalten gemeint ist, wie es beispielsweise bei den Kurztrieben von *Pinus* zu beobachten ist. Die Abrenzungsprobleme zwischen Syllepsis und Prolepsis sensu HALLÉ et al., auf die MUELLER (1985) am Beispiel von *Alstonia scholaris* aufmerksam gemacht hat, würden sich bei Verwendung der klassischen Terminologie garnicht ergeben. Geht man davon aus, daß das Normalverhalten bei *Larix* einschließlich der Kurztriebe ein kataleptisches ist, wäre es vermutlich auch sinnvoller, die von REMPHREY u. POWELL (1985) beschriebenen "sylleptischen" Triebe als proleptisch zu bezeichnen, die man übrigens auch bei kräftigen Jungdouglasien und anderen Koniferen beobachten kann. Die beiden Autoren vermuten aufgrund ihrer Analyse, daß die vorzeitig austreibenden Seitenknospen am oberen Ende des in der Winterknospe schon vorhandenen Sproßabschnittes stehen, während der weitere, zunächst unverzweigt bleibende, Spitzenabschnitt erst nach dem Beginn des Knospenaustriebs gebildet wird.

19 Arten aus der Gattung *Psychotria* aus Panama folgten nach HAMILTON (1985) den Modellen Chamberlain, Leuwenberg und Koriba, wobei einige Arten während der Entwicklung alle drei oder zwei der Verzweigungsmodelle durchliefen. Es zeigt sich dabei, daß die Modelle vom Verzweigungsmodus her gesehen nicht wesentlich sind, vielmehr Phasen im Verlauf der Erstarkung darstellen. Die Zuordnung von Pflanzen zu einem Modell besagt deshalb wenig, vielmehr müßte die Plastizität eines Typus innerhalb einer größeren Verwandtschaft erfaßt werden. Vergleichbare Beobachtungen haben FISHER u. WEEKS (1985) am Verzweigungssystem von *Neea* gemacht, die sich nach den Modellen Prevost und Koriba entwickeln kann. Die Fähigkeit zur Reiteration sensu DE CASTRO E SANTOS (vgl. Progr. Bot. 45: 46) führt in hohem Maß zu opportunistischen Formen. Besonders große Variabilität in der Architektur von Jungpflanzen von *Fagus sylvatica* beobachteten THIEBAUT et al. (1985). Sie fanden allein 30 verschiedene Verzweigungsformen, monopodiale wie sympodiale, und versuchten in den Verjüngungen soziale Beziehungen zwischen diesen Formen herauszuarbeiten. Die monopodialen aufrechten Formen vermochten dabei im Halbschatten wie im vollen Licht durch optimales Längenwachstum bei reduzierter Verzweigung die Dominanz im Bestand zu erreichen.

CREMERS u. SELL (1986) stellten für madagassische *Kalanchoe*-Arten fest, daß sie als "miniaturisierte Baumformen" aufgefaßt werden könnten und den Modellen Holttum, Tomlinson, Chamberlain, Leuwenburg und Rauh folgen. Ihre Infloreszenzen sind geschlossene Rispen, Thyrsen oder Cymoide. Die Ableitung krautiger Pflanzen von akrotonen Bäumen ohne Vermittlung der dank einer überlagerten basitonen Knospenbildung zu basaler Innovation fähigen Strauchformen ist von besonderem Interesse. Die Verzweigungsformen sind bei der Mehrzahl der *Kalanchoe*-Arten akroton. Mesotonie und Basitonie spielen jedoch in der Verwandtschaft ebenfalls eine Rolle, die bei weiteren Untersuchungen dieser Art stärkere Beachtung finden sollten.

Einen umfassenden Bericht über die Entstehung, Phylogenie und Funktion der Holzknollen ("Lignotubers"), die vor allem im mediterranen Ökosystem eine große Rolle spielen, hat JAMES (1984) gegeben. Auch bei diesem Beispiel zeigt sich, daß solide vergleichende und entwicklungsgeschichtliche Arbeiten zu diesem Thema vollständig fehlen.

Wenn sie auch auf eingehende morphologische Analysen verzichten, so haben ANTOS u. ZOBEL (1984) doch die verschiedenen Strategien in der Lebensgemeinschaft von schattigen Wäldern mit Rhizompflanzen *(Clintonia, Smilacina, Arnica, Achlys)*, Ausläuferpflanzen *(Linnea, Rubus)* und Knollenpflanzen *(Tiarella, Valeriana, Erythronium)* herausgearbeitet. Während bei den ersten beiden Gruppen nie Sämlinge gefunden wurden, vermehrt sich die letzte Gruppe ausschließlich über Samen. Ausläuferpflanzen können sich besonders effektiv wechselnden Lichtverhältnissen im Wald anpassen. GIVNISH's (1986b) Versuch, durch Optimierung der Biomechanik am Beispiel von Waldgeophyten (*Podophyllum peltatum, Polygonatum biflorum, Smilacina racemosa* u.a.) Formen verständlich zu machen, mögen eine optimale Konstruktion belegen, ohne vergleichende Morphologie aber die Existenz der verschiedenen Konstruktionstypen nicht zu erklären. Im Fall von *Podophyllum peltatum*, dessen oberirdische Triebe unverzweigt bleiben, werden Blätter mit Sprossen verwechselt. Es ist Tatsache, daß verschiedene Konstruktionstypen miteinander konkurrierend am gleichen Standort optimiert wurden. "Schirmform" *(Podophyllum)* als optimale Form ebener Waldböden und "Bogenform" *(Polygonatum)* als Anpassung an Hanglage zu verstehen, leuchtet nicht ein, wo beide Formen anderswo neben weiteren in beiden Situationen anzutreffen sind. Genau diese Zusammenhänge werden im gleichen Buch im Beitrag von SCHULZE et al. (1986) klar herausgestellt, indem Verzweigungssysteme von Phanerophyten in Hecken analysiert werden. In diesen vom Menschen zwischen seinen Kulturen angebotenen Räumen finden sich alle diejenigen Pflanzen aus der Umgebung ein, die durch ihre opportunistischen Entwicklungsmöglichkeiten von den unterschiedlichsten deterministischen Konstruktionstypen her unter den gegebenen, von den Partnern selbst bestimmten Verhältnissen zu ausreichender Lichternte gelangen können.

Im Vergleich zu den zuvor besprochenen Arbeiten über Wuchsformen sind die Analysen KÄSTNER's (1985, 1986) an *Teucrium* der Sektion *Isotriodon* und *Chamaedrys* auf vergleichend morphologischer Basis als geradezu klassisch zu betrachten. Hier wird ausgehend vom Typus – die Infloreszenzen eingeschlossen – umfassendes Verständnis für die in der Evolution möglichen Strategien bei der Eroberung von Lebensräumen ermöglicht, was insbesondere für das mit unterirdischen ausläuferartigen Zweigsystemen ausgestattete *Teucrium chamaedrys* hervorragend gelungen ist. Ähnliche Gesichtspunkte werden bei der knapp gehaltenen Darstellung der Lebensform einiger psammophiler Pflanzengattungen von LUKASIEWICZ (1985), wie auch einer Analyse des Rhizomsystems von *Aegopodium podagraria* durch CHABOT-JAQUETY (1984) deutlich. Zur Wuchsformenanalyse verschiedener Verwandtschaftsgruppen sind schließlich auch wieder eine Reihe russischer Beiträge erschienen, von denen hier nur einige genannt werden sollen: SEREBRYAKOVA u. PAVLOVA (1986): *Potentilla*, sect. *Potentilla*, SHULKINA (1986): Übersicht über Campanulaceen-Wuchsformen, KALAMBET (1984): *Salvia*.

b) Monocotyledonen

Wuchsform, Erstarkung und Leitbündelsysteme der Rhizomsysteme von *Smilacina* hat LA FRANKIE (1985a,b) untersucht und fand in Übereinstimmung mit *Polygonatum* in den ersten 3 Jahren des Sämlings monopodialen Wuchs, der erst mit dem Erscheinen der ersten Sprosse über dem Boden zum sympodialen Wuchs übergeht. Auch mit Rhizomsystemen von *Maianthemum* hat sich LA FRANKIE (1986a,b) befaßt. Interessant ist die Beschreibung des neuen *Maianthemum paludicolum* durch den gleichen Autor, das in Costa Rica auf Moorflächen in größerer Höhe mit seinen aufgerichteten sympodialen Rhizomen klettert. NESOM u. LA DUKE (1985) haben die Wuchsformen und das Erstarkungswachstum des knollenbildenden *Trillium nivale* dargestellt. Eine ganze Reihe von vergleichenden Untersuchungen ist den Überdauerungsorganen von Liliaceen und Amarylliadaceen gewidmet worden. So hat BARANOWA (1986) bei den Liliaceenzwiebeln den Lilioideen- und Scilloideen-Typus unterschieden und in beiden insgesamt 17 Zwiebelformen aufgefunden. Wuchsform und Knollenentwicklung von *Merendera* sind von OGANESOVA (1986) bearbeitet worden. Bislang unverständliche Wiedersprüche im Bauplan der Knollen von *Colchicum autumnale* hat der Nachweis eines bisher übersehenen adossierten Vorblattes am Blütentrieb geklärt (JAEHN et al. 1985; JAEHN u. ROUX 1986). Ein umfassender Vergleich der Amaryllidaceenzwiebeln ist ARROYO (1984) zu verdanken, bei dem sich gezeigt hat, daß die afrikanischen Arten monopodial und sympodial wachsen, die amerikanischen Arten dagegen sämtlich sympodialen Bau besitzen. Monopodial sind *Amaryllis belladonna* und die Gattung *Nerine*.

Zahlreiche weitere Untersuchungen über die Wuchsformen der Bromelien hat RAUH (1983-1986) publiziert, der mit seinem Werk den vollständigsten Überblick über diese Pflanzenfamilie gegeben hat. Von Interesse ist auch eine Studie von ADAMS III u. MARTIN (1986) über die Entwicklung und Erstarkung von *Tillandsia deppeana*, in der gezeigt wird, daß die Jugendformen dieser großen zisternenbildenden Art den nicht zisternenbildenden Luftformen von Tillandsien entspricht. Da dies auch bei anderen Zisternenbildenden Arten der Fall ist, liegt der Gedanke nahe, die Luftformen dieser Gattung als neoton zu betrachten.

Untersuchungen über die Wuchsformen der feuerresistenten Cyperaceen, die in den Grasländern Zambias vorkommen, haben ergeben, daß sie entweder als Kryptophyten mit unterirdischen Rhizomen geschützt sind, oder aber als Hemikryptophyten Knollen bilden, die gegen Feuer durch die kompakten abgestorbenen Blattreste geschützt sind (MEDWECKA-KORNAS u. KORNAS 1985). Feuer löst unabhängig von der Regenzeit den Austrieb aus, wobei die gespeicherten Wasservorräte ausreichen, die Pflanzen auch während der Trockenzeit zur Blüte zu bringen. Die Wuchsformen von *Ammophila breviligulata* und *Calamovilfa longifolia* wurden von MAUN (1985) bearbeitet, die der ausläuferbildenden *Poa pratensis* ssp. *angustifolia* von ZAUGOLNOVA u. MIKHAILOVA (1986). Die Wuchsformen von *Callisia fragrans (Commelinaceae)* läßt sich mit Holttum's Modell beschreiben (DE SOUZA et al. 1986). Mit der Wuchsform der eigenartigen sympodial verzweigten Palme Ost-Malaysias *Eugeissonia minor*, haben sich HOLBROOK u. PUTZ (1985) befaßt. Die auf hohen Stelzwurzeln stehenden Pflanzen vermehren sich durch Abfaulen der alten Stammteile und Verselbständigen der Zweige vegetativ.

5. Wurzeln

Einen beeindruckenden und sehr anregenden Überblick über die Adventivwurzelbildung und das Spektrum ihrer Formen ist BARLOW (1986) zu verdanken. Darin werden Landpflanzen, Würger, Kletterpflanzen, Epiphyten, Stelzwurzler, Mangroven und selbst ausgefallene Wurzelsysteme wie die der *Psodostemonaceae* behandelt. Einige Bewurzelungstypen, insbesondere die xeromorphen Polsterpflanzen mit tief in den Boden eindringenden "Polwurzeln" (= Primärwurzeln), sowie die Wurzelsysteme von *Chusquea* und einigen Wasserpflanzen aus der argentinischen Flora haben KUTSCHERA-MITTER u. LICHTENEGGER (1986) vorgestellt. Eigenartige endogene Wurzeln fanden GAMM u. WEBERLING (1986) an den unterirdischen Fruchtstielen von *Cardamine chenopodiifolia*.

Naturgemäß ist die Analyse der Wurzelsysteme mit erheblichen Schwierigkeiten verbunden, die eindrucksvoll in den Beiträgen von RICHARDS (1986) und CALDWELL u. RICHARDS (1986) herausgestellt worden sind. In sorgfältigen Untersuchungen haben sie miteinander konkurrierende Wurzelsysteme von tiefwurzelnden Wüstenpflanzen, wie z.B. *Artemisia tridentata* zusammen mit *Agropyron spicatum* und *A. desertorum* untersucht. Die Wurzelsysteme dieser Pflanzen durchmischen sich auf engstem Raum selbst direkt unter den Pflanzen, wobei *Agropyron* bis in 1/2 m Tiefe dominiert, *Artemisia* dagegen in tieferen Bodenschichten. Kosten-Nutzen-Analysen für die aufwendigen Wurzelsysteme sind nach Auffassung der Autoren kaum möglich. Die Anzahl der Wurzelspitzen, die allein absorbieren, Länge und Dicke der Wurzeln als Speicher- und Leitungsorgane und ihrer Widerstände, sowie Absterbe- und Erneuerungsraten sind bei den Wurzeln vergleichsweise zu den kleineren oberirdischen Teilen der Pflanzen kaum zu durchschauen. Mit mathematischen Modellvorstellungen hat FITTER (1985, 1986) das Wurzelsystem von *Trifolium pratense* und anderen Beispielen zu beschreiben versucht, wobei er in Analogie zu Sproßsystemen, die in "Glieder" zwischen zwei Seitenwurzeln (entsprechend den Internodien zwischen zwei Blättern), bzw. die Wurzelspitzen als Endglieder durch ihre Lage im System charakterisiert hat. Da die Entwicklungsgeschichte dieser Einheiten jedoch eine ganz andere ist, als die der Sproßeinheiten, muß zweifelhaft bleiben, ob die Methode das Verständnis der Wurzelsysteme wirklich fördern kann. WILSON u. FIELD (1985) haben bei den Kurzwurzelsystemen von *Pinus radiata* in Kulturgefäßen 3 Typen gefunden, kleine einmalig gabelig verzweigte, gestaucht bleibende Kurzwurzeln und grössere bis zu 10 mal gegabelte "Rhizothamnien", die sich von kräftigen, verlängerten Wurzeln 2. Ordnung herleiten, und schließlich dünnbleibende Wurzeln, deren Gabeläste sich verlängern.

Experimentell hat JERNSTEDT (1984) die Wurzelkontraktion von *Hyacinthus orientalis* überprüft, die sich entsprechend den Vorstellungen RIMBACH's durch bevorzugtes Dickenwachstum der inneren Rindenzellen bei gleichzeitiger Schrumpfung und Faltung der äußeren Rinde vollzieht. IAA verringert die Kontraktion, wobei vermehrt äußere Rindenzellen in den Dickenwachstumsprozeß einbezogen werden. JERNSTEDT schließt aus den Ergebnissen, daß der Schrumpfungsprozeß in der äußeren Rinde wesentlich für den Kontraktionsprozeß ist.

Literatur

ADAMS III, W.W., MARTIN, C.E.: Am. J. Bot. 73, 1207-1214 (1986). - ANTOS, J.A., ZOBEL, D.B.: Bot. Gaz. 145, 508-517 (1984). - ARROYO, S.C.: Kurtziana 17, 55-70 (1984). - AUDUS, L.J.: J. Exp. Bot. 30, 1051-1073 (1979).

BAKER, N.R., DAVIES, W.J., ONG, C.K. (eds.): Control of Leaf Growth. Society for Experimental Biology Seminar Series 27. Cambridge: Cambridge University Press 1985. - BARANOVA, M.V.: Bot. Zh. (Leningr.) 71, 1308-1320 (1986). - BARKER, W.G., HUSSEY, R.B., HORTON, R.F.: Ann. Bot. 58, 397-405 (1986). - BARLOW, P.W.: 67-110 in: siehe JACKSON 1986. - BARLOW, P.W., RATHFELDER, E.L.: Planta 165, 134-141 (1985). - BECK, C.B.: Bot. Rev. 51, 273-294 (1985). - BERGGREN, B.: Nord. J. Bot. 4, 771-789 (1984); - ibid. 5, 475-488 (1985). - BERTAUD, D.S., GANDAR, P.W.: Ann. Bot. 58, 303-320 (1986). - BERTAUD, D.S., GANDAR, P.W., ERICKSON, R.O., OLLIVIER, A.M.: Ann. Bot. 58, 285-301 (1986). - BHAMBIE, S., SHARMA, A.: Proc. Indian Acad. Sci. (Plant Sci.) 95, 295-301 (1985). - BOPP, M. (ed.): Plant Growth Substances. Berlin, Heidelberg, New York: Springer 1985. - BROOKS, D.R., WILEY, E.O.: Evolution as Entropy, Towards a Unified Theory of Biology. Chicago: Univ. of Chicago Press 1986. - BRUCK, D.K., WALKER, D.B.: Bot. Gaz. 146, 188-195 (1985); - Am. J. Bot. 72, 1602-1609 (1985b).

CALDWELL, M.M., RICHARDS, J.H.: 251-273 in: siehe GIVNISH 1986a. - CHABOT-JACQUETY, Y.: Bull. Soc. Bot. Fr. 131, Lett. Bot. 207-221 (1984). - CHAU, R.J.: Am. J. Bot. 72, 1475-1482 (1985). - CHENOU, E., KULIGOWSKI, J., FERRAND, M.: Can. J. Bot. 64, 784-792 (1986). - CHIARETTO, J.A.C.: Hoehnea 11, 47-58 (1984). - CHRETIEN, L., VIETH, J.: Can. J. Bot. 63, 1313-1318 (1985). - CHRISTIANSON, M.L.: Am. J. Bot. 73, 947-958 (1986). - CLOWES, F.A.L.: Ann. Bot. 55, 849-857 (1985). - COONEY-SOVETTS, C., SATTLER, R.: Bot. J. Linn. Soc. 94, 327-371 (1986). - CREMERS, G., SELL, Y.: Bull. Mus. Natl. Hist. Nat., sér. 8, Sect. B., Adansonia No. 1: 63-76 (1986). - CUNNINGHAME, M.E., LYNDON, R.F.: Ann. Bot. 57, 737-746 (1986).

DE MASON, D.A., WILSON, M.A.: Can. J. Bot. 63, 1907-1913 (1985). - DESCHAMP, P.A., COOKE, T.J.: Am. J. Bot. 72, 1377-1387 (1985). - DE SOUZA, J.B., LAURI, P.E., BLANC, P.: Can. J. Bot. 64, 1664-1670 (1986). - DICKINSON, T.A., PARKER, W.H., STRAUSS, R.E.: Taxon 36, 1-20 (1987). - DULIEU, H.: Fac. Sci. Dijon, Trav. Lab. Bot. (1968). - DYER, A.F., PAGE, C.N. (eds.): Biology of Pteridophytes. Proceedings of the Symposium, Edinburgh, Sept. 1983. Edinburgh: Proc. Roy. Soc. Edinburgh, Sect. B. 86, 1985.

EASTMAN, A., PETERSON, R.L.: Bot. Gaz. 146, 44-55 (1985). - EDWARDS, D.S.: Bot. J. Linn. Soc. 93, 173-204 (1986). - ELLENBERG, H.: Flora 176, 169-188 (1985). - EWERS, F.W., SCHMID, R.: Am. J. Bot. 72, 509-513 (1985).

FISHER, J.B.: Bot. Gaz. 147, 84-89 (1986a); - 493-523 in: siehe GIVNISH 1986a. - FISHER, J.B., WEEKS, C.L.: Bull. Mus. Natl. Hist. Nat., Ser. 7, Sect. B. Adansonia No. 4, 385-401 (1985). - FITTER, A.H.: 87-106 in: Ecological Interactions in Soil, eds. A.H. FITTER, D.J. READ, D. ATKINSON, M.B. USHER (Special Publications of the British Ecological Society No. 4). Oxford: Blackwell 1985; - Ann. Bot. 58, 91-101 (1986). - FJELL, J.: Nord. J. Bot. 5, 555-573 (1985). - FONDI, R.: La révolution organiciste, Entretien sur les nouveaux courants scientifiques. Paris: Le Labyrinthe 1986. - FRENCH, J.C., TOMLINSON, P.B.: Am. J. Bot. 71, 1432-1443 (1984).

GAHAN, P.B., RANA, M.A.: Ann. Bot. 56, 437-442 (1985). - GAMM, R., WEBERLING, F.: Flora 178, 19-21 (1986). - GIVNISH, T.J. (ed.): On the Economy of Plant Form and Function. Evolutionary Constraints on Primary Productivity: Adaptive Patterns of Energy Capture in Plants. Cambridge: Cambridge Univ. Press 1986a; - ibid. 525-583, 1986b. - GOMEZ, L.D.: Brenesia 18, 4 (1980). - GONTHIER, R., JACQMARD, A., BERNIER, G.: Planta 165, 288-291 (1985); - ibid. 170, 55-59 (1987). - GOULD, K.S., CUTTER, E.G.: Can. J. Bot. 64, 1268-1276 (1986). - GOYAL, V., GOYAL, S.C., PILLAI, A.: Proc. Indian Acad. Sci. (Plant Sci.) 95, 349-356 (1985). - GRAHAM, L.E.: Am. Sci. 73, 178-186 (1985). - GRAHAM, L.E., TAYLOR, C., III: J. Phycol. (in press) (1986). - GUEDES, M.: Phyton 24, 257-272 (1984).

HAGEMANN, W.: Aufsätze und Reden der Senckenbergischen Naturforschenden Gesellschaft. Bd. 29, 155-158. Frankfurt: Senckenbergische Naturforschende Gesellschaft 1978. - HAMILTON, C.W.: Am. J. Bot. 72, 1081-1088 (1985). - HARTE, C., LINDENMAYER, A.: Biol. Zentralbl. 105, 407-425 (1986). - HAUKE, R.L.: Can. J. Bot. 63, 2430-2438 (1985). - HECKER, M.: Beitr. Biol. Pflanzen 60, 165-184 (1986). - HOLBROOK, N.M., PUTZ, F.E., CHAI, P.: Principes 29, 142-146 (1985). - HOPPE, J.R.: Bot. Jahrb. Syst. 105, 497-581 (1985). - HUGHES, N.F.: Ann. Missouri Bot. Gard. 71, 593-598 (1984).

IMAICHI, R.: J. Jap. Bot. 59, 367-376 pl. VIII-XI (1984); - Bull. Fac. Agric. Tamagawa Univ. 25, 23-33 (1985); - Bot. Mag. Tokyo 99, 309-317 (1986). - IMAICHI, R., NISHIDA, M.: Bot. Mag. Tokyo 99, 85-106 (1986). - IRVINE, D.E.G., JOHN, D.M. (eds.): Systematics of the Green Algae. London, New York: Academic Press 1984.

JACKSON, M.B. (ed.): New Root Formation in Plants and Cuttings. Dordrecht: Nijhoff 1986. - JAEHN, F., ROUX, J.: Bull. Soc. Bot. Fr. 133, Lett. Bot. 225-233 (1986). - JAEHN, F., PFIRSCH, E., ROUX, J.: Beitr. Biol. Pflanzen 60, 303-311 (1985). - JAIN, D.K.: Acta Bot. Indica 12, 26-34 (1984). - JAMES, S.: Bot. Rev. 50, 225-266 (1984). - JERNSTEDT, J.A.: Am. J. Bot. 71, 1080-1089 (1984). - JEUNE, B.: Bull. Mus. Natl. Hist. Nat., Ser. 8, Sect. B. Adansonia 301-323 (1986). - JOEL, D.M., JUNIPER, B.E., DAFUI, A.: New Phytol. 101, 585-593 (1985). - JOHNSON-FLANAGAN, A.M., OWENS, J.N.: Can. J. Bot. 63, 456-462 (1985).

KALAMBET, E.S.: Byull. Mosk. O-Va. Ispyt. Prir. Otd. Biol. 89, 100-114 (1984). - KAPLAN, D.R.: 51-70 in: Cladistics: Perspective on the Reconstruction of Evolutionary History, eds. T. DUNCAN, T. STUESSY. New York: Columbia Press 1984. - KARRFALT, E.: Bot. Gaz. 145, 372-377 (1984). - KÄSTNER, A.: Flora 176, 73-93 (1985); - ibid. 178, 111-138 (1986). - KIRCHOFF, B.K.: Bot. J. Linn. Soc. 89, 37-52 (1984). - KLEKOWSKI, E.J., jr., KAZARINOVA-FUKSHANSKY, N., MOHR, H.: Am. J. Bot.

72, 1788-1800 (1985). - KLOTZ, G.: Flora 176, 189-196 (1985). - KONDRATIEVA-MELVILLE, E.A.: Bot. Zh. (Leningr.) 71, 1321-1330 (1986). - KRAEHENBUEHL, W.: Phyllom-Morphogenese bei *Allium* und *Juncus* unter besonderer Berücksichtigung der Embryogenese. Diss. Bot. 67, Vaduz: Cramer 1983. - KUMAR, V., DEVENDRA, K.A., YERRAMILLI, S.M.: Bot. Mag. Tokyo 97 (1048):437-446 (1984). - KUNZE, H.: Beitr. Biol. Pflanzen 61, 49-77 (1985). - KURTH, E., GIFFORD, E.M., Jr.: Am. J. Bot. 72, 1676-1683 (1985). - KUTSCHERA, L.: 303-322 in: Wurzelökologie und ihre Nutzanwendung/Root Ecology and its Practical Application. Int. Symp. Gumpenstein, Bundesanstalt Gumpenstein (ed.), Irdning (1983). - KUTSCHERA-MITTER, L., LICHTENEGGER, E.: Veröff. Geobot. Inst. ETH, Stiftung Rübel, Zürich 91, 228-250 (1986).

LA FRANKIE, J.V., Jr.: Bot. Gaz. 146, 534-544 (1985a); - Bull. Torr. Bot. Club. 112, 313-317 (1985b); - J. Arnold Arboretum 67, 371-459 (1986a); - Am. J. Bot. 73, 1258-1260 (1986b). - LARSON, P.R.: Can. J. Bot. 63, 2383-2392 (1985); - Am. J. Bot. 73, 1620-1631 (1986a). - ibid. 73, 1632-1641 (1986b). - LUKASIEWICZ, A.: Feddes Rep. 96, 145-156 (1985).

MAGENDANS, J.F.C.: Anatomy of Vein Endings in *Hedera Leaves*. Aspects of Ontogeny. Agricultural University Wageningen Papers 85-5 (1985). - MARTIN, B.F., TUCKER, S.C.: Am. J. Bot. 72, 66-74 (1985). - MATTOX, K.R., STEWART, K.D.: 29-72 in: siehe IRVINE and JOHN 1984. - MAUN, M.A.: Can. J. Bot. 63, 113-124 (1985). - MEDWECKA-KORNAS, A., KORNAS, J.: Flora 176, 61-71 (1985). - MEEUSE, A.D.J.: Anatomy of Morphology. Leiden: Brill 1986. - MEICENHEIMER, R.D.: Am. J. Bot. 73, 1649-1664 (1986). - MELKONIAN, M.: 73-120 in: siehe IRVINE and JOHN 1984. - MERRILL, E.K.: Can. J. Bot. 64, 2645-2649 (1986a); - ibid. 64, 2650-2661 (1986b); - ibid. 64, 2662-2668 (1986c). - MEYEN, S.V.: Bot. Rev. 50, 1-111 (1984). - MIALOUNDAMA, F., PAULET, P.: Can. J. Bot. 64, 208-213 (1986). - MILLER, C.N.: Bot. Rev. 51, 295-318 (1985). - MOORE, R.: Am. J. Bot. 71, 1452-1454 (1984); - ibid. 72, 1048-1053 (1985a); - Ann. Bot. 55:367-373 (1985b); - ibid. 55, 375-380 (1985c); - ibid. 56, 173-187 (1985d); - ibid. 55, 381-385 (1985e); - Am. J. Bot. 72, 733-736 (1985f); - ibid. 72, 785-787 (1985g); - Ann. Bot. 57, 119-131 (1986a); - Am. J. Bot. 73, 417-426 (1986b). - MOORE, R., EVANS, M.L.: Am. J. Bot. 73, 574-587 (1986). - MOORE, R., FONDREN, W.M.: Ann. Bot. 58, 381-387 (1986). - MOORE, R., MC CLELEN, C.E.: Ann. Bot. 56, 73-81 (1985a); - ibid. 56, 83-90 (1985b). - MUELLER, R.J.: Am. J. Bot. 72, 1435-1444 (1985). - MULKEY, T.J., VAUGHAN, M.A.: 241-246 in: siehe BOPP 1985.

NEGBI, M.: Bot. J. Linn. Soc. 93, 247-258 (1986). - NESOM, G.L., LA DUKE, J.C.: Can. J. Bot. 63, 7-14 (1985). - NESTER, J.E.: Bot. Gaz. 146, 510-516 (1985). - NIKLAS, K.J.: Paleobiology 10, 487-506 (1984). - NOUGAREDE, A., LANDRE, P., REMBUR, J., HERNANDEZ, M.N.: Can. J. Bot. 63, 309-323 (1985). - NOZERAN, R., ROSSIGNOL-BANCILHON, L., MANGENOT, G.: Bot. Helv. 94, 199-233 (1984).

OGANESOVA, G.G.: Bot. Zh. (Leningr.) 71, 860-870 (1986). - OROSS, J.W., LUCAS, W.J.: Can. J. Bot. 63, 2295-2304 (1985). - ORR, A.R.: Bot. Gaz. 145, 308-311 (1984); - ibid. 146, 477-482 (1985). - ORR, A.R., SCHONEMAN, T.A.: Am. J. Bot. 73, 161-165 (1986).

PANT, D.D.: NAUTIYAL, D.D., MISRA, D.R.: Gametophytes of Ophioglossaceae. Phyta Monograph No. 1. Allahabad (India): The Society of Indian Plant Taxonomists 1984. - PERIASAMY, K., MURUGANATHAN, E.A.: Proc. Indian. Acad. Sci. (Plant Sci.) 95, 429-436 (1985). - PICKARD, B.G.: 193-281 in: Encyclopedia of Plant Physiology, N.S. Vol. 11, eds. R.P. PHARIS, D.M. REID. Berlin, Heidelberg, New York: Springer 1985. - POETHIG, R.S., SUSSEX, J.M.: Planta 165, 158-169 (1985a); - ibid. 165,

170-184 (1985b). - POSLUSZNY, U., GERRATH, J.M.: Can. J. Bot. 64, 1620-1631 (1986). - PROBST, W.: Biologie der Moos- und Farnpflanzen. UTB 1418. Heidelberg, Wiesbaden: Quelle und Meyer 1986. - PULAWSKA, Z.: Act. Soc. Bot. Pol. 55, 171-179 (1986).

RAGHAVAN, V.: Embryogenesis in Plants. New York: Cambridge University Press 1986. - RASMUSSEN, H.: Taxon 34, 654-657 (1985). - RAUH, W.: Bromelienstudien I. Neue und wenig bekannte Arten aus Peru und anderen Ländern. Mitt. Trop. u. subtrop. Pflanzenwelt, Bd. 41, 42, 43, 50, 52, 53, 58. Mainz: Akad. Wiss. Lit. 1983-1986 - RAVEN, J.A.: 421-492 in: siehe GIVNISH 1986a. - REMBUR, J., NOUGAREDE, A.: Protoplasma 136, 183-190 (1987). - REMPHREY, W.R., POWELL, G.R.: Can. J. Bot. 63, 1296-1302 (1985). - REMY, W., HASS, H.: Argumenta Palaeobotanica, Münster 7, 173-214 (1986). - RICHARDS, J.H., in: Mineral Exploration: Biological Systems and Organic Matter, eds. D. CARLISLE, W.L. BERRY, J.R. WATTERSON, J.R. KAPLAN. Englewood Cliffs, NJ: Prentice-Hall 1986. - RICHARDS, J.H., LEE, D.W.: Am. J. Bot. 73, 1741-1747 (1986). - ROBERTS, D.G., MC COMB, A.J., KUO, J.: New Phytol. 100, 25-36 (1985). - ROTHWELL, G.W.: Bot. Rev. 51, 319-327 (1985). - RUDALL, P.: Bot. Gaz. 145, 524-534 (1984). - RUTH, J., KLEKOWSKI, E.J. Jr., STEIN, O.L.: Am. J. Bot. 72, 1127-1135 (1985). - RUTISHAUSER, R.: Bull. Int. Group for the Study of Mimosoideae 14, 77-108 (1986). - RUTISHAUSER, R., SATTLER, R.: Can. J. Bot. 64, 1987-2019 (1986).

SACHS, T.: 231-235 in: siehe BOPP 1985. - SATTLER, R.: Biophilosophy. Berlin, Heidelberg, New York: Springer 1986. - SAWHNEY, V.K., SEKHAR, K.N.C.: Can. J. Bot. 63, 1564-1566 (1985). - SCHATT, E., LANDRE, P., NOUGAREDE, A.: Can. J. Bot. 63, 2200-2208 (1985). - SCHIAVONE, F.M., COOKE, T.J.: Can. J. Bot. 63, 1573-1578 (1985). - SCHULZE, E.D., KÜPPERS, M., MATYSSEK, R.: 585-602 in: siehe GIVNISH 1986a. - SCHWABE, W.W., CLEWER, A.G.: J. Theor. Biol. 109, 595-620 (1984). - SEILHEAN, V., MICHAUX-FERRIERE, N.: Am. J. Bot. 72, 1089-1094 (1985). - SEKHAR, K.N.C., SAWHNEY, V.K.: Am. J. Bot. 72, 1813-1822 (1985). - SEREBRYAKOVA, T.J., PAVLOVA, N.R.: Bot. Zh. (Leningr.) 71, 154-167 (1986). - SHAH, C.K.: Phytomorphology 33, 62-73 (1983). - SHULKINA, T.V.: Bot. Zh. (Leningr.) 71, 569-579 (1986). - STEINGRAEBER, D.A., FISHER, J.B.: Am. J. Bot. 73, 852-862 (1986). - STERN, W.L., PRIDGEON, A.M.: Taxon 34, 657-658 (1985). - STERN, W.L., PRIDGEON, A.M., LUER, C.A.: Bot. J. Linn. Soc. 91, 457-471 (1985). - STIEBER, J., BERINGER, H.: Bot. Gaz. 145, 465-473 (1984). - SUGIYAMA, M.: Phytomorphology 34, 99-1o9 (1984).

THIEBAUT, B., CUGUEN, J., DUPRE, S.: Can. J. Bot. 63, 2100-2110 (1985). - TILNEY-BASSET, R.A.E.: Plant Chimeras. London: Arnold 1986. - TOMLINSON, P.B.: Taxon 36, 54-65 (1987). - TURLIER, M.F.: Bull. Soc. Bot. Fr. 131, Lett. Bot. 301-314 (1984).

VAN LAMMEREN, A.A.M.: Acta Bot. Neerl.: 35, 169-188 (1986). - VIJAYARAGHAVAN, M.R., KAPOOR, T.: Aquatic Bot. 22, 45-60 (1985). - VISHNU, K., AWASTHI, D.K., MURTY, Y.S.: Bot. Mag. Tokyo 97, 437-446 (1984). - VON ADERKAS, P., GREEN, P.E.J.: Bot. J. Linn. Soc. 93, 307-321 (1986). - VON ADERKAS, P., HICKS, G.: Can. J. Bot. 63, 916-919 (1985). - VON ADERKAS, P., RAGHAVAN, V.: Am. J. Bot. 72, 1067-1073 (1985).

WALKER, J.W., WALKER, A.G.: Ann. Missouri Bot. Gard. 71, 464-521 (1984). - WALLER, D.M., STEINGRAEBER, D.A.: Branching and modular growth: theoretical models and empirical patterns. In: Biology of Clonal Organisms, eds. R. COOK, L. BUSS, J.B.C. JACKSON. New Haven: Yale University Press 1985. - WALTERS, G.A., BARTHOLOMEW, D.P.: Bot. Gaz. 145, 351-357 (1984). - WILLIAMS, E.G., MAHESWARAN, G.: Ann. Bot.

57, 443-462 (1986). - WILSON, E.R.L., FIELD, R.J.: New Phytol. 100, 87-92 (1985). - WOLF, S.D., KUHN SILK, W., PLANT, R.E.: Am. J. Bot. 73, 832-846 (1986).

ZAGORSKA-MAREK, B.: Can. J. Bot. 63, 1844-1854 (1985). - ZAUGOLNOVA, L-B., MIKHAILOVA, N.F.: Bot. Zh. (Leningr.) 71, 292-299 (1986).

Prof. Dr. WOLFGANG HAGEMANN
Institut für Systematische Botanik und
Pflanzengeographie der Universität Heidelberg
Im Neuenheimer Feld 328
D-6900 Heidelberg, FRG

IV. Reproductive Structures of the Flowering Plants

By Rolf Rutishauser

1. Introduction

"Life would be awfully dull without flowers. The love of flowers is common to all people all over the world, no matter how different their cultural patterns may be." (WELLENSIEK in HALEVY 1985)

Reproductive structures, their development and evolution in the flowering plants (angiosperms) are the topics of this report. In surveying the last 2-year period (1985-1986) of publications, it becomes obvious that structural and developmental botany is an active and innovative field of research.

Not reviewed in this chapter are studies on embryology, palynology, and pollination biology, although these topics overlap with that summarized below.

The first four volumes of the superb "CRS Handbook of Flowering" (edited by A.H. HALEVY 1985) have now been published. Fifty scientists contributed new data and literature surveys over the last decades. More than 300 different flowering plants (mainly crops, ornamentals, and forest trees) are featured in separate chapters, with emphasis on floral characteristics, reproductive development and its environmental regulation, sex expression, genetics, and effects of growth regulators. Reproductive morphology is not neglected at all. Many drawings, SEM micrographs, and microtome sections are added. This handbook really deserves to be consulted by all scientists interested in reproductive structures of the flowering plants.

It seems that the "principle of complementarity" is now accepted by various students for interpreting reproductive structure and development. Any interpretation may be complemented by one or more seemingly opposing views, adding other perspectives or facets to our scientific understanding of reproductive structure and development. Interpretations complementing rather than excluding each other were presented at various levels. Depending on the criteria used, the reproductive unit in *Lilaea* may be called flower or inflorescence (POSLUSZNY et al. 1986). Similarly, placentation in *Limnanthes* may be viewed as axial or appendicular (HOFMANN and LUDEWIG 1985). Studying shoot development in *Vitis*, POSLUSZNY and GERRATH (1985) concluded that the shoot may be interpreted as a sympodium as well as a monopodium. This is not due to lack of evidence, but to the partial incongruity of our traditional concepts (models) with reality. A dynamic morphology (including a continuum approach) as proposed by SATTLER (1986) and RUTISHAUSER and SATTLER (1986) may help in overcoming the weaknesses of static concepts by emphasizing developmental processes (programs) rather than structural categories.

2. Reproductive Development (General Aspects)

The flower is a continuously developing system (WILLEMSE in WILLEMSE and VAN WENT 1985:194). During development it shows many functions coordinated for the formation of new offspring. The term "reproductive development" includes initiation and development of both flowers and inflorescences.

To better understand floral induction, a new model emphasizing changes in membrane properties was proposed by GREPPIN et al. (1986). LYNDON (1987) provided evidence for synchronization of cell division just before floral initiation in *Silene*. Timing and sequence of the processes during floral initiation and organogenesis in *Weigela* are described by MALOUPA-IKONOMOU and JACQUES (1986). PHARIS and KING (1985) summarized the role of gibberellins during the various stages of reproductive development from floral initiation until fruit and seed growth. BERNIER's (1986) review is a valuable step toward an integrated, multidisciplinary understanding of the flowering process (induction, evocation, morphogenesis).

Plasticity in Reproductive Development. According to BERNIER (1986), the morphogenetic processes during flower ontogeny are not absolutely fixed, but are plastic to some degree. The morphology of floral organs as well as their number and arrangement may be altered experimentally or spontaneously. As examples of plasticity in floral expression, TRAN THANH VAN (1986) described male-sterile hybrids in *Nicotiana* with naked ovules replacing pollen grains in anthers. Additional backcrosses between *Nicotiana* cultivars even resulted in a hybrid where floral differentiation occurred on leaf blades ("half of a leaf blade was modified into a flower"). Plastic response to temperature was found in *Lycopersicon* (SAWHNEY and POLOWICK 1985). Tomato plants grown under a low temperature regime produced large fruits with 9-13 locules, whereas the fruits produced under a high temperature regime were smaller and consisted of 4-7 locules only. An instructive example showing natural plasticity on floral formation was studied developmentally by POSLUSZNY et al. (1986): *Lilaea scilloides* possesses in its spicate inflorescence five different kinds of "flowers" ranging from complete, bisexual flowers (with tepal, stamen, and carpel) to male flowers (with tepal and stamen) and naked female flowers consisting of one carpel only.

The Juglandaceae provide examples of plasticity with respect to the season for floral initiation. POLITO and LI (1985a,b) studied the phenology of pistillate flower differentiation in several protogynous and protandrous clones of *Juglans regia*, a heterodichogamous species. In the protogynous clones perianth initiation of the pistillate flowers occurred before cessation of organ initiation in summer. By contrast, in protandrous clones tepals did not form until growth resumed the following year. WETZSTEIN and SPARKS (1984) noted a similar phenomenon in the expression of heterodichogamy in *Carya illinoensis*. A protandrous cultivar had well differentiated anther primordia present by mid-summer, while the protogynous types had not attained this stage until the following spring.

Phyllotactic Patterns. Considering phyllotactic patterns in extant primitive angiosperms, ENDRESS (1986a) concluded that phyllotaxis is not yet canalized and restricted to the whorled pattern in primitive flowers. Many different patterns have been shown in Magnoliidae: spiral (e.g., Himantandraceae), whorled (e.g., Lauraceae), more complex symmetrical patterns (e.g., Annonaceae), and unordered (e.g., Winteraceae). According to MOSELEY and UHL (1985), *Nelumbo* has a helical phyllotaxis throughout the entire flower. However, a developmental study is still needed to verify the helical arrangement of the numerous stamens. Although belonging to two different subclasses, *Pereskia* and *Stewartia* possess some similarities in the initiation pattern of their floral appendages (LEINS and SCHWITALLA 1985; ERBAR 1986). In both genera, a spiral (2/5) sequence prevails throughout the flower, and the multistaminate androecium results from five common primordia which show slight developmental differences corresponding to their position within the ontogenetic spiral. In Apiaceae, the initiation patterns of floral appendages vary from a continuous spiral sequence *(Eryngium)* to nearly simultaneous whorls (e.g., *Foeniculum, Levisticum*; ERBAR and LEINS 1985).

Expression of Zygomorphy During Floral Organogenesis. Zygomorphy or bilateral symmetry of the flowers in early developmental stages were found in *Cercidiphyllum*, *Euptelea*, and *Trochodendron* (ENDRESS 1986b). In floral development of *Caesalpinia*, the abaxial stamens arise before the last petals are initiated on the adaxial side of the flower bud (TUCKER et al. 1985). POLOWICK and SAWHNEY (1986) observed a transient stage of zygomorphy during calyx formation in *Brassica*. In the flower of *Wigandia*, zygomorphy is temporarily present in the imbricate aestivation and persistent in the bicarpellate gynoecium. Zygomorphy of the gynoecium is already visible at initiation of the two carpels and is also manifested in different locule depths of the half-inferior ovary (HILGER 1987). In *Oncidium* flowers zygomorphy is also expressed in early development, as shown by TANAKA et al. (1986).

Reproductive Development, Comparative Morphology, and Character State Polarity. Ontogenetic studies may be crucial to some degree for determining the polarity of character states, as discussed in general surveys (DE QUEIROZ 1985; KLUGE and STRAUSS 1985). Similarities in early stages of reproductive development were used as evidence for a closer relationship, e.g., between *Trochodendron*, *Cercidiphyllum*, and *Euptelea* (all Trochodendrales sensu ENDRESS 1986b); between Aristolochiaceae and Alismatidae (LEINS and ERBAR 1985b); and between Caryophyllidae and Dilleniidae (LEINS and SCHWITALLA 1985; ERBAR 1986). Studying floral organogenesis, TUCKER et al. (1985) tried to determine the developmental basis for evolutionary divergence within the genus *Caesalpinia*. The diagnostic differences separating the species studied are manifest late in floral development corroborating TUCKER's hypothesis that differences at the species level are expressed later than differences at higher taxonomic levels. Doubts on the validity of the ontogenetic method were expressed by PHILIPSON (1985) and BARABÉ and CHRÉTIEN (1985b). According to them, observable morphogenesis may give little information on the ancestral state of, e.g., the gynoecium and the number of carpels involved (see below). For PHILIPSON (1985) and BARABÉ and CHRÉTIEN (1985b) anatomical data (e.g., vascularization patterns) seem to be more important than ontogenetic evidence, corresponding to MOSELEY's belief in vascular conservatism (MOSELEY and UHL 1985). Adherents of the ontogenetic method should also rely on the comparative approach ("outgroup comparison"), as done, e.g. by ENDRESS (1986a) and POSLUSZNY et al. (1986). They concluded that bisexual flowers are more primitive than unisexual ones in Chloranthaceae and Lilaeaceae, by deriving them from ancestors near Trimeniaceae and Juncaginaceae, respectively. Polarity of character states may, but should not, be inferred from ontogenetic evidence only. Are unisexual flowers primitive in angiosperms? THOMPSON (1986) avoids a decisive answer because bisexual and unisexual states are readily interchangeable or reversible in many angiosperm taxa.

3. Inflorescences

A new mathematical notation for inflorescence growth in general was proposed by ROBINSON (1986). Beautiful SEM studies were performed on inflorescence development of grasses: *Brachiaria* (STÜR 1986), *Triticum* (GARDNER et al. 1985; PINTHUS in HALEVY 1985), and *Zea diploperennis* (SUNDBERG and ORR 1986). Spikelet initiation in *Brachiaria decumbens* is restricted to one side of the young spike axis (STÜR 1986). In Chinese spring wheat the number of spikelets per spike is correlated with the length of the vernalization period (GARDNER et al. 1985). Evidence from early reproductive development supports the hypothesis that the progenitor of a *Zea mays* ear may be a mixed teosinte inflorescence rather than a pure male tassel or pure female spike (SUNDBERG and ORR 1986).

VAN HEEL et al. (1987) studied reproductive development of *Elaeis guineensis* by means of SEM. Within the normally unisexual spikes the female flower groups develop acropetally as triaxial cincinni, the male units as reduced ones. UHLARZ (1986) described the structure and development of inflorescence and appendix in *Pinellia tripartita*.

Structure and development of marantaceous inflorescences were studied by KUNZE (1985) and KIRCHOFF (1986), mainly using the terminology of TROLL and WEBERLING (see, e.g., WEBERLING 1983). Compared with other Zingiberales, the marantaceous inflorescences appear to be the most complex ones. The polytelic synflorescences with two-flowered coflorescences may be derived from pleiothyrses (cf. *Canna*) through various processes, including terminal ontogenetic abbreviation. The inflorescence of *Thalia geniculata* is among the most complex in Marantaceae (KIRCHOFF 1986). The inflorescence structure of *Orchidantha* (Lowiaceae) shows obvious parallels to the Marantaceae (KUNZE 1986b).

According to WEBERLING (1985), the inflorescences of most Lauraceae are of monotelic character, only a few genera (e.g., *Cassytha*) show polytelic flowering systems with botric or spikelike or umbellate florescences. The inflorescences in the genus *Monimia* (Monimiaceae) appear to represent a reduction series from an indeterminate leafy shoot with simple axillary thyrses to a leafless, determinate thyrse (LORENCE 1985). KAUL (1985, 1986) added again new data on inflorescence architecture of Fagaceae. Inflorescence structure is more specialized in *Quercus* (e.g., showing unisexual spikes and anemophily) when compared with *Lithocarpus*, *Castanopsis*, and *Castanea*. The cupule (scaly or lamellate) of *Quercus* and *Lithocarpus* seems to have "evolved as a structure with its own qualities" (KAUL 1985).

News on Pseudanthia. A careful developmental study on the cyathia glands and their appendages in *Euphorbia*, *Monadenium*, *Synadenium*, and *Pedilanthus* revealed a great variety of developmental modes including the occasional formation of hypopeltate and hyperpeltate structures (HOPPE 1985). CLASSEN (1985) studied the organization and function of the showy bracts in pseudanthia-forming Verbenaceae-Symphoremoideae. The bracts may have evolved first as an aid for wind dispersal, whereas the showiness for pollinator attraction was added afterwards as a new function.

4. Floral Vascular Anatomy

Comparative anatomical studies on mature flowers were undertaken, e.g., in Berberidaceae (TERABAYASHI 1985a,b), Liquidambaroideae (BOGLE 1986), Moringaceae (PERIASAMY and INDIRA 1986), Staphyleaceae (DICKISON 1986), Scrophulariaceae and Bignoniaceae (ARMSTRONG 1985). MOSELEY and UHL (1985) studied floral anatomy of *Nelumbo*. A cortical vascular system accompanies the stelar one in floral and vegetative axes. Bundles of both systems supply sepals, petals, stamens, and carpels. Girdling bundles were found in the floral receptacles of *Nymphaea*, *Brasenia*, and *Cabomba* (ITO 1984, 1986). In *Hydrastis*, the vascular supply to the spirally arranged stamens and carpels is unique when compared with the related Ranunculaceae. The central cylinder gives off fasicle bundles which branch to supply stamens within a given parastichy. Each carpel receives four distinct traces. One becomes the dorsal carpellary bundle. The three remaining traces diverge at higher levels and become the ventral carpellary bundles (TOBE and KEATING 1985). FRENCH (1986b) surveyed stamen vasculature in many Araceae. Depending on the taxon there are one to many bundles entering the stamen, with or without branching or anastomoses.

5. Perianth

In addition to their usual functions (bud protection, optical signals) petals may even serve for secondary pollen presentation, as described by McCONCHIE et al. (1986) in *Acrotriche*. Paleobotanists as well as morphologists studying extant plants added new facets to our understanding of the perianth in angiosperms. WALKER and WALKER (1984) proposed a five-stage model of early angiosperm evolution from Pre-Barremian to Middle Cenomanian. Among other changes, this model includes an early reversal from basic entomophily back to wind pollination in the Barremian - Early Aptian, again followed by a reversal to secondary entomophily. The "basic entomophilous angiosperms" (most Magnoliidae, including Nymphaeales; most monocots) show tepals and tepalar petals which can be derived from floral bracts. In contrast, the "secondary entomophilous angiosperms" (majority of dicots, including Dilleniidae, Rosidae, Asteridae) apparently possess staminodial petals. On the basis of comparative morphology, KUNZE (1986a) interpreted bracts, sepals, and petals (tepals) as ontogenetically abbreviated compared to the adult foliage leaf, e.g., in *Pilea* and *Passiflora*. In addition, the perianth members may form "new structures without a counterpart in the adult leaf", e.g., basal fusion of tepals in *Pilea*, hoods on sepals in *Passiflora*.

The transverse tepal outgrowths (fruit wings) in *Salsola* and *Halothamnus* as well as the corona in the Asclepiadaceae (e.g., *Ceropegia*) may be called emergences on the basis of developmental evidence (FREITAG et al. 1986; HOFMANN and SPECHT 1986). DUBUC-LEBREUX and SATTLER (1985) studied the quantitative distribution of mitotic activity during early corolla development of *Nicotiana*. The petals grow by means of generalized mitotic activity. Terms such as "marginal meristem" and "marginal meristem fusion", therefore, should be dismissed.

6. Androecium

The morphological interpretation and phylogenetic origin of the angiosperm stamen are still doubtful. The stamens of many plants, upon closer inspection, may be accepted as intermediates between phyllomes and caulomes, especially in cases where stamens or stamen fascicles arise in the axil of tepals (RUTISHAUSER and SATTLER 1985; SATTLER 1986:108).

Developmental Modes of Androecia. A basic initiation pattern of the androecium in various Magnoliidae and Alismatidae starts simultaneously with a whorl of six stamens inside a trimerous perianth. This pattern already known from some Alismatidae was observed in *Hydrocharis* (SCRIBAILO and POSLUSZNY 1985b) and also in Aristolochiaceae (LEINS and ERBAR 1985b). In the perianthless flowers of *Anemopsis* the six stamens arise by lateral splitting of three common primordia whose position is equivalent to the three stamens in *Houttuynia* (TUCKER 1985). Androecium development in male *Myristica* flowers starts with three anther primordia among which additional ones are intercalated. All anther primordia (around nine) finally form a synandrium because they remain adnate to the elongating receptacle (ARMSTRONG and TUCKER 1986). In the *Podophyllum* flower, there is a trimerous stamen whorl at first. Additional stamens arise outside and between the three first stamens (= centrifugal inception, DeMAGGIO and WILSON 1986). In *Brassica* the two short stamens (opposite the lateral sepals) are initiated after and basipetal to the long stamens (POLOWICK and SAWHNEY 1986). Though unusual within Dilleniidae, the staminodes in pistillate *Actinidia* flowers seem to arise centripetally (POLITO and GRANT 1984). In *Pereskia* and *Stewartia*, the individual stamens originate centrifugally on five common primordia (LEINS and SCHWITALLA 1985; ERBAR 1986).

Complex Formation Involving the Androecium and Gynoecium. In *Aristolochia* an anther-stigma-tube (gynostemium) is formed above the inferior ovary by common growth under the primordia of the six stamens and the six stigmas which originate on the ventral side of the stamen primordia. According to LEINS and ERBAR (1985b) these commissural stigmas might be viewed as stamen outgrowths or as upper carpel zones. Evolutionary trends in the gynostemium of the Orchidaceae are summarized by BURNS-BALOGH and BERNHARDT (1985). They proposed a progressive change in the position of the anther base relative to the apex of the stigma. SEM micrographs of gynostemium formation in *Oncidium* have been published by TANAKA et al. (1986).

7. Floral Nectaries

SMETS (1986) gave a survey on the distribution and systematic importance of floral nectaries in dicotyledons. Septal nectaries and their phyletic trends in monocotyledons are summarized by SCHMID (1985). Anatomical studies on floral nectaries using transmission electron microscopy were performed on, e.g., *Abutilon* (KRONESTEDT et al. 1986), *Brassica* (DAVIES et al. 1986), *Cynanchum* (CHRIST and SCHNEPF 1985), and *Strelitzia* (KRONESTEDT and WALLES 1986). Anatomical data on the oil-producing part of nectarless *Lysimachia* flowers are given by VOGEL (1986). The epipetalous glands of *Eriocaulon* are formed by the apex of the petal primordia (STÜTZEL 1985a). SCRIBAILO and POSLUSZNY (1985b) found in *Hydrocharis* that the pistillike structure in the center of the male flower is actually nectariferous and probably consists of remnants of staminodial primordia. The nectaries at the sepal bases and petioles of *Ipomoea* probably contribute to antiherbivore defense (KEELER and KAUL 1984).

8. Gynoecium

Syncarpy and Position of Ovary. Various aspects and possible benefits of fusion and epigyny in the angiosperm flower are discussed by ROBINSON (1985). Describing early developmental stages of the inferior gynoecia in Apiaceae, LEINS and ERBAR (1985a) concluded that the conception of congenital fusion of the dorsal carpel flanks with the floral apex or with basal parts of perianth and androecium is a metaphysical one ("ein pures Scheinproblem"). Epigyny in Apiaceae and in many other angiosperms (e.g., *Discaria*) results from an intercalary growth (diffuse meristem) in the receptacle (see similar statements by MEDAN 1985).

According to SCRIBAILO and POSLUSZNY (1985b) syncarpy and epigyny of the female flower of *Hydrocharis* result from intercalary meristematic activity. Various degrees of syncarpy and epigyny were described in *Boykinia* and related genera (GORNALL and BOHM 1985). According to FALLEN (1986), congenital syncarpy is characteristic only of the more primitive part of the Apocynaceae. Similar to the Asclepiadaceae, the more advanced member of the Apocynaceae (70% of the species) are secondarily apocarpous. All these apocarpous species, however, are functionally syncarpous at anthesis due to a temporary postgenital fusion of the carpel apices, forming a style head with vertical compartmentalization into three functional regions: pollen storage, adhesive production, and stigma (i.e., an organization quite similar to that of the Asclepiadaceae). In contrast to other Apocynaceae, *Allamanda* exhibits the "gentianaceous" type of gynoecium development: the two carpel are free at initiation, but fuse completely during development, resulting in a unilocular ovary with parietal placentation (a case of "secondary syncarpy" according to FALLEN 1985). MEDAN (1985) found in *Discaria* a case of "restricted syncarpy": The compitum is restricted to the base of the style, whereas there are three separate pollen tube pathways higher up.

Difficulties in Counting Carpels of Seemingly Syncarpous Gynoecia. Depending on the criteria used, the gynoecium in grasses (e.g., *Bambusa*, *Nardus*) and in *Monodora* consists of either one or more carpels (PHILIPSON 1985; KIRCHER 1986; DEROIN 1985). Comparing gynoecial structure of Restionaceae and Gramineae, KIRCHER (1986) concluded that the pseudomonomerous gynoecium of the grasses corresponds to that of derived Restionaceae. Pseudomonomery in both families must be considered to be derived from a trimerous-syncarpous gynoecium with axile placentation. Sterile lateral chambers as found in the restionaceous genus *Leptocarpus* are no longer formed in grasses. *Bambusa*, however, still shows three stylar rudiments in early gynoecial development (PHILIPSON 1985). According to BARABÉ and CHRÉTIEN (1985b), the pseudomonomery of the gynoecium in various Araceae (e.g., *Spathicarpa*) may be due to the total or partial reduction of a septum, of one carpel, or of both at once. BARABÉ and CHRÉTIEN (1985b) favor the view that the study of ontogenesis does not allow one to determine the number of carpels composing the gynoecium.

DE MAGGIO and WILSON (1986) inferred from vascularization that the second carpel was formerly present in the gynoecium of *Podophyllum*, although the SEM micrographs of gynoecial initiation do not support this view. SINGH and SINGH (1985) used developmental evidence to conclude that the *Cinnamomum* gynoecium primitively consists of three carpels. The triangular shape of the floral apex prior to gynoecial initiation, however, may be due to contact-pressure of the three innermost stamens rather than to an early existence of three carpel primordia (compare SEM micrographs on early gynoecial development in the lauraceous *Pleurothyrium*, ROHWER 1986:27). Although the gynoecium in *Vitis* is initiated as five confluent primordia, it finally becomes a two-loculed ovary by inward growth of two opposite septae from the gynoecial ring (POSLUSZNY and GERRATH 1985). In terms of a dynamic morphology (see Introduction) controversies on carpel numbers and pseudomonomery probably become evanescent!

Stigma Shape and Position. Stigma morphology was used by BOLAND and SEDGLEY (1986) to clarify the systematic position of *Angophora* in relation to the various *Eucalyptus* groups. Commissural stigmas were described in *Aristolochia* (LEINS and ERBAR 1985b, see above) and in Eriocaulaceae-Paepalanthoideae (STÜTZEL 1985c). In *Discaria*, three commissural lobes overgrow the three receptive areas of the stigma (MEDAN 1985).

9. Pistil-Pollen Interactions

Interesting aspects on pistil-pollen interactions were presented in a symposium volume by WILLEMSE and VAN WENT (1985), especially focusing on incompatibility and sterility. The secretory system of the stigma was studied, e.g., in *Medicago* (KREITNER and SORENSEN 1985), *Nicotiana* (CRESTI et al. 1986), and *Oenothera* (HESLOP-HARRISON and HESLOP-HARRISON 1985), also including biochemical and ultrastructural aspects.

Pollen tube growth is controlled by various factors. Self-pollination in incompatible *Citrus* (Orlando tangelo) leads to pollen tube arrest primarily within the stigma (KAHN and DE MASON 1986). Treatment of the stigma with an extract of a compatible pistil overcomes self-incompatibility in *Petunia* (SHARMA and SHIVANNA 1986). *Macadamia* possesses a complex pistil structure which seems to be involved in pollen tube inhibition in the upper style region (SEDGLEY et al. 1985). MULCAHY and MULCAHY (1985) suggested that the ovary plays an important role in pollen-stigma interactions, as indicated by the semi-vivo technique. Late-acting self-incompatibilities (outside or inside the ovule) are reviewed by SEAVEY and BAWA (1986). According to CRUDEN and LYON (1985) there is a positive correlation between style length and pollen grain size in *Polygonum* but not in various members of Apiaceae, Brassicaceae, and *Solanum*. Except for *Polygonum*, these findings are consistent with AMICI's model that pollen tubes receive resources from the stigma and/or style. While studying pollen tube growth in tristylous *Pontederia cordata* ANDERSON and BARRETT (1986) concluded that pollen tube length in styles was positively correlated with pollen size. In most pollinations legitimate pollen tubes grew more rapidly in styles than self-pollen tubes. Inhibition of incompatible pollen tubes was either in the style or ovary. The path of pollen tube growth in *Phaseolus acutifolius* was observed by LORD and KOHORN (1986). Pollen tubes adhere to epidermal secretory cells along the ventral side of the stylar canal and upper ovary, providing a possible site for selection of male gametes. *Raphanus raphanistrum* shows significant variation in the order of fertilization of the linearly arranged ovules within the ovary. The ovules in stylar position are not always the first to be fertilized. HILL and LORD (1986) concluded from these data that the path of pollen tube growth and gynoecial structure deserve greater attention in future studies of gamete competition.

In *Gasteria*, the ultrastructure of the transmitting tissue changes considerably during stylar development from anthesis until passing of the pollen tubes (WILLEMSE and FRANSSEN-VERHEIJEN 1986).

An interesting case of incongruity (sensu HOGENBOOM 1984) was found in *Rhododendron* following interspecific pollination (WILLIAMS et al. 1986). The pollen tube failed to terminate growth within the ovule filling the main body of the embryo sac with a coiled and distorted mass.

NICHOLLS (1985) described an evolutionary breakdown of distyly in *Linum tenuifolium*. The homostylous self-compatible race retains all the floral-morphological traits associated with the long-style (pin) morph of the distylous race. An evolutionary breakdown of tristyly was described in *Eichhornia paniculata* (BARRETT 1985; GLOVER and BARRETT 1986). The NE-Brazil populations are tristylous, whereas the populations on Jamaica are composed exclusively of self-pollinating, semi-homostylous, mid-styled forms.

10. Fruits and Seeds

Ovule and seed development was studied in various small families in order to elucidate their systematic position: e.g., Alzateaceae, Rhynchocalycaceae (TOBE and RAVEN 1984a,b); Datiscaceae, Humiriaceae, Trigoniaceae (BOESEWINKEL 1984, 1985a, 1987); Hydrastidaceae (TOBE and KEATING 1985); Krameriaceae (VERKERKE 1985); Mayacaceae (VENTURELLI and BOUMAN 1986).

Development, morphology, and function of mericarps (partial fruits) were studied in *Limnanthes* (HOFMANN and LUDEWIG 1985), Cynoglosseae and Eritricheae (HILGER 1985) and *Nolana* (BONDESON 1986). The locellar organization of the gynoecium is a common feature of all Nolanaceae. In *Nolana humifusa*, the gynoecium consists of five mericarps with several seeds each, whereas *N. paradoxa* is characterized by a high number of one-seeded mericarps which may be called "carpel-lobes" according to BONDESON (1986).

Morphology and development of fruits and seeds need to be seen in a functional and ecological context. An attractive little book on seed ecology was written by FENNER (1985). Floral architecture and development are better understood when the postfloral stages until fruit or seed dispersal are included in the investigation, as in the Apocynaceae (FALLEN 1986), and in *Discaria* (MEDAN 1985) with its ballistic seed dispersal. Morphological adaptations of fruits and seeds to various dispersal modes were described, e.g. by FOSTER (1986) who studied large seeds of tropical moist forest trees. AUGSPURGER (1986) surveyed wind-dispersed diaspores of neotropical trees. TIFFNEY (1984, 1986) discussed fruit and seed dispersal of early angiosperms and Hamamelidae within an evolutionary framework.

Many botanists investigated surfaces of seeds and fruits, mainly using SEM. Only a few studies can be mentioned here. Considerable intraspecific variation in seed coat ornamentation was found in *Arenaria uniflora* (Wyatt 1984) and *Silene latifolia* (PRENTICE 1986). Seed morphology of the latter species even shows clinal variation which seems to be maintained by climatic selection. In *Portulaca* those North American species with the largest geographical ranges and ecological diversities showed the greatest variation in seed morphology (MATTHEWS and LEVINS 1986). Seed coat ornamentation might be useful in taxonomy in spite of considerable intraspecific variation, as indicated by SAINT-MARTIN (1986) who studied many members of the Papilionoideae by means of SEM. *Cymbalaria* shows unequal seed size (heterospermy) depending on the position within the ovary (SPETA 1986).

11. Origin and Evolution of the Flowering Plants

Fossil evidence throwing light on the early phylogeny of the flowering plants (angiosperms) has increased rapidly over the past decade. Paleobotanists and neobotanists have shown a mutual concern for the origin and early differentiation of the flowering plants (see, e.g., NISHIDA 1985; THOMAS and SPICER 1987). A symposium volume on "Historical perspectives of angiosperm evolution" provides an opportunity to gain some overview of the contemporary field of angiosperm paleobotany (Ann. Missouri Bot. Gard. 71(2):347-630 (1984)).

As an illustrative example the Cercidiphyllaceae may be mentioned. Paleobotanists (CRANE and STOCKEY 1985, 1986) compared inflorescences of extant *Cercidiphyllum* with fossil *Cercidiphyllum*-like plants *(Joffrea, Nyssidium, Trochodendrocarpus)* from the Paleocene. Neobotanists (ENDRESS 1986b; VAN HEEL 1986; MOHANA RAO 1986) studied reproductive structures and development in extant *Cercidiphyllum* and other members of Trochodendrales. A synthetic view of the phylogeny of the Cercidiphyllaceae may be summarized as follows: The condensed female inflorescences of *Cercidiphyllum* evolved by reduction from an elongated form. In contrast to typical extant *Cercidiphyllum*, the fossil taxa may have contained two carpels per flower instead of one, resembling bicarpellate flowers in extant Hamamelidae. Floral elaboration as well as reduction of floral parts played a significant role in evolution of the Hamamelidae, as shown by various contributors to a symposium volume on "Phylogeny of the Hamamelidae" (Ann. Missouri Bot. Gard. 73(2):225-441 (1986)).

NISHIDA (1985) reported a permineralized angiosperm floral axis bearing conduplicate carpels from mid-Cretaceous of Japan. Although similar to extant Monimiaceae and Austrobaileyaceae in respect to its concave receptable it shows several features unknown in any living magnolian taxon, e.g., carpel architecture with a long stigmatic crest and a large number of ovules (similar to *Archaeanthus* as described by DILCHER and CRANE 1984). DILCHER and KOVACH (1986) found another Cretaceous fructification, *Caloda delevoryana*, demonstrating that various putatively primitive angiosperm characters were present by the Cenomanian: e.g., apocarpy, superior gynoecium, conduplicate carpels which are numerous and helically arranged, small seeds. Fossil evidence, therefore, corroborates the classical magnoliid floral theory: Magnoliid-like plants with bisexual flowers bearing many carpels and stamens existed early in angiosperm evolution. CREPET (1984) provided new fossil evidence that insect pollination played an important role in angiosperm success because dipteran, coleopteran, hymenopteran, and lepidopteran pollinators already existed at a time of major angiosperm radiation. Fossils have been useful for determining polarity of reproductive character states in Platanaceae (CRANE et al. 1986; MANCHESTER 1986) and Mimosoideae (CREPET and TAYLOR 1986).

The use of cladistic analysis apparently allows a more objective evaluation of comparative data on reproductive structures and functions of extant and extinct angiosperms (NISHIDA 1985). LAMMERS et al. (1986) have shown that a combination of cladistics, phenetics, and phyletics may help in the elucidation of systematic relationships of primitive angiosperm families (e.g., Lactoridaceae). Using cladistics (phylogenetic analysis) CRANE (1985a,b) suggested that the angiosperms, Gnetales, Bennettitales, and *Pentoxylon* form a monophyletic clade. Taking into account reproductive and vegetative characters, it even seems that the flowering plants are more closely related to the Gnetales than any other group of gymnosperms. DOYLE and DONOGHUE (1986a,b) concluded similarly by using an experimental cladistic approach. To corrobrate a closer relationship of angiosperms and Gnetales, it is also important to intensify research on reproductive structure and development of extant Gnetales.

It is therefore justified to mention here new developmental data on Gnetales. TAKASO (1984, 1985) studied the development of the female strobilus and the early ontogeny of the ovule with its two envelopes in *Ephedra distachya* and *E. equisetina*, using SEM and light microscopy. The two ovule primordia within the female strobilus originate from the axils of the uppermost bract pair. Developmental and anatomical evidence confirms that the outer envelope ("outer integument") resembles vegetative leaves more than the inner one.

12. Systematic Survey on Morphological Literature

The following list contains mainly morphological literature of the years 1985-1986. The letters added indicate interesting topics of each study with respect to reproductive structure and development:

D = reproductive development, including floral initiation (often by means of SEM);
V = vascularization of reproductive structures (mainly flowers);
J = inflorescence morphology;
P,A,G,F,S = noteworthy data on perianth (P), androecium (A), gynoecium (G), fruit (F), and seed (S), respectively;
N = structure of nectaries;
T = pistil-pollen interactions (pollen tube growth, transmitting tissue, incompatibility);
R = reproductive biology considering morphological aspects (e.g., sex distribution, dimorphic flowers, heterostyly);
E = fossil data.

Monocotyledons

Araceae: *Arisaema* J (MURATA 1984), *Monstera*, *Spathicarpa*, *Spathiphyllum* VG (BARABÉ and CHRÉTIEN 1985a,b, 1986), *Orontium* VG (BARABÉ and LABRECQUE 1985), *Pinellia* VJ (UHLARZ 1986), many genera VA (FRENCH 1985, 1986a,b).
Arecaceae: *Elaeis* D (VAN HEEL et al. 1987), *Washingtonia* S (DE MASON 1986).
Cymodoceaceae: *Halodule* F (Bragg and McMILLAN 1986).
Eriocaulaceae: *Eriocaulon* DN, *Tonina* A, *Wurdackia* PG (STÜTZEL 1985a-c).
Gramineae: *Bambusa*, *Nardus* DG (PHILIPSON 1985), *Brachiaria* D (STÜR 1986), *Hordeum* D (DEITZER in HALEVY 1985), *Pennisetum* D (BELLIARD and PERNÈS in HALEVY 1985), *Sorghum* D (EASTIN and LEE in HALEVY 1985), *Triticum* D (GARDNER et al. 1985; PINTHUS in HALEVY 1985), *Zea mays* (HANWAY and RITCHIE in HALEVY 1985), *Zea diploperennis* D (SUNDBERG and ORR 1986).
Hydrocharitaceae: *Hydrocharis* DSR (SCRIBAILO and POSLUSZNY 1984, 1985a,b).
Iridaceae: *Moraea* G (GOLDBLATT 1986).
Lemnaceae: *Lemna* D (KANDELER in HALEVY 1985).
Lilaeaceae (?): *Lilaea* (POSLUSZNY et al. 1986).
Liliaceae: *Erythronium* R (HARDER et al. 1985), *Gasteria* T (WILLEMSE and FRANSSEN-VERHEIJEN 1986).
Lowiaceae: *Orchidantha* DJ (KUNZE 1986b).
Marantaceae: *Thalia* DJ (KIRCHOFF 1986), various genera DJ (KUNZE 1985).
Mayacaceae: *Mayaca* S (VENTURELLI and BOUMAN 1986).
Orchidaceae: *Bulbophyllum* AG (RASMUSSEN 1985), *Oncidium* D (TANAKA et al. 1986).
Pontederiaceae: *Eichhornia* R (BARRETT 1985; GLOVER and BARRETT 1986).
Restionaceae: *Lepyrodia* DG (PHILIPSON 1985), several genera JPG (KIRCHER 1986).
Strelitziaceae: *Strelitzia* GN (KRONESTEDT and WALLES 1986).

Dicotyledons

Actinidiaceae: *Actinidia* D (POLITO and GRANT 1984).
Alzateaceae: *Alzatea* S (TOBE and RAVEN 1984b).
Anacardiaceae: *Rhus* FS (VON TEICHMANN and ROBBERTSE 1986).
Annonaceae: *Isolona*, *Monodora* VG (DEROIN 1985), many genera S (CHRISTMANN 1986).
Apiaceae: *Astrantia*, *Eryngium*, *Foeniculum*, *Hydrocotyle*, *Levisticum* DG (ERBAR and LEINS 1985; LEINS and ERBAR 1985a), European genera P (SCHMITZ and FROEBE 1986).

Apocynaceae: *Allamanda* and many other genera DAG (FALLEN 1985, 1986), Tabernaemontanoideae S (ALLORGE 1985).

Araliaceae: *Arthrophyllum* J (AH-LAN 1985), *Hedera* D (HACKETT and SRINIVASAN in HALEVY 1985).

Aristolochiaceae: *Aristolochia*, *Asarum* D (LEINS and ERBAR 1985b), *Thottea* A (RENUKA and SWARUPANANDAN 1986).

Asclepiadaceae: *Ceropegia*, *Cynanchum*, *Dischidia* PA (HOFMANN and SPECHT 1986), *Cynanchum* N (CHRIST and SCHNEPF 1985).

Asteraceae: *Carlina* PAF (DITTRICH 1985), *Cirsium* R (KAY 1985), *Cosmos* D (MOLDER and OWENS in HALEVY 1985), Cynareae FS (SINGH and PANDEY 1984), *Helianthus* D (SCHUSTER in HALEVY 1985), Israelian genera PF (SHMIDA 1985), several genera S (GRAU and HOPF 1985).

Balsaminaceae: *Impatiens* GS (SHIMIZU and TAKAO 1985), *Impatiens* D (BATTEY and LYNDON 1986).

Begoniaceae: *Begonia* sect. *Solananthera* S (DE LANGE and BOUMAN 1986).

Berberidaceae: *Podophyllum* DV (DE MAGGIO and WILSON 1986), all genera V (TERABAYASHI 1985a,b).

Bignoniaceae: *Kigelia* N (SUBRAMANIAN and INAMDAR 1985), *Schlegelia* and other genera VG (ARMSTRONG 1985).

Boraginaceae: Cynoglosseae, Eritricheae DF, *Mattiastrum* F (HILGER 1985, 1986).

Cactaceae: *Pereskia* D (LEINS and SCHWITALLA 1985).

Callitrichaceae: *Callitriche* T (SCHOTSMAN 1985).

Caprifoliaceae: *Weigela* D (MALOUPA-IKONOMOU and JACQUES 1986).

Caryophyllaceae: *Arenaria* S (WYATT 1984), *Polycarpaea* JGS (RUTISHAUSER and MENDOZA-HEUER 1985), *Silene* D (LYNDON 1987; LYNDON and BATTEY 1985), *Silene* S (PRENTICE 1986).

Celastraceae: *Empleuridium* G (GOLDBLATT et al. 1985).

Cercidiphyllaceae: *Cercidiphyllum* DAG (ENDRESS 1986b), D (VAN HEEL 1986), FS (MOHANA RAO 1986), *Joffrea*, *Nyssidium* JE (CRANE and STOCKEY 1985, 1986).

Chenopodiaceae: *Halothamnus*, *Salsola* PF (FREITAG et al. 1986).

Chloranthaceae: e.g. *Sarcandra* DAG (ENDRESS 1986a).

Convolvulaceae: *Ipomoea* N (KEELER and KAUL 1984).

Cruciferae: *Brassica* D (POLOWICK and SAWHNEY 1986), *Brassica* VN (DAVIS et al. 1986), *Farsetia* S (JONSELL 1986), *Raphanus* T (HILL and LORD 1986).

Cucurbitaceae: e.g. *Citrullus* R (ZAMSKI and RUDICH in HALEVY 1985).

Datiscaceae: *Datisca*, *Octomeles*, *Tetrameles* S (BOESEWINKEL 1984).

Ebenaceae: *Austrodiospyros* E (BASINGER and CHRISTOPHEL 1985).

Epacridaceae: *Acrotriche* P (McCONCHIE et al. 1986).

Ericaceae: *Rhododendron* T (WILLIAMS et al. 1986).

Euphorbiaceae: *Euphorbia*, *Monadenium*, *Synadenium*, *Pedilanthus* DN (HOPPE 1985), *Phyllanthus* J (NOZERAN and ROSSIGNOL-BANCILHON in HALEVY 1985), *Piccodendron* V (HAKKI 1985), *Ricinus* J (ATSMON in HALEVY 1985).

Eupteleaceae: *Euptelea* D (ENDRESS 1986b).

Fagaceae: *Quercus*, *Lithocarpus*, *Castanea*, *Castanopsis* JR (KAUL 1985, 1986, KAUL et al. 1986).

Guttiferae: *Hypericum* S (REYNAUD 1985).

Hamamelidaceae: *Altingia*, *Liquidambar* V (BOGLE 1986), *Noahdendron* PAG (ENDRESS et al. 1985).

Humiriaceae: *Humiria* S (BOESEWINKEL 1985a).

Hydrastidaceae: *Hydrastis* DVS (TOBE and KEATING 1985).

Hydrophyllaceae: *Phacelia* P (LEE 1986), *Pholistoma* GS (BERG 1985), *Wigandia* DVG (HILGER 1987).

Juglandaceae: *Carya* D (WETZSTEIN and SPARKS 1984), *Juglans* D (POLITO and LI 1985a,b).

Krameriaceae: *Krameria* S (VERKERKE 1985).

Lauraceae: *Cinnamomum* D (ENDRESS 1986a; SINGH and SINGH 1985), *Ocotea* s.l. DA (ROHWER 1986; ROHWER and KUBITZKI 1985), several genera J (WEBERLING 1985).

Leguminosae: *Caesalpinia* D (TUCKER et al. 1985), *Kummerowia* J (AKIJAMA and OHBA 1985), *Medicago* T (KREITNER and SORENSEN 1985), *Phaseolus* T (LORD and KOHORN 1986), *Prosopis* S (SERRATO VALENTI 1986), *Stylosanthes* D (ISON and HOPKINSON in HALEVY 1985), Mimosoideae E (CREPET and TAYLOR 1986), Papilionoideae S (SAINT-MARTIN 1986).

Limanthaceae: *Limnanthes* DG (HOFMANN and LUDEWIG 1985).
Linaceae: *Linum* R (NICHOLLS 1985).
Lythraceae: *Alzatea*, *Rhynchocalyx* S (TOBE and RAVEN 1984a,b), *Rotala* and other genera S (PANIGRAHI 1986).
Malvaceae: *Abutilon* N (KRONESTEDT et al. 1986).
Mesembryanthemaceae: *Psilocaulon* FS (IHLENFELDT and BITTRICH 1985).
Monimiaceae: *Monimia*, *Tambourissa* and other genera JG (LORENCE 1985).
Moringaceae: *Moringa* G (PERIASAMY and INDIRA 1986).
Myristicaceae: *Myristica* (ARMSTRONG and TUCKER 1986).
Myrtaceae: *Eucalyptus*, *Angophora* GT (BOLAND and SEDGLEY 1986).
Nelumbonaceae: *Nelumbo* V (MOSELEY and UHL 1985).
Nolanaceae: *Nolana* DF (BONDESON 1986).
Nymphaeaceae: *Nymphaea*, *Brasenia*, *Cabomba* V (ITO 1984, 1986).
Onagraceae: *Oenothera* T (HESLOP-HARRISON and HESLOP-HARRISON 1985), many genera AS (TOBE and RAVEN 1985, 1986).
Oxalidaceae: *Averrhoea* S (BOESEWINKEL 1985b), *Oxalis* R (WELLER 1986).
Papaveraceae: Fumarieae PF (LIDÉN 1986).
Platanaceae: cf. *Platanus* E (CRANE et al. 1986; MANCHESTER 1986).
Plumbaginaceae: *Limonium* DJ (SHILLO and ZAMSKI in HALEVY 1985).
Polygonaceae: *Polygonum* T (CRUDEN and LYON 1985), *Polygonum* F (WOLF and McNEILL 1986).
Portulacaceae: *Portulaca* S (MATTHEWS and LEVINS 1986), *Talinum* GF (GALATI 1986).
Primulaceae: *Lysimachia* N (VOGEL 1986), *Primula* R (PIPER and CHARLESWORTH 1986).
Proteaceae: *Macadamia* D (MONCUR et al. 1985), P (STOREY in HALEVY 1985), T (SEDGLEY et al. 1985).
Ranunculaceae: *Hydrastis* DVS (TOBE and KEATING 1985).
Rhamnaceae: *Discaria* DG (MEDAN 1985).
Rhynchocalycaceae: *Rhynchocalyx* S (TOBE and RAVEN 1984a).
Rosaceae: *Prunus* F (Labrecque et al. 1985a,b).
Rubiaceae: Anthospermeae JFR (PUFF 1986), *Houstonia* S (TERRELL et al. 1986).
Rutaceae: *Citrus* D (LORD and ECKARD 1985), T (KAHN and DE MASON 1986), *Empleuridium* G (GOLDBLATT et al. 1985).
Saururaceae: *Anemopsis* D (TUCKER 1985).
Saxifragaceae: *Boykinia* and other genera JG (GORNALL and BOHM 1985).
Scrophulariaceae: *Cymbalaria* FS (SPETA 1986), *Paulownia* and other genera VG (ARMSTRONG 1985).
Solanaceae: *Capsicum* AG (POLOWICK and SAWHNEY 1985), *Capsicum* T (SHI-YI and LI-YUN 1985), *Lycianthes* P (D'ARCY 1986), *Lycopersicon* G (SAWHNEY and POLOWICK 1985), *Nicotiana* P (DUBUC-LEBREUX and SATTLER 1985), *Nicotiana* DR (MARCOTRIGIANO 1986), *Petunia* T (SHARMA and SHIVANNA 1986).
Staphyleaceae: e.g. *Staphylea*, *Tapiscia* V (DICKISON 1986).
Theaceae: *Stewartia* D (ERBAR 1986).
Trigoniaceae: *Trigonia* S (BOESEWINKEL 1987).
Trochodendraceae: *Trochodendron* DAG, *Tetracentron* AG (ENDRESS 1986b).
Turneraceae: *Turnera* R (BARRETT and SHORE 1985).
Verbenaceae: *Holmskioldia* D (BALA and KUMAR 1986), Symphoremoideae J (CLASSEN 1985).
Vitaceae: *Vitis* D (POSLUSZNY and GERRATH 1985).

Acknowledgments. The author is grateful to Drs. P.K. ENDRESS and L.D. HUFFORD for valuable comments on the manuscript.

References

AH-LAN, L.: Kew Bull. 41, 769-780 (1986). - AKIYAMA, S., OHBA, H.: Bot. Mag. (Tokyo) 98, 137-150 (1985). - ALLORGE, L.: Adansonia, Sér. 4,7, 433-451 (1985). - ANDERSON, J.M., BARRETT, S.C.H.: Can. J. Bot. 64, 2602-2607 (1986). - ARMSTRONG, J.E.: Am. J. Bot. 72, 755-756 (1985). - ARMSTRONG, J.E., TUCKER, S.C.: Am. J. Bot. 73, 1131-1143 (1986). - AUGSPURGER, C.K.: Am. J. Bot. 73, 353-363 (1986).

BALA, R., KUMAR, V.: Acta Bot. Indica 14, 18-23 (1986). - BARABÉ, D., CHRÉTIEN, L.: Can. J. Bot. 63, 1423-1428 (1985a); - Beitr. Biol. Pflanz. 60, 403-412 (1985b); - Can. J. Bot. 64, 1397-1401 (1986). - BARABÉ, D., LABRECQUE, M.: Bull. Soc. Fr. (Lett. Bot. 1985) 132, 133-145 (1985). - BARRETT, S.C.H.: Biol. J. Linn. Soc. 25, 41-60 (1985). - BARRETT, S.C.H., SHORE, J.S.: Ann. Missouri Bot. Gard. 72, 259-263 (1985). - BASINGER, J.F., CHRISTOPHEL, D.C.: Can. J. Bot. 63, 1825-1843 (1985). - BATTEY, N.H., LYNDON, R.F.: Ann. Bot. 58, 333-341 (1986). - BERG, R.Y.: Am. J. Bot. 72, 1775-1787 (1985). - BERNIER, G.: Symp. Soc. Exp. Biol. 40, 257-286 (1986). - BOESEWINKEL, F.D.: Acta Bot. Neerl. 33, 419-429 (1984); - ibid. 34, 183-191 (1985a); - ibid. 34, 413-424 (1985b); - ibid. 36, 81-91 (1987). - BOGLE, A.L.: Ann. Missouri Bot. Gard. 73, 325-347 (1986). - BOLAND, D.J., SEDGLEY, M.: Aust. J. Bot. 34, 569-584 (1986). - BONDESON, W.E.: Nord. J. Bot. 6, 183-198 (1986). - BRAGG, L.H., McMILLAN, C.: Am. J. Bot. 73, 815-821 (1986). - BURNS-BALOGH, P., BERNHARDT, P.: Plant Syst. Evol. 149, 119-134 (1985).

CHRIST, P., SCHNEPF, E.: Israel J. Bot. 34, 79-90 (1985). - CHRISTMAN, M.: Bot. Jahrb. Syst. 106, 379-390 (1986). - CLASSEN, R.: Beitr. Biol. Pflanz. 60, 383-402 (1985). - CRANE, P.R.: Ann. Missouri Bot. Gard. 72, 716-793 (1985a); - Cladistics 1, 329-348 (1985b). - CRANE, P.R., FRIIS, E.M., PEDERSEN, K.R.: Science 232, 852-854 (1986). - CRANE, P.R., STOCKEY, R.A.: Can. J. Bot. 63, 340-364 (1985); - Ann. Missouri Bot. Gard. 73, 382-393 (1986). - CREPET, W.L.: Ann. Missouri Bot. Gard. 71, 607-630 (1984). - CREPET, W.L., TAYLOR, D.W.: Am. J. Bot. 73, 548-563 (1986). - CRESTI, M., KEIJZER, C.J., TIEZZI, A., CIAMPOLINI, F., FOCARDI, S.: Am. J. Bot. 1713-1722 (1986). - CRUDEN, R.W., LYON, D.L.: Bot. Gaz. 146, 143-149 (1985).

D'ARCY, W.G.: Ann. Missouri Bot. Gard. 73, 117-127 (1986). - DAVIS, A.R., PETERSON, R.L., SHUEL, R.W.: Can. J. Bot. 64, 2508-2516 (1986). - DE LANGE, A., BOUMAN, F.: Acta Bot. Neerl. 35, 489-495 (1986). - DE MAGGIO, A.E., WILSON, C.L.: Am. J. Bot. 73, 21-32 (1986). - DE MASON, D.A.: Am. J. Bot. 73, 1332-1340 (1986). - DE QUEIROZ, K.: Syst. Zool. 34, 280-299 (1985). - DEROIN, T.: Adansonia, Sér. 4,7, 167-176 (1985). - DICKISON, W.C.: Bot. Gaz. 147, 312-326 (1986). - DILCHER, D.L., CRANE, P.R.: Ann. Missouri Bot. Gard. 71, 351-381 (1984). - DILCHER, D.L., KOVACH, W.L.: Am. J. Bot. 73, 1230-1237 (1986). - DITTRICH, M.: Bot. Jahrb. Syst. 107, 591-609 (1985). - DOYLE, J.A., DONOGHUE, M.J.: Bot. Rev. 52, 321-431 (1986a); - Relationships of angiosperms and Gnetales: a numerical cladistic analysis, 177-198. In: Systematic and Taxonomic Approaches in Palaeobotany, eds. B.A. THOMAS, R.A. SPICER. Oxford: Oxford University Press 1986b. - DUBUC-LEBREUX, M.-A., SATTLER, R.: Phytomorphology 35, 17-23 (1985).

ENDRESS, P.K.: Plant Syst. Evol. 152, 1-28 (1986a); - Ann. Missouri Bot. Gard. 73, 297-324 (1986b). - ENDRESS, P.K., HYLAND, B.P.M., TRACEY, J.G.: Bot. Jahrb. Syst. 107, 369-378 (1985). - ERBAR, C.: Bot. Jahrb. Syst. 106, 391-407 (1986). - ERBAR, C., LEINS, P.: Bot. Jahrb. Syst. 105, 379-400 (1985).

FALLEN, M.: Am. J. Bot. 72, 572-579 (1985); - Bot. Jahrb. Syst. 106, 245-286 (1986). - FENNER, M.: Seed Ecology, 151 pp. London: Chapman and Hall 1985. - FOSTER, S.A.: Bot. Rev. 52, 260-299 (1986). - FREITAG, H., MAIER-STOLTE, M., STIEGEMEYER, M.: Beitr. Biol. Pflanz. 61, 87-98 (1986). - FRENCH, J.C.: Bot. Gaz. 146, 521-533 (1985); - ibid. 147, 166-179 (1986a); - Am. J. Bot. 73, 434-449 (1986b).

GALATI, B.G.: Parodiana 4, 123-131 (1986). - GARDNER, J.S., HESS, W.M., TRIONE, E.J.: Am. J. Bot. 72, 548-559 (1985). - GLOVER, D.E., BARRETT, S.C.H.: Evolution 40, 1122-1131 (1986). - GOLDBLATT, P.: Ann. Missouri Bot. Gard. 73, 102-116 (1986). - GOLDBLATT, P., TOBE, H., CARLQUIST, S.,

PATEL, V.C.: Ann. Missouri Bot. Gard. 72, 167-183 (1985). - GORNALL, R.J., BOHM, B.A.: Bot. J. Linn. Soc. 90, 1-71 (1985). - GRAU, J., HOPF, H.: Bot. Jahrb. Syst. 107, 251-268 (1985). - GREPPIN, H., AUDERSET, G., BONZON, M., DEGLI AGOSTI, R., LENK, R., PENEL, C.: Saussurea 17, 71-84 (1986).

HAKKI, M.I.: Bot. Jahrb. Syst. 107, 379-394 (1985). - HALEVY, A. (ed.): CRC Handbook of Flowering, Vol. I-IV, 2135 pp. Roca Baton: CRC Press 1985. - HARDER, L.D., THOMSON, J.D., CRUZAN, M.D., UNNASCH, R.S.: Oecologia 67, 286-291 (1985). - HEEL, W.A. VAN: Blumea 31, 273-276 (1986). - HEEL, W.A. VAN, BREURE, C.J., MENENDEZ, T.: Blumea 32, 67-78 (1987). - HESLOP-HARRISON, J., HESLOP-HARRISON, Y.: Israel J. Bot. 34, 187-204 (1985). - HILGER, H.H.: Bot. Jahrb. Syst. 105, 323-378 (1985); - Beitr. Biol. Pflanz. 61, 145-165 (1986); - Am. J. Bot. 74, 250-259 (1987). - HILL, J.P., LORD, E.M.: Evolution 40, 1328-1333 (1986). - HOFMANN, U., LUDEWIG, J.: Bot. Jahrb. Syst. 105, 401-431 (1985). - HOFMANN, U. SPECHT, A.K.: Beitr. Biol. Pflanz. 61, 79-85 (1986). - HOGENBOOM, N.G.: Incongruity: non-functioning of intercellular and intracellular partner relationships through non-matching information, 640-654. In: Encyclopedia of Plant Physiology, N.S., eds. H.F. LINSKENS, J. HESLOP-HARRISON, Vol. 17. Berlin, Heidelberg, New York: Springer 1984. - HOPPE, J.R.: Bot. Jahrb. Syst. 105, 497-581 (1985).

IHLENFELDT, H.-D., BITTRICH, V.: Bot. Jahrb. Syst. 105, 289-322 (1985). - ITO, M.: Acta Phytotax. Geobot. 35, 94-102 (1984); - Bot. Mag. (Tokyo) 99, 169-184 (1986).

JONSELL, B.: Symb. Bot. Upsal. 25(3), 107 pp. (1986)

KAHN, T.L., DE MASON, D.A.: Can. J. Bot. 64, 2548-2555 (1986). - KAUL, R.B.: Am. J. Bot. 72, 1962-1977 (1985); - Ann. Missouri Bot. Gard. 73, 284-296 (1986). - KAUL, R.B., ABBE, E.C., ABBE, L.B.: Biotropica 18, 51-55 (1986). - KAY, Q.O.N.: New Phytol. 100, 457-472 (1985). - KEELER, K.H., KAUL, R.B.: Am. J. Bot. 71, 1364-1372 (1984). - KIRCHER, P.: Diss. Bot. 94, 223 pp. (1986). - KIRCHOFF, B.K.: Can. J. Bot. 64, 859-864 (1986). - KLUGE, A.G., STRAUSS, R.E.: Annu. Rev. Ecol. Syst. 16, 247-268 (1985). - KREITNER, G.L., SORENSEN, E.L.: Can. J. Bot. 63, 813-818 (1985). - KRONESTEDT, E.C., WALLES, B.: Nord. J. Bot. 6, 307-320 (1986). - KRONESTEDT, E.C., ROBARDS, A.W., STARK, M., OLESEN, P.: Nord. J. Bot. 6, 627-639 (1986). - KUNZE, H.: Beitr. Biol. Pflanz. 60, 93-140 (1985); - ibid. 61, 49-77 (1986a); - ibid. 61, 221-234 (1986b).

LABRECQUE, M., BARABÊ, D., VIETH, J.: Can. J. Bot. 63, 242-251 (1985a); - Ann. Sci. Nat. Bot., Sér. 13, 7, 39-53 (1985b). - LAMMERS, T.G., STUESSY, T.F., SILVA, O.M.: Plant Syst. Evol. 152, 243-266 (1986). - LEE, G.J.: Am. J. Bot. 73, 230-235 (1986). - LEINS, P., ERBAR, C.: Bot. Jahrb. Syst. 106, 53-60 (1985a); - ibid. 107, 343-368 (1985b). - LEINS, P., SCHWITALLA, S.: Beitr. Biol. Pflanz. 60, 313-323 (1985). - LIDÊN, M.: Opera Bot. 88, 1-133 (1986). - LORD, E.M., ECKARD, K.J.: Bot. Gaz. 146, 320-326 (1985). - LORD, E.M., KOHORN, L.U.: Am. J. Bot. 73, 70-78 (1986). - LORENCE, D.H.: Ann. Missouri Bot. Gard. 72, 1-165 (1985). - LYNDON, R.F.: Ann. Bot. 59, 67-72 (1987). - LYNDON, R.F., BATTEY, N.H.: Biol. Pl. 27, 339-349 (1985).

MALOUPA-IKONOMOU, E., JACQUES, M.: Ann. Bot. 58, 887-895 (1986). - MANCHESTER, S.R.: Bot. Gaz. 147, 200-226 (1986). - MARCOTRIGIANO, M., Ann. Bot. 57, 435-442 (1986). - MATTHEWS, J.F., LEVINS, P.A.: Syst. Bot. 11, 302-308 (1986). - McCONCHIE, C.A., HOUGH, T., SINGH, M.B., KNOX, R.B.: Ann. Bot. 57, 155-164 (1986). - MEDAN, D.: Bot. Jahrb. Syst. 105, 205-262 (1985). - MOHANA RAO, P.R.: Flora 178, 243-249 (1986). - MONCUR, M.W., STEPHENSON, R.A., TROCHOULIAS, T.: Sci. Hort. 27, 87-96 (1985). - MOSELEY, M.F., UHL, N.W.: Bot. Jahrb. Syst. 106,

61-98 (1985). - MULCAHY, G.B., MULCAHY, D.L.: Am. J. Bot. 72, 1078-1080 (1985). - MURATA, J.: J. Fac. Sci. Univ. Tokyo, Sect. 3, Bot. 13(4), 431-482 (1984).

NICHOLLS, M.S.: Plant Syst. Evol. 150, 291-301 (1985). - NISHIDA, H.: Nature 318, 58-59 (1985).

PANIGRAHI, S.G.: Bot. J. Linn. Soc. 93, 389-403 (1986). - PERIASAMY, K., INDIRA, C.: Ann. Bot. 58, 897-901 (1986). - PHARIS, R.P., KING, R.W.: Annu. Rev. Plant Physiol. 36, 517-568 (1985). - PHILIPSON, W.R.: Am. J. Bot. 72, 1954-1961 (1985). - PIPER, J., CHARLESWORTH, B.: Biol. J. Linn. Soc. 29, 123-137 (1986). - POLITO, V.S., GRANT, J.A.: Sci. Hort. 22, 365-371 (1984). - POLITO, V.S., LI, N.-Y.: Pistillate flower differentiation in *Juglans regia*: a developmental basis for heterodichogamy, 134. In: Sexual Reproduction in Seed Plants, Ferns and Mosses, eds. M.T.M. WILLEMSE, J.L. VAN WENT. Wageningen: Pudoc 1985a; - Sci. Hort. 26, 333-338 (1985b). - POLOWICK, P.L., SAWHNEY, V.K.: Sci. Hort. 25, 117-127 (1985); - Am. J. Bot. 73, 254-263 (1986). - POSLUSZNY, U., GERRATH, J.M.: Can. J. Bot. 64, 1620-1631 (1985). - POSLUSZNY, U., CHARLTON, W.A., JAIN, D.K.: Bot. J. Linn. Soc. 92, 323-342 (1986). - PRENTICE, H.C.: Biol. J. Linn. Soc. 27, 179-189 (1986). - PUFF, C.: A Biosystematic Study of the African and Madagascan Rubiaceae-Anthospermeae. 535 pp. (Plant Syst. Evol., Suppl. 3). Wien, New York: Springer 1986.

RASMUSSEN, F.N.: Bot. J. Linn. Soc. 91, 447-456 (1985). - RENUKA, C., SWARUPANANDAN, K.: Blumea 31, 313-318 (1986). - REYNAUD, C.: Adansonia, Sér. 4, 7, 85-96 (1985). - ROBINSON, D.F.: New Phytol. 103, 587-596 (1986). - ROBINSON, H.: Syst. Bot. 10, 105-109 (1985). - ROHWER, J.G.: Mitt. Inst. Allg. Bot. Hamburg 20, 1-278 (1986). - ROHWER, J.G., KUBITZKI, K.: Bot. Jahrb. Syst. 107, 129-135 (1985). - RUTISHAUSER, R., MENDOZA-HEUER, I.: Candollea 40, 391-402 (1985). - RUTISHAUSER, R., SATTLER, R.: Bot. Jahrb. Syst. 107, 415-455 (1985); - Can. J. Bot. 64, 1987-2019 (1986).

SAINT-MARTIN, M.: Bull. Soc. Bot. Fr. (Lett. Bot. 1986) 133, 137-153 (1986). - SATTLER, R.: Biophilosophy. Analytic and Holistic Perspectives, 300 pp. Berlin, Heidelberg, New York: Springer 1986. - SAWHNEY, V.K., POLOWICK, P.L.: Can. J. Bot. 63, 1031-1034 (1985). - SCHMID, R.: Acta Bot. Neerl. 34, 125-128 (1985). - SCHMITZ, J., FROEBE, H.A.: Bot. Jahrb. Syst. 106, 337-357 (1986). - SCHOTSMAN, H.D.: Adansonia, Sér. 4, 7 (1985). - SCRIBAILO, R.W., POSLUSZNY, U.: Can. J. Bot. 62, 2779-2787 (1984); - ibid. 63, 492-496 (1985a); - Am. J. Bot. 72, 1578-1589 (1985b). - SEAVEY, S.R., BAWA, K.S.: Bot. Rev. 52, 195-219 (1986). - SEDGLEY, M., BLESING, M.A., VITHANAGE, H.I.M.V.: Bot. Gaz. 146, 6-14 (1985). - SERRATO VALENTI, G., MODENESI, P., ROTI-MICHELOZZI, G., BEVILACQUA, L.: Acta Bot. Neerl. 35, 475-487 (1986). - SHARMA, N., SHIVANNA, K.R.: New Phytol. 102, 443-447 (1986). - SHIMIZU, T., TAKAO, S.: Acta Phytotax. Geobot. 36, 97-106 (1985). - SHI-YI, H., LI-YUN, X.: Acta Bot. Sinica 27, 453-454 (1985). - SHMIDA, A.: Ann. Missouri Bot. Gard. 72, 184-186 (1985). - SINGH, R.P., PANDEY, A.K.: Phytomorphology 34, 1-10 (1984). - SINGH, V., SINGH, A.: Phytomorphology 35, 61-67 (1985). - SMETS, E.: Bull. Jard. Bot. Nat. Belg. 56, 51-76 (1986). - SPETA, F.: Phyton (Austria) 26, 23-57 (1986). - STUR, W.W.: Ann. Bot. 58, 569-575 (1986). - STÜTZEL, T.: Beitr. Biol. Pflanz. 60, 271-276 (1985a); - Bot. Jahrb. Syst. 105, 433-438 (1985b); - Flora 177, 335-344 (1985c). - SUBRAMANIAN, R.B., INAMDAR, J.A.: Bot. Mag. (Tokyo) 98, 67-73 (1985). - SUNDBERG, M.D., ORR, A.R.: Am. J. Bot. 73, 1699-1712 (1986).

TAKASO, T.: Acta Bot. Neerl. 33, 257-266 (1984); - ibid. 34, 33-48 (1985). - TANAKA, M., YAMADA, S., GOI, M.: Sci. Hort. 28, 133-146

(1986). - TEICHMANN, I. von, ROBBERTSE, P.J.: Bot. J. Linn. Soc. 93, 291-306 (1986). - TERABAYASHI, S.: Mem. Fac. Sci. Kyoto Univ., Ser. Biol. 10, 73-90 (1985a); - Acta Phytotax. Geobot. 36, 1-13 (1985b). - TERRELL, E.E., LEWIS, W.H., ROBINSON, H., NOWICKE, J.W.: Am. J. Bot. 73, 103-115 (1986). - THOMAS, B.A., SPICER, R.A.: The Evolution and Palaeobiology of Land Plants, 309 pp. London, Sydney: Croom Helm 1987. - THOMPSON, K.: New Phytol. 103, 597-601 (1986). - TIFFNEY, B.H.: Ann. Missouri Bot. Gard. 71, 551-576 (1984); - ibid. 73, 394-416 (1986). - TOBE, H., KEATING, R.C.: Bot. Mag. (Tokyo) 98, 291-316 (1985). - TOBE, H., RAVEN, P.H.: Ann. Missouri Bot. Gard. 71, 836-843 (1984a); - ibid. 71, 844-852 (1984b); - ibid. 72, 451-468 (1985); - Am. J. Bot. 73, 475-488 (1986). - TRAN THANH VAN, K.: Symp. Soc. Exp. Biol. 40, 171-179 (1986). - TUCKER, S.C.: Am. J. Bot. 72, 20-31 (1985). - TUCKER, S.C., STEIN, O.L., DERSTINE, K.S.: Am. J. Bot. 72 , 1424-1434 (1985).

UHLARZ, H.: Beitr. Biol. Pflanz. 61, 241-282 (1986).

VENTURELLI, M., BOUMAN, F.: Acta Bot. Neerl. 35, 497-516 (1986). - VERKERKE, W.: Beitr. Biol. Pflanz. 60, 341-351 (1985). - VOGEL, S.: Akad. Wiss. Lit. Mainz, math.-naturw. Kl., Trop. Subtrop. Pflanzenwelt 54, 168 pp. (1986).

WALKER, J.W., WALKER, A.G.: Ann. Missouri Bot. Gard. 71, 464-521 (1984). - WEBERLING, F.: Bothalia 14, 917-922 (1983); - Bot. Jahrb. Syst. 107, 395-414 (1985). - WELLER, S.G.: Evolution 40, 279-289 (1986). - WETZSTEIN, H.Y., SPARKS, D.: J. Am. Soc. Hort. Sci. 109, 245-252 (1984). WILLEMSE, M.T.M., FRANSSEN-VERHEIJEN, M.A.W.: Acta Bot. Neerl. 35, 297-309 (1986). - WILLEMSE, M.T.M., VAN WENT, J.L. (eds.): Sexual Reproduction in Seed Plants, Ferns and Mosses, 206 pp. Wageningen: Pudoc 1985. - WILLIAMS, E.G., KAUL, V., ROUSE, J.L., PALSER, B.F.: Aust. J. Bot. 34, 413-423 (1986). - WOLF, S.J., McNEILL, J.: Rhodora 88, 457-479 (1986). - WYATT, R.: Syst. Bot. 9, 423-431 (1984).

Dr. ROLF RUTISHAUSER
Botanischer Garten und Institut für
Systematische Botanik der Universität
Zollikerstr. 107
CH-8008 Zürich, Schweiz

B. Physiology

I. Mineral Nutrition: Anions

By ULRICH LÜTTGE and David T. CLARKSON

1. Gradient-Coupled Transport: General Observations

The discovery of the convertibility between the energy of electrochemical proton gradients built up by charge separation at membranes and phosphate-ester-bond energy in ATP (Mitchell hypothesis) has led to a new orthodoxy in transport physiology. In principle all energy-dependent ion fluxes can be explained by secondary coupling to a proton electro-chemical gradient, $\Delta\bar{\mu}_{H^+}$, with

$$\Delta\bar{\mu}_{H^+} = RT \ln \frac{10^{-pH_i}}{10^{-pH_o}} + F\,\Delta E \qquad (1)$$

or a proton-motive force, *pmf*, with

$$pmf = 2.3\,\frac{RT}{F}\,\Delta pH + \Delta E \qquad (2)$$

where R is the universal gas constant, T the absolute temperature, ΔE the transmembrane electrical potential and F the Faraday constant; i and o mark inside and outside compartments respectively, ΔpH is the pH difference between the compartments.

On the basis of this concept a wealth of useful observations on anion-transport mechanisms has been obtained. There are two principle possibilities for approaches, i.e. interference with *pmf* and examination of the consequences for anion transport, or observations of effects of anion transport on *pmf*. Both have been used. Inhibitors and stimulators affecting the membrane mechanisms building up *pmf* were applied. With measurements of transmembrane electrical potentials, known external pH and assumptions on cytosolic pH (e.g. values between pH 7 and 8) thermodynamic considerations (Eqs. 1 and 2) became possible.

In this way comprehensive studies with duckweeds (*Lemna gibba* L., strain G1) demonstrated that all major inorganic anions are co-transported with protons using $\Delta\bar{\mu}_{H^+}$ as the driving force for uptake (see LÜTTGE and CLARKSON 1985). Transient depolarization of the transmembrane electrical potential at the onset of anion uptake by *Lemna* cells showed that anion uptake was associated with a net uptake of positive charge. Thus, more than one H^+ must be co-transported per NO_3^- (ULLRICH and NOVACKY 1981), 2 H^+ are co-transported with each $H_2PO_4^-$ (ULLRICH-EBERIUS et al. 1984), and for sulphate the stoichiometry most likely is 3 H^+/SO_4^{2-} (LASS and ULLRICH-EBERIUS 1984). Fusicoccin, a fungal toxin well known to stimulate active H^+ extrusion from the cells, stimulated uptake of these anions by *L. gibba*.

Progress in Botany, Vol. 49

a) Nitrate Uptake

Studies of nitrate uptake in the unicellular alga *Ankistrodesmus braunii* were performed under conditions where the degree of metabolic utilization of products of NO_3^- reduction varied. Hence, NO_2^- and NH_4^+ were excreted to various extents. All conditions gave good stoichiometries of

$$OH^-_{efflux} = NO_3^- \text{ influx} - NO_2^- \text{ efflux} + NH_4^+ \text{ efflux}$$

(EISELE and ULLRICH 1975). In solutions bubbled with CO_2-free air or nitrogen, uptake of NO_3^- by *A. braunii* was inhibited. Addition of glucose alleviated this inhibition probably via production of a metabolite which is common to photosynthesis and glucose metabolism (EISELE and ULLRICH 1977). In *Chlorella vulgaris* a pre-incubation with glucose led to a five-fold stimulation of NO_3^- uptake. It was proposed that this might be due to induction of the NO_3^- uptake mechanism, which was considered to be a 1 H^+/1 NO_3^- co-transport system (SCHLEE et al. 1985). For *Chara corallina* DEANE-DRUMMOND (1984) argued, however, that there was no obligatory coupling of NO_3^--uptake with H^+ transport and that NO_3^- could be moved across the cell membrane by NO_3^-/HCO_3^- or NO_3^-/OH^- exchange or at high internal NO_3^- concentrations by NO_3^-/NO_3^- exchange.

b) Phosphate Uptake

Phosphate uptake by *A. braunii* showed dependence on both H^+ and Na^+, suggesting an $(H^+ + Na^+)/H_2PO_4^-$ co-transport, where at raised external pH, Na^+ increasingly replaced H^+, while at pH 5.6 Na^+ had no effect and if present gave a low co-transport stoichiometry (ULLRICH and GLASER 1982). Dependence of phosphate uptake on pH by pea-root protoplasts (LEFEBVRE and CLARKSON 1984) and potato-tissue slices (PODER and PENOT 1986) also suggested the possibility of a co-transport with H^+. In the latter work this was further supported by the use of *pmf* inhibitors. In a similar way an H^+/phosphate co-transport (or OH^-/phosphate-antiport) was invoked in maize roots (LIN 1979). With a closer analysis of pH dependence in corn roots SENTENAC and GRIGNON (1985) arrived at the conclusion that $H_2PO_4^-$ is the transported species. In addition, they found that at constant $H_2PO_4^-$ concentrations pH acts by controlling intrinsic kinetic properties of the transport system *or* its possible co-substrates like (H^+, OH^- or HCO_3^- for co- or anti-port respectively). An alternative which they found most difficult to decide.

c) Sulphate Uptake

Sulphate uptake has been extensively studied in cultured tobacco cells (JONES and SMITH 1981 with further references therein). In maize-root protoplasts anion-transport inhibitors, i.e., amino-reactive disulphonic-stilbene derivatives, inhibited SO_4^{2-} and Cl^- uptake, while phosphate uptake was unaffected (LIN 1981). The pH dependence of Cl^- and SO_4^{2-} uptake in these experiments was consistent with the involvement of OH^- exchange or H^+ co-transport mechanisms.

d) Bicarbonate Uptake

The last and most difficult of the metabolized anions to consider is bicarbonate uptake. The particular problems arise from the difficulty of distinguishing between HCO_3^- and OH^- transport, since $HCO_3^- \rightleftharpoons OH^- + CO_2$, and from the fact that it is impossible to distinguish between OH^- transport and H^+ transport in the opposite direction. Moreover,

CO_2 is both the substrate and the product of basic metabolic reaction sequences (photosynthesis, photorespiration, mitochondrial respiration), and CO_2 transport is a possible alternative to HCO_3^- transport. Nevertheless, work on isolated pea-mesophyll protoplasts (VOLOKITA et al. 1981) and extensive studies on aquatic photosynthetic organisms proves that there is a carrier-mediated transport of inorganic carbon which leads to internal CO_2 enrichment (review: LUCAS and BERRY 1985). Sulfhydryl groups are involved in HCO_3^- and OH^- transport at the plasmalemma of *Chara corallina* (LUCAS and ALEXANDER 1980). Experiments with mutants of *Chlamydomonas reinhardii* show that carbonic anhydrase is involved in generating CO_2 from HCO_3^- after uptake (SPALDING et al. 1983 a,b).

In *Chara* and in the leaves of some angiosperm water plants sophistication of the problem is augmented by the formation of separate alkaline and acid zones on the surface. The alkaline-acid-banding phenomenon in *Chara* led to complicated and often contradictory arguments during the long history of its investigation. A fairly recent report by LUCAS (1982) shows that surprisingly high electrical membrane potentials can be built up in *C. corallina*, i.e. -350 to -373 mV between vacuole and medium and -383 to -393 mV between cytoplasm and medium. Energetically this requires the hydrolysis of 1 ATP per H^+ extruded, and the results of this work support a 1 H^+/1 HCO_3^- co-transport into the cells. (The review of LUCAS and BERRY 1985, can also be consulted). In leaves of the water plants *Elodea* and *Potamogeton* acid and alkaline zones are separated on the lower and upper surfaces respectively. H^+ is excreted on the lower surface where K^+ is taken up in exchange. K^+ is transported across the leaves, and K^+ and OH^- extrusion occurs at the upper surface. It is argued that the equilibrium of $HCO_3^- + H^+ \rightleftharpoons CO_2 + H_2O$ is shifted to external CO_2 enrichment on the lower leaf surface and CO_2 is the species taken up (PRINS et al. 1980, 1982).

e) Chloride Uptake

Chloride uptake, i.e. transport of an anion non-metabolized and not directly involved in acid-base equilibria, is more readily discussed in relation to energy-dependent H^+ extrusion and cytoplasmic pH regulation.

Chloride uptake and also $H_2PO_4^-$ uptake were barely affected by fusicoccin (FC) in isolated protoplasts obtained from tobacco cells in suspension culture, although K^+ uptake was stimulated (METTLER and LEONARD 1979). Since FC stimulates energy-dependent H^+ extrusion and *pmf* formation, putative H^+-anion co-transporters should in fact be also stimulated, and cytoplasmic pH changes might be involved. Indeed, internal pH does appear to regulate uptake and release of Cl^- among other anions (HCO_3^-, glycolate) in *Scenedesmus obliquus* (FINDENEGG 1977). Cl^- uptake is inhibited by intracellular acidification in *Chlorella fusca*. In the latter case intracellular acidification was shown to originate from glucose metabolism (i.e. organic acids as products of glucose oxidation; DOBLINGER and TROMBALLA 1982a,b). This is an interesting possibility of a glucose effect alternative or additional to what was discussed above in relation to NO_3^- uptake. Cytoplasmic pH also appears to affect Cl^- fluxes at the plasmalemma of *C. corallina* (SANDERS 1980c).

The expected effect of FC on cytoplasmic pH should be an alkalinization due to stimulated H^+ extrusion. Indeed, many investigations studying distribution and equilibria of weak organic acids (KURKDJAN et al. 1982; MARRÈ et al. 1983, 1986; ROMANI et al. 1985) and determinations of cytoplasmic pH by ^{31}P-nuclear-magnetic resonance (REID et al. 1985a,b) show an FC-dependent alkalinization of the cytoplasm. The issue recently has become controversial, since measurements with pH-sensitive microelectrodes showed an acidification of the cytoplasm after addition of FC (BERTL and FELLE 1985; BRUMMER et al. 1985). For the time being the dispute remains unresolved, but in view of the numerous facets of cytoplasmic pH control, e.g. assimilatory metabolism, organic acid metabolism, energy-dependent H^+ transport at plasmalemma and tonoplast (BELOEIL and LALLEMAND 1987; DAVIES 1986), this is scarcely surprising. It is well not to take too exclusive a view of the process at present. Depending on the physiological situation of the material studied, the different response times of the methods used may result in different observations. SANDERS et al. (1985) also caution that manipulations of the components of $\Delta\mu_{H^+}$ with the aim of discerning the mechanism of transport may lead to erroneous conclusions if the attendant complex regulatory phenomena are ignored. Nevertheless, all authors appear to agree that H^+ extrusion from the cytoplasm is regulated by cytoplasmic pH.

Comprehensive studies on *C. corallina*, i.e. kinetics and current-voltage relations (BEILBY and WALKER 1981), the response of Cl^- uptake to external and internal pH in perfused internodal cells and the stimulation of Cl^- uptake by addition of ATP to the perfusion medium suggest operation of a specific 2 H^+/1 Cl^- co-transport mechanism (SANDERS 1980b, 1981).

f) β-Indolyl-Acetate Transport

A special case of anion transport is that of β-indolyl-acetate (IAA^-). Organic anions often can readily permeate membranes by passive diffusion in the form of the respective undissociated acid. With subsequent dissociation in an alkaline compartment, e.g. the cytoplasm, this may lead to anion accumulation in that compartment (see Sect. 3). The same holds for the phytohormone IAA. Much work now invokes a mechanism of $\Delta\bar{\mu}_{H^+}$-dependent transport, where the undissociated acid HIAA may be taken up into cells by carrier-independent diffusion in the lipid phase of the plasmalemma and the anion IAA^- is extruded by a specific IAA^- carrier. With a polar distribution of the IAA^- carriers, this may also explain polar IAA transport along files of cells in plant organs (e.g. RUBERY and SHELDRAKE 1974; RAVEN 1975; CANDE and RAY 1976; EDWARDS and GOLDSMITH 1980; GOLDSMITH and GOLDSMITH 1981; SUSSMAN and GOLDSMITH 1981; GOLDSMITH 1982; MARTIN and PILET 1986; BENNING 1986; SABATER and SABATER 1986; but see also HASENSTEIN and RAYLE 1984, for deviating views).

In conclusion of this section the observations with more or less intact systems in vivo strongly suggest the occurrence of *pmf*-dependent H^+ anion co-transport mechanisms in plant cells.

2. Transport Proteins

a) Distinct Gene Products?

Varietal and genetically determined differences in transport functions in plants are long known and have been reviewed at regular intervals (recently GERLOFF and GABELMAN 1983). In the context of anion transport the substrate induction of the NO_3^- uptake mechanism dependent on protein synthesis is frequently quoted in favour of genetic control (JACKSON et al. 1973, 1974, 1976; HEIMER 1975; NEYRA and HAGEMANN 1975). But note that DEANE-DRUMMOND and JACOBSEN (1986) suggest caution because work with mutants of *Pisum sativum* shows a requirement of active nitrate reductase for NO_3^- uptake. Differences of transmembrane electric potentials in the order of 10 to 20 mV among maize hybrids suggest genetic control of the energy-dependent H^+ extrusion mechanism (MARRÈ and ROMANI 1982). Thus, it should be accepted that distinct gene products, i.e. concrete transport proteins, are involved both in primary active transport establishing driving forces and in secondary active transport moving anions.

b) Primary Active Transport Mechanisms and the Effects of Anions

Purified plasmalemma and tonoplast fractions are obtained by density-gradient centrifugation which has much stimulated plant-membrane biochemistry as frequently reviewed recently (e.g. LEIGH 1983; SZE 1984, 1985; MARRÈ and BALLARIN-DENTI 1985; BOLLER and WIEMKEN 1986; see also previous reports LÜTTGE 1983; CLARKSON and LÜTTGE 1984). Plasmalemma and tonoplast are also separated by preparative free-flow electrophoresis (AUDERSET et al. 1986; SANDELIUS et al. 1986). Good plasmalemma fractions are obtained by partitioning in aqueous polymer two-phase systems (WINGSTRAND 1985; CLÉMENT et al. 1986; BÉRCZI and MØLLER 1986). Enzymes using Mg-ATP^{2-} as a substrate for primary active electrogenic transport of H^+ have been detected at both membranes and characterized (reviews as cited above). They differ in several properties, e.g. the pH optimum, which is more alkaline for the tonoplast ATPase than for the plasmalemma ATPase. There are further mechanisms of primary active transport at both membranes, i.e. an oxidoreductase at the plasmalemma (see previous report, LÜTTGE and CLARKSON 1985) and a pyrophosphatase at the tonoplast (references below). In the context of anion relations, the effects of anions on these mechanisms shall be considered here.

α) Plasmalemma

i) The ATPase. The ATPase of the plasmalemma is inhibited by vanadate, an effect now so well known that it is already used as a plasmalemma marker (e.g. CHURCHILL and SZE 1983; LEW and SPANSWICK 1984; O'NEILL and SPANSWICK 1984a,b, and reviews cited above). Vanadate is a xenobiotic anion, but its effect is considered to be due to interference with phosphate. In addition to the catalytic site, where ATP is hydrolyzed during functioning of the ATPase, there must be a site where the enzyme protein needs to be phosphorylated to be active.

BRISKIN and LEONARD (1982a,b) discovered an endogenous protein-kinase activity in the plasmalemma fraction of maize roots. A plasmalemma-bound calcium/calmodulin-regulated protein kinase was also found in pea by BLOWERS et al. (1985). BRUMMER and PARISH (1983) separated five specific plasmalemma proteins in corn coleoptiles, two of which were phosphoproteins and could be part of the ATPase. In the preparations of BRISKIN and LEONARD (1982a,b) the protein-kinase activity could be removed by deoxycholate treatment, and this allowed the demonstration of a rapidly turning over covalent phosphorylated intermediate in the plasmalemma fraction. It had a size of 100 kD. Gamma-ray target studies on the red-beet plasmalemma suggested that the ATPase has a molecular weight of 228 kD, while dodecylsulphate gel electrophoresis gave a size of 100 kD for the catalytic subunit, thus the ATPase should be a dimer (BRISKIN et al. 1985b). Estimates obtained by ANTHON and SPANSWICK (1986) come close to these values, giving 90 kD for the catalytic subunit in tomato-root plasmalemma, but these authors suggest that the ATPase is a trimer. Thus, there appears to be little doubt that the 100-kD protein of maize roots studied by BRISKIN and LEONARD was the catalytic subunit of the plasmalemma ATPase. Using red-beet plasmalemma fractions BRISKIN and POOLE further characterized the ATPase phosphorylation. The turnover of the phosphorylated protein is increased by KCl (BRISKIN and POOLE 1983a), where it should be noted that the ATPase requires K^+ for its function. It is an aspartic-acid residue which is phosphorylated (BRISKIN and POOLE 1983b). Formation of the phosphorylated intermediate, but not its breakdown, requires Mg^{2+} (BRISKIN and POOLE 1983c).

The requirement of phosphorylation of the ATPase is an important anion interaction with primary active transport in the plasmalemma. It seems probable that other membrane proteins become phosphorylated by the action of calcium/calmodulin and protein kinases. ZOCCHI (1985) found that "membrane" phosphorylation inhibited not only the vanadate-sensitive plasmalemma ATPase activity, but also the vanadate-insensitive and NO_3^- sensitive tonoplast ATPase activity, although this enzyme is not believed to form a phosphorylated intermediate. This observation needs to be confirmed.

ATP hydrolysis can be measured readily in the inverted (inside-out) microsomes separated by density-gradient centrifugation, but separation by aqueous two-phase partitioning gives sealed "right-side out" plasmalemma vesicles. ATP hydrolysis by such preparations cannot be measured unless they are permeabilized by detergents. This process renders the vesicles useless for transport studies.

Plasmalemma vesicles prepared by gradient centrifugation are frequently leaky when they come from the gradient. This has hampered transport studies on plasmalemma vesicles. However, recently DE MICHELIS and SPANSWICK (1986) obtained tight H^+ transporting plasmalemma vesicles from maize roots. Anions stimulated H^+ transport in relation to their permeability with a sequence of effectiveness of $Br^- > Cl^- = NO_3^- \gg SO_4^{2-}$. This stimulation of H^+ transport was due to a dissipation of $\Delta\bar{\mu}_{H^+}$ by anion transport.

$Mg\text{-}ATP^{2-}$-dependent H^+ transport in sealed plasmalemma vesicles of radish was stimulated by fusicoccin. This stimulation was observed only in the physiological pH range (pH 7.0 to 7.6) and not in the slightly acid pH range (pH 6.0 to 6.6) where the ATPase has its optimum (RASI-CALDOGNO et al. 1986). This suggests that the fusicoccin effects discussed in Section 1 may be due to direct interaction of the toxin with the plasmalemma ATPase.

The plasmalemma ATPase protein itself is insensitive to anions. Generally, no effects of anions like NO_3^-, Cl^- and malate on ATP hydrolysis are observed (reviews as cited above). Only ANTHON and SPANSWICK (1986) reported a weak inhibition by NO_3^- for the plasmalemma ATPase of tomato roots. ATP hydrolysis by the plasmalemma ATPase of red beet was inhibited by free ATP, i.e. an excess of substrate, and by ADP but there was little effect of inorganic phosphate (BENNETT et al. 1985; RAUSCH et al. 1985).

There might also be primary active transport of Cl^-. Physiological evidence by SADDLER (1970, 1971) suggested that *Acetabularia* possesses an electrogenic Cl^- transport at the plasmalemma. GRADMANN et al. (1982) and GOLDFARB et al. (1984a,b) described a Cl^- stimulated, vanadate-sensitive ATPase in a plasmalemma-enriched fraction of *Acetabularia*. This enzyme even seems to be reversible, i.e. capable of forming ATP at the expense of a chloride electrochemical gradient (GOLDFARB et al. 1984b). However, these observations have remained rather isolated.

ii) The Oxidoreductase. The oxidoreductase of the plasmalemma is the second primary active charge-separation system at this membrane. It has been reviewed again (BIENFAIT 1985; MØLLER and LIN 1986) and much new work has been presented since we covered it in a report in this series (LÜTTGE and CLARKSON 1985). Pyridine-nucleotide oxidation has been demonstrated with microsomes and plasmalemma vesicles, and O_2 was shown to be able to serve as an electron acceptor (MØLLER and BÉRCZI 1985, 1986; MACRI and VIANELLO 1986; PUPILLO et al. 1986). Since the system can reduce extracellular Fe compounds, such as the non-permeating ferricyanide, it is generally believed that it does transport electrons and elicit charge separation across the membrane. However, there was one report suggesting that reducing equivalents are transported out of the cells via a malate-oxaloacetate shuttle mechanism and that there were malate dehydrogenase and NADH-Fe^{3+}-reductase activities in isolated cell walls of soybean roots (TIPTON and THOWSEN 1985).

Initially it has been widely accepted that the oxidoreductase transports two electrons and one proton out of the cell and K^+ follows for charge balance. Good stoichiometries supporting this mechanism are obtained under some conditions (see LÜTTGE and CLARKSON 1985). LASS et al. (1986) recently reported stoichiometries of 1.28 H^+ : 0.80 K^+ : 2.00 e^- for iron-deficient and 1.16 H^+ : 1.00 K^+ : 2.00 e^- for non-deficient *Lemna* plants. However, very different stoichiometries are also observed. Several workers maintain that the plasmalemma oxidoreductase transports electrons and protons (BÖTTGER et al. 1985; BÖTTGER and LÜTHEN 1986; BELKOURA et al. 1986). Although vanadate strongly inhibited net H^+ efflux from corn roots in their experiments, vanadate did not prevent the stimulating effect of ferricyanide; the enhancement of net efflux by ferricyanide was the same in vanadate pretreated and in untreated roots (BÖTTGER and LÜTHEN 1986). However, other authors strongly suggest that the oxidoreductase only transfers electrons and that the protons left behind are extruded from the cytoplasm via the ATPase (BRISKIN et al. 1985a; MACRI and VIANELLO 1986; RUBINSTEIN and STERN 1986). This would explain varying stoichiometries, effects of ATPase inhibitors and an initial lag phase of H^+ extrusion after the onset of e^- transport. LASS et al. (1986) suggest that initially H^+ efflux could occur via the redox system but in the steady state probably is brought about by the ATPase.

With a net export of negative charge the plasmalemma oxidoreductase would appear as an ideal system for driving secondary active uptake of anions as stated in our earlier report (LÜTTGE and CLARKSON 1985). However, LASS et al. (1986) found an *inhibition* of anion transport during ferricyanide reduction in *Lemna*.

β) Tonoplast

i) The ATPase. The ATPase of the tonoplast is insensitive to vanadate; it does not form a phosphorylated intermediate (experiments with yeast tonoplasts: LICHKO and OKOROKOV 1985). Dodecyl sulphate-polyacrylamide gel electrophoresis gave two major polypeptides of 72 and 62 kD and gamma-ray inactivation a total size of 400 kD (MANDALA and TAIZ 1985). Thus, the tonoplast ATPase is composed of several polypeptides (also: MANOLSON et al. 1985). In contrast to the plasmalemma ATPase which, apart from the requirement of phosphorylation, is insensitive to anions and needs K^+ for its activity, the tonoplast ATPase is anion-dependent. It is inhibited by nitrate and stimulated by chloride and malate. Hence, it is frequently called an "anion ATPase". Sealed right-side-out tonoplast vesicles, which are at least partially tight to protons, are readily obtained by gradient centrifugation, and both ATP hydrolysis and Mg-ATP^{2-}-dependent *pmf* generation can be studied. Both functions have largely similar properties. The mitochondrial ATPase is also inhibited by nitrate, but at the same time is sensitive to azide. Thus, a vanadate- and azide-resistant, NO_3^--sensitive ATPase activity is now widely accepted as a tonoplast marker. (Reviews as cited above and BENNETT and SPANSWICK 1983a,b, 1984; O'NEILL et al. 1983; JOCHEM et al. 1984; BENNETT et al. 1984; POOLE et al. 1984; TOGNOLI 1985; JOCHEM and LÜTTGE 1987; STRUVE and LÜTTGE 1987).

Several groups have attempted to elucidate the mechanism of anion effects on the tonoplast ATPase. It is essential to distinguish between coupled and uncoupled vesicles. Mg-ATP^{2-}-driven H^+ translocation into the tonoplast vesicles is stalled by the *pmf* built up, when an equilibrium is reached. Hence, in coupled vesicles anions will stimulate the ATPase by increasing its turnover when they are taken up into the vesicles at the expense of *pmf* (see also Sect. 2d). If the vesicles are uncoupled, however, by using detergents (like Triton), NH_4^+ ions or uncouplers (like gramicidin) which dissipate $\Delta\bar{\mu}_{H^+}$, changes of ATP hydrolysis reflect direct effects of anions on the ATPase. In this way inhibition of the vacuolar ATPase by NO_3^- and stimulation by Cl^- and malate were demonstrated and interpreted as a direct interaction of the anions with the ATPase protein (BENNETT and SPANSWICK 1983a; CHURCHILL and SZE 1984; JOCHEM et al. 1984; LEW and SPANSWICK 1985; STRUVE and LÜTTGE 1987). Another tool that was used is the anion-transport blocker 4,4-diisothiocyano-2,2-stilbene disulphonic acid (DIDS) (BENNETT and SPANSWICK 1983a; CHURCHILL and SZE 1984; STRUVE and LÜTTGE 1987).

As GRIFFITH et al. (1986) point out, details of these interactions still need to be elucidated. These authors observed a decrease of the K_m for Mg-ATP^{2-} in response to Cl^- and an increase in response to NO_3^-. Since the tonoplast ATPase needs free SH groups for its catalytic function it can be inhibited by SH reagents like N-ethylmaleimide (NEM). Prior binding of its substrate or product (Mg-ATP^{2-} or Mg-ADP^{2-}) protects the ATPase from NEM inhibition. But Cl^- and NO_3^- have little or no effect on this protection and therefore most likely do not interact at the substrate binding site. NO_3^- is thought to have a chaotropic effect destroying water structures (GRIFFITH et al. 1986) but may also affect the ATPase molecule directly (STRUVE and LÜTTGE 1987).

ii) The Pyrophosphatase. In various membrane fractions of a variety of plant materials, membrane-bound pyrophosphatase activities (PP_i-ases) have been detected, and in several cases their association with the tonoplast and their involvement in establishing a proton electrochemical gradient at the tonoplast was demonstrated (KARLSSON 1975; LEIGH and WALKER 1980; WALKER and LEIGH 1981; DUPONT et al. 1982; CHURCHILL and SZE 1983; WAGNER and MULREADY 1983; BENNETT et al. 1984; BRISKIN et al. 1985c; CHANSON et al. 1985; REA and POOLE 1985, 1986; WANG et al. 1986; MARQUARDT and LÜTTGE 1987). The PP_i-ase constitutes a second $\Delta\bar{\mu}_{H^+}$ generating enzyme at the tonoplast, separate from the ATPase. In red beet the two activities could be solubilized, separated and reconstituted in liposomes where neither of them showed any H^+ transport with the substrate of the other one (REA and POOLE 1986).

The tonoplast PP_i-ase of higher plants has a similar alkaline pH optimum as the ATPase. But otherwise it has very different properties. It requires K^+ and needs a large excess of Mg^{2+} over ATP. The enzyme itself is insensitive to anions. In vitro anions only stimulate PP_i-ase turnover by utilizing *pmf* for movement into the tonoplast vesicles (references as cited above). Pyrophosphate levels in plant cells are between about 5 and 90 µmol kg^{-1} fresh weight (EDWARDS et al. 1984; SMYTH and BLACK 1984; CHANSON et al. 1985; MARQUARDT and LÜTTGE 1987) and thus in the same range as ATP levels. Since in vitro the tonoplast PP_i-ase has about 40 to 50 times higher apparent affinity to Mg-PP_i^{2-} than the ATPase to Mg-ATP^{2-}, the PP_i-ase could contribute considerably to vacuolar H^+ transport in vivo (MARQUARDT and LÜTTGE 1987).

iii) Regulation of the ATPase and the PP_i-ase. Regulation of the ATPase and the PP_i-ase at the tonoplast needs to be elucidated. Apart from the similar pH optimum, the enzymes have little in common. While the ATPase is insensitive to K^+ and effected by anions, the PP_i-ase requires K^+ and does not respond to anions in vitro. Although anion concentrations required to affect the tonoplast ATPase in vitro are rather high, in the lower range of effectiveness they may come close to cytoplasmic concentrations in vivo. Thus, not surprisingly much interest has been taken in these properties in relation to anion-transport processes. This refers both to nitrate and to malate. Regarding nitrate the reasons are obvious (see also Sect. 5b). Malate is most interesting as an anion of general importance in many physiological processes involving transport. During H^+/K^+ exchange at the plasmalemma malic acid is synthesize by CO_2 dark fixation via phosphoenolpyruvate-carboxylase (see

POPP et al. 1982). This produces protons for cytoplasmic pH control preventing undue alkalinization (biochemical pH-stat mechanism; DAVIES 1986). Malate serves for balancing electrical charge of surplus cation uptake. Together with K^+, malate is transported into the vacuoles functioning as an osmoticum in extension growth and in stomatal guard-cell and leaf-pulvinal movements. Malic acid is also synthesized to neutralize the OH^- ions produced during NO_3^- reduction if the OH^- ions cannot be removed by extrusion from the cells (see BLEVINS et al. 1978), and then again malate is accumulated in the vacuoles. Thus, possible interactions of malate with the *pmf*-generating processes at the tonoplast could have important general implications. However, the anion-sensitive ATPase and the anion-insensitive PP_i-ase at the tonoplast may operate in concert, and the PP_i-ase could take over when the conditions for the ATPase are unfavourable (MARQUARDT and LÜTTGE 1987).

γ) Other Membranes

Plant dictyosome membranes have a vanadate-insensitive, partially NO_3^- inhibited H^+ transporting ATPase which is distinct from the plasmalemma and tonoplast ATPases (CHANSON et al. 1984; CHANSON and TAIZ 1985). Interestingly, there are reports that an H^+ translocating ATPase from multivesicular bodies in rat liver is strongly inhibited by NO_3^- and to a lesser extent by SO_4^{2-} and $H_2PO_4^-$ (VAN DYKE 1986). The existence of an anion-binding regulatory site on the H^+-ATPase of endomembranes seems to be a common feature in widely separated taxa.

The endoplasmic reticulum has a vanadate-sensitive Ca^{2+} transporting ATPase like the sacroplasmatic reticulum of animals (BUSH and SZE 1986; LEW et al. 1986).

c) Secondary Active Transport Mechanisms of Anions

α) Plasmalemma

As was pointed out above, sealed plasmalemma vesicles for transport studies are still difficult to obtain. Therefore, little work as yet is available on *pmf*-coupled anion transport at the plasmalemma in vitro.

β) Tonoplast

Elevated concentrations of anions in vacuoles suggest the operation of secondary active transport. A very special case are the sulphuric acid accumulating vacuoles of the marine brown alga *Desmarestia*. With a concentration of 0.2 - 0.4 N H_2SO_4 and a pH of 0.5 to 0.8 inside the vacuoles, in addition to primary active H^+ transport, a powerful sulphate transport mechanism must be present (McCLINTOCK et al. 1982). Another example are the membrane-bound statocysts of *Chara* rhizoids containing solid barium sulphate (SIEVERS and SCHMITZ 1982). Electro-physiological measurements show disequilibrium of Cl^- at the tonoplasts of cells of *Acer pseudoplatanus* (RONA et al. 1982).

The work with microsomes and tonoplast vesicles has much facilitated the development of concepts of anion transport carriers coupled to *pmf* generated by ATPases and PP_i-ases. Thus, Cl^- may move into the vacuoles as a consequence of primary active H^+ transport, and inorganic phosphate may be taken up by an exchange for Cl^- (HAGER and HERMSDORF 1981; HAGER and HELMLE 1981). The increase of ATPase and PP_i-ase turnover in sealed, coupled tonoplast vesicles and the breakdown of transmembrane gradients of pH and electrical charge elicited by anions can be taken as direct evidence for the *pmf*-driven secondary active transport of anions. This clearly suggests the operation of Cl^-, NO_3^- and malate carriers at the tonoplast (see Sect. 2bβi and also BLUMWALD and POOLE 1985; JOCHEM and LÜTTGE 1987; MARQUARDT and LÜTTGE 1987).

Pmf-dependent anion transport was also studied with isolated barley mesophyll vacuoles using the new electro-physiological approach of the patch-clamp technique. An $Mg\text{-}ATP^{2-}$ dependent current of positive charge into the vacuoles was detected, and, assuming that all of this current was carried by protons, thermodynamic considerations allowed the accumulation of 1 $malate^{2-}$ per 2 H^+. A large ion channel was identified possibly carrying both K^+ and malate across the tonoplast driven by *pmf* and their concentration gradients (HEDRICH et al. 1986).

More directly, $Mg\text{-}ATP^{2-}$ driven anion uptake has been demonstrated in isolated vacuoles and tonoplast vesicles, i.e. malate in mesophyll vacuoles of the C_3-plant barley (MARTINOIA et al. 1985) and the crassulacean acid metabolism (CAM) plant *Kalanchoe daigremontiana* (NISHIDA and TOMINAGA 1987) and Cl^- in barley mesophyll vacuoles (MARTINOIA et al. 1986). In the vacuole-like lutoids of *Hevea brasiliensis* *pmf*-dependent accumulation of citrate was observed independent of whether *pmf* was built up by $MgATP^{2-}$ consumption of the ATPase or by nigericin-mediated K^+/H^+ exchange of vesicles preloaded with K^+ (MARIN et al. 1981).

Notwithstanding this very good circumstantial evidence for the extistence of anion carriers as distinct entities, isolation and molecular characterization of anion-transport proteins in plants so far has not been achieved. One preliminary report of an attempt to label plasmalemma peptides during the induction of NO_3^- transport in *Zea mays* is not encouraging since no differential labelling could be detected (McCLURE and BOUTHYETTE 1986). The binding of labelled anion channel blockers such as DIDS suggests an alternative approach.

d) Extrinsic Anion-Binding Proteins at the Plasmalemma

It has been suggested that there is an extrinsic membrane protein in the yeast *Candida* which binds phosphate prior to its transport across the plasma membrane (JEANJEAN and FOURNIER 1979; JEANJEAN et al. 1981, 1982). Loss of this protein by osmotic shock, such as is involved in protoplast formation, leaves cells with only a low affinity phosphate transport system. A protein recovered from the external medium had a K_m for phosphate binding similar to that of the high affinity transport system which had been lost. In protoplasts from roots of *Pisum sativum* the phosphate transport system seemed to retain its high affinity (K_m = 9.9 mmol m^{-3}) although much of its activity was lost during preparation (LEFEBVRE and CLARKSON 1984). Osmotic shock administered to roots of *Zea mays* greatly reduced phosphate uptake capacity but no phosphate-binding protein was released into the outer medium (ATTIA and JEANJEAN 1983). If extrinsic binding proteins are involved in transport by roots they may be much more firmly associated with the plasmalemma than in *Candida* or may be associated with the cell wall. This suggestion was made many years ago by PERSSON (1969) who described metabolically labile binding sites for sulphate in wheat roots. The sulphate binding varied with the S-status of the plants and, at concentrations of < 100 mmol m^{-3} in the external solution, was proportional to entry into the root symplast. He proposed that there were positively charged sites on, or just outside the cell membrane which bound sulphate.

3. Passive Lipid Diffusion

As already stated in Section 1f, passive lipid diffusion of electrically neutral undissociated weak organic acids can indirectly bring about anion transport. This not only is important in IAA transport (Sect. 1f). Another phytohormone, abscisic acid (ABA), is probably transported in this way and trapped as the slowly permeating ABA^- anion in alkaline compartments without any involvement of a carrier (KAISER and HARTUNG 1981). Similarly, the efflux of malate from vacuoles of CAM plants is thought to occur via diffusion of non-dissociated malic acid whose concentration builds up during nocturnal CO_2 fixation, malic-acid synthesis and vacuolar acidification (LÜTTGE and SMITH 1984). Lipid diffusion of weak acids is also often utilized in attempts to manipulate cytoplasmic pH. Weak organic acids supplied to plant tissues and cells diffuse into the cytoplasm, dissociate in this relatively alkaline compartment and tend to acidify it. This affects primary active H^+-transport activities (e.g. ROMANI et al. 1983) and secondary active transport of inorganic ions, including anions, and has become an important tool in transport studies (Sect. 1e).

4. Feedback Inhibition of Anion Uptake

It is long known that anion levels on either side of a membrane affect the anion fluxes across the membrane, and that a feedback inhibition of solutes or anions already accumulated on further anion uptake need not be specific (WEIGL 1970; CRAM 1973, 1980).

Cl^- influx can be controlled directly by the concentration of Cl^- at the inside of the plasmalemma (SANDERS 1980a). According to CRAM (1983) set points and not the rates of uptake determine Cl^- levels attained in a tissue. Rates only affect the time required to reach the steady level.

Phosphorus deficiency increases rates of phosphate uptake (DREW et al. 1984). At least under some conditions cellular phosphate pools can now be readily investigated by using ^{31}P-nuclear-magnetic resonance. Data accumulate which show that vacuolar phosphate pools serve to buffer cytoplasmic phosphate levels (WATERTON et al. 1979; KIME et al. 1982; RÉBEILLÉ et al. 1983; WOODROW et al. 1984), which in pea-root tips were maintained near a concentration of about 18 mM (LEE and RATCLIFFE 1983). This also has important implications for metabolism, e.g. regulation of the free energy available from ATP hydrolysis.

Sulphate transport also is regulated by specific negative feedback mechanisms via intracellular content of soluble sulphate (SMITH 1980; JENSÉN and KÖNIG 1982; LASS and ULLRICH-EBERIUS 1984).

Previous nitrate nutrition restricted NO_3^- influx in corn roots. Even if it is assumed that this is a sole consequence of the N-status of the tissue, the explanation remains very complex (CLARKSON 1986). The effect can be due to a feedback inhibition of plasmalemma transport, to efflux and recycling of NO_3^-, or to inhibition or repression of the uptake system by products of NO_3^- reduction (MacKOWN et al. 1981). A control mechanism can be envisaged by a combination of influx, primarily determined by external NO_3^- concentration and independent of internal NO_3^- concentration, with a variable efflux, directly related to internal NO_3^- concentration (DEANE-DRUMMOND and GLASS 1983a). Also, the well-known inhibitory effect of NH_4^+ on NO_3^- uptake need not be a direct action on the influx. During short-term measurements NO_3^- influx (as labelled using $^{36}ClO_3^-$ as an analogue for NO_3^-) in barley proved to be independent of the presence of NH_4^+ and it was suggested that where net NO_3^- uptake is reduced this occurs as a result of stimulation of efflux (DEANE-DRUMMOND and GLASS 1983b; DEANE-DRUMMOND and JACOBSEN 1986; but see ULLRICH et al. 1984). LEE and CLARKSON (1986) were also unable to find any effect of external NH_4^+ on the efflux of $^{13}NO_3^-$ from the roots of barley plants which had received a steady, sufficient supply of nitrate.

Much recent work is devoted to interaction of different anions with nitrate. Nitrate inhibits phosphate uptake (LAMAZE et al. 1984). The most important interaction appears to be with chloride (DEANE-DRUMMOND 1986). It was suggested from kinetic studies that in barley roots, Cl^- and NO_3^- compete for transport at the tonoplast but not at the plasmalemma (HOOYMANS 1980). In maize roots Cl^- and NO_3^- use different transport systems (McCLURE et al. 1986). A model for the regulation of Cl^- influx in barley developed by GLASS and SIDDIQI (1985) involves both negative feedback effects from vacuolar (NO_3^- plus Cl^-)-concentrations and external NO_3^- inhibition of Cl^- influx at the plasmalemma. These combined effects serve to discriminate against Cl^- accumulation and favour NO_3^- when the latter ion is available.

5. Metabolism and Anion Transport in Whole Plants

a) Sulphate and Phosphate

Observations of feedback controls on the cellular and tissue levels immediately suggest that metabolism and growth must have important effects on anion transport and distribution in intact plants. Indeed, phosphate and sulphate relations of whole plants are determined by nutritional state, long-distance transport and retranslocation and demand by different plant parts (CLARKSON et al. 1978, 1983; DREW and SAKER 1984; SCHJØRRING and JENSÉN 1984a,b). An interesting phenomenon is that circulation of phosphate in whole plants obviously is under the control of zinc (CAKMAK and MARSCHNER 1986; MARSCHNER and CAKMAK 1986). In elongating maize roots SO_4^{2-} uptake along the length of roots was studied in relation to the metabolic activation of SO_4^{2-} by ATP-sulphurylase, and a common regulatory mechanism was suggested (CACCO et al. 1977). Unfortunately, work on sulphate and phosphate is still scarce.

b) Nitrate

A wealth of information was obtained for nitrate in relation to the BENZIONI hypothesis of root-shoot cooperation in NO_3^- nutrition. According to this hypothesis, after uptake by the roots, KNO_3 is transported in the transpiration stream to the shoot, where NO_3^- is reduced and malic acid is synthesized to neutralize the OH^- originating from NO_3^- reduction. Then K_2 malate is transported to the roots via the phloem. Malate is oxidized to HCO_3^- in the roots, where HCO_3^- is exchanged for external NO_3^- . Then K^+ is retranslocated back to the shoots together with NO_3^- (BEN-ZIONI et al. 1971; see RAVEN and SMITH 1976). Support and criticism of the hypothesis were published. Clearly, the hypothesis does not apply under all circumstances. The extent to which NO_3^- is reduced in the shoot relative to NO_3^- reduction in the roots depends on plant species (KIRKBY and ARMSTRONG 1980; ANDREWS 1986a; STEWART et al. 1986), the NO_3^- availability (KIRKBY and KNIGHT 1977; ROBIN et al. 1979; ANDREWS 1986b), the kind of prevailing N-nutrition, i.e. NH_4^+ or NO_3^- (MENGEL et al. 1983), the K^+ availability (FROST et al. 1978; RUFTY et al. 1981; PETTERSSON 1984), the carbohydrate availability and distribution (ASLAM and HUFFAKER 1982), and possibly the capacity for organic acid synthesis (ARNOZIS and FINDENEGG 1986) and other parameters. The N-source of intact plants affects anion levels in whole plants. Analyses with *Ricinus communis* plants and xylem and phloem saps showed little difference in the inorganic cation balance, while NO_3^- and carboxylate were replaced by Cl^- as the dominant anion in NH_4^+ supplied plants as compared to N-nutrition with NO_3^- (ALLEN and SMITH 1986). The degree of NO_3^- reduction in various plant organs also depends on the availability of reduced N from other sources, and hence cycling of reduced N-compounds (e.g. amino acids) through the plant as a whole affects the relation of root vs. shoot reduction of NO_3^- (GOJON et al. 1986).

On the basis of analyses of xylem and phloem sap and the assessment of balance sheets, it was argued that in *Ricinus communis* organic acid retranslocation in the phloem quantiatively is not sufficient for the requirements of the BEN-ZIONI hypothesis and only plays a minor role. Not more than 19 to 24% of the HCO_3^- excretion by the roots arose from oxidative decarboxylation of shoot-borne organic anions (VAN BEUSICHEM et al. 1985). But vacuolar storage and diurnal pattern must be considered. Leaf vacuoles are a major storage pool for nitrate; light stimulates NO_3^- reduction (GRANSTEDT and HUFFAKER 1982). In *Catharanthus roseus* cell cultures, malate accumulation in the vacuoles led to nitrate mobilization and vice versa (MARIGO et al. 1986). Several studies suggest that the BEN-ZIONI mechanism may have a rhythmic pattern. Malate accumulates in the vacuoles of photosynthesizing and NO_3^- reducing leaf cells during the light period and is broken down during the dark period. Amplitudes of day-night changes of malate levels observed in C_3 plants were up to 45 µmol g^{-1} fresh weight day^{-1} in *Moricandia arvensis* (WINTER et al. 1982), about 30 mM day^{-1} in the vacuoles of spinach leaves (GERHARDT and HELDT 1984) and about 10 µmol g^{-1} fresh weight day^{-1} in barley (E. MARTINOIA, personal communication).

6. Root-Soil Interactions

In soils where there is abundant available NO_3^-, the net uptake of anions by roots can considerably exceed that of cations (FINDENEGG et al. 1986). It has long been recognized that electrostatic balance can be maintained under these circumstances by OH^- or HCO_3^- excretion into the rhizosphere (DIJKSHOORN 1962; NYE 1981). Some extremely usefuly techniques have been developed for measuring net H^+ and HCO_3^- fluxes into the rhizosphere; these depend on treating roots with dilute agar containing a suitable pH indicator. The agar slows down the diffusion of H^+ or HCO_3^- from their source and the intensity of colour changes of the indicator can be readily measured. The methods were first developed for roots growing on glass surfaces in water culture (WEISENSEEL et al. 1979) but have been applied more recently to roots growing in soil at an interface with a removable glass sheet (MARSCHNER and RÖMHELD 1983). This technique shows considerable variation in net HCO_3^- or H^+ production (1) at different locations in the same root and (2) between species growing in the same soil. The pattern is strongly influenced by the predominant form of N supplied in the soil. While most species alkalinize the rhizosphere when supplied with NO_3^-, e.g. *Zea mays*, some do not and actually acidify the rhizosphere soil, e.g. *Cicer arietinum*. This observation gains added significance from the fact that the latter species, although a legume, was not nodulated. It is recognized that the rhizosphere of nodulated legumes is usually acidic. All species acidify the rhizosphere when NH_4^+ is the major source of N.

Interest in these phenomena is of two kinds. Firstly, the realization that anions are transported across membranes by H^+ co-transport mechanisms indicates that mildly acid pH values provide optimum conditions for anion absorption. The lack of a suitable pH gradient when rhizosphere conditions are strongly alkalinized may adversely affect the absorption of anions, such as $H_2PO_4^-$, which are present at much lower concentrations than NO_3^-.

Secondly, soil pH can greatly influence the availability of nutrients, particularly phosphate, zinc, iron, manganese and potential toxins such as aluminium. The desorption of phosphate increased ten times when the pH of the rhizosphere soil decreased from 6.2 to 4.5 in *Brassica napus* plants which became deficient in P (GRINSTED et al. 1982). Subsequent analysis revealed that this acidification resulted from disturbed cation/anion balance in the plants (HEDLEY et al. 1982). When growth became limited by P-deficiency the absorption of NO_3^- declined relative to cation inflows (principally those of Ca^{2+} and Mg^{2+}). Net proton release compensated for this imbalance. Experiments in water culture showed that there was an immediate effect on net H^+ efflux by roots of *B. napus*, when NO_3^- was removed from the external solution (MOORBY et al. 1985). Several other workers have noted the interaction between P-deficiency and diminished NO_3^- uptake (e.g. in *Hordeum vulgare*: SCHØRRING 1986); the simplest explanation is that P-limited growth produces a lower demand for N. Since N-intake is closely related to growth (CLARKSON 1986) and the uptake of Ca^{2+} and Mg^{2+} is not regulated by growth, but depends on such factors as transpiration and the membrane potential, the plant takes up excess cations which are compensated for by H^+ excretion. Changes in pH were manifested first and were most marked in the apical zone of lateral roots in *B. napus* deprived of phosphate (MOORBY et al. 1987).

7. Concluding Remarks

We seem to be approaching the point where physiological chemistry can tell us no more about the transport of anions. Further understanding will come when the transport proteins are identified and characterized. When such molecular details are known we can begin to appreciate how anion transport is regulated at the membrane level. We should be encouraged by the considerable successes of colleagues, working with mammalian tissues, who have isolated, purified and reconstituted "band 3" protein from erythrocytes which contains an anion transport (see LOWE and LAMBERT 1983, for a review). The search for membrane proteins associated with anion transport may be assisted by the use of transport mutants or by the considerable alterations in carrier activity accompanying repression and induction of anion transport.

There is now evidence for at least three types of proton translocating reactions, or more correctly water-splitting reactions (GOOD 1986), occurring in the plasmalemma and tonoplast and an additional H^+ translocation from NAD(P)H oxidation. The quantitative significance of each of these in providing the proton gradient remains to be worked out. It is possible that the distribution of anion-transport proteins in membranes is not random, but associated with one or several of the primary active transport mechanisms.

References

ALLEN, S., SMITH, J.A.C.: J. Exp. Bot. 37, 1599-1610 (1986). - ANDREWS, M.: Plant Cell Environ. 9, 511-519 (1986a); - ibid. 9, 605-608 (1986b). - ANTHON, G.E., SPANSWICK, R.M.: Plant Physiol. 81, 1080-1085 (1986). - ARNOZIS, P.A., FINDENEGG, G.R.: Plant Physiol. 125, 441-449 (1986). -

ASLAM, M., HUFFAKER, R.C.: Plant Physiol. 70, 1009-1013 (1982). - ATTIA, A., JEANJEAN, R.: Physiol. Vég. 21, 39-47 (1983). AUDERSET, G., SANDELIUS, A.S., PENEL, C., BRIGHTMAN, A., GREPPIN, H., MORRÉ, D.J.: Physiol. Plant. 68, 1-12 (1986).

BEILBY, M.J., WALKER, N.A.: J. Exp. Bot. 32, 43-54 (1981). - BELKOURA, M., RANJEVA, R., MARIGO, G.: Plant Cell Environ. 9, 653-656 (1986). - BENNETT, A.B., SPANSWICK, R.M.: J. Membr. Biol. 71, 95-107 (1983a); - ibid. 75, 21-31 (1983b); - Plant Physiol. 74, 545-548 (1984). - BENNETT, A.B., O'NEILL, S.D., EILMANN, M., SPANSWICK, R.M.: Plant Physiol. 78, 495-499 (1985). - BENNETT, A.B., O'NEILL, S.D., SPANSWICK, R.M.: Plant Physiol. 74, 538-544 (1984). - BENNING, C.: Planta 169, 228-237 (1986). - BEN-ZIONI, A., VAADIA, Y., LIPS, S.H.: Physiol. Plant 24, 288-290 (1971). - BÉRCZI, A., MØLLER, I.M.: Physiol. Plant. 68, 59-66 (1986). - BERTL, A., FELLE, H.: J. Exp. Bot. 36, 1142-1149 (1985). - BEUSICHEM, M.L. Van, BAAS, R., KIRKBY, E.A., NELEMANS, J.A.: Plant Physiol. 78, 768-773 (1985). - BIENFAIT, H.F.: J. Bioenerg. Biomembr. 17, 73-83 (1985). - BLEVINS, D.G., BARNETT, N.M., FROST, W.B.: Plant Physiol. 62, 784-788 (1978). - BLOWERS, D.P., HETHERINGTON, A., TREWAVAS, A.: Planta 166, 208-215 (1985). - BLUMWALD, E., POOLE, R.J.: Proc. Natl. Acad. Sci. USA 82, 3683-3687 (1985). - BOLLER, T., WIEMKEN, A.: Annu. Rev. Plant Physiol. 37, 137-164 (1986). - BÖTTGER, M., LÜTHEN, H.: J. Exp. Bot. 37, 666-675 (1986). - BÖTTGER, M., BIGDON, M., SELL, H.-J.: Planta 163, 376-380 (1985). - BRISKIN, D.P., LEONARD, R.T.: Plant Physiol. 70, 1459-1464 (1982a); - Proc. Natl. Acad. Sci. USA 79, 6922-6926 (1982b). - BRISKIN, D.P., POOLE, R.J.: Plant Physiol. 71, 507-512 (1983a); - ibid. 72, 1133-1135 (1983b); - ibid. 71, 969-971 (1983c). - BRISKIN, D.P., THORNLEY, W.R., POOLE, R.J.: Arch. Biochem. Biophys. 236, 228-237 (1985a). - BRISKIN, D.P., THORNLEY, W.R., ROTI-ROTI, J.L.: Plant Physiol. 78, 642-644 (1985b). - BRISKIN, D.P., THORNLEY, W.R., WYSE, R.E.: Plant Physiol. 78, 865-870 (1985c). - BRUMMER, B., PARISH, R.W.: Planta 157, 446-453 (1983). - BRUMMER, B., BERTL, A., POTRYKUS, I., FELLE, H., PARISH, R.W.: FEBS-Lett. 189, 109-114 (1985). - BUSH, D.R., SZE, H.: Plant Physiol. 80, 549-555 (1986).

CACCO, G., SACCOMANI, M., FERRARI, G.: Plant Physiol. 60, 582-584 (1977). - CAKMAK, I., MARSCHNER, H.: Physiol. Plant. 68, 483-490 (1986). - CANDE, W.Z., RAY, R.M.: Planta 129, 43-52 (1976). - CHANSON, A., TAIZ, L.: Plant Physiol. 78, 232-230 (1985). - CHANSON, A., FISCHMANN, J., SPEAR, D., TAIZ, L.: Plant Physiol. 79, 159-164 (1985). - CHANSON, A., McNAUGHTON, E., TAIZ, L.: Plant Physiol. 76, 498-507 (1984). - CHURCHILL, K.A., SZE, H.: Plant Physiol. 71, 610-617 (1983); - ibid. 76, 490-497 (1984). - CLARKSON, D.T.: Regulation of the absorption and release of nitrate by plant cells: a review of current ideas and methodology. 3-27 in: Fundamental, Ecological and Agricultural Aspects of Nitrogen Metabolism in Higher Plants, eds. H. LAMBERS, J.J. NEETSON, I. STULEN. The Hague, Nijhoff 1986. - CLARKSON, D.T., LÜTTGE, U.: Progr. Bot. 46, 56-67 (1984). - CLARKSON, D.T., SANDERSON, J., SCATTERGOOD, C.B.: Planta 139, 47-53 (1978). - CLARKSON, D.T., SMITH, F.W., VANDEN BERG, P.J.: J. Exp. Bot. 34, 1463-1483 (1983). - CLÉMENT, J.-D., BLEIN, J.-P., RIGAUD, J., SCALLA, R.: Physiol. Vég. 24, 25-35 (1986). - CRAM, W.J.: J. Exp. Bot. 24, 328-341 (1973); - Plant Physiol. 7, 41-49 (1980); - J. Exp. Bot. 34, 1484-1502 (1983).

DAVIES, D.D.: Physiol. Plant. 67, 702-706 (1986). - DEANE-DRUMMOND, C.E.: Plant Cell Environ. 7, 317-323 (1984); - Physiol. Plant. 66, 115-121 (1986). - DEANE-DRUMMOND, C.E., GLASS, A.D.M.: Plant Physiol. 73, 100-104 (1983a); - ibid. 73, 105-110 (1983b). - DEANE-DRUMMOND, C.E., JACOBSEN, E.: Plant Science 46, 169-173 (1986). - DIJKSHOORN, W.: Nature 194, 165-167 (1962). - DÖBLINGER, R., TROMBALLA, H.W.: Planta 156, 10-15 (1982a); - ibid. 156, 16-20 (1982b). - DREW, M.C., SAKER, L.R.: Planta 160, 500-507 (1984). - DREW, M.C., SAKER, L.R.,

BARBER, S.A., JENKINS, W.: Planta 160, 490-499 (1984). - DuPONT, F.M., GIORGI, D.L., SPANSWICK, R.M.: Plant Physiol. 70, 1694-1699 (1982). - DYKE, R.W. van: J. Biol. Chem. 261, 15941-15948 (1986).

EDWARDS, J., REES, T. ap, WILSON, P.M., MORRELL, S.: Planta 162, 188-191 (1984). - EDWARDS, K.L., GOLDSMITH, M.H.M.: Planta 147, 457-466 (1980). - EISELE, R., ULLRICH, W.R.: Planta 123, 117-123 (1975); - Plant Physiol. 59, 18-21 (1977).

FINDENEGG, G.R.: Planta 135, 33-38 (1977). - FINDENEGG, G.R., BEUSICHEM, M.L. van, KELTJENS, W.G.: Netherl. J. Agric. Sci. 34, 371-379 (1986). - FROST, W.B., BLEVINS, D.G., BARNETT, N.M.: Plant Physiol. 61, 323-326 (1978).

GERHARDT, R., HELDT, H.W.: Plant Physiol. 75, 542-547 (1984). - GERLOFF, G.C., GABELMAN, W.H.: Genetic basis of inorganic plant nutrition, 453-480. In: Inorganic Plant Nutrition. Ecyclopedia of Plant Physiology, N.S. Vol. 15 B, ed. A. LÄUCHLI, R.L. BIELESKI. Berlin, Heidelberg, New York: Springer 1983. - GLASS, A.D.M., SIDDIQI, M.Y.: J. Exp. Bot. 36, 556-566 (1985). - GOJON, A., SOUSSANA, J.-F., PASSAMA, L., ROBIN, P.: Plant Physiol. 82, 254-260 (1986). - GOLDSMITH, M.H.M.: Planta 155, 68-75 (1982). - GOLDSMITH, M.H.M., GOLDSMITH, T.H.: Planta 153, 25-23 (1981). - GOLDFARB, V., SANDERS, D., GRADMANN, D.: J. Exp. Bot. 35, 626-644 (1984a); - ibid. 35, 645-658 (1984b). - GOOD, N.E.: Annu. Rev. Plant Physiol. 37, 1-22 (1986). - GRADMANN, D., TITTOR, J., GOLDFARB, V.: Phil. Trans. R. Soc. London B 299, 447-457 (1982). - GRANSTEDT, R.C., HUFFAKER, R.C.: Plant Physiol. 70, 410-413 (1982). - GRIFFITH, C.J., REA, P.A., BLUMWALD, E., POOLE, R.J.: Plant Physiol. 81, 120-125 (1986). - GRINSTED, M.J., HEDLEY, M.J., WHITE, R.E., NYE, P.H.: New Phytol. 91, 19-29 (1982).

HAGER, A., HELMLE, M.: Z. Naturforsch. 36c, 997-1008 (1981). - HAGER, A., HERMSDORF, P.: Z. Naturforsch. 36c, 1009-1012 (1981). - HASENSTEIN, K.-H., RAYLE, D.: Plant Physiol. 76, 65-67 (1984). - HEDLEY, M.J., NYE, P.H., WHITE, R.E.: New Phytol. 91, 31-44 (1982). - HEDRICH, R., FLÜGGE, U.I., FERNANDEZ, J.M.: FEBS-LETT. 204, 228-232 (1986). - HEIMER, Y.M.: Plant Sci. Lett. 4, 137-139 (1975). - HOOYMANS, J.J.M.: Z. Pflanzenphysiol. 100, 185-187 (1980).

JACKSON, L.A., FLESHER, D., HAGEMAN, R.A.: Plant Physiol. 51, 120-127 (1973). - JACKSON, W.A., JOHNSON, R.E., VOLK, R.J.: Physiol. Plant. 32, 37-42 (1974). - JACKSON, W.A., KWIK, K.D., VOLK, R.J., BUTZ, R.G.: Planta 132, 149-156 (1976). - JEANJEAN, R., FOURNIER, N.: FEBS-Lett. 105, 163-166 (1979). - JEANJEAN, R., ATTIA, A., JARRY, T., COLLE, A.: FEBS-Lett. 125, 69-71 (1981). - JEANJEAN, R., BEDU, S., ATTIA, A., ROCCA-SERRA, J.: Biochem. 64, 75-78 (1982). - JENSÉN, P., KÖNIG, T.: Physiol. Plant. 55, 459-464 (1982). - JOCHEM, P., LÜTTGE, U.: J. Plant Physiol. 129, 251-268 (1987). - JOCHEM, P., RONA, J.-P., SMITH, J.A.C., LÜTTGE, U.: Physiol. Plant. 62, 410-415 (1984). - JONES, S.L., SMITH, I.K.: Plant Physiol. 67, 445-448 (1981).

KAISER, W.M., HARTUNG, W.: Plant Physiol. 68, 202-206 (1981). - KARLSSON, J.: Biochim. Biophys. Acta 399, 356-363 (1975), - KIME, M.J., RATCLIFFE, R.G., LOUGHMAN, B.C.: J. Exp. Bot. 33, 670-681 (1982). - KIRKBY, E.A., ARMSTRONG, M.J.: Plant Physiol. 65, 286-290 (1980). - KIRKBY, E.A., KNIGHT, A.H.: Plant Physiol. 60, 349-353 (1977). - KURKDJAN, A., MATHIEU, Y., GUERN, J.: Plant Sci. Lett. 27, 77-86 (1982).

LAMAZE, T., SENTENAC, H., GRIGNON, C.: Physiol. Vég. 22, 155-161 (1984). - LASS, B., ULLRICH-EBERIUS, C.I.: Planta 161, 53-60 (1984). - LASS, B., THIEL, G., ULLRICH-EBERIUS, C.I.: Planta 169, 251-259

(1986). - LEE, R.B., CLARKSON, D.T.: J. Exp. Bot. 37, 1753-1767 (1986). - LEE, R.B., RATCLIFFE, R.G.: J. Exp. Bot. 34, 1222-1244 (1983). - LEFEBVRE, D.D., CLARKSON, D.T.: J. Exp. Bot. 35, 1265-1276 (1984). - LEIGH, R.: Physiol. Plant. 57, 390-396 (1983). - LEIGH, R.A., WALKER, R.R.: Planta 150, 222-229 (1980). - LEW, R.R., SPANSWICK, R.M.: Pl. Sci. Lett. 36, 187-193 (1984); - Plant Physiol. 77, 352-357 (1985). - LEW, R.R., BRISKIN, D.P., WYSE, R.E.: Plant Physiol. 82, 47-53 (1986). - LICHKO, L.P., OKOROKOV, L.A.: FEBS-Lett. 187, 349-353 (1985). - LIN, W.: Plant Physiol. 63, 952-955 (1979); - ibid. 68, 435-438 (1981). - LUCAS, W.J.: Planta 156, 181-192 (1982). - LUCAS, W.J., ALEXANDER, J.M.: Plant Physiol. 65, 274-280 (1980). - LUCAS, W.J., BERRY, J.A.: Physiol. Plant. 65, 539-543 (1985). - LÜTTGE, U.: Progr. Bot. 45, 76-88 (1983). - LÜTTGE, U., CLARKSON, D.T.: Progr. Bot. 47, 73-86 (1985). - LÜTTGE, U., SMITH, J.A.C.: J. Membr. Biol. 81, 149-158 (1984). - LOWE, A.G., LAMBERT, A.: Biochim. Biophys. Acta 694, 353-374 (1983).

MacKOWN, C.T., VOLK, R.J., JACKSON, W.A.: Plant Physiol. 68, 133-138 (1981). - MACRI, F., VIANELLO, A.: Plant Sci. 43, 25-30 (1986). - MANDALA, S., TAIZ, L.: Plant Physiol. 78, 327-333 (1985). - MANOLSON, M.F., REA, P.A., POOLE, R.J.: J. Biol. Chem. 260, 12273-12279 (1985). - MARRÈ, E., BALLARIN-DENTI, A.: J. Bioenerg. Biomembr. 17, 1-21 (1985). - MARRÈ, M.T., ROMANI, G.: Plant Sci. Lett. 27, 265-273 (1982). - MARRÈ, M.T., ROMANI, G., MARRÈ, E.: Plant Cell Environm. 6, 617-623 (1983). - MARRÈ, M.T., ROMANI, G., BELLANDO, M., MARRÈ, E.: Plant Physiol. 82, 316-323 (1986). - MARIGO, G., BOUYSSON, H., BOUDET, M.: Physiol. Vég. 24, 15-23 (1986). - MARIN, B., SMITH, J.A.C., LÜTTGE, U.: Planta 153, 486-493 (1981). - MARQUARDT, G., LÜTTGE, U.: J. Plant Physiol. 129, 269-286 (1987). - MARSCHNER, H., CAKMAK, I.: Physiol. Plant. 68, 491-496 (1986). - MARSCHNER, H., RÖMHELD, V.: Z. Pflanzenphysiol. 111, 241-251 (1983). - MARTIN, H.V., PILET, P.E.: Plant Physiol. 81, 889-895 (1986). - MARTINOIA, E., FLÜGGE, U.I., KAISER, G., HEBER, U., HELDT, H.W.: Biochim. Biophys. Acta 806, 311-319 (1985). - MARTINOIA, E., SCHRAMM, M.J., KAISER, G., KAISER, W.M., HEBER, U.: Plant Physiol. 80, 895-901 (1986). - McCLINTOCK, M., HIGINBOTHAM, N., URIBE, E.G., CLELAND, R.E.: Plant Physiol. 70, 771-774 (1982). - McCLURE, P.R., BOUTHYETTE, P.Y.: Plant Physiol. 80, Suppl. 89 (1986). - McCLURE, P.R., OMHOLT, T.E., PACE, G.M.: Physiol. Plant. 68, 107-112 (1986). - MENGEL, K., ROBIN, P., SALSAC, L.: Plant Physiol. 74, 618-622 (1983). - METTLER, I.J., LEONARD, R.T.: Plant Physiol. 63, 183-190 (1979). - MICHELIS de M.T., SPANSWICK, R.M.: Plant Physiol. 81, 542-547 (1986). - MØLLER, I.M., BÉRCZI, A.: FEBS-Lett. 193, 180-184 (1985); - Physiol. Plant 68, 67-74 (1986). - MØLLER, I.M., LIN, W.: Plant Physiol. 37, 309-334 (1986). - MOORBY, H., NYE, P.H., WHITE, R.E.: Plant Soil 84, 403-413 (1985). - MOORBY, H., WHITE, R.E., NYE, P.H.: submitted (1987).

NYE, P.H.: Plant Soil 61, 7-26 (1981). - NEYRA, C.A., HAGEMANN, R.H.: Plant Physiol. 56, 692-695 (1975). - NISHIDA, K., TOMINAGA, O.: J. Plant Physiol. 127, 385-393 (1987).

O'NEILL, S.D., BENNETT, A.B., SPANSWICK, R.M.: Plant Physiol. 72, 837-846 (1983). - O'NEILL, S.D., SPANSWICK, R.M.: J. Membr. Biol. 79, 231-243 (1984a); - ibid. 79, 245-256 (1984b).

PERSSON, L.: Physiol. Plant. 22, 959-976 (1969). - PETTERSSON, S.: Physiol. Plant. 61, 663-669 (1984). - PODER, D., PENOT, M.: Physiol. Vég. 24, 551-557 (1986). - POOLE, R.J., BRISKIN, D.P., KŘATKÝ, Z., JOHNSTONE, R.M.: Plant Physiol. 74, 549-556 (1984). - POPP, M., OSMOND, C.B., SUMMONS, R.E.: Plant Physiol. 69, 1289-1292 (1982). - PRINS, H.B.A., SNEL, J.F.H., HELDER, R.J., ZANSTRA, P.E.: Plant Physiol. 66, 818-822 (1980); - Plant Cell Environm. 5, 207-214 (1982). - PUPILLO, P., VALENTI, V., LUCA de L., HERTEL, R.: Plant Physiol. 80, 384-389 (1986).

RASI-CALDOGNO, F., MICHELIS de M.I., PUGLIARELLO, M.C., MARRÊ, E.: Plant Physiol. 82, 121-125 (1986). - RAUSCH, T., ZIEMANN-ROTH, M., HILGENBERG, W.: Plant Physiol. 77, 881-885 (1985). - RAVEN, J.A.: New Phytol. 74, 163-172 (1975). - RAVEN, J.A., SMITH, F.A.: New Phytol. 75, 415-431 (1976). - REA, P.A., POOLE, R.J.: Plant Physiol. 77, 46-52 (1985); - ibid. 81, 126-129 (1986). - RÉBEILLÉ, F., BLIGNY, R., MARTIN, J.B., DOUCE, R.: Arch. Biochem. Biophys. 225, 143-148 (1983). - REID, R.J., DEJAEGERE, R., PITMAN, M.G.: J. Exp. Bot. 36, 535-549 (1985a). - REID, R.J., FIELD, L.D., PITMAN, M.G.: Planta 166, 341-347 (1985b). - ROBIN, P., BLAYAC, D., SALSAC, L.: Phys. Vég. 17, 55-56 (1979). - ROMANI, G., MARRÊ, M.T., MARRÊ, E.: Physiol. Vég. 21, 867-873 (1983). - ROMANI, G., MARRÊ, M.T., BELLANDO, M., ALLOATTI, G., MARRÊ, E.: Plant Physiol. 79, 734-739 (1985). - RONA, J.-P., CORNEL, D., GRIGNON, C., HELLER, R.: Physiol. Vég. 20, 459-463 (1982). - RUBERY, P.A., SHELDRAKE, A.R.: Planta 118, 101-121 (1974). - RUBINSTEIN, B., STERN, A.I.: Plant Physiol. 80, 805-811 (1986). - RUFTY, T.W., JACKSON, W.A., RAPER, C.D.: Plant Physiol. 68, 605-609 (1981).

SABATER, M., SABATER, F.: Planta 167, 76-80 (1986). - SADDLER, A.D.W.: J. Exp. Bot. 21, 345-359 (1970); - ibid. 5, 250-260 (1971). - SANDELIUS, A.S., PENEL, C., AUDERSET, G., BRIGHTMAN, A., MILLARD, M., MORRÉ, D.J.: Plant Physiol. 81, 177-185 (1986). - SANDERS, D.: J. Membr. Biol. 52, 51-60 (1980a); - ibid. 53, 129-141 (1980b); - J. Exp. Bot. 31, 105-118 (1980c); - Plant Physiol. 67, 1113-1118 (1981). - SANDERS, D., SMITH, F.A., WALKER, N.A.: Planta 163, 411-418 (1985). - SCHJØRRING, J.K.: Plant Soil 91, 313-318 (1986). - SCHJØRRING, J.K., JENSÉN, P.: Physiol. Plant 61, 577-583 (1984a); - ibid. 61, 584-590 (1984b). - SCHLEE, J., CHO, B.-H., KOMOR, E.: Plant Sci. 39, 25-30 (1985). - SENTENAC, H., GRIGNON, C.: Plant Physiol. 77, 136-141 (1985). - SIEVERS, A., SCHMITZ, M.: Ber. Dtsch. Bot. Ges. 95, 353-360 (1982). - SMITH, I.K.: Plant Physiol. 66, 877-883 (1980). - SMYTH, D.A., BLACK, C.C.: Plant Physiol. 75, 862-864 (1984). - SPALDING, M.H., SPREITZER, R.J., OGREN, W.L.: Plant Physiol. 73, 268-272 (1983a); - ibid. 73, 273-276 (1983b). - STEWART, G.R., POPP, M., HOLZAPFEL, I., STEWART, J.A., DICKIE-ESKEW, A.: New Phytol. 104, 373-384 (1986). - STRUVE, I., LÜTTGE, U.: Planta 170, 111-120 (1987). - SUSSMAN, M.R., GOLDSMITH, M.H.M.: Planta 150, 15-25 (1981). - SZE, H.: Physiol. Plant. 61, 683-691 (1984); - Annu. Rev. Plant Physiol. 36, 175-208 (1985).

TIPTON, C.L., THOWSEN, J.: Plant Physiol. 79, 432-435 (1985). - TOGNOLI, L.: Eur. J. Biochem. 146, 581-588 (1985).

ULLRICH, W.R., GLASER, E.: Plant Sci. Lett. 27, 155-161 (1982). - ULLRICH, W.R., NOVACKY, A.: Plant Sci. Lett. 22, 211-217 (1981). - ULLRICH, W.R., LARSSON, M., LARSSON, C.-M., LESCH, S., NOVACKY, A.: Physiol. Plant. 61, 369-376 (1984). - ULLRICH-EBERIUS, C.I., NOVACKY, A., BEL, A.J.E. van: Planta 161, 46-52 (1984).

VOLOKITA, M., KAPLAN, A., REINHOLD, L.: Plant Physiol. 67, 1119-1123 (1981).

WAGNER, G.J., MULREADY, P.: Biochim. Biophys. Acta 728, 267-280 (1983). - WALKER, R.R., LEIGH, R.A.: Planta 153, 150-155 (1981). - WANG, Y., LEIGH, R.A., KAESTNER, K.H., SZE, H.: Plant Physiol. 81, 497-502 (1986). - WATERTON, J.C., BRIDGES, I.G., IRVING, M.P.: Biochem. Biophys. Acta 763, 315-320 (1983). - WEIGL, J.: Z. Naturforsch. 25b, 631-636 (1970).

WEISENSEEL, M.H., DORN, A., JAFFE, L.F.: Plant Physiol. 64, 512-518 (1979). - WINGSTRAND, G.: Physiol. Plant. 64, 223-229 (1985). - WINTER, K., USUDA, H., TSUZUKI, M., SCHMITT, M., EDWARDS, G.E., THOMAS, R.J., EVERT, R.F.: Plant Physiol. 70, 616-625 (1982). - WOODROW, I.E., ELLIS, J.R., JELLINGS, A., FOYER, C.H.: Planta 161, 525-530 (1984).

ZOCCHI, G.: Plant Sci. 40, 153-159 (1985).

Prof. Dr. ULRICH LÜTTGE
Institut für Botanik der
Technischen Hochschule Darmstadt
Schnittspahnstraße 3-5
D-6100 Darmstadt, FRG

Prof. Dr. DAVID T. CLARKSON
Long Ashton Research Station
Long Ashton
GB-Bristol, BS 18 9AF, England

II. Photosynthesis: Antenna and Reaction Centers

By J. AMESZ

1. Introduction

Two phases may be distinguished in the primary processes of photosynthesis: (1) the capture of light energy and transfer of this energy to the reaction center and (2) charge separation and electron transport in the reaction center.

Phase (1) occurs in the light-harvesting (also called antenna) system which contains most of the photosynthetic pigments (usually 90% or more). The antenna system consists of one or more pigment-protein complexes. These are normally intrinsic proteins contained in the photosynthetic membrane, but in some groups of organisms a major part of the antenna is located in extramembranous structures. In green photosynthetic bacteria these are the so-called chlorosomes, whereas cyanobacteria and red algae possess phycobilisomes, complex assemblies of red and blue-colored biliproteins which are attached to the thyllakoid membrane.

Energy transfer from the light-harvesting pigments to the reaction center results in the excitation of the primary electron donor, P. In a very short time, this is followed by a charge separation, i.e. the transfer of an electron from excited P to an acceptor molecule. Subsequent electron transfer reactions result in a stabilization of the charge-separated state at the expense of a decrease of the chemical free energy of the reaction products. Electron transport in the reaction center of purple photosynthetic bacteria and in that of photosystem II of plants and algae show a clear similarity, but, as discussed in an earlier review (AMESZ 1983), the reaction center of photosystem I and that of green sulfur bacteria show a different sequence of electron transfer reactions. Like most of the light-harvesting pigments, the components of the reaction center are bound to intrinsic proteins.

In contrast to the primary electron transfer reactions, which appear to proceed in two varieties of reaction center types only, the antenna systems of photosynthetic organisms display a large variety of structures and compositions. Figure 1 illustrates the variety of pigments involved. Together they cover almost the entire spectrum from the near-ultraviolet to wavelengths above 1000 nm in the near infrared region. The most simple antenna systems are those of purple bacteria. In *Rhodospirillum rubrum* only one type of antenna complex is found, the so-called B875 complex, named after the wavelength of maximum absorption of the bacteriochlorophyll (BChl) in the near-infrared. The most extensively studied system is that of *Rhodobacter sphaeroides*[1], which contains two dif-

[1]Formerly called *Rhodopseudomonas sphaeroides*. The taxonomy of the purple photosynthetic bacteria has recently been revised (IMHOFF et al. 1984).

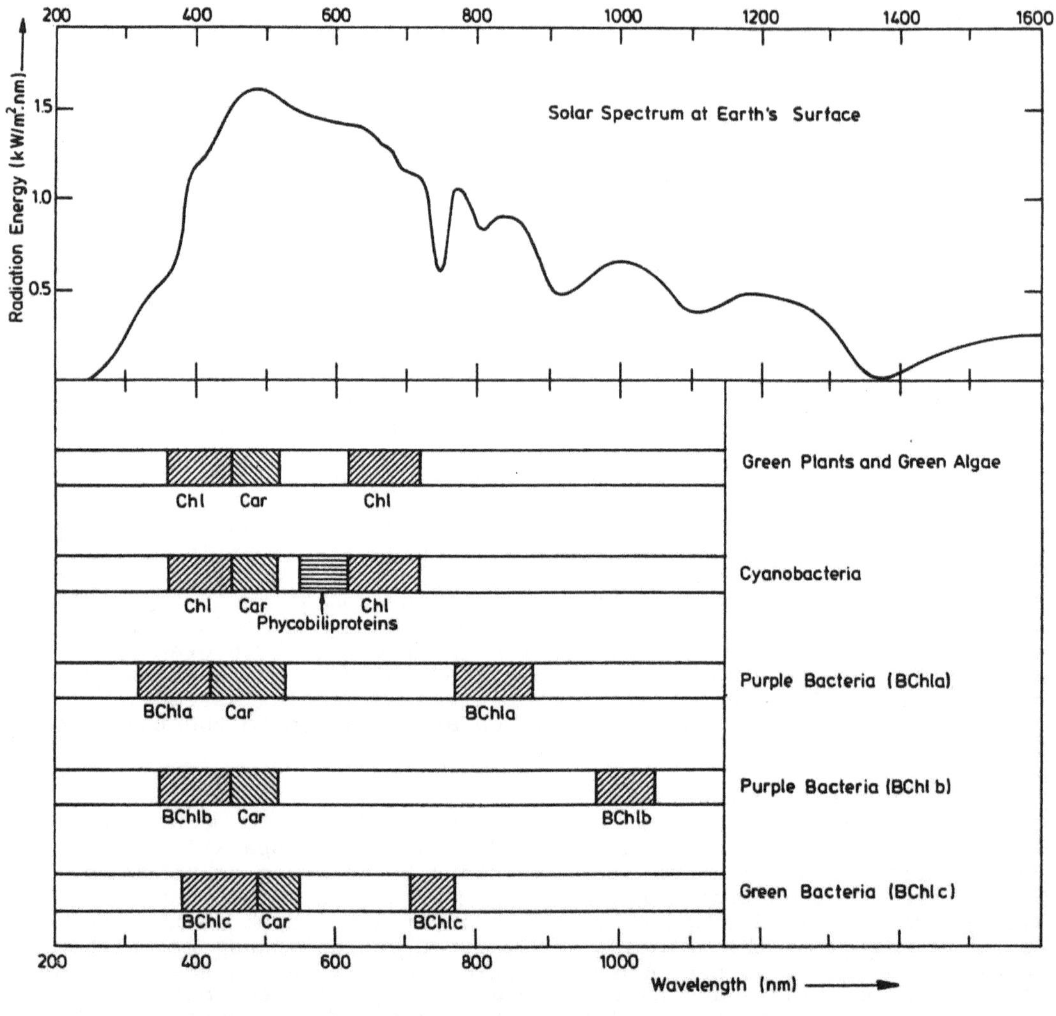

Fig. 1. *Top:* Spectrum of the solar radiation after passage through the atmosphere. *Bottom:* Main absorption bands *(hatched areas)* of antenna complexes of different groups of photosynthetic organisms (VASMEL 1986)

ferent complexes, B800-850 and B875 (see e.g. reviews by AMESZ and KNAFF 1987; ZUBER et al. 1987). The antenna systems of other organisms, like higher plants, algae and green bacteria have a more complicated structure, as will be discussed below.

Our understanding of the primary events in photosynthesis and of the structures in which these processes take place has now become vastly more detailed than could have been envisaged only a few years ago. This knowledge is mainly based on three areas of research, which have been discussed also in earlier reviews in this journal (AMESZ 1983, 1985). First of all, spectroscopic studies of the primary processes, based on advanced optical and electron spin resonance techniques have yielded a wealth of information on the mechanism of electron transfer in the photosynthetic reaction center. Secondly, X-ray diffraction studies of crystalline pigment-protein complexes have yielded a picture of the three-dimensional structure of the reaction center of purple bacteria (DEISENHOFER et al. 1984) and of light-harvesting proteins (C-phycocyanins) of some cyanobacteria (SCHIRMER et al. 1985; ZUBER et al. 1987). A third major advance is due to the efforts of molecular biologists and protein biochemists, who determined the primary structures of almost all of the most important intrinsic proteins that are contained in the light-harvesting and electron transport complexes in plants and bacteria.

The present review will focus on some of these developments and give an overview of the properties of the light-harvesting systems and the reaction center complexes that are contained in, or associated with, the photosynthetic membrane. Emphasis will be given to recent advances that have not been discussed in previous reviews in this series.

2. Purple Bacteria

Traditionally, the photosynthetic bacteria are divided into two groups: the purple and the green photosynthetic bacteria. However, it has become clear now that this division has no taxonomic basis and that most of the major prokaryotic phyla contain both photosynthetic and non-photosynthetic genera (STACKEBRANDT and WOESE 1981; WOESE 1985; see also OLSON and PIERSON 1987), and the recently discovered species *Heliobacterium chlorum* (GEST and FAVINGER 1983) cannot be accomodated in either of the two groups. Nevertheless, for practical purposes we shall retain the traditional division in the present review, but the reader should keep in mind that the only clear distinction between the two groups is their antenna organization, green bacteria having chlorosomes as their major antenna system.

Extensive structural data are available on the antenna system of purple bacteria. These have been discussed in some detail in the previous review (AMESZ 1985). The reader is therefore referred to this review and to more recent ones (AMESZ and KNAFF 1987; ZUBER et al. 1987; AMESZ and VASMEL 1986), for a discussion of the physical and biochemical properties of the antenna complexes. It may suffice here to recall that structural information on these complexes is based on the primary sequence of the constituent polypeptides (which all show a clear homology and are composed of 50 - 55 amino acids with a single α-helical membrane spanning stretch) and on spectroscopic studies. Although many complexes have been obtained in crystalline form now (WOOLLEY et al. 1985; WELTE et al. 1985; WACKER et al. 1986) the X-ray diffraction pattern of these crystals was not well enough resolved for a structural analysis.

Since the previous review was written, the analysis of the structure of the reaction center of *Rhodopseudomonas viridis* has been continuously refined by combining the results of the X-ray studies of the crystalline complex (DEISENHOFER et al. 1984, 1985; MICHEL et al. 1986a) with those obtained from the determination of the primary structure of the L, M and H subunits (MICHEL et al. 1985, 1986b). The L and M subunits are homologous. They span the membrane five times and consist of 273 and 323 amino acids, respectively (MICHEL et al. 1986b). The H subunit has only one α-helical membrane spanning stretch (MICHEL et al. 1985). The primary structures of the L and M subunits show a clear homology to those of *Rb. sphaeroides* and *Rb. capsulatus* (formerly called *Rhodopseudomonas capsulata* (IMHOFF et al. 1984)), which had been determined

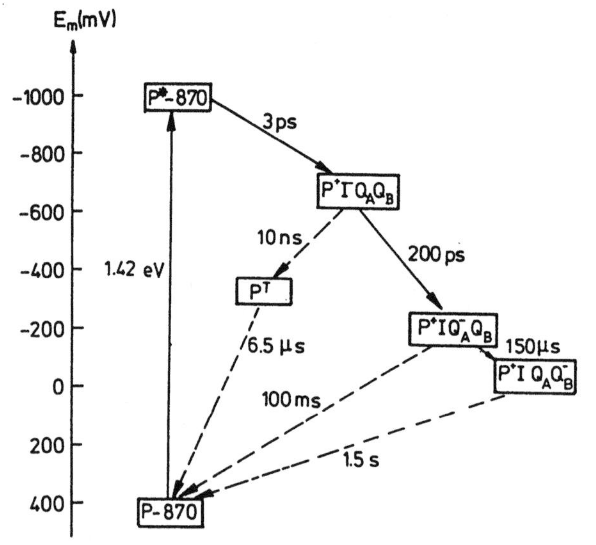

Fig. 2. Energy scheme of electron transfer kinetics in the reaction center of BChl a containing purple bacteria. The excited primary electron donor P 870 is indicated by P^*, excitation by *vertical arrows*. *Downward pointing, solid arrows* indicate main electron transfer pathways; *broken arrows* are "back reactions" that become predominant when main electron transport is blocked. P^T, triplet state of P 870; I, Bph a; Q_A and Q_B, secondary electron acceptors (ubiquinone or menaquinone). The time constants refer to room temperature; they apply to isolated reaction centers of the carotenoidless mutant R-26 of *Rb. sphaeroides*

earlier (WILLIAMS et al. 1983, 1984; YOUVAN et al. 1984; see also AMESZ 1985).

The pigments in the reaction center are arranged in such a way that they form two chains, arranged in an approximately symmetrical fashion, the so-called L and M chains, named after the subunit to which the components of each chain are most closely associated. X-ray diffraction studies of the reaction center of *Rb. sphaeroides*, although at somewhat less resolution, recently indicated a very similar structure for this species (CHANG et al. 1986; ALLEN et al. 1986; DUCRUIX and REISS-HUSSON 1987). The same conclusions was obtained for an analysis of the optical properties (VASMEL et al. 1986a). The reaction center contains six pigment molecules: four BChls (BChl b in *Rps. viridis*; BChl a in *Rb. sphaeroides*) and two bacteriopheophytins (BPh b in *Rps. viridis*; BPh a in *Rb. sphaeroides*). The chains have in common the primary donor P (P 960 in *Rps. viridis*; P 870 in *Rb. sphaeroides*) which is a BChl dimer. In addition to this, each chain consists of a BChl molecule (one of each of the so-called accessory BChls) and a BPh molecule. The first acceptor quinone, Q_A (menaquinone in *Rps. viridis*, ubiquinone in *Rb. sphaeroides*) is associated with the L chain, the second quinone, Q_B (see Fig. 2) with the M chain. The primary charge separation and subsequent electron transport occur only in the L chain, until finally the electron crosses over from Q_A to Q_B in the other chain. In this connection it may be noted that one of the accessory BChls in *Rb. sphaeroides* (presumably that in the M chain) can be converted to BPh and partially removed by treatment with sodium borohydride, essentially without affecting electron transport (MAROTI et al. 1985; SHUVALOV et al. 1986a; SHUVALOV and DUYSENS 1986). Interestingly, the Fe^{2+} ion, located halfway between the two quinones (DEISENHOFER et al. 1984; MICHEL et al. 1986a), has no specific effect on the rate of electron transport between Q_A and Q_B. Its removal slows down the reaction rate by approximately a factor of two, but the rate remains unaffected upon replacement by other divalent ions, like Zn^{2+}, Cu^{2+} or Mn^{2+} (DEBUS et al. 1986). The isolated reaction center of *Rps. viridis*, but not that of *Rb. sphaeroides*, contains a tightly bound cytochrome *c* molecule with four heme groups, which functions as electron donor to P 960.

Electron transfer in the reaction center of purple bacteria was discussed already in earlier reviews (see PARSON 1987 for a recent one).

A scheme of electron transfer is given in Figure 2. The most important new developments concern the mechanism and rate of the primary charge separation. MARTIN et al. (1986) measured the kinetics of the absorbance changes induced by ultra-short (150 fs) laser flashes in isolated reaction centers of *Rb. sphaeroides*. The results indicated that upon excitation of P 870 an electron was transferred in 2.8 ± 0.2 ps to Bph a. Electron transport from P 960 to Pheo b in reaction centers of *Rps. viridis* was found to occur in 6 ps (WASIELEWSKI and TIEDE 1986). The role of the accessory BChl a or b which, according to the X-ray analysis, is situated between P and Bph is still not clear. Experiments of SHUVALOV and co-workers (SHUVALOV and DUYSENS 1986; SHUVALOV et al. 1986b) suggested the formation of an early intermediate state P^+BChl^-, but kinetic evidence for such an intermediate state was not found in the experiments with ultra-short flashes cited above (see also BRETON et al. 1986). The latter results thus would indicate that the accessory BChl may function to facilitate electron transfer from P to Bph, but is not a distinct intermediate in the "classical" sense.

3. Green Bacteria

a) Chlorosomes

The green bacteria are characterized by the presence of chlorosomes, oval-shaped bodies which contain BChl c (or the related pigments BChl d or e), and are attached to the cytoplasmic membrane. On basis of the organization of the cytoplasmic membrane, which contains the reaction center and additional light-harvesting pigments, as well as on basis of their photosynthetic mechanism, two groups of green bacteria may be distinguished: the green sulfur bacteria (Chlorobiaceae) and the green filamentous (or gliding green) bacteria (Chloroflexaceae). The first group are the classical green bacteria known for about a century now; the second group was discovered only in the early 1970s by PIERSON and CASTENHOLZ (1974a,b).

Chlorosomes constitute the main BChl antenna of green bacteria. They also contain most of the carotenoid present in these bacteria (SCHMIDT 1980) and account for about 10 - 20% of the total volume of the cell. The former name "Chlorobium vesicles" has been replaced by the term "chlorosome" (STAEHLIN et al. 1978, 1980) since they are not vesicles in the normal sense of the word and do not contain an inner aqueous phase. The most extensively studied are those of the Chlorobiacea *Chlorobium limicola* and of the Chloroflexacea *Chloroflexus aurantiacus*. For both species, the chlorosome are about 1000 Å long and several hundred Å wide.

Chlorosomes can be detached from the cytoplasmic membrane by mechanical disruption (SCHMIDT 1980) or detergent treatment (FEICK et al. 1982). The chemical composition of chlorosome has been extensively studied by SCHMIDT (1980). Chlorosomes of the Chlorobiacea *Chlorobium limicola* f. *thiosulfatophilum* were found to contain on a dry weight basis, 55% protein, 10% lipid (mainly glycolipids), 32% BChl c and 1% carotenoid. This means that the BChl concentration (on a protein basis) is several times higher than in the antenna of purple bacteria. One chlorosome may contain approximately 10,000 BChl c molecules; experiments of VOS et al. (1987) indicated that the range of excitation transfer extends over the whole chlorosome, which thus serves as a common antenna connecting several reaction centers. The lipid is thought to form a monolayer around the chlorosome (STAEHELIN et al. 1980). The amount of chlorosome material in *Cfl. aurantiacus* is strongly dependent on light intensity (PIERSON and CASTENHOLZ 1974b; SCHMIDT et al. 1980), cells grown at low intensity having a higher BChl c content. A similar, but less pronounced effect has been observed in *C. limicola* (BROCH-DUE et al 1978). Chlorosomes of *Cfl. aurantiacus* grown at low light intensity contained 22% BChl c and 2% carotenoid (SCHMIDT 1980). There is strong evidence now that chlorosome of both species contain small amounts of BChl a, absorbing near 790 - 795 nm (SCHMIDT 1980; BETTI et al. 1982; GEROLA and OLSON 1986; VAN DORSSEN et al. 1986a). The location of the emission band of this BChl a (BETTI et al. 1982; VAN DORSSEN et al. 1986a) as well as its orientation (Van DORSSEN et al. 1986a,b) suggest that it may play an important role in energy transfer from the chlorosome to the cytoplasmic membrane. However, the efficiencies of transfer of absorbed light energy from the chlorosome to the cytoplasmic membrane or to the reaction center are not precisely known (see AMESZ and VASMEL 1986 for a review). The reason for this is the experimental difficulty of obtaining reliable action spectra that cover the spectral region of the chlorosome as well as that of the much weaker BChl a absorption.

Electron micrograph studies indicate that chlorosomes of both species contain rodlike structures that are parallel to the long axis of the chlorosome and may contain a regular array of BChl c protein complexes (STAEHELIN et al. 1978, 1980). A regular arrangement of BChl c is also indicated by measurements of linear dichroism (SWARTHOFF et al. 1980; BETTI et al. 1982; VAN DORSSEN et al. 1986a,b). For *Cfl. aurantiacus*, the primary structure of the polypeptide to which BChl c is bound (FEICK and FULLER 1984) has been determined (WECHSLER et al. 1984). It has a molecular weight of 5.7 kD and may bind 7 BChl c molecules.

b) Green Sulfur Bacteria

Since the last time we discussed this subject (AMESZ 1983) important advances have been made concerning the primary electron transfer reaction in green sulfur bacteria. Because of the large antenna contained in the chlorosomes of these bacteria, all these experiments were done with isolated membranes or with pigment-protein complexes derived from these membranes. These preparations offer the important advantage of a much lower BChl : reaction center ratio (typically about 100 or less) than that of intact cells. Such a ratio is comparable to that of membranes or "chromatophores" of purple bacteria, but whereas chromatophores are easily obtained in large amounts, the preparation of reasonably pure membranes from green sulfur bacteria is laborious and time-consuming, and so far studies of these membranes have been confined to relatively few laboratories. Isolated reaction center complexes, without antenna pigments, have not yet been obtained from green sulfur bacteria, suggesting that the reaction center may be bound to the same peptide as part of the antenna. The smallest complex obtained, the "core complex", contains about 20 BChl a molecules (VASMEL et al. 1983a; HURT and HAUSKA 1984), which appear to be bound to a 68-kD polypeptide, a situation reminiscent of that in photosystem I of plants (Sect. 5).

Membranes or pigment-protein complexes from the green sulfur bacterium *Prosthecochloris aestuarii* upon illumination showed the oxidation of the primary electron donor P 840 (OLSON 1980; SWARTHOFF et al. 1981a) which is probably a dimer of BChl a (OLSON et al. 1977; SWARTHOFF et al. 1981b; WASIELEWSKI et al. 1982). Under reducing conditions, the photoaccumulation of reduced iron-sulfur centers was observed (SWARTHOFF et al. 1981c). Information about the identity of the primary electron acceptor has only recently become available (VAN BOCHOVE et al. 1984; NUIJS et al. 1985a; SHUVALOV et al. 1986c). Absorption difference spectra of mebranes of *P. aestuarii* measured shortly after a 35-ps laser

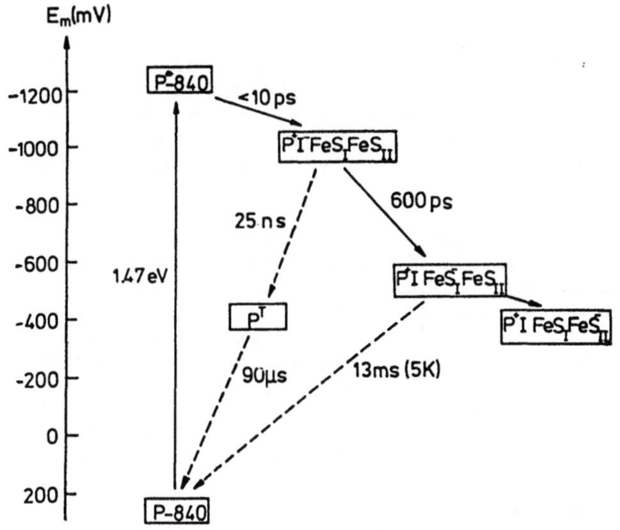

Fig. 3. Energy scheme of electron transfer in the reaction center of green sulfur bacteria. I, BChl c; FeS, iron-sulfur centers. Unless otherwise indicated, the time constants refer to room temperature and to membrane fragments or pigment-protein complexes of *P. aestuarii*. For further explanation, see Fig. 2

flash showed, in addition to the formation of P^+ and of excited states in the antenna, a bleaching of a pigment absorbing near 670 nm. This bleaching was also observed upon illumination at 850 nm, at which wavelength P 840 and the BChl molecules that are associated with the reaction center are directly excited, so that formation of excited states of the 670-nm pigment was avoided (SHUVALOV et al. 1986c). The bleaching was produced in less than 10 ps and disappeared again in 600 - 700 ps, and could be ascribed to reduction of the primary electron acceptor (I) followed by re-oxidation by subsequent electron transfer, presumably to an iron-sulfur center. At first I was tentatively identified as BPh c, but a recent pigment analysis by BRAUMANN and co-workers (1986) demonstrated the presence of a special BChl c in the photosynthetic membrane of *P. aestuarii*. In contrast to the chlorosomal pigment, this BChl c is esterified with phytol (instead of farnesol). Only traces of Bph were detected (BRAUMANN et al. 1986). Thus, one must conclude that the primary electron acceptor in green sulfur bacteria is BChl c. When electron transport to the next acceptor is blocked, a back reaction between P 840^+ and $BChl^-$ c occurs which produces the triplet state of P 840 (SWARTHOFF et al. 1981a,b; VAN BOCHOVE et al. 1984; VASMEL et al. 1984). Figure 3 gives a scheme of electron transport in green sulfur bacteria.

The recently discovered photosynthetic bacterium *Heliobacterium chlorum* (GEST and FAVINGER 1983) contains a "new" bacteriochlorophyll, BChl g, as major pigment (BROCKMANN and LIPINSKI 1983). Its primary donor, P 798 (FULLER et al. 1985; PRINCE et al. 1985; NUIJS et al. 1985b), is probably BChl g too. *H. chlorum* does not contain chlorosomes but flash spectroscopy of isolated membranes indicates that, like in *P. aestuarii*, reduction of the primary acceptor causes a bleaching at 670 nm, suggesting that I is a BChl c-like compounds as in green sulfur bacteria (NUIJS et al. 1985b). Photoaccumulation experiments under reducing conditions showed the reduction of an iron-sulfur center with a midpoint potential below -500 mV (BROK et al. 1986; SMIT et al. 1987). However, it is not clear whether this iron-sulfur center plays a role in the main electron transport chain: recent experiments (H.W.J. SMIT, unpublished) indicate that electron transport may be blocked, with the formation of the triplet of P 798, independent of the redox state of the iron-sulfur center.

c) Green Filamentous Bacteria

Apart from the presence of chlorosomes, the green filamentous bacteria show a clear similarity to the purple bacteria. The photosynthetic membrane of *Cfl. aurantiacus* contains carotenoid and BChl a, but no other bacteriochlorophylls, and the properties of the intrinsic antenna complex, B806-866, and of the reaction center strongly resemble those of e.g. *Rb. sphaeroides*.

The light-harvesting complex, as judged from its absorption spectrum, both in the isolated form (FULLER et al. 1984) and in situ (VASMEL et al. 1986b) contains BChl a 866 and BChl a 806 in a ratio of approximately 2 : 1. This is the same ratio as observed for BChl a 850 and BChl a 800 in the B800-850 complex of *Rb. sphaeroides*. However, the orientation of the porphyrin ring of BChl a 806 with respect to the membrane is different from that of BChl a 800 (VASMEL et al. 1986b), and the primary structures of the constituent apoproteins (WECHSLER et al. 1985; ZUBER et al. 1987) resemble more those of the B875 complex of purple bacteria (THEILER et al. 1985) than of B800-850 (THEILER et al. 1984).

PIERSON and THORNBER (1983) were the first to isolate the reaction center of *Cfl. aurantiacus*. Its pigment composition resembles that of purple bacteria like *Rb. sphaeroides* and *R. rubrum*. However, the *Chloroflexus* reaction center does not contain carotenoid, and the BChl : BPh ratio is 3 : 3 instead of 4 : 2 (BLANKENSHIP et al. 1984), which indicates that one BChl a molecule is replaced by BPh a.

Studies of the photochemical properties of the isolated reaction center, some of which were briefly mentioned in an earlier review (AMESZ 1983), indicate that the mechanism of electron transport is again quite similar to that of purple bacteria (Fig. 2). Difference spectra obtained upon illumination of reaction centers with short laser flashes are in fact easier to interpret than those of purple bacteria, because the component bands are better separated, and indicate the reduction of BPh a together with the oxidation of the primary electron donor, P 865 (KIRMAIER et al. 1983, 1984; BLANKENSHIP et al. 1984). The charge separation appears to be somewheat slower than in purple bacteria (KIRMAIER et al. 1983; SHUVALOV et al. 1986d) and occurs in 13 - 14 ps (SHUVALOV et al. 1986d) when the secondary electron acceptor Q_A, which is menaquinone (VASMEL and AMESZ 1983), is reduced before the flash. Back reaction between the primary reactants produces the triplet of P 865 (DEN BLANKEN et al. 1983). Reduction of BPh a was also observed in mebranes of *Cfl. aurantiacus* upon flash illumination. Transfer of an electron from reduced BPh a to Q_A takes about 400 ps in intact membranes (NUIJS et al. 1986a) and about 300 ps in isolated reaction centers (KIRMAIER et al. 1984).

Analysis of the optical properties of the reaction center of *Cfl. aurantiacus* by means of exciton theory (VASMEL et al. 1986a; SCHERER and FISCHER 1987) indicates that its structure is very similar to that of purple bacteria. If the coordinates obtained from the X-ray analysis of the reaction center of *Rps. viridis* (DEISENHOFER et al. 1984) were used, a satisfactory simulation was obtained of the absorption, linear and circular dichroism spectra of the oxidized and reduced reaction center and of the triplet state if it was assumed that one of the accessory BChls (probably that in the M chain) is replaced by Bph. One of the results of the analysis was that the band near 790 nm, which is observed in the low-temperature spectra (VASMEL et al. 1983b), is essentially the so-called high-energy band of the P 865 dimer. In purple bacteria this band is too weak to be observed (VASMEL et al. 1986a). Since the Chloroflexaceae are probably evolutionarily quite distant from other groups of photosynthetic bacteria, these results indicate that the reaction center of the type found in purple bacteria and Chloroflexaceae has been highly preserved during more than 3 billion years of evolution.

4. Photosystem II

Two years ago (AMESZ 1985) we reviewed the then available evidence concerning the structure of the antenna and the "core" of photosystem II (PS II). To avoid repetition this point will only be briefly discussed here. Spectroscopic (KRAMER et al. 1985) and biochemical evidence (see e.g. NELSON 1987) revealed that the core of PS II has a very similar structure for all oxygen-evolving organisms, including the cyanobacteria. In higher plants and green algae the antenna complex (the well-known Chl a/b complex LHCP II) has been isolated and studied extensively in various laboratories (e.g. GREEN and CAMM 1982; KÜHLBRANDT 1984; CORUZZI et al. 1983; CASHMORE 1984; GÜLEN and KNOX 1984; LI 1985; KARLIN-NEUMANN et al. 1985; ANDERSON and GOODCHILD 1987; ZUBER et al. 1987). In cyanobacteria and red algae the antenna exists in the form of phycobilisomes, complex structures of biliproteins, that are attached to the photosynthetic membrane (see reviews by GLAZER 1984, 1985; ZUBER et al. 1987). During recent years the primary structures of many of the constituent apoproteins have been determined (see ZUBER et al. 1987) and the three-dimensional structures of C-phycocyanins from the cyanobacteria *Mastigocladus laminosus* and *Agmenellum quadruplicatum* have been obtained by X-ray diffraction of the crystalline preparations (SCHIRMER et al. 1985; ZUBER et al. 1987). Light-harvesting complexes containing Chl a and Chl c have been isolated from brown algae (BARRETT and ANDERSON 1977).

Purified PS II core complexes have been obtained from higher plants (DINER and WOLLMAN 1980; TANG and SATOH 1985; GANAGO et al. 1985; IKEUCHI et al. 1985; GHANOTAKIS and YOCUM 1986; VAN DORSSEN et al. 1987; GHANOTAKIS et al. 1987) as well as from cyanobacteria (SATOH et al. 1985). These preparations contain 45 - 60 Chl a molecules per reaction center. Many of these preparations still show a fairly high activity for oxygen evolution. In addition to the intrinsic proteins mentioned in the previous review (AMESZ 1985): CPa-1 (CP-47), CPa-2 (CP-43), the so-called D1 (herbicide-binding) and D2 "33 kD" proteins, and cytochrome b-559, these core complexes contain the extrinsic peptides involved in oxygen evolution (for reviews of the function of these extrinsic peptides and of oxygen evolution in general see MURATA and MIYAO 1985; WITT et al. 1986; YAMASHITA 1986; DEKKER and VAN GORKOM 1987; BABCOCK 1987). It may be recalled that the above numbers used in the designation of the intrinsic peptides refer to the apparent molecular weights as derived from sodium dodecyl-sulfate gel electrophoresis; the true molecular weights based on their primary structures are 20 - 25% higher.

P 680, the primary electron donor of PS II is probably a dimer of Chl a although unequivocal evidence for its dimeric nature is still lacking (RUTHERFORD and HEATHCOTE 1985; MATHIS and RUTHERFORD 1987). So-called photoaccumulation experiments showed the reduction of reduced pheophytin (Phe) a upon illumination of chloroplasts and PS II preparations under reducing conditions (KLIMOV et al. 1977; GANAGO et al. 1985) from which it was concluded that the primary charge separation in PS II would consist of the transfer of an electron from Chl a to Phe a, analogous to what occurs in the reaction center of purple bacteria. This notion has recently been confirmed by flash spectroscopy. NUIJS and co-workers (1986b) obtained flash-induced difference spectra of a partially purified PS II core complex with a time resolution of less than 50 ps. The spectra not only showed the generation and decay of excited Chl a, but also absorbance changes due to $P^{+}680$ and $Phe^{-}a$. The signal due to the latter component decayed after the flash within a time constant of 250 - 300 ps, presumably by electron transfer to Q_A (plastoquinone).

It has previously assumed that the CPa-1 complex (CP-47) would contain the reaction center of PS II (NAKATANI et al. 1984; CAMM and GREEN 1984; DE VITRY et al. 1984; BRICKER et al. 1985; YAMAGISHI and KATOH 1985; SATOH 1986; see also VERMAAS et al. 1986). However, because of the homology between the primary structures of the D1 and D2 complexes and of the L and M subunits of the reaction center of purple bacteria, mentioned in the previous review, it was suggested that D1 and D2, rather than CP-47 might provide the binding sites for the primary reactants (TREBST and DRABER 1986; TREBST 1986). Until recently it was not known whether D1 and D2 contained any pigment at all, but NANBA and SATOH (1987) have now succeeded in isolating a complex of D1, D2 and cytochrome b_{559} which contained bound Chl a, Phe a and carotenoid. The preparation showed photochemical activity as indicated by light-induced absorbance changes which could be attributed to the photoaccumulation of reduced Phe a. Electron transfer from P 680 to Phe a was observed by flash spectroscopy (DANIELUS et al. 1987). The complex contained five Chl a, one β-carotene and two Phe a molecules, one of which could be photochemically reduced, but virtually no plastoquinone (NANBA and SATOH 1987). One of the Chls a appears to be detached from the others as indicated by flash spectroscopy (DANIELUS et al. 1987) and, in fact, preparations obtained by a somewhat different method appear to contain only four chlorophylls (J.J. PLIJTER, unpublished experiments). It thus appears that the reaction center of PS II has a very similar structure and composition as that of purple bacteria. However, the role of the bound cytochrome b_{559} is still unclear (see e.g. GHANOTAKIS and YOCUM 1985; CRAMER et al. 1986); as yet there is no evidence that it would function as an electron donor to P 860 as does the bound cytochrome c of the *Rps. viridis* reaction center (see Sect. 2). The absence of plastoquinone in the complex indicates that the electron acceptors Q_A and Q_B are lost during the isolation procedure. In agreement with this no evidence for secondary electron transport on either the donor or the acceptor side was seen, and the products of the charge separation disappeared again by a back reaction with a time constant of 35 ns (DANIELIUS et al. 1987).

5. Photosystem I

P 700, the primary electron donnor of photosystem I (PS I) of algae and higher plants, is probably a dimer of Chl a (for a discussion see the reviews by RUTHERFORD and HEATHCOTE 1985 and MATHIS and RUTHERFORD 1987) or of an isomer of Chl a (WATANABE et al. 1985). Its midpoint potential is about the same as that of the primary electron donor of purple bacteria (∿ 0.48 V). However, the electron acceptors of PS I are considerably more electronegative than those of purple bacteria or of PS II. ESR and optical studies have shown that PS I contains three bound iron-sulfur centers (EVANS et al. 1972, KE 1973; MALKIN and BEARDEN 1978; SAUER et al. 1979; SWARTHOFF et al. 1982), called F_X, F_A and F_B, with midpoint potentials of about -700, -540 and -590 mV respectively. Most of the data concerning these iron-sulfur centers have been obtained from photoaccumulation experiments in which the reduced forms were trapped at low temperature, and it is still not clear whether the iron-sulfur centers operate in a linear electron transfer chain and in which order. For a discussion we refer to RUTHERFORD and HEATHCOTE (1985) and to MATHIS and RUTHERFORD (1987).

Absorption difference spectroscopy in the ns and sub-ns region indicates that the primary electron acceptor of PS I is a Chl a molecule with an absorption maximum near 693 nm, which is bleached upon reduction. In highly enriched PS I particles prepared by detergent treatment the bleaching has been observed to reverse again in about 200 ps after a flash, presumably by electron transfer to a subsequent acceptor (SHUVALOV et al. 1979), but more recent experiments (NUIJS et al. 1986c; SHUVALOV et al. 1986e) indicate that this reaction is several times faster under more "natural" conditions.

ESR and optical measurements under highly reducing conditions have suggested the possible involvement of two early electron acceptors (GAST et al. 1983; BONNERJEA and EVANS 1982; SWARTHOFF et al. 1982; MANSFIELD and EVANS 1985). The involvement of the first of these, which shows an ESR signal at g = 2.002 and a bleaching at 670 nm ascribed to reduction of Chl a (SWARTHOFF et al. 1982; MANSFIELD and EVANS 1985), was not confirmed by flash spectroscopy (NUIJS et al. 1986c; SHUVALOV et al. 1986e), whereas the identity of the other one has not yet been established with certainty. No absorbance changes were detected in the visible region upon its reduction (MANSFIELD and EVANS 1985), but absorption difference spectra in the ultraviolet region suggest that this electron acceptor, called A_1, may be vitamin K_1 (phylloquinone) (BRETTEL et al. 1986; MANSFIELD and EVANS 1986). Q-band ESR spectra are not incompatible with this suggestion (MANSFIELD and EVANS 1985; THURNAUER and GAST 1985; see also BROADHURST et al. 1986). Vitamin K_1 has long been known to be a constituent of chloroplasts (DAM et al. 1948) and recent evidence has shown that two molecules of vitamin K_1 are associated with PS I (INTERSCHICK-NIEBLER and LICHTENTHALER 1981; TAKAHASHI et al. 1983; MALKIN 1986; SCHROEDER and LOCKAU 1986; LOCKAU et al. 1987). One of these can be easily extracted without impairing the photochemical activity; the other one is more firmly bound (MALKIN 1986).

Purified PS I preparations contain two polypeptides with apparent molecular weights of 60 - 70 kD (VIERLING and ALBERTE 1983; SÉTIF et al. 1980). These peptides are encoded in the chloroplast and from DNA sequencing (FISH et al. 1985) the existence of two homologous polypeptides of slightly different molecular weights (82.5 and 83.2 kD respectively) has been established. Both peptides may span the membrane 11 times. Whereas the primary charge separation presumably occurs on one of the large polypeptides or on a complex of both of them (WASIELEWSKI et al. 1981; MATHIS et al. 1978), smaller polypeptides (15 - 18 kD) have been associated with the iron-sulfur centers F_A and F_B (BENGIS and NELSON 1977; MØLLER et al. 1981; TAKAHASHI and KATOH 1982; LAGOUTTE et al. 1984; BONNERJEA et al. 1985). There is evidence that the early electron acceptors A_1 and F_X are bound to the large polypeptides (INTERSCHICK-NIEBLER and LICHTENTHALER 1981; LAGOUTTE et al. 1984; GOLBECK and CORNELIUS 1986; HIYAMA et al. 1987; LOCKAU et al. 1987; see also HØJ and MØLLER 1986). In addition, the two 60 - 70 kD polypeptides together bind about 60 Chl a molecules (THORNBER et al. 1977; SÉTIF et al. 1980; VIERLING and ALBERTE 1983; MULLET et al. 1980; NECHUSHTAI et al. 1983). Most of these chlorophylls function as antenna pigments only, and since the same polypeptides presumably also bind the reaction center components, it is not surprising that a separate reaction center complex as from purple bacteria and recently also from PS II has not been obtained for PS I. Nevertheless, by selective extraction preparations have been obtained with much lower Chl : P 700 ratios than 60; the lowest ratios obtained so far without loss of photochemical activity are between 7 and 12 (IKEGAMI and KATOH 1975; IKEGAMI and KE 1984).

Most of the chlorophyll associated with PS I is bound to separate light-harvesting complexes. In addition to Chl a, these contain also some

Chl b. Complexes with apparent molecular weights of 20 - 25 kD have been isolated (WOLLMAN and BENNOUN 1982; HAWORTH et al. 1983; LAM et al. 1984a,b) but the primary structures of the polypeptides are as yet unknown.

References

ALLEN, J.P., FEHER, G., YEATES, T.O., REES, D.C., DEISENHOFER, J., MICHEL H., HUBER, R.: Proc. Natl. Acad. Sci. USA 83, 8589-8593 (1986). - AMESZ, J.: Prog. Bot. 45, 89-105 (1983); - ibid. 47, 87-104 (1985). - AMESZ, J., KNAFF, D.B.: Molecular mechanisms of bacterial photosynthesis. In: Biology of Anaerobes, ed. A.J.B. ZEHNDER. New York: Wiley 1987 (in press). - AMESZ, J., VASMEL, H.: Fluorescence properties of photosynthetic bacteria, 423-450. In: Light Emission by Plants and Bacteria, eds. GOVINDJEE, J. AMESZ, D.C. FORK. Orlando, FL: Academic Press 1986. - ANDERSON, J.M., GOODCHILD, D.J.: FEBS Lett. 213, 29-33 (1987).

BABCOCK, G.T.: The photosynthetic oxygen evolving process, 125-158. In: New Comprehensive Biochemistry, Vol. 15, Photosynthesis, ed. J. AMESZ. Amsterdam: Elsevier 1987. - BARRETT, H.J., ANDERSON, J.M.: Plant Sci. Lett. 99, 275-283 (1977). - BENGIS, C., NELSON, N.: J. Biol. Chem. 252, 4564-4569 (1977) - BETTI, J.A., BLANKENSHIP, R.E., NATARAJAN, L.V., DICKINSON, L.C., FULLER, R.C.: Biochim. Biophys. Acta 680, 194-201 (1982). - BLANKENSHIP, R.E., MANCINO, L.J., FEICK, R., FULLER, R.C., MACHNICKI, J., FRANK, H.A., KIRMAIER, C., HOLTEN, D.: Primary photochemistry and pigment composition of reaction centers isolated from the green photosynthetic bacterium *Chloroflexus aurantiacus*, 203-206. In: Advances in Photosynthesis Research, Vol. I, ed. C. SYBESMA. The Hague: Nijhoff/Junk 1984. - BONNERJEA, J., EVANS, M.C.W.: FEBS Lett. 148, 313-316 (1982). - BONNERJEA, J., ORTIZ, W., MALKIN, R.: Arch. Biochem. Biophys. 240, 15-20 (1985). - BRAUMANN, T., VASMEL, H., GRIMME, L.H., AMESZ, J.: Biochim. Biophys. Acta 848, 83-91 (1986). - BRETON, J., MARTIN, J.-L., PETRICH, J., MIGUS, A., ANTONETTI, A.: FEBS Lett. 209, 37-43 (1986). - BRETTEL, K., SÉTIF, P., MATHIS, P.: FEBS Lett. 203, 220-224 (1986). - BRICKER, T.M., PAKRASI, H.B., SHERMAN, L.A.: Arch. Biochem. Biophys. 237, 170-176 (1985). - BROADHURST, R.W., HOFF, A.J., HORE, P.J.: Biochim. Biophys. Acta 852, 106-111 (1986). - BROCH-DUE, M., ORMEROD, J.G., FJERDINGEN, B.S.: Arch. Microbiol. 116, 269-274 (1978). - BROCKMANN, H., LIPINSKI, A.: Arch. Microbiol. 136, 17-19 (1983). - BROK, H., VASMEL, H., HORIKX, J.T.G., HOFF, A.J.: FEBS Lett. 194, 332-326 (1986).

CAMM, E.L., GREEN, B.R.: J. Cell Biochem. 23, 171-179 (1984). - CASHMORE, A.R.: Proc. Natl. Acad. Sci. USA 81, 2960-2964 (1984). - CHANG, C.-H., TIEDE, D., TANG, J., SMITH, U., NORRIS, J., SCHIFFER, M.: FEBS Lett. 205, 82-86 (1986). - CORUZZI, G., BROGLIE, R., CASHMORE, A.R., CHUA, N.-H.: J. Biol. Chem. 258, 1399-1402 (1983). - CRAMER, W.A., THEG, S.M., WIDGER, W.R.: Photosynth. Res. 10, 393-404 (1986).

DAM, H.D., HJORTH, E., KRUSE, I.: Physiol. Plant. 1, 379-381 (1948). - DANIELIUS, R.V., SATOH, K., KAN, P.J.M. VAN, NUIJS, A.M., GORKOM, H.J. VAN: FEBS Lett. 213, 241-244 (1987). - DEBUS, R.J., FEHER, G., OKAMURA, M.Y.: Biochemistry 25, 2276-2287 (1986). - DEISENHOFER, J., EPP, O., MIKI, K., HUBER, R., MICHEL, H.: J. Mol. Biol. 180, 385-398 (1984); - Nature 318, 618-624 (1985). - DEKKER, J.P., GORKOM, H.J. VAN: J. Bioenerg. Biomembr. 19, 125-142 (1987). - DEN BLANKEN, H.J., VASMEL, H., JONGENELIS, A.P.J.M., HOFF, A.J., AMESZ, J.: FEBS Lett. 161, 185-189 (1983). - DE VITRY, C., WOLLMAN, F.-A., DELEPELAIRE, P.: Biochim.

Biophys. Acta 767, 415-422 (1984). - DINER, B.A., WOLLMAN, F.-A.: Eur. J. Biochem. 110, 521-526 (1980). - DUCRUIX, A., REISS-HUSSON, F.: J. Mol. Biol. 193, 419-421 (1987).

EVANS, M.C.W., TELFER, A., LORD, A.V.: Biochim. Biophys. Acta 267, 530-537 (1972).

FEICK, R.G., FULLER, R.C.: Biochemistry 23, 3693-3700 (1984). - FEICK, R.G., FITZPATRICK, H., FULLER, R.C.: J. Bacteriol. 150, 905-915 (1982). - FISH, L.E., KÜCK, U., BOGORAD, L.: J. Biol. Chem. 260, 1413-1421 (1985). - FULLER, R.C., BLANKENSHIP, R.E., FEICK, R.G.: The molecular topography of the photochemical membrane system in the green bacterium *Chloroflexus*, 377-380. In: Advances in Photosynthetic Research, Vol. III, ed. C. SYBESMA. The Hague: Nijhoff/Junk 1984. - FULLER, R.C., SPRAGUE, S.G., GEST, H., BLANKENSHIP, R.E.: FEBS Lett. 182, 345-349 (1985).

GANAGO, I.B., KLIMOV, V.V., KRASNOVSKII, A.A.: Biofizika 30, 811-816 (1985); Translation: Biophysics 30, 882-888 (1985). - GAST, P., SWARTHOFF, T., EBSKAMP, F.C.R., HOFF, A.J.: Biochim. Biophys. Acta 722, 165-175 (1983). - GEROLA, P.D., OLSON, J.M.: Biochim. Biophys. Acta 848, 69-76 (1986). - GEST, H., FAVINGER, J.L.: Arch. Microbiol. 136, 11-16 (1983). - GHANOTAKIS, D.F., YOCUM, C.F.: Photosynth. Res. 7, 97-114 (1985); - FEBS Lett. 197, 244-248 (1986). - GHANOTAKIS, D.F., DEMETRIOU, D.M., YOCUM, C.F.: Biochim. Biophys. Acta 891, 15-21 (1987). - GLAZER, A.N.: Biochim. Biophys. Acta 786, 29-51 (1984); - Annu. Rev. Biophys. Biophys. Chem. 14, 47-77 (1985). - GOLBECK, H., CORNELIUS, J.M.: Biochim. Biophys. Acta 849, 16-24 (1986). - GREEN, B.R., CAMM, E.D.: Biochim. Biophys. Acta 681, 256-262 (1982). - GÜLEN, D., KNOX, R.S.: Photochem. Photobiol. 7, 277-288 (1984).

HAWORTH, P., WATSON, J.L., ARNTZEN, C.J.: Biochim. Biophys. Acta 724, 151-158 (1983). - HIYAMA, T., KATOH, A., SHIMIZU, T., INOUE, K., KUBO, A.: Preparation and properties of photosystem 2 reaction center complex, 45-58. In: Progress in Photosynthesis Research, Vol. 2, ed. J. BIGGINS. Dordrecht: Nijhoff 1987. - HØJ, P.J., MØLLER, B.L.: J. Biol. Chem. 261, 14292-14300 (1986). - HURT, E.C., HAUSKA, G.: FEBS Lett. 168, 149-154 (1984).

IKEGAMI, I., KATOH, S.: Biochim. Biophys. Acta 376, 588-592 (1975). - IKEGAMI, I., KE, B.: Biochim. Biophys. Acta 764, 70-79 (1984). - IKEUCHI, M., YUASA, M., INOUE, Y.: FEBS Lett. 185, 316-322 (1985). - IMHOFF, J.F., TRÜPER, H.G., PFENNIG, N.: Int. J. Syst. Bacteriol. 34, 340-343 (1984). - INTERSCHICK-NIEBLER, E., LICHTENTHALER, H.K.: Z. Naturforsch. 36c, 276-283 (1981).

KARLIN-NEUMANN, G.A., KOHORN, B.D., THORNBER, J.P., TOBIN, E.M.: J. Mol. Appl. Genet. 3, 45-61 (1985). - KE, B.: Biochim. Biophys. Acta 301, 1-33 (1973). - KIRMAIER, C., HOLTEN, D., FEICK, R., BLANKENSHIP, R.E.: FEBS Lett. 158, 73-78 (1983). - KIRMAIER, C., HOLTEN, D., MANCINO, L.J., BLANKENSHIP, R.E.: Biochim. Biophys. Acta 765, 138-146 (1984). - KLIMOV, V.V., KLEVANIK, A.V., SHUVALOV, V.A., KRASNOVSKY, A.A.: FEBS Lett. 82, 183-186 (1977). - KRAMER, H.J.M., WESTERHUIS, W.H.J., AMESZ, J.: Physiol. Végét. 23, 535-543 (1985). - KÜHLBRANDT, W.: Nature 307, 478-480 (1984).

LAGOUTTE, B., SÉTIF, P., DURANTON, J.: FEBS Lett. 174, 24-29 (1984). - LAM, E., ORTIZ, W., MALKIN, R.: FEBS Lett. 168, 10-14 (1984a). - LAM, E., ORTIZ, W., MAYFIELD, S., MALKIN, R.: Plant Physiol. 74, 650-655 (1984b). - LI, J.: Proc. Natl. Acad. Sci. USA 82, 386-390 (1985). - LOCKAU, W., SCHROEDER, H.-U., NITSCHKE, W., ZIEGLER, K.: Bound acceptors of photosystem 1, 37-40. In: Progress in Photosynthesis Research, Vol. 2, ed. J. BIGGINS. Dordrecht: Nijhoff 1987.

MALKIN, R.: FEBS Lett. 208, 343-346 (1986). - MALKIN, R., BEARDEN, A.J.: Biochim. Biophys. Acta 505, 147-181 (1978). - MANSFIELD, R.W., EVANS, M.C.W.: FEBS Lett. 190, 237-241 (1985); - ibid. 203, 225-229 (1986). - MAROTI, P., KIRMAIER, C., WRAIGHT, C., HOLTEN, D., PEARLSTEIN, R.M.: Biochim. Biophys. Acta 810, 132-139 (1985). - MARTIN, J.-L., BRETON, J., HOFF, A.J., MIGUS, A., ANTONETTI, A.: Proc. Natl. Acad. Sci. USA 83, 957-961 (1986). - MATHIS, P., RUTHERFORD, A.W.: The primary reactions of photosystem I and II of algae and higher plants, 63-96. In: New Comprehensive Biochemistry, Vol. 15, Photosynthesis, ed. J. AMESZ. Amsterdam: Elsevier 1987. - MATHIS, P., SAUER, K., RÊMY, R.: FEBS Lett. 88, 275-278 (1978). - MICHEL, H., WEYER, K.A., GRUENBERG, H., LOTTSPEICK, F.: EMBO J. 4, 1667-1672 (1985). - MICHEL, H., EPP, O., DEISENHOFER, J.: EMBO J. 5, 2445-2451 (1986a). - MICHEL, H., WEYER, K.A., GRUENBERG, H., DUNGER, I., OESTERHOLT, D., LOTTSPEICK, F.: EMBO J. 5, 1149-1158 (1986b). - MØLLER, B.L., HØYER-HANSEN, G., HILLER, R.G.: Functional identification of barley thylakoid polypeptides resolved by SDS-polyacrylamide gel electrophoresis, 245-256. In: Photosynthesis, Vol. III, ed. G. AKOYUNOGLOU. Philadelphia: Balaban Int. Science Serv. 1981. - MULLET, J.E., BURKE, J.J., ARNTZEN, C.J.: Plant Physiol. 65, 814-822 (1980). - MURATA, N., MIYAO, M.: Trends Biochem. Sci. 10, 122-124 (1985).

NAKATANI, H.Y., KE, B., DOLAN, E., ARNTZEN, C.J.: Biochim. Biophys. Acta 765, 347-352 (1984). - NANBA, O., SATOH, K.: Proc. Natl. Acad. Sci. USA 84, 109-112 (1987). - NECHUSHTAI, R., MUSTEER, P., BINDER, A., LIVEANU, V., NELSON, N.: Proc. Natl. Acad. Sci. USA 80, 1179-1183 (1983). - NELSON, N.: Structure and function of protein complexes in the photosynthetic membrane, 213-231. In: New Comprehensive Biochemistry, Vol. 15, Photosynthesis, ed. J. AMESZ. Amsterdam: Elsevier 1987. - NUIJS, A.M., BOCHOVE, A.C. VAN, JOPPE, H.L.P., DUYSENS, L.N.M., AMESZ, J.: Biochim. Biophys. Acta 807, 24-34 (1985a). - NUIJS, A.M., DORSSEN, R.J. VAN, DUYSENS, L.N.M., AMESZ, J.: Proc. Natl. Acad. Sci. USA 82, 6856-6868 (1985b). - NUIJS, A.M., VASMEL, H., DUYSENS, L.N.M., AMESZ, J.: Biochim. Biophys. Acta 849, 316-324 (1986a). - NUIJS, A.M., GORKOM, H.J. VAN, PLIJTER, J.J., DUYSENS, L.N.M.: Biochim. Biophys. Acta 848, 167-175 (1986b). - NUIJS, A.M., SHUVALOV, V.A., GORKOM, H.J. VAN, PLIJTER, J.J., DUYSENS, L.N.M.: Biochim. Biophys. Acta 850, 310-318 (1986c).

OLSON, J.M.: Biochim. Biophys. Acta 594, 33-51 (1980). - OLSON, J.M., PIERSON, B.K.: Photosynthetic bacteria, 21-40. In: New Comprehensive Biochemistry, Vol. 15, Photosynthesis, ed. J. AMESZ. Amsterdam: Elsevier 1987. - OLSON, J.M., PRINCE, R.C., BRUNE, D.C.: Brookhaven Symp. Biol. 28, 238-245 (1977).

PARSON, W.W.: The bacterial reaction center, 43-61. In: New Comprehensive Biochemistry, Vol. 15, Photosynthesis, ed. J. AMESZ. Amsterdam: Elsevier 1987. - PIERSON, B.K., CASTENHOLZ, R.W.: Arch. Mikrobiol. 100, 5-24 (1974a); - ibid. 100, 283-305 (1974b). - PIERSON, B.K., THORNBER, J.P.: Proc. Natl. Acad. Sci. USA 80, 80-84 (1983). - PRINCE, R.C., GEST, H., BLANKENSHIP, R.E.: Biochim. Biophys. Acta 810, 377-384 (1985).

RUTHERFORD, A.W., HEATHCOTE, P.: Photosynth. Res. 6, 295-316 (1985).

SATOH, K.: FEBS Lett. 204, 357-362 (1986). - SATOH, K., OHNO, T., KATOH, S.: FEBS Lett. 180, 326-330 (1985). - SAUER, K., MATHIS, P., ACKER, S., BEST, J.A. VAN: Biochim. Biophys. Acta 545, 466-472 (1979). - SCHERER, P.O.J., FISCHER, S.F.: Biochim. Biophys. Acta 891, 157-164 (1987). - SCHIRMER, T., BODE, W., HUBER, R., SIDLER, W., ZUBER, H.: J. Mol. Biol. 184, 257-277 (1985). - SCHMIDT, K.: Arch. Microbiol. 124, 21-31 (1980). - SCHMIDT, K., MAARZAHL, M., MAYER, F.: Arch. Microbiol. 127, 87-97 (1980). - SCHROEDER, H.-U., LOCKAU, W.: FEBS Lett. 199,

23-27 (1986). - SÉTIF, P., ACKER, S., LAGOUTTE, B., DURANTON, J.: Photosynth. Res. 1, 17-27 (1980). - SHUVALOV, V.A., DUYSENS, L.N.M.: Proc. Natl. Acad. Sci. USA 83, 1690-1694 (1986). - SHUVALOV, V.A., KLEVANIK, A.V., SHARKOV, A.V., KRYUKOV, P.G., KE, B.: FEBS Lett. 107, 313-316 (1979). - SHUVALOV, V.A., SHKUROPATOV, A.Y., KULEKOVA, C.M., ISMAILOV, M.A., SHKUROPATOVA, V.A.: Biochim. Biophys. Acta 849, 337-346 (1986a). - SHUVALOV, V.A., AMESZ, J., DUYSENS, L.N.M.: Biochim. Biophys. Acta 851, 327-330 (1986b); - ibid. 851, 1-5 (1986c). - SHUVALOV, V.A., VASMEL, H., AMESZ, J., DUYSENS, L.N.M.: Biochim. Biophys. Acta 851, 361-368 (1986d). - SHUVALOV, V.A., NUIJS, A.M., GORKOM, H.J. VAN, SMIT, H.W.J., DUYSENS, L.N.M.: Biochim. Biophys. Acta 850, 319-323 (1986e). - SMIT, H.W.J., AMESZ, J., HOEVEN, M.F.R. VAN DER, DUYSENS, L.N.M.: Electron transport in *Heliobacterium chlorum*, 189-192. In: Progress in Photosynthesis Research, Vol. 1, ed. J. BIGGINS. Dordrecht: Nijhoff 1987. - STACKEBRANDT, E., WOESE, C.R.: Soc. Gen. Microbiol. Symp. 32, 1-31 (1981). - STAEHELIN, L.A., GOLECKI, J.R., FULLER, R.C., DREWS, G.: Arch. Microbiol. 119, 269-272 (1978). - STAEHELIN, L.A., GOLECKI, J.R., DREWS, G.: Biochim. Biophys. Acta 589, 30-45 (1980). - SWARTHOFF, T., GROOTH, B.G. DE, MEIBURG, R.F., RIJGERSBERG, C.P., AMESZ, J.: Biochim. Biophys. Acta 593, 51-59 (1980). - SWARTHOFF, T., VEEK-HORSLEY, K.M. VAN DER, AMESZ, J.: Biochim. Biophys. Acta 635, 1-12 (1981a). - SWARTHOFF, T., GAST, P., HOFF, A.J.: FEBS Lett. 127, 83-86 (1981b). - SWARTHOFF, T., GAST, P., HOFF, A.J., AMESZ, J.: FEBS Lett. 130, 93-98 (1981c). - SWARTHOFF, T., GAST, P., AMESZ, J., BUISMAN, H.P.: FEBS Lett. 146, 201-207 (1982).

TAKAHASHI, Y., KATOH, S.: Arch. Biochem. Biophys. 219, 219-227 (1982). - TAKAHASHI, Y., HIROTA, K., KATOH, S.: Photosynth. Res. 6, 183-192 (1983). - TANG, X.-S., SATOH, K.: FEBS Lett. 179, 60-64 (1985). - THEILER, R., SUTER, F., ZUBER, H., COGDELL, R.J.: FEBS Lett. 175, 231-237 (1984). - THEILER, R., SUTER, F., PENNOYER, J.D., ZUBER, H., NIEDERMAN, R.A.: FEBS Lett. 184, 231-236 (1985). - THORNBER, J.P., ALBERTE, R.S., HUNTER, F.A., SHIOZAWA, J.A., KAN, K.-S.: Brookhaven Symp. Biol. 28, 133-148 (1977). - THURNAUER, M.C., GAST, P.: Photobiochem. Photobiophys. 9, 29-38 (1985). - TREBST, A.: Z. Naturforsch. 41C, 240-245 (1986). - TREBST, A., DRABER, W.: Photosynth. Res. 10, 381-392 (1986).

VAN BOCHOVE, A.C., SWARTHOFF, T., KINGMA, H., HOF, R.M., GRONDELLE, R. VAN, DUYSENS, L.N.M., AMESZ, J.: Biochim. Biophys. Acta 764, 343-346 (1984). - VAN DORSSEN, R.J., GEROLA, P.D., OLSON, J.M., AMESZ, J.: Biochim. Biophys. Acta 848, 77-82 (1986a). - VAN DORSSEN, R.J., VASMEL, H., AMESZ, J.: Photosynth. Res. 9, 33-45 (1986b). - VAN DORSSEN, R.J., PLIJTER, J.J., DEKKER, J.P., OUDEN, A. DEN, AMESZ, J., GORKOM, H.V. VAN: Biochim. Biophys. Acta 890, 135-143 (1987). - VASMEL, H.: The Photosynthetic Membrane of Green Bacteria. Doctoral thesis, University of Leiden 1986. - VASMEL, H., AMESZ, J.: Biochim. Biophys. Acta 724, 118-122 (1983). - VASMEL, H., SWARTHOFF, T., KRAMER, H.J.M., AMESZ, J.: Biochim. Biophys. Acta 725, 361-367 (1983a). - VASMEL, H., MEIBURG, R.F., KRAMER, H.J.M., VOS, L.J. DE, AMESZ, J.: Biochim. Biophys. Acta 724, 333-339 (1983b). - VASMEL, H., BLANKEN, H.J. DEN, DIJKMAN, J.T., HOFF, A.J., AMESZ, J.: Biochim. Biophys. Acta 767, 200-208 (1984). - VASMEL, H., AMESZ, J., HOFF, A.J.: Biochim. Biophys. Acta 852, 159-168 (1986a). - VASMEL, H., DORSSEN, R.J. VAN, AMESZ, J.: Photosynth. Res. 7, 281-294 (1986b). - VERMAAS, W.F.J., WILLIAMS, J.G.K., RUTHERFORD, A.W., MATHIS, P., ARNTZEN, C.J.: Proc. Natl. Acad. Sci. USA 83, 9474-9477 (1986). - VIERLING, E., ALBERTE, R.S.: Plant Physiol. 72, 625-633 (1983). - VOS, M., NUIJS, A.M., GRONDELLE, R. VAN, DORSSEN, R.J. VAN, GEROLA, P., AMESZ, J.: Biochim. Biophys. Acta 891, 275-285 (1987).

WACKER, T., GAD'ON, N., BECKER, A., MÄNTELE, W., KREUZ, W., DREWS, G., WELTE, W.: FEBS Lett. 197, 267-273 (1986). - WASIELEWSKI, M.R., TIEDE,

D.M.: FEBS Lett. 204, 368-372 (1986). - WASIELEWSKI, M.R., NORRIS, J.R., SHIPMAN, L.L., LIN, C.-P., SVEC, W.A.: Proc. Natl. Acad. Sci. USA 78, 2957-2961 (1981). - WASIELEWSKI, M.R., SMITH, U.H., NORRIS, J.R.: FEBS Lett. 149, 138-140 (1982). - WATANABE, T., KOBAYASHI, M., HONGU, A., NAKAZATO, M., HIYAMA, T., MURATA, N.: FEBS Lett. 191, 252-256 (1985). - WECHSLER, T., SUTER, F., FULLER, R.C., ZUBER, H.: FEBS Lett. 181, 173-178 (1984). - WECHSLER, T., BRUNISHOLZ, R., SUTER, F., FULLER, R.C., ZUBER, H.: FEBS Lett. 191, 34-38 (1985). - WELTE, W., WACKER, T., LEIS, M., KREUTZ, W., SHIOZAWA, J., GAD'ON, N., DREWS, G.: FEBS Lett. 182, 260-264 (1985). - WILLIAMS, J.C., STEINER, L.A., OGDEN, R.C., SIMON, M.I., FEHER, G.: Proc. Natl. Acad. Sci. USA 80, 6505-6509 (1983). - WILLIAMS, J.C., STEINER, L.A., FEHER, G., SIMON, M.I.: Proc. Natl. Acad. Sci. USA 81, 7303-7307 (1984). - WITT, H.T., SCHLODDER, E., BRETTEL, K., SAYGIN, O.: Photosynth. Res. 10, 453-472 (1986). - WOESE, C.R.: Why study evolutionary relationships among bacteria?, 1-30. In: The Evolution of the Prokaryotes, eds. K.H. SCHLEIFER, E. STACKEBRANDT. London: Academic Press 1985. - WOLLMAN, F.-A., BENNOUN, P.: Biochim. Biophys. Acta 680, 352-360 (1982). - WOOLLEY, K.J., MACKENZIE, R.C., COGDELL, R.J., LINDSAY, J.G., MICHEL, H.: Biochem. Soc. Trans. 14, 57 (1985).

YAMAGISHI, A., KATOH, S.: Biochim. Biophys. Acta 807, 74-80 (1985). - YAMASHITA, T.: Photosynth. Res. 10, 473-482 (1986). - YOUVAN, D.C., BYLINA, E.J., ALBERTI, M., BEGUSCH, H., HEARST, J.E.: Cell 37, 949-957 (1984).

ZUBER, H., BRUNISHOLZ, R., SIDLER, W.: Structure and function of light-harvesting pigment protein complexes, 233-271. In: New Comprehensive Biochemistry, Vol. 15, Photosynthesis, ed. J. AMESZ. Amsterdam: Elsevier 1987.

Prof. Dr. J. AMESZ
Department of Biophysics
Huygens Laboratory of the
State University
NL-2300 RA Leiden
The Netherlands

III. Metabolism of Inorganic Nitrogen Compounds

By HERMANN BOTHE

1. Nitrogen Fixation

a) Nitrogen Fixation by Free-Living Microorganisms

A monography worth reading on nitrogen fixation in plants by DIXON and WHEELER (1986) has recently been published which partly replaces the older ones by POSTGATE (1982) or SPRENT (1979). Proceedings from meetings were edited by LUDDEN and BURRIS (1985), EVANS et al. (1985) and SKINNER and UOMALA (1986). Reviews by ORME-JOHNSON (1985) and KÖHLER (1985) should also be mentioned here.

Probably the most exciting point to be mentioned here is the discovery of a new, V-containing nitrogenase. Some years ago already, BISHOP and co-workers observed that several classes of Nif^- mutant strains of *Azotobacter vinelandii* underwent phenotypic reversal under Mo-deficient conditions. The existence of an alternative nitrogenase system in addition to the conventional, Mo-containing enzyme complex was postulated from these observations (BISHOP et al. 1980, 1982). This was not really accepted by others, because explanations other than the existence of an alternative N_2-fixation system were possible. Mutants have now been constructed from *Azotobacter vinelandii* carrying a deletion of the genes nif HDK which are the structural genes of the conventional nitrogenase. These mutants express N_2-fixation activity when the culture medium is molybdenum-deficient (PREMAKUMAR et al. 1984; BISHOP et al. 1986a,b). Work of the Sussex group clarified that the "alternative" nitrogenase has a totally different subunit structure as compared to the "conventional" nitrogenase and contains V instead of Mo in stoichiometric amounts in the larger subunit (ROBSON et al. 1986a). The alternative nitrogenase shows considerably lower C_2H_2-reduction and higher H_2-evolution activities than the conventional enzyme (BISHOP et al. 1986a; ROBSON et al. 1986a; BISHOP et al. 1986b). *Azotobacter* apparently synthesize the Mo enzyme preferentially because it appears to be more stable and does not catalyze the wasteful H_2 evolution to such an extent as the V-nitrogenase. It remains to be shown whether the alternative nitrogenase is expressed in soils deficient in Mo. The two nitrogenases in *Azotobacter* are apparently coded by different sets of genes. The existence of two different and unlinked nif H genes coding for different Fe proteins of nitrogenase in *A. chroococcum* (JONES et al. 1984; ROBSON et al. 1986b) is also established for *Azotobacter vinelandii* (JACOBSON et al. 1986; JOERGER et al. 1986). Recently, evidence for the existence of a second nif K-like sequence has been forwarded for *A. chroococcum* (ROBSON 1986). Up until now, the presence of the alternative nitrogenase has only been demonstrated for some *Azotobacter* strains. Reiterated DNA sequences encoding nif HDK or nif H occur in *Rhizobium phaseoli* (QUINTO et al. 1985) and *Rhodopseudomonas capsulata* (SKOLNIK and HASELKORN 1984). Thus, organisms like these could also possess an alternative nitrogenase. The situation is really complex in *Azotobacter vinelandii* where a third, different Fe protein of nitrogenase (= nitro-

Progress in Botany, Vol. 49

genase reductase) probably exists (CHISNELL and BISHOP 1985; JOERGER et al. 1986).

Electron transport components transferring reducing equivalents to nitrogenase can also occur in multiple forms in N_2-fixing organisms. *Azotobacter vinelandii* ATCC 478 was shown to synthesize three different flavodoxins (KLUGKIST et al. 1986a,b). The physiological functions of these three flavodoxins and of ferredoxin remain to be elucidated. The concentration of flavodoxin II specifically increases under N_2-fixing conditions. Flavodoxin II may, therefore, be the natural electron carrier to nitrogenase in this *Azotobacter* strain (KLUGKIST et al. 1986a,b). Currently, it is not clear whether the presence of three different flavodoxins is restricted to only this specific strain. *Azotobacter* flavodoxin was also reported to contain 2 mol of non-coenzyme phosphorus groups in addition to the tightly bound FMN coenzyme phosphate (EDMONDSON and JAMES 1979). One of these phosphorus residues may be involved in modulating flavodoxin activity and redox properties (EDMONDSON and PELEATO 1987). The occurrence of two distinct ferredoxins was reported for the cyanobacterium *Anabaena variabilis*. One of these may specifically support nitrogenase activity in heterocysts (SCHRAUTEMEIER and BÖHME 1985; BÖHME and SCHRAUTEMEIER 1987).

Nitrogenase proteins from all microorganisms are extremely sensitive to exposure to O_2. The different organisms evolved different strategies to protect their nitrogenase against O_2 damage, including: respiration, oxyhydrogen reaction, leghemoglobin, slime protection and thick cell walls as in heterocysts of cyanobacteria or in the vesicles of actinomycetes (see DIXON and WHEELER 1986). Convincing evidence was forwarded that the cell wall of heterocysts of cyanobacteria provides a barrier that drastically reduces the rate of air entry. The level of N_2 inside the heterocysts is high enough for N_2 fixation to proceed and that of O_2 is sufficiently low to be completely consumed by respiration (WALSBY and SIMPSON 1984). The nonheterocystous, aerobic cyanobacteria apparently protect their nitrogenase by separating photosynthesis and N_2 fixation temporarily. Synchronized *Gloeothece* cells can perform all N_2 fixation in darkness when no O_2 is evolved photosynthetically (GALLON 1981). Experiments with synchronized cells of a marine *Synechococcus* strain suggest that photosynthetic O_2-evolution and N_2-fixation capabilities oscillate inversely to each other (MITSUI et al. 1986, 1987).

Toxicity of O_2 on nitrogenase probably results from the formation of reactive intermediates such as hydrogen peroxide, superoxide radical and hydroxyl radical. Recent experiments indicate that ascorbate and glutathione play an essential role in removing H_2O_2 from nitrogenase by a pathway first described for chloroplasts (NAKANO and ASADA 1981):

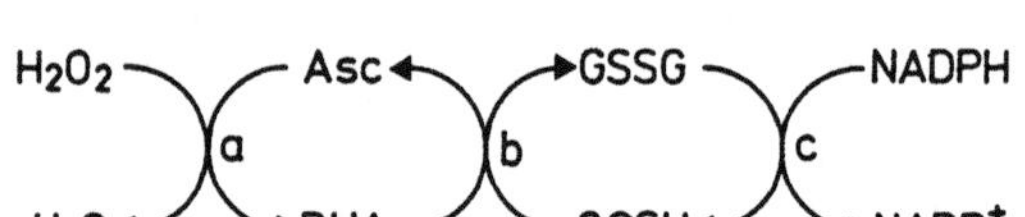

Asc = ascorbate; DHA = dehydroascorbate; GSSG and GSH = oxidized and reduced glutathione; a = ascorbate peroxidase; b = dehydroascorbate reductase; c = glutathione reductase

The enzymes involved are present in cyanobacteria (TEL-OR et al. 1985) and in nodules of soybeans (DALTON et al. 1986), and the concentrations of glutathione and ascorbate suffice for this pathway to operate (with some question mark in cyanobacteria). Catalase catalyzing the cleavage of H_2O_2 to H_2O and 1/2 O_2 is not sufficiently present to eliminate all the H_2O_2 in nodules (HANKS et al. 1983; DALTON et al. 1986) and in cyanobacteria (TEL-OR et al. 1985). It is noteworthy in this context that a positive correlation between N_2-fixation activity and the level of superoxide dismutase scavenging the superoxide anion exists in all organisms investigated so far (PUPPA and RIGAUD 1986; DINGLER and OELZE 1987).

b) Nitrogen Fixation by Symbiotic Microorganisms

Forty-five years after SCHAEDE's book (SCHAEDE 1942), another outstanding monograph on symbiosis has now been published (WERNER 1987).

The molecular basis for the recognition of rhizobia is not yet understood. Attachment studies of *Rhizobium trifolii* to clover root hairs gave positive indications that plant lectins are involved in the attachment of the bacteria. Heterologous rhizobia with a Sym plasmid cured *R. trifolii* (the Sym plasmid contains the nodulation genes) adhered only weakly to clover root hairs as compared to wild strains (DAZZO et al. 1976; DAZZO et al. 1984; ZURKOWSKI 1980). The bacterial surface may contain lectin receptors in the capsular polysaccharides, in extracellular polysaccharides and in the lipopolysaccharide membrane (HRABAK et al. 1981; PLANQUÉ and KIJNE 1977; WOLPERT and ALBERSHEIM 1976). Other investigations indicate that the attachment is not a host-specific process and is not mediated by lectins (BADENOCH-JONES et al. 1985; PULL et al. 1979). SMIT et al. (1986) found a positive correlation between the degree of fibrillation of *R. leguminosarum* cells and the ability to attach to pea root hair tips. In addition to extracellular fibrils (fimbriae), carbon limitation might play an essential role in the recognition process. According to SMIT et al. (1986), expression of host specificity occurs in later steps of the infection process, e.g. in the initiation of the infection thread.

The *Rhizobium*-legume symbiosis is genetically determined both by bacterial genes (the nod genes) and by plant genes (the nodulin genes) (for reviews see KONDOROSI and KONDOROSI 1986; WERNER 1987). Genetic analysis revealed the occurrence of 20-30 nodulin genes encoded by the plant genome. Biochemical functions have been assigned to only a few of them: Glutamine synthetase (CULLIMORE et al. 1984), a nodule-specific form of uricase (NGUYEN et al. 1985), one choline kinase (MELLOR et al. 1986) and four leghemoglobin proteins (APPLEBY 1984). The expression of the different nodulin genes is not induced at the same time during nodule development but successively and is apparently controlled at the post-transcriptional level (GOVERS et al. 1985; MARCKER et al. 1984). It is known that the four leghemoglobin nodulin genes are activated by a signal from *Rhizobium*, most likely by heme which is synthesized by the bacteria (NADLER and AVISSAR 1977). Interestingly, a chimeric gene containing the *Escherichia coli* chloramphenicol acetyl transferase gene inserted to one of the soybean leghemoglobin genes (the Lbc_3 gene) was expressed in yeast, and the expression was controlled at the post-transcriptional level (JENSEN E.O. et al. 1986). This chimeric gene can also be transferred to *Lotus corniculatus* (bird's foot trefoil) using the Ri plasmid of *Agrobacterium rhizogenes* as gene vector (JENSEN J.S. et al. 1986). *Lotus corniculatus* and soybeans are systematically unrelated plants. The chimaeric gene is expressed in fully regenerated plants inoculated with the *Lotus* symbiont, *Rhizobium loti*, at the correct stage of the nodule development (JENSEN J.S. et al. 1986). Therefore, the bacterial induction signal for leghemoglobin expression (heme alone, or heme + additional factors) is conserved within the legume symbiosis.

The bacterial genes required for symbiosis, the nod genes, are located on a very large "Sym" plasmid in most fast-growing rhizobia. An exception to this rule is probably *Rhizobium meliloti*, which carries the genes on the chromosome (CHUA et al. 1985) similar to the slow-growing species of the genus *Bradyrhizobium* (see DIXON and WHEELER 1986; WERNER 1987).

Transposon (Tn 5) mutagenesis indicates the occurrence of four genes nod A, B, C, D which are common to several fast-growing rhizobial species (KONDOROSI et al. 1984; DJORDJEVIC et al. 1985a,b; INNES et al. 1985; DOWNIE et al. 1985; EGELHOFF et al. 1985; JACOBS et al. 1985; MULLIGAN and LONG 1985) and which are functionally interchangeable among the species (KONDOROSI et al. 1984; FISCHER et al. 1985; DJORDJEVIC et al. 1985a). These common nod genes are responsible for the earliest detectable plant responses which are root hair curling and cortical cell division but they are not involved in nodule formation. A region of at least four genes adjacent to the common nod genes controls the host specificity of the symbiosis (KONDOROSI et al. 1984; DJORDJEVIC et al. 1985a,b). In order to study expression, the nod genes were fused to the lac Z gene of *E. coli*. The common nod D gene is expressed constitutively and is subject to autoregulation (ROSSEN et al. 1985). A functional nod D gene product is a prerequisite for the expression of the other common nod genes A, B, C (INNES et al. 1985; MULLIGAN and LONG 1985; ROSSEN et al. 1985). One of the most exciting findings in the last 2 years comes from several groups simultaneously: The expression of the common nod genes A, B, C also requires root exudate (ROSSEN et al. 1985; JACOBS et al. 1985; MULLIGAN et al. 1985). The purified inducing substance is of flavonoid structure in all *Rhizobium* species studied. The most active inducer is luteolin (5, 7, 3', 4' tetrahydroxyflavone) with *Rhizobium meliloti* nodulating alfalfa (Medicago sativa) (PETERS et al. 1986), 7, 4' dihydroxyflavone with *R. trifolii* nodulating white clover *(Trifolium repens)* (REDMOND et al. 1986) and a compound closely related to the flavanone naringenin with *R. leguminosarum* which is the symbiont of *Vicia faba* (ZAAT et al. 1987).

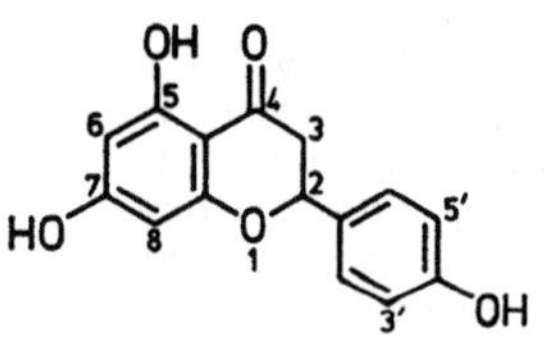

Naringenin
(=5,7,4'-trihydroxyflavanone)
for
R. leguminosarum

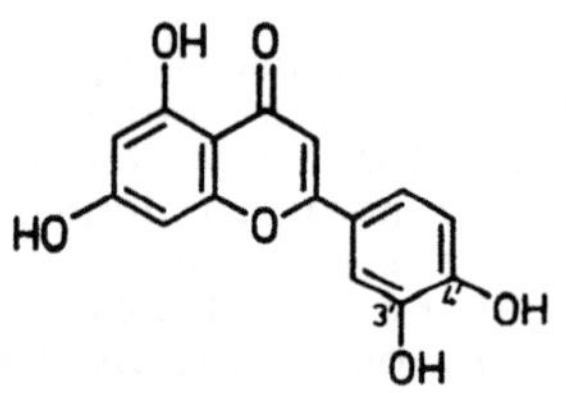

Luteolin
(=5,7,3',4'-tetrahydroxyflavone)
for
R. meliloti

These compounds also induce some of the host-specific genes in a still unidentified way (REDMOND et al. 1986; SHEARMAN et al. 1986). There is no absolute specificity of a given flavonoid acting as an inducer of only one *Rhizobium* species. The transfer of the flavonoid signal to *Rhizobium* and the interaction of the flavonoid with the nod D gene product have not yet been elucidated. Flavonoids are widespread in the plant kingdom and occur outside of the Fabales plant order. Thus, the specificity of the *Rhizobium*-legume interaction cannot be the result of an induction of the nod A, B, C genes by flavonoids. However, a new role has now been assigned to such compounds normally considered as secondary plant metabolites. A similar situation exists in the plant pathogen *Agrobacterium tumefaciens* where the virulence genes are induced by a compound prominent in exudates of wounded plants and recently identified as acetosyringone (STACHEL et al. 1985). Other related phenolic compounds may also serve in the case of *Agrobacterium* (BOLTON et al. 1986). When a part of the nod gene region is transferred from *Rhizobium meliloti* to *Agrobacterium tumefaciens*, the transconjugants can induce nodule like structures, however, not root hair curling and infection thread formation in alfalfa and white clover roots (HIRSCH et al. 1985).

The family of the rhizobiaceae consists of the three genera *Rhizobium*, *Bradyrhizobium* and *Agrobacterium* (JORDAN 1984). In the *Agrobacterium* system, the bacteria redirect plant metabolism to produce opines which are subsequently metabolized by the bacteria. Interestingly, an opine like substance, L-3-O-methyl-scyllo-inosamine, is elicited in the nodules of alfalfa by *R. meliloti* L-5-30 and is a specific growth substance for this strain (TEMPÊ et al. 1982). MURPHY et al. (1987) showed that genes for the synthesis and the catabolism of this compound are closely linked on the Sym plasmid which also carries the symbiotic genes (nif, fix, nod genes). Compounds like methyl-scyllo-inosamine are induced only by a few *R. meliloti* strains and are not universal as the opines in agrobacteria (MURPHY et al. 1987). However, similar compounds, ononitol and O-methyl-scyllo-inositol, were found in pea nodules induced by *Rhizobium leguminosarum* (SKOT and EGSGAARD 1984).

The N_2-fixing nodules of the Ulmaceae *Parasponia* are formed by *Rhizobium*. These nodules contain leghemoglobin as several of the nodules of non-leguminous plants which have actinomycetes of the genus *Frankia* as the symbiont (APPLEBY 1984). The known nodulating, hemoglobin-containing species are widely distributed over several superorders of dicotyledons (DAHLGREN 1980). Legumes contain multigene families coding for the globin part of hemoglobin (HYLDIG-NIELSEN et al. 1982). In contrast, *Parasponia* developed only one hemoglobin locus, which may reflect a relative primitive condition in this plant (LANDSMANN et al. 1986). A cDNA probe of the protein-coding region of the *Parasponia* gene hybridizes at high stringency to hemoglobin genes from *Casuarina* having the symbiont *Frankia* and also to DNA from *Trema*, a non-nodulating Ulmaceae and close relative to *Parasponia*. Therefore, LANDSMANN et al. (1986) suggest that the globin gene family is widespread in modern plants, that it may have a cryptic function in non-nodulating plants and that hemoglobins in plants and animals may have arisen by a vertical evolution from a monophyletic origin.

Although 15 - 30% of the sucrose produced in shoots is transported to the nodules via the phloem (PATE 1975; SCHUBERT 1986), photosynthate is one of the major factors limiting N_2 fixation in legumes (HARDY and HAVELKA 1976; STOUTHAMER 1984). Sucrose catabolism occurs in the nodule cytosol of the plant fraction by glycolysis, the oxidative pentose phosphate cycle and the tricarboxylic acid cycle (STREETER and SALMINEN 1985; STOWERS 1985). In the free-living state, the fast-growing rhizobia can metabolize sucrose and other disaccharides by the ENTNER-DOUDOROFF pathway and partly by the hexosemonophosphate shunt/oxidative pentose phosphate cycle, whereas the slow-growing species are virtually ineffective to do so. Glycolysis is at best poorly expressed and in some strains only (GLENN and DILWORTH 1981; STOWERS 1985). Bacteroids from *Rhizobium japonicum* (now renamed *Bradyrhizobium japonicum*, JORDAN 1984) can metabolize sugars by the pentose phosphate cycle to some extent, but do not express the ENTNER-DOUDOROFF pathway (SALMINEN and STREETER 1987). Mono- and disaccharides are taken up by bacteroids to a limited extent by diffusion (REIBACH and STREETER 1984; SALMINEN and STREETER 1987; STOWERS 1985; but see HOOYMANS and LOGMAN 1984), and the rates of the degradation of these compounds is too slow to support nitrogenase (SALMINEN and STREETER 1987). In contrast, carboxylic acids like succinate, malate, oxoglutarate and pyruvate are taken up actively into the bacteroids of *Bradyrhizobium japonicum* (REIBACH and STREETER 1984; GLENN and DILWORTH 1981; FINAN et al. 1983), and tricarboxylic acid cycle intermediates are likely the sources for energy and reductant for N_2 fixation in the bacteroids (STOWERS 1985). Experiments with mutants showed that rhizobia defective in carbohydrate catabolism form nodules which are effective in N_2 fixation, whereas mutants defective in dicarboxylic acid utilization form ineffective nodules which are, however, otherwise unimpaired in growth (RONSON and ASTWOOD 1985). In the stem nodules of the legume *Sesbania rostrata*, lactate is the principal source for energy and reductant (TRINCHANT and RIGAUD 1987). The degradation of succinate and other carboxylic acids have not been investigated in bacteroids in detail. For the free-living *Azotobacter vinelandii*, recent experiments indicate a direct relationship between nitrogenase activity and the rate of the electron transfer through the respiratory chain (KLUGKIST et al. 1986b). This observation is possibly explained by the findings that electrons from succinate or H_2 can reduce $NAD(P)^+$ by the reversed respiratory electron flow. These NAD(P)H formations in vesicles from both *Azotobacter* (HÄGER and BOTHE 1987) or *Bradyrhizobium japonicum* (HÄGER et al. 1987) are strictly dependent on ATP and on the respiratory complex I. NAD(P)H formation by reversed respiration may function as a valve in *Azotobacter* when the levels of ATP are high and those of NAD(P)H are low (HÄGER and BOTHE 1987). The reversed respiratory electron flow must have an even more important physiological function in *Bradyrhizobium* growing autotrophically with H_2 as the sole source of energy and reductant (EISBRENNER and EVANS 1983; ZUBER et al. 1987).

The root-colonizing *Azospirillum* is still considered as a potential biofertilizer (BODDEY et al. 1986; NAYAK et al. 1986). The biology and ecology of *Azospirillum* has recently been reviewed (ELMERICH 1984; OKON 1985). Four *Azospirillum* species have now been described: *A. brasilense*, *lipoferum*, *amazonense* and recently *A. halopraeferens* (REINHOLD et al. 1987). The nitrogenase system from *A. brasilense* and *A. lipoferum* can be separated into three components: a MoFe protein, an inactive Fe protein and an enzyme which activates the inactive Fe protein (LUDDEN et al. 1978), similar as in *Rhodospirillum rubrum* (LUDDEN and BURRIS 1976). In *R. rubrum*, nitrogenase activity is regulated by covalent modification of the Fe protein (by adenosine diphosphoribosylation of a specific arginyl residue) (POPE et al. 1985). The same is probably true for the Fe protein of *A. brasilense* and *lipoferum* as can be suggested from experiments on the regulation of nitrogenase activity by NH_4^+ (HARTMANN et al. 1986) and O_2 (HARTMANN and BURRIS 1987). In contrast, *A. amazonense* is not subject to regulation by covalent modification (SONG et al. 1985). The organism is acid-tolerant, the nitrogenase is weakly inhibited by NH_4^+ (HARTMANN et al. 1986), and the bacterium is a weak denitrifier (NEUER et al. 1985). *A. amazonense*, therefore, is the preferred *Azospirillum* species for applications currently. *A. halopraeferens*, isolated from roots of Kallar grass in salt marshes *(Leptochloa fusca)* in Pakistan (REINHOLD et al. 1987), has not yet been investigated with respect to applications. Another candidate for applications is the microaerobically N_2-fixing heterotrophic *Pseudomonas stutzeri* recently isolated from a cultivar of *Sorghum nutans* (KROTZKY and WERNER 1987).

2. Nitrate Reduction

a) Assimilatory Nitrate Reduction

Reviews: KLEINHOFS et al. 1985; OAKS and HIREL 1985; Proceedings from a congress: LAMBERS et al. 1985.

New techniques have become available for determining inorganic nitrogen compounds. Nitrate, nitrite (and phosphate) can now easily be measured by reversed phase HPLC (KUCHNICKI et al. 1985), nitrate by a new ion selective electrode (NAMBIAR et al. 1985) and N_2O and NO amperometrically by using a conventional Clark-type O_2 probe (Zimmer et al. 1985).

Significant amounts of NO_x can be produced by nitrate reductase in vivo when plants are tested in the so-called helium-purged in vivo assay. N_2O (MULVANEY and HAGEMAN 1984; DEAN and HARPER 1986) and NO (DEAN and HARPER 1986) are the major products. Acetaldehyde oxime may (MULVANEY and HAGEMAN 1984) or may not be produced (DEAN and HARPER 1986).

The major features of assimilatory NADH: nitrate reductase from higher plants and eukaryotic algae are remarkably similar. The enzyme is a homotetramer which contains one of each of the prosthetic groups: FAD, heme (cytochrome b_{577}) and molybdopterin per subunit. The following sequence of electron transfer is established: NADH → FAD → heme → Mo → NO_3^-. The transfer of electrons from heme to molybdenum is the rate-limiting step. In contrast to the situation with nitrogenase, the reduction of Mo can be monitored by the EPR technique. Mo^V gives a complex signal, whereas Mo^{VI} and Mo^{IV} are silent. The Mo^V signal is abolished by reduction with NADH and reoxidation with nitrate and by reaction of the reduced enzyme with CN^-. EPR titrations revealed midpoint potentials (E_o') for Mo^{VI}/Mo^V = -34 mV and -54 mV for Mo^V/Mo^{IV}. Both couples are apparently involved in catalysis of NADH-dependent nitrate reduction (SALOMONSON and BARBER 1987).

Investigations of the molecular biology are still at infancy with nitrate reductase of higher plants. Nitrate reductase apoprotein may be coded by one gene only (COVE 1979; KLEINHOFS et al. 1985) reflecting that the protein is probably a homotetramer. Work with chlorate-resistant mutants showed that the synthesis and incorporation of the molybdopterin cofactor involves at least three gene loci (NARAYAN et al. 1984; GRAFE and MÜLLER 1983). Regulatory genes involved in nitrate reductase expression are probably numerous (KLEINHOFS et al. 1985).

It is often stated that assimilatory nitrate reduction takes place in the shoots of plants. However, this general statement is wrong. One group of species has the roots as the major site of nitrate reduction, another performs the reaction both in shoots and roots and a third exclusively in shoots (STEWART et al. 1987; WALLACE 1986). Conifers mainly reduce nitrate in roots (STEWART et al. 1987). The differences are genetically determined. A relationship appears to exist between the site of nitrate assimilation and the habitat in which a species occurs. Shade species tend towards root assimilation and sun species towards leaf assimilation, but exceptions to this rule are known (STEWART et al. 1987). In addition, the activity in roots and leaves changes depending on the N-nutritional status. In a N-starved plant, the addition of nitrate almost immediately causes a drastic increase in nitrate uptake and nitrate reduction activity in roots (LEE and DREW 1986; LEE and RUDGE 1986; GOJON et al. 1986; but see DEANE-DRUMMOND and THAYER 1986; GLASS et al. 1985). After this induction process has finished, nitrate is reduced to 75% in shoots of barley and corn seedlings (GOJON et al. 1986). An efflux of nitrate across the plasmalemma amounting up to 40% of the nitrate influx in barley roots may also significantly contribute to the variations in the relative contribution of root and leave nitrate reduction (LEE and CLARKSON 1986).

Immunological investigations revealed that the nitrate reductase from shoots and roots of wheat have a great deal of similarity, but are not identical. Nitrate reductase content is higher in roots, but the specific activity of the active nitrate reductase is 500 times higher in shoot tissues (SOUALMI and CHAMPIGNY 1986). Experiments with tissue cultures from *Silene alba* indicated that NH_4^+-grown cells might contain an enzymatically inactive, antigenic form of nitrate reductase (POULLE and RAMBOUR 1985).

Both the synthesis as well as the activity of the existing nitrate reductase are regulated by light (MALDONADO and APARICIO 1987). The isolated spinach nitrate reductase can be inactivated by incubating with excess of NADH or CN^- and can be reactivated by irradiation with white or blue light. Action spectra indicated that the enzyme-bound FAD moiety acts as the photoreceptor. Photoreaction of nitrate reductase is caused by the triplet state of the flavin. This is able to oxidize a reduced component of the enzyme, probably in the molybdenum domain which leads to the recovery of the enzyme activity (APARICIO et al. 1976; ROLDAN and BUTLER 1980; MAURINO et al. 1983; ARYAN et al. 1983; APARICO et al. 1985). Interestingly, in *Neurospora*, blue light stimulates conidiation, and this photostimulation apparently occurs through photomodulation of the flavin of nitrate reductase of this fungus (KLEMM and NINNEMANN 1979; FRITZ and NINNEMANN 1985; NINNEMANN 1987). Light can also irreversibly damage nitrate reductase, particularly the diaphorase moiety of the enzyme. This photodamage appears to be caused by singlet O_2 generated by the energy transfer from excited triplet flavin to ground triplet O_2 (VARGAS et al. 1982). In long-term experiments, synthesis of nitrate and nitrite reductases is regulated by the phytochrome system in some plants (SHARMA and SOPORY 1984; RAJASEKHAR and SAPORY 1985). In cotyledons of mustards, nitrate and phytochrome (Pfr) act independently of each other in controlling nitrate reductase synthesis. Nitrate is the inducer proper, whereas phytochrome modulates the extent of the nitrate-induced response (RAJASEKHAR and MOHR 1986).

The absorption of iron by many plants requires the reduction of Fe^{3+} to Fe^{2+} as the first step (BROWN 1978). TIPTON and THOWSEN (1985) reported that the reduction of Fe^{3+} to Fe^{2+} is catalyzed by a NADH-dependent enzyme presumably located in the cell wall of soybean roots. This enzyme probably copurifies with nitrate reductase (TIPTON and THOWSEN 1985; CASTIGNETTI and SMARRELLI 1986). Nitrate reductase is known to reduce a number of electron acceptors unspecifically, among which are Fe^{3+} and Fe^{3+}-siderophores (CASTIGNETTI and SMARRELLI 1986; SMARRELLI and CASTIGNETTI 1986). Recently, nitrate reductase was suggested to function physiologically as a siderophore reductase catalyzing the reduction of Fe^{3+}-siderophore to Fe^{2+} + deferrisiderophore (CASTIGNETTI and SMARRELLI 1986). Nitrate reductase may, therefore, be involved in Fe uptake and in nitrate reduction.

b) Dissimilatory Nitrate Reduction

For a review: see PAYNE (1985)

The numbers of enzymes involved in the dissimilatory reduction of nitrate to molecular nitrogen and the conditions for their maximal expression are not yet clear. A NO reductase may occur (as can be deduced from experiments with intact *Pseudomonas perfectomarina*, FRUNZKE and ZUMFT 1986) or NO_2^- may directly be reduced to N_2O by a single NO_2^- reductase (see BRYAN 1981). N_2O reductase in *P. perfectomarina* is inhibited by NO (FRUNZKE and ZUMFT 1986). Therefore, inhibition by NO together with non-optimal expression of NO and/or N_2O reductase may cause the premature termination of denitrification at the level of NO or N_2O in otherwise dinitrogen-evolving bacteria (FRUNZKE and ZUMFT 1986) and may contribute to NO_x formations in soils. In extracts from *Azospirillum*, NO is produced only in soluble nitrite reductase preparations, but not when the enzyme is membrane-bound (DANNEBERG et al. 1985), indicating that NO production may be an artefact of cell-free preparations. However, NO reduction in various denitrifying bacteria is coupled to H^+ translocation and therefore to the generation of a proton motive force (SHAPLEIGH and PAYNE 1985, 1986). N_2O reductase from *Pseudomonas perfectomarina* exists in two catalytically different forms: a pink form obtained by purification under aerobic conditions and a purple form, obtained anaerobically which has higher specific activities (COYLE et al. 1985). Investigations on the molecular biology of dissimilatory nitrate reduction are still at infancy (INGRAHAM 1981; BRYAN et al. 1985). Experiments with mutants of *Pseudomonas perfectomarina* defective in denitrification obtained by transposon mutagenesis indicated that N_2O is the immediate precursor of dinitrogen in denitrification and confirmed that N_2O reductase is a novel type of Cu enzyme (ZUMFT et al. 1985).

It is still an open question for dissimilatory nitrate reduction how the enzymes are orientated on the cytoplasmic membrane. The respiratory nitrate reductases from *E. coli* (INGLEDEW and POOLE 1984; MORPETH and BOXER 1985) and *Paracoccus denitrificans* (CRASKE and FERGUSON 1986) are located at the cytoplasmic side, and their biochemical properties are remarkably similar. The enzyme consists of three subunits with the third subunit being a b-type cytochrome. The branch point between O_2 and NO_3^- respiration is at ubiquinol (CRASKE and FERGUSON 1986). In *P. denitrificans*, nitrite and nitrous oxide reductases are located on the periplasmic side in contrast to nitrate reductase (BOOGERD et al. 1983, 1984). Such an orientation is apparently not valid for all denitrifying organisms. In *Rhodospirillum sphaeroides, f. denitrificans* where photosynthesis and denitrification are closely linked, all three enzymes are located at the periplasmic side (KUNDU and NICHOLAS 1985). For *Azospirillum brasilense*, molar growth yield measurements (ZIMMER et al. 1987) and H^+ translocation experiments (DANNEBERG 1987) indicated that nitrate and nitrous oxide reductases are on the cytoplasmic side and nitrite reductase is on the periplasmic side of the cytoplasmic membrane.

3. Nitrification

Interestingly, new nitrifying microorganisms have been found in cooling towers of a brown coal and furnace gas power plant (KIRSTEIN et al. 1986) and even in chalk stones of historical buildings like the cathedrals of Cologne and Regensburg and the Alte Pinakothek in Munich (BOCK et al. 1987b). The group of BOCK claims that nitrifying bacteria significantly contribute to the corrosion of such buildings by the production of nitric acid.

Probably due to the slow growth of the organisms, publications on the biochemistry and genetics of nitrifying bacteria are still scarce. The current state of knowledge has recently been reviewed (BOCK et al. 1987a). Two groups characterized cytochrome c oxidase from *Nitrosomonas europaea* biochemically (DiSPIRITO et al. 1986; YAMAZAKI et al. 1985; YAMANAKA et al. 1985). The enzyme appears to have 11 nmol copper and 11 nmol heme per mg protein, three different subunits and a molecular weight of approximately 125,000 (DiSPIRITO et al. 1986). As the mitochondrial enzyme, it contains cytochrome aa_3 and not the so-called cytochrome a_1 (DiSPIRITO et al. 1986; YAMAZAKI et al. 1985; YAMANAKA et al. 1985).

Nitrification in soils has attracted special attention recently. Nitrification significantly contributes to losses of fertilizer nitrogen, because the product of nitrification, nitrate, is easily washed out from soils. Nitrification is, therefore, considered as a major cause for the pollution of groundwaters. In addition, nitrification and not denitrification in soils is the major source for the production of N_2O emitted to the atmosphere (CONRAD and SEILER 1985). Over the past 20 years, the atmospheric concentration of N_2O has steadily increased at a rate of approximately 0.2% $year^{-1}$ (WEISS 1981). The use of N-fertilizers is of major concern. Anhydrous ammonia has much greater effects on the emission of N_2O from soils than other commonly used N-fertilizers. The customary method of applying anhydrous ammonia produces alkaline zones of high ammonia-N concentration which favours nitrification and thus N_2O formation (BREITENBACH and BREMNER 1986a,b). Nitrite formed from ammonia during nitrification may be bound to soils by reacting with phenolic compounds present in the humic matter (AZAHR et al. 1986a,b). Various inhibitors of nitrification are available which all do not affect denitrification with the exception of potassium azide and 2,6-diamino-6-trichloromethyl-s-triazine (BREMNER and YEOMANS 1986). Nitrification in neutral and slightly acid soils is carried out by chemoorganotrophic bacteria. In acid forest soils under pH 5.0, heterotrophic fungi like *Verticillum lecanii* significantly contribute to nitrification (LANG and JAGNOW 1986).

References

APARICIO, P.J., BALANDIN, T., MAURINO, S.G., MALDONADO, J.M: Phytochem. Photobiol. 42, 765-770 (1985). - APARICIO, P.J., ROLDAN, J.M., CALERO, F.: Biochem. Biophys. Res. Commun. 70, 1071-1077 (1976). - APPLEBY, C.A.: Annu. Rev. Plant Physiol. 35, 443-478 (1984). - ARYAN, A.P., BATT, R.G., WALLACE, W.: Plant Physiol. 71, 582-587 (1983). - AZAHR, E.S., VERHE, R., PROOT, M., SANDRA, P., VERSTRAETE, W.: Plant Soil 94, 369-382 (1986a). - AZAHR, E.S., VANDENABEELE, J., VERSTRAETE, W.: Plant Soil 94, 383-394 (1986b).

BADENOCH-JONES, J., FLANDERS, D.J., ROLFE, B.G.: Appl. Environ. Microbiol. 49, 1511-1520 (1985). - BISHOP, P.E., JARLENSKI, D.M.L., HETHERINGTON, D.R.: Proc. Natl. Acad. Sci. USA 77, 7342-7346 (1980); - J. Bacteriol. 150, 1244-1251 (1982). - BISHOP, P.E., HAWKINS, M.E., EADY, R.R.: Biochem. J. 238, 437-422 (1986a). - BISHOP, P.E., PREMAKUMAR, R., DEAN, D.D., JACOBSON, M.R., CHISNELL, J.R., RIZOO, T.M., KOPCZYNSKI, J.: Science 232, 92-94 (1986b). - BOCK, E., KOOPS, H.-P., HARMS, H.: 18-38 in: Nitrification, Special Vol. of the Society for General Microbiol. 20, ed. J.I. PROSSER. IRL Press 1987a. - BOCK, E., SAND, W., MEINCKE, M., POHL, M., AHLERS, B., MEYER, C., WOLTERS, B., SAMELUCK, F.: Gemeinsame Frühjahrstagung der Vereinigung für Allgemeine und Angewandte Mikrobiologie und der Deutschen Gesellschaft für Hygiene und Mikrobiologie, Konstanz, F.R.G., 22.-25.03.1987, Abstract V 22 (1987b). - BODDEY, R.M., BALDANI, V.L.D., BALDANI, J.I., DÖBEREINER, J.: Plant Soil 95, 109-121 (1986). - BÖHME, H., SCHRAUTEMEIER, B.E.: Biochim. Biophys. Acta 891, 1-7 (1987). - BOLTON, G.W., NESTER, E.W., GORDON, M.P.: Science 232, 983 (1986). - BOOGERD, F.C., VAN VERSEVELD, H.W., STOUTHAMER, A.H.: Biochim. Biophys. Acta 723, 415-427 (1983). - BOOGERD, F.C., VAN VERSEVELD, H.W., TORENVLIET, D., BRASTER, M., STOUTHAMER, A.H.: Arch. Microbiol. 139, 344-350 (1984). - BREITENBACH, G.A., BREMNER, J.M.: Biol. Fertil. Soil 2, 195-199 (1986a); - ibid. 2, 201-204 (1986b). - BREMNER, J.M., YEOMANS, J.C.: Biol. Fertil. Soil 2, 173-179 (1986). - BROWN, J.C.: Plant Cell Environ. 1, 249-255 (1978). - BRYAN, B.A.: 67-84 in: Denitrification, Nitrification and Atmospheric Nitrous Oxide, ed. C.C. DELWICHE. New York: Wiley 1981. - BRYAN, B.A., JETER, R.M., CARLSON, C.A.: Appl. Environ. Microbiol. 50, 1301-1303 (1985).

CASTIGNETTI, D., SMARRELLI, J.: FEBS Lett. 209, 147-151 (1986). - CHISNELL, J.R., BISHOP, P.E.: 623 in: Nitrogen Fixation Research Progress, eds. H.J. EVANS, P.J., BOTTOMLEY, W.E. NEWTON. Dordrecht: Nijhoff 1985. - CHUA, K.-Y., PANKHURST, C.E., MAC DONALD, P.E., HOPKROFT, D.A., JARRIS, B.D.W., SCOTT, D.B.: J. Bacteriol. 162, 335-343

(1985). - CONRAD, R., SEILER, W.: Soil Biol. Biochem. 17, 893-895 (1985). - COVE, D.J.: Biol. Rev. 54, 291-327 (1979). - COYLE, C.L., ZUMFT, W.G., KRONECK, P.M.H., KÖRNER, H., JAKOBS, W.: Eur. J. Biochem. 153, 459-467 (1985). - CRASKE, A., FERGUSON, S.J.: Eur. J. Biochem. 158, 429-436 (1986). - CULLIMORE, J.V.: J. Mol. Appl. Genet. 2, 589-599 (1984).

DAHLGREN, R.M.T.: Bot. Limn. Soc. 80, 91-124 (1980). - DALTON, D.A., RUSSEL, S.A., HANUS, F.J., PASCOL, G.A., EVANS, H.J.: Proc. Natl. Acad. Sci. USA 83, 3811-3815 (1986). - DANNEBERG, G.: Thesis, Universität Köln, F.R.G. (1987). - DANNEBERG, G., ZIMMER, W., BOTHE, H.: 127-138 in: Azospirillum III Genetics, Physiology, Ecology, ed. W. KLINGMÜLLER. Berlin-Heidelberg-New York: Springer 1985. - DAZZO, F.B., NAPOLI, C.A., HUBBELL, D.H.: Appl. Environ. Microbiol. 32, 166-171 (1976). - DAZZO, F.B., TRUCHET, G.L., SHERWOOD, J.E., HRABAK, E.M., ABE, M., PANKRATZ, S.H.: Appl. Environ. Microbiol. 48, 1140-1150 (1984). - DEAN, J.V., HARPER, J.E.: Plant Physiol. 82, 718-723 (1986). - DEANE-DRUMMOND, C.E., THAYER, J.R.: J. Exp. Bot. 37, 429-439 (1986). - DINGLER, C., OELZE, J.: Arch. Microbiol. 147, 291-294 (1987). - DISPIRITO, A.A., LIPSCOMB, J.D., HOOPER, A.B.: J. Biol. Chem. 261, 17048-17056 (1986). - DIXON, R.O.D., WHEELER, C.T.: Nitrogen Fixation in Plants, 1-157. Glasgow, London: Blackie 1986. - DJORDJEVIC, M.A., SCHOFIELD, P.R., RIDGE, R.W., MORRISON, N.A., BASSAM, B.J., PLAZINSKI, J., WATSON, J.M., ROLFE, B.G.: Plant Mol. Biol. 4, 147-160 (1985a). - DJORDJEVIC, M.A., SCHOFIELD, P.R., ROLFE, B.G.: Mol. Gen. Genet. 200, 463-471 (1985b). - DOWNIE, J.A., KNIGHT, C.D., JOHNSTON, A.W.B., ROSSEN, L.: Mol. Gen. Genet. 198, 255-262 (1985).

EDMONDSON, D.E., JAMES, T.L.: Proc. Natl. Acad. Sci. USA 76, 3786-3789 (1979). - EDMONDSON, D.E. PELEATO, M.L.: 187-189 in: see ref. ULLRICH et al. 1987. - EGELHOFF, T.T., FISHER, R.F., JACOBS, T.W., MULLIGAN, J.T., LONG, S.R.: DNA 4, 242-248 (1985). - EISBRENNER, G., EVANS, H.J.: Annu. Rev. Plant Physiol. 34, 105-136 (1983). - ELMERICH, C.: Biotechnology 2, 967-978 (1984). - EVANS, H.J., BOTTOMLEY, M.J., NEWTON, W.E. (eds.): Nitrogen Fixation Research Progress. Proceedings of the 6th International Symposium on Nitrogen Fixation. Dordrecht: Nijhoff 1985.

FINAN, T.M., WOOD, J.M., JORDAN, D.C.: J. Bacteriol. 154, 1403-1413 (1983). - FISHER, R.F., TU, J.K., LONG, S.R.: Appl. Environ. Microbiol. 49, 1432-1435 (1985). - FRITZ, B.J., NINNEMANN, H.: Photochem. Photobiol. 41, 39-45 (1985). - FRUNZKE, K., ZUMFT, W.G.: Biochim. Biophys. Acta 852, 119-125 (1986).

GALLON, J.R.: Trends Biochem. Sci. 6, 19-23 (1981). - GLASS, A.D.M., THOMPSON, G.R., BORDELEAU, L.: Plant Physiol. 77, 379-381 (1985). - GLENN, A.R., DILWORTH, M.J.: J. Gen. Microbiol. 126, 243-247 (1981). - GOJON, A., SOUSSANA, J.-F., PASSAMA, L., ROBIN, P.: Plant Physiol. 82, 254-260 (1986). - GOVERS, F., GLONDEMANS, T., MOERMAN, M., VAN KAMMAN, A., BISSELING, T.: Embo J. 4, 861-867 (1985). - GRAFE, R. MÜLLER, A.J.: Theor. Appl. Genet. 66, 127-130 (1983).

HÄGER, K.-P., BOTHE, H.: Biochim. Biophys. Acta 892, 213-223 (1987). - HÄGER, K.-P., GÄTZ, A., BOTHE, H.: J. Plant Physiol. 131, 91-100 (1987). - HANKS, J.F., SCHUBERT, K., TOLBERT, N.E.: Plant Physiol. 71, 869-873 (1983). - HARDY, R.W.F., HAVELKA, U.D.: 421-439 in: Symbiotic Nitrogen Fixation in Plants, ed. P.S. NUTMAN. Cambridge: Cambridge University Press 1976. - HARTMANN, A., BURRIS, R.H.: J. Bacteriol. 169, 944-948 (1987). - HARTMANN, A., FU, H., BURRIS, R.H.: J. Bacteriol. 165, 864-870 (1986). - HIRSCH, A.M., DRAKE, D., JACOBS, T.W., LONG, S.R.: J. Bacteriol. 161, 223-230 (1985). - HOOYMANS, J.J.M., LOGMAN, G.J.J.: J. Plant Physiol. 116, 273-277 (1984). - HRABAK, E.M., URBANO, M.R., DAZZO, F.B.: J. Bacteriol. 148, 697-711 (1981). - HYLDIG-NIELSEN, J.J.,

JENSEN, E.O., PALUDAN, K., WIBORY, O., GARRETT, R., JORGENSEN, P., MARCKER, K.A.: Nucleic Acids Res. 10, 689-701 (1982).

INGLEDEW, W.J., POOLE, R.K.: Microbiol. Rev. 48, 222-271 (1983). - INGRAHAM, J.L.: 45-65 in: Denitrification, Nitrification and Atmospheric Nitrous Oxide, ed. C.C. DELWICHE. New York: Wiley 1981. - INNES, R.W., KUEMPEL, P.L., PLAZINSKI, J., CANTER-CREMMERS, H., ROLFE, B.G.: Mol. Gen. Genet. 201, 426-432 (1985).

JACOBS, T.W., EGELHOFF, T.T., LONG, S.R.: J. Bacteriol. 162, 469-476 (1985). - JACOBSON, M.R., PREMAKUMAR, R., BISHOP, P.E.: J. Bacteriol. 167, 480-486 (1986). - JENSEN, E.O., MARCKER, K.A., VILLADSEN, S.: EMBO J. 5, 843-847 (1986). - JENSEN, J.S., MARCKER, K.A., OTTEN, L., SCHELL, J.: Nature 321, 669-674 (1986). - JOERGER, R.D., PREMAKUMAR, R., BISHOP, P.E.: J. Bacteriol. 168, 673-682 (1986). - JONES, R., WOODLEY, P., ROBSON, R.: Mol. Gen. Genet. 197, 318-327 (1984). - JORDAN, D.C.: 242-243 in: Bergey's Manual of Systematic Bacteriology, Vol. 1, eds. N.R. KRIEG, J.G. HOLT. Baltimore, London, Los Angelos, Sidney: Williams & Williams 1984.

KIRSTEIN, K.O.K., STILLER, W., BOCK, E.: Beton Stahlbau 8, 202-205 (1986). - KLEINHOFS, A., WARNER, R.L., NARAYANAN, K.R.: Oxford Surv. Plant Mol. Cell Biol. 2, 91-121 (1985). - KLEMM, E., NINNEMANN, H.: Photochem. Photobiol. 29, 629-632 (1979). - KLUGKIST, J., VOORBERG, J., HAAKER, H., VEEGER, C.: Eur. J. Biochem. 155, 33-40 (1986a). - KLUGKIST, J., HAAKER, H., VEEGER, C.: Eur. J. Biochem. 155, 41-46 (1986b). - KÖHLER, K.-H.: Biol. Rundsch. 23, 137-160 (1985). - KONDOROSI, E., KONDOROSI, A.: Trends Biochem. Sci. 11, 296-299 (1986). - KONDOROSI, E., BANFALVI, Z., KONDOROSI, A.: Mol. Gen. Genet. 193, 445-452 (1984). - KROTZKY, A., WERNER, D.: Arch. Microbiol. 147, 48-57 (1987). - KUCHNICKI, T.C., SARNA, L.P., WEBSTER, G.R.B.: J. Liquid Chromatogr. 8, 1593-1609 (1985). - KUNDU, B., NICHOLAS, D.J.D.: Arch. Microbiol. 140, 358-364 (1985).

LAMBERS, H., NEETESON, J.J., STULEN, I. (eds.): Fundamentals, Ecological and Agricultural, Aspects of Nitrogen Metabolism in Higher Plants. Dordrecht: Nijhoff 1985. - LANDSMANN, J., DENNIS, E.S., HIGGINS, T.J.V., APPLEBY, C.A., KORTT, A.A., PEACOCK, W.J.: Nature 324, 166-168 (1986). - LANG, E., JAGNOW, G.: FEMS Microbiol. Ecology 38, 257-265 (1986). - LEE, R.B., CLARKSON, D.T.: J. Exp. Bot. 37, 1753-1767 (1986). - LEE, R.B., DREW, M.C.: J. Exp. Bot. 37, 1768-1779 (1986). - LEE, R.B., RUDGE, K.A.: Ann. Bot. 57, 471-486 (1986). - LUDDEN, P.W., BURRIS, R.H: Science 194, 424-426 (1976); - (eds): Nitrogen Fixation and CO_2 Metabolism. A Steenbock Symposium in Honor of R.H. Burris. Amsterdam: Elsevier 1985. - LUDDEN, P.W., OKON, Y., BURRIS, R.H.: Biochem. J. 173, 1001-1003 (1978).

MALDONADO, J.M., APARICIO, P.J.: 76-81 in: see ref. ULLRICH et al. 1987. - MARCKER, A., LUND, M., JENSEN, E.O., MARCKER, K.A.: EMBO J. 3, 1691-1695 (1984). - MAURINO, S.G., VARGAS, M.A., APARICIO, P.J., MALDONADO, J.M.: Physiol. Plant 57, 411-416 (1983). - MELLOR, R.B., CHRISTENSEN, T.M.I.E., WERNER, D.: Proc. Natl. Acad. Sci. USA 83, 659-663 (1986). - MITSUI, A., KUMAZAWA, S., TAKAHASHI, A., IKEMOTO, H., CAO, S., ARAI, T.: Nature 323, 720-722 (1986). - MITSUI, A., CAO, S., TAKAHASHI, A., ARAI, T.: Physiol. Plant. 69, 1-8 (1987). - MORPETH, F., BOXER, D.H.: Biochemistry 24, 40-46 (1985). - MULLIGAN, J.T., LONG, S.R.: Proc. Natl. Acad. Sci. USA 82, 6609-6613 (1985). - MULVANEIY, C.S., HAGEMAN, R.H.: Plant Physiol. 78, 118-124 (1984). - MURPHY, P.J., HEYCKE, N., BANFALVI, Z., TATE, M.E., DE BRUIJN, F., KONDOROSI, A., TEMPE, J., SCHELL, J.: Proc. Natl. Acad. Sci. USA 84, 493-497 (1987).

NADLER, K.D., AVISSAR, Y.I.: Plant Physiol. 60, 433-436 (1977). - NAKANO, Y., ASADA, K.: Plant Cell Physiol. 22, 867-880 (1981). - NAMBIAR, O.G.B., WEINZIERL, J., UMLAND, F.: Fresenius Z. Anal. Chem. 322, 327-328 (1985). - NARAYANAN, K.R., MÜLLER, A.J., KLEINHOFS, A., WARNER, R.L.: Mol. Gen. Genet. 197, 358-362 (1984). - NAYAK, D.N., LADHA, J.K., WATANABE, I.: Biol. Fertil. Soils 2, 7-14 (1986). - NEUER, G., KRONENBERG, A., BOTHE, H.: Arch. Microbiol. 141, 354-370 (1985). - NGUJEN, T., ZELECHOWSKA, M., FOSTER, V., BERGMANN, H., VERMA, D.P.S.: Proc. Natl. Acad. Sci. USA 82, 5040-5044 (1985). - NINNEMANN, H.: 82-88 in: see ref. ULLRICH et al. 1987.

OAKS, A., HIREL, B.: Annu. Rev. Plant Physiol. 36, 345-365 (1985). - OKON, Y.: Trends Biotechnol. 3, 223-228 (1985). - ORME-JOHNSON, W.H.: Annu. Rev. Biophys. Chem. 14, 419-459 (1985).

PAYNE, W.J.: 47-65 in: Denitrification in the Nitrogen Cycle, ed. H.L. GOLTERMAN. London: Plenum 1985. - PATE, J.S.: 451-473 in: Encyclopedia of Plant Physiology, N.S., Vol. 1, eds. U. LÜTTGE, M.G. PITMAN. Berlin Heidelberg New York: Springer 1975. - PETERS, N.K., FROST, J.W., LONG, S.R.: Science 233, 977-980 (1986). - PLANQUÉ, K., KIJNE, J.W.: FEBS Lett. 37, 64-66 (1977). - POPE, M.R., MURELL, S.A., LUDDEN, P.W.: Proc. Natl. Acad. Sci. USA 82, 3173-3177 (1985). - POSTGATE, J.R.: The Fundamentals of Nitrogen Fixation, 1-252. Cambridge: Cambridge University Press 1982. - POULLE, M., RAMBOUR, S.: Plant Sci. 40, 111-115 (1985). - PREMAKUMAR, R., LEMOS, E.M., BISHOP, P.E.: Biochim. Biophys. Acta 797, 64-70 (1984). - PULL, S., PUEPPKE, S., HYMOWITZ, T., ORF, J.: Science 200, 1277-1279 (1979). - PUPPO, A., RIGAUD, J.: FEBS Lett. 201, 187-189 (1986).

QUINTO, C., DE LA VEGA, H., FLORES, M., LEMMANS, J., CAVALLOS, M.A., PARDO, M.A., AZIPROZ, R., DE LOURDES GIRARD, M., CALVA, E., PALACIOS, R.: Proc. Natl. Acad. Sci. USA 82, 1170-1174 (1985).

RAJASEKHAR, V.K., MOHR, H.: Planta 168, 369-376 (1986). - RAJASEKHAR, V.K., SOPORY, S.K.: New Phytol. 101, 251-258 (1985). - REDOND, J.W., BATLEY, M., DJORDJEVIC, M.A., INNES, R.W., KNEMPEL, P.L., ROLFE, B.G.: Nature 323, 632-634 (1986). - REIBACH, P.E., STREETER, J.G.: J. Bacteriol. 159, 47-52 (1984). - REINHOLD, B., HUREK, T., FENDRIK, I., POT, B., GILLIS, M., KERSTERS, K., THIELEMANS, S., DE LEY, J.: Int. J. Syst. Bacteriol. 37, 43-51 (1987). - ROBSON, R.L.: Arch. Microbiol. 146, 74-79 (1986). - ROBSON, R.L., EADY, R.R., RICHARDSON, T.H., MILLER, R.W., HAWKINS, M., POSTGATE, J.: Nature 322, 388-390 (1986a). - ROBSON, R.L., WOODLEY, P.R., JONES, R.: EMBO J. 5, 1159-1163 (1986b). - ROLDAN, J.M., BUTLER, W.L.: Photochem. Photobiol. 32, 357-381 (1980). - RONSON, C.W., ASTWOOD, P.M.: 201-207 in: Nitrogen Fixation Research Progress. Proceedings of the 6th International Symposium on Nitrogen Fixation, eds. H.J. EVANS, P.J. BOTTOMLEY, W.E. NEWTON. Dordrecht: Nijhoff 1985. - ROSSEN, L., SHERMAN, C.A., JOHNSTON, A.W.B., DOWNIE, J.A.: EMBO J. 4, 3369-3373 (1985).

SALMINEN, S.O., STREETER, J.G.: Plant Physiol. 83, 535-540 (1987). - SCHAEDE, R.: Die pflanzlichen Symbiosen, 1. Aufl. Jena: Fischer 1942. - SCHRAUTEMEIER, B., BÖHME, H.: FEBS Lett. 184, 304-308 (1985). - SCHUBERT, K.R.: Annu. Rev. Plant Physiol. 37, 539-574 (1986). - SCOLNIK, P.A., HASELKORN, R.: Nature (Lond.) 307, 289-292 (1984). - SHAPLEIGH, J.P., PAYNE, W.J.: J. Bacteriol. 163, 837-840 (1985); - Arch. Biochem. Biophys. 244, 713-718 (1986). - SHARMA, A.K., SOPORY, S.K.: Photochem. Photobiol. 39, 491-493 (1984). - SHEARMAN, C.A., ROSSEN, L., JOHNSTON, A.W.B., DOWNIE, J.A.: EMBO J. 5, 647-652 (1986). - SKINNER, F.A., UOMALA, P. (eds.): Nitrogen Fixation with Nonlegumes. Plant Soil 90, 1-456. Dordrecht: Nijhoff 1986. - SKOT, L., EGSGAAD, H.: Planta 161, 32-36 (1984). - SMARRELLI, J., CASTIGNETTI, D.: Biochim. Biophys. Acta

882, 337-342 (1986). - SMIT, G., KIJNE, J.W., LUGTENBERG, B.J.J.: J. Bacteriol. 168, 821-827 (1986). - SOLOMONSON, L.P., BARBER, J.M.: 71-75 in: see ref. ULLRICH et al. 1987. - SONG, S.-D., HARTMANN, A., BURRIS, R.H.: J. Bacteriol. 164, 1271-1277 (1985). - SOUALMI, K., CHAMPIGNY, M.L.: J. Plant Physiol. 125, 35-45 (1986). - SPRENT, J.I.: The Biology of Nitrogen-Fixing Organisms, 1-196. London: McGraw-Hill 1979. - STACHEL, S.E., MESSENS, E. VAN MONTAGU, M., ZAMBRYSKI, P.: Nature 318, 624-629 (1985). - STEWART, G.R., SUMAR, N., PATEL, M.: 39-44 in: see ref. ULLRICH et al. 1987. - STOUTHAMER, A.H.: 189-197 in: Advances in Nitrogen Fixation Research. Proceedings of the 5th International Symposium on Nitrogen Fixation, eds. C. VEEGER, W.E. NEWTON. The Hague: Nijhoff/Junk/Pudoc 1984. - STOWERS, M.D.: Annu. Rev. Microbiol. 39, 89-108 (1985). - STREETER, J.G., SALMINEN, S.O.: 277-283 in: Nitrogen Fixation Research Progress. Proceedings of the 6th International Symposium on Nitrogen Fixation, eds. H.J. EVANS, D.J. BOTTOMLEY, W.E. NEWTON. The Hague: Nijhoff/Junk/Pudoc 1985.

TEL-OR, E., HUFLEJT, M., PACKER, L.: Biochem. Biophys. Res. Commun. 132, 533-539 (1985). - TEMPÉ, J., PETIT, A., BAUNCROT, H.: C.R. Acad. Sci. Paris (Ser. 3) 295, 413-416 (1982). - TRINCHANT, J.C., RIGAUD, J.: J. Gen. Microbiol. 133, 37-43 (1987). - TIPTON, C.L., THOMPSON, J.: Plant Physiol. 79, 432-435 (1985).

ULLRICH, W.R., APARICIO, P.J., SYRETT, P.J., CASTILLO, F. (eds.): Inorganic Nitrogen Metabolism. Berlin Heidelberg New York: Springer 1987.

VARGAS, M.A., MAURINO, S.G., MALDONADO, J.M., APARICIO, P.J.: Photochem. Photobiol. 36, 223-228 (1982).

WALLACE, W.: Physiol. Plant. 66, 630-636 (1986). - WALSBY, A.E., SIMPSON, J.H.: Proc. R. Soc. Lond. Biol. Sci. 223, 177-196 (1984). - WEISS, R.F.: J. Geophys. Res. 86, 7185-7195 (1981). - WOLPERT, J.S., ALBERSHEIM, P.: Biochem. Biophys. Res. Commun. 70, 729-737 (1976).

YAMANKA, T., FUKUMOR, J., YAMAZAKI, T., KATO, H., NAKAYAMA, K.: J. Inorg. Biochem. 23, 273-277 (1985). - YAMAZAKI, T., FUKUMOR, Y., YAMANAKA, T.: Biochem. Biophys. Acta 810, 174-183 (1985).

ZAAT, S.A., WIJFFELMAN, C.A., SPAINK, H.P., VAN BRUSSEL, A.A.N., OKKER, R.J.H., LUGTENBERG, B.J.J.: J. Bacteriol. 169, 198-204 (1987). - ZIMMER, W., DANNEBERG, G., BOTHE, H.: Current Microbiol. 12, 341-346 (1985). - ZIMMER, W., ROEBEN, K., DANNEBERG, G., BOTHE, H.: 177-182 in: see ref. ULLRICH et al. 1987. - ZUMFT, W.G., DÖHLER, K., KÖRNER, E.: J. Bacteriol. 163, 918-924 (1985). - ZURKOWSKI, W.: Microbios 27, 27-32 (1980).

Prof. Dr. HERMANN BOTHE
Botanische Institut
Universität Köln
Gyrhofstr. 15
D-5000 Köln 41, FRG

IV. Secondary Plant Substances
Aspects of Steroid Biosynthesis

By Horst-Robert Schütte

1. Sterols

Most prokaryotic and eukaryotic cells contain steroids. This type of structure is not only found in sterols but also in corresponding sapogenins, cardiac glycosides and alkaloids. The sterols are preferably concentrated in the plasma membrane (GRUNWALD 1975; HARTMANN-BOUILLON and BENVENISTE 1978; NES and HEFTMANN 1981; HARTMANN et al. 1985) but are also present in lower amounts in other membrane systems. In fungi, the major sterol is ergosterol. In higher plants, sitosterol (X) and stigmasterol (IX) is often constitute more than 80% of the total sterols. The biological meaning of the bulky ethyl group at C-24 is still unknown but could be related to some specific characters of higher plant membranes, such as their high content in phospholipids with highly unsaturated fatty acids (MAZLIAK 1981).

But also cholesterol (IV) is probably present in all plants (HEFTMANN 1984). This is not surprising, since this sterol or one of its precursors is now known to be the starting material from which living organisms produce all the other steroids which they contain. The biosynthesis of steroids in plants, starting from mevalonate via 2,3-oxidosqualene and cycloartenol, as well as their function, has been repeatedly reviewed (Progr. Bot. 37, 133 ff., for earlier references, GRUNWALD 1975, 1980; WILLUHN 1976, 1979; GOODWIN 1977, 1979, 1980, 1981; NES 1977; NES and McKEAN 1977; SCHROEPFER 1981, 1982; HEFTMANN 1983; MUDD et al. 1984; BENVENISTE 1986; CASPI 1986).

Mevinolin and related fungal metabolites are extremely potent inhibitors of the 3-hydroxy-3-methylglutaryl coenzyme A reductase (HMG-CoA reductase) (BACH and LICHTENTHALER 1983a,b; BACH 1985), a key enzyme on the way to mevalonate, and an ammonium analogue of the substrate is an inhibitor of the squalene synthetase (SANDIFER et al. 1982). There is evidence that the HMG-CoA reductase in the latex of *Hevea brasiliensis* is activated by a cytosolic factor and phosphorylation may lower its activity (ISA and SIPAT 1982; SIPAT 1982).

Besides mevalonate as precursor also some amino acids and sugars have been reported to be involved in triterpenoid synthesis in plants (SUGA et al. 1980; GROENEVELD and ROELVINK 1982a,b; GROENEVELD and ELINGS 1984; GROENEVELD and MAHLBERG 1986), e.g., in the laticifers of *Euphorbia lathyris*. Of the amino acids serine tended to be more specifically involved in phytosterol synthesis. Isolated cells and protoplasts from leaves of *Calendula officinalis* synthesize effectively terpenoid compounds from labelled acetate; the levels of sterols, β-amyrine and oleanolic acid were similar in shoots, isolated cells and protoplasts (AUGUSCINSKA and KASPRZYK 1982). The biosynthesis of sterols from mevalonate by a cell-free extract prepared from actively growing tuber portions and leader shoots of *Dioscorea floribunda* has been demonstrated (SELVARAJ and CHANDER 1982).

Progress in Botany, Vol 49

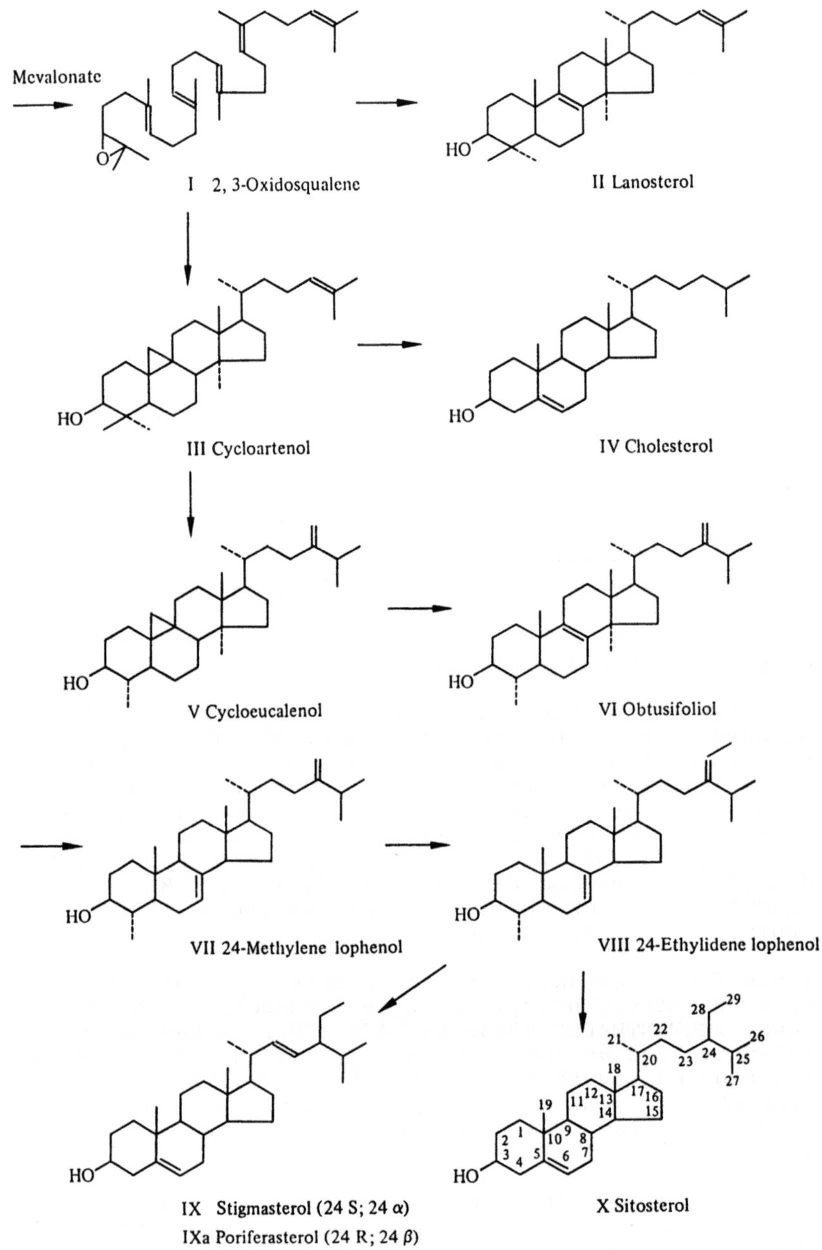

Fig.1. Biosynthesis of sterols

The first cyclization product of 2,3-oxidosqualene in animals is lanosterol, but in plants cycloartenol is usually isolated instead (Fig. 1). The major differences existing between the pathways are (1) the enzymes which cyclize 2,3-oxidosqualene (I) producing lanosterol (II) in non-photosynthetic eukaryotes and cycloartenol (III) in photosynthetic eukaryotes, (2) the existence only in higher plants of cycloeucalenol (V) : obtusifoliol (VI) isomerase, an enzyme which opens the cyclopropane ring; and (3) the presence of enzymes which catalyse the alkylation at C-24 of the lateral chain of sterols and which lead to 24-methyl sterols in most fungi and preferably to 24-ethyl sterols in higher plants. Using D_2O cycloartenol (III) has been confirmed as an obligatory intermediate in the biosynthesis of plant sterols (CASPI and SLIWOWSKI 1975; CASPI 1986). With 2,3-oxidosqualene (I), having a chiral methyl group at C-6, it has been shown using a microsomal preparation from *Ochromonas malhamensis* that the formation of the cyclopropane ring during the 2,3-oxidosqualene cyclase reaction takes place with retention of the configuration (ALTMAN et al. 1978).

Re-investigation of the stereospecific incorporation of mevalonate into sitosterol (X) in *Pisum sativum* and *Calendula officinalis* has shown that the 2proR-protons of mevalonate are located at 1β, 7β, 15α, 22R and 26 positions of the sterol, whereas the 2proS-protons of this precursor are located only at 1α, 22S and 26 positions (CASPI et al. 1975; SLIWOWSKI and CASPI 1976, 1977; CASPI 1986). With (^{13}C, ^{2}H) acetate in *Rabdosia japonica* 1,2-hydride shifts, i.e. 20-H from C-17 and 17-H from C-13, during the biosynthesis of sitosterol (X), have been demonstrated (SEO et al. 1986). 2-Aza-2,3-dihydrosqualene, its N-oxide, a trimethyldecyl-azadecalin as well as several derivatives have been found to be powerful inhibitors of oxidosqualene cyclases in animals and plants (DELPRINO et al. 1983; RAHIER et al. 1983; CERUTTI et al. 1985; DURIATTI et al. 1985; TATON et al. 1986). In vivo, the accumulation of 2,3-oxidosqualene from mevalonate was demonstrated in tobacco tissue treated with AMO-1618 and other growth inhibitors (DOUGLAS and PALEG 1978a,b; NES et al. 1982). It was possible to reverse the growth inhibition by application of exogenous stigmasterol (IX) or GA_3 (DOUGLAS and PALEG 1981).

In most plants the first intermediate without a cyclopropane ring is obtusifoliol (VI), thus alkylation at C-24 and removal of a methyl group from C-4 must occur prior to the opening of the ring at the level of cycloartenol (III). Because the introduction of an alkyl group at C-24 creates a new asymmetric centre, two optical isomers are possible. Whereas the 24-ethyl sterols from most higher plants such as sitosterol (X) and stigmasterol (IX) have the 24α configuration, the 24-methyl sterols occur as a mixture of 24R (24α) and 24S (24β) isomers. The 24S-methyl sterol, dihydrobrassicasterol (XVIII), is frequently isolated together with its 24R epimer, campesterol (XIX), from higher plants. In *Zea mays* 24-ethyl sterol (24R) was synthesized at three to four times the rate of the 24-methyl sterol (RENDELL et al. 1986).

For the Δ^{24}-sterol-C-24-methyl transferase reaction the cyclopropane ring, the 4α-methyl and a free 3β-OH are necessary (BENVENISTE 1986). 25-Azacycloartanol was shown to strongly inhibit the S-adenosyl-methionine : cycloartenol-C-24-methyl transferase (RAHIER et al. 1980, 1983, 1984, 1986; NARULA et al. 1981; SCHMITT et al. 1981, 1982). The alkylation mechanism for the formation of C-29 phytosterols involves the formation of 24-methylene sterol intermediate such as XIV and the migration of a hydrogen from C-24 to C-25 (Fig. 2). The mechanism of the first methylation has been thoroughly studied in *Trebouxia* including the stereochemistry of the hydride shift from C-24 to C-25 (LARGEAU et al. 1977b; MIHAILOVIC 1986).

The incorporation of mevalonic acid and methionine as well as (^{13}C,^{2}H) acetate into sitosterol (X) in barley seedlings, *Zea mays* shoots or tissue cultures of *Rabdosia japonica* is in agreement with the involvement of 24-methylene (XIV) and 24-ethylidene (XIII) intermediates and of isomerization of the 24(28) double bond of isofucosterol (XIII) to the 24(25) double bond (in XVI) prior to the reduction yielding the saturated side chain in sitosterol (X) (LENTON et al. 1975; LARGEAU et al. 1977a; ZAKELJ and GOAD 1983; SEO et al. 1986). 24-Ethylidene lophenol (VIII) was incorporated into sitosterol. The stereochemistry of hydrogen

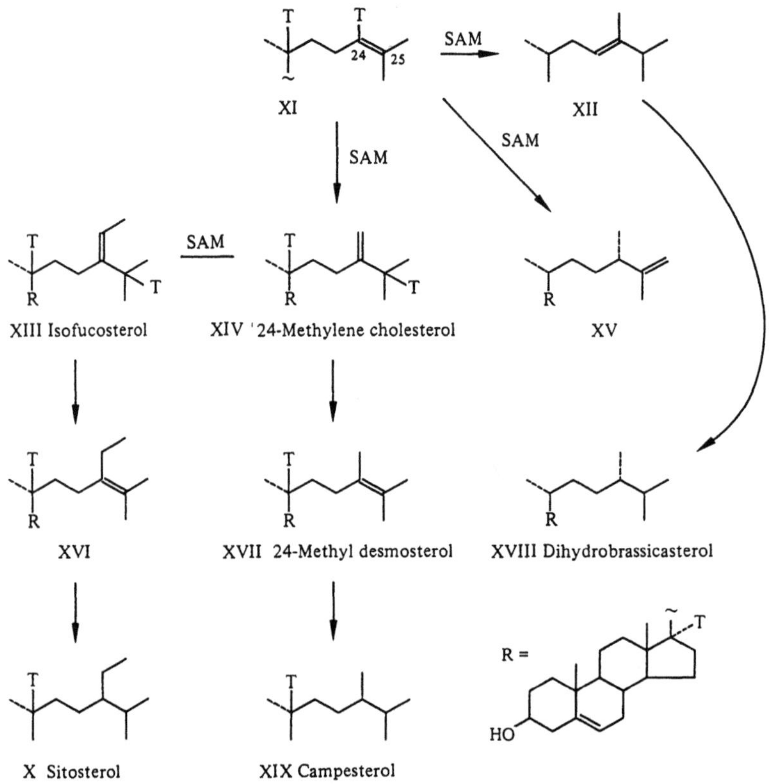

Fig.2. Biosynthetic routes leading to (24 R)-24-methyl and (24 S)-methyl sterols

migration from C-24 to C-25 during alkylation in *Pinus pinea* is such that the C-25 E methyl group of the Δ^{24}-sterol precursor becomes the pro R isopropyl methyl group of isofucosterol (XIII) (NICOTRA et al. 1981). This is the opposite of that found in *Trebouxia* and may indicate that the enzymatic mechanism which leads to 24R-alkyl sterols in vascular plants is different from that leading to 24S-alkyl sterols in cryptogams.

It has been confirmed that preferably the Z-configuration of 24-ethylidene sterols appears to be the intermediate of (24R)-24-ethyl sterol biosynthesis (KNAPP et al. 1977). This is in agreement with the finding of Z-24-ethylidene sterols in suspension cultures of *Rubus fruticosus* cells grown in the presence of fenarimol (SCHMITT and BENVENISTE 1979). Microsomes from suspension cultures of *R. fruticosus* yield in the presence of S-adenosylmethionine,24-methylene cycloartenol (XX) from cycloartenol (III), and 24-ethylidene lophenol (VIII) from 24-methylene lophenol (VII) (FONTENEAU et al. 1977). With cultured cells of *Physalis peruviana* it was demonstrated that the isomerization ($\Delta^{24/(28)}$ to $\Delta^{24/(25)}$) and reduction of the double bond take place stereospecifically leading to sitosterol (X) which has C-26 originating from C-2 and C-27 from C-6 of mevalonate (SEO et al. 1983). With germinating seeds of *Cucurbita maxima* it was shown that both 24R- and 24S-ethyl sterols could be derived from a C-24 ethyl sterol ion without intermediacy of the 24 Z-ethylidene derivative (CATTEL et al. 1980). Furthermore, the conversion of cyclolaudenol (XXII) into ethyl sterols was shown (BALLIANO et al. 1983).

Many reports have been made on the biosynthesis of the mixtures of 24-methyl sterols. Experiments with *Zea mays* seedlings are consistent with

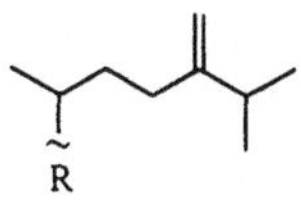

XX 24-Methylene cycloartanol

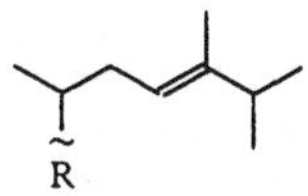

XXI Cyclosadol

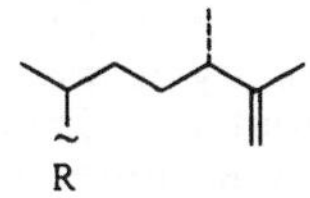

XXII Cyclolaudenol

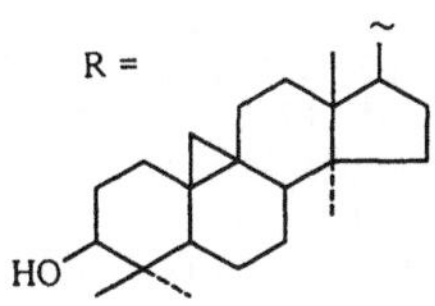

Fig.3. Some cyclopropyl sterols

the hypothesis that two different routes are operating for the elaboration of the 24-methyl cholesterol side chain (McKEAN and NES 1977; ZAKELJ and GOAD 1983). The 24S-methyl sterols such as dihydrobrassicasterol (XVIII) are biosynthesized in this plant by direct C-24 methylation via either a Δ^{25} or Δ^{23} intermediate, whereas the 24R-methyl sterols such as campesterol (XIX) are formed via a 24-methylene intermediate followed by double-bond migration and hydrogenation, e.g. during the formation of the 24R-ethyl sterols (Fig. 2). The investigation of the S-adenosyl-methionine : cycloartenol methyltransferase reaction in a cell-free microsomal preparation of *Z. mays* has shown that besides the major product, 24-methylene cycloartanol (XX), small amounts of the Δ^{23}-sterol, cyclosadol (XXI), and the Δ^{25}-sterol, cyclolaudenol (XXII) were identified (SCHEID et al. 1982; MISSO and GOAD 1983, 1984). Cyclosadol was formed directly by C-24 methylation of cycloartenol and did not originate from isomerization of 24-methylene cycloartanol (Fig. 3). Using (^{13}C)acetate it has been established in cultured *Physalis peruviana* cells that the biosynthetic prochirality at C-25 of dihydrobrassicasterol (XVIII) differs from that of campesterol (XIX) and 24-methylene cholesterol (XIV) (SEO et al. 1984a; UOMORI et al. 1984). Dihydrobrassicasterol has C-26 originating from C-2 of mevalonate, while campesterol has the same carbon derived from C-6 of mevalonate.

Tridemorph, fenpropimorph and related N-alkyl morpholines as well as azadecalin derivatives, which contain structural elements of unstable C-8 and C-9 carbocationic high-energy intermediates occurring during sterol biosynthesis, have been shown to be strong inhibitors of the cycloeucalenol : obtusifoliol isomerase in vitro and in vivo (SCHMITT et al. 1981, 1982; BLADOCHA and BENVENISTE 1983, 1985; RAHIER et al. 1983, 1985, 1986; HOSOKAWA et al. 1984; HARTMANN et al. 1985; TATON et al. 1986, 1987). The accumulated cyclopropylsterols, but no Δ^5-sterols, were composed essentially of 24-methyl sterols such as 24-methyl pollinastanol, whereas in the control 24-ethyl sterols predominate (more than 70% of total Δ^5-sterols). Furthermore, the results suggest that besides the cycloeucalenol : obtusifoliol isomerase also the Δ^7-sterol isomerase and the Δ^{14}-reductase are target enzymes for N-substituted morpholines. Experiments with such tridemorph treated *Zea mays* systems indicate that both (24R) and (24S) epimers originate from a common $\Delta^{24(28)}$ precursor, the 28-CH_3 group contains only two protons of methionine-methyl and the mevalonate incorporation is consistent with the intermediate of a $\Delta^{24(25)}$ sterol (BLADOCHA and BENVENISTE 1985; BENVENISTE 1986). A pathway is suggested where $\Delta^{24(28)}$-cyclopropyl sterols are isomerized to give $\Delta^{24(25)}$-cyclopropyl sterols which in turn would be reduced non-regiospecifically to yield both (24R)- and (24S)-24-methyl pollinastanols. With regards to the results mentioned above with *Zea meays*, there are possibly different pathways in treated plants for cyclopropyl sterols and in control plants for Δ^5-sterols.

The cycloeucalenol : obtusifoliol isomerase catalyzes the opening of the cyclopropane ring of cycloeucalenol (V) giving obtusifoliol (VI) (RAHIER et al. 1977). Results from experiments on this enzyme using ^2H_2O are consistent with a general acid catalyzed cyclopropane ring-opening reaction leading to a carbocationic intermediate. Obtusifoliol is obtained via a trans-antiperiplanar elimination of the 8β-H from such an enzyme-bound intermediate. The cycloeucalenol : obtusifoliol isomerase has been detected in different plant systems and seems to be a biochemical marker of tracheophytes (HEINTZ and BENVENISTE 1974). A 7-oxo-dihydrocycloeucalenol is a potent inhibitor of this isomerase (RAHIER et al. 1982). A membrane fraction rich in endoplasmic reticulum is shown to contain all of the activity of both S-adenosylmethionine : cycloartenol methyltransferase and cycloeucalenol : obtusifoliol isomerase (NES and HEFTMANN 1981).

The later stages in the biosynthetic pathway involve removal of the remaining 4α-methyl group, the conversion of a Δ^7 into a Δ^5 derivative and the insertion of a double bond into the side chain at C-22. The removal of the 14α-methyl group in *Calendula officinalis* involves the formation and reduction of a Δ^{14} double bond during which the 15α-hydrogen is exchanged and the 15β-hydrogen epimerized (CASPI et al. 1975). The involvement of cytochrome P-450 is also highly probable in plants since the 14α-methyl demethylase reaction is strongly inhibited in suspension cultures of *Rubus fruticosus* by fenarimol (SCHMITT and BENVENISTE 1979; SCHMITT et al. 1982), which is capable of binding with cytochrome P-450 (HENRY and SISLER 1984). $\Delta^{8,14}$ sterols, products of the 14α-methyl demethylase reaction, accumulate dramatically in suspension cultures of *R. fruticosus* cells grown in the presence of 15-azasterol, which inhibits the sterol Δ^{14}-reductase (SCHMITT et al. 1980, 1982). The 14α-methyl demethylase is inhibited by several fungicidal chemicals (FRASINEL et al. 1978; SCHMITT and BENVENISTE 1979). During the demethylation of the 4α-methyl group, the conversion of obtusifoliol (VI) into poriferasterol (IXa), the hydrogen in the 4α (axial) position is inverted to the 4α (equatorial) position in the alga *Ochromonas malhamensis* (KNAPP et al. 1977). The conversion of Δ^8- into Δ^5-sterols is catalyzed by the $\Delta^8 \rightarrow \Delta^7$-sterol isomerase. In cell suspension cultures of *R. fruticosus* this enzyme is inhibited by AY 9944, a hypocholesterolemic drug, resulting in an accumulation of Δ^8-sterols (SCHMITT and BENVENISTE 1979). Besides its action on the cycloeucalenol : obtusifoliol isomerase, azadecalins are also potent inhibitors of the $\Delta^8 \rightarrow \Delta^7$-sterol isomerase and of the $\Delta^{8,14}$-sterol reductase (RAHIER et al. 1985; TATON et al. 1987). The latter enzyme is selectively inhibited by the N-oxide of N-benzylazadecalin. The isoquinoline alkaloid pilocereine in *Lophocereus schottii* has been suggested to be responsible for the inhibition of the Δ^{14}-reductase and for the presence in this plant of an interrupted biosynthetic sterol pathway (CAMPBELL and KIRCHER 1980).

Seedlings of tobacco and barley metabolized exogenously supplied sitosterol (X) but neither plant was able to convert it into stigmasterol (IX) (HUANG and GRUNWALD 1986). Both species incorporated mevalonate into sitosterol and stigmasterol. Therefore, it is suggested that the pathway of stigmasterol biosynthesis is not via sitosterol, but through a common precursor which is derived from mevalonate. In other experiments barley apparently converts sitosterol directly into stigmasterol (NAVARI-IZZO and IZZO 1984), but this conversion could not be confirmed with *Nicotiana tabacum* seedlings (GRUNWALD 1985). Exogenous cholesterol significantly inhibited the rate of sterol synthesis in cell cultures of *Solanum dulcamara* (BHATT and BHATT 1984).

2. Sterol Conjugations

One of the most common transformations that sterols undergo in plants is conjugation (GRUNWALD 1975, 1980; MUDD 1980; WOJCIECHOWSKI 1983, 1987; HEFTMANN 1984). They are esterified with fatty acids or conjugated with sugars, which are often also acylated. The common pathway leads from sterols to steryl glycosides and thence to the acylated steryl glycosides. An acetone powder preparation from spinach leaves catalyzes the formation of steryl esters from phytosterols and 1,2-diacylglycerols as acyl donors (GARCIA and MUDD 1978a,b; NES et al. 1982). A similar enzyme preparation from roots of white mustard *(Sinapis alba)* seedlings, which is associated with cell membrane structures, uses preferably triacylglycerols as acyl sources and can be regarded as a specific triacylglycerol : sterol acyltransferase (ZIMOWSKI and WOJCIECHOWSKI 1981a,b).

The glucosyl transferase for glucosylation of phytosterols by glucose transfer from UDP-glucose is tightly membrane-bound, mainly localized in dictyosomes as has been shown with homogenates from cell cultures of *Digitalis purpurea*, from *Calendula officinalis* seedlings, with *Sinapis alba*, onion stems, hypocotyls of *Phaseolus aureus*, cucumber fruits and maize coleoptiles (LERCHER and WOJCIECHOWSKI 1976; BOWLES et al. 1977; HARTMANN-BOUILLON and BENVENISTE 1978; HOPP et al. 1978; HARTMANN-BOUILLON et al. 1979; WOJCIECHOWSKI et al. 1979; YOSHIKAWA and FURUYA 1979; DUPERON et al. 1985). It has been reported that UDPG-sterol glucosyl transferase from different plants is stimulated by some phospholipids suggesting a possible regulatory role of these lipids (BOUVIER-NAVE et al. 1984; ULLMANN et al. 1984). The acylated steryl glucosides are formed by the "low energy" acyl transfer from acyl lipids to steryl glucosides as has been shown with enzymes from marigold leaves, broad bean leaves, carrot roots, white mustard seedlings and potato tubers (EICHENBERGER and SIEGRIST 1975; HEINZ et al. 1975; WOJCIECHOWSKI and ZIMOWSKI 1975; ZIMOWSKI and WOJCHIECHOWSKI 1983; CATZ et al. 1985). The enzyme is membrane-bound and preferably galactosylglycerides are effective fatty acid sources for the acylation of steryl glucosides. White mustard seedlings and germinating barley seeds contain a specific sterol glucoside hydrolase and sterol ester hydrolase with a marked specificity with respect to the length of the acyl chain (KALINOSKA and WOJCIECHOWSKI 1978, 1983, 1986; HUGHES and GOAD 1983).

3. Sterol Synthesis During Seed Germination

Changes in the concentration and composition of free and esterified sterols during seed development and germination have been investigated in several plants (GRUNWALD 1975; DAVIS and PONELEIT 1975; TORRES and GARCIA-OLMEDO 1975). Studies on developing cereal seeds (caryopses) show that there is a manifold increase in sterol concentration as development progresses, thus indicating the existence of an active sterol biosynthetic system. In this seed the rate of sterol synthesis is first regulated by a decline in utilization of 24-methylene cycloartanol (XX), presumably by a loss in C-4 demethylase activity. Accumulation of squalene, cycloartenol (III) and 24-methylene cycloartanol is then alleviated by the loss of 2,3-oxidosqualene : cycloartenol cyclase activity and its replacement by an active 2,3-oxidosqualene : amyrin cyclase, possibly as a result of cyclase synthesis and degradation. It has been suggested that a similar change from cycloartenol to pentacyclic triterpene formation may occur during *Sorghum bicolor* grain development (PALMER and BOWDEN 1977). In *Sorghum bicolor* the total free sterol content of the seed increased during germination. However, as the plant developed (7 - 48 days), there was a logarithmic increase in the leaf blade sterol content, a plateau being reached at the onset of floral differentiation (day 41). Over the next 18 days the sterol content rapidly decreased (HEUPEL et al. 1986). In *Cucurbita maxima* during germination the Δ^5-sterols disappeared, whereas the Δ^7-sterols increased up to seedlings development and maturation (GARG and NES 1985).

An increase in free sterol during germination and seedlings development in part results from hydrolysis of steryl ester reserves in the seeds (TORRES et al. 1976; HUGHES and GOAD 1983). In vitro experiments have demonstrated that a dynamic equilibrium exists between free sterols and their glycoside derivatives in plant tissues (KALINOWSKA and WOJCIECHOWSKI 1984). In *Bryonia dioica* Δ^7-stigmasterol seems to be synthesized only by adult tissue of the plant and α-spinasterol is almost exclusively formed during seed germination (CATTEL et al. 1980). In germinating seeds of *Sinapis alba* a distinct increase in the steryl ester and steryl glucosides can be correlated with a distinctly earlier appearance of the corresponding, synthesizing enzymes (KALINOWSKA and WOJCIECHOWSKI 1984). During the ripening of tomatoes, their stigmasterol (IX) content increased in proportion to a decrease in sitosterol (X) content (CHOW and JEN 1978). Because stigmasterol predominates in shaded leaves of tobacco, basal leaves tend to be richer in stigmasterol, and the apical leaves, which are exposed to higher light intensities, tend to be richer in sitosterol (GRUNWALD 1978). Illumination of etiolated seedlings of *Digitalis purpurea* with red light increased the free sterol and steryl glycoside content and this response was also seen to a lesser extent after exposure to far-red light (JACOBSOHN et al. 1978).

After low temperature hardening an increase in sterol content of wheat was reported (WILLEMOTT 1980). When winter rape *(Brassica napus)* was hardened at low temperatures there was a decrease in the free sterol content of the leaves (SIKORSKA and FARKAS 1982). In an investigation over a 19-month period the sterol content of dandelion leaves *(Taraxacum officinale)* was highest during the winter months (WESTERMAN and RODDICK 1981), which apparently contrasts with the situation in rape plants. An increase in the sitosterol : stigmasterol ratio in *Avena sativa* as temperature decreased was reported (JUSAITIS et al. 1981).

The sterol content of plant tissues appears to increase with age (GRUNWALD 1975) and this may be related to changes in membrane structure which accompany senescence (McKERSIE et al. 1978; LEE and THOMPSON 1980). Long-term exposure of barley stem segments to gibberellic acid caused a reduction in sterol content (JUSAITIS et al. 1981), but in pea plants there was an increase in free sterol and a decrease in steryl glycoside (WESTERMAN and RODDICK 1982). Removal of 2,4-dichlorophenoxyacetic acid from the growth medium of a carrot cell tissue culture resulted in a decrease in mevalonate incorporation into sterol (NISHI and TSURITANI 1983).

4. Cardenolides and Bufadienolides

According to earlier results with cardenolide-producing plants, such as *Digitalis purpurea*, *D. lanata* and *Strophantus kombé*, a sequence is proposed for the biosynthesis of cardenolides in which the formation of the butenolide ring should be produced after hydroxylation at C-14β and at C-21 of a 20-ketopregnane intermediate such as XXVIII (TSCHESCHE 1971, 1972). On the other hand, besides this pregnane route, the condensation of the acetate unit could have been performed on a partially hydroxylated 20-ketopregnane derivative such as pregnenolone, progesterone, 3β-hydroxy-5β-pregnan-20-one, etc. leading to cardenolides by an alter-

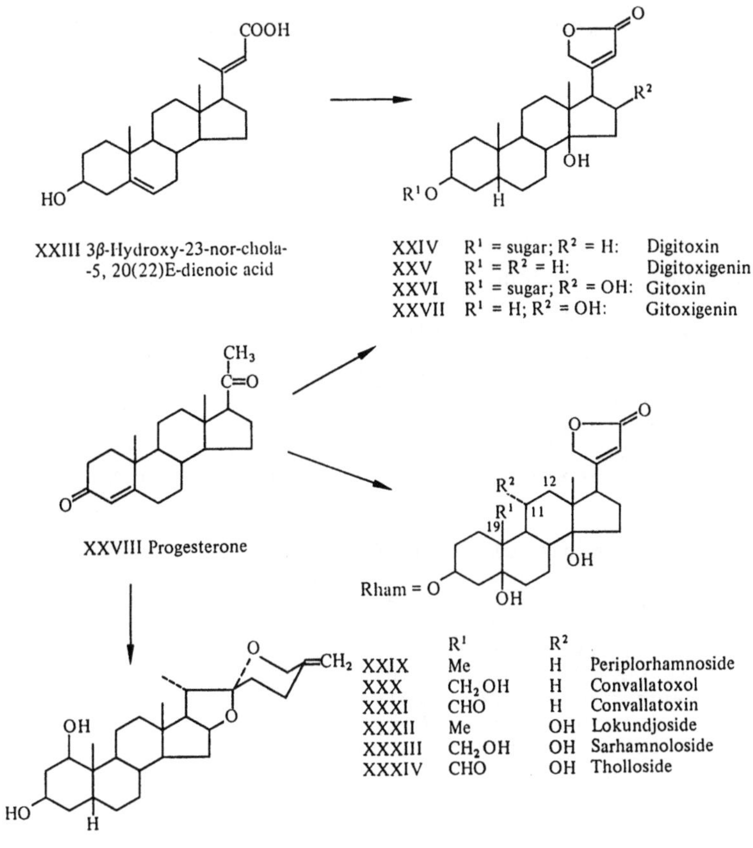

Fig.4. Boisynthesis of cardiac glycosides

native route, which could be called the norcholanoic acid way. Administration of different hydroxynorcholanoic acids such as 3β-hydroxy-23-norchola-5,20(22)E-dienoic acid (XXIII) to *D. purpurea* intact plants produced labelled digitoxin (XXIV) and gitoxin (XXVI), indicating the existence of such an alternative pathway for the biosynthesis of the butenolide ring of cardenolides (Fig. 4) (MAIER et al. 1986). In the biosynthesis of cardenolides, e.g. digitoxigenin (XXV), a 14β-OH compound originates from a 14α-H precursor (ANASTASIA and RONCHETTI 1977).

Convallaria majalis contains about 40 different glycosides of the cardenolide type (KOPP and KUBELKA 1982). Progesterone (XXVIII) which is present in significant amounts in this plant (KOPP and LÖFFELHARDT 1980) has been demonstrated to be a precursor of cardenolide glycosides in leaves only but not in rhizomes (KOPP et al. 1979; SCHRUTKA-RECHTENSTAMM et al. 1985). Afterwards they are transported from leaves into rhizomes (KOPP et al. 1978). The biogenetic pathway of the major cardiac glycosides of *C. majalis* (XXIX-XXXIV) involves different types of hydroxylation reactions at C-19 and C-11, and dehydrogenation at C-19. In leaves of *C. majalis* the cardenolides are subjected to a turnover at least towards the end of the vegetation period (SCHRUTKA-RECHTENSTAMM et al. 1985). Whereas the cardenolides of the underground parts show little variation during growth in the aerial parts, a seasonal change was observed, convallatoxin (XXXI) was the major cardenolide during vegetative growth with convallatoxol (XXX) and convalloside becoming dominant after flowering (KOPP et al. 1981). Using under photosynthet-

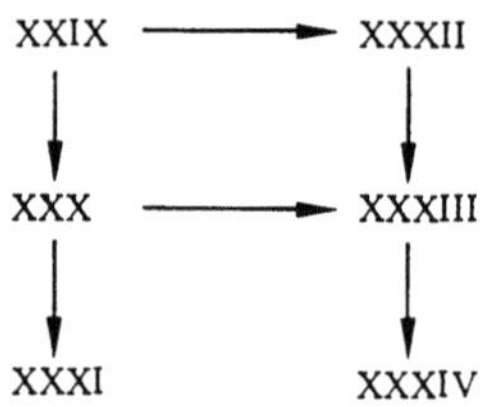

Fig.5. Interconversions of cardenolides of *Convallaria majalis*

ic conditions obtained labelled cardenolides, several interconversions have been observed, e.g. Fig. 5 (KUBELKA et al. 1974, 1977; SCHRUTKA-RECHTENSTAMM et al. 1986). Tholloside (XXXIV) can be considered as the end product of this line of minor glycosides. An NADPH-dependent monooxygenase tentatively localized on mitochondrial membranes converted periplorhamnoside (XXIX) into convallatoxol (XXX) (LÖFFELHARDT et al. 1978). In contrast, the enzyme transforming convallatoxol into convallatoxin (XXXI) could be detected only in the soluble fraction. Desmosterol has been described as precursor of convallamarogenin (XXXV) (TSCHESCHE et al. 1974, 1975).

A stable cardenolide-producing strain of *Digitalis lanata* and *D. purpurea* tissue cultures has been developed (KARTNIG and KOBOSIL 1977; TEWES et al. 1982; DIETTRICH et al. 1986). The accumulation of cardenolides in somatic embryos of these plants was influenced by light (RÜCKER et al. 1976, 1983; GARVE et al. 1980; HAGIMORI et al. 1982; OHLSSON et al. 1983; KUBERSKI et al. 1984; LUCKNER and DIETTRICH 1985). A blue light photoreceptor as well as phytochrome are involved in the regulation of cardenolide biosynthesis and accumulation. Morphological differentiation seems to be an important factor for obtaining cardenolide production. The site of biosynthesis and the accumulation of cardenolides in a lamina of *Digitalis* is mainly in the mesophyll cells (HAGIMORI et al. 1984). The formation of cardenolides by tissue cultures of *D. purpurea* decreases according to the extent of time of cultivation (KARTNIG et al. 1983). The hydroxylation of digitoxin (XXIV) to digoxin (12β-hydroxydigitoxin) is achieved by immobilized cells of *D. lanata* (ALFERMANN et al. 1980).

Cardenolide glycosides such as digitoxin (XXIV) as well as their aglycone such as digitoxigenin (XXV) show an extensive biotransformation in different plant tissues or cell cultures, e.g. in *Digitalis purpurea, D. lanata, Thevetia neriifolia* or *Daucus carota* (REINHARD et al. 1975; DÖLLER et al. 1977; KARTNIG and KOBOSIL 1977; DÖLLER and REINHARD 1979; WEILER 1979; WICHTL and FREIER 1979; HIROTANI and FURUYA 1980; LENKEY et al. 1981; LINLEY and MEKKAWI 1982; ALFERMANN et al. 1985; DOHNAL et al. 1986). Hydroxylation at the positions 12β and 16β as well as acetylation and glycosylation of the sugar moieties were observed. From the leaves of *Nerium oleander* nucleotide-bound D-sarmentose and D-diginose

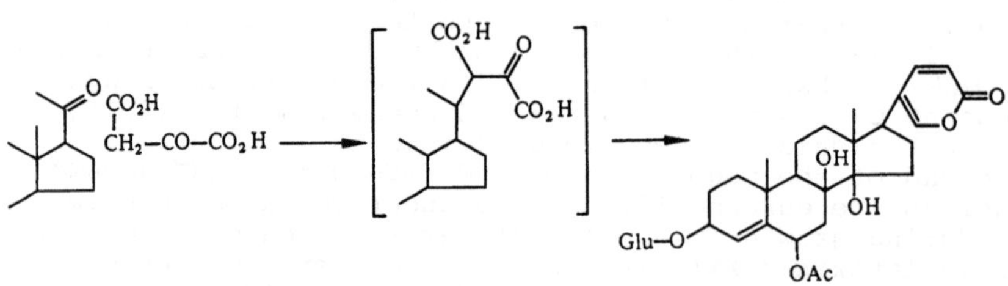

XXXVI Scilliroside

Fig.6. Biosynthesis of scilliroside

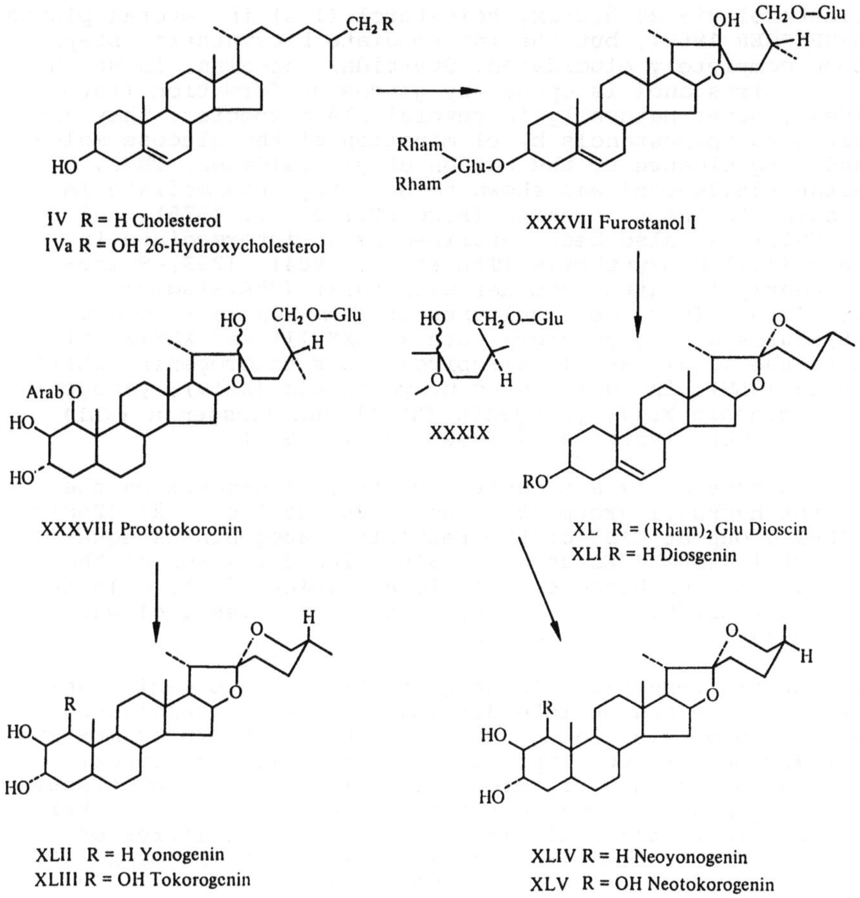

Fig.7. Biosynthesis of sapogenins

were isolated as well as sugar components of the corresponding cardiac glycosides (BAUER et al. 1984).

The biosynthesis of plant bufadienolides resembles that of cardenolides with pregnenolone being a precursor (TSCHESCHE and BRASSAT 1965; PORTO and GROS 1970). Thus, progesterone (XXVIII) was incorporated into bufadienolides of *Urginea maritima*, e.g. proscillaridin A (KOPP and LEEB 1983). Application of bufadienolide shows hydroxylation in 12β and 16β as well as glucosylation. The six-membered lactone ring of the bufadienolides is formed specifically by oxaloacetate as has been shown for scilliroside (XXXVI) in *Scilla maritima* (Fig. 6) (GALAGOVSKY et al. 1982, 1984).

5. Steroidal Saponins

Steroidal saponines, the glycosides of spirostanols, are widely distributed in the plant kingdom (TETENYI 1976; MAHATO et al. 1982). The biosynthesis of steroidal sapogenins has been reviewed earlier (TSCHESCHE and WULFF 1973). In leaves of *Dioscorea deltoidea* mevalonate was incorporated into the aglycone of the steroid saponin deltofolin (PASESHNICHENKO et al. 1984). Spirostanols, e.g. diosgenin (XLI), are synthesized from cholesterol via 26-hydroxycholesterol (IVa) in several plants (Fig. 7) (EICHENBERGER 1982), but the intermediate biosynthetic steps have not yet been completely elucidated. Steroidal saponins, in which the ring F of the spirostanol is opened by glucoside formation (furostanol glycosides), occur naturally in several plant species. They are converted in vitro to spirostanols by elimination of the glucose molecule at C-26 and ring closure by the action of glucosidases. Thus, (25S)-5α-cholestane-3β,26-diol was shown to be a key intermediate in the reaction leading to the sapogenine (RONCHETTI et al. 1975), and furostanol I (XXXVII) has also been described as an intermediate in vivo in diosgenin (XLI) biosynthesis (TAL et al. 1984). (25S)-Spirostanols are frequently isolated together with their (25R)-isomers (MIYAHARA et al. 1983). Thus, cell cultures of *Dioscorea tokoro* contain (25R)- and (25S)-furostanol glycosides such as XXXVIII and XXXIX which are hydrolyzed to the (25R)- and (25S)-spirostanols tokorogenin (XLIII) and neotokorogenin (XLV), in addition to neoyonogenin (XLIV), yamogenin (C-25 isomer of diosgenin XLI), yonogenin (XLII) and diosgenin (XLI) respectively, from other glycosides (UOMORI et al. 1983).

Because C-25 in cholesterol is a prochiral centre, it depends on the introduction of the hydroxyl group at either C-26 (25R) or C-27 (25S) whether the methyl group at C-25 of the resulting sapogenin is equatorial (25α) or axial (25β) (SEO et al. 1984b). The formation of the E and F rings requires, furthermore, functionalization of the C-16 and C-22 (TSCHESCHE and PIESTERT 1975). Thus, 20-hydroxycholesterol was converted by *Digitalis lanata* into tigogenin.

Dioscorea tubers contain reasonably large quantities of diosgenin, which is ideally suited for conversion into 3β-hydroxy-5,16-pregnandien-20-one. Because of its commercial value, the question of obtaining maximum diosgenin yields assumes greater importance. Therefore, some reports have been published on the in vitro production of sapogenins by tissue cultures. Thus, diosgenin (XLI) was obtained either alone or together with solasodine and other steroidal alkaloids by tissue cultures of *Dioscorea deltoidea* and different *Solanum* species (ZACHARIUS and OSMAN 1977; RATHORE and KHARMA 1978; HOSODA and YATAZAWA 1979; HOSODA et al. 1979; KOKATE and RADMAN 1979; UDDIN and CHATURVEDI 1979; JAIN and SAHOO 1981; JAIN et al. 1981; TAL et al. 1983).

6. Steroidal Alkaloids

The biochemistry of steroidal alkaloids has been reviewed (SCHREIBER 1974; RODDICK 1974, 1980; RIPPERGER and SCHREIBER 1981; GROSS et al. 1985). The steroidal alkaloids occur in plants as glycosides, the so-called glycoalkaloids, accompanied by analogous steroidal sapogenins, in the form of saponins, both glycosides containing the same sugars. The incorporation of mevalonate in steroidal alkaloids such as solanidine and tomatidine was confirmed with isolated potato chloroplasts and tomato roots (RODDICK 1974b; RAMASWAMY et al. 1976). Cholesterol (IV) has been demonstrated to be a precursor of tomatidine (L) in *Lycopersicon pimpinellifolium* and solanidine (IL) in *Solanum tuberosum* (Fig. 8) (CANONICA et al. 1977). During the tomatidine formation, the 16β-hydrogen of cholesterol is inserted to the 16α-position, whereas the same hydrogen is lost during solanidine biosynthesis. Probably tomatidine is synthesized in microsomal organelles and accumulated in the vacuole (RODDOCK 1976).

In contrast to the sapogenins, which are isomeric at C-25 but have the same configuration at C-22, the 25-methyl group in the steroidal spiro-

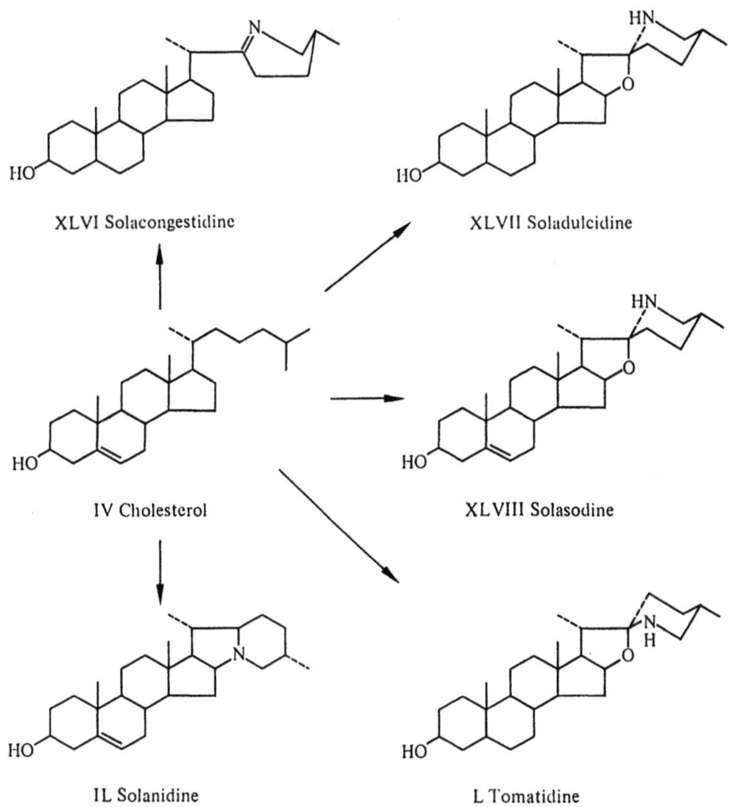

Fig.8. Biosynthesis of steroidal alkaloids

solane alkaloids is always equatorial, whereas the nitrogen may be either in front, as in tomatidine (L), or rear, as in solasodine (XLVIII). Thus, (25S)-5α-cholestane-3β-26-diol was converted into tomatidine (L) (RONCHETTI et al. 1975), while (25R)-5α-cholestane-3β,26-diol was converted into soladulcidine (XLVII) by *Lycopersicon esculentum* (TSCHESCHE et al. 1976). A sapogenin, dormantinol, from *Veratrum grandiflorum*, identified as (25S)-cholest-5-ene-3β,22α,26-triol is discussed as a possible precursor of solanidine (IL) in this plant (KANEKO et al. 1977).

The nitrogen atom is introduced through a simple replacement of the terminal hydroxy group by an amino group. The donor molecule is an amino acid, e.g. arginine for solanidine (IL) in *Veratrum grandiflorum* (KANEKO et al. 1976). The formation of the F ring can, in principle, precede the closure of the E ring. Thus, solacongestidine (XLVI) and dihydrosolacongestidine were converted into soladulcidine (XLVII) and into solasodine (XVLIII by *Solanum laciniatum* (TSCHESCHE and SPINDLER 1978). But the conversion of (25R)-26-aminocholesterol, (25R)-26-amino-16β-hydroxycholesterol, and 26-aminodihydrodiosgenin into solasodine by *S. laciniatum* indicate the opposite line (TSCHESCHE and PIESTERT 1975; TSCHESCHE and BRENNECKE 1980). Verazine was converted into hakurirodine and rubijervine by dormant rhizome slices of *Veratrum grandiflorum* (KANEKO et al. 1975). Like other plant steroids, solanidine is glycosylated by UDP-glucose in the presence of a glucosyltransferase, e.g. from *S. tuberosum* (LAVINTMAN et al. 1977; OSMAN et al. 1980).

Grafting experiments have shown that alkaloid transport between root and shoot does not take place in potato and tomato plants (RODDICK 1982). The alkaloid con-

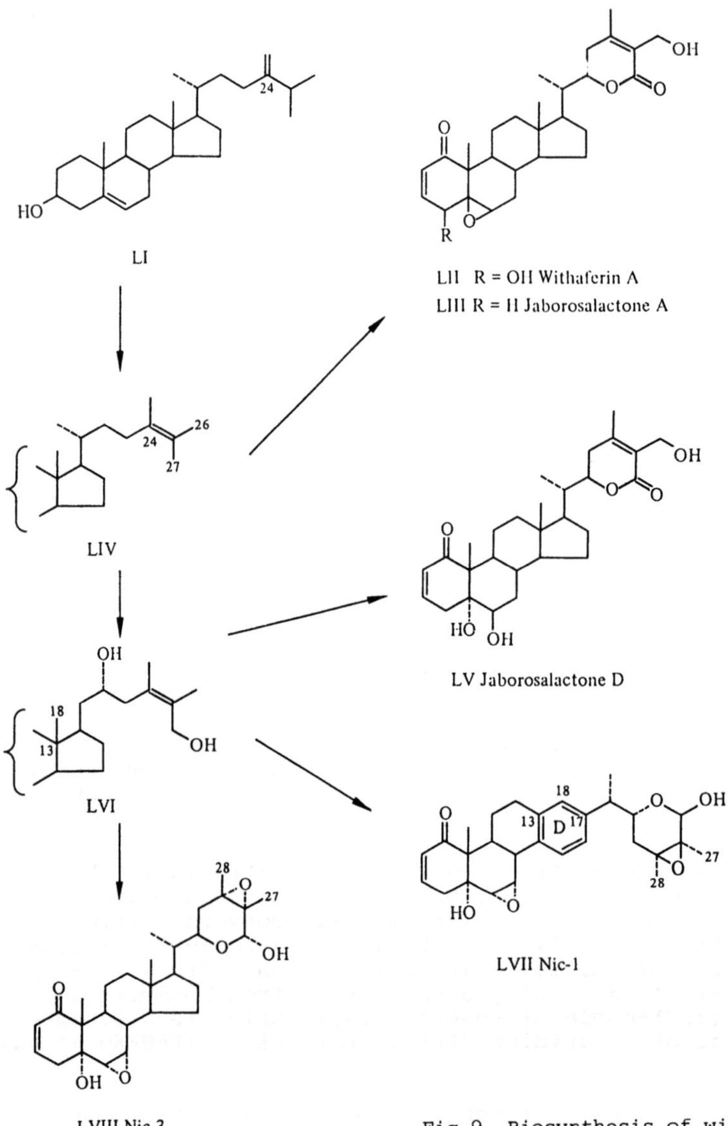

Fig.9. Biosynthesis of withanolides

tent of leaves and fruits of *Lycopersicon esculentum* increases up to a particular developmental stage and thereafter declines (ALI and SCHLÖSSER 1977; ELTAYEB and RODDICK 1984a, 1985). Tomatine is degraded during fruit ripening. Fruit ripeness showed a negative correlation with fruit tomatine (ELTAYEB and RODDICK 1984b). Ethrel treatment of fruits caused a lower tomatine level and fruits in which ripening was retarded have higher alkaloid levels than controls. A reactor containing immobilized cells of *S. aviculare* was designed for the continuous production of glycoalkaloids (JIRKU et al. 1981).

7. Withanolides

The withanolides (KIRSON and GLOTTER 1981) are a large group of 24-methyl plant steroids occurring in the Solanaceae, which are characterized by extensive oxidative modifications. Typical representatives are withaferin A (LII), jaborosalactone A (LIII) and jaborosalactone D (LV). In other withanolides of *Withania somnifera* positions 14 and 17 are hydrolyzed or dehydrogenated (VANDE VELDE and LAVIE 1982; VANDE VELDE et al. 1983).

The incorporation of mevalonate into withaferin A and jaborosalactone A has been described in *Acnistus breviflorus* and *W. somnifera* (LOCKLEY et al. 1976; VELEIRO et al. 1985a,b). C-26 of the withanolides is directly derived from C-2 of mevalonate, suggesting that the 25-pro-R-methyl group of a sterol intermediate had been oxidized to form the lactone ring of the withanolides. The incorporation of 24-methylene cholesterol (LI) as well as the presence in *Withania coagulans* of related compounds suggest such products as precursors of withanolides (Fig. 9). The mode of incorporation is consistent with the hypothesis that a 4-deoxy-5,6-unsaturated withanolide should be the precursor of 4-deoxy-5,6-epoxy withanolides such as jaborosalactone A (LIII) and of 4-hydroxy-5,6-epoxy withanolides such as withaferin A (LII). Jaborosalactone A has been confirmed to be a precursor of withaferin A and jaborosalactone D (LV).

The nicandrenoids (BEGLEY et al. 1976a,b) from *Nicandra physaloides* are related to the withanolides. They contain an unusual epoxylactol side chain and e.g. Nic-1 (LVII), the unique aromatic D ring. The biosynthesis of Nic-1 leads via 24-methylene cholesterol (LI), its 24(25)-double bond isomer (LIV) and the triol (LVI) (Fig. 9) (GILL et al. 1986). A C-26 aldehyde is suggested as intermediate, because experiments with ^{2}H-labelled mevalonate have shown that C-26 in Nic-1 is derived from the mevalonate 3-methyl with some retention of hydrogen. The aromatic D ring of Nic-1 is formed by D-ring expansion in a steroid precursor with oxidative inclusion of the C/D angular methyl (GILL et al. 1986). The co-occurrence of Nic-1 and Nic-3 suggests late stage D-aromatization.

References

ALFERMANN, A.W., SCHULLER, I., REINHARD, E.: Planta Med. 40, 218 (1980). - ALFERMANN, A.W., SPIELER, H., REINHARD, E., 154-163 in: Primary and Secondary Metabolism of Plant Cell Cultures, eds. K.-H. NEUMANN, W. BARZ. Berlin: Springer 1985. - ALI, A., SCHLÖSSER, E.: Angew. Bot. 51, 143-148 (1977). - ALTMAN, L.J., HAN, C.Y., BERTOLINO, A., NANDY, G., LAUNGANI, D., MÜLLER, W., SCHWARTZ, S., SHANKER, D., de WOLF, W.H., YANG, F.: J. Am. Chem. Soc. 100, 3235-3237 (1978). - ANASTASIA, M., RONCHETTI, F.: Phytochemistry 16, 1082-1083 (1977). - AUGUSCINSKA, E., KASPRZYK, Z.: Acta Biochem. Pol. 29, 7-15 (1982).

BACH, T.J.: Plant Science 39, 183-187 (1985). - BACH, T.J., LICHTENTHALER, H.K.: Z. Naturforsch. 38c, 212-219 (1983a); - Physiol. Plant. 59, 50-60 (1983b). - BALLIANO, G., CAPUTO, O., VIOLA, F., DELPRINO, L., CATTEL, L.: Lipids 18, 302-305 (1983). - BAUER, P., KOPP, B., FRANZ, G.: Planta Med. 50, 12-14 (1984). - BEGLEY, M.J., CROMBIE, L., HAM, P.J., WHITTING, D.A.: J. Chem. Soc., Perkin Trans. 1, 296-304 (1976a); - ibid. 1, 304-308 (1976b). - BENVENISTE, P.: Annu. Rev. Plant Physiol. 37, 275-308 (1986). - BHATT, P.N., BHATT, D.P.: J. Exp. Bot. 35, 890-896 (1984). - BLADOCHA, M., BENVENISTE, P.: Plant Physiol. 74, 756-762 (1983); - ibid. 79, 1098-1106 (1985). - BOUVIER-NAVE, P., ULLMANN, P., RIMMELE, D., BENVENISTE, P.: Plant Sci. Lett. 36, 19-27 (1984). - BOWLES, D.J., LEHLE, L., KAUSS, H.: Planta 134, 177-181 (1977).

CAMPBELL, C.E., KIRCHER, H.W.: Phytochemistry 19, 2777-2779 (1980). - CANONICA, L., RONCHETTI, F., RUSSO, G., SPORTOLETTI, G.: Chem. Commun. 286-287 (1977). - CASPI, E., SLIWOWSKI, J.: J. Am. Chem. Soc. 97, 5032-5034 (1975). - CASPI, E., SLIWOWSKI, J., ROBICHAUD, C.S.: J. Am. Chem. Soc. 97, 3820-3822 (1975). - CATTEL, L., BALLIANO, G., CAPUTO, O., DELPRINO, L.: Phytochemistry 19, 465-466 (1980). - CATTEL, L.,

BALLIANO, G., VIOLA F., CAPUTO, O.: Planta Medica 38, 112-120 (1980). - CATZ, D.S., TANDECARZ, J.S., CARDINI, C.E.: J. Exp. Bot. 36, 602-609 (1985). - CERUTTI, M., DELPRINO, L., CATTEL, L., BOUVIER-NAVÉ, P., DURIATTI, A., SCHUBER, F., BENVENISTE, P.: Chem. Commun. 1054-1055 (1985). - CHOW, E.T.S., JEN, J.J.: J. Food Sci. 43, 1424-1426 (1978).

DAVIS, D.L., PONELEIT, C.G.: Phytochemistry 14, 1201-1203 (1975). - DELPRINO, L., BALLIANO, G., CATTEL, L., BENVENISTE, P., BOUVIER, P.: Chem. Commun. 381-382 (1983). - DIETTRICH, B., STEUP, C., NEUMANN, D., SCHEIBNER, H., REINBOTHE, C., LUCKNER, M.: J. Plant Physiol. 124, 441-453 (1986). - DÖLLER, P.C., REINHARD, E.: Planta Med. 37, 277-288 (1979). - DÖLLER, P.C., ALFERMANN, A.W., REINHARD, E.: Planta Med. 31, 1-6 (1977). - DOHNAL, B., ROZKRUTOWA, B., SUPNIEWSKA, J.H.: Herb. Hung. 25, 87-95 (1986). - DOUGLAS, T.J., PALEG, L.G.: Phytochemistry 17, 705-712 (1978); - ibid. 17, 713-718 (1978); - J. Exp. Bot. 32, 59-68 (1981). - DUPERON, P., CHARIFI-KHOUS, M., DUPERON, R.: C.R. Acad. Sci., Ser. 3, 300, 267-370 (1985). - DURIATTI, A., BOUVIER-NAVÉ, P., BENVENISTE, P., SCHUBER, F., DELPRINO, L., BALLIANO, G., CATTEL, L.: Biochem. Pharmacol. 34, 2765-2777 (1985).

EICHENBERGER, W.: Plant Cell Rep. 1, 253-256 (1982). - EICHENBERGER, W., SIEGRIST, H.P.: FEBS Lett. 52, 153-156 (1975). - ELTAYEB, E.A., RODDICK, J.G.: J. Exp. Bot. 35, 252-260 (1984a); - ibid. 35, 261-267 (1984b); - Phytochemistry 24, 253-257 (1985).

FONTENEAU, P., HARTMANN-BOUILLON, M.A., BENVENISTE, P.: Plant Sci. Lett. 40, 147-155 (1977). - FRASINEL, C., PATTERSON, G.W., DUTKY, S.R.: Phytochemistry 17, 1567-1570 (1978).

GALAGOVSKY, L.R., PORTO, A.M., BURTON, G., MAIER, M.S., SELDES, A.M., GROS, E.G.: An. Asoc. Quim. Argent. 70, 327-335 (1982). - GALAGOVSKY, L.R., PORTO, A.M., BURTON, G., GROS, E.G.: Z. Naturforsch. 39c, 38-44 (1984). - GARCIA, R.E., MUDD, J.B.: Plant Physiol. 62, 348-353 (1978a); - Arch. Biochem. Biophys. 190, 315-321 (1978b). - GARG, V.K., NES, W.R.: Lipids 20, 876-883 (1985). - GARVE, R., LUCKNER, M., VOGEL, E., TEWES, A., NOVER, L.: Planta Med. 40, 92-103 (1980). - GILL, H.K., SMITH, R.W., WHITTING, D.A.: Chem. Commun. 1457-1458 (1986a); - ibid. 1459-1460 (1986b). - GOODWIN, T.W.: Biochem. Soc. Trans. 5, 1252-1255 (1977); - Annu. Rev. Plant Physiol. 30, 369-404 (1979); - 485-407 in: The Biochemistry of Plants. A Comprehensive Treatise, Vol. 4, Lipids, Structure and Function, eds. P.K. STUMPF, E.E. CONN. New York: Academic Press 1980. - GOODWIN, T.W., 443-480 in: Biosynthesis of Isoprenoid Compounds, eds. J.W. PORTER, S.L. SURGEON, Vol. 1. New York: Wiley 1981. - GROENEVELD, H.W., ELINGS, J.: Phytochemistry 23, 275-279 (1984). - GROENEVELD, H.W., MAHLBERG, P.G.: Phytochemistry 25, 1111-1113 (1986). - GROENEVELD, H.W., ROELVINK, P.W.: Planta 154, 347-351 (1982a); - ibid. 156, 264-269 (1982b). - GROSS, D., SCHÜTTE, H.R., SCHREIBER, K., 354-384 in: Biochemistry of Alkaloids, eds. K. MOTHES, H.R. SCHÜTTE, M. LUCKNER. Berlin: Deutscher Verlag der Wissenschaften 1985. - GRUNWALD, C.: Annu. Rev. Plant Physiol. 26, 209-236 (1975a); - Phytochemistry 14, 79-82 (1975b); - Plant Physiol. 61, 76-79 (1978); - 221-256 in: Encyclopedia of Plant Physiology. N.S. Vol. 8, Secondary Plant Products, eds. E.A. BELL, B.V. CHARLWOOD. New York, Berlin, Heidelberg: Springer 1980; - Phytochemistry 24, 2915-2918 (1985).

HAGIMORI, M., MATSUMOTO, T., OBI, Y.: Plant Physiol. 69, 653-656 (1980). - HAGIMORI, M., MATSUMOTO, T., MIKAMI, Y.: Plant Cell Physiol. 25, 947-953 (1984). - HARTMANN, M.A., GRANDMOUGIN, A., BENVENISTE, P.: C.R. Acad. Sci., Ser. 3, 301, 601-605 (1985). - HARTMANN-BOUILLON, M.A., BENVENISTE, P.: Phytochemistry 17, 1037-1042 (1978). - HARTMANN-BOUILLON, M.A., BENVENISTE, P., ROLAND, J.C.: Biol. Cellulaire 35,

183-194 (1979). - HEFTMANN, E.: Phytochemistry 22, 1843-1860 (1983); - 487-518 in: Biochemistry and Function of Isopentenoids in Plants, eds. W.D. NES, G. FULLER, L.-S. TSAI. New York: Dekker 1984. - HEINTZ, R., BENVENISTE, P.: J. Biol. Chem. 249, 4267-4274 (1974). - HEINZ, E., DIELER, H.P., RULLKÖTER, J.: Z. Pflanzenphysiol. 75, 78-87 (1975). - HENRY, M.J., SISLER, H.D.: Pestic. Biochem. Physiol. 22, 262-275 (1984). - HEUPEL, R.C., SAUVAIRE, Y., LePHU, H., PARISH, E.J., NES, W.D.: Lipids 21, 69-75 (1986). - HIROTANI, M., FURUYA, T.: Phytochemistry 19, 531-534 (1980). - HOPP, H.E., ROMERO, P.A., DALEO, G.R., PONT LEZICA, R.: Phytochemistry 17, 1049-1052 (1978). - HOSODA, N., YATAZAWA, M.: Agric. Biol. Chem. 43, 821-825 (1979). - HOSODA, N., ITO, H., YATAZAWA, M.: Agric. Biol. Chem. 43, 1745-1748 (1979). - HOSOKAWA, G., PATTERSON, G.W., LUSBY, W.R.: Lipids 19, 449-456 (1984). - HUANG, L.S., GRUNWALD, C.: Phytochemistry 25, 2779-2781 (1986). - HUGHES, M.A., GOAD, L.J.: Biochem. Soc. Trans. 11, 588-589 (1983).

ISA, R.B.M., SIPAT, A.B.: Biochem. Biophys. Res. Commun. 108, 206-212 (1982).

JACOBSOHN, M.K., ORKWISZEWSKI, J.A.J., JACOBSOHN, G.M.: Plant Physiol. 62, 1000-1004 (1978). - JAIN, S.C., SAHOO, S.: Chem. Pharm. Bull. 29, 1765-1767 (1981). - JAIN, S.C., KHANNA, P., SAHOO, S.: J. Nat. Prod. (Lloydia) 44, 125-126 (1981). - JIRKU, V., MACEK, T., VANEK, T, KRUMPHANZL, V., KUBANEK, V.: Biotechnol. Lett. 3, 447-450 (1981). - JUSAITIS, M., PALEG, L.G., ASPINALL, D.: Phytochemistry 20, 417-423 (1981).

KALINOWSKA, M., WOJCIECHOWSKI, Z.A.: Phytochemistry 17, 1533-1537 (1978); - ibid. 22, 59-63 (1983); - ibid. 23, 2485-2488 (1984); - ibid. 25, 45-49 (1986). - KANEKO, K., SETO, H., MOTOKI, C., MITSUHASHI, H.: Phytochemistry 14, 1295-1301 (1975). - KANEKO, K., TANAKA, M.W., MITSUHASHI, H.: Phytochemistry 15, 1391-1393 (1976). - KANEKO, K., TANAKA, M.W., MITSUHASHI, H.: Phytochemistry 16, 1247-1251 (1977). - KARTNIG, T., KOBOSIL, P.: Planta Med. 31, 221-227 (1977). - KARTNIG, T., KUMMER-FUSTINIONI, G., HEYDEL, B.: Planta Med. 47, 247-248 (1983). - KIRSON, I., GLOTTER, E.: J. Nat. Prod. (Lloydia) 44, 633-647 (1981). - KNAPP, F.F., GOAD, L.J., GOODWIN, T.W.: Phytochemistry 16, 1677-1681 (1977a); - ibid. 16, 1683-1688 (1977b). - KOKATE, C.K., RADWAN, S.S.: Z. Naturforsch. 34c, 634-636 (1979). - KOPP, B., KUBELKA, W., JENTZSCH, K.: Sci. Pharm. 49, 265-271 (1981); - KOPP, B., LEEB, R.: Sci. Pharm. 51, 248-258 (1983). - KOPP, B., LÖFFELHARDT, W.: Z. Naturforsch. 35c, 41-44 (1980). - KOPP, B., LÖFFELHARDT, W., KUBELKA, W.: Z. Naturforsch. 33c, 646-650 (1978); - ibid. 34c, 334-339 (1979); - KUBELKA, W., KOPP, B., JENTZSCH, K., RUIS, H.: Phytochemistry 13, 1805-1808 (1974); - ibid. 16, 687-690 (1977). - KUBERSKI, C., SCHEIBNER, H., STEUP, C., DIETTRICH, B., LUCKNER, M.: Phytochemistry 23, 1407-1412 (1984).

LARGEAU, C., GOAD, L.J., GOODWIN, T.W.: Phytochemistry 16, 1925-1930 (1977a); - ibid. 16, 1931-1933 (1977b). - LAVINTMAN, N., TANDECARZ, J., CARDINI, C.E.: Plant Sci. Lett. 8, 65-70 (1977). - LEES, G.L., THOMPSON, J.E.: Physiol. Plant. 49, 215-221 (1980). - LENKEY, B., NANASI, P., TETENYI, P.: Planta Med. 43, 409-411 (1981). - LENTON, J.R., GOAD, L.J., GOODWIN, T.W.: Phytochemistry 14, 1523-1528 (1975). - LERCHER, M., WOJCIECHOWSKI, Z.A.: Plant Sci. Lett. 7, 337-340 (1976). - LINLEY, P.A., MEKKAWI, A.G.: Planta Med. 46, 113-118 (1982). - LOCKLEY, W.J.S., REES, H.H., GOODWIN, T.W.: Phytochemistry 15, 937-939 (1976). - LÖFFELHARDT, W., KOPP, B., KUBELKA, W.: Phytochemistry 17, 1581-1584 (1978). - LUCKNER, M., DIETTRICH, B., 154-163 in: Primary and Secondary Metabolism of Plant Cell Cultures, eds. K.-H. NEUMANN, W. BARZ, E. REINHARD. Berlin, Heidelberg, New York: Springer 1985.

MAHATO, S.B., GANGULY, A.N., SAHU, N.P.: Phytochemistry 21, 959-978 (1982). - MAIER, M.S., SELDES, A.M., GROS, E.G.: Phytochemistry 25, 1327-1329 (1986). - MAZLIAK, P.: Physiol. Vég. 19, 543-563 (1981). - McKEAN, M.L., NES, W.R.: Phytochemistry 16, 683-686 (1977). - McKERSIE, B.D., LEPOCK, J.R., KRUUV, J., THOMPSON, J.E.: Biochim. Biophys. Acta 508, 197-212 (1978). - MIHAILOVIC, M.M.: see BENVENISTE 1986. - MISSO, N.L.A., GOAD, L.J.: Phytochemistry 22, 2473-2479 (1983); - ibid. 23, 73-82 (1984). - MIYAHARA, K., KUDO, K., KAWASAKI, T.: Chem. Pharm. Bull. 31, 348-351 (1983). - MUDD, J.B., 509-534 in: The Biochemistry of Plants. A Comprehensive Treatise, Vol. 4, Lipids, Structure and Function, eds. P.K. STUMPF, E.E. CONN. New York: Academic Press 1980. - MUDD, J.B., MOELLER, C.H., GARCIA, R.E., 487-518 in: Biochemistry and Function of Isopentenoids in Plants, eds. W.D. NES, G. FULLER, L.-S. TSAI. New York: Dekker 1984

NARULA, A.S., RAHIER, A., BENVENISTE, P., SCHUBER, F.: J. Am. Chem. Soc. 103, 2408-2409 (1981). - NAVARI-IZZO, F., IZZO, R.: Phytochemistry 23, 769-772 (1984). - NES, W.R.: Adv. Lipid Res. 15, 233-324 (1977). - NES, W.R., McKEAN, M.L.: Biochemistry of Steroids and Other Isopentenoids. Baltimore: University Park Press 1977. - NES, W.D., HEFTMANN, E.: J. Nat. Prod. (Lloydia) 44, 377-400 (1981). - NES, W.D., DOUGLAS, T.J., LIN, J.-T., HEFTMANN, E., PALEG, L.G.: Phytochemistry 21, 575-579 (1982a). - NES, W.D., YANIV, Z., HEFTMANN, E.: Phytochemistry 21, 581-583 (1982b). - NICOTRA, F., RONCHETTI, F., RUSSO, G., LUGARO, G., CASELLATO, M.: J. Chem. Soc. Perkin Trans. I, 498-502 (1981). - NISHI, A., TSURITANI, I.: Phytochemistry 22, 399-401 (1983).

OHLSSON, A.B., BJÖRK, L., GATENBECK, S.: Phytochemistry 22, 2447-2450 (1983). - OSMAN, S.F., ZACHARIUS, R.M., NAGLAK, D.: Phytochemistry 19, 2599-2601 (1980).

PALMER, M.A., BOWDEN, B.N.: Phytochemistry 16, 459-463 (1977). - PASESHNICHENKO, V.A., GURIELIDZE, K.G., VASILEVA, I.S., GUSEVA, A.R.: Biokhimiya 49, 75-80 (1984). - PORTO, A.M., GROS, E.G.: Experientia 26, 11 (1970).

RAHIER, A., CATTEL, L., BENVENISTE, P.: Phytochemistry 16, 1187-1192 (1977). - RAHIER, A., NARULA, A.S., BENVENISTE, P., SCHMITT, P.: Biochem. Biophys. Res. Commun. 92, 20-25 (1980). - RAHIER, A., SCHMITT, P., BENVENISTE, P.: Phytochemistry 21, 1969-1974 (1982). - RAHIER, A., BOUVIER, P., CATTEL, L., NARULA, A., BENVENISTE, P.: Biochem. Soc. Trans. 11, 537-543 (1983). - RAHIER, A., GENOT, J.C., SCHUBER, F.; BENVENISTE, P., NARULA, A.S.: J. Biol. Chem. 259, 15215-15223 (1984). - RAHIER, A., TATON, M., SCHMITT, P., BENVENISTE, P., PLACE, P., ANDING, C.: Phytochemistry 24, 1223-1232 (1985). - RAHIER, A., SCHMITT, P., HUSS, B., BENVENISTE, P., POMMER, E.H.: Pestic. Biochem. Physiol. 25, 112-124 (1986a). - RAHIER, A., TATON, M., BOUVIER-NAVE, P., SCHMITT, P., BENVENISTE, P., SCHUBER, F., NARULA, A.S., CATTEL, L., ANDING, C., PLACE, P.: Lipids 21, 52-62 (1986b). - RAMASWAMY, N.K., BEHERE, A.G., NAIR, P.M.: Eur. J. Biochem. 67, 275-282 (1976). - RATHORE, A.K., KHANNA, P.: Curr. Sci. 47, 870-871 (1978). - REINHARD, E., BOY, M., KAISER, F.: Planta Med. Suppl. 163-168 (1975). - RENDELL, N., MISSO, N.L.A., GOAD, L.J.: Lipids 21, 63-68 (1986). - RIPPERGER, H., SCHREIBER, K.: The Alkaloids 19, 81-192 (1981). - RODDICK, J.G.: Phytochemistry 13, 9-25 (1974a); - ibid. 13, 1459-1462 (1974b); - ibid. 15, 475-477 (1976); - 167-184 in: Encyclopedia of Plant Physiology, N.S. Vol. 8, Secondary Plant Products, eds. E.A. BELL, B.V. CHARLWOOD. Berlin, Heidelberg, New York: Springer 1980; - Experientia 38, 460-462 (1982). - RONCHETTI, F., RUSSO, G., FERRARA, G., VECCHIO, G.: Phytochemistry 14, 2423-2425 (1975). - RÜCKER, W., JENTZSCH, K., WICHTL, M.: Z. Pflanzenphysiol. 80, 323-335 (1976); - Biochem. Physiol. Pflanzen 178, 91-100 (1983).

SANDIFER, R.M., THOMPSON, M.D., GAUGHAN, R.G., POULTER, C.D.: J. Am. Chem. Soc. 104, 7376-7378 (1982). - SCHEID, F., ROHMER, M., BENVENISTE, P.: Phytochemistry 21, 1959-1967 (1982). - SCHMITT, P., BENVENISTE, P.: Phytochemistry 18, 445-450 (1979a); - ibid. 18, 1659-1665 (1979b). - SCHMITT, P., SCHEID, F., BENVENISTE, P.: Phytochemistry 19, 525-530 (1980). - SCHMITT, P., NARULA, A.S., BENVENISTE, P., RAHIER, A.: Phytochemistry 20, 197-201 (1981a). - SCHMITT, P., BENVENISTE, P., LEROUX, P.: Phytochemistry 20, 2153-2159 (1981b). - SCHMITT, P., RAHIER, A., BENVENISTE, P.: Physiol. Veg. 20, 559-571 (1982). - SCHREIBER, K.: Biochem. Soc. Trans. 2, 1-25 (1974). - SCHROEPFER, G.J.: Annu. Rev. Biochem. 50, 585-621 (1981); - ibid. 51, 555-585 (1982). - SCHRUTKA-RECHTENSTAMM, R., KOPP, B., LÖFFELHARDT, W.: Z. Naturforsch. 40c, 627-630 (1985a); - Planta Med. 387-390 (1985b). - Phytochemistry 25, 1107-1109 (1986). - SELVARAJ, Y., CHANDER, M.S.: Phytochemistry 21, 2137-2138 (1982). - SEO, S., UOMORI, A., YOSHIMURA, Y., TAKEDA, K.: J. Am. Chem. Soc. 105, 6343-6344 (1983); - Chem. Commun. 1174-1176 (1984a). - SEO, S., UOMORI, A., YOSHIMURA, Y., TORKI, K.: J. Chem. Soc. Perkin Trans 1, 869-874 (1984b). - SEO, S.,SANKAWA, U., SETO, H., UOMORI, A., YOSHIMURA, Y., EBIZUKA, Y., NOGUCHI, H., TAKEDA, K.: Chem. Commun. 1139-1141 (1986). - SIKORSKA, E., FARKAS, T.: Physiol. Plant. 56, 349-352 (1982). - SIPAT, A.B.: Phytochemistry 21, 2613-2618 (1982). - SLIWOWSKI, J.K., CASPI, E.: Chem. Commun. 196-197 (1976); - J. Am. Chem. Soc. 99, 4479-4485 (1977). - SUGA, T., TANGE, K., ICCHO, K., HIRATA, T.: Phytochemistry 19, 67-70 (1980).

TAL, B., ROKEM, J.S., GOLDBERG, I.: Plant Cell Rep. 2, 219-222 (1983). - TAL, B., TAMIR, I., ROKEM, J.S., GOLDBERG, I.: Biochem. J. 219, 619-624 (1984). - TATON, M., BENVENISTE, P., RAHIER, A.: Biochem. Biophys. Res. Commun. 138, 764-770 (1986); - Phytochemistry 26, 385-392 (1987). - TETENYI, P.: Herba Hung. 15, 27-48 (1976). - TEWES, A., WAPPLER, A., PESCHKE, E.M., GARVE, R., NOVER, L.: Z. Pflanzenphysiol. 106, 311-324 (1982). - TORRES, J.V., GARCIA-OLMEDO, F.: Biochem. Biophys. Acta 409, 367-375 (1975). - TORRES, J.V., CARBONERO, P., GARCIA-OLMEDO, F.: Phytochemistry 15, 677-680 (1976). - TSCHESCHE, R.: Planta Med. Suppl. 4, 34-39 (1971); - Proc. Roy. Soc. 180B, 187-202 (1972). - TSCHESCHE, R., BRASSAT, B.: Z. Naturforsch. 20b, 707-708 (1965). - TSCHESCHE, R., BRENNECKE, H.R.: Phytochemistry 19, 1449-1451 (1980). - TSCHESCHE, R., PIESTERT, G.: Phytochemistry 14, 435-438 (1975). - TSCHESCHE, R., WULFF, G.: Fortschr. Chem. Org. Naturst. 30, 461-606 (1973). - TSCHESCHE, R., GÜTTEL, H.W., JOSST, G.: Phytochemistry 13, 137-139 (1974). - TSCHESCHE, R., PIESTERT, T., GÜTTEL, H.W.: Phytochemistry 14, 129-131 (1975). - TSCHESCHE, R., GOOSSENS, B., TÖPFER, A.: Phytochemistry 15, 1387-1389 (1976). - TSCHESCHE, R., SPINDLER, M.: Phytochemistry 17, 251-255 (1978).

UDDIN, A., CHATURVEDI, H.C.: Planta Med. 37, 90-92 (1979). - ULLMANN, P., RIMMELE, D., BENVENISTE, P., BOUVIER-NAVE, P.: Plant Sci. Lett. 36, 29-36 (1984). - UOMORI, A., SEO, S., TORI, K., TOMITA, Y.: Phytochemistry 22, 203-206 (1983). - UOMORI, A., SEO, S., YOSHIMURA, Y., TAKEDA, K.: Chem. Comm. 17, 1176-1177 (1984).

VAN DE VELDE, V., LAVIE, D.: Phytochemistry 21, 731-733 (1982). - VAN DE VELDE, V., LAVIE, D., BUDHIRAJA, R.D., SUDHIR, S., GARG, K.N.: Phytochemistry 22, 2253-2257 (1983). - VELEIRO, A.S., BURTON, G., GROS, E.G.: Phytochemistry 24, 2263-2265 (1985a); - ibid. 24, 2573-2575 (1985b).

WEILER, E.W.: Planta Med. 35, 162-166 (1979). - WESTERMAN, L., RODDICK, J.G.: Plant Physiol. 68, 872-875 (1981); - Phytochemistry 21, 1567-1572 (1982). - WICHTL, M., FREIER, R.: Dtsch. Apoth. Ztg. 119, 1124-1129 (1979). - WILLEMOTT, C.: Phytochemistry 19, 1071-1073 (1980). -

WILLUHN, G.: Pharm. Ztg. 121, 1491-1501 (1976); - ibid. 124, 2014-2017 (1979). - WOJCIECHOWSKI, Z.A.: Biochem. Soc. Trans. 11, 565-558 (1983); - 297-307 in: Conjugated Plant Hormones, Structure, Metabolism and Function, eds. K. SCHREIBER, H.R. SCHÜTTE, G. SEMBDNER. Berlin: Deutscher Verlag der Wissenschaften 1987. - WOJCIECHOWSKI, Z.A., ZIMOWSKI, J.: Biochim. Biophys. Acta 398, 111-117 (1975). - WOJCIECHOWSKI, Z.A., ZIMOWSKI, Z., ZIMOWSKI, J.G., LYZNIK, A.: Biochim. Biophys. Acta 570, 363-370 (1979).

YOSHIKAWA, T., FURUYA, T.: Phytochemistry 18, 239-241 (1979).

ZACHARIUS, R.M., OSMAN, S.F.: Plant Sci. Lett. 10, 283-287 (1977). - ZAKELJ, M., GOAD, L.J.: Phytochemistry 22, 1931-1936 (1983). - ZIMOWSKI, J., WOJCIECHOWSKI, Z.A.: Phytochemistry 20, 1795-1798 (1981a); - ibid. 20, 1799-1903 (1981b); - ibid. 22, 2175-2178 (1983).

Prof. Dr. HORST-ROBERT SCHÜTTE
Institut für Biochemie der Pflanzen des
Forschungsbereiches Biowissenschaften und
Medizin der Akademie der Wissenschaften der DDR
DDR-4050 Halle (Saale), Weinberg 3

V. Growth

By GÜNTHER SEMBDNER, WILLIBALD SCHLIEMANN, and GABRIELE HERRMANN

1. General Reviews

With two further volumes in the Encyclopedia of Plant Physiology, New Series, the Hormonal Regulation of Development I - III is now completed. SCOTT (1984) edited Vol. 10 comprehending the function of hormones from the level of the cell to the whole plant, and PHARIS and REID (1985), in Vol. 11, reviewed the role of environmental factors. The chemistry of plant hormones is outlined by TAKAHASHI (1986), and important topics of plant hormone biosynthesis and metabolism were collected by CROZIER and HILLMAN (1984). Plant hormone conjugates, their structures, metabolism and function are reported by SCHREIBER et al. (1987). Receptors of plant hormones and cellular interactions are reviewed by CHADWICK and GARROD (1986). A laboratory guide book on methods in plant hormone analysis has been edited by SEMBDNER et al. (1987) and within the series, Modern Methods of Plant Analysis, valuable contributions appeared on gas chromatography – mass spectrometry, high performance liquid chromatography and immunology (LINSKENS and JACKSON 1986a,b, 1987). The biochemistry and physiology of gibberellins were comprehensively summed up into two volumes edited by CROZIER (1983) and a further gibberellin book was published by MUROMZEV and AGNISTIKOVA (1984). BOPP (1986) edited the Proceedings of the 12th International Conference on Plant Growth Substances. WAREING (1984) gave an interesting survey of the hormonal control of developmental processes. Data on the effectiveness of plant hormones and synthetic plant growth substances practically used in plant production were collected by NICKELL (1983).

2. Gibberellins

a) Chemistry

Chemical synthetic work on gibberellins comprises: (1) total syntheses, (2) partial syntheses of less accessible gibberellins or novel gibberellins as authentic markers, (3) preparation of gibberellin analogues for structure-activity investigation and (4) syntheses of isotopically labelled gibberellins as internal standards in quantitative analyses and for metabolic studies. Total syntheses of GA_3 (DANHEISER 1984; HOOK et al. 1984) are only of theoretical interest. Most of the C_{19}-GAs can now be prepared chemically from GA_3 (TAKAHASHI and YAMAGUCHI 1983; MACMILLAN 1985a). GA_{13} and GA_{36} which are present in relatively high concentrations in the mother liquors after crystallization of GA_3 from industrial fermentation are valuable starting materials for the syntheses of C_{20}-GAs, like GA_{15}, GA_{25}, GA_{37}, GA_{43} and GA_{46} (TAKAHASHI and YAMAGUCHI 1983). In addition to previously published methods, three procedures of general interest have been developed recently: (1) Deoxygenation of 3- and 13-hydroxyl groups by radical reduction of oxalyl esters (DOLAN and MACMILLAN 1985a), giving GA_4, GA_7, GA_9 and GA_{20} starting from GA_3. (2) Catalytic hydrogenolysis of GA_3 or GA_7 followed

Progress in Botany, Vol. 49

by either iodolactonization and then reductive removal of the iodine resulting in high yields of GA_1 and GA_4 respectively, or lactonization by peracids to give 1β,3β-dihydroxy GAs such as GA_{54} and GA_{55} (MACMILLAN 1985a). (3) Use of SeO_2 and t-butylperoxide as oxidative agents yielded 15β-hydroxy GAs, which, when transformed according to the first method, provided GA_{45}, GA_{63}, GA_{67}, GA_{68} and GA_{72} from GA_3 (MACMILLAN 1985a; DOLAN et al. 1985; DOLAN and MACMILLAN 1985b).

VOIGT and ADAM (1983) synthesized GA_{55}, GA_{57} and 1-hydroxylated derivatives of GA_5 and GA_{20} by means of photochemical reactions. Also the synthesis of a C_{20}-GA (GA_{19}) from a C_{19}-GA (GA_3) has been realized in a stereocontrolled manner (DAWE et al. 1985). For studies of structure-activity relationships the preparation of 1- and 2-alkylated derivatives of GA_1 and GA_4 was important (BEALE et al. 1984; MACMILLAN and WILLIS 1985; MACMILLAN and TAYLOR 1985). Several C-16 modified GA_4 derivatives and C-1 modified GA_1 compounds suitable for protein coupling and affinity chromatography have been synthesized (BEALE and MACMILLAN 1985; MACMILLAN 1985a). GA_{53}-7-aldehyde was obtained by a combination of microbial and chemical procedures (DOWN et al. 1983). In order to obtain radiolabelled GAs with high specific activity, $[1\beta,2\beta\text{-}^3H_2]GA_1$ has been prepared from GA_3 using a new method in which the 16,17-double bond is protected as the 16,17-epoxide. Similarly, GA_5 was transformed to $[1\beta,2\beta,3\beta\text{-}^3H_2]GA_{20}$ with a specific activity of 1,1 TBq $mmol^{-1}$ (MACMILLAN 1985a). Also different deuterated GA_1 derivatives were obtained (MACMILLAN and WILLIS 1986), and stereoselective deuteration at C-2 of GAs is possible, too (MACMILLAN and WILLIS 1984). Biochemical synthesis of labelled GA_{12}-7-aldehyde and GA_{12} from mevalonic acid could be achieved using the liquid endosperm of *Cucurbita maxima* (BIRNBERG et al. 1986). The chemical synthesis of gibberellin glucosyl conjugates was comprehensively described by SCHNEIDER (1983) and SCHNEIDER et al. (1984).

Microbial formation of modified GAs was achieved by means of *Gibberella fujikuroi* cultures. The fungus transformed (in the presence of AMO-1618) *ent*-beyer-15-ene to beyergibberellin A_4 and A_9 (FRAGA et al. 1983, 1984) and *ent*-trachylobane to the trachylobagibberellins A_7, A_9, A_{13}, A_{25}, A_{40} and A_{47} (DIAZ et al. 1984). The application of *ent*-12,16-cyclokaurenoic acid to a special mutant, "B1-41a", resulted in the formation of 12,16-cyclo-GA_9 and 12,16-cyclo-GA_{12} (BEALE et al. 1983). The same mutant was used for 13-hydroxylation of 1-methyl-GA_4 (MACMILLAN and WILLIS 1985) and 2,2-dimethyl-GA_4 (BEALE et al. 1984). The formation of 1-methyl GA_7 only from 1β-methyl-GA_4 led to the conclusion that the 1α-H is lost in the dehydrogenation process (MACMILLAN and WILLIS 1985). In the same way, 12α-hydroxylated *ent*-kaur-16-en-19-oic acid yielded 12α-hydroxy derivatives of GA_{12}, GA_{14}, GA_{34} and GA_4 (GASKIN et al. 1984a).

b) Occurrence

α) Isolation, Separation and Determination. The isolation method of GAs based on aqueous methanol extraction followed by solvent partition has changed only little in recent years (CROZIER and DURLEY 1983; HEDDEN 1986; SCHREIBER et al. 1987). Frequently used separation techniques are anion exchange chromatography on DEAE-Sephadex A-25 (CROZIER and DURLEY 1983; SCHNEIDER 1983) and reversed-phase cartridges (HEDDEN 1986). Purification of aqueous methanol extracts from small quantities of plant tissue (<1 g fresh weight) is possible by passing through two C_{18}-reversed-phase Bondapak columns (KOSHIOKA et al. 1983d). Preparative separation of GAs and GA conjugates prior to GC-MS analysis is achieved most commonly by reversed-phase HPLC (KOSHIOKA et al. 1983b). Systematic micromethods for analyzing simultaneously up to four phytohormone groups were established (YAMAGUCHI et al. 1983; RADEMACHER and GRAEBE 1984a). The identification of metabolically formed GA glucoside conjugates was realized by enzymic hydrolysis of GA-O-glucosides and transesterification of GA glucosyl esters with sodium methylate followed by HPLC (TURNBULL et al. 1986).

The only reliable method for unequivocal identification and quantitation of GAs is combined GC-MS (HEDDEN 1986; PARK 1986). The highest stage in this field is reached by capillary gas chromatography-mass spectrometry aided by computer (C-GC-MS-C) (MACMILLAN 1985b). Abbreviated mass spectral data for most of the characterized GAs have been published (CROZIER and DURLEY 1983; HEDDEN 1986).

Immunological methods have been introduced for the detection and quantification of plant hormones (WEILER 1983, 1984, 1986), and their advantages and limitations have been reported (CROZIER et al. 1986; PENGELLY 1986; WANG et al. 1986). Radioimmunoassays with antibodies against GA_1, GA_3 and GA_9 reached a detection limit of 10 - 80 pg GA (ATZORN and WEILER 1983a). An increase in sensitivity was accomplished by a solid-phase enzyme immunoassay with alkaline phosphatase as tracer (ATZORN and WEILER 1983b). By hybridization of spleen cells of Balb/c mice (immunized with GA_{13}-19,20-imide-β-alanine-7-methyl ester bovine serum albumin conjugate) with the myeloma line X63.Ag.8.653 two stable hybridomas were established which produce monoclonal antibodies against GA_4 and related gibberellins (EBERLE et al. 1986a). Recently, a set of monoclonal antibodies were prepared by which different gibberellin epitopes can be specifically recognized without any derivatization (KNOX et al. 1987).

β) Free Gibberellins. Since the last review (Progr. Bot. 45, 136) the number of naturally occurring gibberellins which have been structurally elucidated increased from 62 to 72. The structures can be found in CROZIER (1983), MACMILLAN (1983, 1985a,b) and DATHE (1986). Interestingly, all new GAs are of plant origin.

From immature seeds of *Pyrus communis* and *Malus sylvestris* besides GA_{32} and GA_{45} the 15β-OH-GA_4 (GA_{63}) was isolated and identified (DOLAN et al. 1985) by comparison with authentic synthetic material. Also from embryos and suspensors of *Tropaeolum majus* GA_{63} was identified (PICCIARELLI and ALPI 1987). Immature seeds of *Helianthus annuus* were the source of 15β-OH-GA_{15} (GA_{64}), 15β-OH-GA_{24} (GA_{65}), 15β-OH-GA_{25} (GA_{66}) (HUTCHINSON 1983) and 15β-OH-GA_{20} (GA_{67}) (MACMILLAN 1985b). 15β-OH-GA_7 (GA_{68}) was found in immature seeds of *Malus sylvestris* (MACMILLAN 1985b). Recently, ten GAs have been identified from sporophytes of *Cyathea australis* (YAMANE et al. 1985) including the novel 12β-OH-GA_9 (GA_{69}) 12α-OH-GA_9 (GA_{70}) and 12β-OH-GA_4 (GA_{71}); GA_{69} was first detected in barley grains (GASKIN et al. 1984b; GILMOUR et al. 1984). Additionally, 15β-OH-GA_1 (GA_{72}) was identified in immature seeds of *Helianthus annuus* (MACMILLAN 1985a).

During the last 3 years modern separation and identification techniques have been used in order to analyze the gibberellin status in a number of plants, e.g. immature seeds of *Leucaena leucocephala* (ARIGAYO et al. 1983), *Pharbitis purpurea* (FUJISAWA et al. 1985), *Agrostemma githago* (JONES 1984), seeds of *Ipomoea batatas* and other Convolvulaceae (MATSUO et al. 1984), *Pisum sativum* (GASKIN et al. 1985; SCHWENEN et al. 1985) and *Cucurbita maxima* (BLECHSCHMIDT et al. 1984). Investigation of normal and dwarf-growing *Lycopersicon esculentum* revealed that normal shoots and fruits possessed gibberellin series derived from GA_{12} and GA_{53}, but dwarf mutants contained very low gibberellin amounts (ZEEVAART 1984). Detailed analysis of *Oryza sativa* showed that the major GA in flowering ears is GA_4, whereas in immature seed GA_{34} dominates. The occurrence of 13-deoxygenated GAs in reproductive organs suggest that these GAs may play important roles in the regulation of reproductive growth (KOBAYASHI et al. 1984). The gibberellin spectrum in apical shoot meristems of flowering and vegetative sugarcane plants showed some differences, indicating that metabolism of C_{20}- to C_{19}-GAs may be enhanced

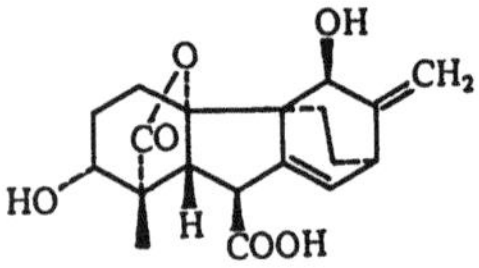

Fig. 1. Structure of antheridic acid

in flowering apices (MOORE et al. 1986). In the weed *Thlaspi arvense* both 13-hydroxylated GAs and 13-non-hydroxylated GAs were found to occur (METZGER and MARDAUS 1986). Several kinds of crown-gall cells and cultured cells derived from non-transformed tissue of *Nicotiana tabacum* contained the same kinds of endogenous gibberellins as the mother plant (PARK et al. 1983).

Pteridophyta also were shown to contain gibberellins; GA_{36} has been identified in sporophytes of *Psilotum nudum* (TAKAHASHI et al. 1984). Structurally related are the antheridiogens (SCHRAUDOLF 1986). The constitution of the antheridium-inducing factor of *Anemia phyllitidis* was elucidated by spectroscopic methods and comparison with synthetic compounds. Antheridiogen An, now renamed as antheridic acid (Fig. 1), contains a rearranged gibban skeleton with a 3α-hydroxyl function. For the parent hydrocarbon the name antheridane was proposed (COREY et al. 1986). The major antheridiogen of *A. mexicana* has been partially characterized as an analogue of antheridic acid, apparently containing only one hydroxyl group (NESTER et al. 1987).

The natural occurrence of GAs in bryophytes, algae and bacteria has not yet been established, but gibberellinlike activity in corresponding extracts has been shown (FURUYA and TAKENO 1983; JACOBS 1986). Gibberellin production by several fungi has been reported. Similar to *Sphaceloma manihoticola*, the culture filtrates of *S. bendentis*, *S. persea* and *S. rhois* contained GA_4 as the major GA (up to 10 mgl^{-1}), and GA_9, GA_{13}, GA_{14} and GA_{24} as minor compounds (RADEMACHER and JUNG 1985). From culture filtrates of *Cercospora rosicola*, which is known to synthesize abscisic acid, GA_3, GA_4, GA_7 and GA_{13} were identified by GC-MS (COOLBAUGH et al. 1985). In *Neurospora crassa* only conidia and mycelia, but not the medium, contained GA_3 (KAWANABE et al. 1983, 1985).

γ) Gibberellin Conjugates. Conjugation is a common, physiologically relevant process in GA metabolism linking covalently GA molecules with other low molecular weight compounds. Among endogenous GA conjugates the O-glucosides and glucosyl esters dominate by far, whilst alkyl ester, acyl derivatives and gibberethione are of limited distribution. The occurrence in plants, chemical synthesis and biochemical formation as well as the biological activity and physiological functions of gibberellin conjugates have been comprehensively summarized (SCHNEIDER 1983; SCHREIBER et al. 1987). SCHNEIDER (1983) described the structurally elucidated native glucosides (2-O-glucosides of GA_8, GA_{26}, GA_{27} and GA_{29}, 3-O-glucosides of GA_1 and GA_3, and 11-O-glucoside of GA_{35}) and glucosyl esters (of GA_1, GA_4, GA_5, GA_9, GA_{37}, GA_{38} and GA_{44}).

Furthermore, a number of structurally not fully identified GA glucosyl conjugates are suggested from the release of free GAs during enzymic hydrolysis of GA conjugate fractions (SCHLIEMANN 1987), e.g. from seeds of *Phaseolus coccineus* (ALBONE et al. 1984), *Sechium edule* (ALBONE et al. 1984; LORENZI and CECCARELLI 1986) and immature grains and seedlings of *Hordeum vulgare* (GASKIN et al. 1984b). Similarly, cellulase treatment of extracts from *Cucurbita maxima* seeds liberated GA_{49}, GA_{58} and 12α-OH-GA_{12}. From the same material the 3-isopentanoyl derivative of GA_{39} was deduced from GC-MS data (BLECHSCHMIDT et al. 1984). MURAKAMI (1985a) checked the distribution of gibberellin glucose conjugates in many plant species.

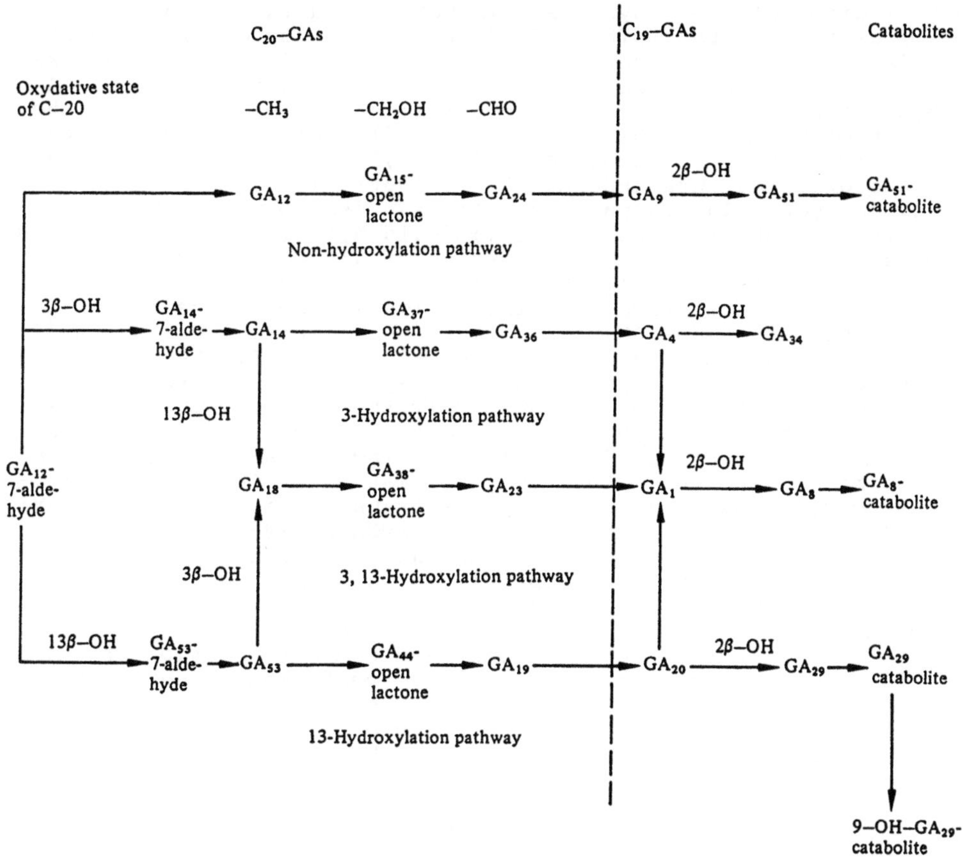

Fig. 2. Simplified pathways of gibberellin biosynthesis in *Gibberella fujikuroi* and in higher plants

c) Biosynthesis and Metabolism

α) Biosynthetic Pathways. Rapid progress was made in gibberellin biosynthesis during the last years, depending mainly on mutant and enzyme studies as well as continued feeding experiments. Recent reviews cover part of these fields (PHINNEY 1984; PHINNEY et al. 1986; GRAEBE 1986).

The early stages in gibberellin biosynthesis up to the GA_{12}-7-aldehyde are rather well known and assumed to proceed likewise in *G. fujikuroi* and higher plants (BEARDER 1983; COOLBAUGH 1983; HANSON 1983; MACMILLAN 1984a). Starting from the common intermediate, GA_{12}-7-aldehyde, four major pathways are operating (Fig. 2) which are connected with one another at different levels. However, only few connecting points are yet clarified. In *G. fujikuroi* only the non-hydroxylation pathway and the early 3-hydroxylation pathway for C_{20}-gibberellins is functioning. On the contrary, in higher plants the early 3-hydroxylation pathway is not known, but the C_{20}-gibberellins are formed on the non-hydroxylation pathway, or the early 13-hydroxylation pathway, or in some cases on the 3,13-hydroxylation pathway (SPONSEL 1983a; DATHE 1986). A further striking difference of the GA metabolism in the fungus and in higher plants is the α-hydroxylation at C-1, C-2, and C-16, which occurs only in the fungus. In higher plants the 2β-hydroxylation of GAs is a typical deactivation process which is followed by glucosylation or catabolite formation (SPONSEL 1983).

The route of gibberellin biosynthesis in maize deduced from detailed analyses of dwarf mutants (PHINNEY 1984) was shown to follow the early 13-hydroxylation pathway (Progr. Bot. 45, 142). The final step, blocked in the d-1 mutant (SPRAY et al. 1984) leads by 3β-hydroxylation of GA_{20} to the physiologically important GA_1 controlling the elongation in the vegetative shoot of maize (PHINNEY et al. 1986). This pathway has been confirmed by feeding of $[^{14}C]GA_{53}$ and $[17-^{13}C,17-^{3}H_2]GA_{20}$ to normal growing maize and dwarf mutants (HEUPEL et al. 1985). In *Pisum sativum* the whole early 13-hydroxylation pathway from GA_{12}-7-aldehyde to GA_{29}-catabolite has been established by $[^{14}C]GA_{12}$-7-aldehyde feeding experiments (DAVIES et al. 1986a) and by identification of all endogenously occurring intermediates (MAKI et al. 1986). The dwarfing genes *le* and *na* in *Pisum sativum* and their relations to gibberellin metabolism have been intensively investigated (INGRAM et al. 1983, 1984, 1985; REID 1983; REID et al. 1983; POTTS and REID 1983; POTTS et al. 1985). Plants carrying the *Le* gene form GA_1, GA_8, GA_8-catabolite, GA_{29} and GA_{29}-catabolite from $[^{13}C,^{3}H]GA_{20}$, whereas in *le* plants only the latter two compounds were found, indicating that the *Le* gene controls the 3β-hydroxylation of GA_{20} to GA_1 (INGRAM et al. 1984). In maturing seeds of *Pisum sativum* cv. Progress No. 9 *(lele-dwarf)* the 2β-hydroxylation of GA_{20} to GA_{29} proceeds in the cotyledons from which GA_{29} is transported to the testa in which the GA_{29}-catabolite formation occurs (SPONSEL 1983). In studies on the biological activity of GA_{20} in dark and red light-grown peas, cv. Progress No. 9, plants responded only in the dark, due to GA_1 formation. The inability of red light-grown Progress No. 9 seedlings to respond to GA_{20} and to hydroxylate GA_{20} in 3β-position to GA_1 is interpreted to be a red light effect on uptake and compartmentation of GA_{20} (SPONSEL 1985, 1986). These results have been confirmed recently (CAMPELL and BONNER 1986) and supplemented with the fact that the red light effects on *lele* plants were largely reversible by far-red irradiation. From this a phytochrome regulation of GA_{20} 3β-hydroxylation was deduced. In the long day plant spinach the non-hydroxylation and the early 13-hydroxylation pathway operate. Results of time course studies indicated that the enzymes oxidizing $GA_{53} \rightarrow GA_{44}$ and $GA_{19} \rightarrow GA_{20}$ are under photoperiodic control, but those oxidizing GA_{44}, GA_{12}, and GA_{15} are not (GILMOUR et al. 1985).

The fate of C-20 in the interconversion of C_{20}-GAs to C_{19}-GAs was newly studied using in vivo ^{13}C-NMR spectroscopy and detecting $^{13}CO_2$ (LEWER and MACMILLAN 1984). The result could be confirmed by enzymic conversion of $[20-^{14}C]GA_{12}$ to GA_{15}, GA_{24}, GA_9, GA_{51} and $^{14}CO_2$ showing that the aldehyde GA_{24} is the ultimate C_{20}-GA (KAMIYA et al. 1986).

β) Enzymic Studies. In vitro investigations contributed much to the progress made in GA biosynthesis. The peculiarities of the cell-free systems from *Marah macrocarpus*, *Cucurbita maxima*, *Pisum sativum*, *Phaseolus coccineus* and *G. fujikuroi*, which are capable of synthesizing GAs from mevalonic acid, have been discussed in detail (HEDDEN 1983; GRAEBE 1984). For the total sequence of the GA biosynthetic pathway from mevalonic acid to a C_{19}-GA about 20 steps are required (Fig. 3).

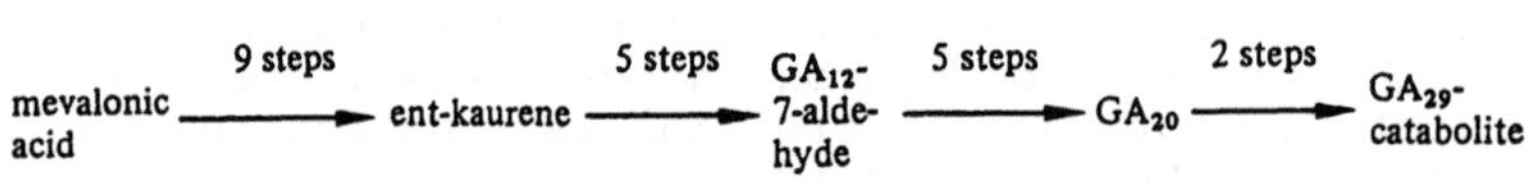

Fig. 3. General pathway of gibberellin biosynthesis

Present evidence suggests that GA_{12}-7-aldehyde is synthesized by a pathway common to all organisms that produce GAs. Lysed chloroplasts from several higher plants synthesized *ent*-kaurene from copalylpyrophosphate, but not from geranyl-geranyl-pyrophosphate. The copalylpyrophosphate transforming activity (so-called B-activity of *ent*-

kaurene synthetase) was relatively stable in plastid lysates from *Pisum sativum*, but not from *Hordeum vulgare*. The bulk of B-activity appeared to reside in the stroma of plastids from *P. sativum*, but required the presence of plastid membranes for maximum activity (RAILTON et al. 1984). In 10-day-old pea plants of both tall and dwarf varieties highest conversion rates of [^{14}C]mevalonic acid to *ent*-kaurene were detected in the youngest internode and leaf (COOLBAUGH 1985; CHUNG and COOLBAUGH 1986). In 4-day-old sunflower seedlings more than 90% of the *ent*-kaurene synthetase activity was found to be localized in cotyledons (SHEN-MILLER and WEST 1985). By storage of these seedlings in liquid nitrogen before homogenization an increase in *ent*-kaurene synthetase activity was observed, obviously caused by partial removal of an inhibitor (SHEN-MILLER and WEST 1984). The cell-free system from liquid endosperm of *Marah oreganus* which converts *ent*-kaurene to *ent*-kaurenol was inhibited by synthetic cytokinin derivatives and dihydrozeation, but not by zeatin (COOLBAUGH 1984).

Present knowledge about the enzymes involved in GA biosynthesis from GA_{12}-7-aldehyde has been summarized by GRAEBE (1986). The reactions from GA_{12}-7-aldehyde on both the non-hydroxylation pathway and the early 13-hydroxylation pathway to GA_{29} (Fig. 2) have been shown in a cell-free system from developing embryos of *Pisum sativum* (KAMIYA and GRAEBE 1983). GA_{12}-7-aldehyde is present in mature pea seeds, too, and becomes converted to GAs already during the earliest stages of germination (GRAEBE 1986). Also by means of a cell-free system from the embryos of immature *Phaseolus vulgaris* seeds nearly all steps of the non-hydroxylation and early 13-hydroxylation pathway have been verified and, after detection of seven additional GAs in immature seeds (GA_9, GA_{12}, GA_{15}, GA_{19}, GA_{23}, GA_{24}, GA_{53}) by gas chromatography – single ion current monitoring mass spectrometry, a tridimensional biosynthesis scheme was established (KAMIYA et al. 1985; TAKAHASHI et al. 1986). Corresponding routes were found using ^{14}C-labelled MVA, *ent*-kaurene and GA_{12}-7-aldehyde with cell-free systems from embryos of *Phaseolus coccineus* seeds (TURNBULL et al. 1985, 1986b).

Dependency upon the pH conditions of the microsomal enzymes from *Cucurbita maxima* endosperm converting GA_{12}-7-aldehyde was studied by HEDDEN et al. (1984). Below pH 6.5 12α-OH-GA_{12}-7-aldehyde and above pH 6.5, GA_{12} are the predominant products. More recently, from the same plant material by isoelectric focussing two soluble enzyme fractions were separated, both belonging to 2-oxoglutarate-dependent dioxygenases (MACMILLAN 1984; VOLGER and GRAEBE 1985). With a pH optimum of 5.0 - 6.0 the conversion of GA_{12}-7-aldehyde to GA_{12} proceeds ten times faster compared to GA_{14}-7-aldehyde. The second fraction (pH optimum 7.0 - 8.0) displayed 3β-hydroxylase activity towards GA_{12}-7-aldehyde, GA_{12}, GA_{53} and GA_{20} (VOLGER and GRAEBE 1985). Whether this early 3β-hydroxylation pathway found in vitro is really operating in vivo has to be confirmed.

Progress was made also in the characterization of enzymes converting C_{19}-GAs. The gibberellin 2β-hydroxylase from *Phaseolus vulgaris* seeds is a 2-oxoglutarate-dependent dioxygenase showing a preference for 3β-hydroxy GAs with relative conversion rates of GA_4 = 100, GA_1 = 70, GA_9 = 4, GA_{20} = 2 (SMITH and MACMILLAN 1984). In dwarf pea seeds a 2β-hydroxylase for 3β-OH-GAs (GA_1, GA_4) (pH optimum: 7.4 - 7.8; K_M: 0.069 µM GA_1) could be distinguished from an enzyme hydroxylating 3-deoxy GAs (GA_{20}, GA_9) (K_M: 1.55 µM GA_{20}). Maximal enzyme activity was observed 22 - 30 days after anthesis and localized in the cotyledons (SMITH and MACMILLAN 1986). A cell-free system from immature seeds of *Phaseolus vulgaris* converted at pH 8.0 GA_{20} simultaneously to GA_1 and GA_5 (7:3 ratio) Neither GA_1 nor GA_{29} served as precursors for GA_5. This enzyme extract showed only weak 2β-hydroxylase activity (KAMIYA et al. 1984).

Industrial research on the development of highly efficient growth retardants acting as inhibitors of gibberellin biosynthesis is continuing. Tetcyclacis

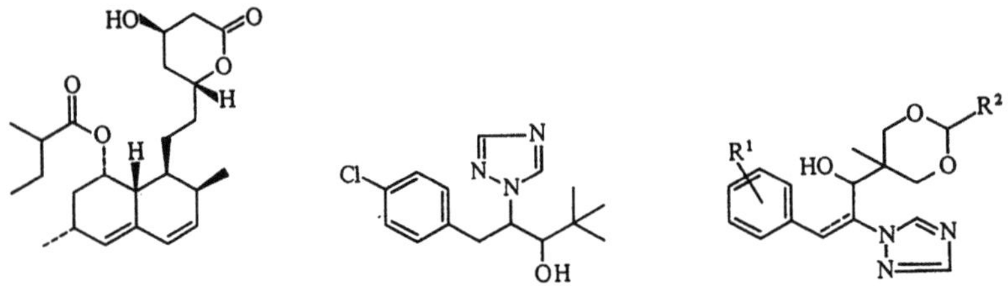

Meviolin (R, S) - Paclobutrazol Dioxanylalk(en)yl triazole type

Fig. 4. Structures of novel growth retardants

(formerly BAS 106...W, Progr. Bot. 45, 139) already introduced to agricultural practice was found to inhibit stem growth and to diminish the endogenous gibberellin formation in *Agrostemma githago* (ZEEVAART 1985). Paclobutrazol inhibited specifically the three steps in the oxidation of *ent*-kaurene to *ent*-kaurenoic acid in a cell-free system from *Cucurbita maxima* endosperm. The K_I values for the separated 2S,3S and 2R,3R enantiomers of paclobutrazol in this system are 2×10^{-8} and 7×10^{-7} M respectively (HEDDEN and GRAEBE 1985). As novel plant growth retardants of high activity (Fig. 4) a series of dioxanylalkyl- and dioxanylalkenyl-triazoles were described and suggested to block the reactions that lead from *ent*-kaurene to *ent*-kaurenoic acid (JUNG et al. 1986).

Some 19-norgibberell-16-enes also act as inhibitors of the oxidative *ent*-kaurene metabolism as shown by a decreased [2-^{14}C]MVA incorporation into GA_3 and GA_{13} in AMO-1618 treated *G. fujikuroi* cultures (BAYNHAM and HANSON 1983). Meviolin, a metabolite from *Aspergillus terreus* cultures, was found to inhibit both shoot growth in *Oryza sativa* and GA_3 formation in *G. fujikuroi* (SHIAO and HAO 1985).

γ) Glucosyl Conjugates: Metabolic Formation and Hydrolysis. Concerning the metabolic formation of gibberellin conjugates after application of radioactively labelled free GAs to plants, it should be pointed out that not all conjugation processes observed necessarily represent physiologically relevant reactions, as many experiments have been performed with readily available rather than endogenous gibberellin substrates and supraoptimal concentrations were applied. In these metabolic studies a great variety of GA conjugates have been detected (SCHNEIDER 1983).

Following the metabolism of [^{3}H]GA_4 in somatic suspension cultures of *Pimpinella anisum* glucosides and glucosyl esters of GA_4 and GA_1 as well as a GA_8 glucose conjugate were detected. [^{3}H]GA_4 glucosyl ester at first increased, and decreased in later stages indicating in vivo hydrolysis (reversible conjugation) (KOSHIOKA et al. 1983a). In carrot cultures the [^{3}H]GA_4 metabolic pattern was similar; additionally, a high polar compound was found and tentatively characterized as [^{3}H]GA_1-3,13-bis-O-glucoside (KOSHIOKA et al. 1983c). Application of [^{3}H]GA_4 to the epicotyl of 7-day-old light-grown *Phaseolus coccineus* seedlings led to the glucoside and glucosyl ester of [^{3}H]GA_4, [^{3}H]GA_8-2-O-glucoside and a [^{3}H]GA_{34}-O-glucoside (TURNBULL et al. 1986a). KOSHIOKA et al. (1985a,b) showed that the formation of GA glucosyl conjugates was similar in developing seeds and cell suspension cultures of *Pharbitis nil* after application of [^{3}H]GA_5. In *Vicia faba* GA glucosyl conjugate-forming capability has been substantiated by the identification of [^{3}H]GA_{20} glucosyl ester and [^{3}H]GA_{20}-13-O-glucoside after application of [^{3}H]GA_{20} to epicotyls of the seedlings (LATTKE and SCHNEIDER 1985). GA_{20}-13-O-glucoside was identified by HPLC and by GC-MS of its permethylated derivative after feeding GA_{20} to derooted seedlings of *Zea mays*. Interestingly, the glucosylation rate of GA_{20} was found to be higher in the dwarf-1 mutant (26%) than in the normal plant (3.6%) (SCHNEIDER et al. 1987). After application of GA_1 and GA_3 to the dwarf-1 mutant, GA_1

yielded the 13-0-glucoside, whereas the non-endogenous GA_3 was converted to its 3-0-glucoside (PHINNEY and SCHNEIDER 1987).

From immature fruits of *Phaseolus coccineus* an enzyme was partially characterized which glucosylates GA_3 (GA_7 and GA_{30} to lower extents) in the presence of UDPG forming the 3-0-glucoside, whereas many other GAs were not accepted (KNÖFEL et al. 1984; SEMBDNER et al. 1985). From cell suspension cultures of *Lycopersicon peruvianum* a cytosolic enzyme was detected which forms glucosyl ester of GA_7 and GA_9 (SEMBDNER et al. 1985; LIEBISCH 1987). Whereas during seed ripening the preferential formation of GA glucose conjugates was found, there are indications that during germination GA conjugates are reconverted into free GAs which may be involved in the promotion of the early stages of seedling growth (SCHNEIDER 1983; ROOD et al. 1983). After feeding of $[^3H]GA_{20}$ to ripening maize cobs, in the seeds during germination, a decrease in the $[^3H]GA$ glucosyl conjugate content and a simultaneous increase of $[^3H]GA_{20}$ and $[^3H]GA_1$ were observed (ROOD et al. 1986; ROOD and PHARIS 1987).

GA glucosyl conjugates are assumed to be biologically inactive per se, and the response of GA conjugates measured in different bioassays is reported to depend on hydrolysis by plant β-glucosidases. This has been confirmed by studies using the dwarf rice bioassay (leaf application) in which GA_3-0-3-glucoside is nearly inactive, whereas the biological activities of GA glucosyl esters and GA_8-2-0-Glc are almost the same as those of the corresponding free GAs. From the etiolated dwarf rice seedlings a β-glucosidase fraction was obtained which catalyzed the hydrolysis of GA_8-2-0-glucoside at a rate 200 times faster than GA_3-3-0-glucoside. Furthermore, this fraction and the major β-glucosidase component showed remarkable GA_3 glucosyl ester hydrolyzing activity (SCHLIEMANN 1984). The GA glucosyl conjugate complex of *Pharbitis nil* seeds was used as a substrate to characterize β-glucosidases from plant and microbial origin. It was pointed out that enzymes from shoots of many species of higher plants are capable of hydrolyzing GA glucosyl esters, but not 3-0-glucosides (MURAKAMI 1985b).

From immature fruits of *Phaseolus coccineus* a soluble β-glucosidase fraction was isolated; and partial purification gave one β-glucosidase specific to the endogenous GA_8-2-0-glucoside accumulating in runner bean fruits at maturity. Concerning its physiological role, this enzyme is reported to catalyze a step important for final GA_8-catabolite formation (SCHLIEMANN 1987).

d) Physiological Roles

Gibberellins are a major hormonal factor in controlling stem elongation. The genetic background has been studied intensively using gibberellin deficient mutants, and remarkable progress was made concerning dwarfing genes in maize and peas (lit. see "Biosynthesis and Metabolism"). It has been shown that GA_1 formation (from GA_{20} is under the control of the maize *dwarf*-1 and pea *le* gene, and GA_1 is the only GA active per se in promoting stem elongation in maize and pea (INGRAM et al. 1986; PHINNEY et al. 1986).

In a gibberellin-insensitive mutant of *Arabidopsis thaliana* the endogenous GAs are not altered compared to the wild type, and the gene is expected to control a step involved in gibberellin action (KOORNEEF et al. 1985). GA-insensitive (Rht 3) wheat showed a decreased $[^3H]GA_1$ metabolism, but in leaves a 12-15 times higher GA level than in the tall control was found (STODDART 1984). On the other hand, the *dwarf*-8 maize (GA-non-responder) contained the same GA level as the normal maize (KATSUMI et al. 1984). Dwarf mutants of cotton (NADZHIMOV et al. 1983) and barley

(HENTRICH et al. 1985) were found to respond sensitively to exogenous GA by stem elongation. Control of plant growth via the GA biosynthetic system has been studied also in apple seedlings (STEFFENS et al. 1985; WANG et al. 1985). Furthermore, the role of gibberellin in stem elongation was analyzed in submerged deep water rice (RASKIN and KENDE 1984; SUGE 1985) and *Thlaspi arvense* (METZGER 1985). Conceptions on the participation of GAs in extension growth were summarized by CROZIER and TURNBULL (1984). Another review deals with plant growth regulators and the regulation of growth under drought (DAVIES et al. 1986b).

The role of gibberellins in reproductive development (flowering and fruiting) has been studied extensively during the last years. A comprehensive review covering the whole field was given by PHARIS and KING (1985), and further reviews deal with the role of gibberellins (and other plant growth regulators) in flower initiation and development of angiosperms (ZEEVAART 1983; VINCE-PRUE 1985; ZIMMERMAN et al. 1985; BERNIER and KINET 1986; CHALAKHYAN 1985; CREVECOEUR et al. 1986) and gymnosperms (DUNBERG and ODEN 1983) as well as in seed and fruit development (BANGERTH 1984, 1986; BULARD and LE PAGE-DEGIVRY 1986).

Among recent investigations a number is directed to induction of flowering in gymnosperms, especially *Picea* species, by application of gibberellins, preferably $GA_{4/7}$ (e.g. MARQUARD and HANOVER 1984, 1985; KRIZO and KORINEKOVA 1985; LONGMAN et al. 1986). Practical techniques for flower and bud formation by GA treatment were developed for *Chamaecyparis obtusa* (KAWAMURA 1985). In apple trees GA_4 promoted flowering, too (LOONEY et al. 1985). Stimulating effects of GA treatment on flower formation in ornamental plants were also described (HENNY 1983; HANKS 1984; PEMBERTON and WILKINS 1985; HALEVY 1986). In poppy GA_3 application enhanced flowering (LEVY et al. 1986) and affected flower differentiation (SLADKY 1986). Investigations on flowering in the classical *Lemna* system were also continued (KANDELER 1984; FUJIOKA et al. 1986). In some studies interactions of applied GAs and stress factors (drought, heat) on flowering are mentioned (MINTER and LORD 1983; PHILIPSON 1983; RAO and MOHAN RAM 1986). Effects of GAs and other plant growth regulators on pollen germination and tube growth have been summarized by MASCARENHAS and CANARY (1985). Additionally, GA_3 was found to affect pollen germination and tube growth in maize (DHINGRA and VARGHESE 1985), sex expression in *Luffa* (GHOSH and BASU 1984) and self-incompatibility of *Raphanus sativus* (MATSUBARA 1984). Male sterility of the tomato mutants "*stamenless*-2" (SAWHNEY 1985) and "*ga*-2" (NESTER and ZEEVAART 1986) could be overcome by GA_3 application. Contrary to the tomato mutants, in pepper and some other species, GA_3 applied at low temperatures can induce male sterility (POLOWICK and SAWHNEY 1985).

Promotion of flowering by gibberellins has been related to effects on photoassimilate partitioning (LEONARD et al. 1983; ROSS et al. 1984). Gibberellins (and other plant growth regulators) affected photosynthate transport also in stems of *Phaseolus vulgaris* (HAYES and PATRIC 1985) and seedling organs of dwarf watermelon (ZACK and LOY 1984). In *Citrus* fruits GA effects on dry matter production and partitioning have been related to abscission (MAUK et al. 1986). The participation of plant hormones in regulation of storage processes during grain filling in cereals was discussed by MICHAEL (1984).

The physiological function of gibberellins in reproductive development cannot only be derived from the effects obtained by exogenous hormone application but has to consider the endogenous hormonal situation, too. A number of publications contribute to the limited progress made in this field. Present knowledge is extensively discussed by DATHE (1987). Correlations between endogenous GA levels and seed development are not yet clear. Seeds in pea fruits cultivated in vitro under the influence of AMO-1618 (inhibitor of GA biosynthesis) reached about 80% of normal size although no endogenous GAs could be detected (GARCIA-MARTINEZ and CARBONELL 1985). The mutant "*ga*-1" of *Arabidopsis thaliana*, deficient in endogenous GAs, is also characterized by normal seed growth (BARENDSE et al. 1986). However, pod growth apparently depends on GAs because both the AMO-1618 treated pea fruits and the *Arabidopsis* mutant showed a remarkable reduction in pericarp. Detailed studies on the role of gibberellins in pod growth and seed development have been done by GARCIA-MARTINEZ et al. (1987), demonstrating that GA_1 is responsive to pericarp growth in normal and parthenocarpic fruits.

The role of gibberellins in seed germination has recently been reviewed and discussed in view of gibberellin-deficient mutants of *Arabidopsis thaliana* (KARSSEN and LACKA 1986). In these mutants environmental control of seed germination occurs through changes both in hormone levels and sensitivity to hormones. The GA requirement of germination depended on the degree of dormancy which was determined by the level of abscisic acid during development. Release from dormancy by dry storage or chilling is accompanied by an increase in GA sensitivity to GA. A further review on plant hormones in seed dormancy and germination, including their modes of action and metabolism, is given by LEWAK (1985).

e) Modes of Action, Mechanisms

Only few studies deal with uptake and translocation of GAs (e.g. SINGH et al. 1984). Evidence was given for translocation of GA_3 and GA-like substances in grafts between normal, *dwarf*-1 and *dwarf*-5 seedlings of maize (KATSUMI et al. 1983). The uptake of GA_1 and GA_3 by suspension-cultured spinach cells has a carrier-mediated component which is not influenced by auxins, ABA and cytokinins (NOUR and RUBERY 1984). The transport of GA_1 in membrane vesicles prepared from hypocotyls of 5-day-old cowpea seedlings was not promoted by Mg-ATP. GA_1 seems to be more permeable than $GA_{\bar{1}}$. Concentrations of GA_1, GA_5 and GA_9 higher than 1 µM led to a collapse of the Mg-ATP generated pH gradient (O'NEILL et al. 1986).

Progress in structure-activity relationships has been reviewed by BEALE et al. (1986). They reported the biological activities of numerous GA_1 and GA_4 derivatives which have substituents at the C-1 and C-2 respectively, using two types of assay, the GA-induced synthesis of α-amylase in aleurone protoplasts of *Avena fatua* and GA-induced stem elongation in dwarf rice and maize. Furthermore, the design of molecular probes was based upon bioactive 17-substituted derivatives; and 17-thiols of varying chain length are further elaborated to promising photoaffinity labels, GA-antigens and affinity columns. HOAD (1983) reviewed former studies on structure-activity relationships using a wide range of bioassays. Further insights in this field were obtained by testing in defined bioassay systems a number of selectively modified GA derivatives, especially C-6 modified GAs (e.g. 6-epi-GA_3, 7-homo-GA_3, 7-nor-GA_3), C-1 substituted GAs (e.g. 1β-chloro-, 1β-bromo- 1β-iodo-, 1β-azido- and 1β-amino-GA_1) and thiogibberellins (e.g. 19 → 10-thiolo-GA_1, 7-thiolo-GA_3) (ADAM and SCHIERHORN 1983; ADAM et al. 1986). SEREBRYAKOV et al. (1984) elaborated a mathematical additive model of structure-activity relationships of gibberellins and confirmed the biological activities of 3-, 13- and 3,13-acetylated GA_3 to be nearly the same as GA_3 and found GA_3-13-0-methyl ether to be most active in all bioassays used.

Gibberellin action is most intensively studied in cereal aleurone cells known to respond to exogenous gibberellins with de novo synthesis and secretion of hydrolytic enzymes, predominantly α-amylase isoenzymes (Progr. Bot. 45, 145). During germination of caryopses the embryo is assumed to provide aleurone cells with gibberellins. This "classical" hormonal regulation conception has been doubted by investigation of the dynamics of GA formation and movement in germinating barley caryopses using immunoassays (ATZORN and WEILER 1983c; ATZORN et al. 1984).

GILMOUR and MACMILLAN (1984) showed that induction of α-amylase in embryoless barley caryopses by GA_1 or GA_4 was not inhibited by the presence of GA biosynthetic inhibitors, like CCC, AMO-1618, ancymidol and paclobutrazol. Further detailed investigations are necessary to understand completely the role of gibberellins in this system. Besides GA levels in the aleurone cells their sensitivity to GAs proved to be decisive for response. Low temperature treatment of genotypically GA_3-insensitive aleurone tissue of wheat produced a significant increase in sensitivity to GA. The low temperature exposure cured or reversed the genetic lesions manifested in the Rht 1-3 genotypes, probably by correcting aberrations in the phospholipid metabolism of the aleurone tissue and, in this way, increasing the hormone receptor sites (SINGH and PALEG 1984a,b,c, 1985a,b).

GA_3-responsive protoplasts, already available from aleurone layers of *Avena fatua* (Progr. Bot. 45, 147) were obtained also from wheat (NORMAN et al. 1983) and barley aleurone cells (JACOBSEN et al. 1985) and have been successfully used in studying GA action.

Present knowledge of GA action in α-amylase synthesis by affecting gene expression is summarized by PARTHIER (1986a,b). It can be concluded that GA controls the synthesis of certain aleurone cell enzymes by enhancing mRNA transcription. Studies using cloned cDNA probes showed the existence of a family of α-amylase genes with at least two active members, both of which are regulated by GA (MUTHUKRISHNAN et al. 1983; ROGERS and MILLIMAN 1984). Abscisic acid counteracts the GA-induced amylase synthesis at the level of mRNA concentration, if assayed by in vitro translation and by RNA blots with the respective cDNA probes (CHANDLER et al. 1984; JACOBSEN and BEACH 1985). Experiments using isolated nuclei from GA-treated aleurone protoplasts gave evidence for a transcriptional control by both hormones (JACOBSEN and BEACH 1985). In addition to transcriptional inhibition, ABA was shown also to inhibit the translation of α-amylase RNA, and this site of action was susceptible to reversal by ethylene (VARTY et al. 1983). By pulse-labelling and DNA-RNA hybridization using cDNA clones containing sequences of RNAs for the high and low isoelectric point (pI) α-amylase, it was established that GA_3 is the principal regulator of α-amylase mRNA accumulation, while CA^{2+} controls high pI α-amylase synthesis at a step after mRNA accumulation and processing (DEIKMAN and JONES 1985, 1986). From further investigations it was concluded that Ca^{2+} influences α-amylase secretion from the aleurone cells (CARBONELL and JONES 1985).

Acid phosphatase isoenzymes are also under the control of GA (Progr. Bot. 45, 148). They have been further studied using either barley endosperm (JONES and CARBONELL 1984; AKIYAMA and SUZUKI 1985) or isolated aleurone protoplasts of *Avena fatua* (HOOLEY 1984a,b) and a differential effect of GA_3 on enhancement and release of the enzyme was found. Carboxypeptidase and protease activities of GA_3-treated barley aleurone layers were investigated using the substrates N-carbobenzoxy-L-Ala-L-Phe and hemoglobin (HAMMERTON and HO 1986). The carboxypeptidase activity develops without hormone application, while the release of this enzyme is enhanced by GA_3. In contrast, GA_3 is required for both the production of protease activity in the aleurone layer and its secretion. GA_3 was shown to increase levels of translatable mRNA encoding $(1 \rightarrow 3, 1 \rightarrow 4)$-β-D-glucanase in barley aleurone, while ABA treatment suppressed the relative abundance of the translatable mRNA (MUNDY and FINCHER 1986). GA_3 also induced release of ferulic acid esters from barley aleurone (GUBLER and ASHFORD 1985).

Mechanisms of GA actions are less intensively studied in systems other than cereal aleurone cells. Modulation of amylolytic activity by GA was also found in a hepatic, *Marchantia polymorpha* (LOOMIS and MARAVOLO 1985). In germinating seeds of *Ricinus communis* the effects of exogenous GA_3 both on the levels of transcripts (MARTIN et al. 1984) and on polysome formation and translation of mRNA have been studied (MARTIN and NORTHCOTE 1983). A number of publications deal with GA_3 action on growth processes in pea. Increase of RNA polymerase activity and stimulation of in vitro RNA synthesis by DNA loosely bound to proteins were found in pea buds (TOMI et al. 1983a,b). In pea epicotyls, growth and DNA synthesis were inhibited by fungicidal inhibitors of ergosterol biosynthesis, and the inhibition could be reversed by GA_3 (BUYSSE et al. 1984). Evidence for the selective replication of dwarf pea DNA under the influence of GA_3 was given by WASILEWSKA et al. (1984). In wheat seedlings GA_3 induced changes in chromatin properties (NOSKOV et al. 1985). In isolated *Zinnia* mesophyll cells GA_3 caused inhibition of cell division and DNA synthesis (IWASAKI et al. 1986). Interactions of GA and light in regulation of phenylalanine ammonia-lyase activity and growth were studied in lettuce (BARNES and JONES 1984). Influences of phytochrome on the amount of GA-like substances were shown in barley leaves and isolated etioplast-enriched suspensions (HILTON 1983, 1984), while the effect of phytochrome on uptake and efflux of GAs was measured in peas (MERTZ 1985).

Effects of GA on microtubule arrangement and stabilization have been detected in onion leaf sheaths (MITA and SHIBAOKA 1984) and the mesocotyl epidermal cells of *dwarf*-5 maize (MITA and KATSUMI 1986).

In gibberellin receptor research, since the previous progress report (Progr. Bot. 45, 144) only few publications appeared, dealing with binding properties of tritium-labelled GA_1 and GA_4 to cytosolic fractions prepared from maize leaf sheaths and cucumber hypocotyls respectively (KEITH et al. 1984; YALPANI and SRIVASTAVA 1985). The evident knowledge on this field has been summarized with competency by STODDART (1986).

Cell elongation in angiosperm shoots is controlled by auxins and GAs. The biophysical and biochemical mechanisms that underlie these auxin and GA responses have been studied intensively during the past decade. An excellent summary of progress made in this field is given by JONES (1986). He stated that the biophysical basis of the growth response to both hormone types is similar, if not identical, and the biochemical processes that lead to changes in cell wall properties may also be similar. Some further surveys of the molecular basis of plant hormone action are given, e.g. by HALL (1984), SACHAR and BERRY (1984) and SEN (1985).

3. Cytokinins

a) Synthesis and Analytical Methods

Chemical syntheses of cytokinins have been done mainly with the aim to follow special structure-activity correlations, like in 6-alkinylpurines (NISHIKAWA et al. 1985, 1986), pyrrolo-[3,2-d]pyrimidines (GREGSON and SHOW 1985) and 9-deazacytokinins of the 7-chloro-imidazo-[1,2-c]pyrimidine type (BRANCA et al. 1987) as well as thiozeatin and related compounds (GREGSON et al. 1985). The urea derivative, thidiazuron, used as a defoliant in cotton, possesses different cytokininlike activities (MOK and MOK 1985; THOMAS and KATTERMAN 1986; YIP and YANG 1986). The synthesized 9-α-anomer of N^6-benzyladenosine showed much lower activity than the β-anomer (HASHIZUME et al. 1985a), formerly found to occur naturally (ERNST et al. 1983a,b). R-(+)- and S-(-)-dihydrozeatin became available by an improved synthesis and resolution of both isomers (CORSE et al. 1983). Some radioactively labelled cytokinins have been synthesized, like *trans*-[8-^{14}C]zeatin (KADIR et al. 1984) and [^{3}H]*trans*- zeatin riboside of high specific activity (LALOUE and FOX 1984).

Gas chromatography (GC) and mass spectrometry (MS) have been further developed and applied to successful cytokinin analysis and combined GC-MS was proved to be the most powerful technique for both qualitative and quantitative analysis of cytokinins and their metabolites. The excellent, comprehensive review given by PALNI et al. (1986) covers all important aspects, like GC and MS instrumentation, derivatization, internal standards, etc. as well as literature data on the mass spectra of 38 native and synthetic cytokinins. Nucleotide identification and quantification requires special care and has been done by direct chemical ionization mass spectrometry (SUMMONS et al. 1983).

Extraction and purification procedures are critical and have to be adapted to the special problem (SCOTT and HORGAN 1984; HASHIZUME et al. 1985c; TAY et al. 1986). High performance liquid chromatography (HPLC) is efficiently and commonly used in purification and separation of cytokinins. Reversed-phase HPLC has been introduced to many procedures for identification and quantitation of cytokinins and their metabolites (ERNST et al. 1983b; ERNSTSEN and JENSEN 1985). By sequential combination of HPLC on ion exchange, gel permeation and reversed-phase columns, a wide range of cytokinin derivatives can be separated (YAMAGUCHI et al. 1983). Enzymes involved in cytokinin metabolism have been assayed by measuring either the formation of the metabolites or the loss of substrate by means of reversed-phase HPLC (CHISM et al. 1984). Combination of HPLC separation and radioimmunoassay detection (RIA) gave further promising results with respect to the quantitation of cytokinins in plant material (BADENOCH-JONES et al. 1984a). Quantitative determination by RIA and GC-MS gave corresponding results but bioassay afforded much lower values (ERNST et al. 1983a). Further progress became available by combination of RIA with GC-MS (ERNST 1986). The available cytokinin RIAs using polyclonal antibodies to isopentenyladenosine and *trans*-zeatin riboside have been reviewed by WEILER (1984). More recently, an RIA for dihydrozeatin and its riboside (HOFMAN et al. 1985) as well as enzyme immunoassays for *trans*-zeatin and its riboside have been described (HANSEN et al. 1984; BARTHE and STEWART 1985). Further progress in sensitivity (femtomol level), specificity and precision is reached by new cytokinin immunoassays using monoclonal antibodies (TRIONE et al. 1985; EBERLE et al. 1986b; WEILER et al. 1986). Purification of polyclonal antibodies to isopentenyladenosine was achieved by affinity chromatography (SENAPATHY et al. 1985), whereas immunoaffinity chromatography was used for rapid and efficient cytokinin separation in conjunction with a combined HPLC-RIA detection (MACDONALD and MORRIS 1985). A promising attempt was made in order to localize cytokinins in root tips by means of immunocytochemistry using fluorescing antibodies to dihydrozeatin riboside (ZAVALA and BRANDON 1983).

b) Biosynthesis and Metabolism

High research intensity in cytokinin biosynthesis and metabolism is documented by the number of reviews (LETHAM and PALNI 1983; MCGAW et al. 1984; WANG et al. 1984; HORGAN 1986; MORRIS 1986; MORRIS et al. 1986; SCHRÖDER et al. 1986).

Outstanding progress was made by the detection and analysis of cytokinin genes in Ti plasmids of the crown-gall pathogen *Agrobacterium tumefaciens*. One gene, tmr, is localized in the T-DNA region (GOLDBERG et al. 1984; LICHTENSTEIN et al. 1984) which is transferred from the bacterium into plant cells (SCHELL et al. 1984). It is responsible for elevated cytokinin content in tumors (AKIYOSHI et al. 1983; JOOS et al. 1983) and, together with two other genes encoding enzymes of an indole-3-acetic acid pathway (SCHRÖDER et al. 1984; INZE et al. 1984), for hormone-independent growth of crown-gall cells (NESTER et al. 1984).

The T-DNA cytokinin gene encodes an isopentenyltransferase (Δ^2-isopentenylpyrophosphate: 5'-AMP-Δ^2-isopentenyl transferase) which attached the isopentenyl side chain to the exocyclic amino group of 5'AMP to give isopentenyladenosine-5'-phosphate (AKIYOSHI et al. 1984; BUCHMANN et al. 1985). A second cytokinin gene, tzs, localized in the vir-region, is expressed primarily in the bacterium (AKIYOSHI et al. 1985). Tzs, exhibiting some sequence homology to tmr, also encodes an isopentenyltransferase. The vir-region-encoded hormone formation might be a factor in the virulence of *Agrobacterium tumefaciens* (MORRIS et al. 1986). Both the tmr and tzs gene have been cloned and transferred to *Escherichia coli*, where they both express isopentenyltransferases and cause zeatin secretion into the medium (MORRIS et al. 1986). Zeatin production in *E. coli* is thought to be due to the presence of a hydroxylase not originating from *Agrobacterium*, which hydroxylates the 4"-methyl group of the isoprenoid side chain to give the *trans*- zeatin series of cytokinins (ribosylzeatin-5'-phosphate, ribosylzeatin, zeatin).

Another plant pathogenic bacterium, *Pseudomonas syringae pv. savastanoi*, also causing plant galls (olive knot disease) has been found to possess cytokinin biosynthesis genes comparable to those of *Agrobacterium tumefaciens* both in structure and function. Comparison of the nucleotide sequences of the coding regions of the *Pseudomonas* and *Agrobacterium* genes showed 50% homology and conservation of the first and second bases (AKIYOSHI et al. 1985; POWELL et al. 1986). The cytokinins secreted have been identified to be isopentenyladenin, its riboside and the novel 1"-methyl-*trans*-ribosylzeatin (SURICO et al. 1985; MACDONALD et al. 1986). Other strains were found to secret cytokinins of the zeatin series only (e.g. SURI and MANDAHAR 1984; MÜLLER and HILGENBERG 1986). Further details on genes specifying cytokinin biosynthesis in *Agrobacterium*, *Pseudomonas* and other phytopathogenes, together with literature data covering surrounding fields, are given by MORRIS (1986) and SCHRÖDER et al. (1986).

Contrary to the outstanding results concerning cytokinin biosynthesis in bacteria there is much less progress in higher plants and the questions of sites and control of cytokinin biosynthesis remain to a large extent unanswered. Indeed, the pathway by which plants produce cytokinins under normal conditions is not fully clear, although it is difficult to imagine a cytokinin biosynthetic pathway that differs from the bacterial route. Present knowledge has been summarized by LETHAM and PALNI (1983), HORGAN (1986) and MORRIS (1986) who gave a general scheme of "direct" biogenesis of cytokinins and their metabolic transformations.

Enzymatic hydroxylation of the isoprenoid side chain has been further studied and confirmed both in vitro (CHEN and LEISNER 1984) and in vivo (PALNI and HORGAN 1983; EINSET 1984), but it is not yet clear whether it occurs at the nucleotide, nucleoside or the free base level. Attempts to localize cytokinin biosynthesis showed in pea plants and carrot roots [8-^{14}C]adenosine incorporation to occur only in the cambium and not in surrounding tissue (CHEN et al. 1985).

In excised maize roots no incorporation of [8-^{14}C]labelled adenosine and adenine could be observed, and the deficiency of a precursor coming from the shoot was postulated (VAN STADEN and FORSYTH 1984). Although there is increasing evidence for de novo biosynthesis, it has not yet proved possible to completely rule out the involvement of tRNA degradation in cytokinin formation. Continued studies by KLÄMBT et al. (1984) indicate cytokinin production by tRNA digestion both in *Lactobacillus acidophilus* and *Phaseolus vulgaris*. In beans tRNA should account for about 50% of the cytokinin formation; further sources are expected to be mRNA, their polyA sequences and degradation products. Re-examination of *cis*-zeatin and *cis*-zeatin riboside formerly reported to occur as free cytokinins in normal (LETHAM and PALNI 1983) and transformed plant cells (GUERIN et al.

1984) failed to find any evidence for them in normal tobacco shoots; thus, they appear to occur only in tRNA (TAY et al. 1986). Formerly, remarkable amounts of the *cis*-isomer exceeding those of the *trans*-forms, had been found in leaves and shoots of alfalfa (HASHIZUME et al. 1985b) and rice (TAKAGI et al. 1985). In tissue cultures of *Nicotiana tabacum* requiring exogenous cytokinins the genes capable of cytokinin biosynthesis are presumably inactive, but habituation to a cytokinin autotrophy can occur. Studies on the epigenetic and genetic changes associated with habituation gave important insights (MEINS et al. 1983; MEINS and FORSTER 1985; MEINS and HANSEN 1986).

Accumulated data on cytokinin metabolism have been comprehensively reviewed and discussed by LETHAM and PALNI (1983), MCGAW et al. (1984) and HORGAN (1986). Exogenously applied cytokinins, and in all probability also the endogenous cytokinins, undergo different metabolic reactions which may be classified as (1) interconversions of bases, nucleosides and nucleotides; (2) irreversible side-chain modification (e.g. hydration, hydroxylation); (3) oxidative side-chain cleavage (catabolism); (4) conjugation with sugars or amino acids; and (5) possible reactivation of conjugates by hydrolysis.

Continued studies on the fate of exogenously applied radioactively labelled cytokinins and cytokinin metabolites in either intact plants, detached organs or cultured cells made some progress in the following selected areas. Although the biologically active zeatin (riboside) and isopentenyladenine (adenosine) are metabolized by all systems, plant species differ in their types and kinetics of metabolite formation. In "glucosidase-type" plants *(Raphanus sativus)* only conjugation to N-glucosides (mainly) and O-glucosides (minor) but no oxidative cleavage of the N^6-side chain takes place, whereas in other species ("oxidase-type") preferably oxidation products and less conjugates are formed. *Phaseolus vulgaris*, like *Zea mays* and *Vinca rosea* crown galls, is characterized by O-glucosylation and lack of N-glucosides (MCGAW and HORGAN 1985).

Differences were found in different organs of *Phaseolus* species with high conformity in embryos (LEE et al. 1985). Nodulated pea roots *(Rhizobium)* fed with zeatin riboside contained a number of known metabolites including oxidation products, O-glucosides and nucleotides (BADENOCH-JONES et al. 1984b). Long-term feeding of [^{3}H]zeatin riboside to the second internode of lupine plants gave intensive metabolization (to known metabolites) and pronounced accumulation in expanding leaves but not in developing seeds (JAMESON et al. 1985a). Also in soybean leaves zeatin riboside was rapidly metabolized to a diversity of known metabolites (NOODEN and LETHAM 1986). In soybean callus long-term metabolism of zeatin led to the degradation product N-(purin-6-yl)glycine (VAN STADEN 1983). In drought-stressed sunflower plants the level of endogenous cytokinins decreased and a zeatin glucosidelike peak appeared indicating an effect of water stress on cytokinin metabolism (HUBICK et al. 1986).

Benzyladenine applied for the initiation of flower bud formation to thin-layer tissue cultures of tobacco is extensively metabolized inside the tissue to its 7-glucoside and presumably the 9-ribosylbenzyladenine mono-, di- and triphosphates (BARENDSE et al. 1987; VAN DER KRIEKEN et al. 1987). In excised embryos of *Picea abies* applied [8-^{14}C]benzyladenine was transformed mainly to nucleotides (VAN STADEN et al. 1986). Benzyladenine can also be metabolized to β-(benzylaminopurin-9-yl)alanine, a conjugate which is analogous to the native lupinic acid (MURAKOSHI et al. 1983; ZHANG et al. 1987). In *Gerbera* shoot cultures [8-^{14}C]-benzyladenine was found to be metabolized to a novel conjugate, tentatively identified as a 9-glucosyl-riboside (HORGAN 1986).

After application of [8-^{14}C]zeatin to embryos and other tissues of *Phaseolus vulgaris*, besides its riboside, ribotide and O-glucosyl derivative, two novel metabolites have been found and characterized as Opentosyl derivatives of zeatin and its riboside respectively (LEE et al. 1985; TURNER et al. 1985). The donor specificity of the conjugate-forming enzyme isolated from cytosol indicates the zeatin derivative to be the O-xylosyl conjugate (TURNER et al. 1986). A cytokinin-specific kinase catalyzing cytokinin nucleotide formation has been isolated and purified from *Phaseolus* callus (MUJER et al. 1986). During studies on cytokinin oxidase in maize kernels new hydroxylated metabolites have been found and identified to be 2-hydroxy-zeatin and 6-(2,3,4,-trihydroxy-3-methylbutylamino)purine (MCGAW and HORGAN 1983a,b). The 6-(Δ^2-isopentenyl-imino)-purine riboside has been isolated as an intermediate in the degradation of isopentenyladenosine to adenosine by a cytokinin oxidase from wheat (LALOUE and FOX 1985).

The studies on the regulation of cytokinin metabolizing enzymes are just at the beginning. Cytokinin oxidase activity in bean callus tissue was shown to be regulated by mechanisms sensitive to cytokinin supply (CHATFIELD and ARMSTRONG 1986). Of high physiological interest is the auxin-induced inhibition of a conjugate-forming enzyme, β-(cytokinin-9-yl)alanine synthetase, already mentioned in this series (Progr. Bot. 45, 155) and which is being further studied. Potent inhibitors of the enzyme are 2,4-D, 5,7-dichloro-IAA and the synthetic cytokinin N-benzyl-N'-phenyl urea (PARKER et al. 1986). In leaves of intact soybean plants 1,7-dimethyl-3-(hex-5-en-1-yl)xanthine suppressed benzyladenine metabolism to the 9-alanine conjugate and in this way enhanced senescence-retarding activity of this cytokinin. With excised leaf discs also 2,4-D strongly inhibited conjugate formation (ZHANG et al. 1987). Another report on the effects of auxins on cytokinin metabolism showed that in artichoke tuber tissue rapid metabolism of zeatin nucleotide to adenine was induced by IAA (PALMER et al. 1984).

Promising in discovering the physiological role of cytokinin metabolism should be the use of suitable mutants. Cytokinin-deficient mutants are not known. However, moss mutants producing higher amounts of endogenous cytokinins (isopentenyl adenine and zeatin) than the wild type have been characterized and used in studies on cytokinin biosynthesis and metabolism (WANG et al. 1984). Hints on the physiological significance of cytokinin metabolism can also be derived from biological activities of the metabolites and relations between metabolic routes and physiological processes. Whether cytokinin bases and/or 9-ribosides are the active form is not yet clear. The 9-ribosides being major cytokinins in xylem and phloem sap are thought to function as translocation forms (LETHAM and PALNI 1983). Nucleotides have often been ignored by workers, because of analytical difficulties (easily hydrolyzed, etc.). However, in xylem exudates of rice (MUROFUSHI et al. 1983) and bean plants (PALMER and WANG 1985) nucleotides of zeatin and dihydrozeatin have been identified as the dominating cytokinin types and interpreted as favoured translocation forms. Other types of conjugates, the 7- and 9-N-glucosides and the 9-alanyl derivatives, are in general biologically inactive (LETHAM et al. 1983, 1984) and remain stable over a long period; they are, therefore, considered to be deactivation or detoxification products (MCGAW and HORGAN 1985). Benzyladenine metabolism studied in relation to leaf senescence in oat leaf segments showed the formation of the 3-β-D-glucopyranosyl conjugate being very active in senescence retardation; all zeatin metabolites, however, formed in this system (side-chain-cleaved compounds and 9-glucoside) were inactive (TAO et al. 1984). The O-glucosides are very active in bioassay (LETHAM et al. 1983) and probably possess a storage role (MCGAW et al. 1985). O-Glucosides apparently are not biologically active per se but can be reactivated by hydrolysis (MCGAW and HORGAN 1985). Oxidative side-chain cleavage gives irreversible deactivation (HORGAN 1986).

c) Occurrence and Physiological Roles

Naturally occurring cytokinins that are structurally elucidated have been reviewed by MCGAW et al. (1984) and PALNI et al. (1986). Some further data are given by PALNI et al. (1983) and SCOTT and HORGAN (1984). Only few novel types have been isolated. These are 1'-hydroxymethyl zeatin isolated from developing wheat caryopses (RADEMACHER and GRAEBE 1984b) as well as 1'-methyl zeatin and its 9-riboside, both isolated from *Pseudomonas syringae pv. savastanoi* (SURICO et al. 1985; EVIDENTE et al. 1986). The latter two compounds also were synthesized and their configurations determined (ITAYA et al. 1986). The 6-benzylamino-purine-9-β-D-riboside was found to occur in anise cell cultures (ERNST et al. 1983). Some novel cytokinin conjugates were already mentioned (Sect. b).

Numerous publications describe the identification and, in part, quantitation of known cytokinins in plant materials. Detailed studies have been done on the cytokinins produced in crown-gall tissues. In *Datura* crown galls 16 cytokinins have been identified (PALNI et al. 1983, 1985) and in *Nicotiana* crown galls 11 cytokinins were found (SCOTT and HORGAN 1984). In crown-gall tissue of *Vinca rosea* the dynamics of cytokinin formation and their accumulation in the culture medium have been followed (PALNI 1984). The endogenous cytokinin spectrum of potato tubers has been identified and quantified by HASHIZUME et al. (1985c). Besides typical plant cytokinins, they detected a novel conjugate of isopentenyladenine, 9-β-D-glucopyranosyl-6-(3-methyl-butenylamino)-purine, and an extremely high amount of *cis*-zeatin riboside which is discussed to be functionally related to the tuberization process. JAMESON et al. (1985b) found an increase in cytokinins (not identified) subsequent to potato tuber initiation, indicating that the tuberization stimulus is unlikely to be cytokininlike in nature. According to combined RIA-HPLC determination zeatin riboside predominates during potato tuber growth and dramatically decreases after harvest. During storage the tubers contained a constant low level but in the buds cytokinin content increased 20-40-fold with the natural break of dormancy (TURNBULL and HANKE 1985a). Application experiments showed a related seasonal pattern of changing sensitivity to cytokinins (TURNBULL and HANKE 1985b). Cytokinins, therefore, are considered as primary factors in the switch from innate dormancy to non-dormancy, but do not control subsequent sprout growth (possibly gibberellins).

The role of cytokinins in flower initiation and development has been further confirmed but not elucidated in detail. Root exudates of short-day induced *Sinapis alba* were much higher in cytokinin content than those of long-day plants (BERNIER and KINET 1986). Flowering in short-day duckweed *(Lemna paucicostata)* was promoted by exogenous 6-benzyladenine more effectively than by zeatin and its riboside (KHURANA and MAHESWARI 1983). In developing flowers of *Cosmos sulphureus* the endogenous cytokinin contents were lower at earlier stages during blooming (SAHA et al. 1985). Cytokinins are also associated with the development of sexually differentiated tissues (DURAND and DURAND 1984).

In *Mercurialis annua* sexual differentiation is controlled by sex-determinating genes and by cytokinins (female) and auxins (male). Feminization of genetic males is possible by cytokinins which induce the mRNA population of female type (DELAIGUE et al. 1984, 1986). In spring sap of male and female plants of *Gymnocladus dioica* identical cytokinins occurred but in different amounts. The sap of females was much higher in zeatin riboside content (HAUTALA et al. 1986).

A classical cytokinin effect is retardation of leaf senescence. Highly active in senescence promotion is methyl jasmonate (Progr. Bot. 47, 148). Counteractions of this native plant growth regulator and cytokinins in triggering senescence processes, like degradation of chlorophyll and ribulose-1,5-bisphosphate carboxylase, have been studied using barley leaf segments (WEIDHASE et al. 1987). In non-senescing cotton leaves cytokinin was found to stimulate ethylene production by increasing both the formation and oxidation of ACC (SUTTLE 1986). The physiological significance of this effect, however, is not clear. Some progress is made concerning the role of cytokinins in senescence control at the whole plant level. Cytokinin control of monocarpic senescence in soybean has been studied and reviewed by NOODEN and LETHAM (1986). The phenomenon is very complex and not yet fully understood; a striking feature is the exchange of regulatory signals between different parts of the plant. The most advanced pods exert "senescence-inducing" signals, the identity of which is unknown; at least some are probably hormonelike (NOODEN 1984). Consequently, mineral and cytokinin fluxes from the roots are being reduced. The diminished cytokinins coming from the root system do not enter the embryo, and the seeds do not compete with the leaves for this supply of cytokinin (NOODEN and LETHAM 1984). Exogenous cytokinin supplied via the xylem of pod-bearing soybean explants delayed but could not prevent foliar yellowing except at concentrations exceeding physiological level (NEUMANN et al. 1983; GARRISON et al. 1984). The role in plant senescence of the endogenous cytokinins in vegetative shoots has been studied by DAVIES et al. (1986b) using G2 peas which become senescent only under long-day conditions. Long-day and short-day grown mature shoots showed no differences in endogenous cytokinins which thus cannot be responsible for apical senescence. Whole plant senescence is initiated by senescence and fading of the apical bud.

The role of cytokinins in stomatal mechanisms has been further studied with regard to cytokinin - ABA interactions (BLACKMAN and DAVIES 1984a) and age-related changes in stomatal responses (BLACKMAN and DAVIES 1984b) as well as dependence upon light conditions (GÖRING et al. 1984). Besides ethylene and ABA, cytokinins may also be involved in plant responses to environmental stress though only little is known at present (MEYER et al. 1987). Changes in endogenous cytokinin levels were followed in water-deficient flax (SEBANEK et al. 1983) and heat-stressed maize seedlings (CAERS et al. 1985). Applied cytokinins might reduce stress injuries (BALASIMHA 1983) and increase stress resistance (TITOV et al. 1986). In *Eucalyptus* plants, infection with *Phytophthora cinnamomi* reduced the cytokinin content in xylem exudate (CAHILL et al. 1986).

In tissues cultures of *Nicotiana tabacum* the endogenous cytokinins have been analyzed and related to the cytokinin requirement and differentiation using different lines (HANSEN et al. 1985; KERBAUY et al. 1986). The role of cytokinins in somatic embryogenesis in anise cell cultures was studied by both analyzing the dynamics of enodgenous cytokinins (ERNST et al. 1984) and application experiments (ERNST and OESTERHELT 1984). Exogenous cytokinins are commonly applied to plant tissue cultures in order to affect cell growth and differentiation. Various publications dealing with those aspects do not give new insights into physiologal roles of cytokinins.

Cytokinins are known to occur not only in higher plants but also in lower plants and microorganisms. Bacterial cytokinin formation has already been mentioned (Sect. b). Occurrence and physiological roles of cytokinins in algae have been summarized by MOONEY and VAN STADEN (1986). In seaweed *trans*-zeatin, dihydrozeatin, isopentenyladenine and their ribosides as well as cytokinin glucosides have been identified and quantified (TAY et al. 1985). BOPP et al. (1986) reviewed the endogenous cytokinins and their metabolism in mosses. Cytokinin control of protonema development in mosses (LEHNERT and BOPP 1983; BOPP 1984) was shown to comprehend both induction of branching of unbranched protonemata and bud formation in *Funaria* (BOPP and JACOB 1986). In this species the mechanism of cytokinin action in bud induction was shown to involve activation and redistribution of plasma-membrane ion channels (SAUNDERS 1986). In *Physcomitella*, induction of bud formation could be induced both on chloronemata and caulonemata by isopentenyladenine (RESKO and ABEL 1985). In the liverwort *Riccia*, cytokinins and auxin affected growth and gametangial formation (KUMRA and CHAPRA 1984).

In *Pterdidophyta* sporophytes isopentenyladenine and its riboside but no zeatin-type cytokinins could be detected until now (YAMANE et al. 1983).

d) Modes of Action

Studies on cytokinin-binding sites were continued less intensively than years ago (Progr. Bot. 43, 112; 45, 115) and made no outstanding progress. Cytokinin-binding proteins from wheat embryos, being studied most frequently, have been further characterized by resolution of the subunit composition (BRINEGAR and FOX 1985) and by following their biosynthesis and degradation during embryogenesis and germination (BRINEGAR et al. 1985).

From tobacco callus cells, by means of fluorescent anticytokinin probes, both a cytokinin-binding protein was isolated from the soluble fraction and a cytokinin-binding site was established on ribosomes (HAMAGUCHI et al. 1985). From a microsomal fraction of beans cytokinin-binding proteins were isolated by means of affinity chromatography (KHARTSCHENKO and TERESCHENKO 1985). This technique has also been successfully used for the isolation of cytokinin-binding proteins from barley leaves (KHARTSCHENKO et al. 1983) and purification of a special zeatin-binding fraction (ROMANOV et al. 1986). Indication of the possible requirement of protein binding in cytokinin action was given by SELIVANKINA et al. (1983, 1985) demonstrating that RNA synthesis in vitro and transcription in barley leaves is activated by the cytokinin-protein complex. IWAMURA et al. (1985) tried to get insights into receptor topology by quantitative structure-activity relationships of cytokinin agonistic and antagonistic pyrido-[2,3-d]-pyrimidines.

Cytokinin action in gene expression is best studied in connection with cytokinin-stimulated plastid development to either etioplasts in dark-grown or chloroplasts in light-grown plants, plant organs and tissue cultures. Effects of light and cytokinin are co-active in nature (TONG et al. 1983). Remarkable results obtained during the last years have been summarized by PARTHIER (1986a,b). Excised pumpkin cotyledons, being deficient in endogenous cytokinins, respond sensitively to exogenous cytokinins with organ enlargement and increase in the activities of various cytoplasmic and plastid enzymes, among them RuBCase (LERBS et al. 1984; PARTHIER et al. 1985). The enzyme activity is increased by de novo synthesis. Cell-free translation of total or poly-A^+ RNA from cytokinin-treated cotyledons in a wheat germ system revealed changes in several polypeptides, among them the increased precursors of the RuBCase small subunit and the chloroplast light-harvesting protein. Translation in a cell-free *E. coli* system of chloroplast RNA (poly-A^- RNA) from cytokinin-treated cotyledons produced no changes in the polypeptide pattern, when compared with water-treated samples, but a three-fold increase in total plastid mRNA. This result was confirmed by quantitative dot hybridization with a cloned RuBCase large subunit gene. The authors concluded that cytokinins do not control plastogenesis via selective transcription of plastid genes but rather through activation of nuclear genes that encode organelle polypeptides. Plastid RNA polymerase seems to be involved in cytokinin-affected processes. While cytokinin is the dominating regulator in etiolated tissues, phytochrome predominantly controls gene expression rates in the light (SHARMA 1985). It is not definitely known whether connections exist between hormonal control and the phytochrome signal system.

The cytokinin-modulated gene expression in excised pumpkin cotyledons has been further studied by CHEN and LEISNER (1985). They found, after cytokinin treatment, numerous alterations in the population of total soluble proteins, among them an increase in hydroxypyruvate reductase, and suggested that cytokinin acts at the transcriptional level. Excised watermelon cotyledons respond to exogenous cytokinin similar to pumpkin cotyledons. [^{3}H]Thymidin incorporation into DNA was stimulated by cytokinin preferably in plastids and nuclei (GALLI 1984). Changes in the protein pattern of plastids included remarkable increase in 32-, 44-, 48-, and 49-kD proteins; especially the 32-kD protein seemed to be cytokinin-dependent (LONGO et al. 1986). In cultured tobacco tissue, cytokinin was found to inhibit the accumulation of a 33-kD protein identified to be a β-1,3-glucanase (FELIX and MEINS 1985).

Tobacco cell suspension cultures respond to exogenous cytokinin by full chloroplast differentiation. The hormone stimulates the appearance of a few prominent plastid proteins, among them the light-harvesting chlorophyll a/b-binding protein (PEAUD-LENOEL and AXELOS 1983; AXELOS et al. 1984). Furthermore, it has been clearly shown that cytokinin modulates the expression of genes encoding the protein of the light-harvesting chlorophyll a/b complex. The expression of the gene activity, in response to exogenous cytokinin, is apparently regulated by the level of mRNA encoding precursor polypeptides of the light-harvesting chlorophyll complex rather than by translational or post-translational controls (TEYSSENDIER DE LA SERVE et al. 1985).

In *Lemna gibba* kept in darkness cytokinin increased the amounts of hybridizable RNA coding for both the major chlorophyll a/b-binding protein and for the small subunit of RuBCase, whereas the levels of mRNA encoding the large subunit of RuBCase and the 32-kD herbicide-binding protein, which are both chloroplast-synthesized messages, are not significantly altered. It is suggested that cytokinin regulates the expression of the two genes for nuclear-encoded chloroplast proteins at a post-transcriptional level, by affecting the stability of the RNA (FLORES and TOBIN 1986).

Like in pumpkin cotyledons, in primary leaf segments of maize seedlings exogenous cytokinin stimulated the activities of RuBCase and some other enzymes (e.g. phosphoenolpyruvate carboxylase, $NADP^+$-malic enzyme, pyruvate orthophosphate dikinase). Effects on plastid ultrastructure depended on the developmental stage (CAERS and VENDRIG 1986). In intact bean leaves benzyladenine treatment induced an increase in DNA content per chloroplast (KINOSHITO and TSUJI 1984).

Some publications deal with cytokinin action at levels other than gene expression. A prospective mode of cytokinin action was found by interaction with lipid-associated oxy-free radicals during senescence (LESHEM 1984). The role of Ca^{2+}/calmodulin in senescence control by cytokinin (and other hormones) has been reviewed by LESHEM et al. (1986). In model membrane systems, cytokinins increased erythritol permeability to phosphatidyl bilayers (STILLWELL et al. 1985).

4. Brassinosteroids

The preceding review in this series (Progr. Bot. 45, 148) recorded brassinolide, castasterone and dolicholide as the only members of this group of native plant growth regulators known to occur endogenously in plants. In the meantime remarkable progress has been made with respect to the discovery of a number of further plant brassinosteroids, chemical syntheses of analogues and studies on physiological activities, including both structure-activity relationships and practical application in agriculture as well as intentions on the mode of action. Despite many published results, the question whether the brassinosteroids function endogenously like plant hormones has to be studied further. Reviews have been published by ADAM and MARQUARDT (1986, 1987), IKEKAWA (1986) and YOKOTA and TAKAHASHI (1986).

a) Chemistry, Analytical Methods, Occurrence, Metabolism

To date 17 native brassinosteroids have been isolated from plants and structurally elucidated; and the number of steroids belonging to this family will be likely to increase in the future. Structurally, part of them exhibit the seven-membered lactone B-ring (brassinolide type), but differ in the substitution at C-24 of the side chain, like homobrassinolide, 28-norbrassinolide (ABE et al. 1983; IKEKAWA and TAKATSUTU 1984; IKEKAWA et al. 1984), dolicholide and homodolicholide (YOKOTA et al. 1983a,b, 1984). Others belong to the castasterone type possessing a normal six-membered B-ring with a 6-oxo function, like brassinone, ethylbrassinone (ABE et al. 1983; MORISHITA et al. 1983; IKEKAWA et al. 1984; SUZUKI et al. 1985; YOKOTA et al. 1987), dolichosterone, homodolichosterone (BABA et al. 1983; ABE et al. 1984a; YOKOTA et al. 1984), 24-epicastasterone (YOKOTA and TAKAHASHI 1986; YOKOTA et al. 1987a), 2-deoxocastasterone (typhasterol) and teasterone (SCHNEIDER et al. 1983; ABE et al. 1984b; YOKOTA et al. 1985). Three further brassinosteroids, 6-deoxocastasterone, 6-deoxodolichosterone (ARIMA et al. 1984; YOKOTA et al. 1984) and 6-deoxohomodolichosterone (YOKOTA and TAKAHASHI 1986) lack the oxygen function at the B-ring. Isolation procedures normally starting with aqueous acetone or methanol extraction have to include extensive purification, e.g. by sequential solvent partition, countercurrent distribution and silica gel column chromatography respectively, followed by HPLC on gel permeation, normal phase and reversed-phase columns before GC-MS-SIM analysis of the derivatized brassinosteroids (preferably bismethaneboranates) can be applied (for literature see ADAM and MARQUARDT 1986; HEDDEN 1986). Each purification step has to be monitored by bioassay, or possibly in the future by immunoassay. To date an enzyme immunoassay (ELISA) has been described using monoclonal antibodies towards the synthetic 24-epibrassinolide; it is suitable for the examiniation of brassinolide (HORGAN et al. 1984).

Brassinosteroids are apparently widely distributed in plants; they have been found to occur in a number of species belonging to different families (dicots, monocots, gymnosperms, algae; details and literature see ADAM and MARQUARDT 1986; YOKOTA and TAKAHASHI 1986). Brassinosteroids may be present in all plant parts, although roots have not yet been studied. The amounts determined vary widely (0.1 ng to 200 µg kg^{-1}); highest levels were found in pollen and immature seeds, whereas vegetative tissues are rather low in their contents. It might be of physiological interest that insect galls of *Castanea crenata* and *Distylium racemosum* contain higher amounts of brassinosteroids than the normal leaf tissue (ARIMA et al. 1984; IKEKAWA et al. 1984).

Conceptions on brassinosteroid biogenesis are still speculative and need experimental confirmation by use of radioactively labelled precursors. Metabolic transformations of brassinosteroids taking place in plant tissues might involve conjugation. Thus, tritiated brassinolide and castasterone fed to rice seedlings were rapidly converted into acidic conjugatelike metabolites; and from *Phaseolus vulgaris* a naturally occurring conjugate has been isolated and structurally elucidated to be 23-O-β-glucopyranosyl-25-methyldolichosterone (YOKOTA et al. 1987b).

b) Physiological Effects, Mode of Action, Agricultural Application

Brassinosteroids exhibit different hormonelike effects in various test systems. Among the bioassays formerly reviewed (Progr. Bot. 45, 148) the rice lamina inclination test is widely used and has been further optimized for microquantitative determination of brassinosteroids and applied to the detection of antibrassinosteroids (WADA et al. 1984).

For the bean first internode curvature bioassay a very sensitive modification has been developed (STRNAD and KAMINEK 1985). Similarities in auxin and brassinosteroid effects observed in several bioassays are interpreted as a synergistic interaction depending on enhancement of sensitivity of plant tissue to auxin (endogenous and/or exogenous) by brassinosteroids, which themselves do not have auxinlike functions (KATSUMI 1985). Brassinosteroids also do not affect auxin uptake, translocation and metabolism (COHEN and MEUDT 1983). In auxin-starved suspension cultured carrot cells brassinosteroids induced cell enlargement, but no cell division (SALA and SALA 1985). Some growth-stimulating effects of brassinosteroids and gibberellins are additive (KATSUMI 1985). Studies on the sequential response to growth regulators in wheat coleoptile and dwarf pea epicotyl segments showed that the peak of brassinosteroid sensitivity lies between those of gibberellin (first) and auxin (SASSE 1985). In the wheat leaf-unrolling test, in which cytokinins are moderately active, but auxins are inhibitory, brassinosteroids showed a remarkable activity (WADA et al. 1985). It has been found that brassinosteroids stimulate the auxin-induced ethylene formation (ARTECA et al. 1983), apparently by increasing the ACC synthase activity (SCHLAGNHAUFER et al. 1984; SCHLAGNHAUFER and ARTECA 1985a,b). Ca^{2+} acts synergistically with brassinosteroid and auxin in stimulating ethylene production (ARTECA 1984).

Studies on structure-activity relationships are being continued intensively (results and literature see ADAM and MARQUARDT 1986; IKEKAWA 1986; YOKOTA and TAKAHASHI 1986) and some structural requirements for a high brassinosteroid activity have been postulated (TAKATSUTO and IKEKAWA 1984; TAKATSUTO et al. 1984; ARTECA et al. 1985; CERNY et al. 1986).

Efforts to explore the mode of action of brassinosteroids and their possible physiological role consider different aspects. Membrane effects have been analyzed in relation to brassinosteroid-stimulated growth in bean epicotyl segments (CERANA et al. 1983a) and maize root segments (CERANA et al. 1983b, 1984, 1985; ROMANI et al. 1983). On the other hand, in brassinolide-treated bean plants enhanced RNA and DNA polymerase activities as well as accumulation of protein, RNA and DNA were found (KALINICH et al. 1985, 1986). A possible physiological function might be derived from the high endogenous contents of brassinosteroids in the pollen mentioned above and the observed stimulation of pollen germination and pollen tube growth by brassinolide (HEWITT et al. 1985). The suggestion that brassinosteroids may display antiecdysteroid activities have been substantiated by the inhibiting effects of seven brassinosteroids on the moult of the dipteran *Phormia terra-novae* (HETRU et al. 1986).

An increasing number of works describe the stimulating effects of brassinosteroids on yield parameters of various crop plants (vegetables, roots, fruits, grains, etc.). Especially some Japanese research groups tested homobrassinolide and brassinolide for agricultural application during the last years and published promising results (in Japanese) which have been summarized by YOKOTA and TAKAHASHI (1986). Stimulating effects of brassinosteroids are most striking in poor-growing plants. Stress injuries, e.g. due to low temperature, salt, herbicides and pathogenic microorganisms, can be protected or alleviated by treatment with brassinosteroids.

Acknowledgement. The authors are indebted to Miss Ines Schulze for help in the preparation of the manuscript.

References

ABE, H., MORISHITA, T., UCHIYAMA, M., TAKATSUTO, S., IKEKAWA, N., IKEDA, M., SASSA, T., KITSUWA, T., MARUMO, S.: Experientia 39, 351-353 (1983). - ABE, H., NAKAMURA, K., MORISHITA, T., UCHIYAMA, M., TAKATSUTO, S., IKEKAWA, N.: Agric. Biol. Chem. 48, 1103-1104 (1984a). - ABE, H., MORISHITA, T., UCHIYAMA, M., TAKATSUTO, S., IKEKAWA, N.: Agric. Biol. Chem. 48, 2171-2172 (1984b). - ADAM, G., MARQUARDT, V.: Phytochemistry 25, 1787-1799 (1986); - Z. Chem. 27, 41-49 (1987). - ADAM, G., SCHIERHORN, A.: Tetrahedron Lett. 24, 351-352 (1983). - ADAM, G., BERGNER, C., LISCHEWSKI, M., SCHIERHORN, A., SEMBDNER, G., VOIGT, B.: Gibberellins - synthesis and biocactivity, 171-189. In: Innovative Approaches in Drug Research, ed. A.F. HARMS. Amsterdam: Elsevier 1986. - AKIYAMA, T., SUZUKI, H.: J. Plant Physiol. 118, 1-5 (1985). - AKIYOSHI, D.E., MORRIS, R.O., HINZ, R., MISCHKE, B.S., KOSUGE, T., GARFINKEL, D.J., GORDON, M.P., NESTER, E.W.: Proc. Natl. Acad. Sci. USA 80, 407-411 (1983). - AKIYOSHI, D.E., KLEE, H., AMASINO, R.M., NESTER, E.W., GORDON, M.P.: Proc. Natl. Acad. Sci. USA 81, 5994-5998 (1984). - AKIYOSHI, D.E., REGIER, D.A., JEN, G., GORDON, M.P.: Nucleic Acid Res. 13, 2773-2788 (1985). - ALBONE, K.S., GASKIN, P., MACMILLAN, J., SPONSEL, V.M.: Planta 162, 560-565 (1984). - ARIGAYO, S., SAKATA, K., FUJISAWA, S., SAKURAI, A., ADISEWOJO, S.S., TAKAHASHI, N.: Agric. Biol. Chem. 47, 2939-2940 (1983). - ARIMA, M., YOKOTA, T., TAKAHASHI, N.: Phytochemistry 23, 1587-1591 (1984). - ARTECA, R.N.: Physiol. Plant. 62, 102-104 (1984). - ARTECA, R.N., TSAI, D.S., SCHLAGNHAUFER, C., MANDAVA, N.B.: Physiol. Plant. 59, 539-544 (1983). - ARTECA, R.N., BACHMANN, J.M., YOPP, I.H., MANDAVA, N.B.: Physiol. Plant. 64, 13-16 (1985). - ATZORN, R., WEILER, E.W.: Planta 159, 1-6 (1983a); - ibid. 159, 7-11 (1983b); - ibid. 159, 289-299 (1983c). - ATZORN, R., GANSZ, A., WEILER, E.W.: Plant Physiol. 75, (Suppl.), 68 (1984). - AXELOS, M., BARBET, J., PEAUD-LENOEL, C.: Plant Sci. Lett. 33, 201-212 (1984).

BABA, J., YOKOTA, T., TAKAHASHI, N.: Agric. Biol. Chem. 47, 659-661 (1983). - BADENOCH-JONES, J., LETHAM, D.S., PARKER, C.W., ROLFE, B.G.: Plant Physiol. 75, 1117-1125 (1984a). - BADENOCH-JONES, J., ROLFE, B.G., LETHAM, D.S.: J. Plant Growth Regul. 3, 41-49 (1984b). - BALASIMHA, D.: Indian J. Plant Physiol. 26, 139-142 (1983). - BANGERTH, F.: Ber. Dtsch. Bot. Ges. 97, 257-267 (1984); - Natural and synthetic plant growth substances in fruit growing, 387-390. In: see ref. BOPP 1986. - BARENDSE, G.W.M., KEPCŽYNSKI, J., KARSSEN, C.M., KOORNEEF, M.: Physiol. Plant. 67, 315-319 (1986). - BARENDSE, G.W.M., CROES, A.F., VAN DER KRIEKEN, W.M., SMULDERS, M.J.M.: Metabolism of hormones in thin-layer tissue cultures from flower stalks of tobacco, 123-127. In: see ref. SCHREIBER et al. 1987. - BARNES, L., JONES, R.L.: Plant Cell Environ. 7, 89-95 (1984). - BARTHE, G.A., STEWART, J.: J. Agric. Food Chem. 33, 293-296 (1985). - BAYNHAM, M.K., HANSON, J.R.: Tetrahedron Lett. 24, 1941-1944 (1983). - BEALE, M.H., MACMILLAN, J.: The synthesis of new gibberellin-macromolecule covalent conjugates. In: see ref. BOPP et al. 1985. - BEALE, M.H., BEARDER, J.R., MACMILLAN, J., MATSUO, A., PHINNEY, B.O.: Phytochemistry 22, 875-881 (1983). - BEALE, M.H., MACMILLAN, J., SPRAY, C.R., TAYLOR, D.A., PHINNEY, B.O.: J. Chem. Soc. Perkin Trans. I, 541-545 (1984). - BEALE, M.H., HOOLEY, R., MACMILLAN, J.: Gibberellins: Structure-activity relationships and the design of molecular probes, 65-73. In: see ref. BOPP 1986. - BEARDER, J.R.: In vivo diterpenoid biosynthesis in *Gibberella fujikuroi*: The pathway after *ent*-kaurene, Vol. I, 251-388. In: see ref. CROZIER 1983. - BERNIER, G., KINET, J.M.: The control of flower initiation and development, 293-302. In: see ref. BOPP 1986. - BIRNBERG, P.R., MAKI, S.L., BRENNER, M.L., DAVIS, G.C., CARNES, M.G.: Analyt. Biochem. 153, 1-8 (1986). - BLACKMAN, P.G., DAVIES, W.J.: J. Exp. Bot. 35, 174-179 (1984); - Ann. Bot. 54, 121-125 (1984). - BLECHSCHMIDT, S., CASTEL, U., GASKIN, P., HEDDEN, P.,

GRAEBE, J.E., MACMILLAN, J.: Phytochemistry 23, 553-558 (1984). - BOPP, M.: Z. Pflanzenphysiol. 113, 435-444 (1984); - (ed.): Plant Growth Substances, Proc. 12th Int. Conf. Plant Growth Substances, Heidelberg 1985. Berlin, Heidelberg, New York: Springer 1986. - BOPP, M., JACOB, H.J.: Planta 169, 462-464 (1986). - BOPP, M., KNOOP, B., RADEMACHER, W.: Abstr. 12th Int. Conf. Plant Growth Substances, Heidelberg 1985. - BOPP, M., GERHÄUSER, D., KESSLER, U.: On the hormonal system of mosses, 263-269. In: see ref. BOPP 1986. - BRANCA, C., MAGGIALI, C.A., MORINI, G., RONCHINI, F., RICCI, D., SCOCCIANTI, V.: Phytochemistry 26, 25-29 (1987). - BRINEGAR, A.C., FOX, J.E.: Biol. Plant. 27, 100-104 (1985). - BRINEGAR, A.C., STEVENS, A., FOX, J.E.: Plant Physiol. 79, 706-710 (1985). - BUCHMANN, J., MARNER, F.J., SCHRÖDER, G., WAFFENSCHMIDT, S., SCHRÖDER, J.: EMBO J. 4, 853-860 (1985). - BULARD, C., LE PAGE-DEGIVRY, M.T.: Phytohormones in seed and fruit development, 308-314. In: see ref. BOPP 1986. - BUYSSE, L., CALLEBAUT, A., DE CRAENE, D.: Meded. Fac. Landbouwwetensch., Rijksuniv. Gent 49, 1019-1024 (1984).

CAERS, M., VENDRIG, J.C.: Physiol. Plant. 66, 685-691 (1986). - CAERS, M., RUDELSHEIM, P., VAN ONCKELEN, H., HEREMANS, S.: Plant Cell Physiol. 26, 52-57 (1985). - CAHILL, D.M., WESTE, G.M., GRANT, B.R.: Plant Physiol. 81, 1103-1109 (1986). - CAMPELL, B.R., BONNER, B.A.: Plant Physiol. 82, 909-915 (1986). - CARBONELL, J., JONES, R.L.: Physiol. Plant. 63, 345-350 (1985). - CERANA, R., BONETTI, A., MARRE, M.T., ROMANI, G., LADO, P., MARRE, E.: Physiol. Plant. 59, 23-27 (1983a). - CERANA, R., COLUMBO, R., LADO, P.: Physiol. Veg. 21, 875-881 (1983b). - CERANA, R., LADO, P., ANASTASIA, M., CIUFFREDA, P., ALLEVI, P.: Z. Pflanzenphysiol. 114, 221-225 (1984). - CERANA, R., SPELTA, M., BONETTI, A., LADO, P.: Plant Sci. 38, 99-105 (1985). - CERNY, V., STRNAD, M., KAMINEK, M.: Coll. Czech. Chem. Commun. 51, 687-697 (1986). - CHADWICK, C.M., GARROD, D.R. (eds.): Hormones, Receptors and Cellular Interactions in Plants. Cambridge: Cambridge University Press 1986. - CHAILAKHYAN, M.K.: Hormonal regulation of plant flowering, 303-307. In: see ref. BOPP 1986. - CHANDLER, P.M., ZWAR, J.A., JACOBSEN, T.J.U., HOGGINS, T.J.V., INGLIS, A.S.: Plant. Mol. Biol. 3, 407-418 (1984). - CHAPMAN, D.J., RAGAN, M.A.: Annu. Rev. Plant Physiol. 31, 639-678 (1980). - CHATFIELD, J.M., ARMSTRONG, D.J.: Plant Physiol. 80, 493-499 (1986). - CHEN, C.M., LEISNER, S.M.: Plant Physiol. 75, 442-446 (1984); - ibid. 77, 99-113 (1985). - CHEN, C.M., ERTL, J.R., LEISNER, S.M., CHANG, C.C.: Plant Physiol. 78, 510-513 (1985). - CHISM, G.W., LONG, A.R., ROLLE, R.: J. Chromatogr. 317, 263-270 (1984). - CHUNG, C.H., COOLBAUGH, R.C.: Plant Physiol. 80, 544-548 (1986). - COHEN, J.D., MEUDT, W.J.: Plant Physiol. 72, 691-694 (1983). - COOLBAUGH, R.C.: Early stages of gibberellin biosynthesis, 53-98. In: see ref. CROZIER 1983; - J. Plant Growth Regul. 3, 97-109 (1984); - Plant Physiol. 78, 655-657 (1985). - COOLBAUGH, R.C., AL-NIMRI, L.F., NESTER, J.E.: Evidence for gibberellins in culture filtrates of *Cercospora rosicola*, 12. In: see ref. BOPP et al. 1985. - COREY, E.J., MYERS, A.G., TAKAHASHI, N., YAMANE, H., SCHRAUDOLF, H.: Tetrahedron Lett. 27, 5083-5084 (1986). - CORSE, J., GRAFFIELD, W., LUNDIN, R.Z.: J. Plant Growth Regul. 2, 47-57 (1983). - CREVECOEUR, M., PENEL, C., GREPPIN, H., GASPAR, T.: J. Exp. Bot. 37, 1218-1224 (1986). - CROZIER, A. (ed.): The Biochemistry and Physiology of Gibberellins, Vol. 1 and 2. New York: Praeger 1983; - The occurrence and structure of gibberellins, Vol. 1, 1-18. In: see ref. CROZIER 1983. - CROZIER, A., DURLEY, R.C.: Modern methods of analysis of gibberellins, Vol. 1, 485-560. In: see ref. CROZIER 1983. - CROZIER, A., HILLMAN, J.R. (eds.): The Biosynthesis and Metabolism of Plant Hormones (Society for Experimental Biology-Seminar series 23). Cambridge: University Press 1984. - CROZIER, A., TURNBULL, C.: What's New in Plant Physiology, Vol. 15, 9-12 (1984). - CROZIER, A., SANDBERG, G., MONTEIRO, A.M., SUNDBERG, B.: The use of immunological techniques in plant hormone analysis, 13-21. In: see ref. BOPP 1986.

DANHEISER, R.L.: The total synthesis of gibberellic acid, 21-70. In: Strategies and Tactics of Organic Synthesis, ed. T. LINDBERG. Orlando: Academic Press 1984. - DATHE, W.: Occurrence of gibberellins in fungi and higher plants in relation to their taxonomical connections, 203-244. In: Hormonal Regulation of Plant Growth and Development, Vol. 3, ed. S.S. PUROHIT. Bikaner: Agro Botanical Publishers (India) 1986; - Rolle der Gibberelline, Abscisinsäure und Jasmonsäure bei ausgewählten physiologischen Prozessen der generativen Entwicklung von *Vicia faba* L. und *Glycine max* (L.) Merr., Dissertation (B), Halle (1987). - DAVIES, P.J., BIRNBERG, P.R., MAKI, S.L., BRENNER, M.L.: Plant Physiol. **81**, 991-996 (1986a). - DAVIES, P.J., METCALFE, T.A., LODGE, T.A., DA COSTA, A.R.: Aust. J. Plant Physiol. **13**, 105-125 (1986b). - DAVIES, P.J., HORGAN, R., HEALD, J.K., MCGAW, B.: Plant Growth Regul. **4**, 311-323 (1986c). - DAWE, R.D., MANDER, L.N., TURNER, J.V.: Tetrahedron Lett. **26**, 363-366 (1985). - DEIKMAN, J., JONES, R.L.: Plant Physiol. **78**, 192-198 (1985); - ibid. **80**, 672-675 (1986). - DELAIGUE, M., POULAIN, T., DURAND, B.: Plant Mol. Biol. **3**, 419-430 (1984). - DELAIGUE, M., POULAIN, T., DABAT, P., CABRE, E., DURAND, R., DURAND, B.: Biol. Plant. **28**, 23-30 (1986). - DHINGRA, H.R., VARGHESE, T.M.: New Phytol. **100**, 563-569 (1985). - DIAZ, C.E., FRAGA, B.M., GONZALEZ, A.G., GONZALEZ, P., HANSON, J.R., HERNANDEZ, M.G.: Phytochemistry **23**, 2813-2816 (1984). - DOLAN, S.C., MACMILLAN, J.: J. Chem. Soc. Chem. Commun. **22**, 1588-1589 (1985a); - J. Chem. Soc. Perkin Trans. I, 2741-2746 (1985b). - DOLAN, S.C., HOLDUP, D.W., HUTCHINSON, M., MACMILLAN, J.: J. Chem. Soc. Perkin Trans. I, 651-654 (1985). - DOWN, G.J., LEE, M., MACMILLAN, J., STAPLES, K.S.: J. Chem. Soc. Perkin Trans. I, 1103-1108 (1983). - DUNBERG, A., ODEN, P.C.: Gibberellins and conifers, 221-296. In: see ref. CROZIER 1983. - DURAND, R., DURAND, B.: Physiol. Plant. **60**, 267-274 (1984).

EBERLE, J., YAMAGUCHI, I., NAKAGAWA, R., TAKAHASHI, N., WEILER, E.W.: FEBS Lett. **202**, 27-31 (1986a). - EBERLE, J., ARNSCHEIDT, A., KLIX, D., WEILER, E.W.: Plant Physiol. **81**, 516-521 (1986b). - EINSET, J.W.: Biochem. Biophys. Res. Commun. **124**, 470-474 (1984). - ERNST, D.: Radioimmunoassay and gas chromatography/mass spectrometry for cytokinin determination, 18-48. In: see ref. LINSKENS and JACKSON 1986b. - ERNST, D., OESTERHELT, D.: Planta **161**, 246-248 (1984). - ERNST, D., SCHÄFER, W., OESTERHELT, D.: Planta **159**, 216-221 (1983a); - ibid. **159**, 222-225 (1983b). - ERNST, D., OESTERHELT, D., SCHÄFER, W.: Planta **161**, 240-245 (1984). - ERNSTSEN, A., JENSEN, E.: J. Liquid Chromatogr. **8**, 369-380 (1985). - EVIDENTE, A., SURICO, G., IACOBELLIS, N.S., RANDAZZO, G.: Phytochemistry **25**, 525-526 (1986).

FELIX, G., MEINS, F.: Planta **164**, 423-428 (1985). - FLORES, S., TOBIN, E.M.: Planta **168**, 340-349 (1986). - FRAGA, B.M., GONZALEZ, A.G., HERNANDEZ, M.G., SAN MARTIN, A., DURI, Z.J., HANSON, J.R.: Tetrahedron Lett. **24**, 1945-1948 (1983). - FRAGA, B.M., GONZALEZ, A.G., HERNANDEZ, M.G., DURI, Z.J., HANSON, J.R.: J. Chem. Soc. Perkin Trans. I, 1105-1107 (1984). - FUJIOKA, S., YAMAGUCHI, I., MUROFUSHI, N., TAKAHASHI, N., KAIHARA, S., TAKIMOTO, A., CLELAND, C.F.: Plant Cell Physiol. **27**, 109-116 (1986). - FUJISAWA, S., YAMAGUCHI, I., PARK, K.-H., KOBAYASHI, M., TAKAHASHI, N.: Agric. Biol. Chem. **49**, 27-33 (1985). - FURUYA, M., TAKENO, K.: Gibberellins in algae and higher cryptogams, Vol. 2, 189-220. In: see ref. CROZIER 1983.

GALLI, M.G.: Planta **160**, 193-199 (1984). - GARCIA-MARTINEZ, J.L., CARBONELL, J.: J. Plant Growth Regul. **4**, 19-27 (1985). - GARCIA-MARTINEZ, J.L., SPONSEL, V.M., GASKIN, P.: Planta **170**, 130-137 (1987). - GARRISON, F.R., BRINKER, A.M., NOODEN, L.D.: Plant Cell Physiol. **25**, 213-219 (1984). - GASKIN, P., HUTCHINSON, M., LEWIS, N., MACMILLAN, J., PHINNEY, B.O.: Phytochemistry **23**, 559-564 (1984a). - GASKIN, P., GILMOUR, S.J., LENTON, J.R., MACMILLAN, J., SPONSEL, V.M.: J. Plant Growth Regul. **2**, 229-242 (1984b). - GASKIN, P., GILMOUR, S.J., MACMILLAN,

J., SPONSEL, V.M.: Planta 163, 283-289 (1985). - GHOSH, S., BASU, P.S.: Biochem. Physiol. Pflanzen 179, 277-287 (1984). - GILMOUR, S.J., MACMILLAN, J.: Planta 162, 89-90 (1984). - GILMOUR, S.J., GASKIN, P., SPONSEL, V.M., MACMILLAN, J.: Planta 161, 186-192 (1984). - GILMOUR, S.J., ZEEVAART, J.A.D., SCHWENEN, L., GRAEBE, J.E.: The effect of photoperiod on gibberellin metabolism in cell-free extracts from spinach, No. 14. In: see ref. BOPP et al. 1985. - GOLDBERG, S.B., FLICK, J.S., ROGERS, S.G.: Nucleic Acid Res. 12, 4665-4677 (1984). - GÖRING, H., KOSHUCHOVA, S., MÜNNICH, H., DIETRICH, M.: Plant Cell Physiol. 25, 683-690 (1984). - GRAEBE, J.E.: Ber. Dtsch. Bot. Ges. 97, 67-74 (1984); - Gibberellin biosynthesis from gibberellin A_{12}-aldehyde, 74-82. In: see ref. BOPP 1986. - GREGSON, S., SHAW, G.: J. Chem. Soc. Perkin Trans. I, 187-190 (1985). - GREGSON, S., SHAW, G., MOK, M.C., MOK, D.W.S.: Plant Sci. 38, 111-114 (1985). - GUBLER, F., ASHFORD, A.E.: Aust. J. Plant Physiol. 12, 297-305 (1985). - GUERIN, B., KAHLEM, G., TELLER, G., DURAND, B.: Plant Physiol. 74, 139-145 (1984).

HALEVY, A.H.: Recent advances in the use of growth substances in ornamental horticulture, 391-398. In: see ref. BOPP 1986. - HALL, M.A.: Hormones and plant development: cellular and molecular aspects, 313-330. In: Developmental Control in Animals and Plants, eds. F. GRAHAM, P.F. WAREING. Oxford: Blackwell 1984. - HAMAGUCHI, N., IWAMURA, H., FUJITA, T.: Eur. J. Biochem. 153, 565-572 (1985). - HAMMERTON, R.W., HO, T.H.: Plant Physiol. 80, 692-697 (1986). - HANKS, G.R.: Sci. Hortic. 23, 379-390 (1984). - HANSEN, C.E., WENZLER, H., MEINS, F.: Plant Physiol. 75, 959-963 (1984). - HANSEN, C.E., MEINS, F., MILANI, A.: Differentiation 29, 1-6 (1985). - HANSON, J.R.: Biochem. Soc. Trans. 11, 522-528 (1983). - HASHIZUME, T., HOSOI, M., SUGIYAMA, T.: Agric. Biol. Chem. 49, 225-227 (1985a). - HASHIZUME, T., KAWAKITA, K., SENDO, S., SUGIYAMA, T.: Agric. Biol. Chem. 49, 3481-3484 (1985b). - HASHIZUME, T., SUYE, S., SUGIYAMA, T.: Agric. Biol. Chem. 49, 387-390 (1985c). - HAUTALA, E., STAFFORD, A., CORSE, J., BARKER, P.A.: J. Chromatogr. 351, 560-565 (1986). - HAYES, P.M., PATRIC, J.W.: Planta 166, 371-379 (1985). - HEDDEN, P.: In vitro metabolism of gibberellins, Vol. 1, 99-150. In: see ref. CROZIER 1983; - The use of combined gas chromatography - mass spectrometry in the analysis of plant growth substances, 1-22. In: see ref. LINSKENS and JACKSON 1986. - HEDDEN, P., GRAEBE, J.E.: J. Plant Growth Regul. 4, 111-122 (1985). - HEDDEN, P., GRAEBE, J.E., BEALE, H., GASKIN, P., MACMILLAN, J.: Phytochemistry 23, 569-574 (1984). - HENNY, R.J.: Hort. Sci. 18, 374 (1983). - HENTRICH, W., BERGNER, C., SEMBDNER, G.: Plant Growth Regul. 3, 103-110 (1985). - HETRU, C., ROUSSEL, J.P., MORI, K., NAKATANI, Y.: C.R. Acad. Sci., Ser. 2, 302, 417-420 (1986). - HEUPEL, R.C., PHINNEY, B.O., SPRAY, C.R., GASKIN, P., MACMILLAN, J., HEDDEN, P., GRAEBE, J.E.: Phytochemistry 24, 47-53 (1985). - HEWITT, F.R., HOUGH, T., O'NEILL, P.O., SASSE, J.M., WILLIAMS, E.G., ROWAN, K.S.: Aust. J. Plant Physiol. 12, 201-211 (1985). - HILTON, J.R.: New Phytol. 95, 545-548 (1983); - ibid. 96, 161-166 (1984). - HOAD, G.V.: Gibberellin bioassays and structure-activity relationships, Vol. 2, 57-94. In: see ref. CROZIER 1983. - HOFMAN, P.J., FORSYTH, C., VAN STADEN, J.: J. Plant Physiol. 121, 1-12 (1985). - HOOK, J.M., MANDER, J.L., URECH, R.: J. Org. Chem. 49, 3250-3260 (1984). - HOOLEY, R.: Planta 161, 355-360 (1984); - J. Exp. Bot. 35, 822-828 (1984). - HORGAN, R.: Cytokinin biosynthesis and metabolism, 92-98. In: see ref. BOPP 1986. - HORGAN, P.A., NAKAGAWA, C.H., IRVIN, R.T.: Can. J. Biochem. Cell Biol. 62, 715-721 (1984). - HUBICK, K.T., TAYLOR, J.S., REID, D.M.: Plant Growth Regul. 4, 139-151 (1986). - HUTCHINSON, M.: Studies on hydroxylated gibberellins. Ph.D. Thesis, University of Bristol 1983.

IKEKAWA, N.: Structure and function of natural sterols, 75-86. In: Natural Products and Biological Activities, eds. H. IMURA, T. GOTO, T. MURACHI, T. NAKAJIMA. Tokyo: University Press 1986. - IKEKAWA, N., TAKATSUTO, S.: Mass Spectroscopy 32, 55-70 (1984). - IKEKAWA, N.,

TAKATSUTO, S., KITSUWA, T., SAITO, H., MORISHITA, T., ABE, H.: J. Chromatogr. 290, 289-302 (1984). - INGRAM, T.J., REID, J.B., POTTS, W.C., MURFET, I.C.: Physiol. Plant. 59, 607-616 (1983). - INGRAM, T.J., REID, J.B., MURFET, I.C., GASKIN, P., WILLIS, C.L., MACMILLAN, J.: Planta 160, 455-463 (1984). - INGRAM, T.J., REID, J.B., MACMILLAN, J.: Planta 164, 429-438 (1985); - ibid. 168, 414-420 (1986). - INZE, D., FOLLIN, A., VAN LIJSEBETTENS, M., SIMOENS, C., GENETELLO, C., VAN MONTAGU, M., SCHELL, J.: Mol. Gen. Genet. 194, 265-274 (1984). - ITAYA, T., FUJII, T., EVIDENTE, A., RANDAZZO, G., SURICO, G., IACOBELLIS, H.S.: Tetrahedron Lett. 27, 6349-6352 (1986). - IWAMURA, H., MURAKAMI, S., KOSHIMIZU, K., MATSUBARA, S.: J. Med. Chem. 28, 577-583 (1985). - IWASAKI, T., FUKUDA, H., SHIBOAKA, H.: Plant Cell Physiol. 27, 717-724 (1986).

JACOBS, W.P.: Are angiosperm hormones present in, and used as hormones by algae?, 249-256. In: see ref. BOPP 1986. - JACOBSEN, J.V., BEACH, L.R.: Nature 316, 275-277 (1985). - JACOBSEN, J.V., ZWAR, J.A., CHANDLER, P.M.: Planta 163, 430-438 (1985). - JAMESON, P.F., ZHANG, R., LETHAM, D.S.: Metabolism and translocation of $[^3H]$zeatin riboside in blue lupin plants, 20. In: see ref. BOPP et al. 1985. - JAMESON, P.F., MCWHA, J.A., HASLEMORE, R.M.: Physiol. Plant. 63, 53-57 (1985b). - JONES, M.G.: Endogenous gibberellins and stem extension, 33-42. In: see ref. MENHENETT and LAWRENCE 1984. - JONES, R.L.: The control of plant cell elongation by auxin, 275-283. In: see ref. BOPP 1986. - JONES, R.L., CARBONELL, J.: Plant Physiol. 76, 213-218 (1984). - JOOS, H., INZE, D., CAPLAN, A., SORMANN, M., VAN MONTAGU, M., SCHELL, J.: Cell 32, 1057-1067 (1983). - JUNG, J., RENTZEA, C., RADEMACHER, W.: J. Plant Growth Regul. 4, 181-188 (1986).

KADIR, K., GREGSON, S., SHAW, G., MOK, M.C., MOK, D.W.S.: J. Chem. Res. 299 (1984). - KALINICH, J.F., MANDAVA, N.B., TODHUNTER, J.A.: J. Plant Physiol. 120, 207-214 (1985); - ibid. 125, 345-353 (1986). - KAMIYA, Y., GRAEBE, J.E.: Phytochemistry 22, 681-689 (1983). - KAMIYA, Y., TAKAHASHI, M., TAKAHASHI, N., GRAEBE, J.E.: Planta 162, 154-158 (1984); - Metabolism of gibberellins in cell-free systems from immature seeds of *Pisum sativum* and *Phaseolus vulgaris*, 9. In: see ref. BOPP et al. 1985; - Planta 169, 524-528 (1986). - KANDELER, R.: Phyton 24, 113-124 (1984). - KARSSEN, C.M., LACKA, E.: A revision of the hormone balance theory of seed dormancy: Studies on gibberellin and/or abscisic acid-deficient mutants of *Arabidopsis thaliana*, 315-323. In: see ref. BOPP 1986. - KATSUMI, M.: Plant Cell Physiol. 26, 615-625 (1985). - KATSUMI, M., FOARD, D.E., PHINNEY, B.O.: Plant Cell Physiol. 24, 379-388 (1983). - KATSUMI, M., YAMANE, H., TAKAHASHI, N., PHINNEY, B.O.: Plant Physiol. 75 (Suppl.), 94 (1984). - KAWAMURA, K.: Forest Breed. 134, 25-27 (1985). - KAWANABE, Y., YAMANE, H., MURAYAMA, T., TAKAHASHI, N., NAKAMURA, T.: Agric. Biol. Chem. 47, 1693-1694 (1983). - KAWANABE, Y., YAMANE, H., TAKAHASHI, N., NAKAMURA, T.: Agric. Biol. Chem. 49, 2447-2450 (1985). - KEITH, N., THOMPSON, R.H., RAPPAPORT, L.: Plant Physiol. 75 (Suppl.), 92 (1984). - KERBAUY, G.B., PETERS, J.A., HELL, K.G.: Plant Sci. 45, 125-132 (1986). - KHARTSCHENKO, V.I., TERESCHENKO, A.F.: Cytokinin-binding proteins from a microsomal fraction of bean hypocotyls (*Phaseolus vulgaris* L.). Int. Symp. Regulation of Plant Integrity, Abstr. 31, Brno 1985. - KHARTSCHENKO, V.I., ROMANKO, E.J., SELIVANKINA, S.J., KULAEVA, O.N.: Fiziol. Rast. 30, 1214-1219 (1983). - KHURANA, J.P., MAHESHWARI, S.C.: Plant Cell Physiol. 24, 913-918 (1983). - KINOSHITA, I., TSUJI, H.: Plant Physiol. 76, 575-578 (1984). - KLÄMBT, D., HOLTZ, J., HELBACH, M., MAAS, H.: Ber. Dtsch. Bot. Ges. 97, 57-65 (1984). - KNÖFEL, H.-D., SCHWARZKOPF, E., MÜLLER, P., SEMBDNER, G.: J. Plant Growth Regul. 3, 127-140 (1984). - KNOX, J.P., BEALE, M.H., BUTCHER, G.W., MACMILLAN, J.: Planta 170, 86-91 (1987). - KOBAYASHI, M., YAMAGUCHI, I., MUROFUSHI, N., OTA, Y., TAKAHASHI, N.: Agric. Biol. Chem. 48, 2725-2729 (1984). - KOORNEEF, M., ELGERSMA, A., HANHART, C.J., VAN LOENEN-MARTINET, E.P.,

VAN RIJN, L., ZEEVAART, J.A.D.: Physiol. Plant. 65, 33-39 (1985). - KOSHIOKA, M., DOUGLAS, T.J., ERNST, D., HUBER, J., PHARIS, R.P.: Phytochemistry 22, 1577-1584 (1983a). - KOSHIOKA, M., HARADA, J., TAKENO, K., NOMA, M., SASSA, T., OGIYAMA, K., TAYLOR, J.S., ROOD, S.B., LEGGE, R.L., PHARIS, R.P.: J. Chromatogr. 256, 101-115 (1983b). - KOSHIOKA, M., JONES, A., KOSHIOKA, M.N., PHARIS, R.P.: Phytochemistry 22, 1585-1590 (1983c). - KOSHIOKA, M., TAKENO, K., BEALL, F.D., PHARIS, R.P.: Plant Physiol. 73, 398-406 (1983d). - KOSHIOKA, M., HISAJIMA, S., PHARIS, R.P., MUROFUSHI, N.: Agric. Biol. Chem. 49, 2627-2631 (1985a). - KOSHIOKA, M., PHARIS, R.P., KING, R.W., MUROFUSHI, N., DURLEY, R.C.: Phytochemistry 24, 663-671 (1985b). - KRIZO, M., KORINEKOVA, M.: Effect of IAA, GA_3 and $GA_{4/7}$ upon the microprothallium formation of some conifers. Int. Symp. Regulation of Plant Integrity, Abstr. 39, Brno 1985. - KUMRA, S., CHAPRA, R.N.: Ann. Bot. 54, 605-620 (1984).

LALOUE, M., FOX, J.E.: Plant Physiol. 75 (Suppl.), 60 (1984); - Characterization of an imine intermediate in the degradation of isopentenylated cytokinins by a cytokinin oxidase from wheat, P 03. In: see ref. BOPP et al. 1985. - LATTKE, P., SCHNEIDER, G.: J. Plant Growth Regul. 4, 71-79 (1985). - LEE, Y.H., MOK, M.C., MOK, D.W.S., GRIFFIN, D.A., SHAW, G.: Plant Physiol. 77, 635-641 (1985). - LEHNERT, B., BOPP, M.: Z. Pflanzenphysiol. 110, 379-391 (1983). - LEONARD, M., KINET, J.-M., BODSON, M., BERNIER, G.: Physiol. Plant. 57, 85-89 (1983). - LERBS, S., LERBS, W., KLYACHKO, N.L., ROMANKO, E.G., KULAEVA, O.N., WOLLGIEHN, R., PARTHIER, B.: Planta 162, 289-298 (1984). - LESHEM, I.: Can. J. Bot. 62, 2943-2949 (1984). - LESHEM, Y., FREUD-SILVERBERG, M., WURZBURGER, J., BARNES, G., MALIK, Z., LANGSAM, Y.: Ca^{2+}: Calmodulin phytohormone-linked plant senescence control, 159-168. In: see ref. BOPP 1986. - LETHAM, D.S., PALNI, L.M.S.: Annu. Rev. Plant Physiol. 34, 163-197 (1983). - LETHAM, D.S., PALNI, L.M.S., TAO, G.Q., GOLLNOW, B.I., BATES, C.M.: J. Plant Growth Regul. 2, 103-115 (1984). - LEVY, A., PALEVITCH, D., MILO, J., LAVIE, D.: Plant Growth Regul. 4, 153-157 (1986). - LEWAK, S.: Adv. Agric. Biotechnol. 16, 95-144 (1985). - LEWER, P., MACMILLAN, J.: Phytochemistry 23, 2803-2811 (1984). - LICHTENSTEIN, C., KLEE, H., MONTAYA, A., GARFINKEL, D., FULLER, S.: J. Mol. Appl. Genet. 2, 354-362 (1984). - LIEBISCH, H.W.: Enzymatic glucosylation of gibberellins, 176-182. In: see ref. SCHREIBER et al. 1987. - LINSKENS, H.F., JACKSON, J.F. (eds.): Gas chromatography/Mass spectrometry. In: Modern Methods of Plant Analysis, N.S. Vol. 3. Berlin Heidelberg New York: Springer 1986a; - Immunology in Plant Sciences. In: Modern Methods of Plant Analysis, N.S. Vol. 4. Berlin Heidelberg New York: Springer 1986b; - High Performance Liquid Chromatography in Plant Sciences. In: Modern Methods of Plant Analysis, N.S. Vol. 5. Berlin Heidelberg New York: Springer 1987. - LONGMAN, K.A., MCPEDICK, J., MUGGLESTONE, M., SMITH, R.J.: Tree Physiol. 1, 101-113 (1986). - LONGO, G.P., BRACALE, M., ROSSI, G., LONGO, C.P.: Physiol. Plant. 68, 684-687 (1986). - LOOMIS, P.D., MARAVOLO, N.C.: Bryologist 88, 65-68 (1985). - LOONEY, N.E., PHARIS, R.P., NOMA, M.: Planta 165, 292-294 (1985). - LORENZI, R., CECCARELLI, N.: Phytochemistry 25, 817-822 (1986).

MACDONALD, E.M.S., MORRIS, R.O.: Methods Enzymol. 110, 347-358 (1985). - MACDONALD, E.M.S., POWELL, G.K., REGIER, D.A., GLASS, N.L., ROBERTO, F., KOSUGE, T., MORRIS, R.O.: Plant Physiol. 82, 742-747 (1986). - MACMILLAN, J.: Biochem. Soc. Trans. 11, 528-534 (1983); - Analysis of plant hormones and metabolism of gibberellins, 1-16. In: see ref. CROZIER and HILLMAN 1984a; - Gibberellin metabolism beyond gibberellin A_{12}-aldehyde: 2-oxoglutarate-coupled oxidases, 13-31. In: see ref. MENHENETT and LAWRENCE 1984b; - Gibberellins: Metabolism and function, 53-66. In: Current Topics in Plant Biochemistry and Physiology, Vol. 4, eds. D.D. RANDALL. Missouri: Columbia Press 1985a; - Biol. Plant. 27, 164-171 (1985b). - MACMILLAN, J., TAYLOR, D.A.: J. Chem. Soc. Perkin Trans. I, 837-842 (1985). - MACMILLAN, J., WILLIS, C.L.: J. Chem. Soc.

Perkin Trans. I, 357-361 (1984); - ibid. 2177-2184 (1985); - ibid. 309-313 (1986). - MAKI, S.L., BRENNER, M.L., BIRNBERG, P.R., DAVIES, P.J., KRICK, T.P.: Plant Physiol. 81, 984-990 (1986). - MARQUARD, R.D., HANOVER, J.W.: Can. J. For. Res. 14, 547-553 (1984); - ibid. 15, 743-746 (1985). - MARTIN, C., NORTHCOTE, D.H.: Planta 158, 16-26 (1983). - MARTIN, C., BEECHING, J.R., NORTHCOTE, D.H.: Planta 162, 68-76 (1984). - MASCARENHAS, J.P., CANARY, D.: Pollen. Symbionts and symbiont-induced structures, 579-598. In: Encyclopedia of Plant Physiology, N.S. Vol. 11, Hormonal Regulation of Development III, eds. R.P. PHARIS, D.M. REID. Berlin Heidelberg New York: Springer 1985. - MATSUBARA, S.: Euphytica 33, 113-122 (1984). - MATSUO, T., ITOO, S., MUROFUSHI, N., TAKAHASHI, N., KOBAYASHI, M., CHISHIKI, T.: Agric. Biol. Chem. 48, 2935-2941 (1984). - MAUK, C.S., BAUSHER, M.G., YELENOSKY, G.: J. Plant Growth Regul. 5, 111-120 (1986). - MCGAW, B.A., HORGAN, R.: Planta 159, 30-37 (1983a); - Phytochemistry 22, 1103-1105 (1983b); - Biol. Plant. 27, 180-187 (1985). - MCGAW, B.A., SCOTT, I.M., HORGAN, R.: Cytokinin biosynthesis and metabolism, 105-133. In: see ref. CROZIER and HILLMAN 1984. - MCGAW, B.A., HORGAN, R., HEALD, J.K.: Phytochemistry 24, 9-13 (1985). - MEINS, F.J., FOSTER, R.: Dev. Biol. 108, 1-5 (1985). - MEINS, F.J., HANSEN, C.E.: Epigenetic and genetic factors regulating the cytokinin requirement of cultured cells, 333-340. In: see ref. BOPP 1986. - MEINS, F.J., FOSTER, R., LUTZ, J.D.: Dev. Genet. 4, 129-141 (1983). - MENHENETT, R., LAWRENCE, D.K. (eds.): Biochemical Aspects of Synthetic and Naturally Occurring Plant Growth Regulators. British Plant Growth Regul. Group - Monogr. 11. Wantage: Wessex Press 1984. - MERTZ, D.: Plant Cell Physiol. 26, 701-707 (1985). - METZGER, J.D.: Plant Physiol. 78, 8-13 (1985). - METZGER, J.D., MARDAUS, M.C.: Plant Physiol. 80, 396-402 (1986). - MEYER, A., MÜLLER, P., SEMBDNER, G.: Biochem. Physiol. Pflanzen 182, 1-21 (1987). - MICHAEL, G.: Ber. Dtsch. Bot. Ges. 97, 151-165 (1984). - MINTER, T.C., LORD, E.M.: Am. J. Bot. 70, 618-624 (1983). - MITA, T., KATSUMI, M.: Plant Cell Physiol. 27, 651-659 (1986). - MITA, T., SHIBAOKA, H.: Protoplasma 119, 100-109 (1984). - MOK, M.C., MOK, D.W.S.: Physiol. Plant. 65, 427-432 (1985). - MOONEY, P.A., VAN STADEN, J.: J. Plant Physiol. 123, 1-21 (1986). - MOORE, P.H., PHARIS, R.P., KOSHIOKA, M.: J. Plant Growth Regul. 5, 101-109 (1986). - MORISHITA, T., ABE, H., UCHIYAMA, M., MARUMO, S., TAKATSUTO, S., IKEKAWA, N.: Phytochemistry 22, 1051-1053 (1983). - MORRIS, R.O.: Annu. Rev. Plant Physiol. 37, 509-538 (1986). - MORRIS, R.O., POWELL, G.K., BEATY, J.S., DURLEY, R.C., HOMMES, N.G., LICA, L., MACDONALD, E.M.S.: Cytokinin biosynthetic genes and enzymes from *Agrobacterium tumefaciens* and other plant-associated prokaryotes, 185-196. In: see ref. BOPP 1986. - MUJER, C.V., MOK, M.C., TURNER, J.E., MOK, D.W.S.: Plant Physiol. 80 (Suppl.), 172 (1986). - MÜLLER, P., HILGENBERG, W.: Physiol. Plant. 66, 245-250 (1986). - MUNDY, J., FINCHER, G.B.: FEBS Lett. 198, 349-352 (1986). - MURAKAMI, Y.: Shokubutso no Kagaku Chosetsu 20, 1-11 (1985a); - Bull. Nat. Inst. Agric. Sci., Ser. D, No. 36, 69-123 (1985b). - MURAKOSHI, I., KOIDE, C., IKEGAMI, F., NASU, K.: Chem. Pharm. Bull. 31, 1777-1779 (1983). - MUROFUSHI, N., INOUE, A., WATANABE, N., OTA, Y., TAKAHASHI, N.: Plant Cell Physiol. 24, 87-92 (1983). - MUROMZEV, G.S., AGNISTIKOVA, W.N.: Gibberellins. Moscow: Nauka 1984. - MUTHUKRISHNAN, S., CHANDRA, G.R., MAXWELL, E.S.: J. Biol. Chem. 258, 2370-2375 (1983).

NADZHIMOV, U.K., LEV, S.V., ABZALOV, M.F., MUSAEV, D.A., KEFELI, V.I.: Fiziol. Rast. 30, 784-787 (1983). - NESTER, E.W., GORDON, M.P., AMASINO, R.M., YANOFSKY, M.F.: Annu. Rev. Plant Physiol. 35, 387-413 (1984). - NESTER, J.E., ZEEVAART, J.A.D.: Plant Physiol. 80 (Suppl.), 7 (1986). - NESTER, J.E., VEYSEY, S., COOLBAUGH, R.C.: Planta 170, 26-33 (1987). - NEUMANN, P.M., TUCKER, A.T., NOODEN, L.D.: Plant Physiol. 72, 182-185 (1983). - NICKELL, L.G. (ed.): Plant Growth Regulating Chemicals, Vol. 1 and 2. Boca Raton, FL: CRC Press 1983. - NISHIKAWA, S., KUMAZAWA, Z., KASHIMURA, N., MIZUTANI, H., KONDO, H.: Agric. Biol.

Chem. 49, 3353-3354 (1985). - NISHIKAWA, S., KUMAZAWA, Z., MITZUTANI, H., KASHIMURA, N.: Agric. Biol. Chem. 50, 1089-1092 (1986). - NOODEN, L.D.: Physiol. Plant. 62, 273-284 (1984). - NOODEN, L.D., LETHAM, D.S.: J. Plant Growth Regul. 2, 265-279 (1984); - Cytokinin control of monocarpic senescence in soybean, 324-332. In: see ref. BOPP 1985. - NORMAN, H.A., BLACK, M., CHAPMAN, J.M.: Planta 158 264-271 (1983). - NOSKOV, V.A., KINZURASHIWILI, L.N., SMIRNOVA, T.A., MANAMSCHJAN, G., KIRJANOV, I., VANJUSHIN, E.F.: Biokhimiya 50, 1901-1908 (1985). - NOUR, J.M., RUBERY, P.H.: Planta 160, 436-443 (1984).

O'NEILL, S.D., KEITH, B., RAPPAPORT, L.: Plant Physiol. 80, 812-817 (1986).

PALMER, M.V., WONG, O.C.: Plant Physiol. 79, 296-298 (1985). - PALMER, M.V., LETHAM, D.S., GUNNING, B.E.S.: J. Plant Growth Regul. 2, 289-298 (1984). - PALNI, L.M.S.: Aust. J. Plant Physiol. 11, 129-136 (1984). - PALNI, L.M.S., HORGAN, R.: Phytochemistry 22, 1597-1601 (1983). - PALNI, L.M.S., SUMMONS, R.E., LETHAM, D.S.: Plant Physiol. 72, 858-863 (1983). - PALNI, L.M.S., TAY, S.A.B., NANDI, S.K., PIANCA, D.J., DE KLERK, G.J.M., WONG, C., LETHAM, D.S., MACLEOD, J.K.: Biol. Plant. 27, 195-203 (1985). - PALNI, L.M.S., TAY, S.A.B., MACLEOD, J.K.: GC-MS methods for cytokinins and metabolites, 214-253. In: see ref. LINKENS and JACKSON 1986a. - PARK, K.-H.: Han'guk Nonghwa Hakhoechi 28, 82-87 (1985). - PARK, K.-H., FUJISAWA, S., SAKURAI, A., YAMAGUCHI, I., TAKAHASHI, N.: Plant Cell Physiol. 24, 1241-1249 (1983). - PARKER, C.W., ENTSCH, B., LETHAM, D.S.: Phytochemistry 25, 303-310 (1986). - PARTHIER, B.: Phytohormones and gene expression, 169-176. In: see ref. BOPP 1986a; - Action of plant growth substances in gene expression, 63-72. In: Selfregulation of Plant Metabolism. Sofia: Bulg. Acad. Sci. 1986b. - PARTHIER, B., LERBS, S., LEHMANN, J., WOLLGIEHN, R.: Biol. Plant. 27, 131-138 (1985). - PEAUD-LENOEL, C., AXELOS, M.: Physiol. Veg. 21, 537-549 (1983). - PEMBERTON, H.B., WILKINS, H.F.: J. Am. Soc. Hort. Sci. 110, 730-735 (1985). - PENGELLY, W.L.: Validation of immunoassays, 35-43. In: see ref. BOPP 1986. - PHARIS, R.P., KING, R.W.: Annu. Rev. Plant Physiol. 36, 517-568 (1985). - PHARIS, R.P., REID, D.M. (eds.): Hormonal Regulation of Development III. Role of Environmental Factors. Encyclopedia of Plant Physiology, N.S. Vol. 11. Berlin Heidelberg New York: Springer 1985. - PHILIPSON, J.J.: J. Exp. Bot. 34, 291-302 (1983). - PHINNEY, B.O.: Gibberellin A_1, dwarfism and the control of shoot elongation in higher plants, 17-41. In: see ref. CROZIER and HILLMAN 1984. - PHINNEY, B.O., SCHNEIDER, G.: Metabolism of GA-conjugates in *Zea mays*, 167-175. In: see ref. SCHREIBER et al. 1987. - PHINNEY, B.O., FREELING, M., ROBERTSON, D.S., SPRAY, C.R., SILVERTHORNE, J.: Dwarf mutants in maize - The gibberellin biosynthetic pathway and its molecular future, 55-64. In: see ref. BOPP 1986. - PICCIARELLI, P., ALPI, A.: Phytochemistry 26, 329-330 (1987). - POLOWICK, P.L., SAWHNEY, V.K.: Sci. Hortic. 25, 117-127 (1985). - POTTS, W.C., REID, J.B.: Physiol. Plant. 57, 448-454 (1983). - POTTS, W.C., REID, J.B., MURFET, J.C.: Physiol. Plant. 63, 357-364 (1985). - POWELL, G.K., MORRIS, R.O.: Nucleic Acid Res. 14, 2555-2565 (1986).

RADEMACHER, W., GRAEBE, J.E.: Biochem. Biophys. Res. Commun. 91, 35-40 (1979); - Ber. Dtsch. Bot. Ges. 97, 75-85 (1984a); - ibid. 97, 167-181 (1984b). - RADEMACHER, W., JUNG, J.: Gibberellin production in different species of *Sphaceloma* and *Elsinoe*, 15. In: see ref. BOPP et al. 1985. - RAILTON, J.D., FELLOWS, B., WEST, C.A.: Phytochemistry 23, 1261-1267 (1984). - RAO, I.V.R., MOHAN RAM, H.Y.: J. Plant Physiol. 122, 181-186 (1986). - RASKIN, I., KENDE, H.: Plant Physiol. 76, 947-950 (1984). - REID, J.B.: Plant Physiol. 72, 759-763 (1983). - REID, J.B., MURFET, I.C., POTTS, W.C.: J. Exp. Bot. 34, 349-364 (1983). - RESKO, R., ABEL, W.O.: Planta 165, 354-358 (1985). - ROGERS, J.C., MILLIMAN, C.: J. Biol. Chem. 259, 12234-12240 (1984). - ROMANI, G., MARRE, M.T., BONETTI,

A., CERANA, R., LADO, P., MARRE, E.: Physiol. Plant. 59, 528-532 (1983). - ROMANOV, G.A., TARAN, V.J., CHVOIKA, L., KULAEVA, O.N.: Fiziol. Rast. 33, 93-104 (1986). - ROOD, S.B., BEALL, F.D., PHARIS, R.P.: Plant Physiol. 80, 448-453 (1986). - ROOD, S.B., PHARIS, R.P.: Evidence for reversible conjugation of gibberellins in higher plants, 183-190. In: see ref. SCHREIBER et al. 1987. - ROOD, S.B., PHARIS, R.P., KOSHIOKA, M.: Plant Physiol. 73, 340-346 (1983). - ROSS, S.D., BOLLMANN, M.P., SWEAT, G.B., PHARIS, R.P.: Plant Physiol. 76, 326-330 (1984).

SACHAR, R.C., BERRY, M.: J. Sci. Ind. Res. 43, 595-614 (1984). - SAHA, S., NAGAR, P.K., SIRCAR, P.K.: Plant Growth Regul. 3, 27-35 (1985). - SALA, C., SALA, F.: Plant Cell Reports 4, 144-147 (1985). - SASSE, J.M.: Physiol. Plant. 63, 303-308 (1985). - SAUNDERS, M.J.: Planta 167, 402-409 (1986). - SAWHNEY, U.K.: Hormonal and temperature control of male-sterility in a tomato mutant, 36-38. In: Proc. 8th Int. Symp. Sexual Reproduction in Seed Plants, Ferms and Mosses. Wageningen 1985. - SCHELL, J., VAN MONTAGU, M., SCHRÖDER, J., SCHRÖDER, G., INZE, D.: Horm. Cell Regul. 8, 245-254 (1984). - SCHLAGNHAUFER, C.D., ARTECA, R.N.: Physiol. Plant. 65, 151-155 (1985a); - Plant Physiol. 78, 300-303 (1985b). - SCHLAGNHAUFER, C., ARTECA, R.N., YOPP, I.H.: Physiol. Plant. 61, 555-558 (1984). - SCHLIEMANN, W.: J. Plant Physiol. 116, 123-132 (1984); - Enzymic hydrolysis of gibberellin conjugates, 191-198. In: see ref. SCHREIBER et al. 1987. - SCHNEIDER, G.: Gibberellin conjugates, Vol. 1, 389-456. In: see ref. CROZIER 1983. - Gibberellin conjugation, 158-166. In: see ref. SCHREIBER et al. 1987. - SCHNEIDER, G., SEMBDNER, G., PHINNEY, B.O.: J. Plant Growth Regul. 3, 207-215 (1984). - SCHNEIDER, G., SCHMIDT, J., PHINNEY, B.O.: J. Plant Growth Regul. 5, 217-223 (1987). - SCHNEIDER, J.A., YOSHIHIRA, K., NAKANISHI, K., KATO, N.: Tetrahedron Lett. 24, 3859-3860 (1983). - SCHRAUDOLF, H.: Phytohormones and Filicinae: Chemical signals triggering morphogenesis in Schizaeaceae, 270-274. In: see ref. BOPP 1986. - SCHREIBER, K., SCHÜTTE, H.R., SEMBDNER, G. (eds.): Conjugated Plant Hormones, Structure, Metabolism and Function. Berlin: Deutscher Verlag der Wissenschaften 1987. - SCHRÖDER, J., BUCHMANN, I., SCHRÖDER, G.: Enzymes of auxin and cytokinin biosynthesis encoded in Ti plasmids, 177-184. In: see ref. BOPP 1986. - SCHRÖDER, G., WAFFENSCHMIDT, S., WEILER, E.W., SCHRÖDER, J.: Eur. J. Biochem. 138, 387-391 (1984). - SCHWENEN, L., HERBST, G., GABRIEL, K., GRAEBE, J.E.: The occurrence of GA_{12}-aldehyde and *ent*-kaurene in mature pea seeds and their utilization during imbibition and early growth, 13. In: see ref. BOPP et al. 1985. - SCOTT, J.M., HORGAN, R.: Planta 161, 345-354 (1984). - SCOTT, T.K. (ed.): Hormonal Regulation of Development II. The Functions of Hormones from the Level of the Cell to the Whole Plant. Encyclopedia of Plant Physiology, N.S. Vol. 10. Berlin Heidelberg New York: Springer 1984. - SEBANEK, J., VITKOVA, H., KLILOVA, C.: Acta Univ. Agr. 1-2, 21-23 (1983). - SELIVANKINA, S.J., ROMANKO, E.G., KARAWAIKO, N.N., KULAEVA, O.N.: Dokl. AN SSSR 272, 761-763 (1983). - SELIVANKINA, S.J., ROMANKO, E.G., OVTSKAROV, A.K.: Sel. Khos. Biologia 5, 123-125 (1985). - SEMBDNER, G., KNÖFEL, H.-D., SCHWARZKOPF, E., LIEBISCH, H.W.: Biol. Plant. 27, 231-236 (1985). - SEMBDNER, G., SCHNEIDER, G., SCHREIBER, K. (eds.) Methoden zur Pflanzenhormonanalyse. Jena: Fischer 1987. - SEN, S.P.: Adv. Agric. Biotechnol. 16, 1-40 (1985). - SENAPATHY, P., JAYABASKARAN, C., JACOB, M.J.: Nucleic Acid Res. 13, 5833-5842 (1985). - SEREBRYAKOV, E.P., EPSTEIN, N.A., YASINSKAYA, N.P., KAPLUA, A.B.: Phytochemistry 23, 1855-1864 (1984). - SHARMA, R.: Photochem. Photobiol. 41, 747-755 (1985). - SHEN-MILLER, J., WEST, C.A.: Plant Physiol. 74, 439-441 (1984); - Phytochemistry 24, 461-464 (1985). - SHIAO, M.S., HAO, Y.Y.: Inhibition of isoprenoid biosynthesis by fungal metabolites, 367-373. In: Trichothecenes and other Mycotoxins, ed. J. LACEY. Proc. Int. Mycotoxin Symp. 1984. Chichester: Wiley 1985. - SINGH, R.K., CHONDHURY, B., MEHTA, S.L., SIVAKAMI, N.: J. Nucl. Agric. Biol. 13, 1-4 (1984). - SINGH, S.P., PALEG, L.G.: Plant Physiol. 76, 139-142

(1984a); - ibid. 76, 143-147 (1984b). - ibid. 74, 437-438 (1984c); - Aust. J. Plant Physiol. 12, 269-275 (1985a); - ibid. 12, 277-289 (1985b). - SLADKY, Z.: Biol. Plant. 28, 31-37 (1986). - SMITH, V.A., MACMILLAN, J.: J. Plant Growth Regul. 2, 251-264 (1984); - Planta 167, 9-18 (1986). - SPONSEL, V.M.: In vivo metabolism of gibberellins in higher plants, Vol. 1, 151-250. In: see ref. CROZIER 1983; - Planta 159, 454-468 (1983a); - Physiol. Plant. 65, 533-538 (1985); - Planta 168, 119-129 (1986). - SPRAY, C., PHINNEY, B.O., GASKIN, P., GILMOUR, S.J., MACMILLAN, J.: Planta 160, 464-468 (1984). - STEFFENS, G.L., BYUN, J.H., WANG, S.Y.: Physiol. Plant. 63, 163-168 (1985). - STILLWELL, W., HESTER, P., BRENGLE, B.: J. Plant Physiol. 118, 105-110 (1985). - STODDART, J.L.: Planta 161, 432-438 (1984); - Gibberellin receptors, 91-114. In: see ref. CHADWICK and GARROD 1986. - STRNAD, M., KAMINEK, M.: Biol. Plant. 27, 209-215 (1985). - SUGE, H.: Plant Cell Physiol. 26, 607-614 (1985). - SUMMONS, R.E., PALNI, L.M.S., LETHAM, D.S.: FEBS Lett. 151, 122-126 (1983). - SURI, R.A., MANDAHAR, C.L.: Experientia 40, 462-463 (1984). - SURICO, G., EVIDENTE, A., IACOBELLIS, N.S., RANDAZZO, G.: Phytochemistry 24, 1499-1502 (1985). - SUTTLE, J.C.: Plant Physiol. 82, 930-938 (1986). - SUZUKI, Y., YAMAGUCHI, I., TAKAHASHI, N.: Agric. Biol. Chem. 49, 49-54 (1985).

TAKAGI, M., YOKOTA, T., MUROFUSHI, N., OTA, Y., TAKAHASHI, N.: Agric. Biol. Chem. 49, 3271-3277 (1985). - TAKAHASHI, N. (ed.): Chemistry of Plant Hormones. Boca Raton: CRC Press 1986. - TAKAHASHI, N., YAMAGUCHI, I.: Preparation and isotopic labeling of gibberellins, Vol. 1, 457-484. In: see ref. CROZIER 1983. - TAKAHASHI, M., KAMIYA, Y., TAKAHASHI, N., GRAEBE, J.E.: Planta 168, 190-199 (1986). - TAKAHASHI, M., YAMANE, H., SATOH, Y., TAKAHASHI, N., IWATSUKI, K.: Phytochemistry 23, 681 (1984). - TAKATSUTO, S., IKEKAWA, N.: J. Chem. Soc. Perkin Trans. I, 439-447 (1984). - TAKATSUTO, S., YAZAWA, N., IKEKAWA, N.: Phytochemistry 23, 525-528 (1984). - TAO, G.-Q., LETHAM, D.S., PALNI, L.M.S., SUMMONS, R.E.: J. Plant Growth Regul. 2, 89-102 (1984). - TAY, S.A.B., MACLEOD, J.K., PALNI, L.M.S., LETHAM, D.S.: Phytochemistry 24, 2611-2614 (1985). - TAY, S.A.B., MACLEOD, J.K., PALNI, L.M.S.: Plant Sci. 43, 131-134 (1986). - TEYSSENDIER DE LA SERVE, B., AXELOS, M., PEAUD-LENOEL, C.: Plant Mol. Biol. 5, 155-163 (1985). - THOMAS, J.C., KATTERMAN, F.R.: Plant Physiol. 81, 681-683 (1986). - TITOV, A.F., DROSDOV, S.N., KRITENKO, S.P., TALANOVA, V.V., SHERUDILO, E.G.: Physiol. Biochem. Kulturpflanzen 18, 64-69 (1986). - TOMI, H., SASAKI, Y., KAMIKUBO, T.: Plant Cell Physiol. 24, 587-592 (1983a); - ibid. 24, 1087-1092 (1983b). - TONG, Z., KASEMIR, H., MOHR, H.: Planta 159, 136-142 (1983). - TRIONE, E.J., KRYGIER, B.B., BANOWETZ, G.M., KATHREIN, I.M.: J. Plant Growth Regul. 4, 101-109 (1985). - TURNBULL, C.G.N., HANKE, D.E.: Planta 165, 359-365 (1985a); - ibid. 165, 359-365 (1985b). - TURNBULL, C.G.N., CROZIER, A., SCHNEIDER, G.: Phytochemistry 25, 1823-1828 (1986a). - TURNBULL, C.G.N., CROZIER, A., SCHWENEN, L., GRAEBE, J.E.: Phytochemistry 25, 97-101 (1986b); - Planta 165, 108-113 (1985). - TURNER, J.E., MOK, M.C., MOK, D.W.S.: Plant Physiol. 79, 321-322 (1985). - TURNER, J.E., MUJER, C.V., MOK, D.W.S., MOK, M.C.: Plant Physiol. 80 (Suppl.), 95 (1986).

VAN DER KRIEKEN, W.M., BARENDSE, G.W.M., CROES, A.F.: BAP metabolism during in vitro flower bud formation in tobacco, 153-157. In: see ref. SCHREIBER et al. 1987. - VAN STADEN, J.: Z. Pflanzenphysiol. 111, 423-428 (1983). - VAN STADEN, J., FORSYTH, C.: Z. Pflanzenphysiol. 114, 27-33 (1984). - VAN STADEN, J., FORSYTH, C., BERGMAN, I., ARNOLD, S.: Physiol. Plant. 66, 427-434 (1986). - VARTY, K., ARREGUIN, B., GOMEZ, M.T., PABLO, P.J.T., GOMEZ, M.A.L.: Plant Physiol. 73, 692-697 (1983). - VINCE-PRUE, D.: Photoperiod and hormones, 308-364. In: see ref. PHARIS and REID 1985. - VOIGT, B., ADAM, G.: Tetrahedron 39, 449-453 (1983). - VOLGER, H., GRAEBE, J.E.: Isolation and characterization of soluble enzymes involved in gibberellin metabolism, 10. In: see ref. BOPP et al. 1985.

WADA, K., MARUMO, S., ABE, H., MORISHITA, T., NAKAMURA, K., UCHIYAMA, M., MORI, K.: Agric. Biol. Chem. 48, 719-726 (1984). - WADA, K., HIROAKI, K., MARUMO, S.: Agric. Biol. Chem. 49, 2249-2251 (1985). - WANG, S.Y., BYUN, J.K., STEFFENS, G.L.: Physiol. Plant. 63, 169-175 (1985). - WANG, T.L., FUTERS, T.S., MCGEARY, F., COVE, D.J.: Moss mutants and the analysis of cytokinin metabolism, 135-164. In: see ref. CROZIER and HILLMAN 1984. - WANG, T.L., GRIGGS, P., COOK, S.: Immunoassays for plant growth regulators - a help or a hindrance?, 26-34. In: see ref. BOPP 1986. - WAREING, P.F.: Hormones and plant development: coordination and control in the whole plant and analogies with insect development, 331-348. In: Developmental Control in Animals and Plants, eds. C.F. GRAHAM, P.F. WAREING. Oxford: Blackwell 1984. - WASILEWSKA, L.D., BRALCZYK, J., MAZURKIEWICZ, J.: Acta Biochim. Pol. 31, 91-102 (1984). - WEIDHASE, R.A., LEHMANN, J., KRAMELL, H., SEMBDNER, G., PARTHIER, B.: Physiol. Plant. 69, 161-166 (1987). - WEILER, E.W.: Biochem. Soc. Trans. 11, 485-495 (1983); - Annu. Rev. Plant Physiol. 35, 85-95 (1984); - Plant hormone immunoassays based on monoclonal and polyclonal antibodies, 1-16. In: see ref. LINSKENS and JACKSON 1986b. - WEILER, E.W., EBERLE, J., MERTENS, R.: Immunoassays for the quantitation of plant growth regulators using monoclonal antibodies, 22-25. In: see ref. BOPP 1986.

YALPANI, N., SRIVASTAVA, L.M.: Plant Physiol. 79, 963-967 (1985). - YAMAGUCHI, I., FUJISAWA, S., TAKAHASHI, N.: Systematic ultramicro analysis of plant growth regulators, 145-150. In: IUPAC Pesticide Chemistry, Human Welfare and the Environment, eds. J. MIYAMOTO, P.C. KEARNEY. Proc. 5th Int. Congr. Pestic. Chem., 1982. Oxford: Pergamon 1983. - YAMANE, H., WATANABE, M., SATOH, Y., TAKAHASHI, N., IWATSUKI, K.: Plant Cell Physiol. 24, 1027-1031 (1983). - YAMANE, H., YAMAGUCHI, I., KOBAYASHI, M., TAKAHASHI, M., SATO, Y., TAKAHASHI, N., IWATSUKI, K., PHINNEY, B.O., SPRAY, C.R., GASKIN, P., MACMILLAN, J.: Plant Physiol. 78, 899-903 (1985). - YIP, W.-K., YANG, S.F.: Plant Physiol. 80, 515-519 (1986). - YOKOTA, T., TAKAHASHI, N.: Chemistry, physiology and agricultural application of brassinolide and related steroids, 129-138. In: see ref. BOPP 1986. - YOKOTA, T., BABA, J., TAKAHASHI, N.: Agric. Biol. Chem. 47, 1409-1411 (1983a). - YOKOTA, T., MORITA, M., TAKAHASHI, N.: Agric. Biol. Chem. 47, 2149-2151 (1983b). - YOKOTA, T., BABA, J., KOBA, S., TAKAHASHI, N.: Agric. Biol. Chem. 48, 2529-2534 (1984). - YOKOTA, T., ARIMA, M., TAKAHASHI, N., CROZIER, A.: Phytochemistry 24, 1333-1335 (1985). - YOKOTA, T., KIM, S.K., FUKUI, Y., TAKAHASHI, N., TAKEUCHI, Y., TAKEMATSU, I.: Phytochemistry 26, 503-506 (1987a). - YOKOTA, T., KIM, S.K., KOSAKA, Y., OGINO, Y., TAKAHASHI, N.: Conjugation of brassinosteroids, 288-296. In: see ref. SCHREIBER et al. 1987b.

ZACK, C.D., LOY, J.B.: J. Plant Growth Regul. 3, 65-73 (1984). - ZAVALA, M.E., BRANDON, D.L.: J. Cell Biol. 97, 1235-1239 (1983). - ZEEVAART, J.A.D.: Gibberellins and flowering, Vol. II, 333-374. In: see ref. CROZIER 1983; - Annu. Report of the MSU-DOE Plant Research Laboratory Michigan, 142 (1984); - Planta 166, 276-279 (1985). - ZEIGLER, R.S., POWELL, L.E., THURSTON, H.D.: Phytopathol. 70, 589-593 (1980). - ZHANG, R., LETHAM, D.S., NOODEN, L.D.: Cytokinin conjugation in relation to soybean leaf senescence, 128-137. In: see ref. SCHREIBER et al. 1987. - ZIMMERMAN, R.H., HACKETT, W.P., PHARIS, R.P.: Hormonal aspects of phase change and precocious flowering, 79-115. In: see ref. PHARIS and REID 1985.

Prof. Dr. GÜNTHER SEMBDNER,
Dr. WILLIBALD SCHLIEMANN and Dr. GABRIELE HERRMANN
Akademie der Wissenschaften der DDR
Forschungsbereich Biowissenschaften und Medizin
Institut für Biochemie der Pflanzen
DDR-4050 Halle (Saale)

VI. Movement of Pulvinated Leaves

By Wolfgang Hensel

1. Introduction

HAUPT, in the 1982 issue of Progress in Botany reviewed the current literature on turgor-mediated movements. But in spite of considerable progress in research since then, the mechanism of leaf movement is still not completely understood. Complexity of the organ and the tissues involved impede the analysis of a structure-function relationship. Besides circadian rhythm several physical and chemical stimuli induce a response of the same motor organs, complicating the evaluation of the regulatory properties of the system. For some recent results on leaf movements as an ecophysiological adaptation, see e.g. LUDLOW and BJÖRKMAN (1984; protective mechanism against heat stress), WERK and EHLERINGER (1984; compass plant *Lactuca serriola* L.) and SCHWARTZ and KOLLER (1986; sun tracking).

2. The Proton/Potassium Exchange Model

In a 1981 published review, SATTER and GALSTON proposed a model focussing on the membrane-located events of the motor tissue, as a substantiated working hypothesis to explain the turgor-mediated movements (Fig. 1).

In this review, the terminology of SATTER is used: extensor cells will gain turgor during opening and lose it during closure, while flexor cells will gain turgor during closure, losing it during opening of the pulvinus.

Potassium is the main cationic osmoticum for the turgor regulation together with anionic chloride and to a lesser extent organic and other anions. The driving force for their diffusional movements across the plasma membrane is an outwardly directed proton pump. A role of calcium was suggested in the light-dependent turgor movements via the phytochrome system. The lower panel of Figure 1 indicates the proposed ion movements and electrophysiological events during the opening movement of the leaf (increase in extensor volume/decrease in flexor volume). The electrogenic proton efflux in extensor cells will drive a potassium influx, hence increasing the turgor. In flexor cells the activity of the proton pump is minimal during cell shrinking. The outwardly directed potassium flux is suggested to be passive, promoted by the depletion of potassium and chloride in the flexor apoplast in favour of apoplastic diffusion directed towards extensor tissue.

Experimental evidence is now in favour of this model (for the older literature see SATTER and GALSTON 1981). In an elemental analysis by atomic spectrometry, ROBLIN and FLEURAT-LESSARD (1983a) found the potassium and chloride content and the capacity to respond to fusicoccin by proton extrusion to be maximal in young internodes and pulvini of *Mimosa*.

Progress in Botany, Vol. 49

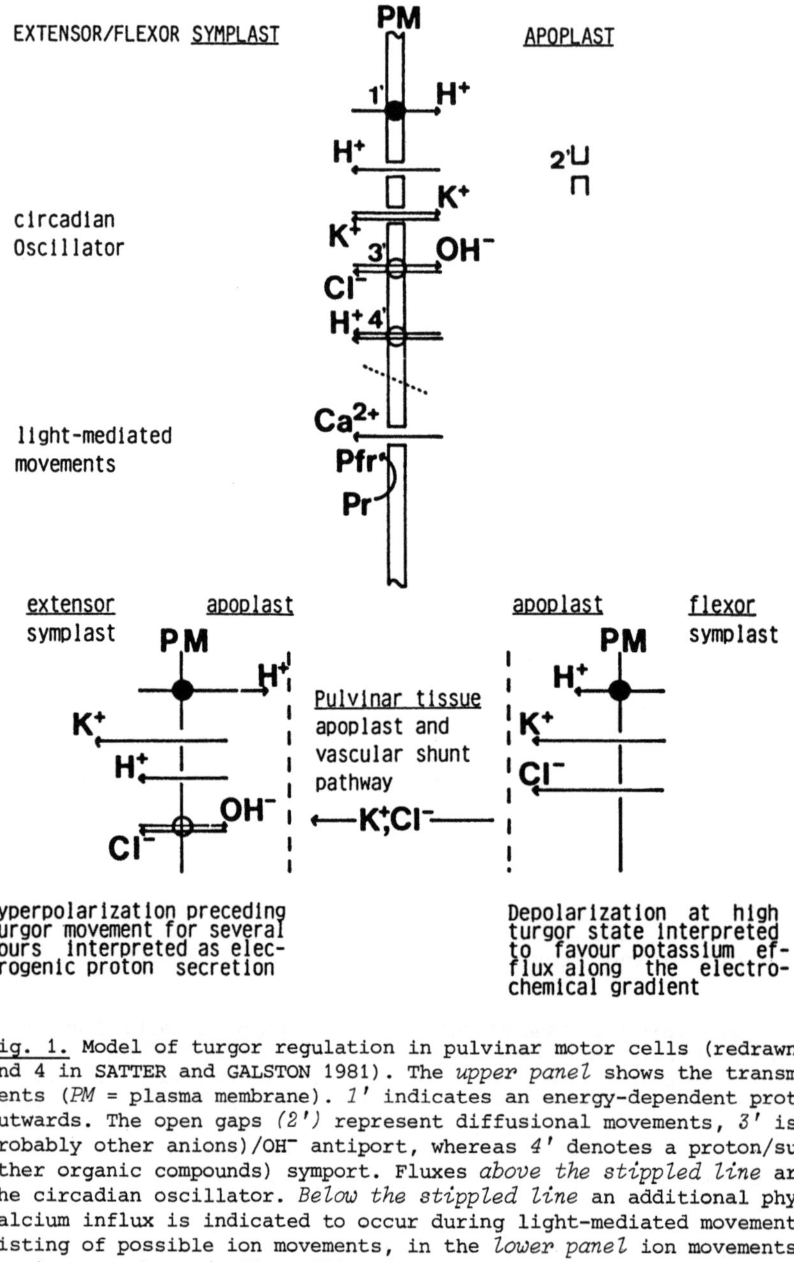

Fig. 1. Model of turgor regulation in pulvinar motor cells (redrawn after Figs. 3 and 4 in SATTER and GALSTON 1981). The *upper panel* shows the transmembrane ion movements (*PM* = plasma membrane). *1'* indicates an energy-dependent proton pump, directed outwards. The open gaps *(2')* represent diffusional movements, *3'* is a chloride (and probably other anions)/OH^- antiport, whereas *4'* denotes a proton/sucrose (and/or other organic compounds) symport. Fluxes *above the stippled line* are influenced by the circadian oscillator. *Below the stippled line* an additional phytochrome-dependent calcium influx is indicated to occur during light-mediated movements. Besides this listing of possible ion movements, in the *lower panel* ion movements during pulvinar opening are shown (cell swelling in the extensor, cell shrinking in the flexor)

In excised tissue strips from extensor and flexor tissue, IGLESIAS and SATTER (1983a) measured the proton/potassium fluxes. Upon illumination, *Samanea* leaflets will open by a turgor increase in the extensor, while the flexor will lose turgor. The energy-dependent acidification of media containing extensor tissue and the alkalization of flexor tissue was found to be correlated with potassium influx in extensor and efflux in flexor tissue. Darkening, which promotes leaflet closure, yielded the opposite results. The proton influx was independent of anoxia treatment. These data were confirmed by measuring the ion fluxes during rhythmic movement of *Samanea* leaflets (IGLESIAS and SATTER 1983b).

Pulvini excised from *Samanea* plants grown in day/night cycles were kept in continuous darkness. Lowering the pH of the bathing solution to 5.5 yielded a proton uptake, however, with rhythmic properties only in the extensor. Lowering the intracellular water content by application of mannitol activated the proton pump, releasing protons into the medium, and induced rhythm in the flexor, while in extensor cells the amplitude of the rhythmic proton efflux/influx increased. The effect of the turgor may indicate an involvement of the cell-turgor state in the feedback mechanism (LEE and SATTER 1987).

In the open-pulvinus state, GORTON and SATTER (1984a,b) isolated protoplasts from the extensor and flexor tissues of *Samanea*. X-ray analysis gained a two-fold higher concentration of potassium and chloride in flexor protoplasts with about the same concentrations of sulphur and phosphorus. The variance to the in situ situation was suggested by the authors to be caused by a leakage of ions during the isolation procedure and by the presence of unidentified osmotica.

Upon stimulation of the *Mimosa* petioles with ice-cold water, the petiole angle decreased transiently by about 10°. In both, extensor and flexor, the extracellular pH, measured in situ, increased transiently from pH 5.1 to 5.7 in the extensor and to 5.4 in the flexor. Hence, the transpulvinar gradient of coupled proton/potassium fluxes fits the model (KUMON and SUDA 1985). From calculations of the potassium and chloride contents in the motor tissue, together with conductivity measurements of the bathing solution, KUMON and SUDA (1984) concluded that (1) about 10% of the ions and the water content of the tissue were lost by efflux into the bathing solution and (2) that the measured fluxes could only be explained, if a bulk flow of solutes (not a diffusional flow) was assumed.

The gradient in the water potential during the open state of *Samanea* leaflets is brought about by the difference in osmotic potential yielding about the same hydrostatic pressure in extensor and flexor. During closure, the hydrostatic pressure in flexor cells increased, caused by an alteration of the osmotic potential (reduced gradient of the water potential), while remaining constant in extensor cells. Re-opening will occur, if the hydrostatic pressure of the flexor decreases again. During both states, hydrostatic pressure remains similar as in extensor cells. In addition, it was shown that upon excision both extensor and flexor increased in length, hence decreasing hydrostatic pressure and the difference in water potential. The restraining effect of the intact tissue thus has to be considered, if data from excised tissue are evaluated (GORTON 1987).

The capacity of *Phaseolus* pulvinar tissue to use different anions, co-redistributing with potassium, was shown by HOSOKAWA and KIYOSAWA (1983b). Depending on the nutrient solution in which the plants were cultivated, the chloride, nitrate or phosphate concentration rose in the pulvini. The diurnal redistribution of potassium, however, was only accompanied by chloride and nitrate. In KH_2PO_4 solution or in distilled water other, probably organic, anions were supposed to balance the potassium movement. In distilled water and, in particular, in KH_2PO_4 solution the actual leaf movements were less pronounced than in KCl or KNO_3 solution.

During the rhythmic movement of *Phaseolus* pulvini, the chloride concentration changed according to the observed turgor changes. In chloride-deprived plants, malate and to a lesser extent citrate were shifted rhythmically between extensor and flexor to compensate for the potassium changes (MAYER et al. 1987). The malic acid concentrations was also shown to change rhythmically in *Phaseolus coccineus* (BIALCZYK and LECHOWSKI 1987).

By means of a different experimental approach, OTSIOGO-OYABI and ROBLIN (1984, 1985a) confirmed the role of proton translocation in light-mediated turgor movements of *Mimosa* leaves. Upon darkening during the day period, the pulvini performed sleep movements (scotonasty), while illumination during the night period caused movement into the day position (photonasty). Glycine-inhibited scotonastic movement and caused alkalization of the medium. Both effects could be overcome by simultaneous fusicoccin application, by stimulating proton efflux, which caused potassium influx in the flexor. ATPase inhibitors decrease the glycine uptake. Together with the depolarizing effect of glycine and the hyperpolarizing effect of fusicoccin (OTSIOGO-OYABI and ROBLIN 1985b), the authors suggested an electrogenic proton/glycine influx.

As known from other tissues, KUMON et al. (1984) found an IAA-induced hyperpolarization of *Mimosa* pulvinar extensor cells coupled to a pH decrease in the bathing solution, indicating an outwardly directed proton pump (however, no IAA effect was found in protoplasts). Since IAA induced opening, brought about by a turgor increase in extensor cells, these data are consistent with the model (see also below). A different viewpont was stressed by FREUDLING et al. (1980; see discussion in SATTER and GALSTON 1981). They determined the membrane properties of cross-sectioned *Phaseolus* pulvinar tissue slices, in particlar the membrane potential, external potassium-induced voltage changes and CN^- induced depolarization. Alterations of the turgor are brought about in their model by an electrogenic anion (probably chloride) influx with an electro-osmotically driven water flux.

In *Phaseolus*, blue light induced small pulvinar movements, superimposed upon the circadian movements (NISHIZAKI 1986). Both flexor and extensor flanks of the pulvini perceived light and responded by a turgor decrease of the illuminated flank, causing a movement towards the light source. Turgor movements were preceded by transient alterations of the membrane potentials of extensor and flexor cells: illumination from the flexor side yielded a positive potential difference between flexor and extensor membrane potentials, i.e., the flexor membrane potential is more negative (hyperpolarized) than that of the extensor cells. Illumination from the extensor side gave the opposite results. As in FREUDLING et al. (1980), these data indicate that hyperpolarization is correlated with the low turgor state of the motor tissue. Movements towards the flexor, i.e. potassium efflux, according to the model of SATTER and GALSTON (1981), should be accompanied by a decreased proton efflux, hence yielding a depolarization rather than the measured hyperpolarization.

3. The Role of Calcium

In the model of SATTER and GALSTON (1981) calcium was suggested to be involved in the regulatory mechanism rather than as a turgor-gaining ion. In favour of this regulatory role are the findings by ROBLIN and FLEURAT-LESSARD (1987) that calcium redistribution during the gravitropic response of *Mimosa* primary pulvini was exactly opposite to the redistribution of turgor-gaining potassium and chloride ions. In an older study by CAMPBELL and THOMSON (1977) lanthanum, a calcium-mimicking but not permeating ion, was found to inhibit closure of seismonastically stimulated *Mimosa* pulvini; opening movement was not affected. On the other hand, EDTA slowed down the opening movement and inhibited closure. By using EGTA, which chelates calcium more specifically, and DMSO, which generally increases membrane permeability, MAYER et al. (1985b) concluded that calcium may act by maintaining the membrane integrity without any effect on contractile elements in the cytoplasm (see literature in MAYER et al. 1985b). Calcium chelation and DMSO led to a passive efflux of solutes from the cells of the primary pulvinus of *Phaseolus*, hence inducing a closure (downward) movement and inhibited opening movement. In tertiary pulvini of *Mimosa* and *Albizzia*, a loss of solutes caused a downward (opening) movement and inhibited the upward (closing) movement. Solute loss-induced movements of both primary and tertiary pulvini were promoted physically by the weight of the leaf and pinnules respectively. The same interpretation appears likely in the DMSO effect on isolated *Cassia* leaflets, described by ROBLIN and FLEURAT-LESSARD (1983b).

At variance are data by ROBLIN and FLEURAT-LESSARD (1984), who found that EGTA inhibited both opening and closure of *Cassia* leaflets, while the calcium ionophore A23187 increased both movements. Generally, the same was found in the gravitropic and seismonastic response of *Mimosa* primary pulvini (ROBLIN and FLEURAT-LESSARD 1987). Calmodulin inhibitors promoted the movement. In a later publication, ROBLIN et al. (1987) used other calmodulin antagonists in *Mimosa* without any effect on the movement. Hence, the contradictory results with calcium ionophores and calmodulin inhibitors await further clarification.

Common to all these studies is the fact that increasing the free intracellular calcium will promote, while reducing it will retard the induced movements.

In animal cells it is now well established that inductor/receptor-dependent calcium regulation of several intracellular responses is triggered by the metabolism of the plasma membrane phosphoinositols. Inositol 1,4,5-triphosphate, released from the plasma membrane by a phosphodiesterase, will trigger calcium efflux from the endoplasmic reticulum (for literature see POOVAIAH et al. 1987). By providing excised secondary pulvini from *Samanea* with labelled inositol, MORSE et al. (1987) were able to demonstrate that the intermediates of the phosphoinositol cycle also occur in motor cells. Further research may shed light on a possible involvement in the regulation of the stimulated movement.

4. Effects of Various Substances

Several publications deal with the effects of different substances on the movement of leaves. In most of these cases, however, the results are rather phenomenological than explanatory for the mechanism of turgor-mediated movement.

Ethylene appears to influence the course of the gravitropic response of *Mimosa*, a re-orientation movement of the primary pulvinus, since the ethylene-gaining compound ethephon promoted and inhibitors of ethylene synthesis reduced the displacement angle. The authors suggested an ethylene effect on the membrane permeability of the motor tissue (ROBLIN and PERRAULT 1985).

TSURUMI and ASAHI (1985) proposed jasmonic acid as a possible endogenous regulator (identified in *Mimosa* pulvinules by mass spectrometry, HPLC[1] and TLC[2]). Authentic jasmonic acid was able, if provided exogenously, to promote closing and inhibit opening.

[1] High-performance liquid chromatography; [2]thin-layer chromatography.

Prostaglandins E_1 and E_2, known to alter the membrane permeability in animal cells, promoted both light-induced opening and closure of *Cassia* leaflets (ROBLIN and BONMORT 1984). The fact that precursors of prostaglandin biosynthesis act accordingly and inhibitors of the biosynthetic pathway also influenced the movements may indicate a role of prostaglandins. However, too little is known about the occurrence and action of prostaglandins in plants, to yield clear-cut interpretations. Although SAEEDI et al. (1984) showed that salicyclic acid and acetylsalicyclic acid inhibited light-induced closure and promoted opening of *Cassia* leaflets, these results do not allow one substantially to suggest a regulatory mechanism. The same holds true for the use of several respiration inhibitors, which yielded differing results, in particular concentration dependency (SAEEDI and ROBLIN 1986). From the data, however, it is obvious that both closure and opening of *Cassia* leaflets are energy-dependent movements, which is in good accordance with the model of an electrogenic proton pump, causing the potassium influx at flexor and extensor cells respectively.

5. Cell Walls

Re-orientation movement of the pulvini requires, besides regulatory mechanisms of ion fluxes at the motor cell plasma membrane, massive ion redistribution from extensor to flexor and vice versa (about 60% of the potassium and chloride content is shifted between the opposite flanks of the motor tissue). On freeze-dried sections SATTER et al. (1982, see also CAMPBELL et al. 1981) determined by X-ray microanalysis the pathways of potassium and chloride redistribution during the pulvinar movements of *Samanea*. They found that ions could not freely diffuse via the apoplastic pathway but a diffusion barrier occurred at the extensor/flexor interface. Hence, at the inner cortex, collenchyma and phloem site a symplastic, plasmodesmatal movement of ions must be considered.

STARRACH et al. (1985) equilibrated isolated extensor cell walls of *Phaseolus* in 1 M KCl and measured the activity of fixed negative charges. They found that cell walls, in particular motor cell walls, effectively bound cations. However, the apoplastic fraction of the tissue was subdivided into a water-free space and about 55% into a Donnan-free space. The capacity of the walls to fix cations is caused mainly by the carboxyl groups of the pectin moiety. Furthermore, it was shown (STARRACH and MAYER 1986) that the cell walls had unequally distributed fixed negative charges. In the petiole, only that wall fraction facing the intercellulars was rich in pectins (middle lamella) and had accordingly a high activity of fixed negative charges, while in the pulvinar tissue the whole cell wall was rich in pectins (ruthenium red positive and digestible by pectinase). Furthermore, the physically upper flanks of the motor tissues (extensor in primary, flexor in secondary pulvini) had a higher content of fixed negative charges than the respective physically lower flanks.

6. Rhythmic Properties

A specific inhibiting effect of lithium chloride on the closure of *Cassia* leaves was described by GAILLOCHET (1981), with lithium being the active component (tested by chloride application together with other cations).

Other alkali ions, in addition to lithium, were tested by RINNAN and JOHNSSON (1986): sodium and potassium had no effect, rubidium shortened and lithium and caesium lengthened the circadian periods. Additionally, rubidium, if not provided continuously, but as a 4 h pulse, resulted in phase shifting by advances.

A dual control system of the leaf position of *Phaseolus* was shown by HOSOKAWA and KIYOSAWA (1983a). Upon inversion, the leaves re-oriented with respect to gravity. If the plants were inverted during the day position (open state), the leaves responded gravitropically negative, while inversion during the night position (closed state) led to a positive gravitropic response (gravinyctitropism). These turgor movements were brought about by an ion redistribution (potassium, chloride, nitrate) as during rhythmic movement. Hence, the plant is able to perceive gravity, the direction of the response, however, is controlled by the actual state of the circadian clock. As an age-dependent response without additional external stimuli, darkened *Albizzia* leaflets performed advanced opening at a larger angle and returned less efficiently to the closed state. While the amplitude of the movement is thus clearly influenced, the duration of the phases remains unaffected by the age of the excised leaves (CHEN et al. 1984; see also SATTER et al. 1987).

An alteration of the period length, however, is caused by affecting the membrane permeability of *Phaseolus* pulvini (MAYER et al. 1985c). Non-permeating osmotica and more pronounced DMSO lengthened the circadian periods of isolated leaves under continuous light, while NaCl and p-hydroxymercuribenzoate shortened the period length. The authors concluded that the circadian clock may be influenced by alterations of the membrane properties and ruled out a more homeostatic regulation of the rhythm.

Exogenous IAA and to a lesser extent gibberellic acid induced the opening of *Mimosa* pulvinules. The response of the tissue, however, strongly depended upon the phase of the circadian cycle and was reversible after washing out the hormones. Hence, the endogenous sensitivity of the motor organs towards the hormones rather than the hormone action itself was suggested to be responsible for the effect (TSURUMI et al. 1985).

A similar result was reported earlier by WATANABE and SIBAOKA (1983), who interpreted the small but nevertheless promotive opening effect of IAA during the night period in favour of an IAA-activated proton pump in the extensor (cf. the model of SATTER and GALSTON 1981). In contrast to light-induced opening, which occurred without time lag, IAA needed 100 min to cause an effect.

On the other hand, if *Mimosa* leaves were excised at the petioles, the primary pulvini, remaining at the stem, maintained their diurnal movements (opening *and* closure, see above) only, if IAA was supplied from the cutting site. Gibberellic acid, kinetin, abscisic acid and the ethylene precursor ACC failed to have a maintaining effect (MORIMOTO et al. 1986). IAA provided as pulses during the light phase, mimicked the control movements of intact leaves best, although concentrations higher than 10^{-6} M induced superimposed irregular movements. Due to the concentration dependency of the response, the authors ruled out that exogenous IAA may simply overcome the senescence effect of the abscission. They proposed instead a role of IAA as an effector of the endogenous clock to regulate the membrane properties (possibly proton pumping) of the motor cells.

7. Structural Aspects

From a mechanistic point of view, pulvinar movement is facilitated by a restriction of rigid (vascular) elements to a narrow cylinder in the central part of the pulvinus (for structural details of the vascular tissue proper, see FLEURAT-LESSARD and BONNEMAIN 1978). In a detailed study, FLEURAT-LESSARD and ROBLIN (1982) extended these structural features. As an "internal control" they used the adjoining petiolus. In contrast to the petiole, the turgor-gaining cortical parenchyma cells (motor cells) of the pulvinus are developed in several layers. Due to the antomical features, in petioles the symplastic connections in transversal direction are rare, transport occurs mainly in a longitudinal direction at the xylem and phloem. The pulvinar cells are characterized additionally by continuous symplastic connections between the central cylinder and the epidermis (cf. the ion-redistribution data of SATTER et al. 1982).

Although in tertiary pulvini of *Mimosa* the flexor cells have thicker cell walls than the extensor (FLEURAT-LESSARD and MILLET 1984) and hence a higher capacity to store ions (see above), this feature does not appear to be strictly necessary, since it is not obvious in primary pulvini (see also FLEURAT-LESSARD and SATTER 1985 for similar data in *Albizzia*). The foldings of the cell walls appear to be more important for the movement of the tertiary pulvini, which will allow a rapid adjustment of the volume to turgor changes.

A different strategy was found in *Phaseolus* motor cells. Reinforcement of the cell walls by cellulose microfibrils facilitates a longitudinal expansion/shrinking of the motor cells during turgor movements. Cellulose microfibrils are oriented normal to the length axes of the pulvini, hence the extensibility of the cell walls is high in a longitudinal and low in a transversal direction (MAYER et al. 1985a).

The occurrence of tannin vacuoles, suggested to have calcium-storing capacity, was confirmed as in other sensitive cells also for *Mimosa* (see literature in FLEURAT-LESSARD and MILLET 1984). The observation of tightly appressed endoplasmic reticulum membranes (FLEURAT-LESSARD 1986) cannot at present be interpreted in the context of a structure-function relationship. Probably the most significant results concerning the regulation of turgor movements are based on a study by CAMPBELL and GARBER (1980 and confirmed by FLEURAT-LESSARD and SATTER 1985). In *Albizzia*, turgor-mediated volume changes of the motor cells are reflected by a different structural organization of the vacuome: during cell shrinking, the large central vacuole splits up into several smaller vacuoles, while during swelling of the cells, the vacuoles fuse and increase in volume. During isolation of protoplasts from motor tissue of *Albizzia* in the open-pulvinus state (GORTON and SATTER 1984a), the extensor cells ajust to the closed pulvinus state be fragmentation of the central vacuole into smaller vacuoles.

In addition to the structural features described above, FLEURAT-LESSARD and SATTER (1985), who found that structural alterations occur mainly in extensor cells of *Albizzia*, discussed an involvement of the cortical cytoskeleton, in particular microtubules, in the movement. In this respect, a report of MUKHERJEE and BISWAS (1982) should be mentioned, who found that colchicine inhibited the seismonastic movement of *Mimosa*. The colchicin-binding protein, less abundant in an insensitive *Mimosa* species, however, was isolated from whole leaf extracts, not from pulvini alone. In contrast, FLEURAT-LESSARD et al. (1985) were not able to confirm these results. They found, instead, an inhibiting effect of cytochalasin B, which indicated an involvement of microfilaments in the seismonastic response.

By means of an elegant method, SAMEJIMA and SIBAOKA (1983) identified cells involved in stimulus transmission in *Mimosa*. They impaled glass electrodes filled with procion yellow into the cells and, if an action potential upon mechanical stimulation was measured, they injected the fluorescent dye. After fixation of the tissue they determined the distribution of the fluorescing cells. The electrical stimulus (propagated action potentials) was transmitted along the protophloem parenchyma, phloem parenchyma and/or companion cells.

Graviperception in *Mimosa* appears to occur at the amyloplast-bearing cells of the starch sheath cells (FLEURAT-LESSARD 1981). In the normal leaf position, amyloplasts were sedimented according to gravity, while the nucleus was located at the apical cell pole. Cell walls facing the collenchyma were thicker and contained more plasmodesmata than the other walls. Upon inversion of the plant, amyloplasts sedimented in the direction of the nucleus, while vacuoles were displaced into that cell site which in the normal position was occupied by the amyloplasts. The nucleus did not change its position, but during gravistimulation, the nuclear envelope became sinuous. In *Cassia*, only the main pulvinus is graviresponsible (GAILLOCHET 1980). Again in the starch sheath cells, amyloplast re-orient with respect to gravity. Removal of the petiolar gland, interpreted by the author to increase the supply of free sugars, led to an increase of amyloplast starch content. Since after removal of the gland, the graviresponse was enhanced, the author suggested a direct effect of amyloplast pressure upon graviperception.

Acknowledgements. Professor Dr. W. HAUPT is thanked for critical reading of the manuscript. The help of colleagues by providing me with reprints and unpublished data is appreciated.

References

BIALCZYK, J., LECHOWSKI, Z.: New Phytol. 105, 469-475 (1987).

CAMPBELL, N.A., GARBER, R.C.: Planta 148, 251-255 (1980). - CAMPBELL, N.A., THOMSON, W.W.: Plant Physiol. 60, 635-639 (1977). - CAMPBELL, N.A., SATTER, R.L., GARBER, R.C.: Proc. Natl. Acad. Sci. USA 78, 2981-2984 (1981). - CHEN, Y.B., LEE, Y., SATTER, R.L.: Plant Physiol. 76, 858-860 (1984).

FLEURAT-LESSARD, P.: Protoplasma 105, 177-184 (1981); - ibid. 130, 1-4 (1986). - FLEURAT-LESSARD, P., BONNEMAIN, J.-L.: Protoplasma 94, 127-143 (1978). - FLEURAT-LESSARD, P., MILLET, B.: J. Exp. Bot. 35, 1332-1341 (1984). - FLEURAT-LESSARD, P., ROBLIN, G.: Ann. Bot. 50, 83-92 (1982). - FLEURAT-LESSARD, P., SATTER, R.L.: Protoplasma 128, 72-79 (1985). - FLEURAT-LESSARD, P., ROBLIN, G., BONMORT, J.: Plant Physiol. 77 (Suppl.), Abstr. 680 (1985). - FREUDLING, C., MAYER, W.F., GRADMANN, D.: Plant Physiol. 65, 966-968 (1980).

GAILLOCHET, J.: J. Exp. Bot. 31, 895-901 (1980); - Planta 151, 544-548 (1981). - GORTON, H.L.: pers. commun., manuscripts submitted to publication (1987). - GORTON, H.L., SATTER, R.L.: Plant Physiol. 76, 680-684 (1984a); - ibid. 76, 685-690 (1984b).

HAUPT, W.: Progr. Bot. 44, 222-229 (1982). - HOSOKAWA, Y., KIYOSAWA, K.: Plant Cell Physiol. 24, 947-951 (1983a); - ibid. 24, 1065-1072 (1983b).

IGLESIAS, A., SATTER, R.L.: Plant Physiol. 72, 564-569 (1983a); - ibid. 72, 570-572 (1983b).

KUMON, K., SUDA, S.: Plant Cell Physiol. 25, 975-979 (1984); - ibid. 26, 375-377 (1985). - KUMON, K., TSURUMI, S., SUDA, S.: Bot. Mag. Tokyo 97, 483-487 (1984).

LEE, Y., SATTER, R.L.: Plant Physiol. 83, in press (1987). - LUDLOW, M.M., BJÖRKMAN, O.: Planta 161, 505-518 (1984).

MAYER, W.-E., FLACH, D., RAJU, M.V.S., STARRACH, N., WIECH, E.: Planta 163, 381-390 (1985a). - MAYER, W.-E., FLACH, D., WIECH, E.: J. Plant Physiol. 118, 79-90 (1985b). - MAYER, W.-E., MAIER, M., FLACH, M.: Chronobiol. Int. 2, 11-17 (1985c). - MAYER, W.-E., RUGE, W.A., STARRACH, N., HAMPP, R.: Z. Naturforsch., in press (1987). - MORIMOTO, N., SHICHIJO, C., WATANABE, S., SUDA, S., HASHIMOTO, T.: Physiol. Plant. 68, 196-200 (1986). - MORSE, M.J., CRAIN, R.C., SATTER, R.L.: Plant Physiol. 83, in press (1987). - MUKHERJEE, J., BISWAS, S.: Phytochem. 21, 1881-1884 (1982).

NISHIZAKI, Y.: Plant Cell Physiol. 27, 155-162 (1986).

OTSIOGO-OYABI, H., ROBLIN, G.: Planta 161, 404-408 (1984); - Physiol. Plant. 65, 40-44 (1985a); - J. Plant Physiol. 119, 19-24 (1985b).

POOVAIAH, B.W., REDDY, A.S.N., McFADDEN, J.J.: Physiol. Plant. 69, 569-573 (1987).

RINNAN, T., JOHNSSON, A.: Physiol. Plant. 66, 139-143 (1986). - ROBLIN, G., BONMORT, J.: Planta 160, 109-112 (1984). - ROBLIN, G., FLEURAT-LESSARD, P.: Ann. Bot. 52, 763-768 (1983a); - Physiol. Plant. 58, 493-496 (1983b); - Plant Cell Physiol. 25, 1495-1499 (1984); - Planta, in press (1987). - ROBLIN, G., PERAULT, J.-M.: Plant Physiol. 77, 922-925 (1985). - ROBLIN, G., FLEURAT-LESSARD, P., SAEEDI, S., BONMORT, J.: Plant Physiol. Suppl., in press (1987).

SAEDI, S., ROBLIN, G.: Plant Physiol. 82, 270-273 (1983). - SAEDI, S., GAILLOCHET, J., BONMORT, J., ROBLIN, G.: Plant Physiol. 76, 851-853 (1984). - SAMEJIMA, M., SIBAOKA, T.: Plant Cell Physiol. 24, 33-39 (1983). - SATTER, R.L., GALSTON, A.W.: Annu. Rev. Plant Physiol. 32, 83-110 (1981). - SATTER, R.L., GARBER, R.C., KHAIRALLAH, L., CHENG, Y.-S.: J. Cell Biol. 95, 893-902 (1982). - SATTER, R.L., CHEN, Y.B., PACALA, S.W., LEE, Y.: In: Plant Senescence: Its Physiology and Biochemistry. Proc. 10th Annu. Symp. in Botany, University of California, Riverside, in press (1987). - SCHWARTZ, A., KOLLER, D.: Plant Physiol. 80, 778-781 (1986). - STARRACH, N., MAYER, W.-E.: J. Plant Physiol. 126, 213-222 (1986). - STARRACH, N., FLACH, D., MAYER, W.-E.: J. Plant Physiol. 120, 441-455 (1985).

TSURUMI, S., ASAHI, Y.: Physiol. Plant. 64, 207-211 (1985). - TSURUMI, S., ASAHI, Y., SUDA, S.: Bot. Mag. Tokyo 98, 89-91 (1985).

WATANABE, S., SIBAOKA, T.: Plant Cell Physiol. 24, 644-647 (1983). - WERK, K.S., EHLERINGER, J.: Plant, Cell Environm. 7, 81-87 (1984).

Dr. WOLFGANG HENSEL
Botanisches Institut
der Universität
Venusbergweg 22
D-5300 Bonn 1, FRG

C. Genetics

I. Replication

By WALTER NAGL

1. Introduction

After a brief review of some new aspects of the initiation and the pattern of DNA replication, this year's essay will emphasize differential DNA replication and DNA endoreduplication in vivo and in vitro, as well as its relation to changes at the chromosome and karyotype levels.

2. General Aspects

An overview over the present understanding of DNA replication in eukaryotes was given by CAMPBELL (1986) and HUBERMAN (1987). BRYANT (1985) reviewed the enzymology of DNA replication in plants, and TAYLOR (1984) the origins of replication. MUKHERJEE et al. (1985) could demonstrate DNA conformational changes at the origin, and BENBOW et al. (1985) discussed self-priming DNA polymerases and suggested an alternative view of chromosomal DNA replication which does not involve replication forks. ORMROD and FRANCIS (1986) could show that a rapid increase in the rate of replication occurs whithin 30 min of the start of flower-inducing, long-day conditions in *Silene*.

Several studies were devoted to the initiation of DNA replication and its regulation. In addition to earlier findings, evidence was obtained for initiation in regions of short, interspersed, repetitive sequences (ANACHKOVA et al. 1984) and identified as Alu elements (ANACHKOVA et al. 1985). In carrot, even a satellite DNA was detected to replicate earlier than the bulk of the DNA (KATO and TANIFUJI 1986).

Among the many factors involved in the control of DNA replication, some new data could be collected, which indicate that pulsed electromagnetic fields can stimulate DNA synthesis (KORENSTEIN et al. 1984; LIBOFF et al. 1984; TAKAHASHI et al. 1986). Additional evidence for the attachment of the replication complex to the nucleoskeleton (or nuclear matrix) was supplemented (JACKSON and COOK 1986; COOK and LANG 1984). The latter authors also discussed the relationship between initiation of replication at the matrix, and the origin of autonomously replicating sequences (ARS; see also OHTANI et al. 1985).

The pattern of replication could be shown to be dynamically related to cell function. Tissue-specific genes (and middle repetitive sequences, probably adjacent to them) replicate early in those cells in which they are potentially expressed (GOLDMAN et al. 1984). An important factor appears to be the position of a gene, rather than its sequence: Rearrangements are accompanied by dramatic effects in the temporal order of replication in mammals (CALZA et al. 1984) and plants (GREENBLATT

Progress in Botany, Vol. 49

1984). Intranuclear ringlike structures of replicating domains that were temporarily organized during the S phase could be shown by immunofluorescent staining with antibody against 5-bromodeoxyuridine (NAKAMURA et al. 1986).

3. Differential DNA Replication in Vivo

The study of differential DNA replication, i.e. underreplication and amplification of certain genes or non-coding sequences, has developed into an important field of genetics and cell biology during the last years. It was shown that these events are involved in differentiation, dedifferentiation, carcinogenesis, resistance to metals, drugs, toxins and antimetabolites, increased and specialized cell activity and evolution (cf. Progr. Bot. 47, p. 155, 1985, but also earlier volumes). The number of publications dealing with DNA amplification and underreplication is too large to be given in a representative list of references, and just a few specific aspects can be selected.

There is increasing evidence that within the same cell type, a certain gene, or group of genes, or repetitive sequences, can be amplified during polyploidization, while another may be underreplicated. The degree of underreplication can vary among sequences, in addition (HAMMOND and LAIRD 1985). Gene amplification is evidently highly regulated to make cells capable of synthesis of high levels of product in a relatively short space of time (KAFATOS et al. 1985). Events of that kind support the concept that the genome is highly flexible and dynamic (NAGL 1985; CAVALLINI et al. 1986). Phenotypic consequences can accompany genomic changes (CULLIS 1986), and the extent of variation in vitro is normally greater than that observed in vivo (e.g. CULLIS and CLEARY 1986). The rapid genomic changes in higher plants with respect to phylogenesis and ontogenesis, genotype and phenotype, nucleus and organelles, were reviewed by WALBOT and CULLIS (1985).

Some of the recent investigations revealed DNA amplification in maize endosperm nuclei during kernel development (KOWLES and PHILIPS 1985), but DNA underreplication in the endosperm and cotyledons of germinating castor bean (GALLI et al. 1986); underreplication was also observed in raphide crystal idioblasts of *Vanilla planifolia* during development (KAUSCH and HORNER 1984), and amplification of GC-rich nuclear satellite DNA during *Vicia faba* root dedifferentiation (NATALI et al. 1986), underreplication of highly repetitive sequences in differentiating root cells of *Vicia* (BASSI et al. 1984), rapid amplification and fixation of sequences in the ribosomal DNA repeat unit in derivatives of a cross between maize and *Tripsacum* (LIN et al. 1985) and amplification of extrachromosomal circular DNA in nuclei of intact bean leaves (KUNISADA et al. 1986). These examples indicate the extremely different events in which differential DNA replication may be involved. New evidence for a role of gene amplification in the resistance of man, animals and plants was also accumulated (Table 1; for a review of this phenomenon in human tumors, see STARK 1986). Amplification and underreplication of specific sequences also seem to be correlated with changes from the normal to the tumorous state in *Nicotiana* tissue cultures (DURANTE et al. 1986).

A phenomenon that is now a bit better understood than a few years ago is karyotype variability in plants, particularly in plant cell and tissue cultures, and its relationship to repetitive DNA and DNA plasticity (amplification, mobile elements, etc.).

Table 1. New examples for resistance phenomena that are related to gene amplification

Resistance to	Amplification of the gene for	Organism, cell lines	References
Antimycin A	Alcohol dehydrogenase 4	Yeast	WALTON et al. (1986)
L-Phosphinotricin (a herbicide)	Glutamine synthetase	Alfalfa	DONN et al. (1984)
Methotrexate	Thymidylate synthetase/ dihydrofolate reductase	*Leishmania*	GARVEY and SANTI (1986)
Quabain	Na,K-ATPase subunit	C^{+} (HeLa)	EMANUEL et al. (1986)
Multidrug	P-glycoprotein et al.	CHO[a]	VAN DER BLIEK et al. (1986)
Multidrug	MDRA[b]	CHL[c]	SCOTTO et al. (1986)

[a]Chinese hamster ovary cells. [b]Multidrug-resistant, associated genes. [c]Chinese hamster lung cells.

4. Chromosome Variation in Vitro

While genetic *stability* is a prerequisite of in vitro plant cloning, genetic *variability* (somaclonal variation) is a prerequisite for somatic breeding of new varieties and improved crops and meliorated ornamental plants (VASIL 1984; DIXON 1985; BAJAJ 1985a,b). The problem lies in the control of the former or the latter during plant tissue growth and regeneration. These aspects were recently reviewed by D'AMATO (1985). The important effect of the initial culture media was emphasized by FISH and KARP (1986). Without going into details one rule and one exception to it will be pointed out. As a rule, we find an inverse relationship between aneuploidization and polyploidization, on the one hand, and both the growth rate and the ability for regeneration, on the other hand. This is valid for plant and animal cell cultures (e.g. ROSEN 1985). In the interdihaploid potato (genotype $H^{2}578$), however, suspension cultures did not survive, if they did not become tetraploid (PIJNACKER et al. 1986). This example shows that not only the culture conditions, but also the genetic constitution, plays a significant role in in vitro development.

Another aspect which has to be taken into consideration comes from the methodical lane. While one part of the studies consists of cytogenetic analyses of cells (karyotype and chromosome studies), the other part consists of histochemical and biochemical techniques (DNA measurements, DNA characterizations). The legitimate interpretations are quite different, and several avenues of investigation should be employed in parallel, in order to understand a system. Figure 1 illustrates that the same variation in nuclear DNA (amount and Tm) can be measured, for instance, although its basis is quite different. In the one case, an aneuploidy (trisomy) leads to DNA changes, in the other case the chromosomal integration of an amplified gene. The results, with respect to the cytometrically measured DNA amounts and the biochemically determined melting temperatures (and GC content), and perhaps, also the reassociation kinetics, are the same, or at least very similar, in both instances.

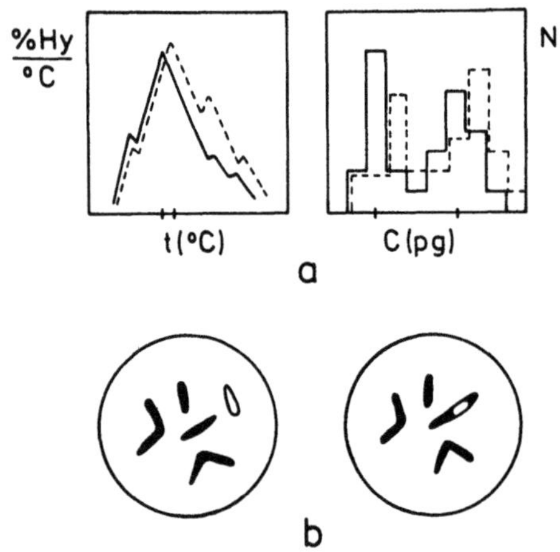

Fig. 1a,b. An illustration of how differences in DNA characterization (e.g. melting temperature) and DNA amount (C values), as shown in *a*, can be caused by either chromosomal aberrations (e.g. a trisomy) or an amplification event (*b*). Therefore, the results obtained at the DNA level can only be adequately interpreted by the aid of analyses at the chromosomal level

In this connection findings should be mentioned, which led to some progress in the understanding of the relationship between changes at the genome (DNA) level and the karyotype (chromosome) level, although only a few such studies have been made in plants so far (e.g. NAGL et al. 1984; PIJNACKER et al. 1986; NAGL and PFEIFER 1987; see also NUTI RONCHI et al. 1986). Most data were obtained in animal cell cultures: MORGAN et al. (1986) reached the conclusion that in hydroxyurea-treated CHO (Chinese hamster ovary) cells, drug-induced chromosomal aberrations are the first step in the process that leads to gene amplification. ROWLAND et al. (1986), among other authors, reported chromosome anomalies to be associated with the amplification of the adenosine deaminase gene (ADA) in deoxycoformycin-resistant rat hepatoma cells. The described, homogeneously staining regions (HSRs) comprise, in some instances, nearly an entire chromosome. KLEINBERGER et al. (1986) suggested that two mechanisms may lead to drug-induced gene amplification. The one might not require direct association between the site of DNA perturbations and the chromosomal location of the amplified regions (induction in *trans*; see also KAFATOS et al. 1985). Alternatively, it is possible that unterminated DNA replication forks induce local amplification, and the latter is, therefore, restricted to the perturbed replicons. In general, certain chromosomal anomalies appear to be specifically associated with the amplification of genes (e.g. ROVIGATTI et al. 1986; WOLOSCHAK et al. 1986).

It should also be remembered that many human tumors can be characterized by specific chromosomal anomalies (numerical and structural), so that they exhibit a tumor-specific karyotype evolution (reviewed by FORMAN and ROWLEY 1982; GERMAN 1983; MITELMAN 1983). Recently, evidence was accumulated that oncogenes are involved in those chromosome aberrations (e.g. VARMUS 1984; KUHN et al. 1985; HALUSKA et al. 1987; HEIM and MITELMAN 1987). Hence, our view of the interrelationships between genes, gene amplification and chromosome aberrations becomes somewhat more complete. Moreover, these findings confirm the existence of a "chromosome field" (LIMA-DE-FARIA 1983), which exhibits an ordering effect on gene and chromosome rearrangements.

5. Endoreduplication and Endopolyploidy

Although differential DNA replication is, in many somatic tissues, restricted to endoreduplicating cells, endocycles with multiplication of the *total* genome will, for systematic reasons, be discussed separately. The great number of new reports on somatic polyploidy in plants shows that endopolyploidization during differentiation is not an exception, but the rule. Several of the recent records are listed in Table 2.

From a functional point of view, different lines of understanding of increased DNA values clearly rule out that endocycles are accidental, or that they are the expression of a pathological condition:

- endopolyploidization is very often related to cell differentiation (in callus cultures of *Zea mays* there appears to be even a dependence of differentiation on polyploidization (K. SCHUMANN, pers. comm.), and it takes place very early in development (DEL NERO BUFFALINO and WITKUS 1984);
- it is often found in reproductive organs (reviewed by D'AMATO 1984), the highest values occurring in cells with evidently high metabolic activity, such as antipodals (cf. BOHDANOWICZ and TURALA-SZYBOWSKA 1985) and the embryo suspensor (see Sect. 6);
- it is clearly related to the storage capacity of cotyledons (LE GAL et al. 1984, 1986) and endosperm (i.e. for instance, the grain weight of wheat; CHOJECKI et al. 1986);
- and it is higher in fast-growing annual plants than in slow-growing perennials and trees (OLSZEWSKA and OSIECKA 1983, 1984).

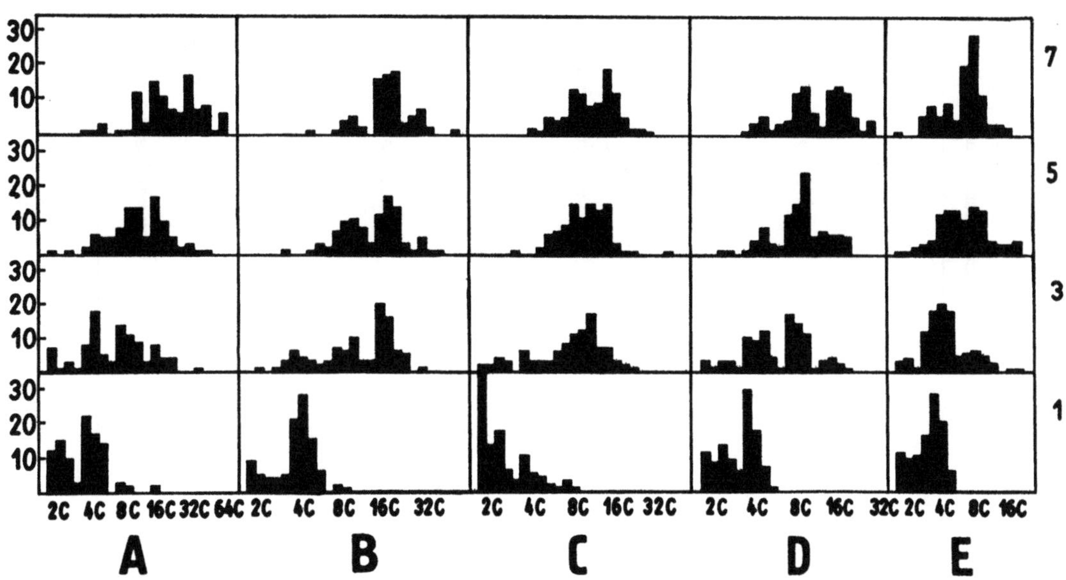

Fig. 2. Characteristic changes in DNA content (ploidy level) in parenchyma cell nuclei during growth and differentiation of roots in some Brassicaceae. *A Lepidium sativum*; *B Raphanus sativus*, cv. Chodowianka; *C R.s.* cv. Tetra Ilowiecka; *D Brassica oleracea*, cv. Amager; *E Armoracia lapatifolia*. *Ordinate*: percent nuclei; *abscissa*: relative DNA content; *1* to *7*: successive root segments (OLSZEWSKA and OSIECKA 1983)

Table 2. New records of endoreduplication and endopolyploidy in different organs of higher plants

Organ	Species	Max. degree (C)	References
Storage organs (seeds)	Various	Not given	GALLI (1984, 1985)
Cotyledons	*Vaccinia pyramidata*	8	BERNHARDT et al. (1984)
Cotyledons	Various Papilionaceae	32	BRYANS and SMITH (1985)
Cotyledons	*Lupinus albus*	64	LE GAL et al. (1984, 1986)
Roots	*Zea mays*, *Z. diploperennis*, hybrids	8	VIDAL (1984)
Root metaxylem	*Triticum aestivum*	16	DEMCHENKO (1984)
Root, hypocotyl	*Spinacia oleraceae*	32	NISHIBAYASHI (1983)
Roots	Various dicots	32	OLSZEWSKA and OSIECKA (1983, 1984)
Leaf epidermis	*Triticum durum*	16	CIONINI et al. (1983)
Various tissues	*Helianthus annuus*	32	CAVALLINI and CIONINI (1986)
Endosperm	*Ottelia alismoides*	12	SHAH and APPARAO (1982)
Endosperm	*Triticum aestivum*	24	CHOJECKI et al. (1986)
Endosperm	*Zea mays*	690	KOWLES and PHILIPS (1985)
Embryo suspensor	*Ottelia alismoides*	16	SHAH and APPARAO (1982)
Antipodals	*Ranunculus baudotii*	1024	WEDZONY (1982)

Figure 2 shows a typical pattern of endopolyploidy in differentiating and growing roots of some Brassicaceae.

For animal tissues, a survey and critical discussion of possible functions of polyploidy and polyteny was given by BRODSKY and URYVAEVA (1985). In mammals and humans, there is, in addition, a positive correlation between nuclear C values and the progress of carcinogenesis (Table 3), which can be used for diagnosis and prognosis. Normal tissues with endopolyploid nuclei are rather rare in higher animals, and those with high degrees of polyploidy, and even polyteny, are restricted to the "cancerlike" trophoblast of the embryo (e.g. in rodents: ZYBINA and ZYBINA 1985; ZYBINA et al. 1984; see also NAGL 1978; in humans: SARTO et al. 1982; see also, for hydatiform moles of the trophoblast: SARTO et al. 1984; THERMAN et al. 1986). As the trophoblast represents a highly active and important organ for the nourishment of the embryo (comparable in some way with the function of the embryo suspensor in some angiosperms: NAGL 1985), endopolyploidization of just this organ

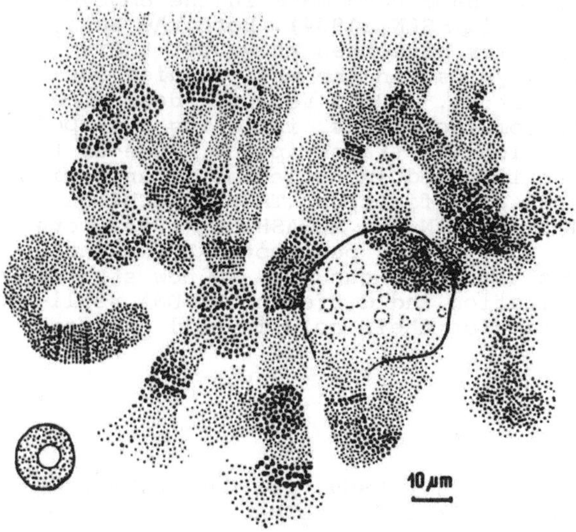

Fig. 3. Polytene chromosomes in an antipodal nucleus of *Ranunculus baudotii* in relation to a diploid nucleus; aceto-carmine (WEDZONY 1982)

Table 3. Records of endopolyploidy in human tumors

Tumor types	References
Brain tumors	DANOVA et al. (1986)
Cervical dysplasia	BIBBO et al. (1985)
Cervix carcinoma	THERMAN et al. (1983, GÖPPINGER et al. (1986)
Gastric carcinoma	HATTORI et al. (1984)
Prostatic cancer	SEPPELT et al. (1986)
Thyroid (Hürthle-cell) tumor	GALERA-DAVIDSON et al. (1986)

is not surprising. But also in other cells, particularly hepatocytes, polyploidization is evidently related to increased function, e.g. after poisoning (BARNI et al. 1985). Also the activity of nucleolar organizers (BACSI and WEJKSNORA 1986) and certain enzymes (VAN NOORDEN et al. 1984), and the electrical charge of the cells (SLIVINSKY et al. 1983), apparently depend on the level of ploidy.

6. Polyteny in Angiosperms

Cells with particularly high degrees of endopolyploidy exhibit the tendency to polyteny, i.e. the formation of giant chromosomes (reviewed by NAGL 1981; CHAPMAN 1985). During the last years, some new examples could be detected, or observations with respect to their function and functional structures could be made. An area of economic interest is the role of polyteny in antipodal nuclei of cereal grain, because they appear to influence endosperm growth (CHAPMAN 1985). Very distinct giant chromosomes were found in the antipodal cells of *Ranunculus baudotii* (WEDZONY 1982; Figure 3). Other records have been made in the embryo suspensor (e.g. of *Psophocarpus tetragonolobus*: SEN, 1984). Both the antipodals and the suspensor of other species have been known for their capability of polytenization for many years. In the well-studied suspensor of *Phaseolus coccineus*, FREDIANI et al. (1986) recently described the functional heterogeneity of ribosomal cistrons, including a temporal correlation between undermethylation and transcription. Some of the condensed chromosome bands which are involved in nucleolar organization have been demonstrated to undergo DNA amplification (TAGLIASACCHI et al. 1984; see also NAGL 1978, 1981). THERMAN and MURASHIGE (1984) were able to induce polytene chromosomes in cultured pea roots. This, and a few earlier attempts (see references therein), can lead to new strategies in the study of DNA endoreduplication and differential DNA replication in vitro, and in the elucidation of their role in cell differentiation and function in vivo.

References

ANACHKOVA, B., TODOROVA, M., VASSILIEV, L., RUSSEV, G.: Eur. J. Biochem. 141, 105-108 (1984). - ANACHKOVA, B., RUSSEV, G., ALTMANN, H.: Biochem. Biophys. Res. Commun. 128, 101-106 (1985).

BACSI, S.G., WEJKSNORA, P.J.: Exp. Cell Res. 165, 283-289 (1986). - BAJAJ, Y.P.S., ed.: Biotechnology in Agriculture and Forestry, Vol. I. Trees I. Berlin, Heidelberg, New York: Springer 1985a; - ed.: Biotechnology in Agriculture and Forestry, Vol. II. Crops I. Berlin, Heidelberg, New York: Springer 1985b. - BARNI, S., BERNOCCHI, G., SCHERINI, E., MARES, V.: Cell Biol. Int. Rep. 9, 993-1001 (1985). - BASSI, P., CIONINI, P.G., CREMONINI, R., SEGHIZZI, P.: Protoplasma 123, 70-77 (1984). - BENBOW, R.M., GAUDETTE, M.F., HINES, P.J., SHIODA, M.: 449-483 in: Control of Animal Cell Proliferation, Vol. 1, eds. A.L. BOYNTON, H.L. LEFFERT. Orlando: Academic Press 1985. - BERNHARDT, D., KÖHLER, K.H., SCHMERDER, B., KNÖSCHE, R.: Biochem. Physiol. Pflanzen 179, 561-571 (1984). - BIBBO, M., BARTELS, P.H., DYTCH, H.E., WIED, G.L.: Anal. Quant. Cytol. Histol. 7, 213-217 (1985). - BOHDANOWICZ, J., TURALA-SZYBOWSKA, K.: Protoplasma 127, 163-170 (1985). - BRODSKY, U.Y., URYVAEVA, I.V.: Genome Multiplication in Growth and Development. Biology of Polyploid and Polytene Cells. Cambridge: Universtiy Press 1985. - BRYANS, C., SMITH, D.L.: Ann. Bot. 55, 225-237 (1985). - BRYANT, J.A.: CRC Crit. Rev. Plant Sci. 3, 169-199 (1985).

CALZA, R.E., ECKHARDT, L.A., DEL GUIDICE, T., SCHILDKRAUT, C.L.: Cell 36, 689-696 (1984). - CAMPBELL, J.L.: Annu. Rev. Biochem. 55, 733-772 (1986). - CAVALLINI, A., CIONINI, P.G.: Protoplasma 130, 91-97 (1986). - CAVALLINI, A., ZOLFINO, C., CIONINI, G., CREMONINI, R., NATALI, L., SASSOLI, O., CIONINI, P.G.: Theor. Appl. Genet. 73, 20-26 (1986). - CHAPMAN, G.P.: Int. Rev. Cytol. 94, 107-126 (1985). - CHOJECKI, A.J.S., BAYLISS, M.W., GALE, M.D.: Ann. Bot. 58, 809-817 (1986). - CIONINI, P.G., CAVALLINI, A., BARONCELLI, S., LERCARI, B., D'AMATO, F.: Protoplasma 118, 36-43 (1983). - COOK, P.R., LANG, J.: Nucl. Acids Res. 12, 1069-1075 (1984). - CULLIS, C.A.: Trends Genet. 2, 307-309 (1986). - CULLIS, C.A., CLEARLY, W.: Can. J. Genet. Cytol. 28, 247-251 (1986).

D'AMATO, F.: 519-566 in: Embryology of Angiosperms, ed. B.M. JOHRI. Berlin, Heidelberg, New York: Springer 1984; - CRC Crit. Rev. Plant. Sci. 3, 73-112 (1985). - DANOVA, M., RICCARDI, A., MAZZINI, G., UCCI, G., MONTECUCCO, C.M., GAETANI, P., SILVANI, V., KNERICH, R., BUTTI, G., ASCARI, E., PAOLETTI, P.: Bas. Appl. Histochem. 30, 175-180 (1986). - DEL NERO BUFFALINO, L., WITKUS, R.: Ann. Bot. 53, 53-58 (1984). - DEMCHENKO, N.P.: Tsitologiya 26, 382-391 (1984). - DIXON, R.A.: Plant Cell Culture. Oxford (USA): IRL Press 1985. - DONN, G.: TISCHER, E., SMITH, J.A., GOODMAN, H.M.: J. Mol. Appl. Genet. 2, 621-635 (1984). - DURANTE, M., GERI, C., BUIATTI, M., CIOMEI, M., CECCHINI, E., MARTINI, G., PARENTI, R., GIORGI, L.: Dev. Genet. 7, 51-64 (1986).

EMANUEL, J.R., GARETZ, S., SCHNEIDER, J., ASH, J.F., BENZ, E.J., jr., LEVENSON, R.: Mol. Cell. Biol. 6, 2476-2481 (1986).

FISH, N., KARP, A.: Theor. Appl. Genet. 72, 405-412 (1986). - FORMAN, D., ROWLEY, J.: Nature 300, 403-404 (1982). - FREDIANI, M., FORINO, L.M.C., TAGLIASACCHI, A.M., CIONINI, P.G., DURANTE, M., AVANZI, S.: Protoplasma 132, 51-57 (1986).

GALERA-DAVIDSON, H., BIBBO, M., BARTELS, P.H., DYTCH, H.E., PULS, J.H., WIED, G.L.: Anal. Quant. Cytol. Histol. 8, 158-167 (1986). - GALLI, M.G.: Planta 160, 193-199 (1984); - J. Exp. Bot. 36, 963-973 (1985). - GALLI, M.G., BALZARETTI, R., SGORBATI, S.: J. Exp. Bot. 37, 1716-1724 (1986). - GARVEY, E.P., SANTI, D.V.: Science 233, 535-540 (1986). - GERMAN, J., ed.: Chromosome Mutation and Neoplasia. New York: Liss 1983. - GOLDMAN, M.A., HOLMQUIST, G.P., GRAY, M., CASTON, L.A., NAG, A.: Science 224, 686-692 (1984). - GÖPPINGER, A., FREUDENBERG, N., ROSS, A., HILLEMANS, H.-G., HILGARTH, M.: Anal. Quant. Cytol. Histol. 8, 148-151 (1986). - GREENBLATT, I.M.: Genetics 108, 471-485 (1984).

HALUSKA, F.G., TSUJIMOTO, Y., CROCE, C.M.: Trends Genet. 3, 11-15 (1987). - HAMMOND, M.P., LAIRD, C.D.: Chromosoma 91, 267-278 (1985). - HATTORI, T., HOSOKAWA, Y., FUKUDA, M., SUGIHARA, H., HAMADA, S., TAKAMATSU, T., NAKANISHI, K., TSUCHIHASHI, Y., KITAMURA, T., FUJITA, S.: Cancer 54, 1591-1597 (1984). - HEIM, S., MITELMAN, F.: Hum. Genet. 75, 70-72 (1987). - HUBERMAN, J.A.: Cell 48, 7-8 (1987).

JACKSON, D.A., COOK, P.R.: EMBO J. 5, 1403-1410 (1986).

KAFATOS, F.C., ORR, W., DELIDAKIS, C.: Trends Genet. 1, 301-306 (1985). - KATO, A., TANIFUJI, S.: Plant Cell Physiol. 27, 261-264 (1986). - KAUSCH, A.P., HORNER, H.T.: Eur. J. Cell Biol. 33, 7-12 (1984). - KLEINBERGER, T., ETKIN, S., LAVI, S.: Mol. Cell. Biol. 6, 1958-1964 (1986). - KORENSTEIN, R., SOMJEN, D., FISCHLER, H., BINDERMAN, I.: Biochim. Biophys. Acta 803, 302-307 (1984): - KOWLES, R.V., PHILIPS, R.L.: Proc. Natl. Acad. Sci. USA 82, 7010-7014 (1985). - KUHN, E.M., THERMAN, E., DENNISTON, C.: Hum. Genet. 70, 1-5 (1985). - KUNISADA, T., YAMAGISHI, H., KINOSHITA, I., TSUJI, H.: Plant Cell Physiol. 27, 355-361 (1986).

LE GAL, M.F., HALLET, J.N., REY, L.: Protoplasma 120, 113-122 (1984). - LE GAL, M.F., LECOCQ, F.M., HALLET, J.N.: Protoplasma 130, 128-137 (1986). - LIBOFF, A.B., WILLIAMS, T., jr., STRONG, D.M., WISTAR, R., jr.: Science 223, 818-820 (1984). - LIMA-DE-FARIA, A.: Molecular Evolution and Organization of the Chromosome. Amsterdam: Elsevier 1983. - LIN, L.-S., HO, T.-H.D., HARLAN, J.R.: Dev. Genet. 6, 101-112 (1985).

MITELMAN, F.: Cytogenet. Cell Genet. 36, 1-515 (1983). - MORGAN, W.F., BODYCOTE, J., FERO, M.L., HAHN, P.J., KAPP, L.N., PANTELIAS, G.E.: Chromosoma 93, 191-196 (1986). - MUKHERJEE, S., PATEL, I., BASTIA, D.: Cell 43, 189-197 (1985).

NAGL, W.: Endopolyploidy and Polyteny in Differentiation and Evolution. Amsterdam: North-Holland 1978; - Int. Rev. Cytol. 73, 21-53 (1981); - 135-172 in: Advances in Chromosome and Cell Genetics, eds. A.K. SHARMA, A. SHARMA. New Delhi: Oxford & IBH Publ. 1985. - NAGL, W., PFEIFER, M.: Plant Syst. Evol., in press (1987). - NAGL, W., RIBICKI, R., MERTLER, H.-O., HEZEL, U., JACOBI, R., BACHMANN, E.: Protoplasma 122, 138-144 (1984). - NAKAMURA, H., MORITA, T., SATO, C.: Exp. Cell Res. 165, 291-297 (1986). - NATALI, L., CAVALLINI, A., CREMONINI, R., BASSI, P., CIONINI, P.G.: Cell Diff. 18, 157-161 (1986). - NISHIBAYASHI, S.: J. Sci. Hiroshima Univ., Ser. B, 18, 191-233 (1983). - NUTI RONCHI, V., BONATTI, S., DURANTE, M., TURCHI, G.: Environ. Exp. Bot. 26, 127-135 (1986).

OHTANI, T., KIYOKAWA, S., OHGAWARA, T., HARADA, H., UCHIMIYA, H.: Plant Mol. Biol. 5, 35-39 (1985). - OLSZEWSKA, M.J., OSIECKA, R.: Biochem. Physiol. Pflanzen 178, 581-599 (1983); - Biochem. Physiol. Pflanzen 179, 641-657 (1984). - ORMROD, J.C., FRANCIS, D.: Protoplasma 130, 206-210 (1986).

PIJNACKER, L.P., HERMELINK, J.H.M., FERWERDA, M.A.: Plant Cell Rep. 5, 43-46 (1986).

ROSEN, E.M., GOLDBERG, I.D., SHAPIRO, H.M., ZOLLER, L.C., MYRICK, K.V., LEVENSON, S.E., HALPIN, P.A.: J. Cell. Physiol. 125, 512-520 (1985). - ROVIGATTI, U., WATSON, D.K., YUNIS, J.J.: Science 232, 398-400 (1986). - ROWLAND, P., III., CHIANG, J., JARGIELLO-JARRETT, P., HOFFEE, P.A.: Cytogenet. Cell Genet. 41, 136-144 (1986).

SARTO, G.E., STUBBLEFIELD, P.A., THERMAN, E.: Hum. Genet. 62, 228-232 (1982). - SARTO, G.E., STUBBLEFIELD, P.A., LURAIN, J., THERMAN, E.: Am. J. Obstet. Gynecol. 148, 1014-1023 (1984). - SCOTTO, K.W., BIEDLER, J.L., MELERA, P.W.: Science 232, 751-755 (1986). - SEN, S.: 59-64 in: Recent Trends in Botanical Research, ed. R.N. GOHIL. Jodhpur (India): Scient. Publ. 1984. - SEPPELT, U., SPRENGER, E., HEDDERICH, J.: Anal. Quant. Cytol. Histol. 8, 152-157 (1986). - SHAH, C.K., APPARAO, B.J.: Ind. Bot. Rep. 1, 144-149 (1982). - SLIVINSKY, G.G., MALKOVSKY, A.D., NARYZHNAYA, I.I.: Stud. Biophys. 97, 205-208 (1983). - STARK, G.R.: Cancer Surv. 5, 1-24 (1986).

TAGLIASACCHI, A.M., FORINO, L.M., CIONINI, P.G., CAVALLINI, A., DURANTE, M., CREMONINI, R., AVANZI, S.: Protoplasma 122, 98-107 (1984). - TAKAHASHI, K., KANEKO, I., DATE, M., FUKUDA, E.: Experientia 42, 185-186 (1986). - TAYLOR, J.H.: Mol. Cell Biochem. 61, 99-109 (1984). - THERMAN, E., MURASHIGE, T.: Plant Syst. Evol. 148, 25-33 (1984). - THERMAN, E., SARTO, G.E., BUCHLER, D.A.: Cancer Genet. Cytogenet. 9, 9-18 (1983). - THERMAN, E., SARTO, G.E., KUHN, E.M.: Cancer Genet. Cytogenet. 19, 301-310 (1986).

VAN DER BLIEK, A.M., VAN DER VELDE-KOERTS, T., LING, V., BORST, P.: Mol. Cell. Biol. 6, 1671-1678 (1986). - VAN NOORDEN, C.J.F., VOGELS, I.M.C., FRONIK, G., BHATTACHARYA, R.D.: Exp. Cell Res. 155, 381-388 (1984). - VARMUS, H.E.: Annu. Rev. Genet. 18, 553-612 (1984). - VASIL, I.K., ed.: Cell Cultures and Somatic Cell Genetics of Plants. Orlando: Academic Press 1984. - VIDAL, B. DE C.: Cell. Mol. Biol. 30, 11-22 (1984).

WALBOT, V., CULLIS, C.A.: Annu. Rev. Plant Physiol. 36, 367-396 (1985). - WALTON, J.D., PAQUIN, C.E., KANEKO, K., WILLIAMSON, V.M.: Cell 46, 857-863 (1986). - WEDZONY, M.: Acta Biol. Cracov., Ser. Bot. 24, 43-62 (1982). - WOLOSCHAK, G.E., DEWALD, G.W., BAHN, R.S., KYLE, R.A., GREIPP, P.R., ASH, R.C.: J. Cell Biochem. 32, 23-34 (1986).

ZYBINA, E.V., ZYBINA, T.G.: Tsitologiya 27, 402-410 (1985). - ZYBINA, E.V., ZYBINA, T.G., STEIN, G.I.: Tsitologiya 26, 525-530 (1984).

Prof. Dr. WALTER NAGL
Abt. Zellbiologie
FB Biologie der Universität
Postfach 3049
D-6750 Kaiserslautern, FRG

II. Recombination: Asexual Recombination in Higher Plants

By WOLFGANG FRIEDT and ULRIKE BRUNE

1. Introduction

Extensive progress has been made in the last couple of years in the fields of plant cell genetics and gene technology. Techniques of asexual recombination of genomes, chromosomes or genes have been improved and applied to more species including cultivars of various crop plants. Scientific areas where particularly large steps forward have been taken are (1) establishment of new hybrids by protoplast fusion, (2) induction of somaclonal variation, (3) elucidation of the structure and function of mobile genetic elements (transposons) and (4) genetic engineering via construction and application of different vector systems.

2. Somatic Hybrids and Cybrids

a) *Solanaceae*

Protoplast regeneration is now a routine technique in the genus *Solanum*, either with or without previous fusion. EHLENFELDT and HELGESON (1987) evaluated two sets of somatic hybrids between *S. brevidens* (2x) and *S. tuberosum* (2x and 4x) for meiotic regularity, male fertility and female fertility. They found tetraploid somatic hybrids from 2x + 2x fusions and hexaploid from 2x + 4x fusions. Pollen stainability (fertility) ranged from 0% to 83% in tetraploids and from 0% to 23% in hexaploids. Meiosis proved to be more regular in tetraploids.

AUSTIN et al. (1986) assessed phenotypic and flowering characteristics of hybrid plants generated by protoplast fusion between a tetraploid *S. tuberosum* line and diploid *S. brevidens* under field conditions. Almost all of the hybrids were hexaploid. A wide range of variation in morphological characters was observed for hybrids and protoclones.

Intraspecific somatic hybridization between amino acid analogue-resistant cell lines of *S. tuberosum* has been carried out by DE VRIES et al. (1987) following electrofusion of protoplasts. Quantitative examination of the free amino acid content revealed characteristics of both parental cell lines in most of the somatic hybrids.

Thirty somatic hybrids between two *Solanum* species have been analyzed for mitochondrial and chloroplast genome rearrangements. No chloroplast DNA alterations but a range of mitochondrial alterations (intra- and intermolecular recombination) were found (KEMBLE et al. 1986). These results show that mtDNA rearrangements can be induced by somatic fusion

Progress in Botany, Vol. 49

of *Solanum* species. Mitochondrial rearrangements are also commonly observed in plants regenerated from individual potato species.

Intergeneric somatic hybrid plants were regenerated by HANDLEY et al. (1986) from fused leaf mesophyll protoplasts of *Lycopersicon esculentum* with suspension culture-derived protoplasts of *Solanum lycopersicoides*. Seven of the obtained eight somatic hybrids had a chromosome number ranging from the expected 2n = 4x = 48 to 2n = 68. Mixoploid root-tip cells containing 48, 53, 54 or 55 chromosomes for two of the hybrids were also observed.

In *Nicotiana*, MEDGYESY et al. (1985a, 1986) described genetic recombination between chloroplasts of two flowering plant species, *N. tabacum* and *N. plumbaginifolia*, after somatic cell fusion. Protoplasts of the two species, differing in three cytoplasmic genetic markers, were fused and calli grown from the fused population were screened for the expression of streptomycin resistance as the result of interspecific chloroplast recombination. A somatic hybrid pt14 expressed a new combination of cytoplasmic genetic markers. In the pt14 chloroplast genome three *N. tabacum* and four *N. plumbaginifolia* parent-specific restriction sites have been identified, indicating that at least six recombination sites are present.

MEDGYESY et al. (1985b) reported also on the use of a cytoplasmic light-sensitive *N. plumbaginifolia* mutant line as a recipient, which made the efficient transfer of chloroplasts from a wild-type donor possible. They fused mesophyll protoplasts of *N. gossei* with such of a *N. plumbaginifolia* line carrying light-sensitive plastids from an *N. tabacum* mutant. The cybrids (new nucleus-cytoplasm combinations) were fully male sterile. This fact and the recombinant-type mitochondrial DNA restriction pattern indicated the co-transfer of chloroplast and mitochondrial traits.

Protoplasts of a chloroplast-defective cultivar of *N. tabacum* were fused with gamma-irradiated protoplasts of *Petunia hybrida* by GLIMELIUS and BONNETT (1986). They regenerated over 100 photoautotrophic plants. From 94 tested plants all but one had *Petunia* chloroplasts. For three of these plants the chloroplast DNA was shown to be identical to that of *Petunia*. Only 30% of the plants showed varying degrees of vegetative and reproductive abnormalities.

The chloroplast and mitochondrial DNAs of *Petunia* somatic hybrid plants derived from the fusion of wild-type *P. parodii* protoplasts with albino *P. inflata* protoplasts were analyzed by CLARK et al. (1986). The mitochondrial genomes of these plants have undergone rearrangements that resulted in mtDNA restriction patterns containing both *P. parodii*- and *P. inflata*-specific fragments as well as novel fragments.

b) *Brassicaceae*

Isolated protoplasts can be used for interspecific hybridization, where both, nuclear genome and the plasmon, from two parental species are combined. Another possibility is to combine only the cytoplasma from one species with the genome of the other species for production of cybrids. A wide range of new somatic hybrids within the *Brassicaceae* and somatic cybrids within *Solanaceae* were produced by GLIMELIUS et al. (1986). The authors used the resynthesis of rape as a model system: they obtained *Brassica napus* by somatic rehybridization of *B. oleracea* and *B. campestris*. As many as 78% of the living population of protoplasts obtained after the protoplast fusion treatment were fusion products and about 60% of the regenerated calli were hybrids. Regarding cytoplasmic male sterility (cms), the involvement and function of mtDNA is of special interest.

CHETRIT et al. (1985) analyzed the mitochondrial genomes of five somatic hybrid plants, combining either *B. napus* chloroplasts and a cytoplasmic male sterility trait from *Raphanus sativus* or chloroplasts of a triazine-resistant *B. campestris* and a cytoplasmic male sterility trait from *R. sativus*. The presence of new bands in the somatic hybrid mitochondrial DNA restriction pattern is evidence of mitochondrial recombination in somatic hybrid cells. Large quantitative variation in a mitochondrial plasmidlike DNA has been observed in both parental and somatic hybrid plants. These results suggest that the genes for male sterility are located in the chromosomal mtDNA and not in the plasmidlike DNA. Additional experiments for genetic improvement of cytoplasmic traits of *Cruciferae* (*Brassicaceae*) through cytoplasmic hybridization were reported by PELLETIER et al. (1986).

TAGUCHI and KAMEYA (1986) fused protoplasts from leaf mesophyll cells of cabbage (*B. oleracea*) and Chinese cabbage (*B. campestris*), which is the first example of a somatic hybrid between heading-type vegetables. An effective procedure for obtaining somatic hybrids between *Brassica* species was recently developed by TERADA et al. (1987). They fused hypocotyl protoplasts of *B. oleracea* with mesophyll protoplasts from three varieties of *B. campestris*; 90% of the regenerated plants were found to be hybrids. All of the hybrids had intermediate leaf morphology and possessed leucine aminopeptidase (LAP) isozymes of both parental species. Chromosome analyses showed a considerable variation in chromosome number of somatic hybrids, which was due to the occurrence of multiple fusions and chromosome loss during culture.

c) Other Species

OHGAWARA et al. (1985) obtained hybrid plants between *Citrus sinensis* and *Poncirus trifoliata* by protoplast fusion which is the first case of asexual hybridization in *Rutaceae*. Protoplasts were isolated from embryogenic cells of *C. sinensis* and from leaves of *P. trifoliata*. Green globular embryoids resulted in the regeneration of trifoliate plants. Morphological characters of these plants were intermediate between both parents. One of the hybrids had 36 chromosomes, which is the sum of the parental chromosome complements ($2n = 18$ each).

3. Somaclonal Variation

a) General

Somaclonal variation technology takes advantage of the genetic variation that appears in plants regenerated from somatic cells grown in tissue culture. The possible causes and origins of somaclonal variation were reviewed by KARP and BRIGHT (1985). They summarized the phenotypic characteristics of somaclonal variation and presented evidence that this variation can be caused by a range of genotypic or epigenetic changes. Genetic changes can be due to e.g. polyploidy, aneuploidy, chromosome deletion, chromosome breakage, late replication of heterochromatin, activation of dormant transposons and cytoplasmic organelle segregation. Many of the obtained gene mutations are recessive.

b) *Triticeae*

It is important to distinguish between true mutations and presumed changes in regenerants due to cross-pollination or admixture.

METAKOVSKY et al. (1987) presented arguments that some so-called somaclonal variants in gliadin-coding genes of wheat (*Triticum aestivum*) are unlikely to be due to mutations since they would have had to involve directed changes at several loci simultaneously. They maintained that simple scoring of changes in the gliadin spectrum is not sufficient to prove that somaclonal variation in gliadin genes has occurred.

Changes in the chromosome complement are a major aspect of the phenomenon of somaclonal variation. A cytological study was carried out by KARP (1986) to determine how much of the variation observed in the wheat plants resulted from changes in the chromosomes. Up to seven cultured immature embryos were selected at random from four wheat cultivars. Chromosomes were studied in up to ten distinct regenerants from each of the cultured embryos. They differed in their morphogenetic response and in their mode of regeneration, i.e. via somatic embryogenesis or from more disorganized structures. On average 70% of the plants retained their normal hexaploid chromosome number and the remaining 30% were aneuploids; the aneuploid frequency appeared to vary non-significantly between cultivars. Structural chromosomal changes including deletions and interchanges were also present at high frequencies amongst the regenerants. One of these structural variants had an altered gliadin pattern in which some protein bands normally associated with chromosome 1B were present in reduced dosage in the structural heterozygote and were absent in a structural homozygote found amongst the selfed progeny. This regenerant was found to carry a deletion of a large part of the long arm of chromosome 1B. For comparison to the allohexaploid wheat somaclones a cytogenetic study was carried out on barley (*Hordeum vulgare*) plants regenerated from cultured, immature embryos. No aneuploids were present amongst 50 regenerants screened and the only chromosome abnormality observed was the occurrence of chromosome breakage in one regenerant.

Furthermore, plants regenerated from embryos and callus derived from embryo culture of wheat cv. Maris Huntsman were investigated for somaclonal variation. Its seed transmission was studied in three successive generations of callus-derived (SC1) plants of wheat (AHLOOWALIA and SHERINGTON 1985). Second, third and fourth generations (SC2-SC4) were compared with those obtained from embryo culture (E1-E4). Genome instability was associated with variation observed in yield. SC4 progeny of a dwarf, partially sterile SC1 variant included aberrant plants, i.e. a haploid, aneuploids, mixoploids, some showing multiploidy, multivalents, spindle abnormalities, chromatin bridges and fragments. The presence of these abnormalities indicated the occurrence of chromosomal aberrations, i.e. translocations and deletions and also the occurrence of chromosome elimination.

As part of another program to investigate somaclonal variation in wheat, field trials have been carried out with selfed progeny generations of the initial regenerants to assess how much of this variation is due to stable genetic changes. MADDOCK and SEMPLE (1986) reported on heritable changes in wheat plants produced by in vitro regeneration from immature embryos or inflorescences. Their results confirmed that stable variation can be induced in wheat. This has already been reported by LARKIN et al. (1984). At a low frequency (<5%) individual lines could be identified that showed stable somaclonal variation after two meiotic cycles. In particular, variation was noted in a number of morphological characters. There are differences in the level of variation between different lines. About 95% of more than 800 lines derived from plants regenerated in tissue culture were uniform, healthy and similar to seed-derived control material; the most noticeable differences were those in height and morphology.

COOPER et al. (1986) could also demonstrate that somaclonal variation in gliadin proteins of wheat plants derived from immature embryo callus culture of the line ND7532 is heritable. The types of variation in gliadin pattern observed in somaclones of ND7532 were similar to those reported for the line Yaqui 50E, but gliadin changes occurred less frequently in ND7532. Four plants produced gliadin patterns different from controls. Three of the four mutant lines showed the presence of a mutant protein of 50 relative mobility units (RMU) and the corresponding loss of a parental protein of 26 RMU.

The progeny of 149 plants regenerated from tissue culture of immature wheat embryos were screened for variation in their grain β-amylase isozyme pattern (RYAN and SCOWCROFT 1987). The special properties of β-amylase isozymes were expected to increase the chances of isolating somaclonal mutants. One regenerant was found to be heterozygous for a variant pattern characterized by the presence of five new isozyme bands. No chromosome abnormalities were evident and no recombination was found between the variant bands. These variant bands were not present in a survey of 111 diverse genotypes nor in developing grain of the parent cultivar; this variant could therefore represent a rare mutation leading to expression of a currently unexpressed locus.

Meiotic karyotypes of microsporocytes and pollen sterility in plants regenerated from immature tassels of selected monosomic and other aneuploid types of *Zea mays* were analyzed by RHODES et al. (1986). At least 40% of 161 plants regenerated from aneuploid cultures had altered karyotypes; this frequency was not related to culture age. Most alterations involved chromosome breakage rather than changes in chromosome number: 18.1% of the alterations were heteromorphic pairs, 12.5% translocations, 10.6% additions, 1.4% losses of chromosomes and 2.8% genomic doubling. Among 115 plants regenerated from euploid cultures there were 49% with alterations: 12.3% translocations, 8.8% heteromorphic pairs, 29.2% genomic doubling, but no single chromosome additions or losses. Karyotypic changes observed in this study rarely involved the monosomic chromosome. Another frequent aberration was lack of cell wall formation after the first and/or second meiotic division (23% of 144 regenerants).

Corresponding somaclonal variation in plants regenerated from immature embryo calluses in rye (*Secale cereale*) was reported by LINACERO and VAZQUEZ (1986). Plants of four rye cultivars regenerated through two different pathways, embryogenesis or organogenesis, were examined. Variation in chromosome modifications, meiotic alterations and in pollen viability was found. Several abnormalities were observed in PMC meiosis, where low chiasmata frequency and univalents were the most frequent abnormality found, but anaphase bridges, laggards, chromosome fragments, stickiness and asynchrony were also noticed. PMCs with different numbers of bivalents were observed in some plants, too.

c) *Solanaceae*

Numerous reports have been given on somaclonal variation in *Solanum* species. Recently, FISH and KARP (1986) regenerated protoplast-derived plants from potato (*S. tuberosum* cv. Maris Bard) on four different initial culture media. Chromosome counts of up to 50 regenerants showed no significant differences between any of the treatments, in the proportions of plants which were euploid, aneuploid at the tetraploid level or aneuploid with higher chromosome numbers. Highly significant differences were found between shoots which rooted quickly (predominantly euploid) and those which rooted only after transfer to a rooting medium (predominantly aneuploid with high chromosome numbers). Overall, more than 60% of the regenerants were normal euploids, which means an improvement of the authors' earlier work with this cultivar (4%).

Amongst 178 plants screened, four regenerants were found with structural chromosome changes.

In agreement with these findings, only some plants with grossly aberrant phenotypes were also found by CREISEN and KARP (1985) from shoot culture-derived protoplasts of potato cv. Majestic. The majority were similar to or indistinguishable from normal controls. Cytological examination showed that 57% of the regenerants had the normal chromosome number ($2n = 4x = 48$). The remainder were aneuploids at about the euploid level ($2n = 44$-49) or plants with higher chromosome numbers (ranging from $2n = 73$ up to $2n = 8x = 96$). Three cases of structural chromosome variation (translocation) were observed.

In another study, 14 different cultivars of potato (*S. tuberosum*) plants regenerated from cultured excised stem, rachis or leaf pieces, morphological differences (e.g. in skin colour) were observed. Still 87% of the regenerated plants contained the euploid chromosome number ($2n = 4x = 48$) (WHEELER et al. 1985). SREE RAMULU et al. (1985) analyzed the pattern of variation in nuclear DNA content and chromosome number in a temporal sequence during in vitro growth of calli and cell suspensions. The results indicated that monoploid, diploid and tetraploid potato lines differed in growth of callus and cell suspension cultures, suggesting genotpyic differences. Both polyploidization and aneuploidy occurred during the initial stages of callus induction in all examined potato genotypes. With further growth of callus, variation increased. The various genotypes differed in their genetic instability; it was concluded that these differences were due to different ploidy levels and genetic make-up of the genotypes.

The large number of reports of phenotypic variation among plants regenerated from tissue culture suggest underlying alterations at the DNA level. Therefore, LANDSMANN and UHRIG (1985) tested the genome of individual, protoplast-derived *Solanum tuberosum* plants at specific sites using Southern hybridization techniques. Two of 12 plants were shown to be variant; thus, both somaclonal variants can be regarded as mutant-deficient in ribosomal RNA genes.

A sample of 50 protoplast-derived plants of *Solanum brevidens* was examined for chromosomal variation by NELSON et al. (1986). Twelve plants were diploid ($2n = 24$), 26 were tetraploid and 12 were aneuploid at the tetraploid level. Phenotypic variation in *S. brevidens* regenerants was restricted to differences in leaf shape and fertility associated with increased ploidy. Tetraploid and aneuploid plants had broader leaves and fewer or no seeds compared to the diploid regenerants.

BUIATTI et al. (1985) screened progenies of the diploid tomato (*Lycopersicon esculentum*) regenerated from cotyledon cultures in vitro for induced mutations. In 17.4% of the progenies they found chlorophyll mutations and other morphological abnormalities segregating 3:1. A multicellular origin of the regenerated plants was suggested by a variegated chimera found in the first generation and by a number of progenies showing single aberrant phenotypes. In 29.9% of the progenies "rogue" plants appeared, an aberrant phenotype showing absence of apical dominance with partial cytoplasmic inheritance.

In tomato, plants resistant to the fungal pathogen *Fusarium oxysporum* were also obtained using in vitro selection (SHAHIN and SPIVEY 1986). Seeds of plants regenerated from toxin-resistant callus of cotyledon-derived protoplasts were rescreened for resistance. Data obtained from this test of R_1 progenies showed a ratio of three resistant to one susceptible line. Further analysis of the R_2 progeny confirmed that resistant plants were either homozygous or heterozygous dominant for the gene conferring the resistance. The nature of this single dominant gene type of resistance is under investigation.

d) Other Species

A proportion of plantlets regenerated from alfalfa (*Medicago sativa*) callus showed obvious morphological variation (BAERTLEIN and McDANIEL 1987), i.e. 9.1% of the somaclones differed significantly from the parent and from the other somaclones judged by quantitative protein pattern variations. Two of the somaclone lineages showed no overlapping with the parental variation which indicated significant molecular divergence of these plantlets as judged by quantitative protein differences.

CULLIS and CLEARY (1986) compared the DNAs from leaves and callus from some flax (*Linum usitatissimum*) genotypes. Abundance of most of the highly repeated DNA sequence families could vary in culture, but the extent of variation depended on the genotroph (i.e. stable, genetically different lines derived after environmental stress treatment). The extent of variation observed between leaf DNA and callus DNA from a single genotroph was greater than that observed between the genotrophs in vivo. Individual progenies from a single inbred regenerated plant also differed.

4. Transposable Elements

a) General

Mobile or transposable elements (transposons, "jumping genes") are DNA segments of varying size which are capable of changing their position in the genome. They are more complex than IS elements (insertion sequences with lengths <2,000 bp). Integration of a transposable element into a functional gene leads to a distortion, i.e. a change of its expression. Insertion of transposons into DNA sequences is another means of separating contiguous genes, besides crossover. By mobile gene segments, genetic material can be transferred from one locus to another. Structures of integrated plant transposable elements are similar to those of prokaryotic elements; both are flanked by a duplication of the target site sequence and often the termini of elements consist of inverted (terminal inverted repeat, TIR) or direct repeat sequences (SAEDLER and NEVERS 1985; PETERSON 1985; see Figure 1).

A molecular model for transposition of plant transposable elements has recently been constructed by SAEDLER and NEVERS (1985). This process may occur via excision and re-integration of the element. Excision generates DNA sequence diversity which suggests the participation of DNA repair enzymes in the healing of the donor molecule (Figure 2). However, the enzymology of transposition is not known in detail as yet.

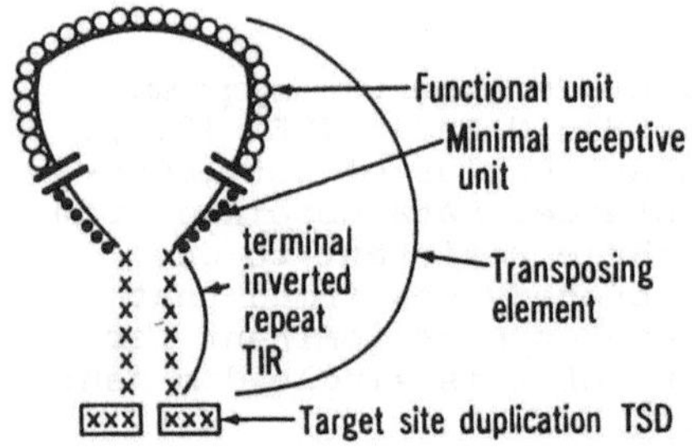

Fig. 1. General configuration of a mobile element. The stem at the terminal end can be of varied length. The very terminal end of the stem at the junction point with the target site contains terminal inverted repeated structure (XXX) (TIR) and is characteristic of the specific system. The body of the element may be varied in size and includes 4.3 kb for *Ac* and 8.4 kb for *En*. The minimal receptive unit and the functional unit are approximations (PETERSON 1985)

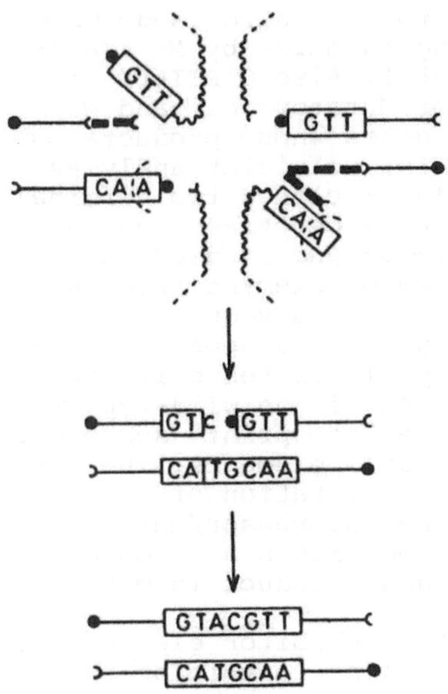

Fig. 2. Possible origin of one somatic excision event in maize. *Wavy lines* represent the termini of the transposable element *Spm*-I8 integrated within the *wx* gene. Single-stranded nicking by transposase has occurred, but the element has not yet been released. The *boxed-in* sequences flanking the element correspond to duplicated target site sequences. *Thin, dashed half-circles* show the extent of exonuclease degradation. Polymerase action is illustrated by a *thick, dashed line*. •- 5'end of the DNA; >- 3'end of the DNA (SAEDLER and NEVERS 1985)

It may be simple with only one element-specific enzyme or enzyme complex, which is called the transposase. Other required enzyme systems (polymerase, exonuclease, ligase) are assumed to be provided by DNA repair systems (SAEDLER and NEVERS 1985). The information needed for enzymes in order to recognize, cut out and transpose a sequence seems to be given by TIR (cf. e.g. FEDOROFF 1984).

For reviews on transposable elements in plants, see e.g. FEDOROFF (1984), WALBOT and CULLIS (1985), DÖRING (1985), PETERSON (1985, 1986), CHARLESWORTH et al. (1986), LILLIS and FREELING (1986) and NEVERS et al. (1986).

b) Maize *(Zea sp.)*

Particularly in maize, it has been shown that mobile elements are pervasive in different genomes where they are extensively distributed as discrete, related units (FEDOROFF 1984; PETERSON 1986). According to PETERSON (1985), mobile elements exist in maize in two broad categories: (1) a type where the insert in a specific gene is autonomously unstable, i.e. instability is inherited with the gene as a single unit [e.g. original *En* (Enhancer)-induced pale green mutable allele]; (2) a second category of two-element control derived from the first (e.g. induced unstable mutants of the *A2* and *C* loci). In the latter case, the gene with the insert is only unstable in the presence of a second factor; this two-element category represents a mobile element system (PETERSON 1985), or a family of related elements (FEDOROFF 1984). Approximately ten such systems exist in maize; each consists of a receptor (a gene with a stable insert) which is unstabilized (excised) in the presence of the second factor (regulatory element) with a transactive function (FEDOROFF 1984; PETERSON 1985).

According to LILLIS and FREELING (1986), there are at least four categories of mobile element systems in maize: (1) controlling elements (eight families), (2) a retrovirallike element, (3) *Mu* and (4) biologically uncharacterized insertions with structures similar to transposons.

Insertion mutants can be very useful as a tool for isolation and cloning of specific genes. Possibilities of gene tagging in maize by *Mu* transposon were described by LILLIS and FREELING (1986). Also O'REILLY et al. (1985a) suggested the use of *Spm* (Suppressor Mutator = *En*) and *Mu* transposable elements as tags for isolation of genes whose products are unknown. Transposons were correspondingly used for molecular analyses of the *A1* and *C2* loci. SCHWARZ-SOMMER et al. (1985) cloned two germinal and 16 somatic reversion events induced by the Enhancer transposable element system at the *wx-8::Spm-I8* allele of *Zea mays* and studied them by sequence analysis. Excision of the *Spm-I8* receptor element from the *wx* gene results in various mutant DNA sequences, i.e. a sequence divergence generated by visitation of a locus by plant transposable elements. Altered gene products which are still capable of restoring the wild-type phenotype are the result. The comparison of the wild-type and the mutant alleles of the *wx* gene showed possible "footprint" sequences that may have arisen by the excision of transposable elements. The authors demonstrated that in the (waxy) *wx-8::Spm-I8* mutation of *Zea mays*, the receptor component, i.e. the inhibitor of the suppressor/mutator (*Spm* or *En*) controlling element system is inserted within an exon of the *wx* gene. In the absence of *En* (*Spm*) the *wx* gene product is missing.

GIERL et al. (1985) described the sequence of the inhibitor element *Spm-I8* (*En*) isolated from the *wx-m8* allele. In a line containing the autonomous *En* element they detected a 2.5-kb transcript hybridizing to *Spm-I8*. Two chimeric transcripts are seen in the *wx-m8* mutant, both are suppressed by the introduction of the autonomous element *En*. The authors also described the cDNA sequence of an *En*-encoded transcript.

A highly mutagenic strain of maize, called Robertson's mutator, carries the *Mu* transposon. ALLEMAN and FREELING (1986) crossed mutator plants with a normal inbred line. The progeny plants had nearly the same number of *Mu* sequences as their mutator parent. One-half of these copies originated from segregation and one-half arose by transposition and are integrated into new positions in the genome. There is an extremely high rate of transposition of the *Mu* element: 10-15 transpositions per gamete per generation. When mutator plants were self-pollinated, the progeny doubled their *Mu* copy number in the first generation, but maintained a constant number of *Mu* sequences after subsequent self-pollinations. Transpositions of *Mu* occur very late in the development of the germ cells but before fertilization. ALLEMAN (1986) studied *Mu* transposition in a series of different alleles of the unstable mutant allele *AdhI1S3034*, resulting from the insertion of a *Mu* transposon into the first intron of *AdhI*. The derivative alleles differed amongst themselves in revertibility and in alcohol dehydrogenase production where several specified an altered polypeptide.

SCHNABLE and PETERSON (1985) compared the *Cy* transposable element system with Robertson's mutator system and were able to demonstrate that *Cy* is a new system subject to a negative relationship to Robertson's mutator.

A newly isolated unstable mutation *wx-844::En-1* of *Zea mays* was also proven to be caused by the insertion of the autonomous transposable element *En* into the *waxy* gene. Molecular analysis of *En-1* revealed that it is 8.4-kb-long, has a 13-bp-long perfect inverted repeat at its termini and generates a 3-bp target site duplication. *En-1* is integrated into an intron located approximately in the middle of the transcribed region of the *wx* gene (PEREIRA et al. 1985).

The *a1* locus of *Zea mays* has been cloned using transposable elements as gene tags for construction of genomic libraries from maize stocks with *a1* mutations induced either by *En(Spm)* or by Robertson's mutator system. Screening of these libraries is carried out either for the *En*-containing mutant or for the *Spm*-induced mutations (O'REILLY et al. 1985b).

Six independent derivatives of the *bz-m13* allele which contains a 2.2-kilobase-pair, defective, suppressor-mutator insertion at the *bronze* locus, have been isolated and analyzed by SCHIEFELBEIN et al. (1985). The derivatives were selected for alterations in the frequency and timing of somatic reversion.

WERR et al. (1985) determined the structure of the *shrunken* gene of *Zea mays* encoding sucrose synthase. The gene is interrupted by 15 introns. The transposable element *Ds* in the mutant *sh-m5933* and *sh-m6233* alleles is inserted in the seventh and first intron respectively. The 3'-untranslated sequence shows two duplications that may have arisen by the insertion and subsequent excision of transposable elements.

The 3-kb *Teo1* DNA insert from *Zea mexicana* teosinte Guerrero is a composite of *Cin2* with an insertion of *Cin3* elements. The integration of one element into a different element has been described already in other organisms including *Drosophila* and some prokaryotes. But in plants such composite structures have not been described until recently when a detailed analysis of *Cin*-DNA insertions of *Zea mays* was reported by BLUMBERG VEL SPALVE et al. (1985, 1986). A composite structure, the *Cin2/Cin3* complex, was detected in teosinte Guerrero by its homology to a unique 5-kb fragment, *LC1* of *Zea mays* line C. DNA sequence comparison between *LC1* and the clone containing *Teo1* revealed that *LC1* still contains one of the direct repeats of *Cin2* flanked by the same duplication. *LC1* or the equivalent clone from teosinte Guerrero is a recombinational derivative of the complex. The *Cin2* structure resembles copia elements of *Drosophila*, *Ty* elements of yeast and retroviruses. Single copies of the long terminal repeats, i.e. (LTR)-like direct repeats, are found. *Cin3* has a similar structure to that of Robertson's mutator. Both are flanked by a 9-bp duplication but the *Mu1* elements have shorter inverted repeats.

Genetic analysis of the behaviour of a multiple gene loss induced by a transposon in maize was reported by HUANG (1985). An exceptional derivative from a frequently mutating allele at the *a* locus showed in addition to this mutation specific chromosome breakage on chromosome 3. The loss includes at least the *a* function, the transposable element *En* and the *Sh2* gene. Heritability of the frequency of an induced loss event was 70%-90%.

There is some evidence that transposition may be initiated or supported by infection with RNA viruses. For example, *Bs-1*, a transposable element that moves into the maize *Adh1* gene following barley stripe mosaic virus infection, was shown by JOHNS et al. (1985) to be present in one to five copies in all maize and teosinte lines. The insertion of *Bs1* is bordered by 304-bp perfect direct repeats different from all other known plant transposons but similar in structure to *Ty1* in yeast, copia and related elements in *Drosophila* and vertebrate pro-retroviruses.

PEREIRA and PETERSON (1985) isolated three transposable element mutations, displaying an unstable phenotype at the *A* locus from lines that originated from maize plants treated with RNA viruses. Each of these three unstable mutants belongs to the *Uq-ruq* transposable element system.

c) Other Species

Mobile genetic elements have also been demonstrated to exist in species other than maize. For example, VERSHININ et al. (1987) studied the occurrence of DNA sequences similar to the *Ds* element of *sh-m5933* maize (*Ds*-like sequences) in other representatives of the *Gramineae*. The number of copies of *Ds*-like sequences in the maize genome is tenfold higher than the content of such sequences found in barley, rye or wheat genomes. No quantitative differences in *Ds*-like sequences between different wheat species with different ploidy levels and genomes as well as between different barley varieties were detected.

It has been shown by RHODES and VODKIN (1985) that the recessive allele for soybean seed lectin results from the insertion of a DNA segment (designated *Tgm1*) into the coding region of the gene. The complete sequence of *Tgm1* contains 3550 base pairs and can be divided into three regions. No large open reading frames were found but an extensive highly structured border with homology to the lectin gene was revealed.

The results of in situ hybridization with an I-rDNA clone from *Vicia faba* (SCHUBERT and WOBUS 1985) strongly support the mobile NOR hypothesis. The NORs of *Allium cepa* and possibly also those of *A. fistulosum* are in all probability able to move between telomeric heterochromatin blocks of different chromosomes.

In *Petunia*, interaction is described between a transregulatory genetic element (*Bi*) and responsive anthocyanin alleles resulting in a variegated flower colour pattern. Crosses segregated white-flowering plants that had lost the *Bi* element. The present study by WIJSMAN (1986) describes the genetic analysis of a two-element system in *Petunia*.

RAMMANA et al. (1985) reported the phenomenon of a breakage-fusion-bridge cycle in some tomato lines with extra chromosomes and subsequently examined the genetic behaviour of these lines in order to test whether the breakage-fusion-bridge cycle can activate transposable elements and cause genetic instability (RAMANNA 1986). Most of the plants of two tomato lines were found to have two to nine extra chromosomes in their somatic cells in addition to the normal set of 24. The extra chromosomes were derived from a single isochromosome of the short arm or nucleolar arm of chromosome 2. Plants with the breakage-fusion-bridge cycle showed a three-fold increase in phenotypically deviating seedlings as compared to the progeny of the control.

d) Importance in Evolution and Breeding

Transposable elements may be involved in significant developmental and evolutionary processes because they can initiate a reconstruction of gene and chromosome structure, i.e. chromosomal rearrangements, such as duplications, deletions, inversions, translocations etc. and, in addition, they can influence tissue specificity, developmental timing and level of gene expression. In particular, transposable elements can be activated under stress conditions. There is evidence reported by PETERSON (1986) that transposable elements are present even in high-bred maize materials. Such high-bred lines show a surprisingly high degree of flexibility allowing the selection for almost any character. However, it remains to be clarified whether mobile genetic elements are really generating "new" genetic variation. However, because of their functions, i.e. inactivating and transferring functional genes, transposable elements can become a very helpful tool for genetic engineering, particularly of cereals.

5. Genetic Engineering

Several ways exist at present to deliver and express foreign genes in plant cells (POTRYKUS et al. 1985a):

1. *Agrobacterium tumefaciens* or *A. rhizogenes* mediated gene transfer;
2. DNA viruses as gene vectors;
3. Liposome fusion with protoplasts;
4. Spheroplast fusion with protoplasts;
5. Microinjection into protoplasts, cells or tissue in vitro.

a) Transformation with Agrobacteria

α) *Agrobacterium tumefaciens.* The most widely used method has been exploited by the natural gene transfer system of the soil bacterium *A. tumefaciens*, the causal agent of crown gall disease. Cells transformed with Ti plasmids of *A. tumefaciens* do not require supplementation with auxins and cytokinins for continued cell growth and division, as most plant tissues usually do.

For many years, *A. tumefaciens* has been the only possible way of introducing foreign genes into plants. This experimental system has been optimized to achieve routine integration of foreign genes into plant genomes and their transfer to sexual offspring. Therefore, it is now a well-established technique, which does not require protoplasts, but obviously shows a limited host range which seems to exclude most gramineous monocots.

Seven plant phenolic compounds widely distributed among angiosperm plants were reported to be able to induce the expression of the *A. tumefaciens* loci needed for virulence (BOLTON et al. 1986). These phenolics in combination induce each transcriptional locus of the *vir* loci.

In order to study the role of the Ti plasmid in the attachment process and to verify the specificity of any observed bacterial binding, KRENS et al. (1985) used several *Agrobacterium* and *Rhizobium* strains, both with and without a Ti plasmid, as well as an *E. coli* strain. Only the Agrobacteria showed attachment. *A. tumefaciens* strains cured of their Ti plasmid were avirulent but still able to attach as were wild-type Agrobacteria lacking the Ti plasmid. Neither the Ti plasmid nor the cryptic plasmid carry all the genes necessary for attachment. Neither bacterial attachment nor transformation was observed for cells without a more or less well-regenerated primary cell wall. The presence of a newly formed primary cell wall was shown to be required for attachment and subsequent transformation of tobacco leaf protoplasts by *A. tumefaciens* in co-cultivation experiments. It is concluded that the primary cell wall was sufficiently well formed so that certain functional receptor molecules were available for attachment of *Agrobacterium* as the first step of a multistep process leading to the transformation of cells. The expression of bacterial functions required for attachment was independent of the presence of a Ti plasmid.

STACHEL et al. (1985) showed that *A. tumefaciens* virulence gene expression is activated in particular by the plant molecules acetosyringone (AS) and α-hydroxyacetosyringone (OH-AS). These molecules induce the entire *vir* regulon in *Agrobacterium* as well as the formation of T-DNA intermediate molecules. AS and OH-AS occur specifically in exudates of wounded and metabolically active plant cells and probably allow *Agrobacterium* to recognize susceptible cells in nature.

A physical and functional map of supervirulent *A. tumefaciens* tumor-inducing plasmid pTiBo542 was prepared by KOMARI et al. (1986). *A. tumefaciens* strains carrying this plasmid induce large, fast-appearing tumors and have an unusually wide host range. A clone bank was made from this plasmid in a wide host-range vector and restriction maps were prepared for BamHI and SalI. The authors were able to localize the virulence genes and the transferred DNA genes, the plasmid incompatibility region and a region that inhibits growth of certain *A. tumefaciens* strains. Genetic organization of this plasmid was compared with that of well-characterized Ti plasmid pTiA6. The six virulence genes and two *tms* genes were highly homologous to the genes of pTiA6, but the *tmr* gene was not.

The involvement of circular intermediates in the transfer of T-DNA from *A. tumefaciens* to plant cells was reported by KOUKOLIKOVA-NICOLA et al. (1985). Co-cultivation of *A. tumefaciens* with plant cells leads to the induction of circular copies of the T-DNA segment of the large tumor-inducing plasmid in bacterial cells. These circular molecules are presumably intermediates in DNA transfer from *A. tumefaciens* to the plant cells. The junction of the T-DNA circles occurs precisely in the 25-base pair terminal sequence involved in T-DNA transfer.

PERALTA and REAM (1985) constructed a deletion removing the three potential T_L right borders (the T_L right border and both T_R borders) in order to determine whether T-DNA transfer and integration require sequences in addition to the 23-bp border repeat (similar 23-bp direct repeats occur at the ends of two adjacent but non-contiguous T-DNAs, T_L- and T_R-DNA in the tumor-inducing plasmid). They also tested the influence of orientation, location and flanking sequences on border repeat function. Therefore, different restriction fragments were introduced, each containing a border repeat, into a Ti plasmid with the right borders deleted and their ability to restore virulence was tested. Experiments defined an 80-bp region that contains a functional right T-DNA border and identified non-essential sequences outside the border repeat that stimulate T-DNA transmission. The right border repeat functioned normally when moved to a new location in the Ti plasmid only when it remained in its wild-type orientation. Since the deletion severely attenuated virulence, the authors re-introduced restriction fragments containing the T_L-right border repeat at a new location to the right of T_L and tested their ability to restore virulence. Fragments that carried the border repeat flanked by at least 67 bp of wild-type Ti plasmid sequences on the left and 1035 bp on the right restored virulence completely. Smaller fragments restored virulence significantly but not fully even though the border repeat remained intact. Therefore, T-region sequences flanking the border repeat in the fully active fragments stimulated T-DNA integration. Fragments that restored virulence fully when inserted in the wild-type orientation, stimulated virulence slightly in the opposite orientation. The right border sequence promotes T-DNA transfer and integration best in one direction.

A cell line which acquired only the T_R-DNA was found among the transformants obtained after co-cultivation of *Nicotiana plumbaginifolia* cells with an octopine-type *A. tumefaciens* strain LBA 4013 (CZAKO and MARTON 1986). The authors demonstrated that transformants which acquired only the T_R-DNA are able to regenerate normal plants. The seed transmission of the T_R-DNA and agropine synthesis has been demonstrated. Agropine was always accompanied by mannopine, but mannopine could also be present alone; agropine could be absent although its precursor mannopine was present in the transformed tissue. It is shown in this chapter that variability exists not only in the T_L- but also in the T_R-DNA in the *N. plumbaginifolia* cell lines recovered in the single cell system after co-culture transformation. The production of agropine and mannopine in some of the transformants (T_R-DNA markers) showed that besides the T_L-region, the T_R-region of the Ti plasmid was also transferred to the plant cells. The authors could for the first time demonstrate inheritance of T_R-DNA in the transformed line 101.

The regenerated plants were tetraploid and segregation data indicated that the T_R-DNA was integrated in a single chromosome; T_R-DNA was transmitted through meiosis without rearrangement. This is evidence that the T_R-region can integrate into the plant genome independently of the T_L-region of the intact octopine-type Ti plasmid. This finding together with the demonstration of Mendelian inheritance of T_R-DNA further contributes to evidence supporting the concept that the T_R-region of octopine-type Ti plasmids functions as a transforming element independently from the T_L-region.

AN et al. (1985) have constructed a set of small vectors based on the tumor-inducing plasmid of *A. tumefaciens* which allows the transfer of exogenous DNA into plant chromosomes. Their simplified transformation system depends on wide host-range vectors which contain T-DNA border sequences flanking multiple restriction sites. Using these vectors any foreign DNA can be cloned in *E. coli*, transferred to *Agrobacterium* and stably maintained in *trans* with the Ti plasmid. The cloned DNA can then be transferred to plant cells employing the natural transformation system of Agrobacteria. Due to the small size of these cosmid vectors, DNA fragments up to 35 kbp can be inserted by an in vitro packaging method in *E. coli*. The ability of these vectors to be stably replicated in both *E. coli* and *A. tumefaciens* allows their subsequent transfer to and maintenance in Agrobacteria without intermediate genetic manipulations. DNA cloned into these vectors in *A. tumefaciens* can efficiently transform plants when in *trans* with a wild-type Ti plasmid which donates the functions necessary for DNA transfer and integration. Only the right border of the T-DNA is necessary for DNA transformation. The authors have constructed a simplified system for the transformation of higher plant cells that utilizes a series of shuttle vectors in combination with the natural transformation system of *A. tumefaciens*. These vectors consist of a wide host-range replicon that contains a chimeric *nos-npt* gene fusion, the T-DNA border sequences and various restriction enzyme sites suitable for the cloning of foreign DNA fragments. The *nos-npt* gene fusion described here yielded high levels of *npt* expression in transformed plant cells. Their vectors do not require homologous recombination for their stable maintenance in *Agrobacterium*; instead, they are maintained in *trans* with the Ti plasmid. The authors have demonstrated that several different Ti plasmids can serve as helpers for the shuttle vectors; however, transformation efficiencies obtained with the different plasmids in co-cultivation were seen to vary widely.

SHAH et al. (1986) constructed a binary vector with a chimeric EPSP synthase gene with the use of CaMV 35S promotor to attain high level expression of EPSP synthase. This chimeric gene was transferred into *Petunia* cells. Transgenic plants, produced by the leaf-disc technique, were tolerant to the herbicide glyphosate, which is a potent inhibitor of the enzyme EPSP.

The CaMV 35S promotor has been shown to direct high level expression of foreign genes into plants. For example, NAGY et al. (1985) have examined the activity of the 35S promotor of CaMV in transgenic tobacco and *Petunia* plants using the bacterial chloramphenicol acetyltransferase coding sequence as a reporter gene. The 35S promotor is approximately 30 times stronger than the nopaline synthase promotor.

SHEWMAKER et al. (1985) introduced a full-length copy of CaMV DNA into the T-DNA of *A. tumefaciens* and integrated the virus into plant genomes. Two transcripts were derived from the integrated CaMV, one 1.9-kb and a larger 7.5-kb transcript. In galls from seven plant species they showed wide variation of both the ratios of the transcripts within one species and of their levels between species.

β) Application of *A. tumefaciens* System. Numerous reports on transformation with *A. tumefaciens* have been given in recent years. BEVAN et al. (1985) isolated the 35S transcript promotor of cauliflower mosaic virus (CaMV) and linked it to a cDNA of the 3' end of tobacco mosaic virus (TMV) that encodes the coat protein of this single-stranded RNA virus. They used a disarmed *A. tumefaciens* binary vector for transformation of this constructed chimeric CaMV 35S promotor-TMV coat protein gene into tobacco plants and they detected viral coat protein synthesis in transformed leaf tissue.

OOMS et al. (1987) isolated derivatives of potato transformed with disarmed T-DNA from genetically engineered *A. tumefaciens* strains. By infection of wounds with mixed cultures of shoot-inducing *A. tumefaciens* strain T37 and either LBA1834 or LBA4404 they induced shoot-forming tumors from which transformed plants were recovered. Plants of cv. Maris Bard transformed with LBA1834 T-DNA were morphologically abnormal and highly aneuploid. Desiree plants transformed with LBA4404 T-DNA were morphologically normal and contained 47 or the euploid 48 chromosomes per cell respectively.

JONES et al. (1985) fused promotor DNA sequences from a *Petunia* chlorophyll a/b binding protein gene to octopine synthase DNA sequences. The resulting chimeric genes were introduced into *Petunia* and tobacco cells using a kanamycin resistance marker. Substantial variation was observed between transformants in the level of chimeric gene expression. In the most actively expressing transformants the amount of mRNA from the introduced chimeric genes was half that of the endogenous wild-type gene. Transcription initiated at the same place in the chimeric and endogenous genes. The authors reported on their investigations into the expression of introduced chimeric genes in a population of transformed plants. They showed that chimeric genes can be introduced to plant cells using *Agrobacterium* and can give rise in regenerated plants to mRNA levels comparable with those derived from the corresponding endogenous gene. Their results showed that careful attention should be paid to the exact sequences at the junctions of chimeric genes which are to be introduced into plants. The authors have investigated the copy number of inserted DNA in 30 transformants and all but one plant contain one to five copies of the introduced DNA. This copy number difference cannot explain the >200-fold variation in expression between transformants.

The development of a gene transfer system for cultivated alfalfa (*Medicago sativa*) using the vector pARC8 and three related vectors in the disarmed *A. tumefaciens* strain LBA4404 were described by SHAHIN et al. (1986a). The disarmed vector was used for introduction of a dominant selectable marker for kanamycin resistance into alfalfa tissues. The marker gene was stably integrated into the plant genome and expressed in selected kanamycin-resistant calli and in regenerated plants. The results demonstrate that foreign DNA can be readily transferred and expressed in alfalfa plants using a binary vector system in disarmed *A. tumefaciens*. The described method has been used successfully to transform alfalfa using two other vectors derived from pARC8 by the insertion of two different fragments of foreign DNA.

SCHÖFFL and BAUMANN (1985) inserted a genomic fragment containing a soybean heat shock gene (hs6871) into the T-DNA region of *A. tumefaciens* pTiC58 plasmid and thereby constructed a T-DNA vector. Then they used a strain carrying the modified Ti plasmid to incite tumors in sunflower hypocotyls. Transcripts from hs6871 are faithfully initiated in sunflower starting at the same site on the DNA as in soybean. This is the first evidence for thermo-induced and faithfully initiated transcription of a soybean *hs* gene in sunflower tumor cells under conditions which also induce the expression of highly homologous sunflower *hs* genes. The above authors have shown for the first time that transcription of plant *hs* genes may occur in heterologous plant cells in a thermo-regulated fashion with faithful initiation.

A novel leaf disc transformation-regeneration method for transformation of *Petunia*, tobacco and tomato plants was described by HORSCH et al. (1985). They inoculated leaf discs with an *A. tumefaciens* strain containing a modified tumor-inducing plasmid (a chimeric gene for kanamycin resistance replaced the phytohormone biosynthetic genes from transferred DNA). The authors have developed an approach to transformation that integrates the gene-transfer capability of *A. tumefaciens* with the simple and general regeneration capability of leaf explants.

Finally, chloroplast transformation by *A. tumefaciens* has been reported by DE BLOCK et al. (1985). They introduced a chimeric gene (*Pnos-cat* gene) by co-cultivation in tobacco protoplasts. Maternal inheritance of a chloramphenicol acetyltransferase gene was observed. The authors showed that the chimeric gene was present in the chloroplasts of chloramphenicol-resistant plants. Furthermore, the chloramphenicol acetyltransferase activity was shown to be associated with the chloroplast fraction. These observations are first proof that *Agrobacterium* Ti plasmid vectors can be used to introduce genes into chloroplasts. These results indicate that direct chloroplast genetic manipulation can be achieved using Agrobacteria (see also HOWE 1985).

ABEL et al. (1986) introduced in tobacco cells a chimeric gene containing a cloned cDNA of the coat protein gene of tobacco mosaic virus (TMV) on a Ti plasmid of *A. tumefaciens* from which tumor-inducing genes had been removed. Plants regenerated from transformed cells expressed TMV mRNA and coat protein as a nuclear trait. The results indicate that plants can be genetically transformed for resistance to virus disease development.

GARDNER and KNAUF (1986) used *A. tumefaciens* strains to transfer complementary DNA copies of a potato spindle tuber viroid (PSTV) to plant cells at a wound site on tomato plant stems. Infectious viroid RNA was found in the leaves of these plants indicating systemic infection. This process utilized the T-DNA transfer mechanism of *A. tumefaciens* since PSTV infection required most virulence genes as well as one of the DNA sequences that flank either side of the T-DNA. The *vir*E gene seems to be directly involved in the integration of foreign DNA into plant chromosomes.

γ) *Agrobacterium rhizogenes*. Like *Agrobacterium tumefaciens*, the related species *A. rhizogenes* can be used for transformation of plants. For details of molecular and genetic analyses of Ri plasmid and T-DNA respectively, compare WHITE et al. (1985), SIMPSON et al. (1986) and BOULANGER et al. (1986).

The structure and expression of Ri T-DNA from *A. rhizogenes* in *Nicotiana tabacum* was reported by DURAND-TARDIF et al. (1985). The transferred DNA (T-DNA) is incorporated into the chromosomal DNA of higher plants and this is correlated with the appearance of a complex phenotype. The transformed geno- and phenotype undergo Mendelian inheritance. A particular part of the foreign DNA has been delineated as the likely source of the transformed phenotype. The transformed phenotype does not appear to be due to structural changes in the Ri T-DNA.

It was demonstrated by COSTANTINO et al. (1984) that the T-DNA of *A. rhizogenes* is transmitted through meiosis to the progeny of tobacco hairy root plants. They analyzed the R_1 generation of self-pollinated plants regenerated from tobacco hairy root callus culture. All of three R_1 plants analyzed contained the same full-length T-DNA as the parental regenerant; only two showed agropine synthesis.

FURNER et al. (1986) reported that the genome of *Nicotiana glauca* contains a large inverted repeat which is homologous to the single copy T_L DNA of *A. rhizogenes*. Related sequences are present in several *Nicotiana* species, but are absent in others. The DNA sequence of the centre of the inverted repeat and the corresponding region from the bacterium both contain open reading frames which could potentially produce similar polypeptides. The data suggest that an infection with Agrobacteria resulted in genetic transformation early in the evolution of the genus *Nicotiana*.

OOMS et al. (1985a) regenerated potato plants from a hairy-root line after infection of a shoot of *S. tuberosum* cv. Desiree with *A. rhizogenes* strain LBA9402. Transformed plants were uniform and had a distinct phenotype and development compared with untransformed controls. Ten of the 42 transformed plants had either 47 or 49 chromosomes instead of the normal 48. In two of these aneuploids structural changes were observed. The hairy-root derived plants showed little phenotypic variation despite some numerical variation (24% were aneuploid) within a narrow range (46-49 chromosomes per cell).

A range of derivatives of four potato cultivars, all transformed with T-DNA from an *A. rhizogenes* plasmid pRi 1855, was isolated by OOMS et al. (1986). They all formed proliferic roots with reduced geotropism. Their leaves were slightly crinkled and their tubers were longer with more abundant and more prominent eyes. The transformed phenotype appeared stable even in a further two years of field trials.

SHAHIN et al. (1986b) genetically transformed tomato using two procedures. Punctured cotyledons were infected with disarmed *A. tumefaciens* or with *A. rhizogenes* each containing the binary vector pARC8. Southern blot analysis showed that only the pARC8 T-DNA was present in the plant genome, thus indicating that the pARC8 T-DNA integrated into the plant genome independently of the pRi A4 T-DNA. Using the second procedure with inverted cotyledons, they recovered transformed plants from *A. rhizogenes*-induced hairy roots. In the genome of the transformed plants both pARC8 T-DNA and the T-DNA of the A4 Ri plasmid were present.

The hairy-root disease caused by *A. rhizogenes* can also be used for rapid in vitro propagation in *Lycopersicon peruvianum* in addition to organ culture and plant regeneration from callus culture. A karyotypic analysis revealed minor structural alterations of chromosomes in roots transformed by *A. rhizogenes* (BANERJEE-CHATTOPADHYAY et al. 1985).

It is also possible to transform *Brassicaceae* like oilseed rape (*B. napus*) using *A. tumefaciens* or *A. rhizogenes* in a frequency of 10%-70%, depending on the genotype. Regeneration of transformed, abnormal but flowering plants was reported; however, these plants failed to form seeds (OOMS et al. 1985b).

A. rhizogenes can also be used successfully to transfer a foreign gene into cucumber plants (*Cucumber sativus*, TRULSON et al. 1986). The transgenic plants looked normal and they all contained vector DNA (T-DNA) but only some of them contained DNA from the Ri plasmid. Selectable marker genes such as kanamycin resistance introduced by this method should facilitate somatic hybridization of cucumbers with other species.

JENSEN et al. (1986) used two intermediate vectors pAR29 and pAR30 to integrate a chimeric gene into *A. rhizogenes* T-DNA. Homologous recombination through the Ri fragments was obtained after conjugation of the two "intermediate vectors" from *E. coli* to the *A. rhizogenes* strain. The complete intermediate cloning vector carries the chimeric soybean leg hemoglobin gene, which was then introduced into the genome of *Lotus corniculatus*, in the T-DNA region of the Ri plasmid. Nodule-specific expression of the chimeric gene was found in root nodules formed on fully regenerated plants inoculated with the *Lotus* microsymbiont.

Genotypic variability of soybean response to Agrobacteria strains harboring the Ti or Ri plasmids was observed by OWENS and CRESS (1985). They tested 24 diverse cultivars of soybean and three lines of its annual wild progenitor for their response to Agrobacteria strains harboring either the Ti plasmid (pTi) from *A. tumefaciens* or the Ri plasmid (pRi) from *A. rhizogenes* following uniform wounding and inoculation. They found three genotypes highly, ten moderately, eleven weakly and two non-susceptible to *A. tumefaciens* strain A348 (pTiA6). Of 26 genotypes inoculated with strain R1000 (pRiA4b) only seven responded in a clearly susceptible fashion by forming small, fleshy roots.

Recently, alfalfa (*Medicago sativa*) was transformed with the agropine type *A. rhizogenes* NCPPB 1855. Transformed plants regenerated via somatic embryogenesis showed phenotypic alterations, mainly affecting the root system (SPANO et al. 1987).

CARDARELLI et al. (1985) identified the genetic locus responsible for non-polar root induction by *A. rhizogenes* 1855. They demonstrated that this locus is actually responsible for the non-polar infectivity of *A. rhizogenes* 1855 by inserting a Tn5 in this region of pRi1855. *A. rhizogenes* can induce root proliferation in carrot discs both on the apical and basal surface or on the apical surface only, depending on the bacterial strain (polarity). The authors correlate polarity with the absence of an Ri plasmid genetic locus. They demonstrate that this locus is responsible for root induction on the basal surface since insertion of a transposon in this region of pRi1855 induces polarity in this strain.

b) Other Transformation Systems

α) Plant Viruses. Gene transfer via a plant DNA virus as was described for the first time for cauliflower mosaic virus (CaMV, BRISSON et al. 1984) has specific merits based on the systemic spread of the virus within an infected plant and the multicopy situation per infected cell. CaMV is a DNA virus with an 8-kb circular double-stranded genome.

Besides the use as a vector itself, CaMV is also frequently used for construction of binary vectors for *Agrobacterium*-mediated transformation. For example, SHAH et al. (1986) constructed a chimeric EPSP synthase gene with the use of CaMV 35S promotor to attain high level expression of EPSP synthase and they introduced this chimeric gene into *Petunia* cells via *Agrobacterium*. Transformed *Petunia* cells as well as transgenic plants were tolerant to the herbicide glyphosate, which is a potent inhibitor of the enzyme EPSP.

The CaMV 35S promotor has been shown to direct high level expression of foreign genes into plants. NAGY et al. (1985) have examined the activity of the 35S promotor of CaMV in transgenic tobacco and *Petunia* plants using the bacterial chloramphenicol acetyltransferase coding sequence as a reporter gene. The 35S promotor is approximately 30 times stronger than the nopaline synthase promotor.

SHEWMAKER et al. (1985) introduced a full-length copy of CaMV into the T-DNA of the *A. tumefaciens* Ti plasmid and integrated the virus into plant genomes. Two transcripts derived from the integrated CaMV: one 1.9-kb and a larger 7.5-kb transcript. In galls from seven plant species they showed wide variation of both the ratios of the transcripts within one species and of their levels between species.

Vectors based on RNA viruses are currently being developed as well. For example, FRENCH et al. (1986) described the efficient expression of a bacterial gene inserted in an engineered RNA virus, brome mosaic virus (BMV), in monocotyledonous plant cells. BMV has a genome of three RNA components. A viral complementary DNA expression system has been used to express wild-type and altered genomic RNAs in barley protoplasts.

β) Direct Gene Transfer. Alternative transformation techniques for plant genetic engineering involve direct transformation accomplished by introducing large amounts of DNA into plant cells. Methods of direct transformation are chemically mediated DNA uptake, electroporation (see MORIKAWA et al. 1986) and microinjection (see CROSSWAY et al. 1986). "Direct gene transfer" describes the phenomenon of uptake of genes into plant cells (protoplasts) through the plasma membrane, independent of a biological vector, and their functional integration into the host genome. It makes use of the fact that DNA can pass plasma membranes and that protoplasts can be totipotent. The procedure requires no more than a gene (under the control of plant expression signals) and protoplasts. Theoretically there should be no host range problem but direct gene transfer of course has its limitations, because it requires naked cells (protoplasts) and totipotent protoplasts are not readily available for every desired plant species.

PASZKOWSKI et al. (1984) introduced into tobacco a plasmid conferring kanamycin resistance. This plasmid contained a gene of the bacterial transposon Tn5 under the control of the 5' and 3' sequences of CaMV gene VI. The tumorous growth caused by *A. tumefaciens* is maintained by the integration into the plant genome of a region (the T-DNA) delineated by short repeated sequences of a bacterial plasmid. Previous attempts at Agrobacteria-independent gene transfer to higher plants used a range of approaches, involving plant organs, cells, protoplasts and pollen treated with naked DNA or sequences packaged in phages or liposomes.

The direct gene transfer experiment in *Nicotiana tabacum* reported above was described in detail by POTRYKUS et al. (1985b). They constructed a hybrid selectable marker gene by placing the protein coding region of the bacterial gene for APH(3')II (from Tn5) under the control of the 5'/3' expression signals from CaMV gene VI. Freshly isolated protoplasts from a shoot culture of *N. tabacum* were incubated with the gene, calf thymus, carrier DNA and PEG, and cultured without selective procedure for one week. This led finally to high levels of stable integrative transformation. The gene is transmitted to sexual offspring and inherited as a Mendelian unit; in most cases it is stably maintained through several sexual cycles. Cases of somatic and meiotic instability were also found. The authors found no host range limitations in transformation experiments with protoplasts from a series of herbaceous dicots and a gramineous monocot. Transferred foreign DNA can be integrated in one or a few places of the genome (SHILLITO et al. 1986). For corresponding transformation experiments in *Brassica*, PASZKOWSKI et al. (1986) constructed a hybrid CaMV genome containing a selectable marker gene (they constructed a non-infectious viral vector molecule based on CaMV DNA by the replacement of the CaMV gene VI coding region with a marker gene). The gene VI coding region was replaced by a special type of gene from Tn5. Stable, genetically transformed cell lines of *B. campestris* var. rapa were obtained after transformation. DNA of the hybrid CaMV genome was found to be integrated into high molecular weight plant genomic DNA. Transformation was achieved only when the hybrid genome was supplied together with wild-type viral DNA. The authors studied the replication and infectivity of this disarmed viral molecule and its possible functional complementation by wild-type CaMV in plant and in a protoplast-transformation system. They recovered protoplast-derived, genetically transformed cell lines of turnip rape and they analyzed them for the form, organization and expression of the transforming viral DNA within the plant cells. The integration pattern of pCaMV6km into the turnip genome is simple. The authors did not observe major DNA rearrangements or integration of separate, small fragments of the transforming molecule.

Direct gene transfer via incubation of protoplasts with PEG and plasmid DNA containing the coding region of aminoglycoside phosphotransferase gene (NPT II) was recently shown to be a simple and efficient method for transformation of moth bean (*Vigna aconitifolia*, KÖHLER et al. 1987a) and tobacco (KÖHLER et al. 1987b). They were able to regenerate kanamycin-resistant transformed plants. However, pronounced effects of plant- and vector-genotypes and significant interactions with regard to transformation rates were observed.

In a similar type of experiment, JUNKER et al. (1986) examined the possibilities of direct DNA transfer for the transformation of maize. They established a protoplast culture system for suspension-derived protoplasts and demonstrated uptake and expression of foreign DNA, as indicated by transient expression studies only.

γ) Electroporation. A general method for electrically stimulating the introduction of DNA into plant cells was developed by FROMM et al. (1985). When an electric pulse is applied to a solution containing DNA and protoplasts gene transfer occurs. To optimize gene-transfer efficiency carrot protoplasts were used as a model system. The efficiency increased with DNA concentration. The results showed that an electric pulse is an effective method for the introduction of genes into monocot and dicot protoplasts; the technique is now available for carrot and tobacco protoplasts.

Subsequently, FROMM et al. (1986) described stable transformation of maize after gene transfer by electroporation. This method uses electrical pulses of high field strength to permeabilize cell membranes reversibly so as to facilitate the transfer of DNA into cells. Electroporation-mediated gene transfer has resulted in stably transformed animal cells and transient gene expression in monocot and dicot plant cells. The authors reported that electroporation-mediated DNA transfer of a chimeric gene encoding neomycin phosphotransferase results in stably transformed maize cells that are resistant to kanamycin.

LANGRIDGE et al. (1986) have developed an electroporation procedure for the transformation of carrot protoplasts with Ti-plasmid DNA from *A. tumefaciens*. The uptake of labelled plasmid DNA 26-200 kbp in size into carrot protoplasts is stimulated by electric pulses. The uptake of plasmid pTic58 resulted in hormone-independent regeneration of carrot protoplasts. After the introduction of TMV RNA into *D. carota* protoplasts approximately 10% of the electrotransfected carrot protoplasts synthesized viral coat protein.

WEBER (1987) reported on DNA uptake by chloroplasts of *Brassica napus* after microirradiation with a focussed laser beam. Chloroplasts were isolated from mesophyll protoplasts of *B. napus*. They were irradiated with a laser beam. Uptake of DNA by individual chloroplasts was monitored using plasmid DNA labelled with a fluorescent dye.

6. Conclusion

Substantial progress has been made recently in developing asexual recombination techniques of genomes and plasmones (somatic hybrids and cybrids), chromosomes (somaclonal variation) or genes (DNA sequences) with the help of gene technology using Agrobacteria, viruses or transposable elements. Particularly the latter group of genetic units has been met with special interest by geneticists, because of their various functions in maize, which have been partly elucidated.

The integration of a transposable element into a functional gene leads to a distortion, i.e. a change of gene expression. Insertion of transposons into DNA sequences can therefore be considered as another means of somatically separating contiguous genes, besides meiotic crossover. Furthermore, by mobile gene segments, genetic material can be transferred from one locus to another, so that tagging and isolation of specific structural genes becomes feasible with the aid of mobile elements. This could become a very interesting alternative to molecular genetic analysis and genetic engineering of monocots, particularly cereals, which do not seem to be susceptible to Agrobacteria and cannot be regenerated from protoplasts as yet.

References

ABEL, P.P., NELSON, R.S., DE, B., HOFFMANN, N., ROGERS, S.G., FRALEY, R.T., BEACHY, R.N.: Science 232, 738-742 (1986). - AHLOOWALIA, B.S., SHERINGTON, J.: Euphytica 34, 525-537 (1985). - ALLEMAN, M.L.: Diss. Abstracts Int. 46, 3708B (1986). - ALLEMAN, M., FREELING, M.: Genetics 112, 107-119 (1986). - AN, G., WATSON, B.D., STACHEL, S., GORDON, M.P., NESTER, E.W.: EMBO J. 4, 277-284 (1985). - AUSTIN, S., EHLENFELDT, M.K., BAER, M.A., HELGESON, J.P.: Theor. Appl. Genet. 71, 682-690 (1986).

BAERTLEIN, D.A., McDANIEL, R.G.: Theor. Appl. Genet. 73, 575-580 (1987). - BANERJEE-CHATTOPADHYAY, S., SCHWEMMIN, A.M., SCHWEMMIN, D.J.: Theor. Appl. Genet. 71, 258-262 (1985). - BEVAN, M.W., MASON, S.E., GOELET, P.: EMBO J. 4, 1921-1926 (1985). - BLUMBERG VEL SPALVE, J., SHEPHERD, N.S., SAEDLER, H.: Maize Genet. Coop. Newslett. 59, 35 (1985); - 771-772 in: see ref. HORN et al. 1986. - BOLTON, G.W., NESTER, E.W., GORDON, M.P.: Science 232, 983-985 (1986). - BOULANGER, F., BERKALOFF, A., RICHAUD, F.: Plant Mol. Biol. 6, 271-279 (1986). - BRISSON, N., PASZKOWSKI, J., PENSWICK, J.R., GRONENBORN, B., POTRYKUS, I., HOHN, T.: Nature 310, 511-514 (1984). - BUIATTI, M., MARCHESCHI, G., TOGNONI, F., LIPUCCI DI PAOLA, M., COLLINA GRENCI, F., MARTINI, G.: Z. Pflanzenzücht. 94, 162-165 (1985).

CARDARELLI, M., SPANO, L., DE PAOLIS, A., MAURO, M.L., VITALI, G., COSTANTINO, P.: Plant Mol. Biol. 5, 385-391 (1985). - CHARLESWORTH, B., LANGLEY, C.H., STEPHAN, W.: Genetics 112, 947-962 (1986). - CHETRIT, P., MATHIEU, C., VEDEL, F., PELLETIER, G., PRIMARD, C.: Theor. Appl. Genet. 69, 361-366 (1985). - CLARK, E., SCHNABELRAUCH, L., HANSON, M.R., SINK, K.C.: Theor. Appl. Genet. 72, 748-755 (1986). - COOPER, D.B., SEARS, R.G., LOOKHART, G.L., JONES, B.L.: Theor. Appl. Genet. 71, 784-790 (1986). - COSTANTINO, P., SPANO, L., POMPONI, M., BENVENUTO, E., ANCORA, G.: J. Mol. Appl. Genet. 2, 465-470 (1984). - CREISSEN, G.P., KARP, A.: Plant Cell Tissue Organ Cult. 4, 171-182 (1985). - CROSSWAY, A., OAKES, J., IRVINE, J., WARD, B., KNAUF, V., SHEWMAKER, C.: Mol. Gen. Genet. 202, 179-185 (1986). - CULLIS, C.A., CLEARY, W.: Can. J. Genet. Cytol. 28, 247-251 (1986). - CZAKO, M., MARTON, L.: Plant Mol. Biol. 6, 101-109 (1986).

DE BLOCK, M., SCHELL, J., VAN MONTAGU, M.: EMBO J. 4, 1367-1372 (1985). - DE VRIES, S.E., JACOBSEN, E., JONES, M.G.K., LOONEN, A.E.H.M., TEMPELAAR, M.J., WIJBRANDI, J., FEENSTRA, W.J.: Theor. Appl. Genet. 73, 451-458 (1987). - DÖRING, H.-P.: BioEssays 3, 164-171 (1985). - DURAND-TARDIF, M., BROGLIE, R., SLIGHTOM, J., TEPFER, D.: J. Mol. Biol. 186, 557-564 (1985).

EHLENFELDT, M.K., HELGESON, J.P.: Theor. Appl. Genet. 73, 395-402 (1987).

FEDOROFF, N.V.: Sci. Am. 250, 65-74 (1984). - FISH, N., KARP, A.: Theor. Appl. Genet. 72, 405-412 (1986). - FRENCH, R., JANDA, M., AHLQUIST, P.: Science 231, 1294-1297 (1986). - FROMM, M.E., TAYLOR, L.P., WALBOT, V.: Proc. Natl. Acad. Sci. USA 82, 5824-5828 (1985); - Nature 319, 791-793 (1986). - FURNER, I.J., HUFFMAN, G.A., AMASINO, R.M., GARFINKEL, D.J., GORDON, M.P., NESTER, E.W.: Nature 319, 422-427 (1986).

GARDNER, R.C., KNAUF, V.C.: Science 231, 725-727 (1986). - GIERL, A., SCHWARZ-SOMMER, Z., SAEDLER, H.: EMBO J. 4, 579-583 (1985). - GLIMELIUS, K., BONNETT, H.T.: Theor. Appl. Genet. 72, 794-798 (1986). - GLIMELIUS, K., FAHLESSON, J., SJÖDIN, C., SUNDBERG, E., DJUPSJÖBACKA, M.: 663-682 in: see ref. HORN et al. 1986.

HANDLEY, L.W., NICKELS, R.L., CAMERON, M.W., MOORE, P.P., SINK, K.C.: Theor. Appl. Genet. 71, 691-697 (1986). - HORN, W., JENSEN, C.J., ODENBACH, W., SCHIEDER, O., eds.: Genetic Manipulation in Plant Breeding. Proc. Int. Symp. EUCARPIA, Berlin. Berlin, New York: de Gruyter 1986. - HORSCH, R.B., FRY, J.E., HOFFMANN, N.L., EICHHOLTZ, D., ROGERS, S.G., FRALEY, R.T.: Science 227, 1229-1231 (1985). - HOWE, C.: Trends Genet. 1, 217-218 (1985). - HUANG, Y.C.: Diss. Abstracts Int. 45, 1968B (1985).

JENSEN, J.S., MARCKER, K.A., OTTEN, L., SCHELL, J.: Nature 321, 669-674 (1986). - JOHNS, M.A., MOTTINGER, J., FREELING, M.: EMBO J. 4, 1093-1102 (1985). - JONES, J.D.G., DUNSMUIR, P., BEDBROOK, J.: EMBO J. 4, 2411-2418 (1985). - JUNKER, B., BAKER, B., LÖRZ, H.: 827-830 in: see ref. HORN et al. 1986.

KARP, A.: 547-554 in: see ref. HORN et al. 1986. - KARP, A., BRIGHT, S.W.J.: Oxford Surveys Plant Mol. Cell Biol. 2, 199-234 (1985). - KEMBLE, R.J., BARSBY, T.L., WONG, R.S.C., SHEPARD, J.F.: Theor. Appl. Genet. 72, 787-793 (1986). - KÖHLER, F., GOLZ, C., EAPEN, S., KOHN, H., SCHIEDER, O.: Plant Cell Rep., in press (1987a). - KÖHLER, F., GOLZ, C., EAPEN, S., SCHIEDER, O.: Plant Sci., in press (1987b). - KOMARI, T., HALPERIN, W., NESTER, E.W.: J. Bacteriol. 166, 88-94 (1986). - KOUKOLIKOVA-NICOLA, Z., SHILLITO, R.D., HOHN, B., WANG, K., VAN MONTAGU, M., ZEMBRYSKI, P.: Nature 313, 191-196 (1985). - KRENS, F.A., MOLENDIJK, L., WULLEMS, G.J., SCHILPEROORT, R.A.: Planta 166, 300-308 (1985).

LANDSMANN, J., UHRIG, H.: Theor. Appl. Genet. 71, 500-505 (1985). - LANGRIDGE, W.H.R., LI, B.J., SZALAY, A.A.: 785-802 in: see ref. HORN et al. 1986. - LARKIN, P.J., RYAN, S.A., BRETTELL, R.I.S., SCOWCROFT, W.R.: Theor. Appl. Genet. 67, 443-455 (1984). - LILLIS, M., FREELING, M.: Trends Genet. 2, 183-188 (1986). - LINACERO, R., VAZQUEZ, A.M.: 479-481 in: see ref. HORN et al. 1986.

MADDOCK, S.E., SEMPLE, J.T.: J. Exp. Bot. 37, 1065-1078 (1986). - MEDGYESY, P., FEJES, E., MALIGA, P.: Proc. Natl. Acad. Sci. USA 82, 6960-6964 (1985a). - MEDGYESY, P., GOLLING, R., NAGY, F.: Theor. Appl.

Genet. 70, 590-594 (1985b). - MEDGYESY, P., FEJES, E., MALIGA, P.: 693-695 in: see ref. HORN et al. 1986. - METAKOVSKY, E.V., NOVOSELSKAYA, A.Y., SOZINOV, A.A.: Theor. Appl. Genet. 73, 764-766 (1987). - MORIKAWA, H., IIDA, A., MATSUI, C., IKEGAMI, M., YAMADA, Y.: Gene 41, 121-124 (1986).

NAGY, F., ODELL, J.T., MORELLI, G., CHUA, N.-H.: 227-235 in: Biotechnology in Plant Science, eds.: M. ZAITLIN, P. DAY, A. HOLLAENDER. New York: Academic Press 1985. - NELSON, R.S., KARP, A., BRIGHT, S.W.J.: J. Exp. Bot. 37, 253-261 (1986). - NEVERS, P., SHEPHERD, N.S., SAEDLER, H.: Adv. Bot. Res. 12, 103-203 (1986).

OHGAWARA, T., KOBAYASHI, S., OHGAWARA, E., UCHIMIYA, H., ISHII, S.: Theor. Appl. Genet. 71, 1-4 (1985). - OOMS, G., KARP, A., BURRELL, M.M., TWELL, D., ROBERTS, J.: Theor. Appl. Genet. 70, 440-446 (1985a). - OOMS, G., BAINS, A., BURRELL, M., KARP, A., TWELL, D., WILCOX, E.: Theor. Appl. Genet. 71, 325-329 (1985b). - OOMS, G., BURRELL, M.M., KARP, A., TWELL, D., ROBERTS, J.: 823-825 in: see ref. HORN et al. 1986. - OOMS, G., BURRELL, M.M., KARP, A., BEVAN, M., HILLE, J.: Theor. Appl. Genet. 73, 744-750 (1987). - O'REILLY, C., SHEPHERD, N.S., PEREIRA, A., SCHWARZ-SOMMER, Z., NIESBACH, U., WEYDEMANN, U., WIENAND, U., ROBERTSON, D.S., PETERSON, P.A., SAEDLER, H.: Maize Genet. Coop. Newslett. 59, 35 (1985a). - O'REILLY, C., SHEPHERD, N.S., PEREIRA, A., SCHWARZ-SOMMER, Z., BERTRAM, I., ROBERTSON, D.S., PETERSON, P.A., SAEDLER, H.: EMBO J. 4, 877-882 (1985b). - OWENS, L.D., CRESS, D.E.: Plant Physiol. 77, 87-94 (1985).

PASZKOWSKI, J., SHILLITO, R.D., SAUL, M., MANDAK, V., HOHN, T., HOHN, B., POTRYKUS, I.: EMBO J. 3, 2717-2722 (1984). - PASZKOWSKI, J., PISAN, B., SHILLITO, R.D., HOHN, T., HOHN, B., POTRYKUS, I.: Plant Mol. Biol. 6, 303-312 (1986). - PELLETIER, G., PRIMARD, C., VEDEL, F., CHETRIT, P.: 653-661 in: see ref. HORN et al. 1986. - PERALTA, E.G., REAM, L.W.: Proc. Natl. Acad. Sci. USA 82, 5112-5116 (1985). - PEREIRA, A., PETERSON, P.A.: Genet. Res. Camb. 46, 219-236 (1985). - PEREIRA, A., SCHWARZ-SOMMER, Z., GIERL, A., BERTRAM, I., PETERSON, P.A., SAEDLER, H.: EMBO J. 4, 17-23 (1985). - PETERSON, P.A.: BioEssays 3, 199-204 (1985). - PETERSON, P.A.: 47-78 in: Genetics, Development and Evolution; 17th Stadler Genetics Symp., eds. J.P. GUSTAFSON, G.L. STEBBINS, F.J. AYALA. New York: Plenum Press 1986. - POTRYKUS, I., SHILLITO, R.D., SAUL, M.W., PASZKOWSKI, J.: Plant Mol. Biol. Rep. 3, 117-128 (1985a). - POTRYKUS, I., PASZKOWSKI, J., SAUL, M., PETRUSKA, I., SHILLITO, R.D.: Mol. Gen. Genet. 199, 169-177 (1985b).

RAMMANA, M.S.: 119-122 in: see ref. HORN et al. 1986. - RAMMANA, M.S., HILLE, J., ZABEL, P.: Theor. Appl. Genet. 71, 145-152 (1985). - RHODES, C.A., PHILLIPS, R.L., GREEN, C.E.: Can. J. Genet. Cytol. 28, 374-384 (1986). - RHODES, P.R., VODKIN, L.O.: Proc. Natl. Acad. Sci. USA 82, 493-497 (1985). - RYAN, S.A., SCOWCROFT, W.R.: Theor. Appl. Genet. 73, 459-464 (1987).

SAEDLER, H., NEVERS, P.: EMBO J. 4, 585-590 (1985). - SCHIEFELBEIN, J.W., RABOY, V., FEDOROFF, N.V., NELSON, O.E., jr.: Proc. Natl. Acad. Sci. USA 82, 4783-4787 (1985). - SCHNABLE, P.S., PETERSON, P.A.: Maize Genet. Coop. Newslett. 59, 4-6 (1985). - SCHÖFFL, F., BAUMANN, G.: EMBO J. 4, 1119-1124 (1985). - SCHUBERT, I., WOBUS, U.: Chromosoma 92, 143-148 (1985). - SCHWARZ-SOMMER, Z., GIERL, A., CUYPERS, H., PETERSON, P.A., SAEDLER, H.: EMBO J. 4, 591-597 (1985). - SHAH, D.M., HORSCH, R.B., KLEE, H.J., KISHORE, G.M., WINTER, J.A., TUMER, N.E., HIRONAKA, C.M., SANDERS, P.R., GASSER, C.S., AYKENT, S., SIEGEL, N.R., ROGERS, S.G., FRALEY, R.T.: Science 233, 478-481 (1986). - SHAHIN, E.A., SPIVEY, R.: Theor. Appl. Genet. 73, 164-169 (1986). - SHAHIN, E.A.,

SPIELMANN, A., SUKHAPINDA, K., SIMPSON, R.B., YASHAR, M.: Crop Sci. 26, 1235-1239 (1986a). - SHAHIN, E.A., SUKHAPINDA, K., SIMPSON, R.B., SPIVEY, R.: Theor. Appl. Genet. 72, 770-777 (1986b). - SHEWMAKER, C.K., CATON, J.R., HOUCK, C.M., GARDNER, R.C.: Virology 140, 281-288 (1985). - SHILLITO, R.D., SAUL, M.W., PASZKOWSKI, J., POTRYKUS, I.: 813-822 in: see ref. HORN et al. 1986. - SIMPSON, R.B., SPIELMANN, A., MARGOSSIAN, L., McKNIGHT, T.D.: Plant Mol. Biol. 6, 403-415 (1986). - SPANO, L., MARIOTTI, B., PEZZOTTI, M., DAMIANI, F., ARCIONI, S.: Theor. Appl. Genet. 73, 523-530 (1987). - SREE RAMULU, K., DIJKHUIS, P., HANISCH TEN CATE, CH.H., DE GROOT, B.: Plant Sci. 41, 69-78 (1985). - STACHEL, S.E., MESSENS, E., VAN MONTAGU, M., ZAMBRYSKI, P.: Nature 318, 624-629 (1985).

TAGUCHI, T., KAMEYA, T.: Jpn. J. Breed. 36, 185-189 (1986). - TERADA, R., YAMASHITA, Y., NISHIBAYASHI, S., SHIMAMOTO, K.: Theor. Appl. Genet. 73, 379-384 (1987). - TRULSON, A.J., SIMPSON, R.B., SHAHIN, E.A.: Theor. Appl. Genet. 73, 11-15 (1986).

VERSHININ, A.V., SALINA, E.A., SVITASHEV, S.K., SHUMNY, V.K.: Theor. Appl. Genet. 73, 428-432 (1987).

WALBOT, V., CULLIS, C.A.: Annu. Rev. Plant Physiol. 36, 367-396 (1985). - WEBER, G., MONAJEMBASHI, S., GREULICH, K.O., WOLFRUM, J.: 7. Wissenschaftl. Tagung der Gesellschaft für Entwicklungsbiologie mit der deutschen Sektion der IAPTC, Berlin, pp. 74. Berlin: Universitäts-Druckerei der FU 1987. - WERR, W., FROMMER, W.-B., MAAS, C., STARLINGER, P.: EMBO J. 4, 1373-1380 (1985). - WHEELER, V.A., EVANS, N.E., FOULGER, D., WEBB, K.J., KARP, A., FRANKLIN, J., BRIGHT, S.W.J.: Ann. Bot. 55, 309-320 (1985). - WHITE, F.F., TAYLOR, B.H., HUFFMAN, G.A., GORDON, M.P., NESTER, E.W.: J. Bacteriol. 164, 33-44 (1985). - WIJSMAN, H.J.W.: Theor. Appl. Genet. 71, 791-796 (1986).

Prof. Dr. WOLFGANG FRIEDT
Dipl. Ing. agr. ULRIKE BRUNE
Lehrstuhl für Pflanzenzüchtung
Inst. f. Pflanzenbau u. Pflanzenzüchtung
Justus-Liebig-Universität
Ludwigstr. 23
D-6300 Gießen, FRG

III. Mutation: Higher Plants

By WERNER GOTTSCHALK

1. Methodological Aspects

The permanently increasing intensity of mutation research requires effective methods for discerning the action of mutant genes. A reliable but time-consuming method in higher plants is the evaluation of mutants, mostly of chlorophyll deficient ones, in the M_2 generation. A considerable advantage in this respect is a test system in which the extent of genetic damage can be evaluated already in M_1. Not generative, but somatic mutations are evaluated. The system has been developed in soybeans in combination with EMS (ethyl methanesulfonate) treatment (FUJII and TANO 1986), but it can certainly also be used in other plant species. Another very efficient system uses the formation of micronuclei from damaged chromosomes as a parameter for discerning the mutagenicity of substances. This system has been developed in *Tradescantia* specifically for testing gaseous or liquid forms of chemical and physical agents. Pollen mother cells in cuttings of young inflorescences, which are maintained in nutrient solution, are cytologically evaluated after treatment with the respective agents. In this way, a great number of different agents can be tested within a short time (Trad-MCN test; MA et al. 1984). A "chromosome image analyzing system", developed by FUKUI (1986), works automatically and digitizes the information on chromosome structure for analyzing them. By means of this method, the karyogram of *Secale cereale* could be obtained within 25 min. This system should also be suited for analyzing chromosome mutations. A new banding technique using hydrochloric acid for identifying somatic chromosomes was successfully used in the genus *Impatiens*. It is supposed that this method will be helpful for studying evolutionary problems (ZINOVEVA-STAHEVITCH and GRANT 1985).

Propane sultone is not only a potent carcinogen but also an effective mutagen. Its effectiveness depends on concentration, temperature and pH, a pH of 4.6 having the highest action (SINGH and SHAMSHERRY 1985). The mutagenic action of EMS and sodium acid was compared in *Avena strigosa* ($2n = 2x = 14$) and *A. sativa* ($2n = 6x = 42$). NaN_3 was found to be an essentially stronger mutagen than EMS. The mutation frequency was considerably higher in the diploid than in the hexaploid species, obviously due to the presence of duplicative loci in the hexaploids (RINES 1985). The effect of presoaking was studied in barley in combination with gamma ray treatment. High mutation rates were obtained after 13, 16 and 19 h of presoaking before redrying. Most of the mutations were induced during S phase but also three distinct stages of G_1 proved to be well suited for mutation induction (YAMAGUCHI 1985). High mutation rates in a distinct locus up to around 9% were obtained in *Poinsettia* after X-irradiation of embryonic cell suspensions. The material used was heterozygous for anthocyanin synthesis. Mutations in that particular gene are easily discernible because anthocyanin-free petioles arise in the developing shoots (KLEFFEL et al. 1986).

Progress in Botany, Vol. 49

2. Environmental Mutagenesis

The problem of mutagens in our environment is being intensively studied in many countries, and the number of substances belonging to this group increases rapidly. Not only more and more pesticides and industrial substances, but also natural compounds of higher plants were found to be mutagenic. In the present chapter, only some examples can be given. A widely used test system is the evaluation of somatic pink mutations in the blue *Tradescantia* flowers (VILLALOBOS-PIETRINI et al. 1986).

The following *pesticides* were found to be mutagenic: Thimet 10-G (PANDITA and KHOSHOO 1984), Dimethoate (TALUKDAR et al. 1985), Prometrin (SOFRADŽIJA and HADŽISELIMOVIĆ 1985), Deltamethrin (CHAUHAN et al. 1986), Methamidophos (AMER and FARAH 1985), Ceresan and Agrosan, two widely used fungicides (NANDI 1985). These substances were tested in root tip meristems of *Allium cepa* or *Vicia faba*. Chromosomal anomalies, such as fragmentations, stickiness, formation of bridges, ring chromosomes, micronuclei or binucleate cells, unequal distribution of the chromatids, polyploidy and other mitotic abnormalities were observed. Eight pesticides more were further tested with regard to their mitotic and meiotic effects on barley, and manifold anomalies and disturbances were found (KAUR and GROVER 1985).

Many *industrial wastes* are tested in cultivated rat or Chinese hamster cells or using the *Salmonella*/microsome assay; however, they are not considered in the present chapter. Some of them, however, have been tested in root tip meristems of *Allium* or *Vicia*. This holds true for dye manufacturing industry waste water (SOMASHEKAR et al. 1985) and for benzene, toluene, hexane, heptane and ethyl acetate (GÓMEZ-ARROYO et al. 1986). In principle, these substances were found to cause the same mitotic anomalies as just mentioned for the pesticides.

Among the *drugs*, a high mutagenic effectiveness was evidenced for pyrolysates of opium and morphine (FRIESEN et al. 1985). Also further medicinal substances were found to induce chromosome aberrations and other cytological anomalies. This holds true for Metamizole, a widely used analgesic antipyretic, spasmolytic and antirheumatic agent in India (REDDY and SUBRAMANYAM 1984; root tips of *Allium cepa*) and for two antihistaminic drugs (SHEHAB and ABO-EL-KHIER 1985; meiosis of *Vicia* and *Allium*). Further investigations on tobacco condensates formed during pipe smoking confirmed the mutagenicity of these compounds (ČERNÁ and ANGELIS 1985; *Salmonella* assay). The lymphocyte chromosomes of Indian betel and betel with tobacco chewers had a higher frequency of sister chromatid exchanges than the control persons (GHOSH and GHOSH 1984). Mitotic anomalies were found by PATNAIK et al. (1984) in *Allium cepa* treated with the extract of processed tobacco used in betel. Ethanol is already known as a weak mutagen (VILLALOBOS-PIETRINI et al. 1986). According to CORTÉS et al. (1986), however, ethanol itself is not mutagenic, it becomes only effective when it is metabolically transformed into acetaldehyde. Rutin has been identified as the major mutagenic substance of red wine (YU et al. 1986; *Salmonella* test). Mutagenic efficiency was furthermore found or confirmed for quercetin, widely distributed in edible plants (NAKAYASU et al. 1986) and for allyl isothiocyanate, a component of mustard oil and of some other human food plants such as cabbage and cauliflower (NEUDECKER and HENSCHLER 1985).

Some plant substances, used in drugs for medicinal purposes, were found to be mutagenic. Fagaronine from roots of *Fagara zanthoxyloides* (Rustaceae), a strong antitumor agent, caused mutations in *Drosophila* (PÉREZ-CHIESA and CINTRÓN 1986). Water extracts of *Ammi majus*, a wild growing Umbelliferae, is used in the treatment of skin diseases and leukodermia; it causes mitotic disturbances (ADAM and RASHAD 1984). This holds also true for yohimbine and vinblastine, two *Vinca* alkaloids used as antidiuretics or anticarcinogenic drugs (SHEHAB 1985). Aqueous fruit extracts of *Emblica officinalis* are a cheap source of ascorbic and gallic acids in India, widely used in indigenous medicine. They were found to cause chromosomal anomalies in root tip cells of *Allium sativum* (YADAV and RATHOE 1984). Similar findings were obtained with rotenone, extracted from the roots of *Derris*, *Lonchocarpus* and *Millettia* species and used in insecticides (AMER and MIKHAEL 1986). Mutagenic efficiency was furthermore found for aristolochic acid from root extracts of *Aristolochia clematitis* (MAIER et al. 1985), for xanthone derivatives from different Gentianaceae (MATSUSHIMA et al. 1985) and for indol-3-acetonitrile from Chinese cabbage. This compound is a mutagen precursor and becomes mutagenic on nitrite treatment (WAKABAYASHI et al. 1985).

3. Antimutagenic Plant Substances

A relatively young field of mutation research is antimutagenesis which, however, is in rapid development. In the present chapter, only those antimutagenic substances are considered which are derived from plants. Furthermore, some compounds are mentioned the antimutagenic action of which has been tested in higher plants.

Extracts of leaves of Japanese green tea (*Camellia sinensis*) were found to reduce the high spontaneous mutation rate in a mutant strain of *Bacillus subtilis*. The effective antimutagenic compound is epigallo-catechin-gallate (KADA et al. 1985). A similar action was found for pyrogallol compounds including gallic acid and for tannic acid (SHIMOI et al. 1985, 1986). The reaction of tannic acid is insofar of additional interest, as it becomes active after UV and nitroquinoline-1-oxide treatment but it does not suppress mutations induced by gamma rays or MNNG (N-methyl-N'-nitro-N-nitrosoguanidine). Different types of antimutagenic action were furthermore found for crude drugs derived from roots of *Paeonia*, *Bupleurum* and *Glycirrhiza* species used in Chinese medicines (SAKAI et al. 1986; Ames test). Potent antimutagens are β-carotene, riboflavin, chlorophyll a and b, and chlorophyllin, a sodium and copper salt of chlorophyll (RENNER 1985; TERWEL and VAN DER HOEVEN 1985; ONG et al. 1986).

The frequency of chromosome aberrations induced by 7,12-dimethylbenz (a) anthracene was significantly reduced under the influence of fresh or boiled juices from cabbage, onion, Welsh onion, egg plant and burdock (ITO et al. 1986; rat bone marrow cells). Similar effects were observed after having used fruit juices of *Phyllanthus emblica* which are a rich source of vitamin C in India. They were tested in male mice in combination with distinct environmental mutagens (GIRI and BANERJEE 1986).

Maleic hydrazide (MH) is a strong mutagen. A significant reduction of the frequency of MH-induced chromatid aberrations in *Vicia faba* mitoses was obtained by pretreatment with hydrazine, hydrogen peroxide or butyl hydroperoxide before applying the mutagen (HEINDORFF et al. 1985a,b). This phenomenon is called "cleistogenic adaptation". The effect was not observed when triethylene melamine (TEM) was used instead of MH. A marked reduction of the yield of chlorophyll and other mutations in *Arabidopsis* was reached by diethyldithiocarbamate and carbon monoxide (GICHNER and VELEMINSKÝ 1984).

It should finally be mentioned that even heat shocks were found to have an antimutagenic effect. A temperature of 40°C, given 10 and 30 min prior to mutagenic treatment of *Vicia faba* root tip meristems, reduced the frequency of chromatid aberrations. The effect was observed in combination with MH or TEM as mutagens but interestingly not when ethanol was used (RIEGER et al. 1985).

4. Gene Mutations

It is impossible to give a detailed survey on hundreds of works in the present review dealing with the behaviour of new, spontaneously arisen or induced mutants of higher plants. I shall only consider genotypes with different meiotic anomalies which are of general botanical or even biological interest and some mutants of practical value. A review on the mutants of *Arabidopsis thaliana*, widely used for studying manifold problems of mutation research, has been given by ESTELLE and SOMERVILLE (1986). The occurrence of periclinal chimeras from frequently mutating unstable alleles and the dependence of these processes on temperature was analyzed in *Petunia* (DOODEMAN and BIANCHI 1985; DOODEMAN et al. 1985). The well-known difficulties to separate mutations of closely linked genes, responsible for single characters, from mutations of pleiotropic genes and from small deficiencies were discussed by PALMER (1984) using two soybean mutants.

The influence of the temperature on the speed of ontogenetic development and the flowering behaviour of 17 *Pisum* mutants and 20 different recombinants was studied under long-day phytotron conditions. Clear differences between the genotypes were found especially with regard to the flowering time. Distinct genes influenced a gene for earliness negatively at low temperature (12°C), whereas high temperatures (25°C) were found to have a strong delaying effect on the flowering of some fasciated mutants in contrast to all the other genotypes studied. A distinct recombinant produced only tiny flower buds at a constantly given temperature of 25°; another one needs the change of low and higher temperatures for normal flowering. Some of our high-yielding fasciated mutants do not flower under short-day conditions. A group of prospective fasciated recombinants was selected in the phytotron which flower under short-day conditions; they will be tested in countries with short-day climates (GOTTSCHALK 1985a, 1986). Interesting gene-ecological findings were obtained in the Sovjet Union in barley after MNU (N-methyl-N-nitrosourea) and DES (diethyl sulfate) treatment. The M_1 plants were grown at an altitude of 2,600 m, whereas the M_2 was grown in two different regions with strongly varying climatic conditions, namely in altitudes of 2,600 m and 3,860 m. A higher frequency of morphological mutants and a wider mutation spectrum was observed at 2,600 m as compared to 3,860 m. Also with regard to the frequency of sterile mutants, different results were obtained at the two altitudes (ABDULOV and NIGMATULLIN 1985).

a) Mutant Genes Affecting the Meiosis of Higher Plants

A review paper on genes controlling meiosis was given by KAUL and MURTHY (1985). In the mutant condition, they cause at least a strong reduction of the fertility, mostly complete sterility. The whole group has been subdivided by the authors into synaptic genes, genes influencing premeiotic and others influencing postmeiotic stages. A voluminous table and many references demonstrate the abundance of findings available in this field.

Asynaptic genes (*as* genes) suppress the pairing of the homologous chromosomes during the early meiotic stages completely or partially. Spontaneous mutants of this type were selected in *Manihot esculenta* (JOS et al. 1984), *Solanum commersonii* (JOHNSTON et al. 1986) and in *Capsicum frutescens* (KUMAR and RAO 1985; this genotype is classified as desynaptic, but according to its meiotic behaviour, it is obviously a partially asynaptic one). New, experimentally induced asynaptic mutants are available in *Hordeum vulgare* (PRASAD and TRIPATHI 1986; DES treatment), *Avena sativa* (RINES 1985; EMS), *Oryza sativa* (KITADA and OMURA 1984b; MNU) and *Allium cepa* (KONVIČKA and FISCHBECK 1984a,b; gamma rays). The last mentioned genotypes were able to produce some seeds when pollinated with genetically normal pollen. In their progenies, not only aneuploid but even polyploid plants arose after formation of restitution nuclei. All the *as* genes mentioned are recessive.

Desynaptic genes (*ds* genes) suppress chiasma formation partially or completely, thus leading to varying numbers of univalents in the meiocytes. A single plant of this type with low chiasma frequency was isolated in *Critesion violaceum* (*Gramineae*; WANG 1984). In mutation experiments, desynaptic mutants were found in *Hordeum vulgare* (PRASAD and TRIPATHI 1986; DES), *Sorghum bicolor* (REDDI and MANI 1985; EMS and gamma rays), *Nigella sativa* (DATTA and BISWAS 1985; EMS) and *Carthamus tinctorius* (PRASAD and PRASAD 1983; gamma rays). All these mutants show a normal pairing behaviour in pachytene; thus, they are really desynaptic and not asynaptic genotypes. The influence of five independent *ds* genes of *Oryza sativa* on the process of recombination was studied by KITADA and OMURA (1984a). Each of these genes was found to reduce the recombination frequency almost to the same extent. The formation of the synaptonemal complex is not affected by them.

Genes for *male sterility* (*ms* genes) become effective during microsporogenesis, whereas they do not affect megasporogenesis. New spontaneous genes of this category were found in *Triticum aestivum* (BING-HUA and JING-YANG 1986), *Brassica napus* (MATHIAS 1985), *Brassica campestris* (ZUBERI and ZUBERI 1983) and *Crotolaria pallida* (ARORA and GUPTA 1984). Experimentally induced male sterile mutants were selected in *Avena sativa* (RINES 1985; EMS) and *Carthamus tinctoria*. The *Ms* gene of this species is dominant; all the male sterile plants of this genotype were found to be heterozygous (JAMBHALE and NERKAR 1985). Dominance was furthermore revealed for the just mentioned *Ms* genes of *Brassica napus* and *Triticum aestivum*. This is insofar remarkable as around 99% of all mutant genes in higher plants are recessive. The ms_1 gene of the soybean has already been under study for several years. Investigations of the megasporogenesis and gametogenesis have shown that more than eight megagametophytes are often formed resulting from failure of postmeiotic cytokinesis. An additional anomaly is the fusion of nuclei; therefore, megagametophytes with up to four egg cells can arise. These findings are in concordance with the occurrence of polyembryonic seedlings and polyploids in the progenies of ms_1 plants. The chromosome number of these seedlings varied from 20 to 200 (KENNELL and HORNER 1985, CHEN et al. 1985). VAN HARTEN et al. (1985) tried to induce cytoplasmic male sterility in *Petunia*. Around 70,000 selfed seeds were treated with mutagenic chemicals but cytoplasmic male sterility was not obtained. In the offspring of the treated material, however, some cases of genic male sterility were observed.

Some meiotic mutants cannot be incorporated into one of the three groups just discussed. An NaN_3-induced barley mutant, for instance, shows an additional chromosome condensation in the late tetrad stage of microsporogenesis but no division and separation of the chromosome follows. The fertility of the plants is not affected (PRINA and HAGBERG 1984). In protoclones of *Solanum tuberosum*, regenerated from mesophyll cells, a recessive gene is present causing spindle anomalies which lead to the formation of restitution and double restitution nuclei. As a result of this anomaly, no haploid pollen grains are produced (GILL et al. 1985). Diploid pollen was also found in a *Pinus* mutant following fusion of spindles (BUTORINA et al. 1985).

b) Mutants of Agronomic Interest

Mutation breeding is intensively carried out in many countries but the proportion of mutants which are of direct agronomic value is extraordinarily low, only around 1% of all the genotypes selected. New results in this field were discussed in two symposia in Japan (Institute of Radiation Breeding; OHMIYA 1982, 1983). A list of mutant varieties with date of release, kind of mutagenic treatment and main improved characters was given by MICKE et al. (1985). Surveys on problems and results of applied mutagenesis were also published by HANNA (1982) for *Pennisetum americanum* and *Sorghum bicolor* and by JARANOWSKI and MICKE (1985) for *Pisum sativum*.

Mutants with quantitatively enhanced resistance to bacterial leaf blight are available in rice (NAKAI et al. 1985; gamma rays, thermal neutrons). A spontaneously arisen shattering-resistant mutant of *Papaver bracteatum* may be of interest to the codeine industry. Under the influence of the dominant mutant gene, the capsules do not form pores of dehiscence. The seed yield of the plants is lower but the thebaine content of the capsules is higher than normal (thebaine is readily converted to codeine; LEVY 1985). X-ray induced mutants of *Chrysanthemum morifolium* were found to tolerate relatively low temperatures, flowering about two weeks earlier than the control plants at 12°C (BROERTJES et al. 1983; BROERTJES and LOCK 1985). A high degree of drought tolerance has a mutant of *Vigna aconitifolia* (HENRY and DAULAY 1983; EMS). The dormancy of hazelnut trees is normally induced by short days. A mutant does not show this reaction. Shoot tips continue to grow and axillary buds begin to grow in December. The non-dormancy is controlled by a recessive gene: already eight cultivars are known to be heterozygous for that gene (THOMPSON et al. 1985). Five floury endosperm mutants of *Oryza sativa* have about twice as much amylose than their parental variety (YANO et al. 1985). The alkaloid content of *Lupinus mutabilis* is strongly reduced under the influence of a mutant gene resulting in "sweet" plants (WILLIAMS et al. 1984; EMS).

The *afila* gene (*af*) of the *Pisum* genome, causing formation of branched tendrils instead of the leaflets, is of increasing importance for pea breeding mainly because of the pronounced lodging resistance of the *afila* genotypes. It has been used for developing several commercial varieties or advanced strains in England, Poland, Czechoslovakia, Italy and USA (JARANOWSKI and MICKE 1985). New findings on the so-called leafless or semileafless peas were published by PYKE and HEDLEY (1983) and by SNOAD et al. (1985). Twenty different recombinant lines, homozygous for *af* and other mutant genes or gene groups of agronomic interest, were studied with regard to positive and negative interactions between *af* and the other genes involved. A favourable cooperation was found between *af* and a gene for earliness, whereas the combination of *af* with genes for stem fasciation gave negative results (GOTTSCHALK and EWEDA 1984).

c) Somaclonal Variation

The phenomenon of somaclonal variation, i.e. the appearance of variability among plants regenerated from tissue cultures without having used any mutagens, is still not yet completely understood. It cannot be excluded that a part of this variation was already present in the plant tissues used for culture initiation. Moreover, not all the morphological changes observed are genetically conditioned. But the number of examples for Mendelian variation, which remains constant in subsequent generations, increases permanently. There is clear evidence that much of this variability is genetically conditioned. The variation includes also gross chromosomal changes such as endopolyploidy, endomitosis and polyteny.

Review papers with many references and examples from sexually as well as asexually propagated species have been published by MEINS (1983), ORTON (1984), SCOWCROFT (1984) and LÖRZ and BROWN (1986).

Some examples may elucidate the present status of this research field. In 17% of the progenies of tomato plants, regenerated from cotyledons in vitro, chlorophyll mutants and plants with distinct morphological anomalies were found; they showed a 3:1 segregation (BUIATTI et al. 1985). Similar findings were obtained in lettuce plants regenerated from mesophyll protoplasts. Ten clones segregated for mutant phenotypes in approximately monohybrid ratios suggesting mutations in single recessive genes (ENGLER and GROGAN 1984). In 10 of 70 progenies following in vitro culture of mung bean de-embryonated cotyledons, mutants for chlorophyll and morphological anomalies appeared similar to those selected in mutagenic treatments (MATHEWS et al. 1986). Based on the experiences available, it is assumed that it may be possible to utilize the material obtained by somaclonal variation as a new genetic resource for crop improvement. More details on the phenomenon of somaclonal variation are found in Part C II (FRIEDT and WENZEL) of the present volume.

5. Genome Mutations

Intensive studies in this field refer to the phenomenon of androgenesis and somatic hybridization. Both of these fields are not only of interest for basic research but to some extent also for applied genetics. Intensive work is also being done in the experimental production of new allopolyploids and in the analysis of the genomic constitution of polyploid plant species. A comprehensive review on genetic, cytological and evolutionary problems of *Gossypium* has been given by ENDRIZZI et al. (1985) including the genomic relations between the species, the origin of the allotetraploids, their diploidization, the behaviour of different groups of aneuploids and the present knowledge on translocations. In another review paper, relevant problems of polyploidy research are discussed (GOTTSCHALK 1985b). Investigations on the relations between ploidy level and climatic conditions were carried out in almost 700 taxa of grasses in the eastern Himalayas. 63.4% of them proved to be polyploid. The incidence of polyploidy in the Indian Himalaya regions decreases with increasing latitudes together with a change from subtropical to subtemperate and temperate climates (SHARMA 1985). A correlation of polyploidy with geographical distribution was also found in the widespread genus *Gladiolus* (OHRI and KHOSHOO 1985b).

a) Haploids

Haploid plants are widely produced in many species via androgenesis from pollen grains. Reviews on methodological aspects and on the prospects and difficulties of the methods used have been given by HEBERLE-BORS (1985) and JENSEN (1986). The problems related to this phenomenon were intensively discussed during an Eucarpia Symposium held in Berlin (HORN et al. 1986).

Investigations in *Nicotiana tabacum* demonstrated once more that the efficiency for inducing pollen embryogenesis depends highly on the genotypic constitution of the material used (SUBHASHINI and VENKATESWARLU 1985). The authors used different F_1 lines. Especially good results were obtained in lines derived from mutants. Around 50% of *Brassica oleracea* plants, regenerated from microspores, proved to be haploids; the remainder were spontaneously doubled haploids (KELLER and ARMSTRONG 1983). In pollen clones of rice:

47.2% - 50.4% haploids;
34.8% - 36.7% dihaploids;
4.8% - 5.3% polyploids; and
8.8% - 11.0% aneuploids

were obtained (CHU et al. 1985).

In distinct, interspecific *Hordeum* crosses, the chromosomes of one of the two parental species are eliminated resulting finally in the production of haploids. This process was found to depend on the temperature, 22°C being more effective than 15°C. This observation, however, was only made in four interspecific *Hordeum* crosses and not following hybridizations between *Hordeum* and *Triticum* or *Secale*, respectively (PICKERING and MORGAN 1985). In the progeny of an interspecific *Pennisetum squamulatum* x *P. americanum* hybrid (2n = 41), a polyhaploid plant with 21 chromosomes was found. It originated probably by parthenogenetic development of a reduced gametophyte and was partially female-fertile but completely male-sterile. The offspring of this plant arose by obligate apomixis (DUJARDIN and HANNA 1986). Subhaploid pollen grains in frequencies of around 80% were present in the anthers of a *Paphiopedilum insigne* cultivar. Its normal chromosome number is n = 13, the abnormal ones ranged between five and 12. This anomaly may be due to segregational irregularities during meiosis, to split spindles or to failure of wall formation (SINGH 1985).

b) Autopolyploids

The spontaneous formation of polyploid plants has been observed in cultivated *Bougainvilleas* via the use of unreduced pollen arising through precocious centromere division at metaphase I. When they are involved in fertilization with haploid or diploid egg cells of other cultivars, triploid or even tetraploid plants can arise (OHRI and ZADOO 1986). Unreduced gametes were also found in many plants of the diploid relatives of alfalfa. It is therefore assumed that *Medicago sativa* (2n = 4x = 32) originated in this way (VERONESI et al. 1986). The tetraploid tuber-bearing species *Solanum kuntzianum* has a high percentage of quadrivalents indicating its autopolyploid origin (SANGOWAWA 1985). Autopolyploid nature is also assumed for the tetra- and hexaploid forms of the Asiatic *Hordeum brevisubulatum* complex (LANDSTRÖM et al. 1984). Thirty-nine cultivars of the autopolyploid *Chrysanthemum morifolium* complex with tetra-, hexa-, octo- and aneuploid forms were meiotically studied. The hexaploids and the higher aneuploids show an obviously genetically controlled diploidlike meiotic behaviour (NAZEER and KHOSHOO 1985).

A high degree of intraplant polyploidy was found in the diploid species *Pennisetum americanum*. Haploid + diploid, diploid + tetraploid, haploid + diploid + aneuploid and diploid + tetraploid + aneuploid pollen mother cells were present in the same anther, floret or spike (PANUGANTI and NIRMALA 1986). A similar variation was observed in PMCs of *Coix gigantea* (2n = 20), the chromosomes of which varied between seven and 29. Associations of three and four chromosomes arose frequently, but even higher configurations with up to ten chromosomes were found (NIRMALA and RAO 1984).

Experimentally produced autotetraploids of *Nicandra physaloides* were found to form exclusively bivalents and showed a strong tendency to reverse to the diploid initial state (GUPTA and ROY 1986). A very high level of ploidy (12n) was reached in *Solanum nigrum* after colchicine treatment. The plants had a diploidlike meiosis and a surprisingly high fertility. This may mainly be due to the fact that the initial material was already hexaploid (SINGH and ROY 1985). Twenty-five percent of *Solanum tuberosum* protoclones (2n = 4x = 48), regenerated from mesophyll protoplasts, were octoploids or near-octoploids (2n = 92 - 96); another 3% were hexaploids and 12% were mono- and trisomics possibly due to endomitosis and non-disjunction (GILL et al. 1986).

c) Allopolyploids

α) Natural Allopolyploids. Allotriploid plants as natural hybrids between the tetraploid *Solanum acaule* and the diploid *S. infundibuliforme*, two tuber-bearing species, are constituents of the North Argentinian flora. They agree morphologically with the artificially produced hybrids and are completely male sterile (= *Solanum* x *viirsooi* hybr. nov.; OKADA and CLAUSEN 1985). The subspecies *somniferum* of *Papaver somniferum* (2n = 22) is possibly a triploid hybrid between taxa closely related to *P. glaucum* (2n = 14) and others related to *P. gracile* (2n = 28). Based on morphological and hybridological investigations, the ssp. *setigerum* (2n = 44) could have arisen as a hexaploid form from ssp. *somniferum* (KADEREIT 1986). Populations of natural triploid hybrids between the diploid species *Antennaria microphylla* and the two tetraploid species *A. rosea* and *umbrinella* were found in western Canada completing the euploid series 2x to 8x of the genus (CHINNAPPA 1984).

Genome analyses in the genus *Hordeum* have revealed that the hexaploid species *H. procerum* and *lechleri* have obviously two genomes in common. *Hordeum procerum* has furthermore two common genomes with the hexaploid species *H. parodi* and one common genome with the tetraploid species *Elymus virginicus* (REDDY and SUBRAHMANYAM 1985; GUPTA and FEDAK 1986). Common genomes were also found in:

Papaver bracteatum (2x) and *P. orientale* (4x) (MILO et al. 1986);
Papaver bracteatum (2x) and *P. pseudo-orientale* (4x) (MILO et al. 1986);
Stellaria longifolia (2x) and *S. longipes* (4x) (CHINNAPPA 1985);
Ocimum gratissimum (AABB) and *O. viride* (AACC) (KHOSLA and SOBTI 1986).

The *Scilla scilloides* complex has been intensively studied for many years in East Asia with regard to the genomic constitutions of different representatives. Comprehensive material has been recently analyzed in Korea involving 3,384 plants from 40 populations at 35 different localities. The following variation was found:

AA:	19.6%
AABB:	72.4%
ABBB:	4.3%
AABBB:	3.0%
aneuploids:	0.8%

The A genome of the polyploids was found to be identical with A of the Japanese allopolyploids of the complex, whereas the A genome of the diploids is karyotypically different from A in the allopolyploids (ARAKI 1985).

The groundnut (*Arachis hypogaea*) is considered to be an allotetraploid species AABB, the A genome deriving from *A. cardenasii* and the B genome from *A. batizocoi*. According to pachytene studies of interspecific hybrids, *Arachis correntina* could be the donor of the A genome (MURTY and JAHNAVI 1986). Seven diploid wild relatives of the groundnut were again found to agree with each other with regard to the A genome (SINGH 1986b). *Lotus corniculatus* was one of the best examples for the autotetraploid origin of a plant species derived from the closely related diploid species *L. tenuis* or *alpinus*. According to new morphological, cytogenetic, but preferably biochemical findings, however, this hypothesis can obviously not be maintained. The species seems to be an allotetraploid form derived from *L. alpinus* and/or *tenuis* and *L. uliginosus* (ROSS and JONES 1985).

β) Experimentally Produced Allopolyploids. New allopolyploids have been mainly produced in different Gramineae in which the difficulties of interspecific or even intergeneric hybridizations are less pronounced than in dicots. The following new amphiploids derived from species of the same genus are available:

Avena sativa/magna (AAAACCCCDD, 2n = 10x = 70). Univalents to quadrivalents were present in metaphase I; seed set and pollen fertility showed great variation (PREMACHANDRAN et al. 1986).

Avena abyssinica/sativa (2n = 10x = 70). All the decaploid plants proved to be somatically instable. In their progenies, chromosome elimination occurred (FRITZ and SORRELLS 1985).

Bromus inermis/pumpellianus (8x). The octoploid plants, derived from two tetraploid species, had a seed set of around 50%, a high value for this high ploidy level (ARMSTRONG 1985).

Nicotiana knightiana (2n = 24)/*umbratica* (2n = 46). The amphiploids with 70 chromosomes showed regular bivalent formation. They were fully male fertile but female sterile due to disintegration of the embryo sacs (GANGADEVI et al. 1985).

The following amphiploids were produced via intergeneric hybridizations:

Hordeum chilense/Triticum turgidum conv. *durum* ($AABBH^{ch}H^{ch}$ = "*Tritordeum*"). There was no pairing between the chromosomes of the two parental species (FERNANDEZ et al. 1985).

Hordeum pubiflorum/Secale africanum. Over 90% of the pollen proved to be stainable but only 40% seed set was obtained (FEDAK 1985).

Aegilops ventricosa/Secale cereale. A high diminution of chromosome pairing between the three genomes was observed in the hexaploid plants (CERMEÑO et al. 1985). This work is remarkable because of the excellent photomicrographs of the C-banded chromosomes.

Triticum turgidum/Dasypyrum villosum (AABBVV; JAN et al. 1986).

Triticum aestivum/Dasypyrum villosum (AABBDDVV). The octoploid plants had poor vigour and low fertility (JAN et al. 1986).

A survey on the improvement of the synthetic genus *Triticale* by means of chromosome manipulation and modification was given by KALTSIKES et al. (1984) including the prospects of further development and the problem of meiotic instability of the representatives of different ploidy levels. Another review, mainly on breeding aspects of *Triticale*, was published by FEDOROVA (1985).

The immense variation of garden gladioli, around 30,000 cultivars are available at present all having 2n = 4x = 60 chromosomes, has been obtained from only eight elemental species. The chromosome numbers of the 180 *Gladiolus* species known range between 2n = 30 and 180. They are crossable not only at the same but also at different ploidy levels. Thus, many new garden forms with different genomic constitutions can be expected in the near future (OHRI and KHOSHOO 1985a).

γ) Somatic Hybridization. The production of allopolyploids via protoplast fusion of somatic cells, i.e. without using gametes for the hybridizations, is a fascinating method, but it is still too early to judge its potentialities for producing new plant species or even genera. Review papers on the whole complex including methodological aspects were published by GLIMELIUS (1985) and SCHIEDER et al. (1986). Problems of somatic hybridization were also discussed at an Eucarpia Meeting held in Berlin (HORN et al. 1986). Some examples may demonstrate not only the potentialities of the methods used but also the difficulties in getting fully functioning allopolyploid somatic hybrids.

Nicotiana plumbaginifolia/N. tabacum
The protoplast fusion was electrically induced. The regenerated plantlets were true hybrids as confirmed by esterase isozyme patterns (BATES and HASENKAMPF 1985).
Nicotiana rustica/N. tabacum
Of the 17 independently arisen somatic hybrids, three proved to be partially self-fertile, whereas the other ones did not set seed. The level of self-fertility could be increased by recurrent selection in the progenies (HAMILL et al. 1985).
Solanum melongena/S. sisymbriifolium
The two species are sexually incompatible. All the 26 somatic hybrids obtained were aneuploid and sterile (GLEDDIE et al. 1986).
Solanum tuberosum/S. brevidens
Almost all the hybrids obtained were hexaploid as expected. Most of them flowered, had viable pollen and produced tubers. With regard to their morphological characters, great variability was observed (AUSTIN et al. 1986).
Lycopersicon esculentum/Solanum rickii
The plants tested regenerated from the same callus. The chromosome numbers of three plants varied around the expected number of 48 ($2n = 4x = 44 - 52$). Two other plants had chromosome numbers ranging between 60 and 64; they were approximately hexaploid. In another plant, numbers of $2n = 67$, 71 and 130 were found in different root tips (O'CONNELL and HANSON 1986).
Lycopersicon esculentum/Solanum lycopersicoides
The chromosome numbers of seven somatic hybrids ranged from the expected $2n = 4x = 48$ to 68. Four plants of this group produced some fruits (HANDLEY et al. 1986).
Citrus sinensis/Poncirus trifoliata
The somatic hybrids had the amphidiploid chromosome number of $2n = 4x = 36$. The leaves were trifoliate; other morphological characters were intermediate between both parents (OHGAWARA et al. 1985).

An interesting method was used in the genus *Nicotiana* for producing allotriploid plants. Diploid somatic protoplasts of *Nicotiana tabacum* ($2n = 4x = 48$) were fused with haploid gametic protoplasts of *N. glutinosa* ($n = 12$) resulting in five "gametosomatic" hybrid plants. They had the expected chromosome number of $2n = 5x = 60$ (PIRRIE and POWER 1986).

d) Aneuploids

In pollen clones of *Oryza sativa*, derived from anther culture, many aneuploids were selected including primary and double trisomics, tetrasomics, monosomics and nullisomics (CHU et al. 1985). A series of different aneuploids was isolated from a wild population of *Coix gigantea* in India. Altogether, 96 aneuploids were found ranging from nullisomics to hexasomics. Penta- and hexasomics are for the first time reported in higher plants. In 8% of the normal disomic plants of the population, non-disjunction was observed being obviously the mechanism of this high degree of chromosomal diversity (SAPRE and BARVE 1984, 1985a,b). The 21 possible monosomic lines of *Avena byzantina* are now available. They are morphologically scarcely distinguishable from each other and from the disomics. Their identification was not only carried out by means of cytogenetic but also biochemical methods (MORIKAWA 1985).

The trisomics are a very heterogeneous group. A comprehensive survey on the present knowledge of rye genetics and cytogenetics was given by SCHLEGEL et al. (1986). The importance of trisomics and telotrisomics as a new standard tester material is discussed in this paper. New single and double trisomics were selected in *Chrysanthemum coronarium* (GUPTA and GILL 1984), *Plantago lagopus* (SHARMA et al. 1985) and *Hordeum vulgare* (PRASAD and TRIPATHI 1985). A tetrasomic and a heptasomic plant was found in *Plantago lagopus* (SHARMA et al. 1985) and *Capsicum annuum* (SADANANDAM and SUBHASH 1985).

Monotelotrisomics are aneuploids which contain a telocentric chromosome in addition to the normal complement. Seven plants of this category appeared in the progeny of primary rye trisomics. They were identified by means of Giemsa banding patterns. Their vitality and fertility were better than those of the primary trisomics (MELZ and SCHLEGEL 1985). Tertiary trisomics have a translocated chromosome in addition to the normal karyotype. They appear in progenies of interchange heterozygotes and have been meiotically studied in rye (JANSE 1985), pearl millet (KUMAR et al. 1985) and in the garden pea (MERCYKUTTY and KUMAR 1985). A complicated chromosomal structure has been found in the telotertiary compensating trisomics: a normal chromosome is lacking and is replaced by a telocentric and a translocated chromosome. Plants of this type have been isolated in *Secale cereale* (DE VRIES 1985).

References

ABDULOV, I.A., NIGMATULLIN, F.G.: Genetika USSR 21, 1210-1214 (1985). - ADAM, Z.M., RASHAD, TH.: Cytologia 49, 265-271 (1984). - AMER, S.M., FARAH, O.R.: Cytologia 50, 521-526 (1985). - AMER, S.M., MIKHAEL, E.: Cytologia 51, 171-176 (1986). - ARAKI, H.: Genetica 66, 3-10 (1985). - ARMSTRONG, K.C.: Can. J. Genet. Cytol. 27, 538-541 (1985). - ARORA, D.K., GUPTA, P.K.: Cytologia 49, 685-690 (1984). - AUSTIN, S., EHLENFELDT, M.K., BAER, M.A., HELGESON, J.P.: Theor. Appl. Genet. 71, 682-690 (1986).

BATES, G.W., HASENKAMPF, C.A.: Theor. Appl. Genet. 70, 227-233 (1985). - BING-HUA, L., JING-YANG, L.: Z. Pflanzenzücht. 97, 204-209 (1986). - BROERTJES, C., LOCK, C.A.M.: Euphytica 34, 97-103 (1985). - BROERTJES, C., KOENE, P., PRONK, TH.: Euphytica 32, 97-101 (1983). - BUIATTI, M., MARCHESCHI, G., TOGNONI, F., LIPUCCI DI PAOLA, M., GRENCI, F.C., MARTINI, G.: Z. Pflanzenzücht. 94, 162-165 (1985). - BUTORINA, A.K., MURAYA, A.K., MASHKINA, L.S., VISOTSKY, A.A.: Genetika USSR 21, 103-111 (1985).

CERMEÑO, M.C., CUÑADO, N., ORELLANA, J.: Theor. Appl. Genet. 70, 679-683 (1985). - ČERNÁ, M., ANGELIS, K.J.: Mutation Res. 143, 161-164 (1985). - CHAUHAN, L.K.S., DIKSHITH, T.S.S., SUNDARARAMAN, V.: Mutation Res. 171, 25-30 (1986). - CHEN, L.-F.O., HEER, H.E., PALMER, R.G.: Theor. Appl. Genet. 69, 271-277 (1985). - CHINNAPPA, C.C.: Can. J. Genet. Cytol. 26, 133-136 (1984); - Can. J. Genet. Cytol. 27, 318-321 (1985). - CHU, Q., ZHANG, Z., GAO, Y.: Theor. Appl. Genet. 71, 506-512 (1985). - CORTÉS, F., MATEOS, S., ESCALZA, P.: Mutation Res. 171, 139-143 (1986).

DATTA, A.K., BISWAS, A.K.: Cytologia 50, 535-543 (1985). - DE VRIES, J.N.: Genetica 68, 47-58 (1985). - DOODEMAN, M., BIANCHI, F.: Theor. Appl. Genet. 69, 297-304 (1985). - DOODEMAN, M., BINO, R.J., UYTEWAAL, B., BIANCHI, F.: Theor. Appl. Genet. 69, 489-495 (1985). - DUJARDIN, M., HANNA, W.: Theor. Appl. Genet. 72, 33-36 (1986).

ENDRIZZI, J.E., TURCOTTE, E.L., KOHEL, R.J.: Adv. Genet. 23, 271-375 (1985). - ENGLER, D.E., GROGAN, R.G.: J. Hered. 75, 426-430 (1984). - ESTELLE, M.A., SOMERVILLE, C.R.: Trends Genet. 2, 89-93 (1986).

FEDAK, G.: Can. J. Genet. Cytol. 27, 1-5 (1985). - FEDOROVA, T.N.: Genetika USSR 21, 181-190 (1985). - FERNANDEZ, J.A., GONZALES, J.M., JOUVE, N.: Theor. Appl. Genet. 70, 85-91 (1985). - FRIESEN, M., O'NEILL, I.K., MALAVEILLE, C., GARREN, L., HAUTEFEUILLE, A., CABRAL, J.R.P., GALENDO, D., LASNE, C., SALA, M., COUROULINKOV, I., MOHR, U., TURUSOV, V., DAY, N.E., BARTSCH, H.: Mutation Res. 150, 177-191 (1985). - FRITZ, S.E., SORRELLS, M.E.: Can. J. Genet. Cytol. 27, 399-404 (1985). - FUJII, T., TANO, S.: Environ. Exp. Bot. 26, 191-195 (1986). - FUKUI, K.: Theor. Appl. Genet. 72, 27-32 (1986).

GANGADEVI, T., RAO, P.N., RAO, B.H., SATYANARAYANA, K.V.: Theor. Appl. Genet. 70, 330-332 (1985). - GHOSH, R., GHOSH, P.K.: Mutation Res. 139, 79-81 (1984). - GICHNER, T., VELEMINSKY, J.: Mutation Res. 139, 29-33 (1984). - GILL, B.S., KAM-MORGAN, L.N.W., SHEPARD, J.F.: J. Hered. 76, 17-20 (1985); - ibid. 77, 13-16 (1986). - GIRI, A.K., BANERJEE, T.S.: Cytologia 51, 375-380 (1986). - GLEDDIE, S., KELLER, W.A., SETTERFIELD, G.: Theor. Appl. Genet. 71, 613-621 (1986). - GLIMELIUS, K.: Hereditas Suppl. 3, 41-47 (1985). - GÓMEZ-ARROYO, S., CASTILLO-RUIZ, P., VILLALOBOS-PIETRINI, R.: Cytologia 51, 133-142 (1986). - GOTTSCHALK, W.: Theor. Appl. Genet. 70, 207-212 (1985a); - 1-43 in: Advances in Chromosome and Cell Genetics, eds. A.K. SHARMA, A. SHARMA, 1985b; - Proc. 5th Int. SABRAO Congress Bangkok, 75-83 (1986). - GOTTSCHALK, W., EWEDA, M.: Genetika (Beograd) 16, 245-261 (1984). - GUPTA, P.K., FEDAK, G.: Can. J. Genet. Cytol. 28, 416-419 (1986). - GUPTA, R.C., Gill, B.S.: Cytologia 49, 789-795 (1984). - GUPTA, S.K., ROY, S.K.: Cytologia 51, 319-324 (1986).

HAMILL, J.D., PENTAL, D., COCKING, E.C.: Theor. Appl. Genet. 71, 486-490 (1985). - HANDLEY, L.W., NICKELS, R.L., CAMERON, M.W., MOORE, P.P., SINK, K.C.: Theor. Appl. Genet. 71, 691-697 (1986). - HANNA, W.W.: Mutat. Breed. Rev. 1, 1-13 (1982). - HEBERLE-BORS, E.: Theor. Appl. Genet. 71, 361-374 (1985). - HEINDORFF, K., AURICH, O., RIEGER, R., MICHAELIS, A.: Mutat. Res. 142, 183-186 (1985a). - HEINDORFF, K., MICHAELIS, A., AURICH, O., RIEGER, R.: Mutat. Res. 142, 23-27 (1985b). - HENRY, A., DAULEY, H.S.: Indian J. Genet. Plant Breed. 43, 342-344 (1983). - HORN, W., JENSEN, C.J., ODENBACH, W., SCHIEDER, O. (eds.): Genetic Manipulation in Plant Breeding, 217-263 (1986).

Institute of Radiation Breeding, Ohmiya-machi/Japan: Gamma Field Symp. 21, 131 pp. (1982); - ibid. 22, 1-105 (1983). - ITO, Y., MAEDA, S., SUSIYAMA, T.: Mutat. Res. 172, 55-60 (1986).

JAMBHALE, N.D., NERKAR, Y.S.: Z. Pflanzenzücht. 95, 185-188 (1985). - JAN, C.C., DE PACE, C., McGUIRE, P.E., QUALSET, C.O.: Z. Pflanzenzücht. 96, 97-106 (1986). - JANSE, J.: Can. J. Genet. Cytol. 27, 393-398 (1985). - JARANOWSKI, J., MICKE, A.: Mutat. Breed. Rev. (IAEA Vienna) 2, 23 pp. (1985). - JENSEN, C.J.: 231-256 in: see ref. Horn et al. 1986. - JOHNSTON, S.A. RUHDE, R.W., EHLENFELDT, M.K., HANNEMAN, R.E.: Can. J. Genet. Cytol. 28, 520-524 (1986). - JOS, J.S., BAI, K.V., NAIR, R.B.: Cytologia 49, 273-277 (1984).

KADA, T., KANEKO, K., MATSUZAKI, S., MATSUZAKI, T., HARA, Y.: Mutat. Res. 150, 127-132 (1985). - KADEREIT, J.W.: Bot. Jahrb. Syst. 106, 221-244 (1986). - KALTSIKES, P.J., GUSTAFSON, J.P., LUKASZEWSKI, A.J.: Can. J. Genet. Cytol. 26, 105-110 (1984). - KAUL, M.L.H., MURTHY, T.G.K.: Theor. Appl. Genet. 70, 449-466 (1985). - KAUR, P., GROVER, I.S.: Cytologia 50, 187-197 and 199-211 (1985). - KELLER, W.A., ARMSTRONG, K.C.: Euphytica 32, 151-159 (1983). - KENNELL, J.C., HORNER, H.T.: Can. J. Genet. Cytol. 27, 200-209 (1985). - KHOSLA, M.K., SOBTI, S.N.: Cytologia 51, 225-234 (1986). - KITADA, K., OMURA, T.: Genetics 108, 697-706 (1984a); - Can. J. Genet. Cytol. 26, 264-271 (1984b). - KLEFFEL, B., WALTHER, F., PREIL, W.: 113-120 in: Nuclear Techniques and in vitro Culture for Plant Improvement (IAEA Vienna) (1986). - KONVIČKA, O., FISCHBECK, G.: Cytologia 49, 771-779 (1984a); - Allium Newsl. 1, 42-55 (1984b). - KUMAR, O.A., RAO, K.G.R.: Cytologia 50, 69-73 (1985). - KUMAR, R.S., SINGH, U.P., SINGH. R.M., SINGH, R.B.: Cytologia 50, 433-443 (1985).

LANDSTRÖM, T., VON BOTHMER, R., DEWEY, D.R.: Can. J. Genet. Cytol. 26, 569-577 (1984). - LEVY, A.: Euphytica 34, 811-815 (1985). - LÖRZ, H., BROWN, P.T.H.: 513-534 in: see ref. Horn et al. 1986.

MA, T.-H., HARRIS, M.M., VAN ANDERSON, A., AHMED, I., MOHAMMAD, K., BARC, J.L., LIN, G.: Mutat. Res. 138, 157-167 (1984). - MAIER, P., SCHAWALDER, H.P., WEIBEL, B., ZBINDEN, G.: Mutat. Res. 143, 143-148 (1985). - MATHEWS, V.H., RAO, P.S., BHATIA, C.R.: Z. Pflanzenzücht. 96, 169-173 (1986). - MATHIAS, R.: Z. Pflanzenzücht. 94, 170-173 (1985). - MATSUSHIMA, T., ARAKI, A., YAGAME, O., MURAMASSU, M., KOYAMA, K., OHSAWA, K., NATORI, S., TOMIMORI, H.: Mutat. Res. 150, 141-146 (1985). - MEINS, F.: Annu. Rev. Plant Physiol. 34, 327-346 (1983). - MELZ, G., SCHLEGEL, R.: Euphytica 34, 361-366 (1985). - MERCYKUTTY, V.C., KUMAR, H.: Cytologia 50, 859-863 (1985). - MICKE, A., MALUSZINSKI, M., DONINI, B.: Mutat. Breed. Rev. 3, 92 pp. (1985). - MILO, J., LEVY, A., LADIZINSKY, G., PALEVITCH, D.: Theor. Appl. Genet. 72, 524-529 (1986). - MORIKAWA, T.: Theor. Appl. Genet. 70, 271-278 (1985). - MURTY, U.R., JAHNAVI, M.R.: Cytologia 51, 241-250 (1986).

NAKAI, H., KOBAYASHI, M., SAITO, M.: Euphytica 34, 577-585 (1985). - NAKAYASU, M., SAKAMOTO, H., TERADA, M., NAGAO, M., SUGIMURA, T.: Mutat. Res. 174, 79-83 (1986). - NANDI, S.: Cytologia 50, 921-926 (1985). - NAZEER, M.A., KHOSHOO, T.N.: Nucleus 28, 35-41 (1985). - NEUDECKER, T., HENSCHLER, D : Mutat. Res. 156, 33-37 (1985). - NIRMALA, A., RAO, P.N.: Can. J. Genet. Cytol. 26, 334-338 (1984).

O'CONNELL, M.A., HANSON, M.R.: Theor. Appl. Genet. 72, 59-65 (1986). - OHGAWARA, T., KOBAYASHI, S., OHGAWARA, E., UCHIMIYA, H., ISHII, S.: Theor. Appl. Genet. 71, 1-4 (1985). - OHRI, D., KHOSHOO, T.N.: Nucleus 28, 216-221 (1985a); - Cytologia 50, 213-231 (1985b). - OHRI, D., ZADOO, S.N.: Z. Pflanzenzücht. 97, 227-231 (1986). - OKADA, K.A., CLAUSEN, A.M.: Euphytica 34, 219-231 (1985). - ONG, T.-M., WHONG, W.-Z., STEWART, J., BROCKMAN, H.E.: Mutat. Res. 173, 111-115 (1986). - ORTON, T.J.: 427-468 in: Gene Manipulation in Plant Improvement, ed. J.P. GUSTAFSSON, Plenum Press, New York, London 1984.

PALMER, R.G.: J. Hered. 75, 445-447 (1984). - PANDITA, T.K., KHOSHOO, T.N.: Nucleus 27, 168-171 (1984). - PANUGANTI, N., RAO, P.N., NIRMALA, A.: Can. J. Genet. Cytol. 28, 203-206 (1986). - PATNAIK, S., SARAN, B.L., PATNAIK, S.N.: Cytologia 49, 807-814 (1984). - PÉREZ-CHIESA, Y., CINTRÓN, C.N.: Mutat. Res. 173, 193-196 (1986). - PICKERING, R.A., MORGAN, P.W.: Theor. Appl. Genet. 70, 199-206 (1985). - PIRRIE, A., POWER, J.B.: Theor. Appl. Genet. 72, 48-52 (1986). - PRASAD, G., PRASAD, J.: Indian J. Genet. Plant Breed. 43, 276-278 (1983). - PRASAD, G., TRIPATHI, D.K.: Cytologia 50, 837-840 (1985). - ibid. 51, 11-19 (1986). - PREMACHANDRAN, M.N., CHOUBEY, R.N., GUPTA, S.K.: Z. Pflanzenzücht. 97, 268-271 (1986). - PRINA, A.R., HAGBERG, A.: Hereditas 101, 261-263 (1984). - PYKE, K.A., HEDLEY, C.L.: Euphytica 32, 193-203 (1983).

REDDI, V.R., MANI, N.S.: Cytologia 50, 247-251 (1985). - REDDY, G.A., SUBRAMANYAM, S.: Egypt. J. Genet. Cytol. 13, 189-206 (1984). - REDDY, M.K., SUBRAHMANYAM, N.C.: Genetica 66, 53-61 (1985). - RENNER, H.W.: Mutat. Res. 144, 251-256 (1985). - RIEGER, R., MICHAELIS, A., SCHUBERT, I.: Mutat. Res. 143, 79-82 (1985). - RINES, H.W.: Environ. Exp. Bot.

25, 7-16 (1985). - ROSS, M.D., JONES, W.T.: Theor. Appl. Genet. 71, 284-288 (1985).

SADANANDAM, A., SUBHASH, K.: Cytologia 50, 125-128 (1985). - SAKAI, Y., NAGASE, H., OSE, Y., SATO, T., YAMADA, A., HIBI, M., YAMADA, F.: Mutat. Res. 174, 1-4 (1986). - SANGOWAWA, B.G.: Nucleus 28, 152-158 (1985). - SAPRE, A.B., BARVE, S.S.: Cytologia 49, 345-349 (1984); - Genetica 66, 73-80 (1985a); - J. Hered. 76, 387-389 (1985b). - SCHIEDER, O., HEIN, T., KOHN, H.: 641-651 in: see ref. Horn et al. 1986. - SCHLEGEL, R., MELZ, G., METTIN, D.: Theor. Appl. Genet. 72, 721-734 (1986). - SCOWCROFT, W.R.: 35-48 in: Genetics, New Frontiers IV, eds. V.L. CHOPRA, B.C. JOSHI, R.P. SHARMA, H.S. BANSAL. Oxford and IHB Publ. Comp. New Delhi 1984. - SHARMA, M.L.: Cytologia 50, 483-486 (1985). - SHARMA, P.K., LANGER, A., KOUL, A.K.: Genetica 67, 131-135 (1985). - SHEHAB, A.S.: Cytologia 50, 427-431 (1985). - SHEHAB, A.S., ABO-EL-KHIER, Z.A.: Cytologia 50, 789-795 (1985). - SHIMOI, K., NAKAMURA, Y., TOMITA, I., KADA, T.: Mutat. Res. 149, 17-23 (1985). - SHIMOI, K., NAKAMURA, Y., TOMITA, I., HARA, Y., KADA, T.: Mutat. Res. 173, 239-244 (1986). - SINGH, A.K.: Theor. Appl. Genet. 72, 433-439 (1986). - SINGH, C., SHAMSHERRY, A.P.: Cytologia 50, 603-610 (1985). - SINGH, F.: Genetica 67, 57-61 (1985). - SINGH, R.N., ROY, S.K.: Cytologia 50, 59-68 (1985). - SNOAD, B., FRUSCIANTE, L., MONTI, L.M.: Theor. Appl. Genet. 70, 322-329 (1985). - SOFRADŽIJA, A., HADŽISELIMOVIĆ, R.: Genetika (Beograd) 17, 13-16 (1985). - SOMASHEKAR, R.K., GURUDEV, M.R., RAMIAH, S.: Cytologia 50, 129-134 (1985). - SUBHASHINI, U., VENKATESWARLU, T.: Theor. Appl. Genet. 70, 225-226 (1985). - SUKHADEV, P., JANAKE, P.S., MURTHY, J.V.V.S.N., RAO, M.V.S., MANGA, V.: Cytologia 50, 347-350 (1985).

TALUKDAR, A.R., SHARMA, A., TALUKDER, G.: Nucleus 28, 243-259 (1985). - TERWEL, L., VAN DER HOEVEN, J.C.M.: Mutat. Res. 152, 1-4 (1985). - THOMPSON, M.M., SMITH, D.C., BURGESS, J.E.: Theor. Appl. Genet. 70, 687-692 (1985).

VAN HARTEN, A.M., VAN DER SPEK, D.W., BAL, E.C.J.: Euphytica 34, 769-778 (1985). - VERONESI, F., MARIANI, A., BINGHAM, E.T.: Theor. Appl. Genet. 72, 37-41 (1986). - VILLALOBOS-PIETRINI, R., HERNÁNDES, R., GUADARRAMA, A., GÓMEZ-ARROYO, S.: Cytologia 51, 211-218 (1986).

WAKABAYASHI, K., NAGAO, M., OCHIAI, M., TAHIRA, T., YAMAIZUMI, Z., SUGIMURA, T.: Mutat. Res. 143, 17-21 (1985). - WANG, R.R.-C.: Can. J. Genet. Cytol. 26, 532-537 (1984). - WILLIAMS, W., HARRISON, J.E.M., JAYASEKERA, S.: Euphytica 33, 811-817 (1984).

YADAV, S.K., RATHOE, J.S.: Egypt. J. Genet. Cytol. 13, 181-188 (1984). - YAMAGUCHI, I.: Acta Radiobot. Genetica 6, 1-66 (1985). - YANO, M., OKUNO, K., KAWAKAMI, J., SATOH, H., OMURA, T.: Theor. Appl. Genet. 69, 253-257 (1985). - YU, C.-L., SWAMINATHAN, B., BUTLER, L.G., PRATT, D.E.: Mutat. Res. 170, 103-113 (1986).

ZINOVEVA-STAHEVITCH, A.E., GRANT, W.F.: Caryologia 38, 139-145 (1985). - ZUBERI, M.I., ZUBERI, S.: Indian J. Genet. Plant Breed. 43, 438-440 (1983).

Prof. Dr. WERNER GOTTSCHALK
Institut für Genetik
der Universität Bonn
Kirschallee 1
D-5300 Bonn-1, FRG

IV. Function of Genetic Material. Ribulose-Bisphosphate-Carboxylase as a Tool for the Study of Gene Function in Plants

By Rolf Blaich

1. Introduction

Ribulose bisphosphate carboxylase (RBC) (3-phospho-D-glycerate carboxylyase, EC 4.1.1.39) is a key enzyme in the photosynthetic process. Literature on its function and on its catalytic properties has been reviewed recently in this series by KELLY and LATZKO (1984) and HOLTUM et al. (1986) and will not be discussed in the following survey.

Two different forms are known from prokaryotes. The first is composed of identical subunits, form II contains large and small ones and seems homologous to the RBC of eukaryotic plants. Although RBC of eukaryotic plants is present exclusively in the chloroplast, it is composed of subunits (SU) encoded both in the genomes of the chloroplast (large subunit, LSU) and the nucleus (small subunit, SSU). A precursor of the SSU is imported into the chloroplast (after being bound to its surface: CLINE et al. 1985) where it is processed in two steps (ROBINSON and ELLIS 1984; MISHKIND et al. 1985). The subunits then combine to form an oligomeric structure.

These properties make RBC particularly suited for studies on the interactions of different cellular compartments. The analyses of the function of the respective genes have been facilitated by the fact that the expression of the SSU is regulated by a light-sensitive promoter (for literature see CORUZZI et al. 1984; MORELLI et al. 1985). Among others this structure has been used to demonstrate the activity of chimeric genes in alien plants (VAN DEN BROEK et al. 1985; SCHREIER et al. 1985; for literature see KAULEN et al. 1986).

This review is restricted to works which contribute to an understanding of gene function, whereas other interesting aspects can only be mentioned briefly: The detailed knowledge of both the RBC genes and their products in a large number of plants allows their utilization as markers to test the differential activity of genomes after fusion of protoplasts (for literature see MELCHERS et al. 1978; SCHILLER et al. 1982; GLEDDIE et al. 1986; KUMASHIRO and KUBO 1986). It also yields a base for studies on the evolution of genes, selection mechanisms and phylogeny (for more literature, see PICHERSKY et al. 1986; MARTIN and DOWD 1984).

Progress in Botany, Vol. 49

2. Maps and Sequencing of RBC Genes

a) Prokaryotes

In prokaryotes there exist at least two forms of RBC. For *Rhodopseudomonas capsulata* SHIVELEY et al. (1984) described an intermediate (MW 300,000, SUs of only one type, MW 64,000) and a large form (MW 542,000, two types of SUs with 58,000 and 10,000, respectively). The two types are depressed under different conditions.

The LSU gene from the cyanobacterium *Anabaena azolla* was isolated and sequenced by CURTIS and HASELKORN (1983). Sequence analysis of *Anacystis nidulans*-LSU was published by SHINOZAKI et al. (1983). HERRERO and WOLK (1986) localized the RBC genes of the cyanobacterium *Anabaena variabilis* in close proximity on one of 40 linkage groups based on cosmid hybridization. This corroborates earlier results of NIERZWICKI-BAUER et al. (1984) who in an *Anabaena* strain found cotranscription of both genes, which are separated by 545 bp (base pair) spacer without an open reading frame. Hybridization experiments with probes containing the genes for the LSU and the SSU from *Anacystis nidulans* with a gene bank of the cyanobacterium *Chlorogloeopsis fritschii* indicated a near linkage also in *C. fritschii* (VAKERIA et al. 1986).

FORNARI and KAPLAN identified the LSU gene in *Rhodopseudomonas sphaeroides* in 1983. MULLER et al. (1985) cloned the form II RBC gene of this organism and achieved in vitro transcription-translation. LSU and SSU genes from *Spirulina platensis* were cloned by TIBONI et al. (1984) and expressed in *E. coli*. Analogous work for the RBC of the purple bacterium *Rhodospirillum rubrum* has been published by SOMERVILLE and SOMERVILLE (1984). NARGANG et al. (1984) sequenced this gene.

b) SSU Genes in Eukaryotic Plants

The SSU in higher plants is encoded by different numbers of gene families, each with one or more genes.

Chlamydomonas reinhardtii contains two SSU genes which have been sequenced by GOLDSCHMIDT-CLERMONT and RAHIRE (1986). DEAN et al. (1985) obtained clones from five nonoverlapping regions of *Petunia* DNA that carry eight SSU genes which, based on homology tests, can be divided into three gene families. Two of them contain single genes and the third six genes, four of which are closely linked in the genome. A family of five genes is known to encode the SSU in pea (FLUHR et al. 1986), the nucleotide sequence of which shows two introns. The coding regions are conserved but the 3'- untranslated portions show considerable differences. *Glycine max* (soybean) also contains a multigene family encoding the SSU. In 1982 BERRY-LOWE et al. sequenced one of at least ten members. The gene had two introns; the first exon encodes the entire transit peptide (55 AAs) and two AAs (amino acid) of the mature sequence. The whole precursor consists of 178 AAs. Two members were now analyzed in more detail (GRANDBASTIEN et al. 1986). Regulatory sequences are conserved between the two genes, two introns are present, the three exons which encode the 178 amino acid precursor peptide differ in only one amino acid.

In leaves of *Nicotiana silvestris* four transcripts with identical nucleotide sequences of the noncoding regions and three differences in the region coding for the mature peptide are present (PINCK et al. 1984, 1986). The analysis of restriction fragments of tomato (VALLEJOS et al. 1986) revealed three loci with SSU sequences: *Rbcs-1* on chromosome 2,

Rbcs-2 and *Rbcs-3* on chromosome 3 and 2, respectively. The latter seems to contain more than one gene copy. GREENLAND et al. (1987) report only one SSU gene for cucumber.

c) LSU Genes in Eukaryotic Plants

A detailed review on the plastid genome including figures of physical maps have been published in this series by HAGEMANN et al. (1985) and in this volume, therefore only some supplementary or more recent papers will be mentioned here.

The typical chloroplast genome containing the LSU gene is about 150 kbp (kilobase pairs) in size and has an inverted repeat of about 20 kbp. In *Euglena gracilis* it is a split gene containing nine introns (KOLLER et al. 1984); in vascular plants no introns have been reported so far. Earlier results on the localization of this gene were corroborated by FREYSSINET et al. (1985) using hybridizations with mutants. The homology of the enzyme protein to that of prokaryotes is about 70%.

In *Nicotiana acuminata* the inverted repeats are larger as compared to other *Nicotiana* spp. Except in the extra segment the restriction maps are, however, identical (SHEN et al. 1982). According to KISHIMA et al. (1986) the LSU gene of *Beta vulgaris* is located in the larger of two single copy regions (84.3 kbp and 23,2 kbp), the total length of the ptDNA being 148.5 kbp with two inverted repeat sequences of 20.5 kbp. A restriction map for *Beta vulgaris* (sugar beet) is also presented by BREARS et al. (1986). In both sugar beet and *Zea mays* (maize) the inverted repeats recombine.

Mapping, nucleotide sequencing and expression in *E. coli* minicells are reported for the LSU gene of *Petunia hybrida* by BOVENBERG et al. (1984). He found similarity of the untranslated 3'-end of the gene with that of *Nicotiana tabacum*. Stem and loop structures possibly involved in termination and transcription can be formed. In 1984 ZHU et al. had already demonstrated that both *Chlamydomonas* and *Nicotiana otophora* LSU genes could be expressed in *E. coli* and *Bacillus subtilis*.

Differences between the genomes of chloroplasts and chromoplasts of *Capsicum annuum* (pepper) could not be found by GOUNARIS et al. (1986), the location of the genes being identical to those in other Solanaceae and higher plants in general. Already THOMPSON (1981) had reported the identity of plastid DNA in *Narcissus pseudonarcissus* (daffodil). According to REITH and CATTOLICO (1986) *Olisthodiscus luteus* has a plastid genome of 150 kbp with an inverted repeat of 22 kbp. The isolation of circular DNA from chloroplasts of the liverwort *Marchantia polymorpha* has been reported by OHYAMA et al. (1982).

3. Evolutionary Studies

The study of RDC in different groups of organisms allows the study of evolutionary aspects of both sequence and function of these genes. At least the LSU is a highly conserved protein. From comparison of the LSU of *Anacystis* with three other plants, a rate of amino acid substitution of 0.4×10^{-9} per year per site is calculated, a value which is but a third of that of Hb *a* (SHINOZAKI et al. 1983).

REICHELT and DELANEY (1983) compared the LSU gene from the cyanobacterium *Synechococcus* to those of spinach and maize. About 70% homology was found in the coding region and even more in the ribosome-binding site (two 11-bp sequences at the 5'-flanking region). GATENBY et al.

(1981), GATENBY (1984) and GATENBY et al. (1981, 1982) fused *rbcL* from *Zea mays* with the lambda *N* operon. Thus, its transcription was being controlled by the phage promoter PL. At 32°C transcription was repressed by the presence of a temperature sensitive *E. coli* repressor. Unmoderated transcription from PL occurred at 45°C, yielding a polypeptide the same size as that synthesized in maize in amounts up to 2% of the total *E. coli* cell proteins.

According to BERRY-LOWE (1982) who compared the 123 AAs SSU1 polypeptide of soybean to those of spinach and pea, there are 34 and 30 AA replacements in the SSU1 polypeptide. Comparison of soybean and pea SSU sequences suggested that the two genes diverged long before the divergence of pea and soybean. Detailed comparative studies on the *Hordeum* and *Zea* LSU genes are presented by ZURAWSKI et al. (1984). McKNIGHT et al. (1986) have sequenced two cDNA clones encoding the SSU of tomato. The coding regions were 85% conserved, the untranslated 3'-region was only 55% conserved. Comparison of nucleotide and amino acid sequences of the *Anabaena azolla* RBC with those of *Zea mays*, *Spinacia*, *Chlamydomonas* and *Synechococcus* genes revealed homology of about 75% and 80% at the nucleotide and amino acid level, respectively. Conservation of sequence is lost immediately outside the coding region (CURTIS and HASELKORN 1983).

An LSU probe for *Chlamydomonas reinhardtii* hybridized with DNA of *Chlorogloeopsis fritschii* but no homology could be found using the LSU probes from *Zea mays* or the SSU probe from pea (VAKERIA et al. 1986). The two SSU genes *Chlamydomonas reinhardtii* which have been sequenced by GOLDSCHMIDT-CLERMONT and RAHIRE (1986) yield subunits differing by four amino acid residues. Although the sequence is homologous to those of higher plants and of cyanobacteria, there are differences in number and location of intervening sequences and in the length at the amino terminus.

Although the comparative studies on SSU amino acid sequences of numerous flowering plants made by the group of MARTIN are not immediately related to gene function, they seem to be a valuable data base for geneticists and should be mentioned in this context (for literature, see MARTIN and DOWD 1984).

4. Translation and Processing of Transcripts

a) Small Subunit

KARLIN-NEUMANN and TOBIN (1986) demonstrated the existence of three major homology blocks which are shared by the transit peptides of the SSU and the light-harvesting chlorophyll a/b protein II (LHPC II) of the thylakoid membrane in several plant species. The homology blocks lie at the beginning, middle and end of the transit sequence and are separated by sequences unshared in both proteins. Functions common for all transit peptides are proposed for the shared sequences, while the interblocks might be dispensable or might confer protein-specific properties.

A gene encoding the SSU in wheat was transferred to tobacco where it is not expressed under the control of its own promoter (KEITH and CHUA 1986). Several wheat transcripts (both spliced and unspliced) accumulate if a viral promoter is used. Inefficient splicing was also observed with an intron from the maize *Adh-1* gene. In contrast to earlier results (e.g. MISHKIND et al. 1985), these findings seem to indicate that even monocot and dicot plants require different RNA sequences for processing.

HURT et al. (1986) showed that both the LSU and the SSU of *Chlamydomonas* could be imported into isolated yeast mitochondria if the presequence of a protein transported into the yeast mitochondrial matrix had been attached. The SUs, however, did not assemble with each other inside the yeast mitochondria.

STEINMÜLLER et al. (1983) synthesized LSU and SSU of the two red algae *Cyanidium caldarum* and *Porphyridium aerugineum*. They found that, in contrast to green algae and higher plants, the SSUs are not made in the form of larger precursor molecules but are synthesized with MWs in the range of the mature form.

b) Large Subunit

A long and a short transcript (200 bases) of the LSU gene are formed in barley, the short one being predominant in dark-grown seedlings. The long one contains completely conserved prokaryotic-type transcription promoters. Comparisons to genes of maize, spinach and tobacco revealed that the two monocot species differ by an insertion of about 140 bp in the mRNA leader sequence (POULSEN 1984).

Characterization of the LSU gene of *Pisum sativum* and its transcription has been published by OISHI and TEWARI (1983). According to BLOOM et al. (1984) the LSU gene of *Zea mays* has two translation start sites. They are 18 bp apart, each being preceded by a suitable ribosome-binding region.

A cloned plastid DNA template of *Zea mays* containing 172 bp of the LSU protein-coding region and 791 bp of upstream sequences was transcribed in vitro (HANLEY-BOWDOIN et al. 1985). Three RNA species were obtained. The shortest (-63), where the 5'-terminus is positioned 63 nucleotides upstream of the protein-coding region, is a processed form of a primary (-300) RNA containing the promoter. ERION (1985) isolated two transcripts from maize leaves. The smaller one was supposed to be a processed form of the larger primary transcript.

5. Assembly and Structure of the Enzyme

Studies on the assembly of the enzyme monomers have been carried out by ROY et al. (1982), who showed that the LSU is able to form larger complexes which later dissociate to combine with SSUs forming the RBC molecule. BARRACLOUGH and ELLIS (1980) found that newly synthesized LSUs are not incorporated into pre-existing RBC holoenzymes nor are they precipitated by antibodies to holoenzymes or to LSUs prepared from this. In addition, there is a time lag between the onset of LSU synthesis and the onset of assembly. A protein of subunit MW 60,000 is found to be associated with newly synthesized LSUs which he thought to be involved in the assembly of RBC.

VAN DER VRIES et al. (1986) produced a functional hybrid enzyme using LSUs of the cyanobacterium *Synechococcus* PCC 6301 and SSUs of wheat in an *E. coli* system. The activity of the hybrid RBC was reduced to about 10% as compared to the homologous bacterial enzyme. This system allowed interesting studies on the assembly mechanisms: Bacterial LSUs as well as SSUs expressed in *E. coli* assembled into inactive oligomeric structures if the corresponding SSUs/LSUs were absent. If these were introduced by infection with an M13 recombinant phage carrying the gene, enzyme activity was restored. The specific activity of the enzyme could be modulated by changing the intracellular pool size of the SSU.

There are contradictory reports on the tertiary structure of the active enzyme. It seems to be different for various species or due to differing analytical procedures. A hexadecameric model is now most widely accepted.

EM investigations of BARCENA and SHAW (1985), corroborating the analysis by small angle neutron scattering of DONNELLY et al. (1984), suggest a model of eight SSUs arranged equatorially around a core of eight LSUs in *Spinacia oleracea*. According to HURT et al. (1986) the RBC of *Chlamydomonas* contains eight large and eight small subunits and ANDREWS and LORIMER (1985) combined successfully the LSU octamer of the cyanobacterium *Synechococcus* and the SSU of spinach to produce a catalytically active enzyme.

GRAY et al. (1978), however, found two SSUs and three LSUs in tobacco and DADAY et al. (1986) reported that only one LSU polypeptide seems to be characteristic for higher plants: The RBC of *Medicago sativa* and *M. falcata* consists of one LSU and two or three SSUs, depending on the genotype. According to peptide maps, amino acid composition and isoelectric points, the LSUs of different genotypes seem to be identical, whereas the SSUs show heterogeneity. *Spinacia oleracea* and *Nicotiana spp.* also have variable SSUs and a single LSU polypeptide but with differing isoelectric points.

The active forms of RBC of the bacterium *Rhodopseudomonas sphaeroides* have been described to be tetrameric or hexameric in an in vitro system (CHORY et al. 1985). The genes for RBC (form I) of the same organism were isolated by GIBSON and TABITA (1986) and expressed in *E. coli*. In this case they seem to be hexadecameric. According to DONNELLY et al. (1984), the RBC of *Rhodospirillum rubrum* is dimeric.

6. Regulation of Gene Activity

Studies on the control of RBC activity are numerous and new ones are produced continuously. Preliminary studies usually are more or less descriptive analyses of enzyme activity under different conditions. To reduce the number of cited papers, the following text includes only studies which give some indications of the underlying genetic mechanisms. Table 1 gives an overview. Mainly recent studies will be discussed in more detail.

a) Light Induction of RBC

Due to the fact that RBC is light-inducible, a large number of papers deals with the control of enzyme activity by different light regimes. Both transcriptional and post-transcriptional mechanisms are involved in control, the first ones being responsible for a long-term adjustment of gene expression to environmental changes, while translation can be controlled rapidly. The phytochrome system is involved in the regulation of the SSU, a red-light induction being reversible by a far-red light given immediately after the red-light pulse (for older literature, see LINKS 1982; STIEKEMA et al. 1983; SILVERTHORNE and TOBIN 1984).

The competence of the gene towards phytochrome regulation depends on endogenous factors (for literature, see SCHMIDT et al. 1987). A prerequisite for the expression of the SSU nuclear genes is the integrity of the plastids (OELMÜLLER and MOHR 1986; SIMPSON et al. 1986), a continuous signal from the plastid being necessary which is released only

Table 1. Control of ribose-bisphosphate-carboxylase activity

Species	Subject of investigations	Author(s)
Algae		
Chlorella fusca	Enzyme turnover, mutant	BULLMANN and GALLING (1986)
Chlorogonium elongatum	Hetero-autotrophic conditions	WESTHOFF et al. (1981)
Spirodela oligorrhiza	Light-regulation	FROMM et al. (1985)
Vascular plants		
Amaranthus hypochondriacus	Light: transcription, post-transcription	BERRY et al. (1985)
	Light: translation	BERRY et al. (1986)
Cucumis sativus	Hormone(benzyladenine), light	OHYA et al. (1986), GREENLAND et al. (1987)
Cucurbita pepo	Light and cytokinin influence	LERBS et al. (1984)
Glycine max	Light: transcription of LSU	BERRY-LOWE and MEAGHER (1985)
Gossypium hirsutum	Expression during embryogenesis	BORROTO and DURE (1986)
Hordeum vulgare	Light development	VIRO and KLIPPSTECH (1983)
	Light induction	POULSEN (1983)
	Plastid factor: transcription	BATSCHAUER et al. (1986)
Lemna gibba	Phytochrome: transcription regulation	SILVERTHORNE and TOBIN (1984), STIEKEMA et al. (1983)
	Hormone(benzyladenine)	FLORES and TOBIN (1986)
Lolium perenne	Di/tetraploidy effects	McNEIL et al. (1981)
	Di/tetraploidy effects	REJDA et al. (1981)
Medicago sativa	Polyploidy effects	MEYERS et al. (1982), MOLIN et al. (1982)
Nicotiana sylvestris	Temperature influence	LETT et al. (1980)
	Expression during differentiation	VERNET et al. (1982)
Petunia sp.	Light, transgenic plants	FLUHR and CHUA (1986), FLUHR et al. (1986a,b)
Pisum sativum	Tissue specificity, light	CORUZZI et al. (1984)
	Light regulation	BENNETT et al. (1984)
	Sequence for light sensitivity	MORELLI et al. (1985)
	Effects of amanitin on mRNA	SASAKI (1986)
	Light, phytochrome development	FLUHR and CHUA (1986)

Table 1 (continued)

Species	Subject of investigations	Author(s)
Pisum sativum (continued)	Phytochrome development	KAUFMAN et al. (1986)
	Expression in transgenic tobacco, light, tissue specificity	SIMPSON et al. (1986)
	Light: translation of LSU	INAMINE et al. (1985)
Raphanus sativus	Phytochrome, red light	FOURCROY et al. (1985)
	Phytochrome, red light	FOURCROY (1986)
Sinapis alba	Phytochrome: control of mRNA	LINK (1982)
	Photo-oxidative damage	OELMÜLLER and MOHR (1986)
	Influence of inhibitors	OELMÜLLER et al. (1986a)
	Phytochrome regulation	OELMÜLLER et al. (1986b)
	Competence of genes to phytochrome regulation	SCHMIDT et al. (1987)
Solanum tuberosum	Organ specificity	ECKES et al. (1985)
Spinacia oleracea	Localization of translation	MINAMI and WATANABE (1984)
Triticum aestivum	Control amount of enzyme ageing, decline of synthesis	JELLINGS et al. (1983), BRADY (1981), SPEARS and BRADY (1981)
Zea mays	Light induction	BEDBROOK et al. (1978)
	Two different LSU-mRNAs	CROSSLAND et al. (1984)
	Supercoiling	STIRDIVANT et al. (1985)
	Cell-type specificity	LINK et al. (1987), MARTINEAU and TAYLOR (1986)
	Mapping of photoactive sites on plastid genome	RODERMEL and BOGORAD (1985)

during a limited time span (OELMÜLLER et al. 1986a,b). MORELLI et al. (1985) introduced derivatives of a *Petunia* SSU gene into pea plants and found a conserved 33-bp sequence close to the "TATA" box of the SSU gene to be involved in the light sensitivity of the gene. STIRDEVANT et al. (1985) proposed changes in DNA conformation (supercoiling) as a possible mechanism for differential regulation of RBC and another gene.

FLUHR and CHUA (1986) found that changes in some SSU transcripts are regulated by phytochrome in etiolated primary leaves of *Pisum satium*, whereas in mature leaves an additional blue-light photoreceptor seems to be involved. These responses are the same in transgenic *Petunia*. Depending on the gene, o.4 kb and 2 kb of upstream sequences are sufficient for the photoresponses. An additional receptor is also demonstrated by SIMPSON et al. (1986).

A detailed analysis of the phytochrome control of 12 different mRNAs in developing pea plants was published by KAUFMAN et al. (1986). There are different time courses for accumulation of the transcripts after the red-light pulse. The transcript for the SSU accumulates at a linear rate during 24 h in darkness, the reversibility by far-red light is lost within a short time, whereas responses for some other transcripts are retained at least 7 h. SIMPSON et al. (1986) reported analoguous findings with chimeric genes (5'-flanking sequences of pea in transgenic tobacco).

BERRY et al. (1985, 1986) found in *Amaranthus* that light-induced accumulation of LSU and SSU polypeptides was not coordinated. Their results indicate that the SSU transcription is positively regulated by white light, while far-red has a negative influence which is a classic phytochrome-linked response (BERRY-LOWE and MEAGHER 1985). A similar, rapid control by red/far-red light of the SSU in radish is reported by FOURCROY (1986). INAMINE et al. (1985) found that light increases the copy number of the LSU gene but the primary regulation of LSU synthesis is at the level of translation. These results are also supported by inhibitor studies with amanitin (SASAKI 1986).

b) Influence of Hormones and Growth Conditions

The action of growth hormones has been studied by FLORES and TOBIN (1986). Cytokinin seems to regulate expression of mRNA at a post-transcriptional level, possibly by influencing the stability of the mRNA. The studies of OHYA et al. (1986) indicated that BA in the dark can replace the light to a certain extent.

Another possibility to induce RBC activity is the transfer of heterotrophically grown cells to autotrophic conditions, which has been demonstrated by WESTHOFF et al. (1981) for *Chlorogonium elongatum*. In this case a tight cosynthesis of LSUs and SSUs was found which was not disturbed by different inhibitors of protein and RNA synthesis.

c) Organ Specificity of Gene Expression

FLUHR et al. (1986a,b) used the 3'-sequence difference (see Sect. 2b) of different pea SSU genes to compare their expression levels in different organs. Some transcripts are underrepresented in seeds and petals. In leaves ten-fold differences were observed. Studies with light, blue-light and phytochrome induction during development of etiolated and green pea leaves showed that the genes are differentially activated.

A successful application of hybridization to mRNA in photosynthetic plant tissue sections was reported by MARTINEAU and TAYLOR (1986). Comparing bundle sheath and mesophyll, they could corroborate earlier reports based on the analysis of extracts (LINK et al. 1978) that LSU mRNA accumulates only in bundle sheath cells, whereas phosphoenolpyruvate carboxylase is restricted to mesophyll cells. ECKES et al. (1985) reported that organ-specific expression of SSU in *Solanum tuberosum* (100-500-fold higher in leaves than in roots or tubers) is controlled primarily at the level of transcription and depends also on RNA polymerase II. BERRY et al. (1985) compared the RBC activities of the cotyledons of 1- to 8-day-old etiolated and light-grown *Amaranthus* seedlings. In darkness the rate of synthesis of both SUs was much lower than in light, it occurred as a burst between 2-5 days; however, mRNAs were present which were functional in vitro despite their inactivity in vivo after day 5.

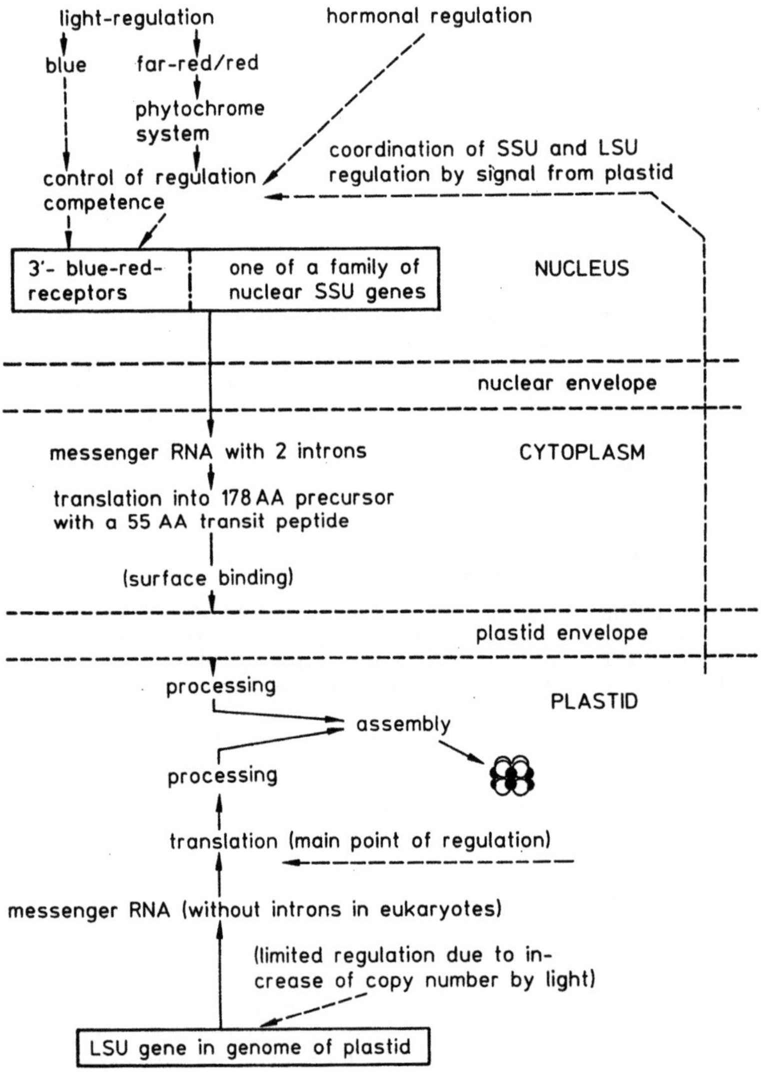

Fig. 1. Simplified scheme of the RBC system in eukaryotes

d) Influence of the Genotype on RBC Activity

The specific activity of RBC in alloplasmic lines of wheat fell into high or low categories if measured in vitro. EVANS and AUSTIN (1986) showed that the activity depends on the type of cytoplasm and is not influenced by SSUs which were introduced by substituting different (hexaploid) nuclear genomes into different cytoplasms.

RHODES et al. (1980) tested RBC from *Nicotiana sylvestris* and from cultivars and nuclear substitution lines of *N. tabacum*. They found two types of enzymes: Either with identical SSUs and variable LSUs or identical LSUs and variable SSUs. The ratios of carboxygenase to oxygenase activity were influenced by the composition of the enzymes.

Several groups tried without success to verify earlier reports on qualitative influences of tetra- or polyploidy on RBC (REIJDA at al. 1981; McNEIL et al. 1981; MEYERS et al. 1982; MOLIN et al. 1982).

7. Concluding Remarks

To give an overview, some results are summarized in Figure 1. For clarity a number of interesting findings were deliberately omitted.

At the moment about 20 papers dealing with RBC both in prokaryotes and in eukaryotes are being published monthly. Only part of them, however, is related to genetic problems, most of them referring to the catalytic function of the enzyme in photosynthesis or to its molecular properties at a pure biochemical level.

Nevertheless, detailed knowledge of the enzyme enhances the value of genetic results, the most valuable of which have been obtained only recently by the application of modern methods of molecular genetics. More important results are therefore to be expected in the near future. Certainly the RBC system will be one of the first really comprehensive models for a better understanding of the function of genetic material.

References

ANDREWS, T.J., LORIMER, G.H.: J. Biol. Chem. 260, 4632-4636 (1985).

BARCENA, J.A., SHAW, P.J.: Planta 163, 141-144 (1985). - BARRACLOUGH, R., ELLIS, R.J.: Biochim. Biophys. Acta 608, 19-31 (1980). - BATSCHAUER, A., MOESINGER, E., KREUZ, K., DÖRR, I., APEL, K.: Eur. J. Biochem. 154, 625-643 (1986). - BEDBROOK, J.R., LINK, G., COEN, D.M., BOGORAD, L., RICH, A.: Proc. Natl. Acad. Sci. USA 75, 3060-3064 (1978). - BENNETT, J.L., PHILLIPS, J.N.: Z. Naturforsch. 39, 623-626 (1984). - BERRY, J.O., NIKOLAU, B.J., CARR, J.P., KLESSIG, D.F.: Mol. Cell. Biol. 5, 2238-2246 (1985); - ibid. 6, 2347-2353 (1986). - BERRY-LOWE, S.L., MEAGHER, R.B.: Mol. Cell. Biol. 5, 1910-1917 (1985). - BERRY-LOWE, S.L., McKNIGHT, T.D., SHAH, D.M., MEAGHER, R.B.: J. Mol. Appl. Genet. 1, 483-498 (1982). - BLOOM, M., BROT, N., COHEN, B.N., WEISSBACH, H.: Plant Mol. Biol. 3, 403-406 (1984). - BORROTO, K.E., DURE, L.(II): Plant Mol. Biol. 7, 105-114 (1986). - BOVENBERG, W.A., KOES, R.E., KOOL, A.J., NIJKAMP, H.J.J.: Curr. Genet. 8, 231-241 (1984). - BRADY, C.J.: Aust. J. Plant Physiol. 8, 591-602 (1981). - BREARS, T., SCHARDL, C.L., LONSDALE, D.M.: Plant Mol. Biol. 6, 171-178 (1986). - BULLMANN, J., GALLING, G.: 583-586 in: Plant Biology, Vol. 2, eds. G. AKOYUNOGLOU, H. SENGER. New York: Liss 1986.

CHORY, J., MULLER, E.D., KAPLAN, S.: J. Bacteriol. 16, 307-313 (1985). - CLINE, K., WERNER-WASHBURNE, M., LUBBEN, T.H., KEEGSTRA, K.: J. Biol. Chem. 260, 3691-3696 (1985). - CORUZZI, G., BROGLIE, R., EDWARDS, C., NAN-HAI CHUA: EMBO 3, 1671-1680 (1984). - CROSSLAND, L.D., RODERMEL, S.R., BOGORAD, L.: Proc. Natl. Acad. Sci. USA 81, 4060-4064 (1984). - CURTIS, S.E., HASELKORN, R.: Proc. Natl. Acad. Sci. 80, 1835-1839 (1983).

DADAY, H.V., WHITECROSS, M.I., SHAW, D.C.: Theor. Appl. Genet. 71, 708-715 (1986). - DANIELL, H., KRISHNAN, M., BAI, U.U., GNANAM, A.: Biochem. Biophys. Res. Comm. 135, 248-255 (1986). - DEAN, C., VAN DEN ELZEN, P., TAMAKI, S., DUNSMUIR, P., BEDBROOK, J.: Proc. Natl. Acad. Sci. USA 82, 4964-4968 (1985). - DONNELLY, M.I., HARTMANN, F.C., RAMAKRISHNAN, V.: J. Biol. Chem. 259, 406-411 (1984).

ECKES, P., SCHELL, J., WILMITZER, L.: Mol. Gen. Genet. 199, 216-224 (1985). - ERION, J.L.: Plant Mol. Biol. 4, 169-180 (1985). - EVANS, J.R., AUSITN, R.B.: Planta 167, 344-350 (1986).

FLORES, S., TOBIN, E.M.: Planta 168, 340-349 (1986). - FLUHR, R., CHUA, N.H.: Proc. Natl. Acad. Sci. USA 83, 2358-2362 (1986). - FLUHR, R., KUHLEMEIER, C., NAGY, F., CHUA, N.H.: Science 232, 1106-1112 (1986a). - FLUHR, R., MOSES, P., MORELLI, G., CORUZZI, G., CHUA, N.H.: EMBO 5, 2063-2072 (1986b). - FORNARI, C.S., KAPLAN, S.: Gene 25, 291-300 (1983). - FOURCROY, P.: Plant Sci. 44, 183-190 (1986). - FOURCROY, P., KLEIN-EUDE, D., LAMBERT, C.: Plant Sci. Lett. 37, 235-244 (1985). - FREYSSINET, G., FREYSSINET, M., LEBRUN, M.: Plant Sci. Lett. 37, 245-250 (1985). - FROMM, H., DEVIC, M., FLUHR, R., EDELMAN, M.: EMBO 4, 291-296 (1985).

GATENBY, A.A.: Eur. J. Biochem. 144, 361-366 (1984). - GATENBY, A.A., CASTLETON, J.A., SAUL, M.W.: Nature 291, 117-121 (1981); - Mol. Gen. Genet. 185, 424-429 (1982). - GIBSON, J.L., TABITA, F.R.: Gene 44, 271-278 (1986). - GOLDSCHMIDT-CLERMONT, M., RAHIRE, M.: J. Mol. Biol. 191, 421-432 (1986). - GLEDDIE, S., KELLER, W.A., SETTERFIELD, G.: Theor. Appl. Genet. 71, 613-621 (1986). - GOUNARIS, I., MICHALOWSKI, C.B., BOHNERT, H.J., PRICE, C.A.: Curr. Genet. 11, 7-16 (1986). - GRANDBASTIEN, M.A., BERRY-LOWE, S., SHIRLEY, B.W., MEAGHER, R.B.: Plant Mol. Biol. 7, 451-466 (1986). - GRAY, J.C., KUNG, S.D., WILDMAN, S.G.: Arch. Biochem. Biophys. 185, 272-281 (1978). - GREENLAND, A.J., THOMAS, M.V., WILDEN, R.M.: Planta 170, 99-110 (1987).

HAGEMANN, R., LINDENHAHN, M.M., METZLAFF, M.: Prog. Botany 47, 208-227 (1985). - HANLEY-BOWDOIN, L., OROZCO, E.M., CHUA, N.H.: Mol. Cell Biol. 5, 2733-2745 (1985). - HERRERO, A., WOLK, C.P.: J. Biol. Chem. 261, 7748-7754 (1986). - HOLTUM, J.A., KELLY, G.J., LATZKO, E.: Progr. Botany 48, 97-123 (1986). - HURT, E.C., GOLDSCHMIDT-CLERMONT, M., PESOLT-HURT, B., ROCHAIX, J.D., SCHATZ, D.: J. Biol. Chem. 261, 11440-11443 (1986).

INAMINE, G., NASH, B., WEISSBACH, H., BROT, N.: Proc. Natl. Acad. Sci. USA 82, 5690-5694 (1985).

JELLINGS, A.J., LEESE, B.M., LEECH, R.M.: Mol. Gen. Genet. 192, 272-274 (1983).

KARLIN-NEUMANN, G.A., TOBIN, E.M.: EMBO 5, 9-14 (1986). - KAUFMAN, L.S., ROBERTS, L.L., BRIGGS, W.R., THOMPSON, W.F.: Plant Physiol. 81, 1033-1038 (1986). - KAULEN, H., SCHELL, J., KREUZALER, F.: EMBO 5, 1-8 (1986). - KEITH, B., CHUA, N.H.: EMBO 5, 2419-2426 (1986). - KELLY, G., LATZKO, E.: Progr. Botany 46, 68-93 (1984). - KISHIMA, Y., MIKAMI, T., HARADA, T., SHINOZAKI, K., SUGIURA, M., KINOSHITA, T.: Plant Mol. Biol. 7, 201-206 (1986). - KOLLER, B., GINGRICH, J.C., STIEGLER, G.L.,

FARLEY, M.A., DELIUS, H., HALLICK, R.B.: Cell 36, 545-554 (1984). - KUMASHIRO, T., KUBO, T.: Jpn. J. Breed. 36, 39-48 (1986).

LERBS, S., LERBS, W., KLYACHKO, N.L., ROMANKO, E.G., KULAEVA, O.N., WOLLGIEHN, N.L., PARTHIER, B.: Planta 162, 289-298 (1984). - LETT, M.C., FLECK, J., DURR, A., HIRTH, L.: Planta 148, 211-216 (1980). - LIN, C.M., KUNG, S.D.: Theor. Appl. Genet. 68, 213-218 (1984). - LINK, G.: Planta 154, 81-86 (1982). - LINK, G., COEN, D.M., BOGORAD, L.: Cell 15, 725-732 (1978).

MARTIN, P.G., DOWD, J.M.: Aust. J. Bot. 11, 301-310 (1984). - MARTINEAU, B., TAYLOR, W.C.: Plant Physiol. 82, 613-618 (1986). - McKNIGHT, T.D., ALEXANDER, D.C., BABCOCK, M.S., SIMPSON, R.B.: Gene 48, 23-32 (1986). - McNEIL, P.H., FOYER, C.H., WALKER, D.A., BIRD, I.F., CORNELIUS, M.J., KEYES, A.J.: Plant Physiol. 67, 530-534 (1981). - MELCHERS, G., SACRISTAN, M.D., HOLDER, A.A.: Carlsberg Res. Commun. 43, 203-218 (1978). - MEYERS, S.P., NICHOLS, S.L., BAER, G.R., MOLIN, W.T., SCHRADER, L.E.: Plant Physiol. 70, 1704-1709 (1982). - MISHKIND, M.L., WESSLER, S.L., SCHMIDT, G.W.: J. Cell Biol. 100, 226-234 (1985). - MINAMI, E.I., WATANABE, A.: Arch. Biochem. Biophys. 235, 562-570 (1984). - MOLIN, W.T., MEYERS, S.P., BAER, G.R., SCHRADER, L.E.: Plant Physiol. 70, 1710-1714 (1982). - MORELLI, G., NAGY, F., FRALEY, R.T., ROGERS, S.G., CHUA, N.H.: Nature 315, 200-204 (1985). - MULLER, E.D., CHORY, J., KAPLAN, S.: 319-324 in: Molecular Biology of the Photosynthetic Apparatus, eds. K.E. STEINBECK, S. KAPLAN. NATO Advanced Workshop. New York: Cold Spring Harbor 1985.

NARGANG, F., McINTOSH, L., SOMERVILLE, C.: Mol. Gen. Genet. 193, 220-224 (1984). - NIERZWICKI-BAUER, S.A., CURTIS, S.E., HASELKORN, R.: Proc. Natl. Acad. Sci. USA 81, 5961-5965 (1984).

OELMÜLLER, R., MOHR, H.: Planta 167, 106-113 (1986). - OELMÜLLER, R., LEVITAN, I., BERGFELD, R., RAJASEKHAR, V.K., MOHR, H.: Planta 168, 482-492 (1986a). - OELMÜLLER, R., DIETRICH, G., LINK, G., MOHR, H.: Planta 169, 260-266 (1986b). - OHYA, T., KOJIMA, S., NAITO, K., SUZUKI, H.: J. Plant Physiol. 125, 115-122 (1986). - OHYAMA, K., WETTER, L.R., YAMANO, Y., FUKUZAWA, H., KOMANO, T.: Agric. Biol. Chem. 46, 237-242 (1982). - OISHI, K.K., TEWARI, K.K.: Mol. Cell Biol. 3, 587-595 (1983).

PICHERSKY, E., BERNATZKY, R., TANKSLEY, S.D., CASHMORE, A.R.: Proc. Natl. Acad. Sci. USA 83, 3880-3884 (1986). - PINCK, M., GUILLEY, E., DURR, A., HOFF, M., PINCK. L., FLECK, J.: Biochimie 66, 539-546 (1984). - PINCK, M., DORE, J.M., GUILLEY, E., DURR, A., PINCK. L., HIRTH, L., FLECK, J.: Plant Mol. Biol. 7, 301-310 (1986). - POULSEN, C.: Carlsberg Res. Commun. 48, 57-80 (1983) and 49, 89-104 (1984).

REICHELT, B.Y., DELANEY, S.F.: DNA 2, 121-130 (1983). - REITH, M., CATTOLICO, R.A.: Proc. Natl. Acad. Sci. USA 83, 8599-8603 (1986). - REJDA, J.M., JOHAL, S., CHOLLET, R.: Arch. Biochim. Biophys. 210, 617-624 (1981). - RHODES, P.R., KUNG, S.D., MARSHO, T.V.: Plant Physiol. 65, 69-73 (1980). - ROBINSON, C., ELLIS, R.J.: Eur. J. Biochem. 142, 343-346 (1984). - RODERMEL, S.R., BOGORAD, L.: J. Cell Biol. 100, 463-476 (1985). - ROY, H., BLOOM, M., MILOS, P., MONROE, M.: J. Cell Biol. 94, 20-27 (1982).

SASAKI, Y.: FEBS Lett. 204, 279-282 (1986). - SCHILLER, B., HERRMANN, R.G., MELCHERS, G.: Mol. Gen. Genet. 186, 453-459 (1982). - SCHMIDT, S., DRUMM-HERREL, H., OELMÜLLER, R., MOHR, H.: Planta 170, 400-407 (1987). - SCHREIER, P.H., SEFTOR, E.A., SCHELL, J., BOHNERT, H.J.: EMBO J. 4, 25-32 (1985). - SHEN, G.F., CHEN, K., WU, M., KUNG, S.D.: Genet. 187, 12-18 (1982). - SHINOZAKI, K., YAMADA, C., NAOYUKI, T.,

SUGIURA, M.: Proc. Natl. Acad. Sci. USA 80, 4050-4054 (1983). - SHIVELY, J.M., DAVIDSON, E., MARRS, B.L.: Arch. Microbiol. 138, 233-236 (1984). - SILVERTHORNE, J., TOBIN, E.M.: Proc. Natl. Acad. Sci. USA 81, 1112-1116 (1984). - SIMPSON, J., VAN MONTAGU, M., HERRERA-ESTRELLA, L.: Science 233, 34-38 (1986). - SOMERVILLE, C.R., SOMERVILLE, S.C.: Mol. Gen. Genet. 193, 214-219 (1984). - SPEIRS, J., BRADY, C.J.: Aust. J. Plant Physiol. 8, 603-618 (1981). - STEINMÜLLER, K., KALING, M., ZETSCHE, K.: Planta 159, 308-313 (1983). - STIEKEMA, W.J., WIMPEE, C.F., SILVERTHORNE, J., TOBIN, E.M.: Plant Physiol. 72, 717-724 (1983). - STIRDIVANT, S.M., CROSSLAND, L.D., BOGORAD, L.: Proc. Natl. Acad. Sci. USA 82, 4886-4890 (1985).

THOMPSON, J.A.: Z. Naturforsch. Biosci. 35, 1101-1103 (1980). - TIBONI, O., DI PASQUALE, G., CIFERRI, O.: Biochim. Biophys. Acta 783, 258-264 (1984).

VAKERIA, D., CODD, G.A., HAWTHORNTHWAITE, A.M., STEWART, W.D.P.: Arch. Microbiol. 145, 228-233 (1986). - VALLEJOS, C.E., TANKSLEY, S.D., BERNATZKY, R.: Genetics 112, 93-106 (1986). - VAN DEN BROEK, G., TIMKO, M.P., KAUSCH, A.P., CASHMORE, A.R., VAN MONTAGU, M., HERRERA-ESTRELLA, L.: Nature 313, 358-363 (1985). - VAN DER VRIES, S.M., BRADLEY, D., GATENBY, A.A.: EMBO 5, 2439-2444 (1986). - VERNET, T., FLECK, J., DURR, A., FRITSCH, C., PINCK, M., HIRTH, L.: Eur. J. Biochem. 126, 489-494 (1982). - VIRO, M., KLIPPSTECH, K.: Planta 157, 202-208 (1983).

WESTHOFF, P., ZIMMERMANN, K., BOEGE, F., ZETSCHE, K.: Z. Naturforsch. Biosci. 36, 11-12 (1981).

ZHU, Y.S., LOVETT, P.S., WILLIAMS, D.M., KUNG, S.D.: Theor. Appl. Genet. 67, 333-336 (1984). - ZURAWSKI, G., CLEGG, M.T., BROWN, A.H.D.: Genetics 106, 735-750 (1984).

Prof. Dr. ROLF BLAICH
Bundesforschungsanstalt
für Rebenzüchtung
Geilweilerhof
D-6741 Siebeldingen, FRG

V. Extranuclear Inheritance: Plastid Genetics

By RUDOLF HAGEMANN, MONIKA M. HAGEMANN, and MICHAEL METZLAFF

1. Introduction

This review is a continuation of our articles in Progress in Botany 43, 45 and 47 (1981, 1983, 1985).

We shall first report on the complete sequencing of the plastid DNAs of a liverwort (*Marchantia polymorpha*) and a dicotyledonous angiosperm (*Nicotiana tabacum*) and shall characterize the coding capacity of their plastid DNAs by listing the genes so far identified. Subsequently, the presence of introns in plastidal genes and interesting splicing mechanisms are reported. Mutational rearrangements and drastic size reductions of plastid DNA which were (surprisingly) found as a consequence of anther culture of cereals will be briefly described. Evolutionary discussions, based on findings of intercompartmental gene transfer between the plastids and the nucleus, will also be summarized.

During the past two years several books have been published which contain relevant data on our topic. Regulation of Chloroplast Differentiation (edit. G. AKOYUNOGLOU and H. SENGER, 1986) and Chloroplast Biogenesis (edit. N.R. BAKER and J. BARBER, 1984) deal with many aspects of genetics, molecular biology, biochemistry and developmental biology of plastids. In addition, the reviews by PALMER (1985) and DYER (1984, 1985), which appeared parallel to our report in Progress in Botany, 1985, contain abundant information on the plastid genome and its organization.

2. The Complete Sequencing of the Plastid DNAs of *Marchantia* and *Nicotiana*

In 1986 an exciting breakthrough was achieved in the field of molecular biology and genetics of plastids: The plastid DNA molecules of two different plant species were completely sequenced by two Japanese research groups. OHYAMA and co-workers reported the total nucleotide sequence of the chloroplast DNA molecule of the liverwort *Marchantia polymorpha* (OHYAMA et al. 1986) and the research group of SUGIURA and SHINOZAKI sequenced the chloroplast DNA molecule of the higher plant species *Nicotiana tabacum* (SHINOZAKI et al. 1986c). The plastid DNA of *Marchantia* consists of 121,024 base pairs, the *Nicotiana* plastid DNA contains 155,844 base pairs.

Both the liverwort and the tobacco chloroplast genomes are remarkably similar with regards to their general structure, gene content and gene organization (Figs. 1 and 2). The circular DNA molecules consist of a "large single copy region", a "small single copy region" and two "inverted repeats". The size of the single copy regions is very similar.

Progress in Botany, Vol. 49

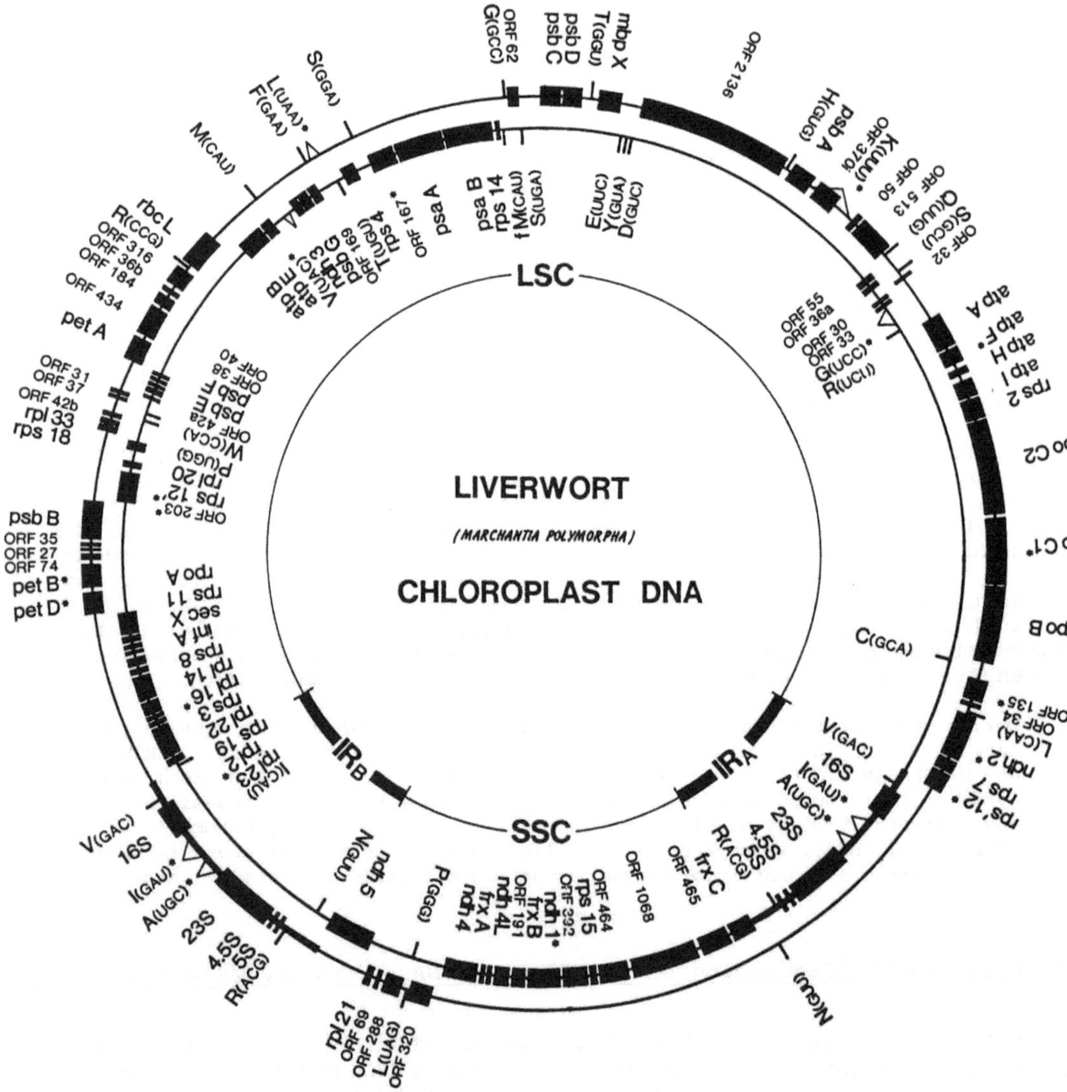

Fig. 1. Physical map of the plastid DNA of *Marchantia polymorpha* (OHYAMA et al. 1986). For gene symbols see Table 1. LSC large single copy region; SSC small single copy region; IR_A, IR_B inverted repeats; ORF open reading frame. Symbols outside the circle designate genes with counterclockwise direction of transcription, symbols inside the circle designate genes with clockwise direction of transcription. Asterisks mark genes containing introns

In the large single copy region there is a sequence rearrangement between the two species: an inversion of about 30,000 base pairs. Such sequence rearrangements have been found in other plant species, too, e.g. in *Oenothera* (HERRMANN et al. 1983), *Triticum aestivum* (HOWE et al. 1983), *Zea mays* (PALMER and THOMPSON 1982), *Vigna radiata* and *Lactuca sativa* (PALMER 1985). With regards to the size of the inverted repeats there is a remarkable difference: In *Marchantia polymorpha* the inverted

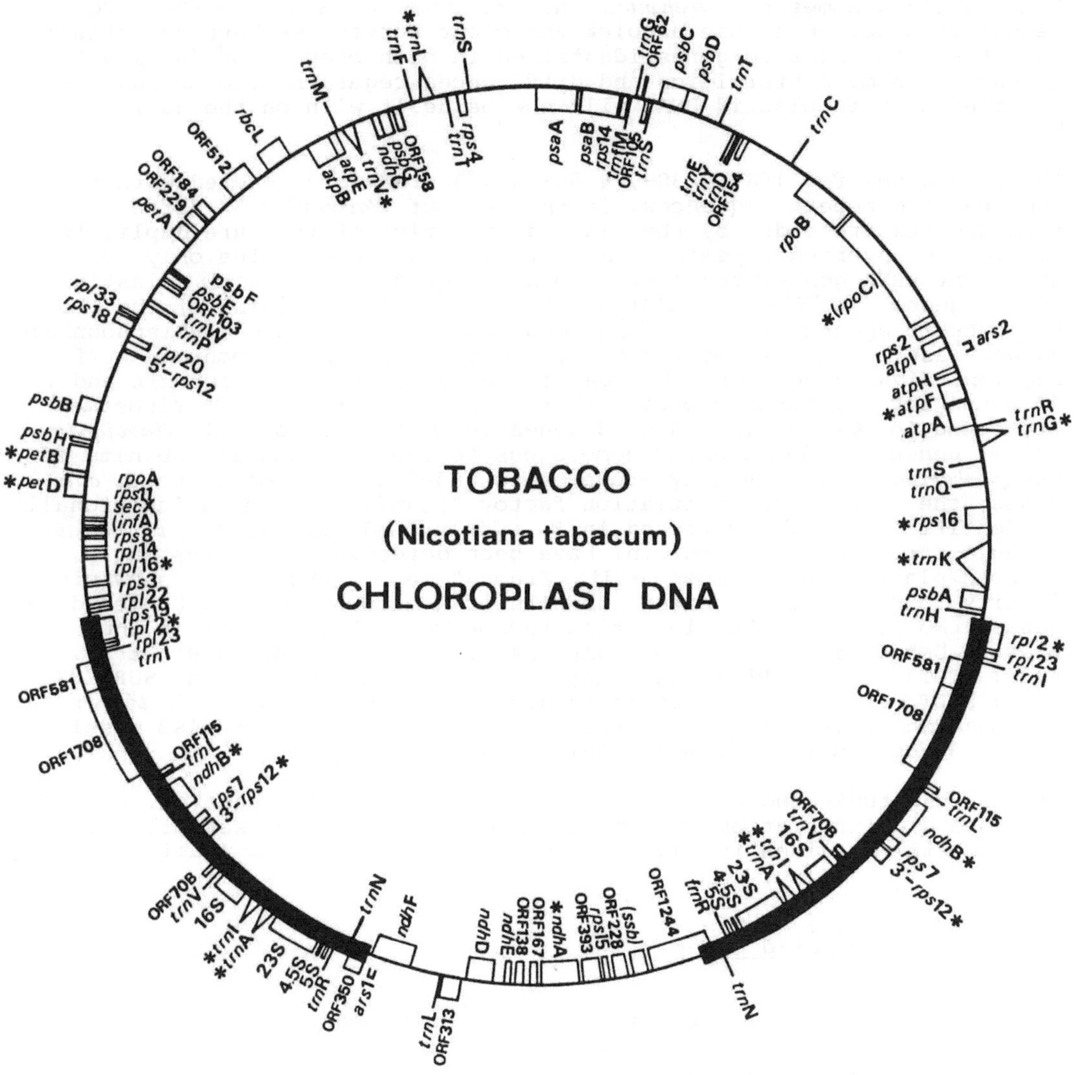

Fig. 2. Physical map of the plastid DNA of *Nicotiana tabacum* (SHINOZAKI et al. 1986c). For gene symbols see Table 1. Symbols outside the circle designate genes with counterclockwise direction of transcription, symbols inside the circle designate genes with clockwise direction of transcription. Asterisks mark genes containing introns

repeat consists of only 10,058 base pairs, whereas in the case of *Nicotiana tabacum* an inverted repeat has 25,339 base pairs.

In both plant species rRNAs, tRNAs and a large number of proteins are encoded by the plastid DNAs and synthesized within the plastids. Another part of the proteins is coded by chromosomal DNA within the nucleus and synthesized on 80S ribosomes of the cytoplasm. Those proteins are transported through the double membrane of the plastids.

Most of the plastid genes coding for RNAs and proteins have been localized and identified over the last years (cf. HAGEMANN et al. 1985).

The plastid genomes of *Marchantia* and *Nicotiana* contain more than 120 genes. In order to enable a quick and clear survey, we have summarized in Table 1 the plastid genes identified in both species or (at least) in one of them. Conformities and differences regarding the coding capacities of both plastid DNAs will now be dealt with on the basis of Table 1.

The four genes for 16S-, 23S-, 4,5S- and 5S-rRNA are located within the inverted repeat sequences. In the case of *Marchantia* 32 different tRNA species are coded by the plastid DNA, five of them are duplicated within the inverted repeats. *Nicotiana* plastid DNA contains only 30 different tRNA genes, seven of them duplicated. In *Marchantia* plastid DNA 19 genes (in *Nicotiana*: 20) are found encoding proteins of the photosynthetic apparatus, including components of ribulose-1,5-bisphosphate carboxylase, photosystems I and II, ATP synthase and cytochrome b_6/f complex. Eight genes for ribosomal proteins of the large subunit and 11 genes for ribosomal proteins of the small subunit of 70S ribosomes are found in *Marchantia*, 7 and 12 genes in *Nicotiana*. Four (in *Nicotiana*: three) genes encode subunits homologous to the α, β and β' subunits of the *Escherichia coli* RNA polymerase. Also based upon homology to *E. coli* genes, the genes *infA* (initiation factor 1) and *secX* (for a functionally unidentified protein expressed in *E. coli*) as well as, in *Nicotiana*, *ssb* (single-strand binding protein) have been determined. Interestingly, in both plant species "open reading frames" were found, which show similarity to six genes encoding subunits of a mitochondrial NADH dehydrogenase. No gene for the elongation factor Tu (*tufA*), present in the plastid DNA of *Euglena gracilis* (HALLICK et al. 1983; PASSAVANT et al. 1983; STUTZ et al. 1984) and *Chlamydomonas reinhardii* (WATSON and SURZYCKI 1982; STUTZ et al. 1984), could be determined so far in *Marchantia* or in *Nicotiana*. A gene for the σ-factor of RNA polymerase is also missing. Nearly 30 open reading frames remain unidentified.

Some of the tRNA- and protein-coding genes contain introns (cf. Table 1, Footnote b). Information on introns in plastid genes of several plant species will be given in the next section in more detail (Table 2).

3. Introns in Plastid Genes

For several years many introns from different types of genes have been studied. They can be assigned to three general groups (SHARP 1985; CECH 1986; NEUHAUS and LINK 1987).

Introns of nuclear mRNA precursors with invariant dinucleotides (GU and AG) at their boundary sequences form the first group of introns (*group 1*). A second group includes introns of nuclear tRNA genes (*group 2*) and a third group (*group 3*) comprises introns in mitochondrial and chloroplast genes as well as introns in the nuclear rRNA genes of *Tetrahymena* and *Physarum*. These introns can form conserved stem and loop structures by intramolecular pairing.

In 1983 MICHEL and DUJON published a paper in which they subdivided the *group 3* introns containing mitochondrial-, chloroplast- and nuclear-encoded members into two classes (MICHEL and DUJON 1983). They proposed a general core structure for *class I* introns including four base-paired stems and two short, highly conserved sequence stretches. *Class II* introns comprise a core with six base-paired stems, which may bring the 5' and 3' intron-exon junctions into relatively close proximity. For *class I* introns open reading frames encoding "maturase"-like proteins have been found also in chloroplast introns of the *Chlamydomonas reinhardii* 23S rRNA

Table 1. Coding capacity of the plastid DNAs of *Marchantia polymorpha* and *Nicotiana tabacum* (presence of rRNA, tRNA and protein coding genes)

Genes	*Marchantia polymorpha*[a]	*Nicotiana tabacum*[a]
rRNA genes		
16S rDNA	+	+
23S rDNA	+	+
4,5S rDNA	+	+
5S rDNA	+	+
tRNA genes		
trnA (UGC)	+[b,c]	+[b,c]
trnC (GCA)	+	+
trnD (GUC)	+	+
trnE (UUC)	+	+
trnF (GAA)	+	+
trnG (GCC)	+	+
trnG (UCC)	+[b]	+[b]
trnH (GUG)	+	+
trnI (CAU)	+	+[c]
trnI (GAU)	+[b,c]	+[b,c]
trnK (UUU)	+[b]	+[b]
trnL (CAA)	+	+[c]
trnL (UAA)	+[b]	+[b]
trnL (UAG)	+	+
trnfM (CAU)	+	+
trnM (CAU)	+	+
trnN (GUU)	+[c]	+[c]
trnP (GGG)	+	-
trnP (UGG)	+	+
trnQ (UUG)	+	+
trnR (ACG)	+[c]	+[c]
trnR (CCG)	+	-
trnR (UCU)	+	+
trnS (GCU)	+	+
trnS (GGA)	+	+
trnS (UGA)	+	+
trnT (GGU)	+	+
trnT (UGU)	+	+

Table 1 (continued)

Genes	*Marchantia polymorpha*[a]	*Nicotiana tabacum*[a]
tRNA genes		
trnV (GAC)	+[c]	+[c]
trnV (UAC)	+[b]	+[b]
trnW (CCA)	+	+
trnY (GUA)	+	+
RNA polymerase genes		
rpoA	+	+
rpoB	+	+
rpoC1	+[b]	+ *rpoC*
rpoC2	+	
Genes for photosynthesis		
rbcL	+	+
psaA	+	+
psaB	+	+
psbA	+	+
psbB	+	+
psbC	+	+
psbD	+	+
psbE	+	+
psbF	+	+
psbG	+	+
psbH	-	+
atpA	+	+
atpB	+	+
atpE	+	+
atpF	+[b]	+[b]
atpH	+	+
atpI	+	+
petA	+	+
petB	+[b]	+[b]
petD	+[b]	+
Ribosomal protein genes		
rpl2	+[b]	+[b]
rpl14	+	+
rpl16	+[b]	+[b]

Table 1 (continued)

Genes	*Marchantia polymorpha*[a]	*Nicotiana tabacum*[a]
Ribosomal protein genes		
rpl20	+	+
rpl21	+	–
rpl22	+	+
rpl23	+	+
rpl33	+	+
rps2	+	+
rps3	+	+
rps4	+	+
rps7	+	+
rps8	+	+
rps11	+	+
rps12	+[b]	+[b]
rps14	+	+
rps15	+	+
rps16	–	+[b]
rps18	+	+
rps19	+	+
Other genes		
infA	+	+
secX	+	+
ssb	–	+
Genes predicted by amino acid sequence homology		
ndh1	+[b] *ndhA*	+[b]
ndh2	+[b] *ndhB*	+[b]
ndh3	+ *ndhC*	+
ndh4	+ *ndhD*	+
ndh4L	+ *ndhE*	+
ndh5	+ *ndhF*	+

[a] + = Genes present; – = genes not determined.

[b] Presence of introns.

[c] Genes present twice.

tRNAs are symbolized by the one letter code of the accepting amino acids and their anticodon.

gene and *psbA* (ROCHAIX et al. 1985; ERICKSON et al. 1984). Other *class I* introns in plastid genes of higher plants (*trnL-UAA*) are postulated to be spliced autocatalytically as known from the *Tetrahymena* rRNA precursor (CECH 1983; BONNARD et al. 1984; MA and DOEBLEY 1986; YAMADA et al. 1986). Viroids and virusoids might also be related to *class I* introns (DINTER-GOTTLIEB 1986).

The introns of some protein-coding genes and tRNA genes belong to *class II* introns according to MICHEL and DUJON (1983). The intron in the gene *rps12* of *Nicotiana* (FROMM et al. 1986) may be folded into a secondary structure model that follows the rules for *class II* introns. The same is true for *rpl16* of *Spirodela oligorhiza* (POSNO et al. 1986a). Highly conserved loop structures at the 3' end sharing similarities to *class II* introns are also found in *atpF* from *Triticum aestivum* (BIRD et al. 1985) and *Spinacia oleracea* (HENNIG and HERRMANN 1986) as well as *trnG-UCC* of *Hordeum vulgare* (OLIVER and POULSEN 1984), *trnA-UGC* and *trnI-GAU* of *Nicotiana tabacum* (TAKAIWA and SUGIURA 1982) and *Zea mays* (KOCH et al. 1981). The nine introns of *rbcL* of *Euglena gracilis* (GINGRICH and HALLICK 1985) contain highly conserved boundary sequences also typical of *class II* introns. None of these introns contain internal open reading frames; thus, it is unlikely that they code for any polypeptide. In contrast, the intron of the *Euglena gracilis psbC* gene (MONTANDON et al. 1986) contains an open reading frame of 458 codons coding for a protein of 54 kD of yet unknown function. The intron boundary sequences are very similar to the consensus sequences proposed for intron *class II*. The intron of *trnK-UUU* of *Sinapis alba* (NEUHAUS and LINK 1987) showing characteristic features of *class II* introns has a long open reading frame coding for 524 amino acids. The derived polypeptide near its carboxyl terminus appears to have structural relations to maturases. In the gene *trnG-UCC* a *class II* intron has been described, based on sequence and structural homologies (QUIGLEY and WEIL 1985). Three of the four introns in *psbA* of *Euglena gracilis* have sequence similarities with *class II* introns, but their sequences could not be folded into the characteristic secondary structure. Only the fourth intron shows the typical *class II* core structure. For the first time it was shown that the excised intron RNAs from a plastid gene were circularized (KOLLER et al. 1985).

In 1985/86 SUGITA et al. (1985) and SHINOZAKI et al. (1986a) proposed the subdivision of *group 3* introns not only into two classes, but into three classes with special regard to introns in chloroplast genes. *Class I* introns should be those which can be folded into a secondary structure similar to that of the intron of the autosplicable rRNA precursor of *Tetrahymena* (cf. BONNARD et al. 1984). They might be spliced by an autocatalytic reaction or by "maturases". *Class II* introns show a complex secondary structure for excision (cf. MICHEL and DUJON 1983). *Class III* introns have conserved boundary sequences resembling those of nuclear gene introns (cf. CECH 1983).

According to SHINOZAKI et al. (1986a), introns described in some previous papers as *class II* introns now belong to *class III*, e.g. introns of *trnG-UCC* (QUIGLEY and WEIL 1985) and *atpF* (BIRD et al. 1985) of *Triticum aestivum*. On the basis of the conserved boundary sequences, the introns of the genes *rps12*, *rps16*, *rpl12*, *rpl16*, *trnK-UUU* and *atpF* of *Nicotiana tabacum* (TORAZAWA et al. 1986; SHINOZAKI et al. 1986a; TANAKA et al. 1986; SUGITA et al. 1985; SHINOZAKI et al. 1986b), *rps12* of *Spirodela oligorhiza* (POSNO et al. 1986b) and in the gene for 23S rRNA of *Chlorella ellipsoidea* (YAMADA and SHIMAJI 1987) are described as *class III* introns. The *Chlorella* intron, according to the model of WARING and DAVIES (1984), might be an example for a self-splicing intron.

Table 2. Intronss in plastid genes

Species	Gene	Number of introns	Size of introns (bp)	Class	Reference
Chlamydomonas reinhardii	*23S rDNA*	1	888	I	ROCHAIX and MALNOE (1978) ROCHAIX et al. (1985)
Chlorella ellipsoidea	*23S rDNA*	1	243	III	YAMADA and SHIMAJI (1987)
Marchantia polymorpha	*trnA*-UGC				OHYAMA et al. (1986)
Nicotiana tabacum	*trnA*-UGC	1	710	II	TAKAIWA and SUGIURA (1982)
Zea mays	*trnA*-UGC	1	806	II	KOCH et al. (1981)
Hordeum vulgare	*trnG*-UCC	1	677	II	OLIVER and POULSEN (1984)
Marchantia polymorpha	*trnG*-UCC				OHYAMA et al. (1986)
Nicotiana tabacum	*trnG*-UCC	1	691	III	DENO and SUGIURA (1984)
Triticum aestivum	*trnG*-UCC	1	676	II III	QUIGLEY and WEIL (1985) SHINOZAKI et al. (1986a)
Marchantia polymorpha	*trnI*-GAU				OHYAMA et al. (1986)
Nicotiana tabacum	*trnI*-GAU	1	707	II	TAKAIWA and SUGIURA (1982)
Zea mays	*trnI*-GAU	1	949	II	KOCH et al. (1981)
Marchantia polymorpha	*trnK*-UUU				OHYAMA et al. (1986)
Nicotiana tabacum	*trnK*-UUU	1	2526	III	SUGITA et al. (1985)
Sinapis alba	*trnK*-UUU	1	2574	II	NEUHAUS and LINK (1987)
Marchantia polymorpha	*trnL*-UAA				OHYAMA et al. (1986)
Nicotiana tabacum	*trnL*-UAA	1	503	I	YAMADA et al. (1986)
Sorghum bicolor	*trnL*-UAA	1	450	I	MA and DOEBLEY (1986)
Vicia faba	*trnL*-UAA	1	451	I	BONNARD et al. (1984)
Zea mays	*trnL*-UAA	1	458		STEINMETZ et al. (1982)
Hordeum vulgare	*trnV*-UAC	1	597	III	ZURAWSKI and CLEGG (1984)
Marchantia polymorpha	*trnV*-UAC				OHYAMA et al. (1986)

Table 2 (continued)

Species	Gene	Number of introns	Size of introns (bp)	Class	Reference
Nicotiana tabacum	*trnV*-UAC	1	571	III	DENO et al. (1982)
					SUGITA and SUGIURA (1983)
Zea mays	*trnV*-UAC	1	603	III	KREBBERS et al. (1984)
Marchantia polymorpha	*rpoC1*				OHYAMA et al. (1986)
Euglena gracilis	*rbcL*	9	404		KOLLER et al. (1984)
			514	II	GINGRICH and HALLICK (1985)
			513		
			568		
			413		
			479		
			382		
			420		
			441		
Euglena gracilis	*psaA*	1	214		MANZARA et al. (1987)
Chlamydomonas reinhardii	*psbA*	4	1350	I	ERICKSON et al. (1984)
			1400		
			1100		
			1800		
Euglena gracilis	*psbA*	4	435		KARABIN et al. (1984)
			443		KELLER and STUTZ (1984)
			434	II	KELLER and MICHEL (1985)
			617		
Euglena gracilis	*psbC*	1	1604	II	MONTANDON et al. (1986)
Marchantia polymorpha	*atpF*				OHYAMA et al. (1986)
Nicotiana tabacum	*atpF*	1	695	III	SHINOZAKI et al. (1986b)
Spinacia oleracea	*atpF*	1	764	II	HENNIG and HERRMANN (1986)
Triticum aestivum	*atpF*	1	823	II	BIRD et al. (1985)
				III	SHINOZAKI et al. (1986a)
Marchantia polymorpha	*petB*				OHYAMA et al. (1986)

Species	Gene	Number of introns	Size of introns (bp)	Class	Reference
Nicotiana tabacum	*petB*	1	753		SHINOZAKI et al. (1986c)
Marchantia polymorpha	*petD*				OHYAMA et al. (1986)
Marchantia polymorpha	*rpl2*				OHYAMA et al. (1986)
Nicotiana debneyi	*rpl2*	1	666	III	ZURAWSKI et al. (1984) SHINOZAKI et al. (1986a)
Nicotiana tabacum	*rpl2*	1	666	III	TANAKA et al. (1986)
Spirodela oligorhiza	*rpl2*	1	661		POSNO et al. (1986c)
Marchantia polymorpha	*rpl16*				OHYAMA et al. (1986)
Nicotiana tabacum	*rpl16*	1	1020	III	TANAKA et al. (1986)
Spirodela oligorhiza	*rpl16*	1	1411	II	POSNO et al. (1986a)
Marchantia polymorpha	*rps12*				FUKUZAWA et al. (1986)
Nicotiana tabacum	*rps12*	1	536	III	TORAZAWA et al. (1986) ZAITA et al. (1987)
Nicotiana tabacum line 92 (ptDNA: *N. undulata*)	*rps12*	1	540	II	FROMM et al. (1986)
Spirodela oligorhiza	*rps12*	1	542	III	POSNO et al. (1986b)
Nicotiana tabacum	*rps16*	1	860	III	SHINOZAKI et al. (1986a)
Euglena gracilis	*tufA*	2	103 110		MONTANDON and STUTZ (1983)
Marchantia polymorpha	*ndh1*				OHYAMA et al. (1986)
Nicotiana tabacum	*ndhA*	1	1242		SHINOZAKI et al. (1986c)
Marchantia polymorpha	*ndh2*				OHYAMA et al. (1986)
Nicotiana tabacum	*ndhB*	1	757		SHINOZAKI et al. (1986c)

Finally, a special type of splicing, trans-splicing, should be mentioned: The gene *rps12* is split into different DNA strands in *Marchantia polymorpha* and *Nicotiana tabacum* (FUKUZAWA et al. 1986; TORAZAWA et al. 1986; ZAITA et al. 1987). In the case of *Marchantia* the gene consists of three exons: Exon 1 is located on the same DNA strand as *atpB* and *psaA*, exons 2 and 3 on the opposite strand like *rbcL* and *psbA*. Exons 2 and 3 are split by a 0,4 kb intron (FUKUZAWA et al. 1986). Nearly the same situation is found in *Nicotiana tabacum*: The gene *rps12* also consists of three exons, but exons 2 and 3 are present twice, located within the inverted repeat sequences. There is a 536-bp intron between exons 2 and 3. In both plant species trans-splicing is needed to join exons 1 and 2. ZAITA et al. (1987) proposed that the 3'-flanking sequence of exon 1 be designated as "transon" 1 and the 5'-flanking sequence of exon 2 as "transon" 2. Both transons fit the conserved boundary sequences of most chloroplast introns and are able to form a long, complementary structure (Fig. 3).

4. DNA Rearrangements and Drastic Size Reduction of Plastid DNA as a Consequence of Anther Culture in Cereals

Anther culture and culture of microspores are nowadays widespread methods in plant cell biology and somatic cell genetics. They can be successfully applied in several genera, e.g. *Nicotiana*, *Datura*, *Brassica*, *Petunia* (REINERT and BAJAJ 1977). However, cell biologists working with cereals have been confronted with two severe difficulties. First, the amount of haploids regenerating from cereal anthers and microspores, respectively, is very low in comparison with standard objects, like tobacco. This difficulty could partly be overcome by special treatments (cold treatment of cereal spikes, high temperature treatment or new culture procedures) (HU 1984; HUANG and SUNDERLAND 1982). Second, among the regenerated haploids the great majority are albinos, and only few haploids are green (HU 1984; CLAPHAM 1977; WENZEL et al. 1977).

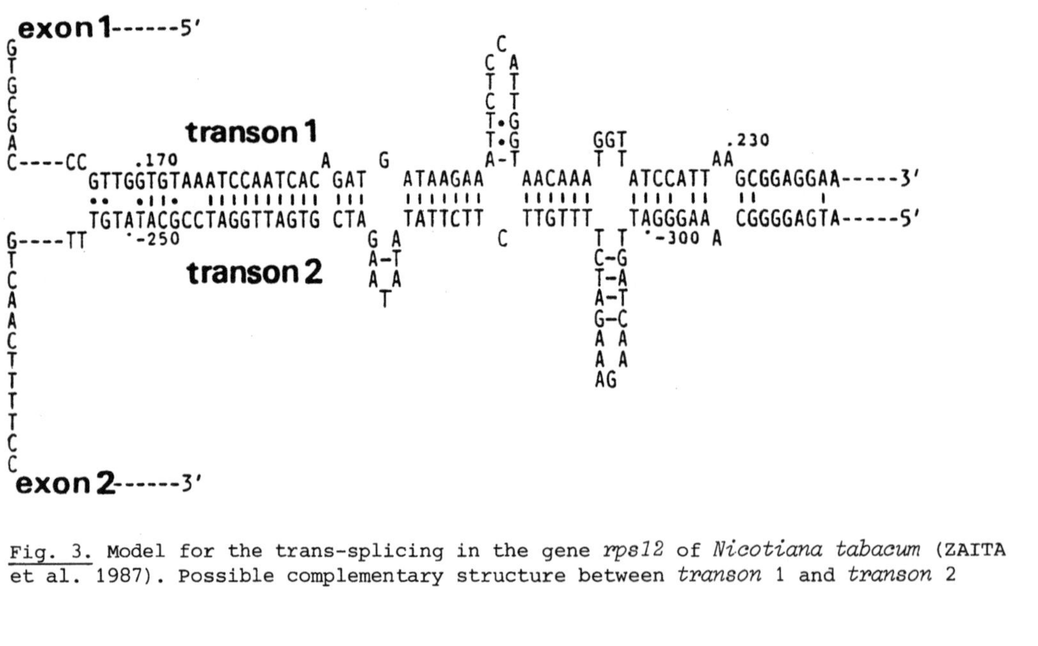

Fig. 3. Model for the trans-splicing in the gene *rps12* of *Nicotiana tabacum* (ZAITA et al. 1987). Possible complementary structure between *transon* 1 and *transon* 2

There has been a discussion over several years about the reasons for the appearance of albino plantlets, derived from anther culture, and about the structure of their plastid DNA. Recent studies by DAY and ELLIS (1984, 1985) as well as ELLIS and DAY (1986) led to the surprising result that the plastid genome of albino plants has obviously undergone a series of gross DNA rearrangements, most of which are deletions. The plastid DNA of several of the albino variants of wheat and barley has been carefully analyzed. Two deletion mutants of the plastid DNA of wheat are likely to have arisen as a consequence of single recombination events between two single copy regions leading to a deletion (DAY and ELLIS 1984, 1985).

In a thorough study of a barley albino plant ELLIS and DAY (1986) characterized its plastid DNA: The white plastids contain a linear double-stranded DNA of only 21 kb with a hairpin structure. Monomers and dimers of this molecule have been found. This result leads to the assumption that a deletion of (about) 100 kb of the wild-type plastid DNA of barley gave rise to the monomeric form.

These findings seem to indicate that the stress conditions of an anther culture are able to cause drastic DNA rearrangements and deletions in the plastid DNA of the microspores treated.

5. Sequence Homologies Between Plastid DNA and the DNAs of the Nucleus and Mitochondria

The eukaryotic plant cell originated by endosymbiosis; it contains three semi-autonomous genetic compartments which harbour their own specific DNA. It is well known that many protein subunits, essential for plastidal and mitochondrial biogenesis, are encoded by nuclear DNA, transcribed in the nucleus, translated on cytoplasmic 80S ribosomes and transported as precursor polypeptides into the organelles. There is a complex interaction between the three genetic compartments at the protein level. However, it is still not completely understood how this interaction is regulated.

The use of methods of modern molecular genetics, like gene cloning, hybridization techniques and DNA sequencing, for the analysis of plant DNA species has increased the knowledge on the structure and function of all three plant DNAs impressively in less than ten years. In the early 1980s it seemed certain that the three plant DNA species (nucDNA, mtDNA, ptDNA) were distinct from each other and for their specific compartment. Thus, it was even more surprising when STERN and LONSDALE (1982) published the occurrence of a 12-kb plastid DNA sequence in the mitochondrial DNA of maize. Within this sequence the genes for the plastidal 16S rRNA and two tRNAs have been localized. By restriction mapping of the surrounding mitochondrial DNA regions it has been convincingly proved by the authors that the location of a plastid DNA sequence in the mitochondrial DNA is not due to an artificial cloning event.

After this first discovery of sequence homology between DNAs of two different plant organelles, an intense search for sequence homologies was started. Today a number of examples exists for sequence homologies between all three compartments (for review, see GELLISSEN 1987; BÖRNER 1984). In 1982 ELLIS coined the term "promiscuous DNA" for all DNA sequences which are located in more than one compartment of eukaryotic cells. In this review only the data of sequence homologies in connection with plastid DNA will be summarized.

a) Sequence Homologies Between Plastid and Mitochondrial DNA

The occurrence of a 12-kb plastid DNA sequence within the mitochondrial DNA of maize was already mentioned. It is interesting that male fertile and male sterile maize plants differ, in addition, in many other changes also in this integrated DNA. However, it is unlikely that the integrated chloroplast genes are transcribed in mitochondria. In 1983 LONSDALE et al. published the finding of a further plastid sequence homology in maize mitochondrial DNA: 20,000 bp away from the sequence with homology to the plastid 16S rRNA gene, a sequence was found which shows strong homology to the plastid gene for the large subunit of ribulose-1,5-bisphosphate carboxylase/oxygenase (= *rubisco*). In this case the sequence is identical in fertile and sterile maize plants. The transcription/translation of the cloned gene in an *E. coli* cell-free system resulted in a 21,000 dalton polypeptide which is much smaller than the native maize plastidal large subunit of *rubisco* of 52,000 dalton. Also, these findings do not confirm the expression of the integrated plastid gene in maize mitochondria.

STERN and PALMER (1984) reported widespread homologies between plastid and mitochondrial DNA in several higher plants. Cloned mung bean and spinach plastid DNA fragments hybridize with a number of mitochondrial DNA fragments of mung bean and spinach and of two other plant species, corn and pea. The strongest hybridizations were found for cloned fragments containing sequences of the plastid inverted repeats. Fragments with the gene for the large subunit of *rubisco* also hybridized with corn mitochondrial DNA. For mung bean mitochondrial DNA, a strong hybridization signal was found for the cloned β subunit of the plastid ATPase. However, in addition to the strong hybridization signals for all tested plastid DNA fragments at least one fragment with sequence homology was found in mitochondrial DNA. Most of the homologies were confirmed by using cloned spinach mitochondrial DNA fragments in reciprocal hybridizations. In general, the hybridization between plastid and mitochondrial DNA fragments of the same plant species was stronger than between different species. From these findings the authors concluded that the observed sequence homologies are a result of relatively recent DNA transfer events which took place after the evolutionary separation of plant species.

DRON et al. (1985) reported the integration of the plastid tRNA Leu2 gene into the mitochondrial DNA of *Brassica oleracea*. The sequences of the native plastid gene are completely homologous, with the exception of an insertion of 5 bp in the T-loop of the mitochondrial copy. Two distinct gene conversion events, which occurred between two sets of short directed repeats inside and close to the gene, caused the insertion which resulted in an inactivation of the integrated plastid gene.

One hypothesis for the mechanism of gene transfer events in eukaryotic cells is based on the concept of the autonomous replication of separate DNA fragments containing "autonomously replicating sequences" (*ars*). SEDEROFF et al. (1986) found that the linear mitochondrial S-1 DNA molecule from maize contains a 420-bp fragment which is homologous with the plastid photosystem II gene *psbA*. This is the only report of a plastid sequence in plasmidlike or viruslike molecules.

So far, no report has been published on the presence of a mitochondrial DNA sequence in the plastid DNA, although many physical maps and sequences of plastid DNAs from many plant species have been determined and compared with other sequences.

b) Sequence Homologies Between Plastid and Nuclear DNA

WHISSON and SCOTT (1985) reported that a plastid DNA fragment with the gene for a P_{700} chlorophyll a apoprotein shares homology not only with the mitochondrial DNA but also with the nuclear DNA of spinach. The nuclear DNA contains a number of regions which hybridize with this plastid DNA fragment. This seems to be the first example of a promiscuous DNA sequence, which has been found in all three genetic compartments of the eukaryotic cell. Five years ago in the same laboratory hybridization experiments were performed with plastid DNA probes from a clone bank harbouring the entire plastid DNA of spinach (TIMMIS and SCOTT 1983). Each tested clone hybridized with fragments of *Eco*RI and *Bam*HI digested nuclear DNA of spinach. The amount of DNA, homologous with plastid DNA within the nuclear DNA of spinach, was estimated to be about three to six copies of the plastome per haploid genome. However, long continuous tracts of homologous DNA are very rare in the recent nuclear genome; moreover, each sequence with homology to plastid DNA occurs in more than one nuclear location.

These findings can be explained by two alternative hypotheses: (1) The sequences, identified by these hybridization experiments, were originally DNA sequences of the plastids and the progenitors of the plastids (e.g. cyanobacteria), respectively, and the promiscuous plastid DNA sequences were altered extensively after their integration into nuclear DNA. (2) The alternative interpretation would be that in the large amount of nuclear DNA (unique, middle repetitive and highly repetitive sequences) there are DNA tracts which hybridize with relatively short plastid DNA sequences. These nuclear DNA sequences may be partly promiscuous DNA, but they may partly represent sequences with accidental homology.

c) Hypotheses for the Mechanisms of Intercompartmental Gene Transfer

Today the "endosymbiotic hypothesis" is generally accepted to explain the evolutionary origin of plastids and mitochondria. The plastids originated most likely from free-living photosynthetic cyanobacteria. In fact, among the cyanobacteria there are several genera (e.g. *Anabaena*, *Microcystis*, *Spirulina*) which have the genes for the large subunit (*rbcL*) and the small subunit (*rbcS*) of *rubisco* closely linked on the cyanobacterial genome (GUREVITZ et al. 1985; TIBONI et al. 1984). The homology between the *rbcL* genes of higher plants and cyanobacteria is about 75% at the nucleotide sequence level and about 85% at the protein sequence level (CURTIS and CLEGG 1984; MIZIORKO and LORIMER 1983; GUREVITZ et al. 1985). Also in cyanelles, present as endosymbionts in the unicellular biflagellate *Cyanophora paradoxa*, the genes for the large and the small subunit of *rubisco* are located on the same DNA molecule (LÖFFELHARDT et al. 1985; HEINHORST and SHIVELEY 1983). In contrast, in all plastid genomes of eukaryotic plants the gene for the large subunit of *rubisco* (*rbcL*) is present and the gene for the small subunit (*rbcS*) is absent, but present in the nuclear DNA. Thus, a gene transfer from the symbionts to the eukaryotic nucleus must have taken place during evolution.

In the literature, two main hypothetical mechanisms for this transfer have been discussed:

1. A direct physical contact of compartments by membrane continuities and associations might have facilitated a DNA transfer and intermolecular recombination of DNAs from different organelles. Such direct physical contacts have been reported for several species like barley, corn, tobacco and *Euglena* (STERN and PALMER 1984).

2. Sequence transfer events might have occurred also without physical contact of organelles. Thus, it should be possible that broken or lysed plastids can release plastid DNA sequences which are taken up randomly by the nucleus or mitochondria. Specific sequences, like short recombination sites, transposonlike elements or autonomously replicating sequences (*ars*) may enhance transfer events.

The experimental data for the verification of these hypotheses are still very limited. But there are data available that gene transfer between compartments is an ongoing process in recent organisms. In the fungi *Podospora* and *Neurospora* sequences are transferred between the nucleus and mitochondria during senescence (ESSER et al. 1986). Thus, the occurrence of promiscuous DNA demonstrates impressively that the evolution of the eukaryotic cell is a complex intercompartmental process.

6. Outlook

Many genetic and molecular biological aspects of biogenesis and the function of chloroplasts are being investigated at present. For this review, we could only select some topics in which scientific progress has been particularly quick and outstanding.

Thus, we had to omit several other interesting studies, whose results will be dealt with in one of the following volumes of Progress in Botany.

In Section 5c we mentioned the homologies between plastidal, cyanobacterial and cyanelle genes. These results have to be summarized in some detail. Moreover, the high degree of homology between plastidal and cytobacterial genes has been the basis of successful experiments on site-specific mutagenesis. They aim at the induction of directed changes in the amino acid sequence of the active centres of enzymes, e.g. *rubisco*, in cyanobacteria (*Anabaena* 7120: GUREVITZ et al. 1985) and in eubacteria (*Rhodospirillum rubrum*: ESTELLE et al. 1985). The results, obtained and expected, have direct consequences for the study of homologous plastid genes and the polypeptides, which they encode.

References

AKOYUNOGLOU, G., SENGER, H. (eds.) Regulation of Chloroplast Differentiation (Plant Biology, Vol. 2). New York: Liss 1986.

BAKER, N.R., BARBER, J. (eds.): Chloroplast Biogenesis. Amsterdam: Elsevier 1984. - BIRD, C.R., KOLLER, B., AUFFRET, A.D., HUTTLY, A.K., HOWE, C.J., DYER, T.A., GRAY, J.C.: EMBO J. 4, 1381-1388 (1985). - BONNARD, G., MICHEL, F., WEIL, J.H., STEINMETZ, A.: Mol. Gen. Genet. 194, 330-336 (1984). - BÖRNER, T.: Biol. Zentralbl. 103, 485-493 (1984).

CECH, T.R.: Cell 34, 713-716 (1983); - ibid. 44, 207-210 (1986). - CLAPHAM, D.H.: Haploid induction in cereals, 279-298. In: Plant Cell, Tissue and Organ Culture, eds. J. REINERT, Y.P.S. BAJAJ. Berlin, Heidelberg, New York: Springer 1977. - CURTIS, S.E., CLEGG, M.T.: Mol. Biol. Evol. 4, 291-301 (1984).

DAY, A., ELLIS, T.H.N.: Cell 39, 359-368 (1984); - Curr. Genet. 9, 671-678 (1985). - DENO, H., SUGIURA, M.: Proc. Natl. Acad. Sci. USA 81, 405-408 (1984). - DENO, H., KATO, A., SHINOZAKI, K., SUGIURA, M.:

Nucl. Acids Res. 10, 7511-7520 (1982). - DINTER-GOTTLIEB, G.: Proc. Natl. Acad. Sci. USA 83, 6250-6254 (1986). - DRON, M., HARTMANN, C., RODE, A., SEVIGNAC, M.: Nucl. Acids Res. 13, 8603-8610 (1985). - DYER, T.A.: The chloroplast genome: its nature and role in development, 23-70. In: Chloroplast Biogenesis, eds. N.R. BAKER, J. BARBER. Amsterdam: Elsevier 1984; - Oxford Surv. Plant Mol. Cell Biol. 2, 147-177 (1985).

ELLIS, J.: Nature 299, 678-679 (1982). - ELLIS, T.H.N., DAY, A.: EMBO J. 5, 2769-2774 (1986). - ERICKSON, J.M., RAHIRE, M., ROCHAIX, J.-D.: EMBO J. 3, 2753-2762 (1984). - ESSER, K., KÜCK, U., LANG-HINRICHS, C., LEMKE, P., OSIEWACZ, H.D., STAHL, U., TUDZYNSKI, P.: Plasmids of Eukaryotes. Berlin, Heidelberg, New York: Springer 1986. - ESTELLE, M., HANKS, J., McINTOSH, L., SOMERVILLE, C.: J. Biol. Chem. 260, 9523-9526 (1985).

FROMM, H., EDELMAN, M., KOLLER, B., GOLOUBINOFF, P., GALUN, E.: Nucl. Acids Res. 14, 883-898 (1986). - FUKUZAWA, H., KOHCHI, T., SHIRAI, H., OHYAMA, K., UMESONO, K., INOKUCHI, H., OZEKI, H.: FEBS Lett. 198, 11-15 (1986).

GELLISSEN, G.: Biologie uns. Zeit 17, 15-20 (1987). - GINGRICH, J.C., HALLICK, R.B.: J. Biol. Chem. 260, 16156-16161 (1985). - GUREVITZ, M., SOMERVILLE, C.R., McINTOSH, L.: Proc. Natl. Acad. Sci. USA 82, 6546-6550 (1985).

HAGEMANN, R., LINDENHAHN, M.M., METZLAFF, M.: Progr. Botany 47, 208-228 (1985). - HALLICK, R., GREENBERG, B.M., GRUISSEM, W., HOLLINGSWORTH, M.J., KARABIN, G.D., NARITA, J.O., NICKOLOFF, J.A., PASSAVANT, C.W., STIEGLER, G.L.: Organization and expression of the chloroplast genome of *Euglena gracilis*, 155-166. In: Structure and Function of Plant Genomes, eds. O. CIFERRI, L. DURE III. New York, London: Plenum Press 1983. - HEINHORST, S., SHIVELEY, J.M.: Nature 304, 373-374 (1983). - HENNIG, J., HERRMANN, R.G.: Mol. Gen. Genet. 203, 117-128 (1986). - HERRMANN, R.G., WESTHOFF, P., ALT, J., WINTER, P., TITTGEN, J., BISANZ, C., SEARS, B.B., NELSON, N., HURT, E., HAUSKA, G., VIEBROCK, A., SEBALD, W.: Identification and characterization of genes for polypeptides of the thylakoid membrane, 143-153. In: Structure and Function of Plant Genomes, eds. O. CIFERRI, L. DURE III. New York, London: Plenum Press 1983. - HOWE, C.J., BOWMAN, C.M., DYER, T.A., GRAY, J.C.: Mol. Gen. Genet. 190, 51-55 (1983). - HU, H.: Crop improvement by anther culture, 77-84. In: Genetics, New Frontiers IV. Proc. XV. Int. Congr. Gen. 1984. - HUANG, B., SUNDERLAND, N.: Ann. Bot. 49, 77-88 (1982).

KARABIN, G.D., FARLAY, M., HALLICK, R.B.: Nucl. Acids Res. 12, 5801-5812 (1984). - KELLER, M., MICHEL, F.: FEBS Lett. 179, 69-73 (1985). - KELLER, M., STUTZ, E.: FEBS Lett. 175, 173-177 (1984). - KELLER, W.: Cell 39, 423-425 (1984). - KOCH, W., EDWARDS, K., KÖSSEL, H.: Cell 25, 203-213 (1981). - KOLLER, B., CLARKE, J., DELIUS, H.: EMBO J. 4, 2445-2450 (1985). - KOLLER, B., GINGRICH, J.C., STIEGLER, G.L., FARLAY, M.A., DELIUS, H., HALLICK, R.B.: Cell 36, 545-553 (1984). - KREBBERS, E., STEINMETZ, A., BOGORAD, L.: Plant Mol. Biol. 3, 13-20 (1984).

LÖFFELHARDT, W., MICHALOWSKI, CH., BOHNERT, H.J.: Endocyt. C. Res. 2, 213-231 (1985). - LONSDALE, D.M., HODGE, T.P., HOWE, C.J., STERN, D.B.: Cell 34, 1007-1014 (1983).

MA, D.P., DOEBLEY, J.: Gene 43, 169-174 (1986). - MANZARA, T., HU, J.-X., PRICE, C.A., HALLICK, R.B.: Plant Mol. Biol. 8, 327-336 (1987). - MICHEL, F., DUJON, B.: EMBO J. 2, 33-38 (1983). - MICHEL, F., Lang, F.: Nature 316, 641-643 (1985). - MIZIORKO, H.M., LORIMER, G.H.: Annu. Rev. Biochem. 52, 507-535 (1983). - MONTANDON, P.-E., STUTZ, E.: Nucl. Acids

Res. 11, 5877-5892 (1983). - MONTANDON, P.-E., VASSEROT, A., STUTZ, E.: Curr. Genet. 11, 35-39 (1986).

NEUHAUS, H., LINK, G.: Curr. Genet. 11, 251-257 (1987).

OHYAMA, K., FUKUZAWA, H., KOHCHI, T., SHIRAI, H., SANO, T., SANO, S., UMESONO, K., SHIKI, Y., TAKEUCHI, M., CHANG, Z., AOTA, S., INOKUCHI, H., OZEKI, H.: Nature 322, 572-574 (1986). - OLIVER, R.P., POULSEN, C.: Carlsberg Res. Commun. 49, 647-673 (1984).

PALMER, J.D.: Annu. Rev. Genet. 19, 325-354 (1985). - PALMER, J.D., THOMPSON, W.F.: Proc. Natl. Acad. Sci. USA 78, 5533-5537 (1981); - Cell 29, 537-550 (1982). - PASSAVANT, C.W., STIEGLER, G.L., HALLICK, R.B.: J. Biol. Chem. 258, 693-695 (1983). - POSNO, M., VAN VLIET, A., GROOT, G.S.P.: Nucl. Acids Res. 14, 3181-3195 (1986a). - POSNO, M., VERWEIJ, W.R., DEKKER, J.C., DE WAARD, P.M., GROOT, G.S.P.: Curr. Genet. 11, 25-34 (1986b). - POSNO, M., VAN VLIET, A., GROOT, G.S.P.: Curr. Genet. 10, 923-930 (1986c).

QUIGLEY, F., WEIL, J.H.: Curr. Genet. 9, 495-503 (1985).

REINERT, J., BAJAJ, Y.P.S. (eds.): Plant cell, tissue and organ culture. Berlin, Heidelberg, New York: Springer 1977. - ROCHAIX, J.D., MALNOE, P.: 135-142. In: Genetic Engineering, eds. H.W. BOYER, S. NICOSIA. Amsterdam: Elsevier 1978. - ROCHAIX, J.D., RAHIRE, M., MICHEL, F.: Nucl. Acids Res. 13, 975-984 (1985).

SEDEROFF, R.R., RONALD, P., BEDINGER, P., RIVIN, C., WALBOT, V., BLAND, M., LEVINGS III, C.S.: Genetics 113, 469-482 (1986). - SHARP, P.: Cell 42, 397-400 (1985). - SHINOZAKI, K., DENO, H., SUGITA, M., KURAMITSU, B., SUGIURA, M.: Mol. Gen. Genet. 202, 1-5 (1986a). - SHINOZAKI, K., DENO, H., WAKASUGI, T., SUGIURA, M.: Curr. Genet. 10, 421-423 (1986b). - SHINOZAKI, K., OHME, M., TANAKA, M., WAKASUGI, T., HAYASHIDA, N., MATSUBAYASHI, T., ZAITA, N., CHUN-WONGSE, J., OBOKATA, J., YAMAGUCHI-SHINOZAKI, K., OHTO, C., TORAZAWA, K., MENG, B.Y., SUGITA, M., DENO, H., KAMOGASHIRA, T., YAMADA, K., KUSUDA, J., TAKAIWA, F., KATO, A., TOHDOH, N., SHIMADA, H., SUGIURA, M.: EMBO J. 5, 2043-2049 (1986c). - STEINMETZ, A., GUBBINS, E.J., BOGORAD, L.: Nucl. Acids Res. 10, 3027-3037 (1982). - STERN, D.B., LONSDALE, D.M.: Nature 299, 698-702 (1982). - STERN, D.B., PALMER, J.D.: Proc. Natl. Acad. Sci. USA 81, 1946-1950 (1984). - STUTZ, E., MONTANDON, P.E., ROUX, E., RUTTI, B., SCHLUNEGGER, B.: Organization and capacity of the chloroplast genome in algae, 11-22. In: Compartments in Algal Cells in their Interaction, eds. W. WIESSNER, D. ROBINSON, R.C. STARR. Berlin, Heidelberg, New York: Springer 1984. - SUGITA, M., SUGIURA, M.: Nucl. Acids Res. 11, 1913-1918 (1983). - SUGITA, M., SHINOZAKI, K., SUGIURA, M.: Proc. Natl. Acad. Sci. USA 82, 3557-3561 (1985).

TAKAIWA, F., SUGIURA, M.: Nucl. Acids Res. 10, 2665-2676 (1982). - TANAKA, M., WAKASUGI, T., SUGITA, M., SHINOZAKI, K., SUGIURA, M.: Proc. Natl. Acad. Sci. USA 83, 6030-6034 (1986). - TIBONI, O., DiPASQUALE, G., CIFERRI, O.: Biochim. Biophys. Acta 783, 258-264 (1984). - TIMMIS, J.N., SCOTT, N.S.: Nature 305, 65-67 (1983). - TORAZAWA, K., HAYASHIDA, N., OBOKATA, J., SHINOZAKI, K., SUGIURA, M.: Nucl. Acids Res. 14, 3143 (1986).

WARING, R.B., DAVIES, R.W.: Gene 28, 277-291 (1984). - WATSON, J.C., SURZYCKI, S.J.: Proc. Natl. Acad. Sci. USA 79, 2264-2267 (1982). - WENZEL, G., HOFFMANN, F., THOMAS, E.: Theor. Appl. Genet. 51, 81-86 (1977). - WHISSON, D.L., SCOTT, N.S.: Plant Mol. Biol. 4, 267-273 (1985).

YAMADA, K., SHINOZAKI, K., SUGIURA, M.: Plant Mol. Biol. 6, 193-199 (1986). - YAMADA, T., SHIMAJI, M.: Curr. Genet. 11, 347-352 (1987).

ZAITA, N., TORAZAWA, K., SHINOZAKI, K., SUGIURA, M.: FEBS Lett. 210, 153-156 (1987). - ZURAWSKI, G., BOTTOMLEY, W., WHITFELD, P.R.: Nucl. Acids Res. 12, 6547-6558 (1984). - ZURAWSKI, G., CLEGG, M.T.: Nucl. Acids Res. 12, 2549-2559 (1984).

Prof. Dr. RUDOLF HAGEMANN
Dr. MONIKA M. HAGEMANN
Dr. MICHAEL METZLAFF
Sektion Biowissenschaften
Wissenschaftsbereich Genetik
der Martin-Luther-Universität
Domplatz 1
DDR-4020 Halle (Saale)

VI. Population Genetics

By KLAUS WÖHRMANN and VOLKER LOESCHCKE

1. Introduction

During the last years rapid development has occurred in two fields of population genetics. New techniques in molecular biology made data available in prokaryotes and eukaryotes which could be analyzed at the population level. After about 20 years of enzyme variability research interest switched to DNA sequence variation, its origin and change within and between populations. The new techniques led to the discovery of interrupted genes, pseudogenes, transposed elements, nonhomologous recombination, sequence expansions and multigene families. This new information needs interpretation and incorporation into our knowledge about evolution.

On the other hand, population geneticists became aware again of the organism's phenotype as a unit which is based on the interaction of a great number of genes in which expression might be modified by exogenous and endogenous factors. Studies on fitness components, morphological, physiological and behavioral traits have appeared. These traits are generally quantitative traits determined by an unknown number of genes of which the contribution of each single gene to the trait is very small. Methods from "quantitative genetics" were improved and applied to tackle problems in population biology and evolution.

In this review we will concentrate on those publications from molecular biology and quantitative genetics which are of relevance to population and evolutionary genetics. For other recent reviews of interest in this connection see: BARKER and THOMAS (1987), DEVILLERS (1985), DÖRING and STARLINGER (1986), HARTL and DYKHUIZEN (1984), KRIEBER and ROSE (1986), LEWONTIN (1985), MITCHELL-OLDS and RUTLEDGE (1986), NEVERS et al. (1985), PALMER (1985), REANNEY and CHAMBAN (1985), SHAPIRO (1983), SYLVANEN (1984), WALBOT and CULLIS (1985).

2. DNA Variation and Population Genetics

a) Transposable Elements

Not long ago it was believed that the genome was a stable unit in which all genes are arranged on chromosomes in a defined order. The sequence of the genes could be altered only by chromosomal rearrangements and the gene itself by mutations leading to new alleles. Mutation and recombination were considered as the only source generating genetic variability and were believed to be adaptive. Recently, so-called transposable elements were described which generate further genetic varia-

Progress in Botany, Vol. 49

bility within the genome. A transposable element is a sequence of DNA which can move freely within the genome. It might be excised from one site and inserted at another site of the same or another chromosome. Transposable elements can promote their own multiplication at the expense of the fitness of the host organism. This observation led to the concept of "selfish DNA" (DOOLITTLE and SOPIENZA 1980; ORGEL and CRICK 1980). The elements do not need to provide a relative advantage to the host. If the replication rate of the transposon is greater than that of the genome, it becomes fixed in sexual reproducing populations. Mathematical models have shown that this might even happen if the element reduces the fitness of a host (HICKEY 1982). There is another hypothesis which is based on the assumption that hosts that give rise to successful variants will be favored. As transposable elements operate as a complex molecular apparatus during speciation they must in turn be favored (SYVANEN 1984).

Transposable elements may cause inactivation or activation of a gene, transpositions within the genome, structural changes on the chromosomes (duplications, deletions, insertions and translocations) and/or changes in the expression of genes, tissue specificity and developmental timing (FEDEROFF 1984; SCHWARZ-SOMMER et al. 1985). It is speculated whether the variability caused by transposons give rise to gross morphological changes which would lead to rapid macroevolutionary changes and thus support the saltationists view of evolution (GOULD and ELDRIDGE 1977).

The existence of transposable elements was first deduced in the 1940s by B. McCLINTOCK, after analyzing unusual phenotypic variation in the color of maize seeds. Only recently could these elements be isolated by the techniques of molecular biology, and the analysis of the DNA structure and sequences be conducted. Transposable elements are found in many plant species (NEVERS et al. 1985) and many of them have been biochemically identified, especially in *Zea mays* and, in addition, in *Zea mays* subsp. *parviglumis*, *Tripsacum dactyloides*, *Antirrhinum majus* and *Glycine maximus*. None has yet been described in the genetically well-known tomato (*Lycopersicum esculentum*) (DÖRING and STARLINGER 1986).

But not only plants may carry transposable elements. They were found in yeast, bacteria and animals (e.g. in *Drosophila*). From *Drosophila* we have obtained up to now the most extensive knowledge on the behavior of transposable elements and an impression of their importance at the population level. Therefore, a short review on the elements found in *Drosophila* will be given. Transposons could be shown in laboratory strains as well as in natural populations (ANANIEV et al. 1984; MONTGOMERY and LANGLEY 1983). Many of the transposons isolated in *D. melanogaster* were also found in other species. Sequence differences allowed conclusions on the phylogenetic origin (MARTIN et al. 1983; HUNT et al. 1984). DANIELS and STRAUNSBAUGH (1986) discussed in addition a horizontal transfer. Three types of transposable elements are distinguished by their intrinsic structure: (1) "copia", (2) "foldback" and (3) P-elements. The latter are of most interest because of their epidemic spread and their ability to cause hybrid dysgenesis which might be an important factor in speciation. As the transposition activity of P-elements is restricted to germ line cells, a rapid spread within the gene pool is possible without damage to the host by random mutations (ENGELS 1983). Comparative investigations of laboratory strains with natural populations showed no P-elements in the laboratory strains sampled before 1950. This observation was explained by a "random loss" (ENGELS 1981) during cultivation. However, KIDWELL (1983) postulated the "rapid invasion" hypothesis, which is supported by experiments of ANXOLABEHERE et al. (1986) and, moreover, by the distribution patterns found in natural populations. The investigation on natural populations led to the conclusion that the spreading of P-elements started in North America about 35 years ago where they were found everywhere, whereas the invasion into European populations is still ongoing. It is assumed that the P-elements invaded only recently *D. melanogaster* and *D. simulans* by horizontal gene transfer (BROOKFIELD et al. 1984). P-homologous sequences have also been found in the *D. obscura* group (ANXOLABEHERE et al. 1985).

Data comparable to natural populations of *Drosophila* are not yet available in plants. The role of transposable elements in the molecular evolution of organisms depends on the frequency of insertions by mobile elements into the genome and is limited by integration and excision. In maize the frequency of germinal and somatic excisions for the EN-I transposable element system is for instance 1% and 20%, respectively, and is assumed to be not rate-limiting for the generation of mutations. No information is available concerning the frequency of insertions. The possibility to obtain unbiased estimates of inserted frequencies is discussed by SCHWARZ-SOMMER et al. (1985). They proposed to estimate the frequencies of sequence alterations by using seeds stored for many years in comparison with fresh seeds of the same line.

Differences exist in evaluating the importance of transposons to the fitness of the host, which are related to a more "selectionistic" and "neutralistic" view (such as the hypothesis of "selfish DNA"). In *E. coli* at least one insertion sequence (IS10 and IS50) provides a selective advantage in growth chamber competition experiments (CHAO et al. 1983; CHAO and McBROOM 1985). The advantage of IS50R mutants is explained by the fact that they possess a direct growth advantage only under certain conditions, e.g. immediately after transfer into glucose-limiting chemostats (BIEL and HARTL 1983; HARTL et al. 1983).

For constructing models simplifying assumptions have to be made. These assumptions include the number of sites in the genome which can accept a transposon, the frequency of at least one element type, the transposition frequency of an element and the function at occupied sites with a deleterious effect on host fitness. Of course, the results depend upon the assumptions made. Several papers have been published on models assuming one or the other condition mentioned above and on data from prokaryotes and eukaryotes (e.g. *Drosophila*). Models on the population dynamics and evolution of transposons (CHARLESWORTH and CHARLESWORTH 1983, 1985; OHTA 1986; KAPLAN et al. 1985; LANGLEY et al. 1983), conditions under which self-regulated rates of transpositions can evolve in populations of transposable elements infecting sexually reproducing hosts (CHARLESWORTH and LANGLEY 1986) and methods estimating the parameters in the frequency spectrum for the evolution of transposable elements in finite Mendelian populations were proposed (KAPLAN and BROOKFIELD 1983). SLATKIN (1985) developed a model to predict the extent of genetic differentiation in a family of transposons. All the former papers refer to the populations of transposons alone in a Mendelian population. GOLDING et al. (1986) are interested in the invasion of panmictic sexual reproducing populations by a newly introduced transposon and described the time at which such an invasion will occur. A model of KING (1985) is concerned with the role of transposons in the evolution of eukaryotes.

b) Multigene Families

OHTA (1980) defined multigene families as a group of genes with the properties of "multiplicity, close linkage, sequence homology and related or overlapping functions". Multigene families are also called "gene clusters" or "gene complexes". Gene families may consist of highly uniform members, e.g. ribosomal RNA or histone genes, or those which are composed of quite diverse gene copies, e.g. immunoglobin genes.

They are known to be fairly widespread in the genomes of higher organisms. Gene families exhibit a so-called concerted evolution or coincidental evolution. In such families the members evolve in unison. They may be caused by the continued occurrence of unequal crossovers, gene conversions, duplications or transpositions. Multigene families may provide efficient means of accumulating genetic information for the evolution of complexity.

For a population geneticist and evolutionist the evolution and variation within and between families is of interest and, moreover, the influence on the variability of the host. A large number of mainly theoretical publications on this important topic exists. The reader is referred for further information to OHTA (1980, 1985, 1986a,b), the review by CLEGG and EPPERSON (1985) and for a different point of view to DOVER (1986).

c) Extranuclear Variation

Genetic information is not only located in the cell nucleus, but also in the organelles of the cell. Most of the works on organelle genome variation deal with qualitative differences in the mitochondrial DNA (mtDNA) and chloroplast DNA (cpDNA) between species and populations rather than with quantitative differences (PALMER 1985b).

The plant cpDNA seems to be more conservative in its structural organization than the mtDNA. The largest survey of intrapopulational variation for cpDNA in plants yields a maximum level of sequence divergence of 0.3% in barley (CLEGG et al. 1984). As a whole there exists only relatively little information on the variability in DNA sequences within and between plant populations. In barley (*Hordeum*) CLEGG et al. (1986) showed that cpDNA diversity has been restricted during domestication after comparing 11 lines of *Hordeum spontaneum* and nine lines of *H. vulgare*. Moreover, barley has a comparatively low level of cpDNA variation at the intraspecific level. PALMER et al. (1985a) compared cpDNA variation in wild and culture lines in the genus *Pisum*. From these and previous data, the authors conclude that the evolution of the cpDNA is much slower in comparison to the nuclear DNA. Two major lineages could be defined by cpDNA restriction site mutations: one includes all tested accessions of *P. fulonium* and the other includes cultivated lines of the garden pea, *P. sativum*, and a wild population of *P. humile*. On the basis of their data the authors speculate on the evolution of cultivated *P. sativum*.

In a theoretical study TAKAHATA and PALUMBI (1985) considered extranuclear differences and gene flow in a finite island model, using sequence information from extranuclear genomes. This allows estimates of relative interclone gene flow (extranuclear genomes are divided into a finite number of demes) when deme size is constant and genetic variation is selectively neutral. TAKAHATA (1984) investigated a single locus model of extranuclear genomes assuming within-generation drift caused by random transmission of multiple copy genomes during cell division.

d) Adaptive Value of DNA Variation

Since the first formulation of the neutral theory by KIMURA in 1968, it is still a matter of discussion whether nucleotide substitutions are of significance for the organism's fitness. KIMURA postulated that if the then known rate of substitutions was extrapolated to the entire genome, HALDANE's principle of the cost of natural selection could not be valid. The discussion of the neutral theory divided population geneticists and evolutionary geneticists into two groups, the "neutralists" and the "selectionists". Since its first formulation the neutral theory itself has evolved. Today, different allelic forms are no longer assumed to be strictly neutral. The theory should be renamed "mutation random-drift theory" or more precisely "the theory that the molecular level evolutionary changes and polymorphism are mainly due to mutations that are nearly enough neutral with respect to natural selection that their behavior and fate are mainly determined by mutation and random drift" (KIMURA 1983). Today, the theory is widely accepted and is able to explain most of the observed variation at the molecular level. However, the relationship between the molecular and morphological evolution is not yet understood. The apparent discrepancy between evolution on both levels is discussed by KING (1984). He assumes in a polygenic system that the background genotype interacts with the locus considered in a positive-feedback manner. Selective effects per locus are probably so small that they are unimportant in determining the fate of individual alleles. Polygenic variation is not in equilibrium but in a steady state of conditional turnover.

Several contributions on the problem of the selective importance of DNA substitutions have been published. MAYO and NANCOCK (1985) found that constant evolutionary rates are expected only in very large populations. Protein sequence data are not compatible with a molecular clock model based on the Poisson process. Using a Cox process, a Poisson process with randomly varying rates, it appears that the variance in the rates of molecular evolution is an episodic process. It is argued that this finding is more easily accounted for by the action of natural selection than by the neutral allele model (GILLESPIE 1986).

A method for quantitatively estimating the influence of point mutations and selection on the frequency of codons and amino acids (FOEMMEL and HOLZHUETTER 1985), stochastic models for treating geographically structured populations (MARUYAMA 1983) and a study on the estimation of the fraction of selective neutral alleles among newly arisen mutations (KIMURA 1983) have been published. Rates lower than 5×10^{-9} substitutions per site per year indicate the presence of constraints imposed by negative (natural) selection (JUKES and KIMURA 1984). The evolution of selective neutrality is discussed by HARTL et al. (1985).

e) DNA and Protein Variability in Populations of Microorganisms

Although bacteria have been a main object of molecular genetics for several decades, only few scientists were engaged in their genetics at the population level. It is generally accepted that genetic variability is a prerequisite for evolution. Bacteria are asexual organisms and have developed special types of gene transfer. These are: (1) transduction: the transfer of genetic information from one bacterium to another by a phage, (2) transformation: the conversion of the genotype of one bacterium by exposure of the cell to DNA isolated from bacteria having a different genotype and (3) conjugation: unidirectional transfer of DNA from a donor cell to a recipient cell. In the context of evolution it is of importance to know to which extent genetic variability can be maintained in bacterial populations.

More extended investigations on natural populations of bacteria were first conducted by MILKMAN (1973) using enzymatic variation. It was found that an unexpectedly high amount of average genic diversity exists in bacteria (0.23 over five enzyme loci). An extension of this work resulted in an estimated diversity of 0.47 (SELANDER and LEVIN 1980). Recently, this work was extended to species like *Shigella* and *Legionella* where, in addition, serotypes and outer membrane proteins were considered. The investigations showed clearly the clonal structure of the species and an extraordinary amount of genetic diversity (SELANDER 1985; MILKMAN 1985; REANNERY et al. 1983; SLATER 1985). Although the serotypes are generally different between humans and other animals, some serotypes can be found in both groups. Interclonal similarities between human and nonhuman isolates have also been found in electrophoretic studies (HARTL and DYKHUIZEN 1984).

E. coli consists of numerous, more or less independent clones. Recombination is observed, however, with a very low frequency. This could nevertheless have a very important effect on an evolutionary time scale and in rapidly changing environments. In addition, plasmids and transposons may be transferred among unrelated clones. An extensive migration rate is observed. Some clones appear to be distributed worldwide, whereas others are geographically more restricted. Although periodic selection is assumed to exist in natural populations, its relative importance to random genetic drift cannot be evaluated on the basis of the available data. In most cases electrophoretic alleles from natural populations have been found to be selectively neutral or nearly neutral, but sometimes the selective value was shown to depend on the environmental conditions and the genetic background (HARTL and DYKHUIZEN 1985a,b). Corresponding results have already been found for yeast by STROBEL and WÖHRMANN (1977).

A tremendous amount of variability and a rapid change of nucleotides have been reported by PALESE (1986) for influenza A-virus which changed most rapidly in comparison to types B and C. Isolates obtained since 1933 exhibit an accumulation of 0.22–0.34% nucleotide changes per year in their NS genes. This means an evolutionary rate for influenza A-virus of $2.2–3.4 \times 10^{-3}$ substitutions per site per year. This is 10^6-fold higher than values found for eukaryotic genes such as globulin.

3. The Impact of Release of Genetically Manipulated Organisms

a) The Problem

At the Cordon Research Conference on Nucleic Acids in 1973 the potential risk resulting from the release of genetically engineered organisms into the environment was first discussed before a broader audience. Since that time the controversy has continued among those who are engaged in genetic engineering and therefore who are more optimistic with regards to the impact of a release than those who are afraid of damage to the existing ecological systems. The actual state of the discussion is well reflected by the contributions to an interdisciplinary symposium in Philadelphia on the "Engineered Organisms in the Environment" (HALVORSON et al. 1985). In one point the contributors to the symposium agreed: requests for environmental release should be based on a benefit-risk analysis in a process of case-by-case consideration. However, the definitions of benefit and risk seem more likely to depend on individual assessment than on objective criteria. To forecast the fate of a released, engineered organism, one has to have fundamental knowledge of the processes going on in the course of evolution. But "natural selection is perhaps the least understood process in evolutionary biology" (HARTL 1985) and, similarly, ENDLER (1986) pointed out that we are "most remarkably ignorant about ecological aspects of natural selection".

The same is true for the definition of risk. Is it, e.g. a risk, if a cultivated species like maize or beet becomes a weed? Or must a hazard be much worse than plagues such as bubonic plague, smallpox or acquired immune deficiency? Indeed, the risk is very limited if one advocates the view that if "the very worst and most implausible outcome occurs in which a released organism becomes a potential danger, the situation would surely be amenable to containment or control by using the methods that have already been developed in the public health and agriculture for the control of natural parasites and pests" (HARTL 1985). On the background of our poor knowledge on the factors which are involved in evolution or which are effective in the ecosystem, it is at least difficult to predict the outcome of a release of genetically engineered organisms into nature. So scientists can argue only on the basis of past experiences, because up to now an engineered organism has never been released (KOWARIK and SUKOPP 1986).

There is a long tradition behind crop plant engineering, i.e. plant breeding, which goes back to primitive man's effort at preserving seeds to keep them for planting the following season. Since the turn of the century, simple selection methods have been supplemented by techniques such as hybridization (to recombine genetic material) and mutagenesis (to induce new genetic variability). So far, modern technology of gene engineering is just a new technique to combine genetic information from different organisms. But there is a fundamental difference: in the past the combined material was derived from individuals of the same or at least from very closely related species within the same genus. New technology, however, enables plant engineers to combine genetic material from unrelated genera, and gene combinations might arise which have never existed before. As an organism represents more than the sum of its genes, and because genes interact with one another and with the environment in a complex way (HARTL 1985), it is a matter of believe whether in each case "the addition of a few characterized genes to a plant results in properties that are relatively easy to predict" (BRILL 1985a,b).

b) Some Goals of Genetic Engineering

Scientists at the University of California, Berkeley, e.g. have developed recombinant DNA up to the point where it could be field tested. They found bacteria which act as "nucleation centers" for the formation of crystals such that they prevent the freezing of plant tissues. The genes responsible in nucleation have been transferred into *Pseudomonas syringae*. It is hoped that the altered bacteria sprayed onto the plant can compete with and replace the natural varieties to prevent damage to plants by frost (LINDOW 1985). For a long time, farmers have used preparations of *Bacillus thuringensis* to protect plants (cabbage, cotton, etc.) against attacks by lepidopteran insects. To save repeated spraying the gene responsible for the organism's "toxicity" was transferred to *Pseudomonas fluorescens*, which is in contact with plant roots. Applied to seed or soil at the time of planting, the engineered organism could thus provide long-term protection against soilborne pests (WATRUD et al. 1985). The possibility to transfer genes taken from influenza, hepatitis B and other viruses into the vaccinia virus genome which was used for immunization against smallpox is under examination. This would allow the fabrication of "polyvalent" vaccines. One shot would confer immunity against several diseases (MOSS and BULLER 1985). Another goal is concerned with the engineering of "leaching bacteria" which are modified or selected microbes. Introduced into orebodies or mine wastes, they are able to recover metals (BRIERLEY 1985).

There are, of course, numerous potential applications (listed in ANONYMOUS 1985) for genetic engineering in changing the genetic constitution of organisms. These include the introduction of new quality characters such as endosperm storage proteins and increased tolerance to environmental stresses such as high and low temperature, salinity, drought or resistance to pests caused by animals or microorganisms. However, in some cases the potential developments seem to be limited by our ability to develop delivery systems for the genetic information and methods for isolating and controlling the expression of the DNA for various characters.

One of the perhaps most controversial applications of genetic engineering is the transfer of herbicide-resistant genes into crops. Herbicides have come to play an important role in modern agriculture. Plant species differ in their sensitivity to phytotoxic chemicals and therefore in their response to herbicide treatments. This makes herbicides "useful" because of their differential selection of weeds and crops. Herbicides, however, create also a selection process on the weed population, whereby resistant weed genotypes are selected. Consequently, the same problems arise as are known from the treatment of pest populations by fungicides and insecticides (see below). Furthermore, several herbicides sprayed in the USA corn belt persist sufficiently long in the soil to affect, e.g. soybeans sown the next year. A solution is proposed by the development of herbicide-tolerant crops using recombinant DNA techniques. Genes for herbicide resistance could come from a far wider range of donors. There are efforts to transform plants (tobacco, tomato) with DNA from a mutant of the *Salmonella typhimurium* that carries a resistant form of the target enzyme EPSP synthase (GOODMAN and NEWEL 1985).

c) Discussion of Some Arguments

It is not aimed here to discuss whether the described goals of genetic engineering are convenient or significant. Attention shall rather be directed to problems in which population genetics is involved.

α) The Niche. Some ecologists are afraid of the release of engineered organisms because of the potential risk that these organisms will get out of man's control and become established in the ecosystem causing damage to the environment. Gene engineers argue that the new organisms are very poor competitors and, moreover, all available niches are already occupied or "full".

The term niche is a verbally formulated ecological concept and was first used 1917 by GRINNELL in an attempt to explain the geographical distribution of the Californian thrasher as due to adjustments of the bird to specific environmental conditions (LOESCHCKE 1987). Subsequently, the term has been used in a variety of contexts and the definitions remained vague until HUTCHINSON (1957) gave a more mathematical, tractable definition: "The fundamental niche may be regarded as a set of points in an abstract n-dimensional space". A realized niche is then that part of the fundamental niche in which a population is not excluded by other species. Later, it was defined as a fitness measure in an environmental space (LEVINS 1968) in a hypervolume where each point corresponds to the probability of survival and reproduction. This definition of a niche made the concept more dynamic. Niches could now be modified by natural selection, and adaptive changes could be considered in terms of changes in the biotic and abiotic environment. However, a niche is still a theoretical concept, and in practice a niche can only be recognized with reference to an existing species. The rate of evolution is limited by ecological opportunities in the form of new niches (WRIGHT 1978). If, however, all niches are "full", no evolution can occur and the spread of species after introduction into a new environment, as has occurred very often in the past, would not be possible (SIMBERLOFF 1985; PIMENTEL 1985). There are many well-known and often cited examples. The spread of rabbits in Australia after introduction by British immigrants, the case of prickly pear and cactus on the same continent, the oyster problem and the damage caused by the salt cedar in the USA are only a few examples. For further information, see KOWACK and SUKOPP (1986) and SUKOPP and TREPL (1987). All these organisms found a suitable "niche", i.e. a set of favorable environmental conditions to which they became particularly well adapted.

β) Introgression. Ecologists fear that due to horizontal gene flow engineered genes might introgress into the wild relatives of our crop plants and, e.g. in the case of herbicide resistance, become resistant, too, and escape control by herbicides. The introgression of genes into wild species presumes a horizontal gene flow between populations and species. SHAPIRO (1985) claimed that bacteria live like eukaryotes in organized populations and that genetic interactions are part of intercellular communications. Consequently, transfer and reorganization of DNA molecules are viewed as an operation of adaptive mechanisms. It is necessary to think of bacterial populations as biological units. But knowledge is still insufficient on how bacteria interact with one another and with their natural environment. The temporal and spatial contact between microbial species may result in lower rates of transfer of particular genes between different species. Nevertheless, the case can be made that if a gene can be transferred between individuals under laboratory conditions, then somewhere, at some time, a similar event may occur in nature (SLATER 1985).

Interspecific crosses have occurred in the evolution of plants and have resulted in amphidiploid bastards. The crops maize and wheat, for instance, are such species. Most crops have weed relatives and weed races. These are known from annuals. But also weedy forms of perennial crops are not rare (e.g. *Citrus*, *Punica*, *Musa*, *Prunus*, etc.). In certain areas (east and southern Africa) hybridization occurs between wild and cultivated races of *Sorghum bicolor*. The hybrid weeds are very aggressive and cause serious crop losses. There is a tendency for local weed to "mimic" the particular race of cultivated sorghum with respect to the type of inflorescences that it infests (DEWET et al. 1976). The mimicry is partly due to gene flow and partly due to selective pressure applied by weed control. *Sorghum halepense*, which can be a very aggressive and troublesome weed in the USA, is assumed to have evolved by means of introgression of cultivated diploid sorghum into *S. halepense* introduced from the Mediterranean area to the USA. In its homeland *S. halepense* is a species which lacks vigor and aggressiveness (HARLAN 1983).

Pearl millet (*Pennisetum typhoides*) is an outcrosser. Its cultivated and its spontaneous form coexist under normal field conditions in Africa. Crosses between both forms are very easily conducted (JOLY and SARR 1985). There are mimetic races of maize, wheat, barley, oats, rye, pearl millet (*Pennisetum glaucum*) and finger millet (*Eleusine coracana*). Genetic interactions between wild weed and cultivated races are practically universal (HARLAN 1983). Crop weeds are by no means only confined to cereals (e.g. *Rhaphanus*, *Daucus*, *Gossipium*, *Glycine*, *Solanum*, etc.).

γ) Herbicide Resistance. The use of herbicides became extremely important in agriculture because it reduces the frequency and thus the cost of tillage to control weeds. A crop should be resistant to the herbicide and the weed should be susceptible. Fortunately, a variation exists in the response to herbicides within and between species to allow a selective treatment in the field. Triazine is one of the herbicides used in agriculture and several species exhibit tolerant genotypes for the chemical, e.g. *Amaranthus hybridis*, *A. powellii*, *A. retroflexus* (AHRENS et al. 1981), *Poa annua* (DARMENCY and GASQUEZ 1981), *Senecio vulgaris* (WARWICK 1981), and *Chenopodium album* (AL MOUEMAR and GASQUEZ 1983). Variation for atrazine tolerance was observed in *Chenopodium album* and *Ch. strictum* (WARWICK and BLACK 1981), *Amaranthus retroflexus* (SOLYMOSI and KOSTYAL 1985), for diazinon in *Lucilia caprina* (McKENZIE and PURVIS 1984), for simazine in *Senecio vulgaris* (HOLLIDAY and PUTWAIN 1981; SCOTT and PUTWAIN 1981) and, last but not least, for erodicane in *Zea mays* (LANDI et al. 1981). Steady application of herbicides leads to tolerant plant populations (HOLLIDAY and PUTWAIN 1981; JANA and NAYLOR 1982; THAI et al. 1985; SOLYMOSI and KOSTYAL 1985). An increase in herbicide tolerance by continuous spraying was accompanied by a reduction of enzyme polymorphism in *Chenopodium album* (AL MOUEMAR and GASQUEZ 1983). For the spread of tolerance genes in a population, the outcrossing rate of the weed is of importance. *Poa annua* exhibits an outcrossing rate of 13%. The inheritance of resistance is maternal. In only 0.1% of cases was transmission via pollen observed. Paternal transmission is therefore not important for the flow of tolerant genes in *P. annua* (DARMENCY and GASQUEZ 1981). Maternal inheritance of simazine resistance in *Senecio vulgaris* was reported by SCOTT and PUTWAIN (1981). In maize no reciprocal differences were found for erodicane resistance. Maternal inheritance is supposed to increase the spread of resistance genes in a population. Modifications in the genetic background produce changes in the fitness of genotypes at the diazinon resistance locus in *Lucilia caprina* (McKENZIE and PURVIS 1984).

4. Quantitative Genetics

Ample evidence exists of genetic variation for fitness components (survivorship, time to age at first reproduction, longevity and fertility), and of morphological, physiological and behavioral traits in natural populations (for references, see reviews by MITCHELL-OLDS and RUTLEDGE 1986 and by BARKER and THOMAS 1987). Although some of this variation is thought to be due to one or a few loci (GOTTLIEB 1984, but see COYNE and LANDE 1985 and MORTON 1982), most of the variation is probably polygenic. Recently, evolutionary biologists have focussed much attention to explain the organization and maintenance of quantitative genetic variation (LANDE 1976, 1980; BULMER 1980; TURELLI 1984, 1985, 1986; MITCHELL-OLDS and RUTLEDGE 1986; CLARK 1987; among others). Here, we will focus on the actual debate concerning the maintenance of quantitative genetic variation. Furthermore, a few other aspects connected to quantitative genetics are considered.

a) Maintenance of Quantitative Genetic Variation

There is some agreement that the neutral theory of molecular evolution does not apply to the evolution of morphological traits or of life-history characters (NEY and GRAUR 1984), and some type of balancing selection is generally assumed to maintain quantitative genetic variation (BARKER and THOMAS 1987).

α) Mutation-Selection Balance. LANDE (1976, 1980) analyzed mutation (stabilizing)-selection models to explain observed levels of quantitative genetic variation in natural populations in combination with a review of the data (LANDE 1976). His contributions received much attention and initiated a renewal of interest in theoretical quantitative genetics (which now was considered as a potential link between genetics, ecology and evolution). The type of models LANDE analyzed is now generally referred to as K-L-F models (TURELLI 1984) after

KIMURA (1965), LANDE (1976) and FLEMING (1979). TURELLI (1984, 1985, 1986) started the critical examination of assumptions in the K-L-F models and constructed mutation-selection models which he considered as being more biologically plausible. The assumptions of the K-L-F models are listed carefully by CLEGG and EPPERSON (1985) and are discussed in detail by BARKER and THOMAS (1987). Controversial opinions concern among others per locus mutation rates, effects of new mutations and the nature and intensity of selection (BARKER and THOMAS 1987). The K-L-F models as well as those of TURELLI all predict high levels of genetic variation at equilibrium by mutation-selection balance.

β) Other Evolutionary Factors. Mutation and selection seem to be obvious factors to be taken into account when explaining the maintenance of quantitative genetic variation. However, also other models, not including these factors, predict the maintenance of quantitative genetic variation. In the context of life-history evolution, ROSE et al. (1987) compared mutation-selection balance models with antagonistic pleiotropy and optimization models to explain the maintenance of variation in life-history characters or fitness components and their evolution. They concluded that both mutation-selection balance and antagonistic pleiotropy may play a considerable role in these processes. GILLESPIE (1984), ignoring mutation, suggests that pleiotropic overdominance provides an alternative to the mutation-selection balance hypothesis in explaining the maintenance of quantitative genetic variation. LYNCH and HILL (1986) derived a general model for predicting the genetic variance within populations for quantitative traits as a generalization of the neutral theory to the phenotypic level. They assume no selection and the joint action of drift and mutation.

γ) Genetic Variation in Heterogeneous Environments. The importance of heterogeneous environments in maintaining genetic polymorphism is still under discussion (for review, see HEDRICK 1986 and ENNOS 1983), and it is still far from clear what proportion of polymorphism is maintained by the heterogeneity of the environment. The literature, with regards to this topic, refers more to single locus variation than to quantitative variation. However, due to the greater simplicity of the system, the genetics is much more specific in these kinds of investigations and some of the results of these studies might have implications that should be recognized when investigating quantitative variation.

δ) Experimental Evaluation. In general, it can be said that there is no lack of possible explanations for the maintenance of quantitative genetic variation in the form of models, but rather a need for their experimental evaluation and assessment of relative importance (BARKER and THOMAS 1987). The experimental evaluation, however, is difficult to attain, as available techniques in quantitative genetics cannot be applied in natural populations without perturbations (but see ROACH 1986). When organisms are brought to the laboratory or greenhouse, new gene by environment interactions may result. One way around the problem of inducing gene by environment interactions would be to rear organisms under various environmental conditions to obtain the norm of reaction: i.e. the range of potential phenotypes, a single genotype could develop if exposed to a range of environmental conditions (WOLTERECK 1909; see also GUPTA and LEWONTIN 1982 and STEARNS and KOELLA 1986). This, however, creates a new problem. In quantitative genetics the environment has usually been treated as discrete, whereas most environmental variations experienced by organisms are probably continuous (for a detailed description of this problem, see VAN NOORDWIJK and GEBHARDT 1987).

b) Quantitative Variation and the Analysis of Individual Loci

A correlation between heterozygosity and quantitative traits has been reported in a number of studies (MITTON and GRANT 1984; ZOUROS and FOLTZ 1984; among others). These observations are often explained with reference to the concept of developmental homeostasis (LERNER 1954).

Population differences in quantitative characters may not be necessarily connected to differences in gene frequencies at individual loci (LEWONTIN 1984 and discussion by ROGERS 1986; FELSENSTEIN 1986 and LEWONTIN 1986). If the variation is neutral at both levels, population differentiation is expected to be more easily detected at the level of individual loci (LEWONTIN 1986).

c) Phenotypic Plasticity

A change in the average phenotype expressed by a given genotype in different environments is called phenotypic plasticity. The modification of the phenotype in response to different environmental conditions is thought to provide an advantage over a constant phenotype. Phenotypic plasticity can evolve so that the adaptation to variable environments increases (for review, see SCHLICHTING 1986). MARSHALL et al. (1986), for instance, showed that different, closely related *Sesbania* species (Leguminoseae) differed in the amount and pattern of plasticity in yield components, whereas within species half-sib families differed only in the amount, but not in the kind of plastic response. Quantitative genetic models on the evolution of phenotypic plasticity are presented by VIA and LANDE (1985) and VIA (1987).

d) Genetic Correlations

Genetic connections between characters were recognized long ago. When selecting for one character, another responds in a predictable manner:

"Hairless dogs have imperfect teeth; long-haired and coarse-haired animals are apt to have, as is asserted, long or many horns; pigeons with feathered feet have skin between their toes... Hence if man goes on selecting, and thus augmenting, any peculiarity, he will almost certainly modify unintentionally other parts of the structure, owing to the mysterious laws of correlation" (CH. DARWIN 1859).

Genetic correlations can be due to linkage and to pleiotropy. However, which of these factors is responsible in a given case will not be revealed in a typical quantitative genetic analysis (CLARK 1987). The evolutionary consequences of the two causes may be quite distinct. Genetic correlations are neither fixed nor absolute: there are genes that act on several characters independently, and there are genes that act in the same or in opposite directions. Constructional features in the developmental system may impose constraints on the selection response (SCHARLOO 1987, see also MAYNARD SMITH et al. 1985).

During the last years a large number of studies has appeared that concern genetic and phenotypic covariance structures (see VIA 1984 for a study on natural populations and MITCHELL-OLDS and RUTLEDGE 1986 for a recent review).

References

AHRENS, W.H., WAX, L.M., STOLLER, E.M.: Weed Sci. 29, 345-348 (1981). - AL MOUEMAR, A., GASQUEZ, J.: Weed Res. 23, 141-150 (1983). - ANANIEV, E.V., BARSKY, U.E., ILGIN, Y.V.: Chromosoma 90, 366-377 (1984). - ANONYMOUS: Newswatch 5, 7-8 (1985). - ANXOLABEHERE, D., NOUAUD, D., PERIQUET, G., TCHEN, P.: Proc. Natl. Acad. Sci. USA 82, 5418-5422 (1985). - ANXOLABEHERE, D., NOUAUD, D., PERIQUET, G., RONSSERAY, S.: Genetica 69, 81-95 (1986).

BARKER, J.S.F., THOMAS, R.H.: A quantitative genetic perspective on adaptive evolution, 3-23. In: Genetic Constraints on Adaptive Evolution, ed. V. LOESCHCKE. Berlin, Heidelberg, New York: Springer 1987. - BIEL, S.W., HARTL, D.L.: Genetics 103, 581-592 (1983). - BRIERLEY, J.A.: Use of microorganisms for mining metals, 141-146. In: see ref. HALVORSON et al. 1985. - BRILL, W.J.: Science 227, 381-384 (1985a); - ibid. 229, 115-117 (1985b). - BROOKFIELD, J.F.Y.: Genetics 112, 392-408 (1986). - BROOKFIELD, J.F.Y., MONTGOMERY, E., LANGLEY, C.H.: Nature (Lond.) 310, 330-332 (1984). - BULMER, M.G.: The Mathematical Theory of Quantitative Genetics. Oxford: Clarendon Press 1980.

CAMPBELL, A.: Transposons and their evolutionary significance, 258-279. In: Evolution of Genes and Proteins, eds. M. NEI, R.K. KOEHN. Sunderland, MA: Sinauer 1983. - CHAO, L., McBROOM, S.M.: Mol. Biol. Evol. 2, 359-369 (1985). - CHAO, L., VARGAS, C., SPEAR, B.B., COX, E.C.: Nature (Lond.) 303, 633-635 (1983). - CHARLESWORTH, B.: Genetics of transposable elements. In: see ref. OHTA and AOKI 1985. - CHARLESWORTH, B., CHARLESWORTH, D.: Genet. Res. 42, 1-28 (1983). - CHARLESWORTH, B., LANGLEY, C.H.: Genetics 112, 359-383 (1986). - CLARKE, A.G.: Genetic correlations: the quantitative genetics of evolutionary constraints, 25-45. In: Genetic Constraints on Adaptive Evolution, ed. V. LOESCHCKE. Berlin, Heidelberg, New York: Springer 1987. - CLEGG, M.T., EPPERSON, D.K.: Adv. Genet. 3, 235-269 (1985). - CLEGG, M.T., BROWN, A.H.D., WHITFIELD, P.R.: Genet. Res. 43, 339-343 (1984). - CLEGG, M.T., RITLAND, K., ZURAWSKI, G.: Processes of chloroplast DNA evolution, 275-294. In: see ref. KARLIN and NEVO 1986. - COLWELL, R.K., NORSE, E.A., PIMENTEL, D., SHARPLES, F.E., SIMBERLOFF, D.: Science 229, 111-115 (1985). - COYNE, J.A., LANDE, R.: Am. Nat. 126, 141-145 (1985).

DANIELS, S.B., STRAUSBAUGH, L.: J. Mol. Evol. 23, 138-148 (1986). - DARMENCY, H., GASQUEZ, J.: New Phytol. 89, 487-494 (1981). - DARWIN, CH.: The origin of species, quoted after Mentor Ed. New York: Times Mirror 1859. - DEVILLERS, C.: Annee Biol. 24, 153-178 (1985). - DEWET, J.M.J., HARLAN, J.R., PRICE, E.G.: Variability in *Sorghum bicolor*, 453-464. In: Origins of African Plant Domestications, eds. J.R. HARLAN, J.M.J. DEWET, A.B.L. STEMLER. The Hague: Mounton 1976. - DÖRING, H.P., STARLINGER, P.: Annu. Rev. Genet. 20, 175-200 (1986). - DOOLITTLE, W.F., SOPIENZA, C.: Nature (Lond.) 284, 601-603 (1980). - DOVER, G.A.: The spread and success of Non-Darwinian novelties, 199-238. In: see ref. KARLIN and NEVO 1986. - DROPPA, M., HORVATH, G., DEMETER, S.: Z. Pflanzenphysiol. 105, 75-80 (1981).

ENDLER, J.A.: Natural Selection in the Wild. Princeton: Princeton Univ. Press 1986. - ENGELS, W.R.: Cold Spring Harbor Symp. Quant. Biol. 45, 561-565 (1981); - Annu. Rev. Genet. 17, 315-344 (1983). - ENNOS, R.A.: Evol. Biol. 16, 129-155 (1983).

FEDEROFF, N.: Sci. AM. 250, 64-74 (1984). - FELSENSTEIN, J.: Am. Nat. 127, 731-732 (1986). - FENCHEL, T.M.: Oecologia (Berl.) 20, 19-32 (1975). - FLEMING, W.H.: SIAM J. Appl. Math. 36, 148-168 (1979). - FOEMMEL, C., HOLZHUETTER, H.G.: J. Mol. Evol. 21, 233-257 (1985).

GASQUEZ, J.: Breeding systems and genetic structure of *Chenopodium album* populations according to crop and herbicide rotation, 56-66. In: Genetic Differentiation and Dispersal in Plants, eds. P. JACQUARD, G. HEIM, J. ANTONOVICS. Berlin, Heidelberg, New York: Springer 1986. - GILLESPIE, J.H.: Genetics 107, 321-330 (1984); - Statistical aspects of the molecular clock, 255-274. In: see ref. KARLIN and NEVO 1986. - GINZBURG, L.R., BINGHAM, P.M., YOO, S.: Genetics 107, 331-341 (1984). - GOLDING, G.B., AQUADRO, C.F., LANGLEY, C.H.: Proc. Natl. Acad. Sci. USA 83, 427-431 (1986). - GOODMAN, R.M., NEWELL, N.: Genetic engineering of plants for herbicide resistance: Status and prospect, 47-53. In: see ref. HALVORSON et al. 1985. - GOTTLIEB, L.D.: Am. Nat. 123, 681-709 (1984). - GOULD, S.J., ELDREDGE, N.: Paleobiology 3, 115-151 (1977).

HALVORSON, H., PRAMER, D., ROGAL, M., eds.: Engineered Organisms in the Environment: Scientific issues. Washington, DC: Acad. Soc. Microbiology 1985. - HARLAN, J.R.: Some merging of plant populations, 267-276. In: Genetics and Conservation, eds. CH.M. SCHONEWALD-COX, ST.M. CHAMBERS, B. McBRYDE, W.L. THOMAS. London: Benjamin 1983. - HARTL, D.L.: Engineered organisms in the environment: Interference from population genetics, 83-87. In: see ref. HALVORSON et al. 1985. - HARTL, D.L., DYKHUIZEN, D.E.: Annu. Rev. Genet. 18, 31-68 (1984); - The neutral theory and the molecular basis of preadaptation, 107-124. In: see ref. OHTA and AOKI 1985. - HARTL, D.L., DYKHUIZEN, D.E., MILL, R.D., GREEN, L., DEFRAMOND, J.: Cell 35, 503-510 (1983). - HARTL, D.L., DYKHUIZEN, D.E., DEAN, A.M.: Genetics 111, 655-674 (1985). - HEDRICK, PH.W.: Annu. Rev. Ecol. Syst. 17, 535-366 (1986). - HICKEY, D.A.: Genetics 101, 519-531 (1982). - HOLLIDAY, R.J., PUTWAIN, P.D.: J. Appl. Ecol. 17, 779-792 (1981). - HUNT, J.A., BISHOP, J.G., CARSON, H.L.: Proc. Natl. Acad. USA 81, 7146-7150 (1984). - HUTCHINSON, E.G.: Cold Spring Harbor Symp. Quant. Biol. 22, 415-427 (1957).

JANA, S., NAYLOR, J.M.: Can. J. Bot. 60, 1611-1617 (1982). - JOLY, H., SARR, A.: Preferential associations among characters in crosses between Pearl millet (*Pennisetum typhoides*) and its wild relatives, 95-111. In: Genetic differentiation and dispersal in plants, eds. :. JACQUARD, G. HEIM, J. ANTONOVICS. Berlin, Heidelberg, New York: Springer 1985. - JUKES, T.H., KIMURA, M.: J. Mol. Evol. 21, 99-92 (1984).

KAPLAN, N.L., BROOKFIELD, J.F.Y.: Genetics 104, 485-496 (1983). - KAPLAN, N., DARDEN, T., LANGLEY, C.H.: Genetics 109, 459-480 (1985). - KARLIN, S., NEVO, E., eds.: Evolutionary Processes and Theory. New York: Academic Press 1986. - KIDWELL, M.C.: Proc. Natl. Acad. Sci. USA 80, 1655-1659 (1983). - KIMURA, M.: Proc. Natl. Acad. Sci. USA 54, 731-736 (1965); - Nature (Lond.) 217, 624-626 (1968); - The Neutral Theory of Molecular Evolution. Cambridge: Cambridge Univ. Press 1983a; - Mol. Biol. Evol. 1, 84-93 (1983b). - KING, C.C.: J. Theor. Biol. 114, 447-462 (1985). - KING, J.L.: Evol. Theor. 7, 73-79 (1984). - KOWARIK, I., SUKOPP, H.: Ökologische Folgen der Einführung neuer Pflanzenarten, 111-135. In: Die ungeklärten Gefahrenpotentiale der Gentechnologie, eds. R. KOLLECK, B. TAPPESER, G. ALTNER. München: Schweizer 1986. - KRIEBER, M., ROSE, M.R.: Annu. Rev. Ecol. Syst. 17, 465-485 (1986).

LANDE, R.: Genet. Res. 26, 221-235 (1976); - Genetics 94, 203-215 (1980). - LANDI, P., CATIZONE, P., CONTI, S.: Riv. Agron. 15, 85-94 (1981). - LANGLEY, C.H., BROOKFIELD, J.F.Y., KAPLAN, N.: Genetics 104, 457-472 (1983). - LERNER, I.M.: Genetic Homeostasis. Edinburgh: Oliver and Boyd 1954, - LEVINS, R.: Evolution in Changing Environments. Princeton: Princeton Univ. Press 1968. - LEWONTIN, R.C.: Am. Nat. 123, 115-124 (1984); - Annu. Rev. Genet. 19, 81-102 (1985); - Am. Nat. 127, 733-734 (1986). - LINDOW, S.E.: Ecology of *Pseudomonas syringae* relevant to the field use of ice deletion mutants constructed in vitro for plant frost control, 23-35. In: see ref. HALVORSON et al. 1985. - LOESCHCKE,

V.: Niche structure and evolution in ecosystems, 320-332. In: Ecological Studies, Vol. 61, eds. E.D. SCHULZE, H. ZWÖLFER. Berlin, Heidelberg, New York: Springer 1987a. - LYNCH, M., HILL, W.G.: Evolution 40, 915-935 (1986).

MARSHAL, D.L., LEVIN, D.A., FOWLER, N.L.: Am. Nat. 127, 508-521 (1986). - MARTIN, G., WIERNASZ, D., SCHEDL, P.: J. Mol. Evol. 19, 203-213 (1983). - MARUYAMA, T.: Bull. Math. Biol. 45, 521-554 (1983). - MAYO, O., HANCOCK, T.W.: Experientia 41, 1080-1081 (1985). - McKENZIE, J.A., PURVIS, A.: Heredity 53, 625-634 (1985). - MILKMAN, R.: Science 182, 1024-1026 (1973); - Two elements of unified theory of population genetics and molecular evolution, 65-84. In: see ref. OHTA and AOKI 1985. - MITCHELL-OLDS, T., RUTLEDGE, J.J.: Am. Nat. 127, 379-402 (1986). - MITTON, J.B., GRANT, M.C.: Annu. Rev. Ecol. Syst. 15, 479-499 (1984). - MONTGOMERY, E.A., LANGLEY, C.: Genetics 104, 473-483 (1983). - MOSS, B., BULLER, M.L.: Vaccina virus vectors: Potential use as live recombinant virus vaccines, 36-39. In: see ref. HALVORSON et al. 1985. - MUKAI, T.: Experimental verification of the neutral theory, 125-146. In: see ref. OHTA and AOKI 1985.

NEI, M., GRAUR, D.: Evol. Biol. 17, 73-118 (1984). - NEVERS, P., SHEPHERD, N.S., SAEDLER, H.: Adv. Bot. Res. 12, 102-203 (1985).

OHTA, T.: Evolution and Variation of Multigene Families. Berlin, Heidelberg, New York: Springer 1980; - Genetic variation of multigene families, 233-242. In: see ref. OHTA and AOKI 1985; - Population genetics theory of multigene families with emphasis on genetic variation contained in the family, 239-254. In: see ref. KARLIN and NEVO 1986a; - Genetics 113, 145-160 (1986b). - OHTA, T., AOKI, K., eds.: Population Genetics and Molecular Evolution. Berlin, Heidelberg, New York: Springer 1985. - ORGEL, L.E., CRICK, F.H.C.: Nature (Lond.) 284, 604-607 (1980).

PALESE, P.: Rapid evolution of human influenza viruses, 53-68. In: see ref. KARLIN and NEVO 1986. - PALMER, J.D.: Annu. Rev. Genet. 19, 325-354 (1985). - PALMER, J.D., JORGENSEN, R.A., THOMSON, W.F.: Genetics 109, 195-213 (1985). - PIMENTEL, D.: Using genetic engineering for biological control, 129-140. In: see ref. HALVORSON et al. 1985.

REANNEY, D.C., CHAMBON, P.: Int. Rev. Cytol. 93 (1985). - REANNEY, D.C., GOWLAND, P.C., SLATER, J.H.: Genetic interactions between microbial communities, 379-421. In: Microbes in Their Natural Environments, eds. J.H. SLATER, R. WHITTENBURY, J.W.T. WIMPENNY. Cambridge: Cambridge Univ. Press 1983. - ROACH, D.A.: Am. Nat. 128, 47-57 (1986). ROGERS, A.R.: Am. Nat. 127, 729-730 (1986). - ROSE, M.R., SERVICE, P.M., HUTCHINSON, E.W.: Three approaches to trade-offs in life-history evolution, 91-105. In: Genetic Constraints on Adaptive Evolution, ed. V. LOESCHCKE. Berlin, Heidelberg, New York: Springer 1987.

SCHARLOO, W.: Constraints in selection response, 125-149. In: Genetic Constraints on Adaptive Evolution, ed. V. LOESCHCKE. Berlin, Heidelberg, New York: Springer 1987. - SCHLICHTING, C.D.: Annu. Rev. Ecol. Syst. 17, 667-693 (1986). - SCHWARZ-SOMMER, Z., GIERL, A., CUYPERS, H., PETERSON, P., SAEDLER, H.: EMBO J. 4, 591-597 (1985). - SCOTT, K.R., PUTWAIN, P.D.: Weed Res. 21, 137-140 (1981). - SELANDER, R.K.: Protein polymorphism and the genetic structure of natural populations of bacteria, 85-106. In: Population Genetics and Molecular Evolution, eds. T. OHTA, K. AOKI. Berlin, Heidelberg, New York: Springer 1985. - SELANDER, R.K., LEVIN, B.R.: Science 210, 545-547 (1980). - SHAPIRO, J.A., ed.: Mobile Genetic Elements. New York: Academic Press 1983; - Communication and genetic change in bacterial populations, 63-69. In: see ref. HALVORSON et al. 1985. - SIMBERLOFF, D.: Predicting ecological effects of novel entities, 152-161. In: see ref. HALVORSON et al. 1985. -

SLATER, J.H.: Gene transfer in microbial communications, 89-98. In: see ref. HALVORSON et al. 1985. - SLATKIN, M.: Genetics 110, 145-158 (1985). - SOLYMOSI, P., KOSTYAL, S.: Weed Res. 25, 411-414 (1985). - STACEY, G.: The rhizobium experience, 109-121. In: see ref. HALVORSON et al. 1985. - STEARNS, S.C., KOELLA, J.C.: Evolution 40, 893-913 (1986). - STROBEL, R., WÖHRMANN, K.: Biochem. Genet. 15, 1015-1021 (1977). - SUKOPP, H., TREPL, L.: Extinction and naturalization of plant species as related to ecosystem structur and function, 245-276. In: Potentials and Limitations of Ecosystem Analysis, eds. E.D. SCHULZE, H. ZWÖLFER. Berlin, Heidelberg, New York: Springer 1987. - SYVANEN, M.: Annu. Rev. Genet. 18, 271-293 (1984).

TAKAHATA, N.: Genet. Res. 44, 109-116 (1984). - TAKAHATA, N., PALUMBI, S.R.: Genetics 109, 441-457 (1985). - THAI, K.M., JANA, S., NAYLOR, J.M.: Weed Science 33, 829-835 (1985). - TURELLI, M.: Theor. Pop. Biol. 25, 138-193 (1984); - Genetics 111, 165-195 (1985); - Gaussian versus non-Gaussian genetic analyses of polygenic mutation-selection balance, 607-628. In: see ref. KARLIN and NEVO 1986.

VAN NOORDWIJK, A.J., GEBHARDT, M.: Reflections on the genetics of quantitative traits with continuous environmental variation, 73-90. In: Genetic Constraints on Adaptive Evolution, ed. V. LOESCHCKE. Berlin, Heidelberg, New York: Springer 1987. - VIA, S.: Evolution 38, 896-905 (1984); - Genetic constraints on the evolution of phenotypic plasticity, 47-71. In: Genetic Constraints on Adaptive Evolution, ed. V. LOESCHCKE. Berlin, Heidelberg, New York: Springer 1987. - VIA, S., LANDE, R.: Evolution 39, 505-522 (1985).

WALBOT, V., CULLIS, CH.A.: Annu. Rev. Plant Physiol. 36, 367-396 (1985). - WARWICK, S.I.: Weed Res. 20, 299-304 (1981). - WARWICK, S.I., BLACK, L.: Can. J. Bot. 59, 689-693 (1981). - WATRUD, S., PERLAK, F.J., TRAN, M.-T., KUSANO, K., MAYER, E.J., MILLER-WIDEMAN, M.A., OBUKOWICZ, M.G., NELSON, D.R., KREITINGER, J.P., KAUFMAN, R.J.: Cloning of *Bacillus thuringiensis* subsp. *kurstaki* delta-endotoxin into *Pseudomonas fluorescens*: Molecular biology and ecology of an engineered microbial pesticide, 40-46. In: see ref. HALVORSON et al. 1985. - WOLTERECK, R.: Verh. Dtsch. Zool. Ges. 9, 110-172 (1909). - WRIGHT, S.: Evolution and the Genetics of Populations, Vol. 4. Variability within and among Populations. Chicago: University of Chicago Press 1978.

ZOUROS, E., FOLTZ, D.W.: Malacologia 25, 583-591 (1984).

Prof. Dr. KLAUS WÖHRMANN
Institut für Biologie II
Universität Tübingen
D-7400 Tübingen 1, FRG

Dr. VOLKER LOESCHCKE
Institute of Ecology and Genetics
University of Aarhus
DK-8000 Aarhus C, Denmark

D. Taxonomy

I. Systematics of the Bryophytes

By JAN-PETER FRAHM

1. Introduction

There are two antagonistic trends in current bryophyte systematics: on the one hand, numerous new species are described. In the years 1984 and 1985 approximately 5 new genera and 70 species of hepatics have been described and about 6 genera and 50 species of mosses. On the other hand, monographs and revisions reduce the number of species consistently. Especially in tropical genera and families, reductions of 50 - 70% of the number of species is normal, eliminating dozens of species described as "endemic" from small ranges and thus preparing a base for further phytogeographic considerations. The description of new genera in mosses especially concerns the pleurocarpous mosses (*Harveya*, Hookeriaceae: CRUM 1985a, *Podperaea*, Hypnaceae: IWATZUKI and GLIME 1984).

Special attention is paid to the bryology of the tropics. In contrast to the long-lasting period of pure collection and the phase of checklists in the last decades, now comprehensive floras have been compiled by teams of specialists. The "Manual of Mexican Mosses", which will be completed in the near future, presents numerous new species and even genera from Mexico, which are published for the first time (CRUM 1985, 1985a,b, CRUM and SHARP 1984, IRELAND 1984). For part of Papua New Guinea, from an illustrated, comprehensive bryoflora edited by KOPONEN and NORRIS (1983a), about 250 pages have been published in 11 parts in the course of only 2 years (KOPONEN and NORRIS 1983b, 1984, 1985b; PIIPPO 1984, 1984a, 1985, 1985a,b; GROLLE and PIIPPO 1984; NORRIS and KOPONEN 1985; FRAHM et al. 1985).

Nomenclatural activities concern lectotypifications of hence untypified species, especially from *Hedwigian* species or species described from the first half of the last century. An increasing number of publications deal with "theoretical bryology", especially in North America. Such discussions concern species concepts, systematics and evolution (for literature see Sect. 3a).

The state of knowledge in bryology is summarized in the proceedings of congresses in Tokyo (First World Conference of Bryology, 1983) and Brighton (Third International Congress of Systematic and Evolutionary Biology, 1985).

2. Handbooks, General

Two comprehensive manuals are now available, dealing with all aspects of bryology and giving an updated overview. SCHOFIELD (1985) treats the major groups of mosses, hepatics and anthocerotes as well as cy-

Progress in Botany, Vol. 49

tology, genetics, physiology, ecology and geography and is specifically useful for bryophyte courses or courses in cryptogamic botany. Fifty years after the publication of the first "Manual of Bryology" edited by F. VERDOORN, SCHUSTER (1983, 1984) in collaboration with several specialists edited a "New Manual of Bryology" in two volumes, presenting on 1300 pages in 21 chapters on different subjects of bryology, a most comprehensive and modern overview over the present state of knowledge. A second volume of the "Advances in Bryology" has been published (SCHULTZE-MOTEL 1984) dealing with mitosis, desiccation and ultrastructure, applied bryology and bryophytes as indicators for pollution.

"Recent progress" in bryology has been summarized by several authors in a manual devoted to Cryptogamic Botany by the University of Lucknow (BHARADWAJ 1982). Current research projects, herbaria and addressses of bryologists are listed in a compendium (VITT et al. 1985).

3. Taxonomy

a) General

The basic concept of systematics of pleurocarpous mosses goes back to the systematic treatment of Fleischer, wich was followed by Brotherus in "Die Natürlichen Pflanzenfamilien". For 60 years, no major changes have been made in this concept. Recently neglected characters such as paraphyllia or SEM structures of the peristome have been studied as well as cluster analysis to calculate the relationships of species and genera. For the Hypnaceae, a preliminary new circumscription and subdivision are presented by NISHIMURA et al. (1984). The Rhytidiaceae are now included in the Hylocomiaceae (ROHRER 1985). New ideas concerning the phyletic relationships of the Bryopsida are treated in detail by VITT (1984) including a historical summary, an evaluation of the taxonomic criteria and considerations of modern taxonomic characters such as peristome criteria and chromosome numbers.

A discussion of the species concept in bryophytes was presented at the 35. Annual Meeting of the American Institute of Biological Sciences (MILLER 1985). Theoretical approaches (recommendations for a phylogentic alternative: DONOGHUE 1985; input from population biology: WYATT 1985; reproductive biology: STARK 1985a; relevance of ecology: SHAW 1985b; relevance of morphological and developmental facts: MISHLER 1985, 1985)a) were faced with practical concepts developed during taxonomic studies (ZANDER 1985; CRUM 1985d), showing the discrepancies between theoretical perfectionism and practical usefulness or the impossibility to apply theoretical techniques fully to practical problems. The discussion of this problem shows that there is no common comcept but different author-orientated concepts. It becomes thus evident that different taxonomic groups require the application of different concepts.

Reviews of modern methods of analyzing taxon relationships are given by HORTON (1984), a review of the literature on the systematics of bryophytes by WYATT and STONEBURGER (1984).

b) Monographs, Revisions

Anthocerotae: Revision of Japanese Anthocerotae: HASEGAWA (1984).

Hepatics: *Archilejeunea:* GRADSTEIN and BUSKES (1985); *Blepharolejeunea:* SLAGEREN and KRUIJT (1985); *Caudalejeunea:* VANDEN BERGHEN (1984); *Dicranolejeunea:* KRUIJT

(1985); Geocalycaceae subfamily. Lophocoleoideae: ENGEL and SCHUSTER (1984). *Marchantia* in the New World: BISCHLER (1984); *Metzgeria* sect. *Biseria:* KUWAHARA (1984); *Plagiochila* in SE Asia: INOUE (1984); *Schiffneriolejeunea:* GRADSTEIN and VANDEN BERGHEN (1985); *Stictolejeunea:* GRADSTEIN (1985), *Symbiezidum:* GRADSTEIN and VAN BEEK (1985); *Teleranea* subg. *Neolepidozia:* ENGEL and SCHUSTER (1983).

Mosses: *Atractylocarpus* (worldwide): PADBERG and FRAHM (1985); Bartramiaceae in Patagonia and the Antarctic: MATTERI (1984); review of *Brotherobryum:* ALLEN (1984); *Campylopodiella:* FRAHM (1984); Campylopodioideae of Papua New Guinea: FRAHM et al. (1985); *Campylopus* subg. *Thysanomitrion:* FRAHM (1984); *Campylopus* in Africa: FRAHM (1985); *Campylopus* in Sri Lanka: FRAHM (1984b); Grimmiaceae (*Indusiella, Aligrimmia, Coscinodontella*)*:* MURRAY (1984). *Homaliodendron* in Indochina: NINH (1984); *Orthotrichum* in Australasia: LEWINSKY (1984b); *Orthotrichum* in South America: LEWINSKY (1984), Phyllogoniaceae: LIN (1984); *Pohlia* in North and Central America: SHAW (1983); *Plagiothecium* in North America: IRELAND (1985). Ptychomniaceae: HATTAWAY (1984); *Racomitrium-canescens*-group: FRISVOLL (1983). *Rhodobryum* in Asia: MOHAMED (1984). Revision of *Schimperella* (Brachytheciaceae): BUCK (1985a): Sphagnaceae in North America: CRUM (1984b). African Sphagnales: EDDY (1985). *Timmia* sect. *Timmia:* BRASSARD (1984).

c) Biosystematic and Experimental Study

Biosystematic analysis of *Frullania tamarisci:* STOTLER et al. (1985). Cultural studies in the *Fissidens bryoides* complex: CONKLIN (1984). Quantitative study of *Epipterygium:* SHAW (1984). Chromosomal, morphological, chemical and geographical studies on *Frullania jackii* and *F. davurica:* MUES et al. (1984). Comparative study of *Brachythecium starkei* and *B. curtum:* VYSOTSKAYA (1984). Statistical analysis of *Didymodon:* DÜLL-HERMANNS (1984). SEM separation of *Braunia* and *Hedwigidium:* SHARP et al. (1978). Quantitative taxonomic study of *Epipterygium:* SHAW (1984b). Distinction of *Fontinalis dalecarlica* and *F. flaccida:* SEE and GLIME (1984).

d) Chemosystematics

The relevance of chemical data for taxonomy is discussed by MUES (1985). Relationships between algae, bryophytes and pteridophytes based on chemical data are discussed by ASAKAWA (1985). A general survey of occurrence and distribution of chemical compounds in mosses, hepatics and anthocerotes is given by HUNECK (1983), including a discussion of the biological activities and chemotaxonomic value.

Flavonoid chemistry of ten species of *Radula:* MUES (1984). Terpenoids of *Jungermannia vulcanicola:* YOKOUCHI et al. (1984), in Lejeuneaceae: GRADSTEIN et al. (1985). Structure determination of a glycoside from *Radula complanata:* MARKHAM and MUES (1984). Sesquiterpenes in Taiwanese liverworts: WU (1984). Seasonal changes in the compounds of *Bazzania trilobata:* HUNECK (1984), HUNECK et al. (1984). Chemical constituents of *Marchantia* species: ASAKAWA et al. (1984); of Chilean *Plagiochila* species: ASAKAWA and INOUE (1984); of Chilean liverworts: ASAKAWA and INOUE (1984a). Cytotoxic, antifungal, antimicrobial, antifeedant and piscicidal compounds from hepatics: ASAKAWA (1984). Fatty acids from Japanese mosses: ICHIKAWA et al. (1983, 1984).

e) Cytosystematics

The knowledge of the cytosystematics of bryophytes has considerably increased over the last 30 years. Up until 1950, chromosome numbers of 340 taxa had been published. Between 1950 - 1980, the number of taxa, for which counts have been published, has increased to 2020 (ANDERSON 1984). However, as pointed out by ANDERSON (1984) the usefulness of chromosome data for bryophyte taxonomy is limited. Because infraspecific polyploidy and aneuploidy is so frequent, the recognition of evolutionary trends may be difficult. Accordingly, there are contrary opinions even on the basic numbers of chromosomes (4 vs 9 in the Hepaticopsida, 7 vs 11 in the Bryopsida). Karyotype studies, C-banding or measurements of DNA amounts are not yet widely used for bryophytes, since these techniques are difficult to perform with bryophytes, especially with mosses.

Basic reviews of the cytology are given for mosses by RAMSAY (1983) and hepatics and hornworts by NEWTON (1983) and by NEWTON (1984) for bryophytes in general. Newton discusses especially karyotype characters such as chromosome size, centromere position, heterochromatin, accessory chromosomes and polyploidy. The evolution of sex chromosomes in bryophytes is briefly reviewed by NEWTON (1985a). ANDERSON (1984) gives a critical assessment of the abundant data accumulated over the past decades.

Karyological studies have been undertaken for *Riccia* from SW Africa (BORNEFELD 1984), for species of *Orthotrichum* from Australia (RAMSAY and LEWINSKY 1984), for mosses from the Antarctic (PRZYWARA et al. 1984a), British mosses (NEWTON 1984a), Polish bryophytes (KUTA et al. 1984, including the first records of the genera *Myurella* and *Gymnomitrion*; FRITSCH 1984; PRZYWARA et al. 1984b), mosses from the Himalayas (KUMAR and VERMA 1984), the USSR (DANILKIV et al. 1984) and Pennsylvania (LIN 1983) as well as for species of Thuidiaceae (INOUE 1984d) and Japanese species of *Fissidens* (IWATSUKI and INOUE 1984).

f) Special Taxonomic Treatments

A n t h o c e r o t a e : Taxonomy of Anthocerotales, discussion of the taxonomic significance of morphological characters, description of the new family Phaeocerotaceae: BHARADWAJ (1981a). Key to Japanese species of Anthocerotae: HASEGAWA (1985). Anthocerotae of Bangladesh: HADIUZZAMAN and CHAKRAVARTY (1984). Classification of Anthocerotae, proposes *Aspiromitus* for *Anthoceros* and *Anthoceros* for *Phaeoceros:* SCHUSTER (1984a). *Megaceros* in New Zealand: CAMPBELL (1984).

H e p a t i c a e : Balantiopsidaceae: *Acroscyphus* gen. nov. from New Caledonia: KITAGAWA (1984). Calypogeiaceae: in Papua New Guinea: PIIPPO (1984a). Cephaloziaceae: subfam. Schiffneroideae in Papua New Guinea: GROLLE and PIIPPO (1984). Condoniaceae: New species of *Fossombronia* from Australia: SCOTT and PIKE (1984). Haplomitriaceae: in Papua New Guinea: PIIPPO (1984). Jubulaceae: new definitions for subgenera of *Frullania:* SCHUSTER (1985). *Frullania* species of Latin America: YUZAWA (1984). *Frullania* subg. *Meteoriopsis* in the Neotropics: KRON (1985). New taxa of *Frullania* from Queensland: HATTORI (1984a). Frullaniaceae from New Caledonia: HATTORI (1984). *Frullania tamarisci* subspecies complex: STOTLER et al. (1985). *Frullania* species of Switzerland: BISANG (1985). Jungermanniaceae: in Taiwan: VANA and INOUE (1983). Lepidoziaceae: Zoopsoideae from Papua New Guinea: GROLLE and PIIPPO (1984). Synopsis of African Lepidoziaceae: POCS (1984). *Bazzania* in New Caledonia: TIXIER (1985). Austral *Telaranea:* ENGEL and SCHUSTER (1983). Leujeuneaceae: *Austrolejeunea bidentata* sp. nov. from Australia: THIERS (1985). Leujeuneaceae in Australia: THIERS (1985a). New and little known Lejeuneaceae from Africa: JONES (1985a). *Cololejeuna kashyapii* sp. nov. from India: UDAR and SRIVASTAVA (1985). New species of *Marchesinia* and *Cheilolejeunea* from Rêunion: VANDEN BERGHEN (1984). Review of *Blepharoleujeunea:* SLAGEREN and KRUIJT (1985). Taxonomic characters of *Brachiolejeunea:* SLAGEREN (1984). Lejeuneaceae from Sri Lanka: ONRAEDT (1985). Subfamilies of Leujeuneaceae: MIZUTANI (1985). *Cololejeunea schaeferi* sp. nov. from the Canary Islands: GROLLE (1985a). *Cyclolejeunea* in Cuba: GROLLE (1984a): Review of *Aphanolejeunea:* POCS (1984a). *Aphanoleujeunea* in tropical Africa: POCS (1984a). *Colelejeunea* in Japan: MIZUTANI (1984, 1984a). *Plagiolejeunea* gen. nov. MIZUTANI (1984b). *Kymatoleujeunea*, a new genus from New Zealand: GROLLE (1983). *Schiffnerolejeunea* in Africa: GRADSTEIN and VANDEN BERGHEN (1985). *Schiffnerolejeunea amazonica* sp. nov.: GRADSTEIN (1985a). *Lopholejeunea erugata* nom. nov. for *Ptychocoleus inermis:* THIERS (1984). *Lindigianthus* gen. nov. from tropical America: KRUIJT and GRADSTEIN (1985). *Chonecolea ruwenzoriensis* sp. nov. from Uganda: JONES (1985). Marchantiaceae: *Marchantiopsis* gen. nov. from China: GAO and CHANG (1983). Metzgeriaceae: *Metzgeria* sect. *Biseria:* KUWAHARA (1984). Plagiochilaceae: *Plagiochila lenis* sp. nov. from China: INOUE (1984). Three new species of *Plagiochila* from Chile: INOUE (1984a). Radulaceae: several new species of *Radula* from New Caledonia: YAMADA (1985). Tasmanian species of *Radula:* YAMADA (1984a). Three new species of *Radula* from Papua New Guinea: YAMADA (1985a). Three new species of *Radula* from Queensland: YAMADA (1984). Systematics of Indian Radulaceae: KUMAR and UDAR (1985). Ricciaceae: *Riccia parvo-areolata* sp. nov. from the Cape: VOLK and PEROLD (1984). *Riccia gemmifera* sp. nov. from SW Africa: VOLK (1984). New classification of *Riccia:* VOLK (1983). Schistochilaceae in Australia: SCHUSTER and ENGEL (1985).

M u s c i : Amblystegiaceae: *Crassicosta* gen. nov. from Mexico: CRUM and SHARP (1984). Two new species of *Hygrohypnum* from Mexico: CRUM (1985b). *Sciaromium lacustre* sp. nov. from Lake Titicaca: RICHARDS (1984). Archidiaceae: *Archidium minutissimum* sp. nov. from Queensland: STONE (1985). *Archidium thalliferum* sp. nov.: STONE (1985a). Bartramiaceae: *Breutelia* in Brazil: GRIFFIN (1984). *Breutelia subarcuata* and *B. chrysea* in Latin America: GRIFFIN (1984a). The *Breutelia subarcuata* complex in Colombia: GRIFFIN (1984b). Brachytheciaceae: *Brachythecium* sub. *Chamberlainia* stat. nov.: VOHRA (1983). Bryaceae: Two new species of *Anomobryum:* SHAW and FIFE (1984). New combinations in *Mielichhoferia:* SHAW and CRUM (1984). Bryaceae from Papua New Guinea: KOPONEN and NORRIS (1984). *Rhodobryum ontariense* has to replace *R. spathulatum:* TOUW (1984). Australasian Bryoideae: OCHI (1984). *Pohlia andrewsii* in the European Alps: NORDHORN-RICHTER (1984). Calymperaceae: Nomenclatural changes in *Calymperes*: REESE (1984a,b). *Leucophanes* in the Americas: SALAZAR (1985). *Calymperes* and *Syrrhopodon* in Brazil: REESE (1984). Calymperaceae in China: LIN (1984). *Calymperes* in Bangladesh: HADIUZZAMAN (1984). Dicranaceae: *Campylopus* in Sri Lanka: FRAHM (1984b). Review of *Campylopodiella*: FRAHM (1984a). Dicranaceae from Peru, *Campylopus bryotropii* sp. nov.: FRAHM (1984c). *Campylopus* of Australia: CATCHESIDE and FRAHM (1985). *Campylopus* subg. *Campylopidulum* subg. nov.: VITAL (1984a). Two new species of *Trematodon* from Japan: AKIYAMA (1984). *Dicranum tubulifolium* sp. nov. from Mexico: IRELAND (1984). Ditrichaceae: *Eccremidium* in Brazil: VITAL and EGUNYOMI (1984). Entodontaceae: *Entodon buckii* sp. nov. from China: LIN (1984b). Fissidentaceae: *Fissidens* in Bangladesh: HADIUZZAMAN (1984a). *Fissidens elegans* complex in the Neotropics: PURSELL (1984). *Fissidens henryae* sp. nov. from Queensland: STONE (1984). *Fissidens* sect. *Pachylomidium* in Europe, *F. coacervatus* sp. nov.: BRUGGEMAN-NANNENGA (1985). Funariaceae: *Entosthodon steerei* sp. nov. from Ecuador: FIFE (1984). Grimmiaceae: Grimmiaceae from Patagonia: DEGUCHI (1984). Taxonomic notes on *Racomitrium crispulum*: FRISVOLL (1984). *Racomitrium crumianum* sp. nov. from New Zealand: FIFE (1984a). *Racomitrium heterostichum* group: FRISVOLL (1984a). Lectotypification of several species of *Racomitrium*: FRISVOLL (1984b). *Schistidium obtusifolum* sp. nov. from Canada: IRELAND and CRUM (1984). *Grimmia* and *Schistidium* in South Georgia: BELL (1984). *Grimmia chenii* nom. nov. from China: LIN (1984c). Helicophyllaceae: *Helicophyllum torquatum* in Brazil: YANO (1984a). Hookeriaceae: *Harveya* gen. nov. from Mexico: CRUM (1985a). Hypnaceae: Subdivision of Hypnaceae: NISHIMURA et al. (1984). *Hypnum bizotii* nom. nov. in Africa: OCHYRA (1984). *Platygrium repens* in Japan: HIGUCHI and NISHIMURA (1984). *Ptychophyllum* congeneric with *Extropothecium*: NISHIMURA (1984a). *Caribeohypnum* gen. nov. for *Hypnum polypterum*: ANDO and HIGUCHI (1984). *Podperaea* gen. nov.: IWATSUKI and GLIME (1984). *Gollania philippinensis* in China: HIGUCHI and LIN (1984). West Indian Hypnaceae: BUCK (1984a). Leptodontaceae: *Forstroemia* in North America: STARK (1985). Meesiaceae: *Neomeesia* gen. nov. from southern South America: DEGUCHI (1983). Phyllogoniaceae: reclassification: LIN (1984a). Pottiaceae: *Crossidium asirense* sp. nov. from Saudi Arabia: FREY and KÜRSCHNER (1984). *Barbula jacksharpii* sp. nov. from Mexico: CRUM (1984a). Systematic studies in *Tortula*: MISHLER (1984). *Didymodon* in Europe: DÜLL (1984). *Leptodontium gem-*

mascens in France: ROGEON and SCHUMACKER (1984). *Uleobryum curtisii* sp. nov. from Queensland: STONE (1984b). *Tortula* in Germany: HÄUSSLER (1984). *Tortula* in New Zealand: LIGHTOWLERS (1985). *Tortula* in South Georgia: LIGHTOWLERS (1985a). *Gymnostomiella orcutii* in Brazil: VITAL (1984). *Weissia* in the Faroes: LEWINSKY (1984a). Racopilaceae: *Racopilum* in Brazil: YANO (1984b). Regmatodontaceae: South American species: SCHULTZE-MOTEL (1984a). Sematophyllaceae: *Pylaisiadelpha* replaces *Brotherella*: BUCK (1984). *Pylaisiadelpha duellii* sp. nov. from Mexico: CRUM (1985). Sphagnaceae: *Sphagnum* in Svalbard: FLATBERG and FRISVOLL (1984). *Sphagnum arcticum* sp. nov.: FLATBERG and FRISVOLL (1984a). New *Sphagnum* species from western North America: CHRISTY (1984). New Sphagna from Brazil: CRUM (1985). Two new species of *Sphagnum* from Costa Rica: CRUM (1984c). Trachypodiaceae: placement of *Bryowijkia*: VITT (1984).

4. Nomenclature

Index Hepaticarum vol. 10: GEISSLER and BISCHLER (1985). Index Hepaticarum Supplement 1978-79: ENGEL (1984). Neotypification of *Plagiochasma rupestre*: BISCHLER and SERGIO (1984). Recommendations for the citation of taxa: FRAHM and GRADSTEIN (1984). Different types of nomina nuda (published and unpublished): TIXIER (1984). Proposal for conservation of *Jubula*, *Mannia*, *Pellia* and *Calypogeia*: ZIJLSTRA (1985), *Papillaria*: ISOVIITA (1985), *Rhodobryum*: ISOVIITA and KOPONEN (1984). Nomenclatural changes in the Mielichhoferioideae: SHAW (1985a). Lectotypification of *Ditrichum flexicaule*: FRISVOLL (1985a). Various taxonomic changes in hepatics: GROLLE (1984, 1985). Proposal to amend *Papillaria*: ISOVIITA (1985). Neotypification of *Neckera angustifolia*: SMITH (1984). Validation of numerous taxa of hepatics. SCHUSTER (1984b).

5. Floristics

a) Checklists

Australia, Lord Howe Island: RAMSAY (1984); New South Wales (RAMSAY 1984a); Brazil: YANO (1984); Greek: PRESTON (1984, 1984a); Sri Lanka: ABEYWICKRAMA and JANSEN (1978); Cap Verde Islands: MUHLE (1982); Paraguay: BUCK (1985).

b) Books for Identification

Australia, liverworts: SCOTT (1985); Germany: DÜLL (1985); North America, Sphagna: CRUM (1984), Great Lakes Forest: CRUM (1983); Rumania: MOHAN (1984); Ungarn: ORBAN and VAJDA (1983).

c) Floristic Changes

The changes in the floristic composition of bryophytes during the last decade or even centuries have been studied by several authors. PENTECOST and ROSE (1985) analyzed the changes in the cryptogam flora of sandrocks in Great Britain between 1688 and 1984. In the Netherlands, BOELE and VAN ZANTREN (1984) studied the regression of epilithic bryophytes on megalithic cairns. TOVONEN (1985) reported on the changes in the pleustonic macrophyte flora in Finnish lakes over 30 years and the expansion of *Ricciocarpus natans*, which is credited to eutrophication. The surprisingly long survival of tubers of the moss *Dicranella staphylina* in soil for nearly 50 years was observed by WHITEHOUSE (1984).

6. Phylogeny, Evolution

A new approach by cladistics has enriched the discussion on phylogeny (MISHLER and CHURCHILL (1984). Based on 51 character states this analysis revealed that bryophytes s. lat. are paraphyletic and mosses seem to be derived later from tracheophytes than hornworts or liverworts. The discussion on the evolution of bryophytes is, however, mostly hypothetical and restrained by the fact that evolutionary processes in bryophytes are very slow (SZWEYKOWSKI 1984), as indicated by large geographical ranges (species ranges of bryophytes are comparable to genera and family ranges of flowering plants) and the fact that fossil bryophytes, even the oldest from the Permian, are very similar to extant forms.

General survey: ANDO (1982, 1983), KUMAR (1984). Morphogenetic approach: CRANDALL-STOTLER (1985). Ultrastructural approach: DUCKETT and CAROTHERS (1985). Cladistic approach: MISHLER (1985), MISHLER and CHURCHILL (1984, 1985), ROBINSON (1984, 1985). Chemosystematic approach: ASAKAWA (1985); cytological approach: NEWTON (1985). The use of evolutionary distance analysis of RNA sequences reveals that bryophytes are evolved from pteridophytes by degeneration: HORI et al. (1985). Similarity indices of RNA sequences show high identity with phanerogams and Chlorophyceae: KATOH et al. (1984). Based on anatomical, ontogenetic, cytological and biochemical facts it is hypothesized that mosses, hepatics and anthocerotes are phylogenetically distinct: CRANDALL-STOTLER (1984). Phylogenetic classification scheme of hepatics: SCHUSTER (1984). Origin of anacrogynous and thalloid hepatics from leafy acrogynous hepatics: MEHRA (1982).

Evolution of the moss operculum: ROBINSON and SHAW (1985), of the moss peristomes: SHAW and ROBINSON (1984). Evolution of sex chromosomes: NEWTON (1985a). Induction of phenovariants to elucidate modes of evolution in the branching and perianth development of hepatics: BASILE and BASILE (1984). Differences in isoenzymes as a cause for polymorphism and evolutionary processes. SZWEYKOWSKI (1984a).

7. Bryogeography

A general survey on the phytogeography of bryophytes is given by SCHUSTER (1983), who deals in detail with fossils and tectonic events, distribution patterns and dispersibility factors. As a result of intensive, worldwide taxonomic studies it becomes evident that many bryophytes have large ranges. Especially in the tropics, numerous Afro-American or Afro-SE-Asian disjunctions at the species level have been recognized. GRADSTEIN et al. (1983) give 50 examples of such macro-disjunctions. BUCK and GRIFFIN show that *Trachyphyllum* hitherto known only from Africa occurs also in South America. STONE (1984a) gives examples for disjunctions between Australia and South Africa. FRAHM (1985) recognized that *Campylopus shawii* known from the British Isles occurs also on the Azores and in the Caribbean. The possibilities for long-distance dispersal were experimentally tested by VAN ZANTEN (1983, 1984), who showed that many of the small-sized spores are drought- and freeze-resistant, and are thus able to survive aerial, long-range dispersal. Larger spores of hepatics and not the drought-resistant spores of tropical hepatics, however, support the theory that disjunctions must in part also be explained by continental drift.

Special surveys on the ranges of several Mediterranean mosses are given by HEBRARD (1984). The heavy metal moss *Scopelophila cataractae* hitherto known from scattered localities in tropical, montane regions has been found new to Europe in the British Isles (CORLEY and PERRY 1985), and the North American

Didymodon reedii is reported as new to England by APPLEYARD (1985). Several bryogeographical treatments are given from special regions such as Taiwan (LIN and TSAI 1984), CSSR (POSPISIL 1984), Patagonia (HÄSSEL et al. 1984), Pacific North America (SCHOFIELD 1984), the Sahel (TIXIER 1984a), Macaronesia (SERGIO 1984; DÜLL 1984a), Mexico (SHARP 1984; DELGADILLO 1984), Newfoundland (BRASSARD 1984a), Japan (DEGUCHI and IWATSUKI 1984; HASEGAWA 1984), Hispaniola (STEERE 1984), Norway (STOERMER 1984), Philippines (TAN 1984) and New South Wales (RAMSEY 1984a).

Distribution maps of several mosses in Poland have been prepared by OCHYRA and SMAJDA (1983) and part of the German Democratic Republic by MEINUNGER (1984). The ranges of the European species of *Campylopus* have been mapped by FRAHM (1984d). An account of the progress of the European bryophyte mapping project is given by SCHUMACKER (1984).

8. Morphology, Anatomy

a) General

A comprehensive review of the morphological structures found in liverworts is given by SCHUSTER (1984c), including branching and inserting patterns, leaf and stem anatomy and ontogeny, embryological patterns, as well as evolutionary hypotheses derived from these facts and the construction of a prototype of all hepatics. SCHUSTER (1984) gives a separate treatment of the Anthocerotae, which are regarded as an isolated division, lacking affinities to other bryophytes. Further studies on oil bodies have been published on *Radula* (YAMADA 1985c), hepatics from southern Chile (INOUE 1984b), Andean Aneuraceae (MEENKS and DE JONG 1985) and hepatics of the western Himalayas (UDAR and SHAHEEN 1983). Branching patterns of Leujeuneaceae are used as a tool for taxonomy by THIERS (1984a,b). Ramification pattern, branch type, inflorescence positioning and leaf sequences are used as reliable characters for the delimination of subfamilies and allow suggestions on the interrelationships of suprageneric categories. The techniques to study the branching patterns of bryophyte systematics are discussed separately (THIERS 1985c). With the objective of selecting taxonomically useful characters, a survey is given of the lobule characters of some European and African species of Lejeuneoideae (JONES 1984). From Lepidoziaceae, bulbils are described by HÄSSEL DE MENENDEZ (1984), which are supposed to function as a nutrient reserve.

Studies on the peristome structures of mosses led to systematic reinterpretations. Thus, *Pseudoditrichum mirabile* is shifted from the Funariales to the Bryales (SHAW 1984c) and the isolated position of *Mittenia* is supported by the fact that no other peristomal structure is known homologous to the inner peristome of *Mittenia* (SHAW 1985c). A general account of the interrelationships of moss peristomes is given by EDWARDS (1984), SHAW and ROBINSON (1984) and SHAW and ROHRER (1984). The latter give an exhaustive survey of the endostome structures of 70 species of mosses, which, however, revealed no new information on classification at the familial or generic level. The significance of peristome structures for the taxonomy of Bryaceae has been recognized by SHAW and CRUM (1984) and SHAW and FIFE (1984). SHAW and ROBINSON (1984), in a new survey on the development of peristomes, recognized six basic types, two nematodontous (*Tetraphis*, *Polytrichum*) and four arthrodontous (*Buxbaumia*, *Dicranum*, both haplolepideous; *Bryum* and *Funaria*, both diplolepideous). Nine possible phylogenetic developments of the arthrodontous peristomes are discussed, but there is no evidence favouring one of them.

A review of the moss spore sturcture is given by MOGENSEN (1983). Morphology and anatomy of the moss gametophyte in general is treated by SCHOFIELD and HEBANT (1984).

A textbook on the morphology, growth and differentiation of bryophytes was published by PURI (1981), who gives a review of the classical morphology, morphogenesis, regeneration, and reproductive biology.

Recently, and probably due to phanerogamists, the reproductive biology of bryophytes has received much attention. LONGTON and SCHUSTER (1983) discuss in detail asexual and sexual reproduction, change of both methods due to climatic changes, effects of sperm and spore dispersal and evolutionary trends. Several "strategies" (although plants seem merely to react to varying ecological conditions) are given for hepatics. Examples for the role of asexual reproduction of bryophytes are given by SELKIRK (1984) for subantarctic mosses colonizing snow-free areas on Macquarie Island, by EGUNYOMI (1984) for tropical mosses in Nigeria and by REESE (1984c) for *Syrrhopodon texanus*.

b) Ultrastructures

SEM studies on the ultrastructure of spores have been performed by INOUE and HIBINO (1984) on *Metzgeriales*, STEINKAMP and DOYLE (1984) and CASTALDO-COBIANCHI et al. (1982) on *Fossombronia*, GAO and GAO (1984) and MCQUEEN (1985), on *Sphagnum*, UDAR and SRIVASTAVA (1983) on *Targionia* and *Cyathodium*, RUSHIN (1984) on *Bruchia* and UDAR and SRIVASTAVA (1984) on eight species of Indian liverworts. Separation of the genera *Braunia* and *Hedwigigidium* by the SEM structure of the leaf surface was possible by studies of SHARP et al. (1978).

c) Fungal Parasites and Mycorrhizas, Endophytic Bacteria and Symbiontic Blue-Green Algae

The symbiontic relationships between fungi and hepatics have been studied intensively. Swollen rhizoid tips which were found in 33 species of British hepatics were always associated with fungal hyphae (POCOCK and DUCKETT 1985, 1985a). It is postulated that the fungi serve for nutrient uptake. These relationships between liverworts and endophytic fungi are discussed by POCOCK and DUCKETT (1985b). Such endophytic fungi were also observed in leafless flagellar branches of Lepidoziaceae from Malaysia (POCOCK et al. 1984). Comparisons between the carbohydrates of axenically cultured hepatics with those collected in the field revealed that both show identical carbohydrates, which therefore are not synthesized by fungi (CHRISTIE et al. 1985).

New bryophilous ascomycetes have been found in Bavaria by DÖBBELER (1984).

Cyanophyaceae associated with tropical bryophytes from Minas Gerais, Brazil, have been studied by SANT ANNA (1984). It may be supposed that these Cyanophyceae play an important role for bryophytes in such unsuitable habitats such as tropical lowland forests or deserts. This may be supported by the observation that nitrogen-fixing Cyanophyceae have been found on the ventral side of *Porella navicularis*.

An influence of bacteria on the growth and development of mosses has been observed by SPIESS et al. (1984, 1984a).

9. Applied Bryology

A review of all applications of bryophytes, such as for horticultural and decorative use, for medicines and biologically active substances, for biomonitoring and controlling erosion, is given by ANDO and MATSUO (1984). Biologically active substances in bryophytes are summarized by ASAKAWA et al. (1978). The absorption of metals by aquatic mosses is used as an indicator for prospecting heavy metals (SHACKLETTE 1984). Publications of WATANABE (1978) and INOUE (1979) deal with Japanese moss gardens.

10. Bibliographies

The fourth continuation of the catalogue of Russian and foreign bryological publications with 1686 titles, covering the period 1976-80 and arranged alphabetically according to authors, has been published by TSEVETKOVA et al. (1983). For Japan annual reviews of bryological literature are given by IWATSUKI (1978, 1979) and IWATZUKI and HIGUCHI (1985). The bryological titles of the American botanical literature are indexed by BUCK et al. (1984, 1985, 1985a). The bryological references for Spain are compiled by CASAS et al. (1984). All current bryological journals are listed by O'SHEA (1984).

References

ABEYWICKRAMA, B.A., JANSEN, M.A.B.: 277-281 in: UNESCO, Man and the Biosphere National Committee for Sri Lanka (1978). - AKIYAMA, H.: Acta Phytotaxon. Geobot. 35, 21-29 (1984). - ALLEN, B.H.: Lindbergia 10, 111-120 (1984). - ANDERSON, L.E.: J. Hattori Bot. Lab. 55, 187-197 (1984). - ANDO, H.: Proc. Japan. Soc. Plant Taxonom. 4(6), 2-5 (1982); - Heredity 37(5), 39-45 (1983). - ANDO, H., HIGUCHI, M.: Cryptogamie Bryol. Lich. 5, 7-14 (1984). - ANDO, H., MATSUO, A.: Applied bryology, 132-22. In: Advances in Bryology, Vol. 2, ed. W. SCHULTZE-MOTEL. Vaduz: Cramer 1984. - APPLEYARD, J.: J. Bryol. 13, 319-321 (1985). - ASAKAWA, Y.: J. Hattori Bot. Lab. 56, 215-219 (1984); - Abstract. Third International Congress of Systematic and Evolutionary Biology, 9 (1985). - ASAKAWA, Y., INOUE, H.: Chemical constituents of selected Chilean liverworts, 109-115. In: Studies on Cryptogams of Southern Chile, ed. H. INOUE. Tokyo: Kenseisha 1984; - Chemical constituents of Chilean Plagiochila species, 117-124. Ibid. 1984a. - ASAKAWA, Y., TOYOTA, M., TAKEMOTO, T.: Proc. Bryol. Soc. Japan 2, 78-80 (1978). - ASAKAWA, Y., TOYOTA, M., BISCHLER, H., CAMPBELL, E.O., HATTORI, S.: J. Hattori Botanical Laboratory 57, 383-389 (1984).

BASILE, D.V., BASILE, R.M.: J. Hattori Bot. Lab. 55, 173-185 (1984). - BELL, B.G.: Bull. Brit. Antarct. Surv. 67, 71-109 (1984). - BHARADWAJ, D.C. (ed.): Recent Advances in Cryptogamic Botany. Lucknow 1982. - BHARADWAJ, D.C.: Taxonomy of Anthocerotales, 132-151. In: see ref. BHARADWAJ 1982, 1982a. - BISANG, I.: Botanica Helvetica 95, 247-278 (1985). - BISCHLER, H.: Marchantia L. The New World Species. Bryophytorum Bibliotheca, Vol. 26, 228 pp. Vaduz: Cramer 1984. - BISCHLER, H., SERGIO, C.: Cryptogamie Bryol. Lich. 5, 173-174 (1984). - BOELE, C., VAN ZANTEN, B.O.: Lindbergia 10, 187-189 (1984). - BORNEFELD, T.: Nova Hedwigia 40, 313-328 (1984). - BRASSARD, G.R.: Lindbergia 10, 33-40 (1984); - Bryologist 87, 56-65 (1984a). - BRUGGEMAN-NANNENGA, M.A.:

Proc. Kon. Nederlandse Ak. Wet. C, 88, 183-207 (1985). - BUCK, W.R.: Yushania 1, 11-13 (1984). - Brittonia 36, 178-183 (1984a); - Candollea 40, 201-209 (1985); - Brittonia 47, 36-40 (1985a). - BUCK, W.R., BRAKO, L., SLAVICK, A.D., WETTER, M.A.: Bull. Torrey Bot. Club 111, 241-272, 393-427, 511-539 (1984). - BUCK, W.R., BRAKO, L., LEONARDI, L., WETTER, M.A.: Bull. Torrey Bot. Club 112, 95-126 (1985). - BUCK, W.R., BRAKO, L., RICHARDSON, P.M.: Bull. Torrey Bot. Club 112, 204-236 (1985b). - BUCK, W.R., GRIFFIN, D.: J. Nat. Hist. 18, 63-69 (1984).

CAMPBELL, E.O.: Tuatara 27, 105-120 (1984). - CASAS, C.M., BRUGES, M., CROS, R.M.: Treballs Inst. Bot. Barcelona 9, 2-24 (1984). - CASTALDO-COBIANCHI, R., GIORDANO, S., PERILLO, R.: Giornale Bot. Italiana Suppl. 1, 116, 13-14 (1982). - CHRISTY, J.A.: Bull. Native Plant Soc. Oregon 17, 11 (1984). - CHRISTY, A., POCOCK, K., LEWIS, D.H., DUCKETT, J.D.: J. Bryol. 13, 417-421 (1985). - CONKLIN, E.L.: Abstracts, Botanical Society of America, 5-6 (1984). - CORLEY, M.F.V., PERRY, A.R.: J. Bryol. 13, 323-328 (1985). - CRANDALL-STOTLER, B.: Musci, hepatics and anthocerotes - An essay on analogues, 1093-1129. In: see ref. SCHUSTER 1984; - Abstract. Third International Congress of Systematic and Evolutionary Biology, 41 (1985). - CRUM, H.: Mooses of the Great Lakes Forest, 3rd ed. 417 pp. Ann Arbor: University Microfilms 1983; - Sphagnopsida, Sphagnaceae. North American Flora, Series II part 11. New York 1984; - Bryologist 87, 203-216 (1984a). - Sphagnaceae. In: North American Flora ser. II part 11, 180 pp, 1984b; - Cryptogamie Bryol. Lich. 5, 293-297 (1984c); - ibid. 6, 181-184 (1985); - Bryologist 88, 23-24 (1985a); - ibid. 88, 22-23 (1985b); - ibid. 88, 27-28 (1985c); - ibid. 88, 221-222 (1985d). - CRUM, H., SHARP, A.J.: Bryologist 87, 149-150 (1984).

DALTON, D.A., CHATFIELD, J.M.: Am. J. Bot. 72, 781-784 (1985). - DANILKIV, I.S., LESNYAK, E.N., VYSOTSKOVA, E.I.: Botaniceskij Zurnal 69, 1209-1212 (1984). - DEGUCHI, H.: Bull. National Sci. Mus. Tokyo ser. B 9, 143-148 (1983). - DEGUCHI, H.: 17-22 in: Studies on Cryptogams in Southern Chile, ed. H. INOUE. Tokyo: 1984. - DEGUCHI, H., IWATSUKI, Z.: J. Hattori Bot. Lab. 55, 1-11 (1984). - DELGADILLO, C.: Bryologist 87, 12-16 (1984). - DÖBBELER, P.: Ber. Bayer. Bot. Ges. 55, 79-84 (1984). - DONOGHUE, M.J.: Bryologist 88, 172-181 (1985). - DUCKETT, J.G., CAROTHERS, Z.B.: Abstract. Third International Crongress of Systematic and Evolutionary Biology, 49 (1985). - DÜLL, R.: J. Hattori Bot. Lab. 55, 259-266 (1984); - Proc. Third Meeting Bryologists Central and East Europe, Praha, 177-190 (1984a); - Exkursionstaschenbuch der wichtigsten Moose Deutschlands, 273 pp. Rheurdt: IDH 1985. - DÜLL-HERMANNS, I.: J. Hattori Bot. Lab. 55, 253-257 (1984).

EDDY, A.: Bull. Brit. Mus. 12, 77-162 (1985). - EDWARDS, S.R.: Homologies and inter-relationships of moss peristomes, chapter 12 in: see ref. SCHUSTER 1984. - EGUNYOMI, A.: J. Hattori Bot. Lab. 56, 115-121 (1984). - ENGEL, J.J.: Taxon 33, 761-779 (1984). - ENGEL, J.J., SCHUSTER, R.M.: Fieldiana Bot. N.S. 14, 1-7 (1983); - Nova Hedwigia 39, 385-463 (1984).

FIFE, A.J.: Bryologist 87, 154-157 (1984); - N.Z. J. Bot. 22, 1-6 (1984a). - FLATBERG, K.I., FRISVOLL, A.A.: J. Hattori Bot. Lab. 56, 287-319 (1984); - Bryologist 87, 143-148 (1984a). - FRAHM, J.-P.: Nova Hedwigia 39, 585-621 (1984); - Bryologist 87, 249-250 (1984a); - J. Bryol. 13, 163-192 (1984b); - Nova Hedwigia 39, 145-167 (1984c); - Proc. Third Meeting Bryologists Central and East Europe, Praha, 191-212 (1984d); - J. Bryol. 13, 329-332 (1985). - FRAHM, J.-P., GRADSTEIN, S.R.: Bryol. Times 28, 1-2 (1984). - FRAHM, J.-P., GIESE, M., PADBERG, M., KOPONEN, T., NORRIS, D.H.: Acta Bot. Fenn. 161, 63-88 (1985). - FREY, W., KÜRSCHNER, H.: J. Bryol. 13, 25-31 (1984). - FRISVOLL, A.A.: J. Bryol. 13, 285-290 (1984); - ibid. 13, 291-292 (1984a); - J. Hattori

Bot. Lab. _57_, 299-318 (1984b); - Gunneria _41_, 1-181 (1983); - Bryologist _88_, 31-40 (1985a). - FRITSCH, R.: Lindbergia _9_, 160-162 (1984).

GAO, C., CHANG, K.: Bull. Bot. Res. _2_, 113-122 (1983). - GAO, C., GAO, T.: Bull. Bot. Res. _4_, 129-136 (1984). - GEISSLER, P., BISCHLER, H.: Indes Hepaticarum, Vol. 10, Lembidium to Mytilopsis, 352 pp. Vaduz: Cramer 1985. - GRADSTEIN, S.R.: Nova Hedwigia Beih. _80_, 195-220 (1985); - ibid. _80_, 13-29 (1985a). - GRADSTEIN, S.R., BUSKES, G.M.C.: Nova Hedwigia Beih. _80_, 89-112 (1985). - GRADSTEIN, S.R., VAN BEEK, J.: Nova Hedwigia Beih. _80_, 221-248 (1985). - GRADSTEIN, S.R., VANDEN BERGHEN, C.: Nova Hedwigia Beih. _80_, 173-193 (1985). - GRADSTEIN, S.R., MATSUDA, R., ASAKAWA, Y.: Nova Hedwigia Beih. _80_, 63-86 (1985). - GRADSTEIN, S.R., POCS, T., VANA, J.: Acta Bot. Hungarica _29_, 127-171 (1983). - GRIFFIN, D.: J. Hattori Bot. Lab. _57_, 83-95 (1984); - Bryologist _87_, 233-237 (1984a); - Acta Bot. Neerl. _33_, 275-282 (1984b). - GROLLE, R.: Wiss. Z. Friedrich Schiller Univ. Jena, Math. Naturwiss. Reihe _32_, 1005-1012 (1983); - J. Hattori Bot. Lab. _55_, 501-511 (1984); - Wiss. Z. Friedrich Schiller Univ. Jena, Math. Naturwiss. Reihe _33_, 759-764 (1984a); - J. Hattori Bot. Lab. _58_, 197-202 (1985); - J. Bryol. _13_, 487-495 (1985a). - GROLLE, R., PIIPPO, S.: Ann. Bot. Fenn. _21_, 299-307 (1984).

HADIUZZAMAN, S.: Bangladesh J. Bot. _13_, 161-166 (1984); - ibid. _13_, 147-160 (1984a). - HADIUZZAMAN, S., CHAKRAVARTY, A.: Bangladesh J. Bot. _13_, 101-107 (1984). - HÄSSEL DE MENENDEZ, G.G.: J. Hattori Bot. Lab. _56_, 105-114 (1984). - HÄSSEL DE MENENDEZ, G.G., GREENE, S.W., MATTERI, C.M.: J. Hattori Bot. Lab. _55_, 45-64 (1984). - HASEGAWA, J.: J. Hattori Bot. Lab. _56_, 21-28 (1984). - HATTAWAY, R.A.: A Monograph of the Ptychomniaceae, 311 pp. Ann Arbor: University Microfilms 1984. - HATTORI, S.: J. Hattori Bot. Lab. _57_, 405-426 (1984); - Cryptogam. Bryol. Lichenol. _5_, 177-189 (1984a). - HÄUSSLER, M.: Bryol. Beitr. _3_, 1-22 (1984). - HEBRARD, J.P.: Cryptogam. Bryol. Lichenol. _5_, 33-45 (1984). - HIGUCHI, M., LIN, P.J.: Proc. Bryol. Soc. Japan _3_, 193 (1984). - HIGUCHI, M., NISHIMURA, N.: Hikobia _9_, 51-55 (1984). - HORI, H., LIM, B.L., OSAWA, S.: Proc. Nat. Acad. Sci. USA _82_, 820-823 (1985). - HORTON, D.G.: J. Hattori Bot. Lab. _55_, 199-208 (1984). - HUNECK, S.: Chemistry and biochemistry of bryophytes, 1-116. In: see ref. SCHUSTER 1983; - Proc. Third Meeting Bryologists from Central and East Europe, 25-36 (1984). - HUNECK, S., JÄNICKE, S., MEINUNGER, L., SNATZKE, C., CONNOLLY, J.D., ASAKAWA, Y.: J. Hattori Bot. Lab. _57_, 337-342 (1984).

ICHIKAWA, T., NAMIKAWA, M., YAMADA, K., SAKAI, K., KONDO, K.: Tetrahedron Lett. _24_, 3337-3340 (1983). - ICHIKAWA, T., YAMADA, M., NAMIKAWA, K., SAKAI, K., KONDO, K.: J. Hattori Bot. Lab. _56_, 209-213. - INOUE, H.: Proc. Bryol. Soc. Japan _2_, 101 (1979); - The Genus Plagiochila in Southeast Asia, 142 pp. Tokyo: Academic Scientific Book 1984; - Notes on some species of Plagiochila from Southern Chile and Argentina, 97-108. In: Studies on Cryptogams in Southern Chile, ed. H. INOUE. Tokyo: 1984a; - Studies on oil bodies and chromosomes of some hepatics from southern Chile, 73-96. In: Studies on Cryptogams in Southern Chile, ed. H. INOUE. Tokyo: 1984b; - J. Jap. Bot. _59_, 344-349 (1984c). - J. Hattori Bot. Lab. _57_, 443-454 (1984d). - INOUE, H., HIBINO, R.: Bull. Natl. Sci. Mus. Tokyo Ser. B _10_, 177-189 (1984). - IRELAND, R.R.: Bryologist _87_, 355-360 (1984); - IRELAND, R.R.: Evansia _2_, 4-9 (1985). - IRELAND, R.R., CRUM, H.: Bryologist _87_, 371-373 (1984). - ISOVIITA, P.: Taxon _34_, 534-536 (1985). - ISOVIITA, P., KOPONEN, T.: Taxon _33_, 736-739 (1984). - IWATSUKI, Z.: Proc. Bryol. Soc. Japan _2_, 70 (1978); - ibid. _2_, 95-98 (1979). - IWATSUKI, Z., GLIME, J.M.: J. Hattori Bot. Lab. _55_, 495-500 (1984). - IWATSUKI, Z., HIGUCHI, M.: Proc. Bryol. Soc. Japan _4_, 23-28 (1985). - IWATSUKI, Z., INOUE, S.: J. Hattori Bot. Lab. _57_, 343-362 (1984).

JONES, E.W.: Cryptogam. Bryol. Lichenol. 5, 159-171 (1984); - J. Bryol. 13, 385-398 (1985); - ibid. 13, 497-508 (1985a).

KATO, K., HORI, H., OSAWA, S.: J. Hattori Bot. Lab. 56, 133-138 (1984). - KITAGAWA, N.: Acta Phytotaxon. Geobot. 35, 1-7 (1984). - KOPONEN, T., NORRIS, D.H.: Ann. Bot. Fennici 20, 15-29 (1983); - ibid. 20, 31-40 (1983b); - ibid. 21, 265-290 (1984); - ibid. 131, 1-51 (1985); - ibid. 131, 97-127 (1985b). - KRON, K.A.: Am. J. Bot. 72, 794 (1985). - KRUIJT, R.CH.: Nova Hedwigia Beih. 80, 155-163 (1985). - KRUIJT, R.CH., GRADSTEIN, S.R.: Nova Hedwigia Beih. 80, 165-172 (1985). - KUMAR, D., UDAR, R.: Third Int. Congress Syst. Evol. Bot., 103 (1985). - KUMAR, S.S.: J. Hattori Bot. Lab. 55, 219-226 (1984). - KUMAR, S.S., VERMA, S.K.: Hikobia 9, 65-70 (1984). - KUTA, E.L., PRZYWARA, L., OCHYRA, R.: Bryol. Beitr. 3, 28-45 (1984). - KUWAHARA, Y.: Hikobia 9, 31-42 (1984).

LEWINSKY, J.: Lindbergia 9, 200-203 (1984a); - J. Hattori Bot. Lab. 56, 359-460 (1984b); - Lindbergia 10, 65-94 (1985). - LIGHTOWLERS, P.J.: J. Bryol. 13, 369-375 (1985); - British Antarct. Surv. Bull. 67, 41-77 (1985a). - LIN, P.-J.: J. Hattori Bot. Lab. 55, 267-272 (1984). - LIN, S.H.: Biol. Bull. Tunghai Univ. 59, 1-54 (1984); - J. Hattori Bot. Lab. 55, 295-301 (1984a); - Yushania 1, 1-6 (1984b); - Biol. Bull. Tunghai Univ. 60, 747-748 (1984c). - LINS, S.H., TSAI, C.C.: J. Taiwan Mus. 37, 81-100 (1984). - LONGTON, R.E., SCHUSTER, R.M.: Reproductive biology, 386-462. In: see ref. SCHUSTER 1983.

MARKHAM, K.R., MUES, R.: Z. Naturforsch. Sect. C 39, 309-310 (1984). - MATTERI, C.M.: Darwinia 25, 143-162 (1984). - MCQUEEN, C.B.: Bryologist 88, 1-4 (1985). - MEENKS, J.L.D., DE JONG, C.: Cryptogam. Bryol. Lichenol. 6, 1-24 (1985). - MEHRA, P.N.: Abstract. Recent Advances in Cryptogamic Botany, 152. Lucknow: Paleobotanical Society 1982. - MEINUNGER, L.: Proc. Third Meeting Bryologists Central and Eastern Europe, Praha, 143-166 (1984). - MILLER, N.G.: Bryologist 88, 171 (1985). - MISHLER, B.D.: Abstract. Botanical Society of America, 1-2 (1984); - Systematic studies in the genus Tortula. Thesis Univ. Ann. Arbor, 236 pp. (1984a); - Abstract. Third International Congress of Systematic and Evolutionary Biology, 128 (1985); - Bryologist 88, 207-214 (1985a). - MISHLER, B.D., CHURCHILL, S.P.: Brittonia 36, 406-424 (1984); - ibid. 37, 282-285 (1985). - MIZUTANI, M.: J. Hattori Bot. Lab. 57, 427-442 (1984); - ibid. 57, 153-170 (1984a); - ibid. 56, 331-338 (1984b); - Proc. Bryol. Soc. Japan 4, 3-4 (1985). - MOGENSEN, G.S.: The spore, 325-342. In: see SCHUSTER 1983. - MOHAMED, M.A.H.: J. Hattori Bot. Lab. 55, 281-293 (1984). - MOHAN, G.: Briofite, Determinator Illustrate al Brioflorei Romaniei, 643 pp. Bucarest: 1984. - MUES, R.: Proc. Third Meeting Bryologists Central and East Europe 37-66 (1984); - Bryol. Times 33, 1-3 (1985). - MUES, R., HATTORI, S., ASAKAWA, Y., GROLLE, R.: J. Hattori Bot. Lab. 56, 227-239 (1984). - MUHLE, H.: Cour. Forsch. Inst. Senckenberg 52, 277-281 (1982). - MURRAY, B.M.: Bryologist 87, 24-36 (1984).

NEWTON, M.E.: Cytology of the Hepaticae and Anthocerotae, 117-148. In: see ref. SCHUSTER 1983; - The cytogenetic of bryophytes, 65-96. In: The Experimental Biology of Bryophytes, eds. A.F. DYER, J.G. DUCKETT. London: 1984; - J. Bryol. 13, 263-268 (1984a); - Abstract. Third International Congress of Systematic and Evolutionary Biology, 134 (1985); - Bull. Brit. Bryol. Soc. 45, 9 (1985a). - NINH, T.: J. Hattori Bot. Lab. 57, 1-39 (1984). - NISHIMURA, N.: Bull. Natl. Sci. Mus. Tokyo Ser. B 10, 153-158 (1984). - NISHIMURA, N., HIGUCHI, M., SEKI, T., ANDO, H.: J. Hattori Bot. Lab. 55, 227-234 (1984). - NORDHORN-RICHTER, G.: J. Bryol. 13, 15-18 (1984). - NORRIS, D.H., KOPONEN, T.: Acta Bot. Fennica 131, 1-51 (1985).

OCHI, H.: N.Z. J. Bot. 22, 179-182 (1984). - OCHYRA, R., SZMAJDA, P.: Series V Musci Part I. In: Atlas of Geographical Distribution of Spore Plants in Poland, eds. J. SZWEYKOWSKI, T. WOJTERSKI. Warszawa: 1983. - ONRAEDT, M.: Cryptogam. Bryol. Lichenol. 6, 151-175 (1985). - ORBAN, S., VAJDA, L.: Magyarorszag Mohaflorajanak Kezikönyve, 518 pp. Budapest: Akademiai Kiado 1983. - O SHEA, B.J.: Brit. Bryol. Soc. 44, 35-39 (1984).

PENTECOST, A., ROSE, F.: Bot. J. Linn. Soc. 90, 217-230 (1985). - PIIPPO, S.: Ann. Bot. Fenn. 21, 21-48 (1984); - ibid. 21, 309-335 (1984a); - ibid. 131, 89-96 (1985); - ibid. 131, 128-167 (1985a); - ibid. 131, 168-179 (1985b). - POCOCK, K., DUCKETT, J.G.: Bull. Brit. Bryol. Soc. 45, 10-11 (1985); - Bryol. Times 31, 2-3 (1985a); - New Phytologist 99, 281-304 (1985). - POCOCK, K., DUCKETT, J.G., GROLLE, R., MOHAMED, M.A.H., PANG, W.C.: J. Bryol. 13, 241-246 (1984). - POCS, T.: Proc. Third Meeting Bryologists Central and East Europe, 107-119 (1984); - Cryptogam. Bryol. Lichenol. 5, 239-267 (1984a); - J. Hattori Bot. Lab. 55, 307-313 (1984b). - POSPISIL, V.: Proc. Third Meeting Bryologists Central and East Europe, Praha, 165-175 (1985). - PRESTON, C.D.: J. Bryol. 13, 43-95 (1984); - ibid. 13, 97-100 (1984a). - PURI, P.: Bryophytes: Morphology, Growth and Differentiation, 310 pp. Delhi: 1981. - PRZYWARA, L., KUTA, E., OCHYRA, R.: J. Hattori Bot. Lab. 57, 127-137 (1984a). - PRZYWARE, L., OCHYRA, R., KUTA, E.: Lindbergia 9, 178-185 (1984b). - PURSELL, R.A.: J. Hattori Bot. Lab. 55, 235-252 (1984).

RAMSAY, H.P.: Cytology of mosses. Chapter 2 in: see ref. SCHUSTER 1983; - Telopea 2, 455-533 (1984); - ibid. 2, 535-547 (1984a). - RAMSAY, H.P., LEWINSKY, J.: N.Z. J. Bot. 22, 345-351 (1984). - REESE, W.D.: Lindbergia 10, 123-126 (1984); - Phytologia 56, 306 (1984a); - ibid. 56, 305 (1984b); - Bryologist 87, 217-222 (1984c). - RICHARDS, P.W.: J. Bryol. 13, 33-38 (1984). - ROBINSON, H.: Abstracts, Botanical Society of America, 9-10 (1984); - Brittonia 37, 279-281 (1985). - ROBINSON, H., SHAW, J.: Bryologist 88, 293-296 (1985). - ROGEON, M.A., SCHUMACKER, R.: Bull. Soc. Bot. Centre-Oeust N.S. 15, 81-102 (1984). - RUSHING, A.E.: Am. J. Bot. 72, 75-86 (1984).

SALAZAR, A.N.: Am. J. Bot. 72, 796 (1985). - SANT ANNA, C.L.: Rickia 11, 129-142 (1984). - SCHOFIELD, W.B.: J. Hattori Bot. Lab. 55, 35-43 (1984); - Introduction to Bryology, 431 pp. New York: Macmillan 1985. - SCHOFIELD, W.B., HEBANT, C.: The morphology and anatomy of the moss gametophyte. Chapter 11 in: see ref. SCHUSTER 1984. - SCHULTZE-MOTEL, W. (ed.): Advances in Bryology, Vol. 2, 224 pp. Vaduz: CRAMER 1984. - J. Hattori Bot. Lab. 55, 303-305 (1984a). - SCHUMACKER, R.: Proc. Third Meeting Bryologists Central and East Europe, Praha, 135-141 (1984). - SCHUSTER, R.M.: Phytogeography of the Bryophyta, 463-626. In: see ref. SCHUSTER 1983; - (ed.): New Manual of Bryology, Vol. 1. Nichinan: 1983. - Evolution, phylogeny and classification of the Hepaticae, 892-1070. In: see ref. SCHUSTER 1984; - Morphology, phylogeny and classification of the Anthocerotae, 1071-1092. In: see ref. SCHUSTER 1984 (1984a); - Phytologia 56, 65-74 (1984b); - Comparative anatomy and morphology of the Hepaticae, 760-891. In: see ref. SCHUSTER 1984 (1984c); - Phytologia 57, 369-373 (1985). - SCHUSTER, R.M., ENGEL, J.J.: J. Hattori Bot. Lab. 58, 255-539 (1985). - SCOTT, G.A.M.: Southern Australian Liverworts, 216 pp. Canberra: Australien Government Publishing Service 1985. - SCOTT, G.A.M., PIKE, D.C.: J. Hattori Bot. Lab. 56, 339-349 (1984). - SEE, A., GLIME, J.M.: Cryptogamie Bryol. Lich. 5, 79-85 . - SELKIRK, P.M.: J. Hattori Bot. Lab. 55, 105-111 (1984). - SERGIO, C.: J. Hattori Bot. Lab. 56, 7-13 (1984). - SHACKLETTE, H.T.: J. Geochem. Explor. 21, 89-93 (1984). - SHARP, A.J.: J. Hattori Bot. Lab. 56, 15-17 (1984). - SHARP, A.J., GARDNER, R., GIESY, R.M.: Phytomorphology 28, 329-331 (1978). - SHAW, A.J.: Phylogeny and Ecology of *Pohlia* Hedw. with a revision of the Species of North and Central America, 272 pp. Ann Arbor: University Microfilms 1983. - SHAW, J.: Bryologist 87, 132-142 (1984); - ibid. 87, 314-318 (1984a); - ibid. 88, 28-30 (1985); - ibid. 88, 199-206 (1985a); - ibid. System. Bot. 10, 224-233 (1985b). - SHAW, J., CRUM, H.: J. Hattori Bot. Lab. 57, 363-381 (1984). - SHAW, J., FIFE, A.J.: J. Hattori Bot. Lab. 57, 285-298 (1984). - SHAW, J., ROBINSON, H.: J. Hattori Bot. Lab. 57, 319-335 (1984). - SHAW, J., ROHRER, J.R.: J. Hattori Bot. Lab. 57, 41-61 (1984). - SLAGEREN, M.v.: Proc. Third Meeting Bryologists Central and East Europe, 121-130 (1984). - SLAGEREN, M.W., KRUIJT, R.CH.: Nova Hedwigia Beih. 80, 113-154 (1985). - SMITH, D.K.: J. Hattori Bot. Lab. 56, 281-286 (1984). - SPIESS, L.D., LIPPINCOTT, B.B., LIPPINCOTT, J.A.: J. Hattori Bot. Lab. 55, 67-77 (1984); - Bot. Gaz. 145, 302-307 (1984a). - STARK, L.R.: Am. J. Bot. 71, 2 (1984); - ibid. 72, 797-798 (1985). - STEERE, W.C.: J. Hattori Bot. Lab. 56, 19-20 (1984). - STEINKAMP, M.P., DOYLE, W.T.: Can. J. Bot. 62, 1871-1879 (1984). - STOERMER, P.: Cryptogam. Bryol. Lichenol. 5, 135-141 (1984). - STONE, I.G.: J. Bryol. 13, 159-162 (1984); - ibid. 13, 129-130 (1984a); - ibid. 13, 19-24 (1984b); - ibid. 13, 353-357 (1985); - ibid. 13, 345-352 (1985b). - STOTLER, R.E., CRANDALL-STOTLER, B., GEISSLER, P.: Am. J. Bot. 72, 798 (1985). - SZWEYKOWSKI, J.: Species problems and taxonomic methods in bryophytes, 1130-1171. In: see ref. SCHUSTER 1984; - J. Hattori Bot. Lab. 55, 209-218 (1984).

TAN, B.C.: J. Hattori Bot. Lab. 55, 13-22 (1984). - THIERS, B.M.: Brittonia 36, 174-177 (1984); - Lindbergia 10, 4-18 (1984a); - Bryologist 88, 350-352 (1985); - Am. J. Bot. 72, 798-799 (1985a); - Nova Hedwigia Beih. 80, 31-61 (1985b); - Evansia 2, 39-41 (1985c). - TIXIER, P.: Bryol. Times 29, 3 (1984); - C. R. Seances Soc. Biogéogr. 60, 131-134 (1984a); - Cryptogam. Bryol. Lichenol. 6, 177-180 (1985). - TOUW, A.: Lindbergia 9, 151-152 (1984). - TOVONEN, H.: Ann. Bot. Fenn. 22, 37-44 (1985). - TSVETKOVA, N.N., KRYUKOVA, Z.F., CHERNAYA, T.A., CHERNOBAEVA, M.B.: Literature Catalog of Bryophytes 1976-1980, 230 pp. Leningrad 1983.

UDAR, R., SHAHEEN, F.: J. Indian Bot. Soc. 62, 151-158 (1983). - UDAR, R., SRIVASTAVA, G.: J. Indian Bot. Soc. 62, 434-436 (1983); - Geophytology 15, 64-66 (1985). - UDAR, R., SRIVASTAVA, S.C.: J. Hattori Bot. Lab. 56, 97-103 (1984).

VANA, J., INOUE, H.: Bull. Natl. Sci. Mus. Tokyo Ser. B 9, 125-142 (1983). - VANDEN BERGHEN, C.: Bull. Jard. Bot. Bat. Belg. 54, 7-14 (1984). - VITAL, D.M.: Bryologist 87, 244-245 (1984); - Cryptogam. Bryol. Lichenol. 5, 15-26 (1984a). - VITAL, D.M., EGUNYOMI, A.: J. Bryol. 13, 39-41 (1984). - VITT, D.H.: Classification of the Bryopsida, 696-759. In: see ref. SCHUSTER 1984. - VITT, D.H., BUCK, W.R.: Brittonia 36, 300-306 (1984). - VITT, D.H., GRADSTEIN, S.R., IWATSUKI, Z.: Compendium of Bryology, 355 pp. Vaduz: Cramer 1985. - VOHRA, J.N.: Rec. Bot. Surv. India 23, 336 pp. (1983). - VOLK, O.H.: Mitt. Bot. Staatssamml. München 19, 453-465 (1983); - Nova Hedwigia 39, 117-143 (1984). - VOLK, O.H., PEROLS, S.M.: Bothalia 15, 117-124 (1984). - VYSOTSKAYA, E.I.: Ukr. Bot. Zh. 41, 29-33 (1984).

WATANABE, R.: Proc. Jap. Bryol. Soc. 2, 66-68 (19878). - WHITEHOUSE, H.L.K.: J. Bryol. 13, 131-133 (1984). - WU, C.L.: J. Hattori Bot. Lab. 56, 221-226 (1984). - WYATT, R.: Bryologist 88, 182-189 (1985). - WYATT, R., STONEBURGER, A.: Biosystematics of bryophytes, 519-542. In: Plant Biosystematics, ed. W.F. GRANT. New York: Academic Press 1984.

YAMADA, K.: Cryptogam. Bryol. Lichenol. 5 191-199 (1984); - J. Jap. Bot. 59, 91-96 (1984a); - J. Hattori Bot. Lab. 58, 111-130 (1985); -

J. Jap. Bot. 60, 257-264 (1985a); - Proc. Bryol. Soc. Japan 4, 22-23 (1985b). - YANO, O.: J. Hattori Bot. Lab. 56, 481-548 (1984); - Rickia 11, 35-42 (1984a); - Rev. Bras. Bot. 7, 57-63 (1984b). - YOKOUCHI, Y., SATAKE, K., AMBE, Y.: Bryologist 87, 323-326 (1984). - YUZAWA, Y.: J. Hattori Bot. Lat. 57, 391-403 (1984).

ZANDER, R.: Bryologist 88, 215-220 (1985). - ZANTEN, B.O. van: Naturwiss. Verein Hamburg, Sonderband 7, 49-64 (1983); - Acta Bot. Neerl. 33, 231-232 (1984). - ZIJLSTRA, G.: Bryol. Times 32, 3 (1985).

Prof. Dr. JAN-PETER FRAHM
Universität - Gesamthochschule - Duisburg
Fachbereich 6, Botanik
Postfach 101629
D-4100 Duisburg, FRG

II. Evolution and Classification of Seed Plants

By Hans-Helmut Poppendieck

1. Introduction

Seed plant systematics and economic botany have formed closer links in recent years. This is reflected not only by joint symposia of mutual interest (e.g. infraspecific classification), but also the increase of respective publications may well be an indication of funding favouring applied rather than basic research. The study of seed plant phylogeny has greatly benefited from the close integration of palaeontological and neontological data, and from their critical analysis. Cladistic methodology has proven to be a valuable research tool once the narrow classificatory approach is abandoned. Comparative DNA analysis has been used to resolve speciation events and phylogenies, but it is questionable whether such methods will serve as a panacea for all cases of conflicting evidence (if there is any at all). Among the groups in focus, Myrtales (p. 320). Hamamelidales (p. 319), Rubiaceae and Bromeliaceae may be mentioned. Biosystematics has proceeded further towards becoming a multidiscipline to be mastered only by research teams of increasing size. But still, progress in systematics relies considerably on comprehensive studies by individuals: Burmanniaceae (RÜBSAMEN 1986) and *Lysimachia* (VOGEL 1986) provide exemplary cases.

The value of traditional systematics for applied research has been impressively demonstrated by SPJUT (1986): A screening programme for anticancer products had to be put on a taxonomic basis after earlier programmes with an ethnomedical or chemotaxonomical bias had turned out to be unsatisfactory. For reasons of cost efficiency, and to avoid duplication of work, the samples will have to be taken from as many independent evolutionary lines as possible. Plant genera and phytogeographic regions provide the guidelines for a random screening programme which eventually will cover 85,000 species. The genus is considered to be the lowest category to correspond with chemical diversity. For the non-systematists, this will illustrate the practical predictive value of the discipline, and may give an idea that controversies on classificatory concepts are not purely academic. The plant taxonomist will find many interesting details: The "hollow curve phenomenon" is used as a decision module when the problem arises whether to discontinue unsuccessful screening of a genus; the list of the worlds most common plants will be a complementary result of this programme, and SPJUT (1986) proposes to produce a manual on these plants which should greatly facilitate plant identification.

PLUCKNETT and SMITH (1986) stress the importance of traditional disciplines for crop research. DUVICK (1986) points to the promises of biotechnology and the concomitant task to preserve genetic diversity. FARNSWORTH and SOEJARTO (1986) calculated the value of a pharmaceutical species threatened with extinction: It amounts to 203 million dollars! HEYWOOD (1985) gives a succinct analysis of the present situation of botanical gardens and outlines a concerted conservation strategy.

Progress in Botany, Vol. 49

Finally, JANZEN's (1986) thoughtful and provocative essay on the future of tropical ecology should be widely read.

2. Methodological Aspects

Introductory textbook: SIVARAJAN (1984). Taxonomic concepts in relation to computers, models and data bases: ABBOTT et al. (1986).

a) Phylogenetic Reconstructions and Evolutionary Processes

This review will concentrate on some points where the methodological advances are manifest; for current discussions on cladistics, see the new journal Cladistics, and those mentioned in Progr. Bot. 47, 240. In an increasing number of studies, cladograms are not the ultimate goal of the investigation, but rather a provisional result which is further used for critical character analysis, for comparing evolutionary hypotheses with geological and ecological data, and for a re-evaluation of the taxonomic structure (see Table 1); it is here that cladistics has its undisputed merits.

α) Exemplary Studies in Seed Plant Phylogeny. The close integration of palaeontological and neontological studies in recent years is a highly welcomed trend. Both CRANE (1985) and DOYLE and DONOGHUE (1986a,b) have presented cladistic studies comprising both recent and fossil groups.

Table 1. Some methodologically exemplary cladistic studies

Seed plants (DOYLE and DONOGHUE 1986b)	See p. 298. Re-evaluation of character evolution, retranslation into ideas on structural evolution
Seed plants (CRANE 1985; BREMER 1985)	High amount of homoplasies
Primitive angiosperms (DAHLGREN and BREMER 1985)	Excessive homoplasy suggests poly- or paraphyly within most lineages
Caprifoliaceae s.l., pollen (DONOGHUE 1985)	Detection of small but consistent differences previously overlooked confirms parallel evolution
Orchidaceae (BURNS-BALOGH and FUNK 1986)	Reconsideration of character evolution from cladogram
Pontederiaceae (ECKENWALDER and BARRETT 1986)	No unambiguous tree generated, subparsimonious trees accepted; identification of four independent dispersal events
Myrtales (JOHNSON and BRIGGS 1984)	Methodological parsimony may imply biologically impossible reversals
Harpullia, pollen: MULLER (1985), revision: LEENHOUTS (1985)	Pollen-based cladogram checked against geological and ecological evidence; detailed evolutionary narrative; difficult to express clear phylogenetic subdivisions taxonomically
Phyllanthoideae (G.A.L. LEVIN 1986a,b)	Homoplasy in leaf characters, but restricted even in this plastic character; phenetics unable to detect convergences
Phytolaccaceae (BROWN and VARADARAJAN 1985)	Probably paraphyletic; *Stegnosperma* and other "anomalous" genera to be kept out

DOYLE and DONOGHUE (1986b) have made a major step forward, not because they have finally solved the abominable mystery of angiosperm origin – we are still far from that! – but because they have attained an exemplary level for the discussion of both their subject and their methodology. There are indications that angiosperms, Bennettitales, *Pentoxylon* and Gnetales form a monophyletic group bridging the gap between angiosperms and gymnosperms. Within the likewise monophyletic Gnetales, *Gnetum* and *Welwitschia* are bound together by additional advances such as vessels, simple stamens, embryological and chemical (see p. 307) features. Though often cited as evidence for a close relationship to angiosperms, they probably represent parallelisms.

On the basis of their most parsimonious trees, DOYLE and DONOGHUE (1986b) offer various interpretations for the evolution of angiosperms. Although the angiosperm lineage may date back to the Triassic, this does not apply for all of its modern features. Alternatively, a Cretaceous derivation from Bennettitales seems only slightly less probable. In any case, many key adaptations such as insect pollination and rapid reproduction may have been inherited from gymnosperm precursors. DNA sequencing of extant seed plant groups is proposed as a means to test these ideas.

The results from CRANE (1985) have been incorporated into BREMER's (1985) summary of green plant classification. DAHLGREN and BREMER (1985) presented a first tentative cladogram for angiosperms, restricted to monocotyledons, centrosperms and magnolioid dicotyledons; the want for synapomorphies in most clades is taken to indicate their paraphyletic or polyphyletic nature, and it is hoped that a redifinition of most terminal taxa will make a more meaningful analysis possible. The most difficult obstacle to a satisfactory analysis of seed plants is the high number of homoplasies encountered, a trend which culminated in angiosperms (see below).

β) Limits and Prospects of Numerical Cladistic Studies. The advantage of numerical cladistic methods is their use in processing abundant data and avoiding biases derived from the concentration on a few favoured characters. Many critical studies, however, have provided somewhat disappointing results. This is due to the fact that parallelisms, reversals or horizontal gene transfers (e.g. by hybridization) can only be identified as long as they are comparatively rare; if they become frequent, it will be increasingly difficult to distinguish meaningful phylogenetic information from evolutionary noise. As in the cladistic studies cited above, the results are numerous most parsimonious cladograms, and even more supraminimal trees which may be equally acceptable or unacceptable (JOHNSON and BRIGGS 1984). For example, DOYLE and DONOGHUE (1986b) obtained no less than 36 most parsimonious trees with a step length of 123, which, for the 62 characters coded, means an average of one reversal or convergence per character. When the data were rearranged on the basis of widely different morphological concepts (e.g. those of MEYEN, BECK, WETTSTEIN), the resulting trees differed minimally from each other and the most parsimonious ones as far as step length was concerned. At this stage, the confidence put into these results can only be low, and the goal of obtaining a single, undisputed phylogenetic tree has to be discarded. It is here that phenetic and cladistic studies converge, but it is uncertain whether they converge on a true evolutionary relationship. JOHNSON and BRIGGS (1984) emphasized that the present methodology is inappropriate for these cases, and new ways for assessing the confidence of phylogenies have to be sought (e.g. FELSENSTEIN 1985c). The situation is aggrevated by the fact that many angiosperm taxa are ill-defined, and poly- or paraphyletic (STEVENS 1986; DAHLGREN and BREMER 1985); note that it took DNA sequence analysis to detect hidden paraphyly in *Heterogaura* (SYTSMA and GOTTLIEB 1986a,b, see p. 310).

Table 2. Some cases where evolutionary "noise" has been interpreted in a more or less unorthodox way

Nucleotide sequence basically similar in legume and non-legume hemoglobin (LANDSMAN et al. 1986)	Cryptic expression in non-symbiotic tissue, detectable only in N_2-fixing nodules which occur erratically in plants
Saccate pollen of Lactoridaceae similar to Mesozoic pteridosperms (ZAVADA and TAYLOR 1986)	Primitive character indicating the gradational nature of pollen characters in seed plants
Exineless pollen in some angiosperms (KRESS 1986)	Primivite character expressed only in some terminal taxa
Various characters in Labiatae (CANTINO 1985)	"Non-universally derived states" (earlier termed "apomorphic tendencies"), may be regarded as an indication of phylogenetic affinity because of evolution from a common basis
Reduction trends in the embryology of orchids and Burmanniaceae (RÜBSAMEN 1986)	Parallel evolutionary trends due to a similar genetic programme rather than convergence or synapomorphies
Various similarities between angiosperms and Gnetales (embryology, vessels, etc.) (DOYLE and DONOGHUE 1986)	Parallelisms, indicating close phylogenetic relationships, but not synapomorphies
Homeria-type flower evolving several times independently in *Moraea* (GOLDBLATT 1986b)	Involving only a few genic mutations (similar cases in crop plant evolution)
Trimery in monocots and magnolioid dicots (KUBITZKI 1987)	Phylogenetic constraints on ontogenetic development
3-4 styles in *Stipa* (FREITAG 1985)	Reversal to primitive condition
Geographical convergences: divaricate habit in New Zealand shrubs, phyllodes in Australia, etc. (BARTHELEMY 1983)	Horizontal gene transfer by "infectious" inheritance

At the same time, the value of cladistics as a research tool becomes strikingly apparent: Perhaps because it places emphasis on ordered patterns of evolution, it has brought those processes into focus which are at variance with them, such as homoiologies, parallelisms, reversals horizontal gene transfer and phylogenetic constraints upon ontogeny (see Table 2). It should be a challenge for evolutionary biology to provide interpretations for these phenomena in the light of present genetic knowledge. It should also be rewarding to investigate whether the notoriously difficult seed plant groups have common evolutionary features; these groups include angiosperms as a whole, grasses and "older" legume genera such as *Baphia* (SOLADOYE 1985), and *Ocotea* (ROHWER 1986). Note that in cases where characters appear to be combined at random, irreversible processes like polyploidization or the switch to dioecism gain importance (BULL and CHARNOV 1985).

γ) Further Aspects. Symposium on phylogenetic systematics and comparative biology: Ann. Missouri Bot. Gard. 72, 591-822 (1985). Introductions to cladistics: FUNK and CRACRAFT (1985), FUNK (1986). Key to common errors of cladistics: ROBINSON (1986). Thoughtful, traditionally oriented comments: BUCK (1986). Symposium on character weighting: Syst. Zool. 35, 100-134 (1985): a priori weighting, the only non-circular approach (NEFF 1986); a posteriori weighting, the only workable solution (WHEELER 1985). Important statistical problems underlying phlyogenetic reconstructions: FELSENSTEIN (1985a,b). Review in defence of phenetics: SOKAL (1986). Even if the phylogeny, as in hypothetical organisms used for simulation studies, is fully known, various methods show deficiencies in reconstructing these phylogenies: FIALA and SOKAL (1985), SOKAL (1985), SOKAL et al. (1984).

b) Infraspecific Classification of Wild and Cultivated Plants

For the expression of variation at the infraspecific level in a formal system of classification, no universally accepted solution has been brought forward so far. Judging from a symposium volume on this topic (Syst. Ass. Spec. Vol. 29, 1-435, 1986), it will probably never be: Continuous variation and incongruent sets of data may force one to make arbitrary decisions (PRENTICE 1986b); cultivar classification, requiring great homogeneity, is at variance with the concept of land races (PICKERSGILL 1986), and both can be brought into agreement with botanical nomenclature only in a highly resolved and complicated scheme (LEWIS 1986); legal issues involved in plant breeding pose quite unexpected problems for the taxonomist, in particular the determination of the amount of genetic change necessary to make a variety distinct (LANGTON 1986). An open and multiple system of classification is favoured by many authors (e.g. BRANDENBURG 1986; HANELT 1986). For the operation of a multipurpose classification, the only workable means will be a multidimensional system based on electronic data processing, as is already applied to several groups of cultivated plants; the EDS data base (DERRICK et al. 1986) aims at a system subservient to all formal, informal and anecdotal (i.e. by tables, graphs) approaches to infraspecific taxonomy.

Note how different the situations may be: In forestry, the 200-year-old concept of provenances will have to be replaced because of the increasing domestication of forest trees; the challenge is to develop a system based on the genetic history of populations (TURNBULL and GRIFFIN 1986). In grasses, the classification scheme may relate to the date of flowering under different conditions (e.g. in *Lolium*) and requires likewise computer storing of multidimensional genecological studies (BREESE and TYLER 1986); economic importance forbids reduction into broad specific groups, as is often the solution to complex situations in wild plants (e.g. *Poa secunda*-complex: KELLOGG 1985). Classification of cultivated plants is much more user-oriented than that of wild plants (HEYWOOD 1986).

The lack of critical and illustrated manuals for the non-specialist has been pointed out by PARKER (1986); apparently, no funding is available for this important task. Modules for deciding on infraspecific subdivisions are discussed by LEWIS (1986). There is no need for cultivar names for varieties which are only marketed locally (LESTER et al. 1986).

c) Concepts of Classification

A central philosophical problem underlying the classificatory process is to what extent the derived patterns reflect the way in which we see the world rather than patterns in the world. WALTERS (1986), who first pointed out the socio-cultural European bias in angiosperm classification, and STEVENS (1985, 1986) should be read for recent contributions to a long-standing debate; they do not differ in their analyses, but rather in value judgements. While WALTERS (1986) advocates classification based on "epigenetic choices" (TURNER 1985), which means that they should be based as much as possible on previous ones, STEVENS (1985) argues equally convincingly that stability is bought at the price of loss of biological information.

Since systematics is an intellectual activity shaped by individuals (TURNER 1985), it is not surprising that museum taxonomists and teaching botanists with urgent obligations to students or the general public differ on generic concepts, the former being mostly "splitters", the latter "lumpers" (BARKLEY 1985). WIEGLEB's (1986) review of data analysis in plant ecology should be profitably studied by plant taxonomists. He reminds us that rigorism is a sign of immaturity in a scientific discipline. The studies cited above are distinguished by their balanced view. While the necessity as well as the intellectual fascination of these issues are obvious to him, the reviewer is uncertain as to their impact on the general scientific community: Perhaps systematists should be aware that while they discuss principles of classification, the results of good systematic research are increasingly being labelled as ecology or even physiology (WILSON 1985).

BHAT (1984) proposed another system of combined description and classification for the plant kingdom.

d) Institutional Bases and Means of Communication

Role of museum collections in biological research, symposium: MILLER (1985). Importance of herbaria: AMMANN (1986). Pest control in herbaria: BARROSO and VIERA (1985). Importance of living collections for cytological research: BÜCHLER (1986). History of botany and botanical gardens 1730-1840, bibliography: JOHNSON (1985). Background information on the systematic community: EDWARDS et al. (1985). Systematic resources of the Pacific basin: SOHMER (1985).

The Australian data bank on Gramineae generated a comprehensive classification down to supertribes (WATSON et al. 1985), a detailed description of 728 grass genera based on 300 characters (WATSON et al. 1986a), and an automated output of keys in English and French for Canadian genera (WATSON et al. 1986b). Ultimate goal is the development of on-line procedures, making printed keys superfluous, as have been presented by WILSON and PARTRIDGE (1986) and, for trees of Guiana, by FORGET et al. (1986). Computer-aided routines for Flora of Ecuador project: FROST-OLSEN and HOLM-NIELSEN (1986). Combined polyclave and dichotomous key: WESTFALL et al. (1986).

JANZEN (1986a) strongly advises against the burial of data in computers: For the tropics, printing has a much greater legitimacy, and is the only way of ensuring flow of information to the field worker.

3. Morphology

a) General Aspects

MEEUSE (1986) presents a summary of his controversial but stimulating ideas on plant morphology.

Characters overlooked because of their simplicity may, shrewdly employed, provide quite important data: In Umbelliferae, the shape of petals, including the distal lobulum inflexum, is very constant and may be used for keys and diagnoses in floristic works (SCHMITZ and FROEBE 1986). The ratio between stomata and trichome number in Bromeliaceae provides an evolutionary and ecological advancement index and gives clues for the history of migration in *Tillandsia* (WINKLER 1986).

Emendations and alterations have been proposed for several morphological classifications, mostly because similar structures can be produced by different ontogenies: This applies to seed coat anatomy (SCHMID 1986) and leaf venation architecture (CARR et al. 1986). Exemplary for this are the leguminous tribes Amorpheae and Psoraleae, where neither glandular dots in the leaves (TURNER 1986) nor floral structures, such as the so-called discontinuity plate (GRIMES 1986), may be homologous despite a first glance similarity. Further discussion on stomata classification: BARANOVA (1985), INAMDAR et al. (1986).

RUTISHAUSER and SATTLER (1985) stress the heuristic value of complementary morphological concepts. For the phyllode-stipule whorls of *Acacia verticillata*, juvenile pinnate leaves, phyllodes and stipules are interpreted as parts of a gradual and at times saltatory morphocline which connects structural categories nonhomologously to each other (RUTISHAUSER and SATTLER 1986); a similar interpretation is offered for the leaf whorls of Rubiaceae (e.g. *Galium*) by RUTISHAUSER (1984).

An interesting, though as yet not fully convincing hypothesis has been brought forward by HAINES and LYE (1986): A l l apocarpy in angiosperms, from Magnoliaceae to Ochnaceae and Sterculiaceae, should be considered as being secondarily derived from a primitive syncapous (capsular?) state. Secondary apocarpy, associated with a switch from animal to wind dispersal, has been confirmed for Apocynaceae by FALLEN (1986) and TRAN DINH LY (1986). Furthermore, the unilocular ovary of *Allamanda* may even be a case of secondary syncarpy derived from secondary apocarpy (FALLEN 1985).

Highly complicated fruits with specialized ejection mechanisms may be especially prone to parallelism, as could be shown for Mesembryanthemaceae (HARTMANN and LIEDE 1986) and for *Geranium* (YEO 1984), where seed ejection has evolved several times independently from *Erodium*-like ancestors.

b) Vegetative Characters

The significance of drip tips in tropical plants has been controversial for a long time. Based on extensive evidence, and resorting to ideas developed in the beginning of this century, ELLENBERG (1985) rightly dismisses most adaptive interpretations as being uncritical: Drip tips can only be formed where hot and humid conditions prevail and leaves are produced rapidly and are unprotected by buds. Tips are formed in early ontogeny of most leaves, but obscured by slow development. The ascribed drainage function (e.g. LIGHTBODY 1985) is at best an exaptation, a term recently coined by GOULD and VRBA (1982) and synonymous with the older German "Ausnutzung" (e.g. GOEBEL 1920). As in grass-herbivore relationships (COUGHENOUR 1985), it is often difficult to tell exaptations from adaptations.

JÜRGENS (1985) postulated that epidermal storage tissues, as found in various unrelated families (e.g. Piperaceae, Gesneriaceae, Mesembryanthemaceae and even Polypodiaceae), denote the starting point for the evolution of more advanced types of leaf succulence, as found in Crassulaceae and Mesembryanthemaceae. In the latter family, a clear correlation between the type of succulence and climatic conditions has been found (JÜRGENS 1986). IHLENFELDT (1985) differentiates between all-cell succulents, where water storage and photosynthetic activity take place in the same tissue, and more specialized storage succulents.

RURY and DICKISON (1984) compared the hydrovascular system of *Hibbertia* and *Erythroxylon* and concluded that the foliage is its most significant component; but also stem succulence, habit or root system exert a buffering influence on xylem. Discussions on the ecological properties of xylem may be ill-advised in the absence of correlative leaf structural data. Interestingly, semisucculent cane-like stems of Begoniaceae and

Piperaceae, show some ecologically correlated similarities such as vessels with scalariform pitting, little alterations of rays during ontogeny, etc. (CARLQUIST 1985).

Primary and secondary thickening meristems in Iridaceae are sequential developmental phases of the same meristem; secondary thickening may have evolved either anew from primary thickening, or may have been retained as a primitive character (RUDALL 1984). Variation in bulb structure in Amaryllidaceae (ARROYO 1984) and in Liliaceae (BARANOVA 1986) points to polyphyly in both families.

Systematic and ecological aspects of xylem anatomy: see Progr. Bot. 48, 393-395 (1986). Symposium on growth and form of modular organisms with readable essays on the "plant as metapopulation" theme: HARPER et al. (1986). Highly variable architecture in *Psychotria* makes application of the descriptive system of HALLE and OLDEMAN difficult: HAMILTON (1985).

Micromorphological characters: The supposedly primitive P-type sieve-tube plastids apparently have evolved independently de novo in Acanthaceae (BEHNKE 1986b). Eucryphiaceae should be dismembered, *Eucryphia* with P-type plastids is rightly placed near Connaraceae and Cunoniaceae, *Paracryphia* with S-type plastids is close to Dilleniaceae/Actinidiaceae/Theaceae (BEHNKE 1985). Stigma structures provide a wealth of data in Bromeliaceae (BROWN and GILMARTIN 1984). Floral nectaries in *Biscutella*: OLOWOKUDEJO (1986). Stylar morphology in *Eucalyptus*: BOLAND and SEDGLEY. Carpopodium (=abscission zone) in Compositae: HAQUE and GODWARD (1984). Corolla venation patterns in *Phacelia*: LEE (1986). Stegmata (= silica cells) in orchids: MØLLER and RASMUSSEN (1984). Stigmatic surfaces in angiosperms, Ericaceae and Liliaceae are heterogeneous: SCHILL et al. (1985).

c) Palynology and Embryology

Symposia and composite works: Pollen form and function with emphasis on systematics and evolution: BLACKMORE and FERGUSON (1986); biotechnology and ecology of pollen, partly interesting for systematics because of evolutionary aspects of pollen tube competition: MULCAHY et al. (1986); comprehensive textbook covering all aspects of angiosperm embryology: JOHRI (1984). Didactically well done and illustrated survey of embryological aspects of phylogeny and homology: FAVRE-DUCHARTRE (1984).

The intellectual fascination of pollen may be its ready availability as an intact cellular system in flowering plants, which otherwise are multicellular organisms (KNOX 1984); and the fact that phylogenetic, ontogenetic and functional constraints and their interactions are especially evident (CRANE 1986). The importance of the reproductive system for economic botany has furthermore led to increased studies on pollination dynamics.

CHALONER (1986) reviews the electrostatic properties of exine ornamentation which help to deposit the grain on the stigma, and add to the multiple functions of the exine. CRANE (1986) reviews adaptations for wind dispersal, BLACKMORE and BARNES (1986) harmomegathic devices, and THANIKAIMONI (1986) evolutionary trends of apertures.

While WALKER and WALKER (1984) present new versions of their ideas on the evolution of angiosperm pollen (see Progr. Bot. 37, 276 and 47, 300) which, among others, consider granular walls to be primitive, TAYLOR and ZAVADA (1986) point out that these types occur only in isolated groups, and suggest that alveolar types are more probably as candidates for angiosperm forerunners.

In *Erythrina*, a broad correlation exists between pollen type and hummingbird and passerine pollination respectively, but many open questions remain, e.g. when is selection pressure exerted by the most effective pollinator strong enough to disrupt established patterns of exine ornamentation? (HEMSLEY and FERGUSON). Similar results have been obtained for Neotropical species of *Bauhinia* (FERGUSON and PEARCE 1986). In Araceae, a psilate exine may indicate a weak adaptation to beetle pollination (GRAYUM 1986b). Broad correlations have also been found between climate and exine sculpture, harmomegathic devices and polyad grain number in *Acacia*; as for the latter, reduction in reproductive output seems to be vaguely associated with the increase in species number for the SW Australian region (GUINET 1986). Apparently, it is easier to interpret structures contributing to the overall fitness of pollen grains, such as harmomegathy, rehydration or energy storage, than to find correlations between individual characters and environmental variables. The close integration of the various functions of pollen may be responsible for this.

Spiroaperature pollen enables germination on numerous sites, and has evolved repeatedly in angiosperms (FURNESS 1985). The exineless pollen found in numerous angiosperm families shows little correlation to the pollination system but should be viewed in terms of the phylogeny of the respective groups; for Zingiberales, KRESS (1986) even assumes that this is a primitive character expressed only in some terminal taxa.

Phylogenetic specialization of the echinolophate grains in *Scorzonera* is reflected by the involvement of the callose wall to a greater extent than in the more simple echinate pollen of *Cosmos* (BARNES and BLACKMORE 1986). For Saxifragaceae, columella-derived structures are assumed to be more primitive than tectum-derived ones (HIDEUX and ABADIE 1986).

Orchidaceae: Compact pollinia are formed due to neotenic processes, with exine formation occurring successively earlier in ontogeny (WOLTER and SCHILL 1986), elastoviscin is homologous to pollenkitt (SCHILL and WOLTER 1986); easy degradation of pollen and pollinia (e.g. in acetolysis) may explain their absence from the fossil record (WOLTER and SCHILL 1985). Annonaceae: Review of pollen evolution, starting from simple grains to compound ones: LE THOMAS et al. (1986); GOTTSBERGER and SILBERBAUER-GOTTSBERGER (1985) maintain that both single grains and polyads may be basal for the family. Progression from non-columellate to columellate exine from *Malmea* to *Ephedranthus*: WAHA (1985). Coherence of Neotropical and Asian *Anaxagorea* is stressed: HESSE et al. (1985). Zonoaperture pollen of *Guatteria* unique in angiosperms, some resemblances in Eupomatiaceae, Nymphaeaceae, Rapateaceae: MORAWETZ and WAHA (1985). Araceae: Repeated evolution of trinucleate pollen: GRAYUM (1986a). Frequent reversals to and from small/starchless and large/starchy grains: GRAYUM (1985). Various groups: Buzz-pollinated Solanaceae have pollen exceptional for its protein and nitrogen content: BUCHMANN (1986). Threadlike exine connections in *Jaqueshuberia* are not viscin threads, but are "functionally homologous": PATEL et al. (1985).

A scutellum which grows into and through the endosperm helps to exploit the nutrients more rapidly, and adds to the competitive success of *Avena* and some non-cereal Gramineae (NEGBI and SARGENT 1986). The similarity of the ramified protocorm of *Zostera* to the grass embryo is taken to interpret Zosteraceae as reduced aquatic Gramineae (JACQUES-FELIX 1985). In Compositae, endosperm has only a protective and no nutritive function; generally, reserve polysaccharides may be instructive for the systematics of dicotyledons (GRAU and HOPF 1985). Additions to the descriptive system of endosperm: DI FULVIO and COCCUCCI (1986).

Evolutionary trends for the tapetum in land plants include the intrusion of tapetal cells between the spores, the loss of tapetal cell walls and increased nutrition through direct contact in narrower anthers; the secretory tapetum with its rather ineffective nutritive system is primitive, various other types can be derived (PACINI et al. 1985). HESSE (1986) stresses the homology between tapetum and sporogenous tissue which contain nearly identical genetic information for sporopollenin production; Ubisch bodies are concretions homologous to exinous sporopollenin, without a functional role. HIDEUX and ABADIE (1986) suggest an intricate regulatory mechanism between the two types of surfaces to be responsible for exine formation. Secretory tapetum in *Viburnum* and *Sambucus* stresses their isolation from Caprifoliaceae which is characterized by various forms of an amoeboid tapetum (WEBERLING and HILDENBRAND 1986).

4. Kayology and Cytogenetics

There is growing evidence that the formation of triploid-derived lineages is more common than usually assumed, not only in vegetatively reproducing taxa (*Amorphophallus*: CHAUHAN and BRANDHAM 1985b), but also in tropical trees and in anuals: KADEREIT (1986c) presents evidence that *Papaver somniferum* (2n=22) and its 4n subspecies *P. setigerum* (morphologically and chemically distinct: LA VALVA et al. 1985) may have arisen via a triploid hybrid from a cross between *P. glaucum* (2n=14) and *P. gracile* (2n=28), which thus confirms the idea of OJALA and ROUSI (1985), who suspected an ancient allopolyploid origin from observations on chromosomal behaviour of intersectional hybrids. KADEREIT (1986c) points out that infragenomic pairing, still rudimentarily observable in *P. somniferum*, which is the precondition for the formation of fertile triploid hybrids, has been repeatedly reported recently. Since n=11 occurs in unrelated groups of Papaveraceae, whose base number is x=7, the family may possess a predisposition for forming triploids. Triploid formation and subsequent diploidization is also hypothesized for some woody tropical groups by MORAWETZ (1986b); another general result of his wide ranging study is that there is no fundamental difference to temperate groups in karyotype evolution (MORAWETZ 1986a), polyploidy being rather frequent, particularly in the Neotropics.

Further results from MORAWETZ (1985, 1986a,b): Since Annonaceae are nearly the only diploid members of the Magnoliidae, and x=8 ist most frequent here, it is considered to be the base number for Laurales and Magnoliales; karyomorphological schemes for various groups of this complex show how the same chromosome number can be reached by different means; Aristolochiaceae and Annonaceae are clearly related; the erratic occurrence of a karyotype with unusual condensing behaviour is taken to indicate its primitive status; the higher incidence of polyploidy in Neotropical Cochlospermeaceae, Annonaceae and Myristicaceae as compared to Palaeo-tropical members may be due to a more diverse ecological history.

In *Viola* sect. *Melanium*, ascending or descending dysploidy from basal x=11 to n=17 or n=3, has been found. Backcrosses between diploids (mainly Balkan peninsula) and polyploids (mainly Greece) give rise to aneuploids which are either stabilized by elimination or double splitting of univalents, or eliminated because of lower viability (ERBEN 1985).

Chromosome numbers are known to vary considerably within species, populations, individuals and organs in different groups of *Ornithogalum*, where vegetative multiplication allows the diffusion of a quasi-random chromosomal variability. In the *O. divergens* group this may range from 2n=11 to 2n=108, but the intra-individual variation may facilitate crosses between species with a different chromosomal formula (COUDERC et al. 1985). In *O. angustifolium*, aneuploidy in seedlings of 2n=27 progeny ranges from 26 to 36; selection favours euploid numbers, the odd-numbered polyploids being fertile and even-numbered ones sterile (RAAMSDONK 1985a). Crossing experiments reveal the *O. umbellatum* group to be coherent (RAAMSDONK 1985b), its evolution being shaped by hybridization and polyploidization in relation to glaciation events; karyotypes may give hints to the history of introduction into northern Europe (RAAMSDONK 1986). Swiss *Mercurialis perennis* populations consist of an (auto-) polyploid/aneuploid complex ranging from 2n=48 to 2n=101; the males are mostly fertile, no sex having a particular chromosome number (KRÄHENBÜHL 1984).

Salix × *hegetschweileri* (2n=76,95,114) forms a triple hybrid complex where according to ploidy level either *S. phylicifolia* and *S. nigricans* (2n=114) or the 2n=28 *S. hastata* shows stronger introgression (BÜCHLER 1986). In the *Tormentillae* group of *Potentilla*, the diploid level is extinct, the two 4n species intergrade towards the 6n and 8n level leading to the 8n *P. anglica* (IETSWAART and KLIPHUIS 1985). *Betula murrayana* originated by repeated combinations of reduced and unreduced gametes from the respective parents (BARNES and DANCIK 1985).

Hybridogenous apomictic and polyploid groups have continued to receive attention, such as the *Arnica frigida* complex (DOWNIE and DENFORD 1986). *Euphrasia minima* (VITEK 1986), *Festuca vivipara* (PILS 1985) and various complexes within *Antennaria* (BAYER 1985a,b; BAYER and CRAWFORD 1986). For the North Atlantic region, the neopolyploids are taken to define a distinct phytoelement (KNABEN 1985). In Asia, polyploids of *Eupatorium* are restricted to recent man-made habitats, while speciation has occurred mainly at the diploid level; this contrasts with America, where apomictic polyploids are widespread, and diploids relict (WATANABE 1984, 1986). Since polyploidy and parthogenesis are often associated in plants, it is difficult to say which factor is responsible for the observed ecological effects (BIERZYCHUDEK 1985). In *Antennaria*, apomictic races have been found to produce more seeds, but their survivorship is lower than that of sexual species (MICHAELS and BAZZAZ 1986). Polyploids are not intrinsically better adapted than diploids, and their superior fitness is due to increased heterozygosity (lacking, of course, in autotetraploids); this has been confirmed again from a long-term study by STEBBINS (1985). Whereas in northern hemisphere groups glaciation events have shaped these phenomena, adaptations to xeric climates seem to have had similar effects for the evolution of Australian species of *Bulbine* (WATSON 1986a,b). In addition to intersectional hybridization (lacking in other regions) and apomixis, interference with introduced as well as cultivated species complicate the taxonomy of South African *Rubus* (SPIESS and DU PLESSIS 1985).

Somatic aneuploidy in roots of bromeliads prevails especially among epiphytic species where roots have lost most of their physiological functions; meiotic counts nearly exclusively give n=25,50,75, etc. based on x=25 (BROWN and GILMARTIN 1986).

Increased heterozygosity and allozyme multiplicity have been found in well-documented autotetraploids in *Tolmiea*: SOLTIS and RIESEBERG (1986). Meiotic irregularities point to the recent allotetraploid origin of *Paeonia officinalis*: SCHWARZACHER-ROBINSON (1986). Autosyndetic pairing confers fertility to tetraploid *Gibasis* hybrids, the diploid ones being highly sterile: KENTON and JONES (1985).

Oenothera wolfii evolved from its homozygous ancestor recently by accumulation of reciprocal translocations and acquisition of balanced lethals; systems of suppressing homozygotes present in structural or complex heterozygotes with the same basic genotype may have given rise to the present hybrid species with fundamentally different complexes (WASMUND and STUBBE 1986). Translocation heterozygosity is "pumped" into the seedbank and acts against the deleterious effects of inbreeding in *Clarkia williamsii* whose populations fluctuate drastically over the years (S.C. PRICE et al. 1985). Chromosome repatterning leading to redundancy of translocations and concomitant self-incompatibility in Hawaiian Madiinae (Compositae) make this group an exception among the Hawaiian flora where, after initial colonization, self-compatibility prevails (G.D. CARR and KYHOS 1986; G.D. CARR et al. 1986; monograph: G.D. CARR 1985).

5. Chemical Systematics and Evolution

Supplement volume of chemotaxonomic treatise: HEGNAUER (1986); biology and chemistry of alkaloids: MOTHES et al. (1985), of isoquinoline alkaloids: PHILIPPSON et al. (1985), of plant phenolics: SUMERE and LEA (1985); polyphenols in secondary metabolism and as plant defences, review: BEAST et al. (1985); German textbook of ecological biochemistry: SCHLEE (1986); concise introduction (in German) to principles of allelochemistry in plants: HARTMANN (1985). WATERMAN (1986) reports the dual importance of phytochemical work in tropical regions in ecosystem research and in training. Symposium on plant allelochemistry: COOPER-DRIVER et al. (1985). Community patterns and mechanism in plant-herbivorous insect interactions: STRONG et al. (1984).

a) Phytochemical Systematics

VALANT-VETSCHERA (1985) stresses the necessity to compare chemical data with other chemical data prior to comparison with morphological data in order to avoid overestimation of congruence. Biogenetic pathways, accumulation tendencies, substitution trends and co-occurrence of different flavonoid types should be used to asses C-gylcosylflavone evolution in a critical manner comparable to that used in morphology.

The leguminous Tephrosieae are unique in accumulating non-protein guanidino derivates in seeds, and are suggested to include Lonchocarpinae; some chemical groupings coincide with morphological ones. A phylogenetic analysis of these compounds may be fruitful once the biogenetic relationships have been clarified (EVANS et al. 1985). Non-protein amino acids are highly toxic and taxonomically restricted: Hence, their importance for deterring generalist feeders (*Vigna*: BIRCH et al. 1985) and their utility for systematic studies (*Diphysa*: LAVIN 1986).

The detection of gnetins in *Welwitschia* (and some erratic occurrences in angiosperms) adds to the list of characters common between this genus and *Gnetum* (LINS et al. 1986). A certain chemical similarity to the Rosiflorean block of angiosperms (GOTTLIEB and KUBITZKI 1984), however, should be viewed as a parallel development in the field of transition between angiosperms and gymnosperms.

Evolutionary advancement is coupled with simplification of flavonoid patterns in *Circaea* (AVERETT and BOUFFORD 1985) and in *Rhododendron* subgen. *Vireya* (HARBORNE 1986b). In *Robinsonia* on Juan Fernandez, however, adaptive radiation is associated with a gain in flavonoid complexity (PACHECO et al. 1985).

Introductory review of ecology of cyanogenesis: NAHRSTEDT (1985). Cyanogen profiles in Turneraceae (SPENCER et al. 1985b), Malesherbiaceae (SPENCER and SEIGLER 1985a) and primitive tribes of Flacourtiaceae stress the relationship and give clues for the origin of Passifloraceae (SPENCER and SEIGLER 1985).

b) Evolutionary Aspects of Plant Allelochemicals

An attractive hypothesis maintains that chemically mediated interactions between plants and herbivores shaped the evolution of both groups, up to the origination of major clades through the acquisition of novel allelochemicals or detoxification mechanisms. But it still remains difficult to relate the complex situations encountered in case studies to the few and necessarily simplistic models so far proposed. ABRAMS (1986) questions the arms race analogy which seems to be more suitable for prey-predator than for plant-herbivore relations. COLEY et al. (1985) find resource availability to be crucial: Slow-growing plants on limited resources, mostly with long-lived leaves, will have quantitative defences such as tannins; fast-growing ones will resort to initially more costly qualitative defences in a rapid turnover. Thus, caffein content in *Coffea* leaves is initially high and decreases as mechanical protection of the leaves takes over (FRISCHKNECHT et al. 1986), the metabolic costs of caffein production may be up to 15% of the carbon used for respiration. Caffeic acid functions as a sink for surplus cinnamic acid, but is by no means ubiquitous in dicotyledons (MØLGAARD 1985): It is concentrated in advanced groups where proanthocyanidin has lost its importance and little lignin is formed, e.g. Ranunculales, Asterales, or in Caryophyllidae, as its derivative, ferulic acid.

Resistance of *Pastinaca sativa* to the apparently co-evolved parsnip webworm is mediated by furanocoumarins whose amount and composition are variable between plants and plant parts, but genetically controlled. Since resistance seems to be linked to disadvantageous growth characteristics, which lessen the competitive ability of *Pastinaca* in the absence of herbivory, both partners appear to have reached a stalemate condition in the co-evolutionary arms race (BERENBAUM et al. 1986; BERENBAUM and ZANGERL 1986). In other cases, a large number of highly diverse plants is associated with a few insect herbivores, so that plant natural product diversity cannot be responsible for patterns of host specificity, e.g. in the *Passiflora/Heliconius* alliance (SMILEY and WISDOM 1985). In Solanaceae this is taken to indicate recent (ithominiid butterflies: DRUMMOND 1986; VASCONCELLOS-NETO 1986) or incipient colonization (chrysomelid beetles: HSIAO 1986).

RHOADES (1985) adds to his concept of apparency (see Progr. Bot. 43, 201) by classifying herbivores into two types: stealthy ones have low, relatively invariant population levels and minimize damage and defensive responses of their food plant; opportunistic ones exert stress on plants by mass attack and take advantage of the lowered resistance thus induced.

Methodological aspects: The influence of the natural environment is often neglected in laboratory studies (MADDOX and CAPPUCINO 1986). A high variation of secondary metabolites has been found in grasses both within seasons and over the years (LINDROTH et al. 1986). Leaf monoterpenes are variable only in young leaves of *Sequoia* (HALL and LANGENHEIM 1986).

The life history of a rare butterfly feeding on *Zamia floridana* is an example for the risks of an aposematic life-style: ROTHSCHILD et al. (1986). Chemical races in *Chaerophyllum aureum* correspond to phytosociological status: GONNET (1986). Herbivory lessens resin production which is at present difficult to explain: JOHNSON and BRAIN (1985). Cultivation of early harvested potato cultivars induces the Colorado bettle to switch to weedy *Solanum* species: HARE and KENNEDY (1986). Influence of rapidly induced chemical changes on herbivore populations is questioned: FOWLER and LAWTON (1985). Trichome length influences plant susceptibility to aphid attack: MORAN (1986). Herbivory on aberrant colour morphs in *Byttneria* increases with morph frequency: A.P. SMITH (1986).

c) Allelochemistry of Wild and Cultivated Plants

Apart from alkaloids, Solanaceae possess a wide range of allelochemicals which, however, have been studied mostly in cultivated plants. HARBORNE (1986) points out the necessity of screening wild Solanaceae for useful characters such as aphid resistance. LUNDGREN et al. (1985) found a dramatically narrow range of leaf volatiles in cultivated tomatoes when compared with wild species; this points to a complicated genetic background within the latter. Even if these are specialized on a single main insecticidal component, they will be less apt to undergo gene-for-gene contradaptations than single-resistance gene cultivars. In *Lupinus*, the loss of quinolizidine alkaloids can only be compensated by the application of insecticides (WINK 1985). Generally, domestication seems to be associated with selection against the plant's own protective properties, such as glycolalkaloids in the early history of the potato (JOHNS and OSMAN 1986), or high amounts of wax or lignin products in tropical crops which influence their palatability; in the latter case this seems to be compensated by an increase in phytoalexine responses (BRAGA et al. 1986). Also in *Vigna* and *Phaseolus*, resistance mediated by non-protein amino acids is enforced by synergistically acting factors; differential toxicity to predators from different continents shows that some co-evolution must have taken place (BIRCH et al. 1986).

d) Selected Defence Compounds

Sesquiterpene lactones react with thiol groups of vitally important compounds, such as enzymes, but no further major generalization between chemical structure and biological effects seems to be possible (review: PICMAN 1986a). Among the many activities of these compounds, which range from anti-tumorgenic to growth regulation, some studies on pest resistance (*Cichorium*: REES and HARBORNE 1985), herbivore deterrence (*Hymenaea* and *Copaifera*: NASCIMENTO and LANGENHEIM 1986; LANGENHEIM et al. 1986a,b) and germination inhibition (of Amaranthaceae and Chenopodiaceae: PICMAN 1986b) may be mentioned. The evolution of sesquiterpene lactones within angiosperms is characterized by diversification of structural types and amplification of oxidation status (EMERENCIANO et al. 1985); in Compositae, however, this trend has been reversed (EMERENCIANO et al. 1986).

Phototoxic polyacetylenes in Compositae derive their efficiency from excited state chemistry which makes a wide diversity of high energy reactions possible. Since the energy costs for the plants are relatively low, some genera such as *Tagetes* (thiophene compounds) possess a few other defence chemicals (McLACHLAN et al. 1986). Phototoxic benzopyranes and benzofuranes are prevented from being autotoxic by storage in resin ducts, and attack DNA of pathogens when a plant is wounded, as in *Encelia* (PROKSCH 1985).

6) Tracing Evolution

a) General Aspects

The state of the art of biosystematics is reviewed in a symposium volume edited by GRANT (1984). Forty years after CAMP and GILLY (1943), the discipline is characterized by a still growing number of subdisciplines. Consequently, (1) team research is inevitable because no individual can master all the respective skills, (2) some groups, such as *Claytonia*, continue to be studied with new techniques and (3) it becomes more and more difficult to arrive at a unifying picture (SEMPLE 1984). The incongruence of data sets has become a recurrent theme (PRENTICE 1986b). VICKERY (1984) demonstrated how different approaches "see" the situation from their own perspective, and add to our understanding in a complementary way. Translated into higher order systematics, this problem emerges as mosaic evolution (STEBBINS 1984) or heterobathmy (SOLO'EVA et al. 1985).

As early as 1971, WENT proposed infectious "horizontal" inheritance to be responsible for otherwise barely explicable phenomena, such as the accumulation of divaricate shrubs in New Zealand, or phyllodes in Australia; the idea is strengthened by the fact that only certain developmental phases are involved. BARTHELEMY (1983) re-examines this hypothesis in the light of current knowledge: There are growing indications that transposable elements, which may play some part in transferring genes between otherwise isolated lineages, makes species barriers permeable (KRIEBER and ROSE 1986; SCHWARZ-SOMMER et al. 1985; NEVERS et al. 1986). For the cases cited above, homeotic genes which determine general form development may be responsible. Though pollen is used experimentally to introduce exotic DNA into a recipient genome, it is unknown how horizontal gene transfer will act in nature; perhaps, though, in analogy to the counterfeit hybrids described by HARLAN and DE WET (1986). Within genera, divaricate habit and heteroblasty in New Zealand spread apparently by hybridization (GODLEY 1985).

In a study undertaken to widen the host range of *Agrobacterium tumefaciens* for genetic engineering, BOGANI et al. (1985) found the behaviour of *Nicotiana* species in hormone-free tissue culture to be correlated with their systematics. As this complements the well-known effects of plant hormones on morphogenesis (e.g. *Streptocarpus*: ROSENBLUM and BASILE 1984), STEBBINS and BASILE (1985) see a promising line of research in phyletic phenocopies: The application of hormones or proteins with key roles for morphogenesis may reveal the genetic and developmental basis of evolutionary change.

Since reproductive isolation and adaptation to the environment are operational with different impacts in plants and animals, it is not surprising that botanists and zoologists may disagree on the nature of speciation. CARSON (1985) stressed that in all cases, a disorganization of the genome under relaxed selection regimes is the starting point. This is in line with ELDREDGE's (1986) theory of evolution as restoring organismic complexity after collapse.

b) Phylogenetic Analysis at the Molecular Level

The analysis of chloroplast DNA restriction sites in *Clarkia* provided an unambiguous and detailed genealogical history for the section *Peripetasma* and verified previous models based on genetic studies (gene duplication); these results are partly at variance with morphological data or those inferred from reproductive isolation. For example, the morphologically distinct monotypic genus *Heterogaura* fits genetically well into *Clarkia*, where its sister species can easily be identified by molecular methods (SYTSMA and GOTTLIEB 1986a,b). Congruence between isozyme and other genetic data, and incongruence to morphological data have also been found in *Cucumis* (PERL-TREVES and GALUN 1985; PERL-TREVES et al. 1985) and in *Lisianthus* (SYTSMA and SCHAAL 1985). This approach has been applied successfully to the study of relationships in cultivated plants, e.g. *Lycopersicon* (PALMER and ZAMIR 1982) or *Pisum* (PALMER et al. 1985).

The elegance of these methods lies in the fact that numerous and individually independent mutations can be used as markers, and, since DNA structures are persistent and long-lived, each species can be supposed to bear its own "fossil" record. However, several aspects remain to be resolved: For chloroplast DNA, it must be known if the plastom is inherited in a matroclinous way or, as in some Onagraceae, biparentally (SYTSMA and GOTTLIEB 1986a). As to the organelle analyzed, chloroplast, mitochondrial and ribosomal nucleic acids evolve at different speeds (*Hordeum*: HOLWERDA et al. 1986), which does not necessarily invalidate the results, but puts the theoretical foundations of the molecular clock in question (SYTSMA and SCHAAL 1985; AYALA 1986). As to the method, these investigations require the application of highly refined numerical cladistics, to which the problems discussed on p. 298 are relevant. Statistical problems associated with inferring phylogenies from gene frequencies are mostly overlooked (FELSENSTEIN 1985). As pointed out by PERL-TREVES and GALUN (1985), a high degree of back mutations and horizontal gene transfer (e.g. by hybridization) invalidates many of the underlying assumptions. Thus, while the traditional taxonomist may prefer to ignore cladistic methodology for some time, the molecular evolutionist cannot.

A DNA sequence study in Cucurbitaceae revealed large genomic diversity which may account for sterility barriers despite morphological and cytological similarity: BHAVE et al. (1986). Comparison of ribosomal RNA genes in Cucurbitaceae: GAMAL and HEMLEBEN (1986). Ribosomal RNA patterns show populations of *Ferula communis* to diverge: OLMEDILLA et al. (1985). Note that for frogs, RNA data have been used to elucidate evolutionary history for well over 50 million years: HILLIS and DAVIS (1986). Continuation of phylogenesis based on the molecular clock for angiosperms (MARTIN et al. 1985), for some monocotyledons and gymnosperms (MARTIN and DOWD 1986b) and for phytogeography of *Solanum* (MARTIN et al. 1986).

Since the effects of change in nuclear DNA amount on chromosome homology and pairing behaviour are surprisingly small (REES 1984), and the mechanisms for amplification or reduction have been shown to be heritable into the F_2 generation within *Microseris* (H.J. PRICE et al. 1985), it is not unexpected to find these changes associated with small-scale adaptations to special environmental regimes. Within a local British flora (!), GRIME et al. (1985) found low DNA content to be correlated with late and fast production of new leaves, and high DNA content with early and slow development. This suggests that nuclear DNA measurements may provide an index of temporal niche differentiation in plant communities. Also in *Sisyrinchium* (KENTON et al. 1986) and in maize (RAYBURN et al. 1985) nuclear DNA content proved to be associated with phenological strategies, e.g. reduction within early maturating cultivars of the latter due to selection for a shorter mitotic cycle. Nuclear DNA measurements could be profitably applied to the many studies on phenology (see p. 314).

Extraordinarily high DNA variation in *Bulnesia* correlated with trends of specialization: POGGIO and HUNZIKER (1986). Nuclear DNA content of gymnosperms is generally higher but less variable than in angiosperms: OHRI and KHOSHOO (1986).

c) Evolution at the Population and Species Level

Symposia on plant population ecology: DIRZO and SARUKHAN (1984), JACQUARD et al. (1986). Population biology and evolution of clonal organisms: JACKSON et al. (1985). Review of phenotypic plasticity: SCHLICHTING (1986).

WATKINSON and WHITE (1986) stress the importance of somatic mutations for variation and evolution in plants and other modular organisms, which is operative at different levels of organization. In long-lived trees, it is decisive for their large intraplant variability (e.g.

Pinus ponderosa : MAZE et al. 1986), and, according to a mathematical model by ANTOLIN and STROBECK (1985), sufficient to deter herbivores; in herbaceous groups with vegetative reproduction (e.g. *Limonium*: INGROUILLE and STACE 1985; *Taraxacum*: FORD 1985; *Caladium* cultivars: SARKAR 1986), it may be the decisive evolutionary process.

The rapid evolution of weedy ecotypes has been studied in a number of cases with respect to different phenological strategies. In many cases, phenotypical plasticity adds to the flexible response of weeds to their sites. Since modification varies and seems to be controlled by selection, it is suggested that phenotypical plasticity is a character in itself which may evolve independently to the character of which it is the expression (SCHLICHTING and LEVIN 1986). In *Senecio vulgaris*, an intense weeding regime has been found to select for races with fast vegetative development; due to resource limitations imposed by a lesser leaf area, these may have less reproductive output (KADEREIT and BRIGGS 1985). In such cases where competitive interaction between individuals is important, group (=strain) selection may be more important than individual selection (experimental study: GOODNIGHT 1985).

Weedy ecotypes: *Panicum miliaceum* in Canada: BOUGH et al. (1986); *Abutilon theophrasti* in North America: WARWICK and BLACK (1986); *Echium* (WOOD and DEGRABIELE 1985) and *Cirsium* in Australia: FORCELLA and WOOD (1986). Additional influence of phenotypical plasticity: *Galium aparine*: GROLL and MAHN (1986); *Capsella*: STEINMEYER et al. (1985), NEUFFER and HURKA (1986).

WARWICK and GOTTLIEB (1985) postulated that speciation by gradual divergence should be reflected in a high divergence in structural genes coding for enzymes, and rapid progenitor-descendent relationships by a low divergence; this has been confirmed for various groups by LOWREY and CRAWFORD (1985), GOTTLIEB et al. (1985), HELENURM and GANDERS (1985) and CRAWFORD et al. (1985). The adaptive value of allozyme polymorphism has become apparent in *Hordeum spontaneum* (NEVO et al. 1986) and in Northern Flint races of maize (DOEBLEY et al. 1986), where morphological uniformity contrasts with genetic variability. In general, the genus *Zea* is characterized by an exceptionally high degree of genetic variation of a magnitude comparable to long-lived forest trees (DOEBLEY et al. 1986). How difficult it is to generalize these findings has been shown by HEYWOOD and LEVIN (1985): Only one of the four varieties of *Gaillarda pulchella* showed allozyme/soil covariation. ASHTON et al. (1984) concluded from enzyme analysis that speciation in the tropics is usually allopatric, being further shaped by re-immigration and differentiation of sympatric species.

Further selected isozyme studies: Relict populations of *Pinus* in Scotland: KINLOCH et al. (1986), of *Washingtonia*: McCLENAGHAN (1986); genetic diversity and conservation in forest trees: CHEN et al. (1986), KNOWLES and GRANT (1985); clonal population structure in *Alnus*: HUENNEKE (1985), density-dependent selection regimes in *Salvia*: SHAW (1986).

Hybrid zones and their dynamics are shaped by gene flow and selection, and hence may vary from narrow and recent (*Phlox drummondii* : LEVIN and SCHMIDT 1985) to very old and stable (*Populus*: ROOD et al. 1986; *Eucalyptus*: POTTS 1986; general aspects: BARTON and HEWITT 1985). Phenological effects determine the gene flow in mixed populations of *Rumex obtusifolius* and *R. crispus* where hybridization is believed to have detrimental effects on both parent species (ZIBURSKI et al. 1986). Recent studies on character displacement indicate problems of interpretation which arise when the characters involved are functionally or genetically correlated (*Phlox*: LEVIN 1985; *Dalechampia*: ARMBRUSTER 1985, 1986); further complications may arise from convergence due to common selection regimes (*Gaillardia*: J.S. HEYWOOD 1986).

Apart from their theoretical interest, transplant experiments and reciprocal sowins have practical importance for the management of rare and endangered species (e.g. GIVEN 1985). As for the latter, the results are discouraging, since native populations have proven to be finely adjusted to their habitat and have a competitive advantage to cospecific provenances coming from a distance as small as 50 m (*Spartina*: SILANDER 1985; *Anthoxanthum odoratum*: SCHMITT and ANTONOVICS 1986a,b; *Phlox drummondii*: SCHMIDT and LEVIN 1985; HEYWOOD and LEVIN 1986; CLAY and LEVIN 1986). This also applies to germination responses, so that probably gene flow by seed dispersal may be minimal (CLAY and LEVIN 1986). In transplant experiments, GROENENDAHL (1985) suggested to supplement them by sowing in order to correctly assess all aspects of environmental interactions. The responses of populations in these experiments may be shaped by predators and pathogens (SCHMITT and ANTONOVICS 1986b; PARKER 1985), and mediated by phenotypic plasticity (GARBUTT et al. 1985).

d) Reproductive Biology

Textbook on plant breeding systems: RICHARDS (1986). Review of self-incompatibility mechanisms in angiosperms: GIBBS (1986), of reproductive strategies: URBANSKA (1984, 1985), of gametophytic apomixis: NOGLER (1984a), of dichogamy: LLOYD and WEBB (1986), of herkogamy: WEBB and LLOYD (1986), of sex allocation in plants: GOLDMAN and WILLSON (1986). Symposium on the evolution of sex: Experientia 41(10), 1231-1310 (1985). List of gynodioecious plants, 543 species in 178 genera and 50 families: DEMYANOVA (1985). Longevity of flowers, breeding system and gene flow: PRIMACK (1985).

In *Ranunculus auricomus*, apospory is due to a dominant factor which favours aposporous embryo sacs to outcompete meiotically formed ones; at the same time, this factor is a homozygotic lethal which cannot be inherited by monoploid gametes. This explains the absence of apomictic diploids in nature. Apomixis is a blind alley at the diploid level, but not at the polyploid level, especially since the influence of the apospory factor may weaken (NOGLER 1984b). Population-specific apospory varies between 19 and 72% in *Ranunculus parnassifolius* subsp. *heterocarpus*, a tetra- and pentaploid species (VUILLE and KÜPFER 1985). Apospory is unknown for the native North American range of *Aster novae-angliae*, but has been found to disturb megasporogenesis to a high degree in Poland, where it occurs as an alien; this may be due to the original introduction of hybrid material, and to long-lasting breeding (KLUSKA 1986). Further studies including aspects of pseudocompatibility pertain to *Rubus* (NYBOM 1985) and *Ranunculus acris* (ØSTERBY 1986). MULCAHY (1984) argues that the continuous trend from self-incompatibility via pseudocompatibility to self-compatibility is brought about by an increasing competitive advantage of self-pollen on the stigma; since non-self pollen will be outcompeted, this process is mostly irreversible.

Although plants producing both cleistogamous and chasmogamous flowers have been widely studied, few generalizations on the effects of inbreeding vs. outbreeding emerge. Cleistogamous progeny has been found to be both more and less homogeneous than chasmogamic progeny in different species (CLAY and ANTONOVICS 1985; MITCHELL-OLDS and WALLER 1986); the demography of the different types of offspring has been studied by ELLSTRAND and ANTONOVICS (1985) and BELL and QUINN (1985). Since only cleistogamous flowers produce seeds vigorous enough for the development of chasmogamous plants, a regular change of the different forms has been found in *Amphicarpea bracteate* (SCHNEE and WALLER 1986). Phenotypic changes in the breeding system have been found in *Ipomopsis rubra*, where male sterile flowers are formed after self-, but not after cross-pollination (BENJAMIN and HAINSWORTH 1986), and in *Campsis radicans*, where the acceptability of foreign pollen increases over the vegetation period and blurs the distinction between resource limitation and pollen limitation for seed set (BERTIN 1985). Apparently, LIEBIG's law of the minimum is applicable also to this problem (HAINSWORTH 1985).

Studies in American *Crataegus* confirm that outbred progeny is more variable than that derived uniparentally by selfing or from apomixis (DICKINSON 1986; DICKINSON and PHIPPS 1985, 1986; MUNIYAMMA and PHIPPS 1985).

The pollen/stigma system and its importance for both rapid selection and control of gene flow has received wide attention. MULCAHY (1986) stresses the analogy of pollen to microorganisms: Haploidy and large numbers ensures quick selection; for example, if plants heterozygous for metal tolerance are experimentally grown on metal-enriched soil, their pollen germinates successfully on stigmata of metal-tolerant plants (SEARCY and MULCAHY 1985). CRUDEN and LYON (1985) have found a correlation between pollen size and the distance the pollen tube has to grow to reach the transmission tissue, if lack of exogenous supply is the reason for crossing failure, it may be overcome, e.g. by the application of exogenous sugar (CALLAN and THOMPSON 1986). SEAVY and BAWA (1986) interpret the rejection of fertilized ovules as late-acting self-incompatibility, and a refined way of "testing" the progeny.

THOMPSON (1986) confirms that no overall interpretation for the basal sexual state in angiosperms is possible; although unisexual flowers emerge as being primitive in SPORNE's calculations, this may be due to their association with woodiness. Since the numbers of stamens are fixed in most families, andromonoecism is the commonest way to increase pollen production and increase male fitness (WEBB 1984; COUVET et al. 1985; WHALEN and COSTICH 1986). Unusual cryptic dioecism has been reported from *Solanum appendiculatum* (LEVINE and ANDERSON 1986).

NIKLAS (1985), in a summary of his studies on the aerodynamics of wind pollination, calls attention to the fact that many plant lineages have a fossil history of consolidating individual reproductive organs into compound structures, e.g. cones, with increased pollen-catching ability.

e) Ecological Aspects

Epiphytes. They make up 10% of the vascular plant species (KRESS 1986a), but due to their diverse taxonomic origins and heterogeneity, they lack a unifying profile in vegetative characters (BENZING 1986) as well as for pollination (ACKERMANN 1986) and dispersal (ROCKWOOD 1985; Melastomataceae: RENNER 1986). Epiphytes choose specialized strategies already existing in terrestrial plants. This suggests recent specialization, as in hemi-epiphytes, where the epiphytic and terrestrial phases vary strongly in water economy (PUTZ and HOLBROOK 1986). CORNER (1985a), however, holds the strangling habit to be primitive, being a mean of a leptocaul to invade the pachycaul forest.

Review of ecophysiology of epiphytes: LÜTTGE (1985). Symposium on biology of tropical epiphytes: Selbyana 9(1), 1-125 (1986).

Phenology. Review: RATHCKE and LACEY (1985). Phenological patterns indicate incomplete adaptations on the border of the Mediterranean region: BIANCO and SCHIVONE (1985). Savanna species change their leaves in the dry season which is supposed to prevent leaching of young and little cutinized leaves: SARMIENTO et al. (1985).

Primary Production: Review of resource limitation in plants in analogy to economics: BLOOM et al. (1985). Evolutionary increase in photosynthetic efficiency by concentrating leaves below the inflorescences in *Carlina* (KÖRNER and MEUSEL 1986) and by leaf orientation in *Lactuca* (WERK and EHLERINGER 1986).

Plant-Animal Mutualisms: Long-spurred orchids of the *Angraecum* type apparently have not co-evolved with long-tongued Sphingidae, but entered an already existing sphingid-plant relationship on Madagascar (NILSSON et al. 1985). In contrast, *Ficus* evolution into subgenera preceded the differentiation of fig wasps (CORNER 1985b). The balanced evolutionary equilibrium between style length in *Ficus* and ovipositor length in fig wasps is upheld by controlled fruit abortion if the number of larvae is too high for optimal seed set (MURRAY 1985). A continuous progression towards elaborate ornithophily has been found in *Deplanchea* culminating in the formation of a platform on top of the inflorescence (WEBER and VOGEL 1986). Highly coordinated floral events ensure pollination in *Philodendron selloum* with scattered distribution (GOTTSBERGER and AMARAL 1984). Discrepancies between ecological and evolutionary equilibrium are pointed out again by JANZEN (1986c) for dense stands of succulents, an extension of the extinct megafauna theme (see Progr. Bot. 45, 262).

Textbooks on ant-plant mutualisms: BEATTIE (1985), JOLIVET (1986). Only ascidia-forming *Dischidia* species are ant-plants proper, others have only recently invaded already existing relations between ants and host trees (WEIR and KIEW 1986). Phylogenetic inertia, or some unknown mutualism are invoked to explain extrafloral nectaries in Hawaiian plants despite the lack of native ants (KEELER 1985).

f) Selected Monographs and Studies in Special Groups

Lysimachia (Primulaceae): Floral attractants suggest a diphyletic origin not only of the genus, but of Primulaceae as a whole: Subgenus *Idiophyton* with predominant pollen flowers has closer links to Myrsinaceae, subgenus *Lysimachia* with oil flowers is connected by forms with hydrophilous floral trichomes to Theophrastaceae. Since the areas of *Macropis* bees and oil-flowered *Lysimachias* are congruent except for some orphaned populations, a close co-evolution of both dating back to Late Tertiary is suggested (VOGEL 1986).

Lisianthus (Gentianaceae): Repeated founder events and genetic drift in small and isolated populations is responsible for the high genetic diversity, which has been studied by a wide array of methods including chloroplast and nuclear DNA analysis and isozyme study (SYTSMA and SCHAAL 1985).

Ocotea (Lauraceae): A catch-all genus, impossible to divide because of numerous intermediate forms, with characters deliberately combined and all combinations, if not present, then extinct or to be expected. However, stamen characters give clues to the phylogeny: Dioecism evolved independently in several lines via gynodioecism, and is predominant in Amazonian rainforest. This suggests on Old World origin and migration from North to South America, and radiation from the Andes into lowland habitats (ROHWER 1986a; ROHWER and KUBITZKI 1985; synopsis with keys: ROHWER 1986b).

Silene (Caryophyllaceae): Summary of the multidisciplinary study of *Silene pratensis* (see Progr. Bot. 47, 257) concludes that the races originated by selective forces, but their present-day distribution in Central Europe is the result of migration, and not of selection (MASTENBROEK and BREDERODE 1986); some characters, such as seed morphology, are shaped by both selection and migration (PRENTICE 1986a). Subspecies have differentiated polytypically with the opening of habitats after glaciation in *Silene vulgaris* (AESCHIMANN 1985a,b).

Microseris (Compositae): Further studies on the genetics of the pappus part system concentrated on the analysis of dispersal events and of polygeny, and revealed two independent duplications of the ten-part determining gene (BACHMANN et al. 1985a,b; ZENTGRAF et al. 1985; MAUTHE et al. 1985; see Progr. Bot. 47, 250).

Grasses: Early hybridization and polyploidy is responsible for the reticulate evolutionary patterns (numerical study: HILU 1985); the high percentage of polyploids is due to the fluctuations of their habitats after glaciation, which brought about cycles of separation and new contact (STEBBINS 1985). In the *Poa secunda* complex, many characters usually employed to discern sections may vary even within the offspring of self-pollinated plants; neither concordant characters nor evolutionary marks could be found, and the pattern of variation does not reflect micro-evolution (KELLOGG 1985). Likewise in *Hordeum*, the discrepancy of the groups derived from enzyme analysis with the existing morphologically based classification may be due to the prevalence of allopolyploidy, with *H. murinum* as a frequent partner; a Cretaceous origin, and two centres in SW Asia and South America may be probable (JØRGENSEN 1986); the American diploids are closely related (BOTHMER and JACOBSEN 1986). Within the *Hordeum-murinum* complex, three discrete lines emerge which have been taken to show the difference of approach for pheno-, clado- and biospecies (GILES and LEFKOVITCH 1986). Symposium on biological adaptation in grasses: Ann. Missouri Bot. Gard. 72, 823-863 (1985).

7. Plants and Man

A major event in this field was the publication of the second edition of MANSFELD's dictionary (SCHULTZE-MOTEL 1986). It contains nomenclature, bibliography and notes on uses for almost any plant known to be cultivated except for ornamentals. This list should stabilize the use of names for economically important plants even if they are not formally conserved.

Glossary for horticultural crops, also useful for the general botanist: SOULE (1985). Medicinal plants of China: DUKE and AYENSU (1985). Guide to chief honey-producing plants: CRANE et al. (1984). Wide-ranging symposium on plants for arid lands: WICKENS et al. (1985). Symposium on prospects for economic botany: Econ. Bot. 40, 261-309 (1986).

A series of lectures on plant genetic resources by CHANG (1985) gives a very readable introduction to the subject. Concerning rice, the exceptional genetic diversity has been conserved mainly because (1) plant breeding began late for this crop, (2) conservation efforts started before the Green Revolution and (3) many ecological niches require the planting of specialized land races. CHANG (1985) concludes that even with limited resources genetic conservation can be successful as long as workers are interested and willing to strive. The importance of traditional farming practice for the maintenance of genetic diversity is again pointed out by CLAWSON (1985): The rationale for growing varietal mixtures is to achieve harvest security by spreading the risk over several biotypes, which at the same time allows a staggering of harvests throughout the year. As a consequence, cultivars for this intraspecific polyculture have been selected for perceptual distinction, as in manioc (BOSTER 1985).

Review of progress in securing crop genetic resources: HOLDEN and WILLIAMS (1984). Introductory text on gene banks and their biological background: FORD-LLOYD and JACKSON (1986).

Various aspects of neotropical ethnobotany: FRANCE and KALLUNKI (1984). Essays on Andean ethnobotany: HEISER (1985). Uses of *Coca*: PLOWMAN (1984). Importance of contributions by native collectors: BERLIN (1984). Diversity of plant products on the local tropical market, now declining: M.E. v.d. BERG (1984).

SMALL (1984) shows how repeated cycles of hybridization between wild, cultivated and weedy biotypes within species contribute to the complexity of these groups. HAMMER (1984) stresses that the domestication syndrome pertains to many plants of the agro-ecosystem, not just the cultivated ones. In seed-crop plants, as in *Lens* (LADIZINSKY 1985; PINKAS et al. 1985) or bitter vetch (LADIZINSKY and VAN OSS 1984), the initial step in the evolution is usually a change in dispersal traits, often governed by a single gene. Subsequent gene flow tends to be from cultivated to wild species, bringing about a genetic bottleneck due to founder effects. Breeding programmes with wild relatives may restore the genetic variability. Polytypic inculturation, however, may be responsible for the unusually high allozyme variability in cultivated lentil (PINKAS et al. 1985). In *Setaria*, gene flow is unidirectional from crop species *S. italica* to weedy *S. verticillata* (POIRIER-HAMON and PERNES 1986). The weedy *Hordeum vulgare* subsp. *agriocrithon* evolved from cultivated barley by repeated mutations, the wild barley was not involved (SCHOLZ 1986). HAMMER et al. (1985) propose that fields where gene flow between wild and cultivated plants occurs should be conserved in order not to disturb the evolutionary process.

a) Systematic Studies in Various Groups

Solanaceae. See also D'ARCY (1986). Hybrid origin of *Solanum tuberosum* subsp. *andigena* confirmed by artificial crosses of putative ancestors and subsequent morphological and chemical analysis: CRIBB and HAWKES (1986). Origin of scarlet eggplant, *Solanum aethiopicum*: LESTER and NIAKAN (1986). Monograph of domesticated *Capsicum* peppers: ANDREWS (1984). Cultivated *Physalis philadelphica* varies little from wild progenitor: HUDSON (1986). *Jaltomata* persists as week in primitive agriculture because of its edible fruits: DAVIS IV (1986). Biosystematics of weedy *Solanum* species in native South America and Europe: EDMONDS (1986). Maize. History of biosystematics, nice synthetic diagrams: ILTIS and DOEBLEY (1984). Inflorescence development confirms recent evolutionary hypothesis: SUNDBERG and ORR (1986). No recent introgression between maize and teosinte: DOEBLEY (1984), J.C.S. SMITH et al. (1985). Barley. Genetic diversity in *Hordeum spontaneum* is high and correlated to environmental factors; this may form the basis of a screening programme: NEVO et al. (1985b, 1986a,b). Ancient divergence of naked barley from covered barley, persisted in regions with unfavourable climate for other cereals as in Nepal, which forms a secondary centre: WHITCOMBE and MURPHY (1986). Wheat. Allozyme markers used to predict stipe resistance: NEVO et al. (1985a). C-banding reveals common ancestry of *Triticum timopheevii* and *T. aestivum*, but numerous substitutions and translocations prevent breeding programme: BADAEVA et al. (1986). Survey of Ethiopian land races: NEGASSA (1986). Cucurbits. Allozyme study in wild and cultivated *Cucurbita*: KIRKPATRICK et al. (1985). Bananas. Epicuticular waxes (FREEMAN and TURNER 1985) and isozymes are helpful for identification and elucidation of parentage in cultivars; the lack of germ plasm collection is pointed out by JARRET and LITZ (1986). Sesame: Phenetic study (BEDIGIAN et al. 1986), anthropological and archaeological investigations (BEDIGIAN and HARLAN 1986) and lignan chemistry are taken to indicate that *Sesamum indicum* did not originate from *S. latifolium* from E-Africa, but from *S. orientale* var. *malabaricum* from India (BEDIGIAN et al. 1985). Vigna. *V. mungo* (urid bean) and *V. radiata* (mung bean) were domesticated from different lineages: CHANDEL et al. (1984). Onions. Wild species of *Allium* sect. *Cepa* endangered in the wild: HANELT (1985). Ornamental plants: Continuation of European Garden Flora: WALTERS et al. (1985). Horticultural monograph of hardy *Geraniums*: YEO (1985). New species of *Impatiens* for breeding programme of *I. walleriana*: MERLIN and GRANT (1986).

Acer buergerianum 'Jaka Kaede' turned out to be the verbenaceous (!) *Premna japonica* (DELENDICK 1985); confusion arose due to mixing of vernacular and cultivar name, the Japanese practice to remove floral buds in foliaceous ornamentals, and the reluctance of taxonomists to work with cultivated material.

8. Evolution and Classification of Larger Groups

a) Families of Dicotyledons

A very useful book on the families of dicotyledons has been presented by GOLDBERG (1986). It contains descriptions and illustrations of all families, a comparison of the family placement in different systems of classification, and a tentative phylogenetic tree with a basal position for Trochodendrales and Hamamelidales. Furthermore, there is a list of evolutionary trends together with an assessment of their reversibility. This immense compilation is quite an unusual achievement for an individual today.

b) Reproductive Biology of Primitive Angiosperms

Starting point for the evolution of zoophily in angiosperms may have been flowers which formed mating sites for insects, and were thus preferred because they offer readily accessible high-quality food (ovules and pollen); thus, the relationship between the primitive vesselless *Zygogynum* (Winteraceae) and a likewise primitive Micropterigidae moth may be epitomizing an ancient association from which more advanced zoophily has arisen several times independently. Fine tuning may have occurred by floral fragrances supplementing insect pheromones; feeding deterrents may simultaneously act as floral attractants to specialist pollinators (THIEN et al. 1985; PELLMYR and THIEN 1986). Perhaps this is also the clue for the otherwise barely explicable occurrence of pollinator attractant compounds, such as benzoates or salicylates, in male and female strobili of Pinaceae, strangely enough in different spectra (BORG-KARLSON et al. 1985). Presence of petal sculpturing in mimosoid and papilionoid flowers from the Paleocene-Eocene boundary gives evidence of highly specialized bee pollination which is taken to be responsible for the major radiation of angiosperms (CREPET 1984; CREPET and TAYLOR 1985).

The correlations between seed size and ecological parameters, originally established by SALISBURY (1942), have been re-investigated for tropical (FOSTER 1986) and temperate communities (HODGSON and MACKEY 1986). These correlations form the basis for TIFFNEY's (1984) stimulating, though speculative ideas on the ecological history of angiosperms: They are supposed to have evolved from small-seeded plants of disturbed habitats to large-seeded climax trees in Tertiary when their major radiation was coupled with that of their major dispersers (birds and mammals). Since then, further radiation produced a wide range, from very small to very large propagules. For Hamamelidiae (TIFFNEY 1986), the importance of biotic dispersal for fostering diversity becomes evident, since animal-dispersed families such as Fagaceae, Moraceae and Cepropiaceae have a much higher species number than wind-dispersed ones. An important conclusion is that the primitive dispersal state for a given group can only be inadequately inferred from extant material; the question remains whether the ecological correlations observed today can be adequately applied to Tertiary or Cretaceous conditions.

c) Magnoliidae

The Lactoridaceae are characterized by saccate pollen quite similar to that of Mesozoic pteridosperms, while pollen wall formation suggests closeness to *Magnolia*; a further indication of the gradational nature of pollen characters in seed plants (ZAVADA and TAYLOR 1986). Leaf flavonols are concordant with a placement in archaic Magnoliidae (CRAWFORD et al. 1986), but neither cladistic nor phenetic analyses are conclusive for their exact position (LAMMERS et al. 1986). The latter authors encounter the same difficulties for their study as DAHLGREN and BREMER (1985) (see p. 298), and they try to explain the extensive character changes by the long evolutionary history of Magnoliidae.

This explanation gains some plausibility through the finding of secondary entomophily in the juglandaceous *Platycarya*, whose anemophilous ancestors probably derived from entomophilous Cretaceous forerunners (ENDRESS 1986c); this author points to important concepts which were derived from the study of extant plants later to be confirmed by palaeobotanists, such as the monophyly of angiosperms (ENDRESS 1986a). He also pleas for more comparative structural investigations rather than undertaking phenetic or cladistic studies based only on the literature (ENDRESS 1986b).

HUBER (1985) presents a new infrafamilial classification for the Aristolochiaceae based on seed morphology: *Asarum* should be placed into a separate family, while *Aristolochia* is more heterogeneous than the specialized floral adaptations suggest. Staminodes in *Thottea* are suggestive of a reduction series in stamen number, ranging from 46 to 6 (RENUKA and SWARUPANADAN 1986). A spiral androeceum following a trimerous perianth may be taken as a further indication of a relationship between this family and Alismatidae, and further of magnolioid dicots and monocotyledons (LEINS and ERBAR 1985b). The occurrence of trimerous perianths in both groups is held to be the result of a unidirectional and irreversible switch from spiral to whorled anthotaxy, still observable in extant magnolioids but fixed in monocots (KUBITZKY 1987), while the erratic distribution of trimery in higher dicots is due to the oligomerization of pentamerous and tetramerous whorls.

RÜFFLE (1985) interprets the floral structure of Monimiaceae in the light of different morphological theories and stresses that some of the structures resemble to those found already in the Palaeozoic. S-type sieve-tube plastids suggest placement of Austrobaileya either near Myristicaceae and Winteraceae, or Monimiaceae; the sieve elements of gymnosperms stand apart from the rest of vascular plants (BEHNKE 1986a). Epidermal characters put *Degeneria* closer to Illiciales than to Magnoliales (HARVEY 1985). Gametogenesis of *Galbulimima* suggests links to Annonaceae and Degeneriaceae (PRAKASH et al. 1984).

d) Hamamelididae

A comparative study of pollen, analyzed both phenetically and cladistically, confirms a previous study by BARABÉ (1981, 1982) and identifies three reasonably well-defined groups, viz. Hamamelidales to Trochodendrales, "Amentiferae" and Urticales, and Fagaceae to Eucommiaceae (ZAVADA and DILCHER 1986). The Hamamelididae lack "exotic" micromolecules which is taken to indicate biochemical conservatism, as compared to their putative magnoliid ancestors (GIANNASI 1986), but see different interpretations by GOTTLIEB (1982). - The Trochodendrales form a homogeneous group between Magnoliidae and Hamamalididae, but are not ancestral to the latter; ENDRESS (1986b) draws attention to parallel trends in floral evolution with Hamamelidales. A continuous line of evolution from Palaeocene multicarpellate to extant bi- or tricarpellate infructescens of *Cercidiphyllum* has been demonstrated by CRANE and STOCKEY (1986), the reduced nature of which is stressed by the detection of a growing point by HEEL (1986). Fruit and seed anatomy of *Cercidiphyllum* stress its isolation (MOHANA RAO 1986). In Fagaceae, various transitional stages from perfect, entomophilous flowers to unisexual anemophilous ones, and from mixed to unisexual spikes can be found; in Betulaceae and Juglandaceae, the separation of sexes is well established (KAUL 1986). Nothofagaceae stand apart (ROMERO 1986) and may be closer to Betulaceae, Trigonobalanoideae are heterogeneous (JONES 1986).

Symposium on morphology, systematics and evolution of Hamamelididae: Ann. Missouri Bot. Gard. 73, 225-441 (1986). Serology confirms relations between Fagaceae and Juglandaceae (POLECHKO and CLARKSON 1986), and suggests Moraceae to be closer to Simaroubaceae or Anacardiaceae than to Hamamelididae (PETERSEN and FAIRBROTHERS 1985).

e) Myrtales

Symposium: Ann. Missouri Bot. Gard. 71, 631-970 (1984).

The Myrtales are apparently one of the most clearly defined orders of angiosperms. Diagnostic characters are, among others, intra-axillary phloem (anatomical review: VLIET and BAAS 1984) and pseudocolpi (PATEL et al. 1984) which may be lacking in some groups but are nearly restricted to the order. Both cladistic analysis (JOHNSON and BRIGGS 1984, see also p. 297) and a survey of character distributions by DAHLGREN and THORNE (1984) are in a broad agreement. *Rhynchocalyx* (TOBE and RAVEN 1984a) and *Alzatea* (JOHNSON and BRIGGS 1984; GRAHAM 1984; TOBE and RAVEN 1984b) are placed in families of their own, mostly from embryological evidence. There are some relationships of Myrtales to Rosales and, more faintly, to Gentianales, Cornales and Theales (DAHLGREN and THORNE 1984). The position of the Thymelaeaceae, however, remains controversial: CRONQUIST (1984) makes a plea to include them in Myrtales for want of a better placement within a general system of classification, and suggests that the lack of tannins, which is the main reason for their exclusion, is secondarily derived. He is supported by MARTIN and DOWD (1986a) on the basis of rubisco protein sequence analysis. However, it must be noted here that Thymelaeaceae agree with Euphorbiaceae chemically in many respects, e.g. by sharing unusual diterpenes.

The Myrtales comprise now Onagraceae, Trapaceae, Lythraceae, Oliniaceae, Alzateaceae, Rhynchocalycaceae, Pennaeceae, Combretaceae, Melastomataceae, Memecyclaceae, Psiloxylaceae, Heteropyxidaceae, Crypteroniaceae and Myrtaceae (RAVEN 1984).

Sieve-tube plastids in Myrtales: BEHNKE (1984). Leaf histology: KEATING (1984). Base number for Lythraceae is x = 8 and x = 5 for *Peplis* and *Lythrum*: TOBE et al. (1986). Different lines of embryo evolution in Myrtaceae related to starch-storing properties: LANDRUM and STEVENSON (1986). Histogenesis of integuments provides good generic characters for Onagraceae: TOBE and RAVEN (1985). Parallelisms characterize evolution of septation in sporogenous tissue in Onagraceae: TOBE and RAVEN (1986). Flavonoid evolution in *Fuchsia*: AVERETT et al. (1986).

f) Monocotyledons

The embryological similarities between Burmanniaceae and orchids, such as tetrad pollen, reduction in megasporogenesis or abbreviation of embryogenesis, should be considered as parallel evolutionary trends due to a similar genetic programme rather than as synapomorphies or convergences. For saprophytic monocotyledons, where reduction has led to a nearly total loss of information on vegetative characters, gametophytic traits alone provide data on their relationship. The main results of RÜBSAMEN's (1986) exhaustive study include (1) Corsiaceae and Burmanniaceae are distinct within the order Burmanniales, (2) show strong differences to Triuridaceae and *Petrosavia* and (3) affinities to Philydrales and to Orchidales, but interestingly only weak ones to the supposedly primitive Apostasioideae.

A detailed study of floral morphology in the Restionaceae reveals numerous similarities between the primitive *Leptocarpus* and the bamboo *Streptochaeta* (KIRCHER 1986), which are taken to confirm the derivation of the grass flower from a typical diplochlamydeic monocot ground plan, and of the pseudomonomerous gynoeceum from a trimerous-syncarpous one (also confirmed by PHILIPSON 1985). Palynology, too, favours that Restionaceae and Poaceae be grouped together (LINDER and FERGUSON 1985). The similarities extend to such characters as the presence of a swelling tissue formed by inner perianth (= lodiculae), and the dispersal of fruits together with the tepals. In grasses, the shape of these anthoecia is suggested to have evolved under insect and herbivore pressure: hence, its limited taxonomic value (THOMASSON 1985).

Overview of systematic anatomy of Cyclanthaceae, related families keyed out by anatomical characters: TOMLINSON and WILDER (1984). Compound vascular bundles in monocots: FRENCH and TOMLINSON (1986). Inflorescence morphology of Cannaceae and Zingiberaceae (KUNZE 1985) and of Lowiaceae (KUNZE 1986) shows pleiothyrus to be basal for the group; *Orchidantha* is similar to Musaceae in floral morphology, but inflorescence structure parallels Marantaceae. In a cladistic analysis of flavonoids of Commelinales, Juncaceae emerges as an isolated line (DEL PERO MARTINEZ 1985). *Ripogonum* is only distantly related to Smilacaceae or Dioscoreaceae, at best some links with Luzuriagaceae (CONRAN and CLIFFORD 1985). Embryology stresses the isolated position of *Eriospermum* (LU 1985). The New Caledonian endemic *Campynemanthe* should be placed in (DAHLGREN and LU 1985) or near (GOLDBLATT 1986a; GOLDBLATT et al. 1984) Melanthiaceae, as is stressed in the latter study on types of crystals in Iridaceae and related families.

The abrupt change from a vegetative to generative apex in *Lilaea* suggests that the "inflorescence" is the fundamental floral unit here (POSLUSZNY et al. 1986). Since Alismatidae show all degrees of organization from prefloral to near-floral stages, neither the classical concepts nor the concepts of BURGER (1977) or MEEUSE (1975 et seq.) seem to be adequate for this group

9. Studies at the Family Level and Below

Gymnosperms

Carotenoids in gymnosperms, survey: CZECZUGA (1986). Cycadaceae. Evolutionary trends in *Dioon*, reduction to relict centre in Mexico in Tertiary, later radiation into xeric habitats; key to species: SABATO and DE LUCA (1985). Cupressaceae. Biflavones at variance with current groupings, closer links between southern and northern hemisphere genera: GADEK and QUINN (1985). Ephedraceae. Karyotype evolution, x = 7 base number: CHOUDRY (1984). Pinaceae. Colour dimorphism of *Abies concolor* cones yet inexplicable: FARRIS and MITTON (1985). Florin rings (i.e. wax rings around stomata) aid *Pinus* taxonomy: YOSHI and SAKAI (1985). Serotaxonomy of European *Pinus*: PRUS-GLOWACKI et al. (1985). Studies of terpenes for biosystematics: YATAGAI and SATO (1986), ZAVARIN and SNAJBERG (1986), POLLACK and DANCIK (1985). *Pinus pseudostrobus* group: STEAD and STYLES (1984). Podocarpaceae. Flavonoids stress isolation of *Phyllocladus*: MARKHAM et al. (1985). *Podocarpus*, key: LAUBENFELS (1985). Taxodiaceae. Karyology favours separation of *Sciadopitys*: SCHLARBAUM and TSUCHIYA (1985). Welwitschiaceae. Biology: WILLERT (1985).

Monocotyledons

Aponogetonaceae. Monograph of *Aponogeton*, interesting diagram of morphological diversification: BRUGGEN (1985). Amaryllidaceae. South African *Cyrtanthus*: REID and DYER (1984). Strumariinae, synopsis: MÜLLER-DOBLIES and MÜLLER-DOBLIES (1985). Araceae. Pollen see p. 304. *Pinellia tripartita* may be anemogamous: UHLARZ (1985). Stamen vasculature and endothecial wall thickening: FRENCH (1985, 1986). Blastogeny of *Acorus* so different that a separate family is proposed: TILLICH (1985). *Arisaema* in Africa, with notable diagrams: MAYO and GILBERT (1986). *Cryptocoryne* in Borneo: JACOBSEN (1985), *Anthurium* in Panama: CROAT (1985). Bromeliaceae. Stigma structure: BROWN and GILMARTIN (1984). Chromosomes: BROWN and GILMARTIN (1986). Evolutionary ecology of trichomes: BENZING et al. (1985). *Nidularium* subgen. *Canistropsis*: PEREIRA and LEME (1986). *Tillandsia* classification highly artificial: GARDNER (1986); cladistics of *Tillandsia* subgen. *Phytarrhiza*: GILMARTIN and BROWN (1986). Burmanniaceae (see p. 320). Flora Neotropica revision: MAAS et al. (1986). Colchicaceae: *Wurmbaea* in South Africa: NORDENSTAM (1986). Commelinaceae. Flavonoid data and inflorescence morphology support separation into Commelineae and Tradescantieae, origin of family may lie in Africa: DEL PERO MARTINEZ and SWAIN (1985). *Gibasis*: HUNT (1986). Convallariaceae. *Maianthemum* species with upright stems in Central America discovered: LA FRANKIE (1986a). Revision for New World: LA FRANKIE (1986b). Cyperaceae. Inflorescence typology, continuation: KUKKONEN (1986). Nomenclature of suprageneric taxa: GOETGHEBEUR (1985). Vegetative shoots in *Carex*: REZNICEK and CATLING (1986). *Bulbostylis* in Central Africa: GOETGHEBEUR and COUDIJZER (1985). *Scleria* in South Africa: HENNESSY (1985). *Rhynchospora* sect. *Dichromena*, monograph, Wagner tree, insect pollination: THOMAS (1984). Eriocaulaceae. Morphology of flowers and inflorescences: STÜTZEL (1984). Position of *Wurdackia*: STÜTZEL (1985). Gramineae. Poaceae data bank (see p. 301); evolutionary studies (see p. 316); economic aspects (see p. 318 f.). Grasses of Bahia: RENVOIZE (1984). Unique *Steyermarkochloa* from white sand riverine soils put into separate tribe of arundinoid affinity: DAVIDSE and ELLIS (1984). Phytochemical phenetics of Chinese bamboos: CHOU et al. (1984), CHOU and HWANG (1984). Fleshy fruited bamboos embryologically distinct: HAN GOPAL and MOHANA RAM (1985). Re-alignments in perennial Triticeae: DEWEY (1984), key to genomically devised genera: BARKWORTH and DEWEY (1985). Artificial hybridization points to close genomic relationships between *Vulpia*, *Festuca* and *Lolium*: BARKER and STACE (1986). Paniceae, key: ZULVAGA and SODERSTROM (1985). Occasional reversals to sexuality and backcrosses may lead to triploid races in *Festuca vivipara* group: PILS (1985). Seed peptide analysis more conclusive than morphology in *Briza*, reveals two groups, South American and Eurasian: KING (1986). *Stipa* in Australia: VICKERY et al. (1986), in SW and S-Asia: FREITAG (1985), in Italy: MORALDO (1986). *Neurachne*, cytogeography: PRENDERGAST and HATTERSLEY (1985). *Paspalum* in Malesia: KONING and SOSEF (1985). *Poa* in New Zealand: EDGAR (1986). *Rottboellia*, circumscription: VELDKAMP et al. (1986). *Zizania aquatica* biosystematics for breeding programme: ARCHIBOLD and WEICHEL (1985). Hydrocharitaceae. *Elodea*: COOK and URMI-KÖNIG (1985), *Apalanthe*: COOK (1985). Hypoxidaceae. *Hypoxis* in E-Africa: NORDAL et al. (1985). Iridaceae. Phenolic patterns highly diversified and heterogeneous, few links to other families: WILLIAMS et al. (1986). Leaf anatomy for Australasian members: RUDALL (1986). Tertiary African diversification and subsequent Pliocene and Pleistocene radiation in Cape area in *Geissorhiza*; very narrow endemism points to conservation problems; cladistics of sections (monograph: GOLDBLATT 1985). Low base numbers ($x = 8$ or 9) suggest South American origin for polyploid North American species of *Sisyrhinchium* (monograph: RODRIGUEZ RIOS 1986). Aristea in South Africa: VINCENT (1985). *Hesperanthera* in Natal: HILLIARD and BURTT (1986). Juncaceae. *Juncus bufonius* s.str. is a variable polyploid originating from some diploid W-European species: COPE and STACE (1985). Liliaceae. Monosomy in *Cardiocrinum* suggests that this genus and *Lilium* with 2n = 24 are allopolyploids from x = 6 ancestor: CHAUHAN and BRANDHAM (1985a). Chromosomes in *Hemerocallis* remarkably similar (JIN XIOBAI 1986), but C-banding of satellites reveals patterns of evolution: NOGUCHI (1986). *Ornithogalum* (see p. 305). Marantaceae. *Maranta* subgen. *Maranta*: ANDERSSON (1986). Orchidaceae. 20,000 to 23,000 may be a realistic estimate of species number, 73% of which are epiphytic: ATWOOD (1986). Comprehensive cladistic study, key to tribes: BURNS-BALOGH and FUNK (1986). Androecium evolution: BURNS-BALOGH and BERNHARDT (1985). Statistical (NILSSON 1985) and enzyme electrophoresis based study on European orchid hybrids: STEINBRÜCK et al. (1986). Karyology and generic limits in *Orchis* and related genera: CAUWET-MARC and BALAYER (1984). *Corybas* in W-Malesia and Asia: DRANSFIELD et al. (1986). *Cymbidium* sect. *Iridorchis*, phenetics: DU PUY et al. (1985). *Dendrobium* sect. *Spatulata*: CRIBB (1986). *Dimerandra*: SIEGERIST (1986). Distribution of anacheiloside (a dihydroflavonol glycoside) within flowers suggests re-alignments in *Epidendrum* and related genera: FERREIRA et al. (1986). *Paphiopedilum*, synopsis and key: KARASAWA and SAITO (1982). Stem structure and taxonomy of *Pleurothallis*: STERN et al. (1985). Palmae. Entomophily prevails in extant palms, cantharophily may be basal, two groups of beetles are clearly co-evolved: HENDERSON (1986). Indigenous palms of New Caledonia: MOORE and UHL (1984). Philydraceae. Pollen ultrastructure stresses monophyly: SIMPSON (1985b). Pontederiaceae. Cladistic analysis reveals South American origin and four independent dispersal events in Africa: ECKENWALDER and BARRETT (1986). Heterostyly in *Eichhornia* (BARRETT 1985; GLOVER and BARRETT 1986) and in *Pontederia*: BARRETT and GLOVER (1985). Restionaceae (see p. 321). Conspectus for Africa: LINDER (1985). Sparganiaceae. *Sparganium*, first part of monograph: COOK and NICHOLLS (1986). Tecophilaeaceae. Palynology: SIMPSON (1985a). Triuridaceae: Flora Neotropica revision: MAAS and RÜBSAMEN (1986). Xyridaceae. *Abolboda*, embryology, floral morphology, systematics: TIEMANN (1985). Zingiberaceae. *Costus speciosus* with di-, tri- and tetraploid races: SHARMA and CHATTOPADHYAY (1984). Alpinieae in Borneo: R.M. SMITH (1985, 1986).

D i c o t y l e d o n s

Acanthaceae. Revision for Australia, 28 genera: BARKER (1986). Alseuosmiaceae. Wood anatomy stresses relation to Saxifragaceae: DICKISON (1986a). Amaranthaceae. *Gomphrena* in Brazil: SIQUIERA (1985). Annonaceae. Palynology (see p. 304); karyology (see p. 305). Histology of seeds: CHRISTMANN (1986). Leaf anatomy for Neotropical members: SETTEN and KOEK-NOORMAN (1986). Apocynaceae. Survey of palynology: NILSSON (1986). Tabernaemontanoideae: ALLORGE (1985a,b). *Voacanga*: LEEUWENBERG (1985b). Series of smaller revisions: LEEUWENBERG (1983, 1985a). Aristolochiaceae (see p. 319). *Asarum*, conspectus, position: TREPLIAKOVA (1985). Asclepiadaceae. Corona formed by androecial appendages: HOFMANN and SPECHT (1986). *Stapelia*, revision: LEACH (1985). Avicenniaceae. Iridoids stress relation to Verbenaceae: KÖNIG and RIMPLER (1985). Balanophoraceae. Revision for tropical Africa: HANSEN (1986). Begoniaceae. (Artificial) key to all species of *Begonia*, the first full treatment since 1864, illustrated: L.B. SMITH et al. (1986). Placentation in *Begonia* advances from parietal to central, derivation of septae from carpellary or sterile placental tissue may vary in the same ovary: REITSMA (1983); alternative theory: BARABÉ et al. (1985). Palynology, careful discussion, *Hillebrandia* close to primitive *Begonias*: R.G.v.d. BERG (1983, 1985). *Begonia* sect. *Gireoudia*: BURT-UTLEY (1985). Berberidaceae. Two lines of evolution from elongate receptacle with sepals in many whorls to reduced structures; conspectus, based mainly on floral anatomy: TERABAYASHI (1985a,b). *Berberis* sect. *Wallichianae*, synopsis, key: CHAMBERLAIN and HU (1985). Bombaceae. Systematic wood anatomy for Neotropics: DÉTIENNE et al. (1983). Boraginaceae. Carpological studies: VOYTENKO and OPARINA (1985), HILGER (1985), HILGER et al. (1985). Outcrossing rates in hetero- and homostylous *Amsinckia*: GANDERS et al. (1985). *Cordia* subgen. *Varronia*: TARODA and GIBBS (1986). Brunnelliaceae. Supplement to monograph: CUATRECASAS (1985). Buxaceae. *Sarcococca*: SEALY (1986). Cactaceae. Consensus list worked out by a team of specialists enumerates 86 genera with braodly drawn generic limits; a lack of data impedes the phylogenetic analysis of cereoid cacti: IOS WORKING PARTY (1986). Growth form study: LOUP (1983). Campanulaceae.

Tropical members with hard-walled capsules have relict occurrences in Eurasian mountain areas, from where more specialized types of capsular dehiscence evolved: KOLAKOWSKI (1986). Carpology, with nice diagrams: KOLAKOWSKI (1985). Three possible lines of derivation for porate pollen from polycolpate pollen: AVETISJAN (1986). Capparaceae. Myrosin cells suggests placement of monotypic *Oceanopapaver* from New Caledonia: SCHMID et al. (1984). Caprifoliaceae. Palynology in line with separation of *Sambucus*, *Viburnum* and *Adoxa*; independent evolution of retiplate or pilate grains: DONOGHUE (1985). Caryophyllaceae. *Silene* (see p. 315); *S.* sect. *Siphonomorpha* in W-Europe: JEANMONOD (1985). Distribution maps for subspecies of *S. vulgaris* generated by computer: AESCHIMANN (1985). *Uebelinia*: OUSTED (1985). Celastraceae. *Maytenus* in tropical Africa and Madagascar: SEBSEBE (1985). Cephalotaceae. Closer to Saxifragaceae than to Crassulaceae as far as flavonoid chemistry is concerned: NICHOLLS et al. (1985). Adaptations for carnivory: PARKES and HALLAM (1984). Ceratophyllaceae. Fruit spine variation: LES (1986). Plumule morphology: LES (1985). Complex discontinuous leaf and fruit variation, two species with several varieties in *Ceratophyllum*: WILMOT-DEAR (1985). Chenopodiaceae. Only two species in European *Salicornia*, the diploid one being more variable than the tetraploid one: HUISKES et al. (1985). Chrysobalanaceae. Flavonoid patterns point to African origin: CORADIN et al. (1985). Clusiaceae. Palynology: SEETHARAM (1985). Cneoraceae. Palynology reveals cospecificity of European and Cuban species of *Cneorum*: LOBREAU-CALLEN and JÉRÉMIE (1986). Compositae. Symposium on generic concepts: Taxon 34, 5-88 (1985). Ecological and evolutionary aspects of inconsistently deciduous pappus: SHMIDA (1985). Continuous hybridization in *Aster ascendens* group: ALLEN (1985). *Acmella*: JANSEN (1986). *Brachycome*, outbreeders with wider distribution: WATANABE et al. *Bubonium* in Macaronesia: HALVORSEN and BORGEN (1986). Calenduleae, embryology: AHLSTRAND (1985). *Carlina*, floral morphology and carpology: DITTRICH (1985). *Centaurea* sect. *Chamaecyanus*: FERNANDEZ CASAS and SERNA (1985). *Chuquiraga*: EZCURRA (1985). *Gutierrezia*: LANE (1985). Phloroglucin derivates in *Helichrysum*: JAKUPOVIC et al. (1986). Liabeae, palynology: ROBINSON and MARTICORENA (1986). Hawaiian Madiinae, monograph: G.D. CARR (1985). *Moscharia* has simple capitulum, no complex one: HELLWIG (1985). *Senecio* in Bolivia: CABRERA (1985). Biosystematics of Australian *Senecio*: LAWRENCE (1985). Senecioneae in tropical Africa: JEFFREY (1986). Reproductive biology of Finnish *Taraxaca*: ROUSI et al. (1985). *Verbesina* sect. *Pseumontanoa*: TURNER (1985a), sect. *Ochractinia*: OLSEN (1985). Convolvulaceae. *Calycobolus* in tropical Africa: LEJOLY and LISOWSKI (1985). Cornaceae. Inflorescence morphology suggests separation of e.g. *Hellwingia*, *Corokia*, *Nyssa* and *Davidia*: JAHNKE (1986). Crassulaceae. Architecture and inflorescence morphology of *Kalanchoë*: CREMERS (1986). Cruciferae. Flavonoid spectra suggest that primitive sections of *Crambe* remained at disjunct ends of former area, while the more recent taxa evolved in the centre of the area: AGUINA-GALDE and GOMEZ-CAMPO (1984). Didymelaceae. Wood anatomy: TAKHTAJAN et al. (1986). Dipsacaceae. Systematic and evolutionary study, central Asian origin with subsequent differentiation in Middle East, accompanied by descending dysploidy, exine elaboration and fruit radiation: VERLAQUE (1985, 1986). Dipterocarpaceae. First report of agamospermy: KAUR et al. (1986). Droseraceae. Cytotaxonomy for Australia, aneuploid series, $x = 5$ probable base number: KONDO and LAVARACK (1984). Elaeocarpaceae. *Vallea* and *Aristotelia*: COODE (1985). Ericaceae. *Cassiope*, wood anatomy: WALLACE (1986). *Pernettya*, monograph: SLEUMER (1985), leaf anatomy: BAAS (1985). Second *Rhododendron* conference: Notes Royal Bot. Gard. Edinburgh 43(1), 1-187 (1985). *Rhododendron*, revision of several subgenera: PHILIPSON and PHILIPSON (1986). Euphorbiaceae. Morphology of cyathia: HOPPE (1985). *Phyllanthus*, architecture and evolutionary trends: ROSSIGNOL and ROSSIGNOL (1985). Cladistics of foliar characters: G.A. LEVIN (1985a,b). Fagaceae. *Quercus* in North America: MILLER and LAMB (1985). Cupule morphology of *Quercus*: KAUL (1985). Gentianaceae. *Lisianthus* (see p. 315). Cytogeography of *Centaurium pulchellum*, 4n species in N-Europe, 6n in Mediterranean area, only one 2n population based on $x = 9$ found: ZELTNER (1985). Secoiridoid glucosides stress isolation of *Centaurium pulchellum*: SLUIS (1985). Neotropical saprophytic *Voyria* and *Voyriella*: MAAS and RUYTERS (1986). Geraniaceae. Horticultural monograph of hardy *Geraniums*, exemplary treatment: YEO (1985). *Geranium* in South Africa: HILLIARD and BURTT (1985). *Pelargonium* sect. *Pelargonium*: WALT (1985),

P. sect. *Eumorpha*, karyotypes: GIBBY and WESTFOLD (1986). Gesneriaceae. Wood anatomy: CARLQUIST and HOEKMAN (1986). Phenolics: KVIST and PEDERSEN (1986). Gunneraceae. P-type sieve-tube plastids not in line with placement near Haloragales: BEHNKE (1986c). Gyrostemonaceae. *Waltheranthus*, new monotypic genus: KEIGHERY (1985). Haloragaceae. Myriophyllum in Australia: ORCHARD (1985). Hamamelidaceae. Gynoeceum in *Liquidambar* very primitive: BOGLE (1986). Humiriaceae. Ovule and seed confirm closeness to Malphigiales and Linales: BOESEWINKEL (1985). Hydrangeaceae. Flavonoids similar to herbaceous Saxifragaceae except for low frequency of myricetin: BOHM et al. (1985). Hydrastidaceae. High chromosome number, floral structure and leaf anatomy favour separate family status: TOBE and KEATING (1985). Hypericaceae. Diploids centred in Mediterranean area, x = 10 base number for palaeopolyploids, x = 8 for neopolyploids: REYNAUD (1986). Juglandaceae. Fossil and recent fruits stress close relation between *Carya* and *Platycarya*: SCHAARSCHMIDT (1985). Krameriaceae. Crassinucellate bitegmic seeds of malphigiaceous affinity, no links to Leguminosae or Polygalaceae: VERKERKE (1985b). Labiatae. Re-investigation favours ERDTMAN's classification over BRIQUET's: CANTINO and SANDERS (1986). *Aeollanthus*: RYDING (1986). *Scutellaria albida* group, biosystematics: BOTHMER (1985). Karyotypes, pollen fertility, seed proteins and nuclear DNA value elucidate phylogeny of *Sideritis*: FERNÁNDEZ-PERALTA and CONZÁLEZ-AGUILERA (1986). Flavonoids (HARBORNE et al. 1986) and growth forms in *Teucrium*: KÄSTNER (1985). Cytoplasmatically transmitted male sterility in *Thymus* decreases after colonization: COUVET et al. (1985a). Lauraceae. Both monotelic and polytelic inflorescences found: WEBERLING (1985). Neotropical member found in otherwise Asian *Caryodaphnopsis*: WERF and RICHTER (1985). *Cinnamomum*: KOSTERMANS (1986). *Ocotea* (see p. 315). Leguminosae. Ontogeny of floral symmetry is in line with HAECKEL's rule: TUCKER (1984). Caesalpinioideae. Cotyledon venation: SMITH and SCOTT (1985). Mimosoideae. Fruits and seeds, keys, illustrations, descriptions: GUNN (1984); ditto for Brazil: LIMA (1985). Evolutionary trends and biogeography of *Acacia*: PEDLEY (1986). *Archidendron*: NIELSEN et al. (1984). *Inga* in Guyana: PONCY (1985). *Mimosa*, palynology: CACCAVARI (1985). *Parkia*, palynology: FEUER et al. (1985). Papilionoideae. Seed coat sculpturing: SAINT-MARTIN (1986). Biochemical characters, *Rhizobium* specifity, etc. suggest origin of *Lotus corniculatus* from male *L. uliginosus* and female *L. tenuis* or *L. alpinus*: ROSS and JONES (1985). 7S seed proteins, parallelisms taken to indicate strong selection constraint, useful for solving relations of Phaseoleae: J.J. DOYLE et al. (1985). Seedling anthocyan pigmentation in *Vigna* and *Phaseolus*: NOZZOLILLO and McNEILL (1985). *Psoralea* should be split on chemical evidence: BOARDLEY et al. (1986). Some major revisions: *Cajanus*: MAESEN (1985a). *Milletia* in China: WEI ZHI (1985). *Pueraria*: MAESEN (1985b). *Spatholobus*: RIDDER-NUMANN and WIRIADINATA (1985). *Taverniera*: THULIN (1985). *Baphia*, Revision: SOLADOYE (1985), palynology: SOLADOYE and CRANE (1985). Lennoaceae. revision: YATSKIEVYCH and MASON (1986). Lentibulariaceae. *Utricularia* in Brazil: FROMM-TRINTA (1985). Sympatric speciation in *U. alpina*: JÉRÉMIE and JEUNE (1985). Limnanthaceae. Neither morphology nor phytochemistry conclusive for position: HOFMANN and LUDEWIG (1985). Breeding system and population structure of *Limnanthes*: KESSELI and JAIN (1985). Linaceae. Breeding system in *Linum tenuifolium*: NICHOLLS (1985, 1986). Loganiaceae. *Strychnos* in Vietnam, evolutionary trends: KHAN (1985, 1986). Loranthaceae. Palynology of small flowered Neotropical group: FEUER (1985). Lythraceae. Palynology: GRAHAM et al. (1985). *Ammannia*: GRAHAM (1985). Magnoliaceae. Vascular system: UEDA (1986) Malvaceae. Palynology, 3-zonocolporate pollen primitive, origin probably Gondwanaland: CHRISTENSEN (1986). *Sida*, key: FRYXELL (1985). Melastomataceae. Key to Asiatic genera based on foliar glands: HANSEN (1985). Memecycleae of Madagascar: JACQUES-FÉLIX (1984/1985). Meliaceae. Biology of South African groups: WHITE (1986). Mesembryanthemaceae. Generic monograph of Mesembryanthemoideae, basal position in family, South African origin, distinct from Aizoaceae s.str., but reconsideration of systematics of Aizoaceae s.l. will be necessary: BITTRICH (1986). *Psilocaulon* and related genera: IHLENFELDT and BITTRICH (1985). *Astridia*, *Acrodon Ebracteola*: GLEN (1986). Monimiaceae. Monograph for Malagasy region; *Diptera* and specialized beetles feeding on leaves are main pollinators: LORENCE (1985). *Kibara* in Malesia: PHILIPSON (1985). Moraceae. Essay on *Ficus* biology: CORNER (1985), key for Brazil: C.C. BERG et al. (1984). Moringaceae. Synopsis: VERDCOURT

(1985). Palynology inconclusive for position: FERGUSON (1985). Myricaceae. Palynology: SUNDBERG (1985). Myristicaceae. *Myristica* flower basically trimerous: ARMSTRONG and TUCKER (1986). *Horsfieldia*: WILDE (1984/1985). Myrsinaceae. *Rapanea* in New Guinea: SLEUMER (1986). Nolanaceae. Uniovulate placentae in pluriovulate carpels basic innovation in the family: BONDESON (1986). Ochnaceae. Ouratineae in Africa, key, evolutionary trends: FARRON (1985). Onagraceae. Trend towards deciduousness in *Fuchsia* sect. *Hemsleyella*: BERRY (1985). Alternative scheme for *Oenothera* subsect. *Oenothera*: ROSTAŃSKI (1985). *Oenothera caespitosa*-aggr.: WAGNER et al. (1985). *Gaura*, biosystematics: CARR et al. (1986). Oncothecaceae: Floral morphology and palynology inconclusive for position: DICKISON (1986b). Papaveraceae. Synopsis of Fumarioideae, widely based monograph of Fumarieae: LIDEN (1986). Carpology stresses isolation of *Pteridophyllum*: BRÜCKNER (1985). Revisions in Eurasian *Papaver*: KADEREIT (1986a,b). Pedaliaceae. *Sesamum* (see p. 317). Petiolary nectaries are aborted flowers: MONOD (1986). Phytolaccaceae. Careful cladistic analysis: BROWN and VARADARAJAN (1985). Picrodendraceae. Floral morphology suggests closeness to Euphorbiaceae rather than to Rutales: HAKKI (1985). Piperaceae. x = 11 base number, *Peperomia* distinct: OKADA (1986). Pittosporaceae. Separate order close to both Araliales and Cornales proposed based on fatty acids in fruits and seeds, etc.: STUHLFAUTH et al. (1985). Polygalaceae. Evolutionary trends for fruits and seeds: VERKERKE (1985a). Polygonaceae. *Triplaris*: BRANDBYGE (1986). Primulaceae. *Dionysia*, xeromorphic features: BOKHARI and WENDELBO (1985). *Lysimachia*, evolution: VOGEL (1986), in mainland SE Asia: HU CHI-MING (1985). Proteaceae. Pollen evolution: FEUER (1986). Pyrolaceae. Trend from verrucate to rugulate and psilate pollen in *Pyrola*: TAKAHASHI (1986). Rafflesiaceae. Trend from monocolpate or monosulcate to tricolnate pollen, should be split into several families: TAKHTAJAN et al. (1985). Ranunculaceae. Pollinator guilds in *Cimicifuga*: PELLMYR (1986). *Ranunculus auricomus* in Bavaria: BORCHERS-KOLB (1985). Retziaceae. Wood anatomy favours merger with Stilbaceae: CARLQUIST (1986). Rhabdodendraceae. Ellagic acid supports placement in Myrtiflorae: WOLTER FILHO (1985). Rhamnaceae. Floral morphology and fruit dehiscence in amphi-Pacific *Discaria*: MEDAN (1985). Rosaceae. Apospory and pseudogamy in *Amelanchier*: CAMPBELL et al. (1985). Synopsis of European *Pyrus*: TERPO (1985). Growth from (SEREBRYAKOVA and PAVLOVA 1986) and hybridogenous species in *Potentilla*: SOJAK (1986). Rubiaceae. Monumental biological and systematic study of the exclusively wind-pollinated Anthospermeae: PUFF (1986). Pollen and stomata of Gardenieae from Gabon, including cladistic study of *Rothmannia*: KEDDAN-MALPLANCHE (1985). Wide-ranging survey of morphology, floral biology and distribution of Gardenieae, new classification for the ca. 100 genera proposed: ROBBRECHT and PUFF (1986). *Houstonia*, palynology and seed morphology: TERRELL et al. (1986). *Keetia*: BRIDSON (1986). *Psydrax*: BRIDSON (1985). Salicaceae. *Salix* in SE-USA: ARGUS (1986), in Italy: MARTINI and PAIERO (1984). Sapindaceae. Highly important studies in *Harpullia*: Functional pollen morphology and ecological and geological data combined into cladogram: MULLER (1985); revision and discussion of different taxonomic approaches: LEENHOUTS (1985). *Nephelium*: LEENHOUTS (1986). Saxifragaceae. *Boykinia* and related genera: GORNALL and BOHM (1985). *Ribes* sect. *Grossularia* in North America: SINNOTT (1985). *Saxifraga*, trichome morphology: GORNALL (1986). Scrophulariaceae. Schlegelieae clearly placed here, based on floral characters: ARMSTRONG (1985). *Leucophyllum* and *Eremogeton*: HENRICKSON and FLYR (1985). Maurandynae migrated from North America to S-Mexico, cladistics, monograph: ELISENS (1985). Complicated gland hairs in *Odontites*: BOLLIGER (1985). *Parentucciella* has vesicular-arbuscular mycorrhiza and stresses the idea that root parasitism is derived from mycorrhiza: ALEXANDER and WEBER (1985). Seed proteins establish relationships between European and North American *Rhinanthus* where morphology is inconclusive: VAN HULST (1986). *Striga*, biosystematics: MUSSELMAN and AYENSU (1984), geopyrophytic *Striga baumannii*: RAYNAL-ROQUES (1985), biology and control: AYENSU (1984). Solanaceae. Second Solanaceae conference with stress on chemistry, ecology and economy: D'ARCY (1986); see also p. 309 and p. 317. Serotaxonomy useful in assessing relationships in *Solanum*, where hybridization experiments fail; the genus should be split: LESTER and ROBERTS (1986). *Solanum* in New Guinea has relations to Asia, Australia and America: SYMON (1985b). Solanaceae of New Guinea: SYMON (1985a). *Solanum* subgen. *Lepto-*

stemonum: WHALEN (1984). Flavonol ring substitution patterns useful for sectional classification in *Solanum*: STEINHARTER et al. (1986). Extrafloral nectaries, rare in Solanaceae, detected in Australian *Solanum*: ANDERSON and SYMON (1985). Evolutionary tendencies in fusion of lateral veins in sympetalous calyces discussed from *Lycianthes*: D'ARCY (1986b). Staphyleaceae. Cunonialean and saxifragalean affinity confirmed by floral morphology: DICKISON (1986c). Symplocaceae. Biogeography in West Indies: MAI (1986). Theaceae. Genetic drift and migration in *Camellia japonica* populations: WENDEL and PARKS (1985). Thymelaeaceae. Palynology suggests separation of *Aetoxylon*, *Amyxa* and *Gonystylus*: NOWICKE et al. (1985). Tiliaceae. *Luehea*: CUNHA (1985). Umbelliferae. Ontogeny of flower and fruit: LEINS and ERBAR (1985a), ERBAR and LEINS (1985). Descending dysploidy in *Bunium*: VASIL'EVA et al. (1985). *Deverra*: PFISTERER and PODLECH (1986). *Lilaeopsis*: AFFOLTER (1985). Urticaceae. *Sphaerotylos*, new primitive Chinese genus: CHEN CHIA-JUI (1985). Vacciniaceae. *Symphysia* should be kept out of *Vaccinium*: VANDER KLOET (1985). Valerianaceae. Phylogenetic aspects of axis organization: LÖRCHER and WEBERLING (1984). *Phyllactis* and *Valeriana* sect. *Bracteata*: LARSEN (1986). Verbenaceae. Bracts act as pseudocorollas and dispersal organs in Symphoremoideae: CLASSEN (1985). Winteraceae. *Exospermum* and *Bubbia* sunk into *Zygogynum*: VINK (1985).

Genera incertae sedis. *Blepharocarya* biflavonols suggest relation to Anacardiaceae: WANNAN et al. (1985). *Calyptrotheca* belongs to Portulaccaceae and not to Capparidaceae; morphology, leaf flavonoids, palynology, seed morphology: NYANANYO (1986). Pollen tetrads question position of *Carpodetus* in Escalloniaceae (Saxifragales) and suggest affinities to Ericaceae: PRAGLOWSKI and GRAFSTRÖM (1985). *Empleuridium*, a monotypic and scarcely known Cape genus, should be placed in Celastraceae, and not in Rutaceae: GOLDBLATT et al. (1985). KOSTERMANS (1985) insists on family status for *Pakaraimea* with Monotoideae, close to Tiliaceae. *Stegnosperma* embryology suggests bridging position between Caryophyllaceae and Phytolaccaceae: KAMELINA and PROSKURIANA (1985), NARAYANA and NARAYANA (1986).

References

ABBOTT, L.A., BISBY, F.A., ROGERS, D.J.: Taxonomic Analysis in Biology: Computers, Models and Databases, 336 pp. New York: Columbia University Press 1986. - ABRAMS, P.A.: Evolution 40, 1229-1247 (1986). - ACKERMAN, J.D.: Selbyana 9, 52-60 (1986). - AESCHIMANN, D.: Candollea 40, 67-98 (1985a); - ibid. 40, 57-65 (1985b). - AFFOLTER, J.M.: Syst. Bot. Monogr. 6, 1-140 (1985). - AGUINAGALDE, I., GOMEZ-CAMPO, C.: Bot. J. Linn. Soc. 89, 277-288 (1984). - AHLSTRAND, L.: Nord. J. Bot. 5, 79-97 (1985). - ALEXANDER, T., WEBER, H.C.: Beitr. Biol. Pflanz. 60, 23-34 (1985). - ALLEN, G.: Am. J. Bot. 72, 268-277 (1985). - ALLORGE, L.: Mem. Mus. Natl. Hist. Nat., N.S., Sér. B., Bot. 30, 1-216 (1985a); - Bull. Mus. Natl. Hist. Nat., Sect. B, Adansonia 7, 433-451 (1985b). - AMMANN, K.: Bot. Helv. 96, 109-132 (1986). - AMOUROUX-PEZAS, C.: Bull. Soc. Bot. France Actual. Bot. 132, 81-87 (1985). - ANDERSON, G.J., SYMON, D.E.: Biotropica 17, 40-45 (1985). - ANDERSSON, L.: Nord. J. Bot. 6, 729-765 (1986). - ANDREWS, J.: Peppers, the Domesticated Capsicums, 170 pp. Austin: University of Texas Press 1984. - ANTOLIN, M., STROBECK, C.: Am. Nat. 126, 52-62 (1985). - ARCHIBOLD, O.W., WEICHEL, B.J.: Can. J. Bot. 64, 1204-1211 (1985). - ARGUS, W.: Syst. Bot. Monogr. 9, 1-123 (1986). - ARMBRUSTER, W.S.: Evolution 39, 733-752 (1985); - Ecology 67, 522-533 (1986). - ARMSTRONG, J.E.: Am. J. Bot. 72, 755-766 (1985). - ARMSTRONG, J.E., TUCKER, S.C.: Am. J. Bot. 73, 1131-1143 (1986). - ARROYO, S.C.: Kurtziana 17, 55-70 (1984). - ASHTON, P.S., GAN YIK-YUEN, ROBERTSON, F.W.: Bot. J. Linn. Soc. 89, 293-304 (1984). - ATWOOD, J.T.: Selbyana 9, 171-186 (1986). - AVERETT, J.E., BOUFFORD, D.E.: Syst. Bot. 10, 363-373 (1985). - AVERETT, J.E., HAHN, W.J., BERRY, P.E., RAVEN, P.H.: Am. J. Bot. 73, 1525-1534 (1986). - AVETISJAN, E.M.: Bot. Zh.

71, 1003-1010 (1986). - AYALA, F.J.: J. Heredity 77, 226-235 (1986). - AYENSU, E.S. (ed.): *Striga*: Biology and Control, 216 pp. Paris: International Council of Scientific Unions 1984.

BAAS, P.: Bot. Jahrb. Syst. 105, 481-495 (1985). - BACHMANN, K., HEUDSEN, A.W. Van, CHAMBERS, K.L., PRICE, H.J.: Experientia 41, 1348-1350 (1985a). - BACHMANN, K., CHAMBERS, K.L., GRAU, J., PRICE, H.J.: Beitr. Biol. Pflanz. 60, 51-88 (1985b). - BADAEVA, E.D., SHKUTINA, F.M., BOGDEVICH, I.N., BADAEV, N.S.: Plant. Syst. Evol. 154, 183-194 (1986). - BARABÉ, D., BROUILLET, L., BERTRAND, C.: Bull. Mus. Natl. Hist. Nat., Sect. B, Adansonia 7, 403-414 (1985). - BARANOVA, M.V.: Bot. Zh. 70, 1585-1594 (1985); - ibid. 71, 1308-1320 (1986). - BARKER, C.M., STACE, C.A.: Nord. J. Bot. 6, 1-10 (1986). - BARKER, R.M.: J. Adelaide Bot. Gard. 9, 1-286 (1986). - BARKLEY, T.M.: Taxon 34, 17-21 (1985). - BARKWORTH, M.E., DEWEY, D.R.: Am. J. Bot. 72, 767-776 (1985). - BARNES, B.V., DANCIK, B.P.: Can. J. Bot. 63, 223-226 (1985). - BARNES, S.H., BLACKMORE, S.: Linn. Soc. Symp. Ser. 12, 71-80 (1986). - BARRETT, S.C.H.: Biol. J. Linn. Soc. 25, 41-60 (1985). - BARRETT, S.C.H., GLOVER, D.E.: Evolution 39, 766-774 (1985). - BARROSO, M. de F., VIERA, M.H.S.A.: Anu. Soc. Broteriana 51, 19 (1985). - BARTHELEMY, D.: Convergences géographiques: l'hypothese d'une transmission d'information génétique par de vois non sexuelles, 84 pp. Montpellier: Thèse 1983. - BARTON, N.H., HEWITT, G.M.: Annu. Rev. Ecol. Syst. 16, 113-148 (1985). - BAYER, R.J.: Plant Syst. Evol. 150, 143-163 (1985a); - Rhodora 87, 321-339 (1985b). - BAYER, R.J., CRAWFORD, D.J.: Am. J. Bot. 73, 287-296 (1986). - BEART, J.E., LILLEY, T.H., HASLAM, E.: Phytochemistry 24, 33-38 (1985). - BEATTIE, A.J.: The Evolutionary Ecology of Ant-plant Mutualisms. Cambridge: Cambridge University Press 1985. - BEDIGIAN, D., HARLAND, J.R.: Econ. Bot. 40, 137-154 (1986). - BEDIGAN, D., SEIGLER, D.S., HARLAN, J.R.: Biochem. Syst. Ecol. 13, 133-139 (1985). - BEDIGAN, D., SMYTH, C.A., HARLAN, J.R.: Econ. Bot. 40, 353-365 (1986). - BEHNKE, H.-D.: Ann. Missouri Bot. Gard. 71, 824-831 (1984); - Taxon 34, 607-610 (1985); - Plant Syst. Evol. 152, 101-121 (1986a); - Bot. Jahrb. Syst. 106, 499-510 (1986b); - Plant Syst. Evol. 151, 215-222 (1986c). - BELL, T.J., QUINN, J.A.: Bot. Gaz. 146, 252-258 (1985). - BENJAMIN, R.B., HAINSWORTH, F.R.: Evolution 40, 843-855 (1986). - BENZING, D.H.: Selbyana 9, 23-43 (1986). - BENZING, D.H., GIVNISH, T.J., BERMUDES, D.: Syst. Bot. 10, 81-91 (1985). - BERENBAUM, M.R., ZANGERL, A.R.: Phytochemistry 25, 659-661 (1986). - BERENBAUM, M.R., ZANGERL, A.R., NITAO, J.K.: Evolution 40, 1215-1228 (1986). - BERG, C.C., VÁSQUEZ AVILA, M., KOOY, F.: Acta Amazonica 14, 159-194 (1984). - BERG, M.E. v.d.: Ver-o-peso: The ethnobotany of an Amazonian market, 140-149. In: see ref. PRANCE and KALLUNKI 1984. - BERG, R.G. v.d.: Meded. Landbouwhoogesch. Wageningen 83-9, 55-66 (1983); - Agric. Univ. Wageningen Papers 84-3, 1-94 (1985). - BERLIN, B.: Contribution of native American collectores to the ethnobotany of the Neotropics, 24-33. In: see ref. PRANCE and KALLUNKI 1984. - BERRY, P.E.: Ann. Missouri Bot. Gard. 72, 213-251 (1985). - BERTIN, R.I.: Am. Nat. 126, 750-759 (1985). - BHAT, K.P.: Phytomorphology 34, 78-81 (1984). - BHAVE, M.R., GUPTA, V.S., RANJEKAR, P.K.: Plant Syst. Evol. 152, 133-151 (1986). - BIANCO, P., SCHIVONE, B.: Taxon 34, 436-439 (1985). - BIERZYCHUDEK, P.: Experientia 41, 1255-1264 (1985). - BIRCH, A.N.E., FELLOWS, L.E., EVANS, S.V., DOHERTY, K.: Phytochemistry 25, 2745-2749 (1986). - BITTRICH, V.: Mitt. Inst. Allg. Bot. Hamburg 21, 5-116 (1986). - BLACKMORE, S., BARNES, S.H.: Linn. Soc. Symp. Ser. 12, 137-149 (1986). BLACKMORE, S., FERGUSON, I.K.: Pollen and Spores. Form and Function, 443 pp. London: Academic Press (= Linn. Soc. Symp. Ser. 12) 1986. - BLOOM, A.J., CHAPIN, F.S., MOONEY, H.A.: Annu. Rev. Ecol. Syst. 16, 363-392 (1985). - BOARDLEY, M., STIRTON, C.H., HARBORNE, J.B.: Biochem. Syst. Ecol. 14, 603-613 (1986). - BOESEWINKEL, F.D.: Acta Bot. Neerl. 34, 183-191 (1985). - BOGANI, P., BUIATTI, M., TEGLI, S., PELLEGRINI, M.G., BETTINI, P., SCALA, A.: Plant Syst. Evol. 151, 19-29 (1985). -

BOGLE, A.L.: Ann. Missouri Bot. Gard. 73, 325-347 (1986). - BOHM, B.A., NICHOLLS, K.W., BHAT, U.G.: Biochem. Syst. Ecol. 13, 441-445 (1985). - BOKHARI, M.H., WENDELBO, P.: Notes R. Bot. Gard. Edinb. 42, 327-345 (1985). - BOLAND, D.J., SEDGLEY, M.: Aust. J. Bot. 34, 569-584 (1986). - BOLLIGER, M.: Bot. Jahrb. Syst. 107, 153-175 (1985). - BONDESON, W.E.: Nord. J. Bot. 6, 183-198 (1986). - BORCHERS-KOLB, E.: Mitt. Bot. Staatssamml. München 21, 49-300 (1985). - BORG-KARLSON, A.-K., EIDMANN, H.H., LINDSTRÖM, M., NORIN, T., WIERSMA, N.: Phytochemistry 24, 455-456 (1985). - BOSTER, J.S.: Econ. Bot. 39, 310-325 (1985). - BOTHMER, R. von: Nord. J. Bot. 5, 421-439 (1985). - BOTHMER, R. von, JACOBSEN, N.: Plant Syst. Evol. 153, 49-64 (1986). - BOUGH, M., COLOSI, J.C., CAVERS, P.B.: Can. J. Bot. 64, 1188-1198 (1986). - BRAGA, M.R., YOUNG, M.C.M., PONTE, J.V.A., DIETRICH, S.M.C., EMERENCIANO, V. de P., GOTTLIEB, O.R.: Biochem. Syst. Ecol. 14, 507-514 (1986). - BRANDBYGE, J.: Nord. J. Bot. 6, 545-570 (1986). - BRANDENBURG, W.A.: Syst. Assoc. Spec. Vol. 29, 87-98 (1986). - BREESE, E.L., TYLER, B.F.: Syst. Assoc. Spec. Vol. 29, 53-69 (1986). - BREMER, K.: Cladistics 1, 369-385 (1985). - BRIDSON, D.M.: Kew Bull. 40, 687-725 (1985); - ibid. 41, 965-994 (1986). - BROWN, G.K., GILMARTIN, A.J.: Brittonia 36, 364-374 (1984); - Selbyana 9, 88-93 (1986). - BROWN, G.K., VARADARAJAN, G.S.: Syst. Bot. 10, 49-63 (1985). - BRÜCKNER, C.: Feddes Rep. 96, 199-213 (1985). - BRUGGEN, H.W.E. van: Bibl. Bot. 137, 1-76 (1985). - BÜCHLER, W.: Bot. Helv. 95, 165-175 (1985); - Bot. Helv. 96, 135-143 (1986). - BUCHMANN, S.L.: Vibratile pollination in *Solanum* and *Lycopersicon*: a look at pollen chemistry, 237-252. In: see ref. D'ARCY 1986. - BUCK, W.R.: Taxon 35, 306-309 (1986). - BULL, J.J., CHARNOV, E.L.: Evolution 39, 1149-1155 (1985). - BURNS-BALOGH, P., BERNHARDT, P.: Plant Syst. Evol. 149, 119-134 (1985). - BURNS-BALOGH, P., FUNK, V.: Smithson. Contrib. Bot. 61, 1-79 (1986). - BURT-UTLEY, K.: Tulasne Stud. Zool. Bot. 25, 1-131 (1985).

CABRERA, A.: Darwiniana 26, 79-217 (1985). - CACCAVARI, M.A.: Pollen Spores 28, 29-42 (1985). - CALLAN, N.W., THOMPSON, M.M.: Theor. Appl. Genet. 71, 657-661 (1986). - CAMP, W.H., GILLY, G.L.: Brittonia 4, 323-385 (1943). - CAMPBELL, C.S., GREENE, C.W., NEUBAUER, B.F., HIGGINS, J.M.: Am. J. Bot. 72, 1397-1403 (1985). - CANTINO, P.D.: Syst. Bot. 10, 119-122 (1985). - CANTINO, P.D., SANDERS, R.W.: Syst. Bot. 10, 163-185 (1986). - CARLQUIST, S.: Bull. Torrey Bot. Club 112, 59-69 (1985); - Aliso 11, 299-316 (1986). - CARLQUIST, S., HOCKMAN, D.A.: Aliso 11, 279-297 (1986). - CARR, B.L., GREGORY, D.P., RAVEN, P.H., TAI, W.: Syst. Bot. 10, 98-111 (1986). - CARR, D.J., CARR, S.G.M., LENZ, J.R.: Aust. J. Bot. 34, 53-62 (1986). - CARR, G.D.: Allertonia 4, 1-123 (1985). - CARR, G.D., KYHOS, D.W.: Evolution 40, 959-976 (1986). - CARR, G.D., POWELL, E.A., KYHOS, D.W.: Evolution 40, 430-434 (1986). - CARSON, H.L.: Syst. Bot. 10, 380-390 (1985). - CAUWET-MARC, A.M., BALAYER, M.: Bot. Helv. 94, 391-406 (1984). - CHALONER, W.G.: Linn. Soc. Symp. Ser. 12, 103-108 (1986). - CHAMBERLAIN, D.F., HU, C.-M.: Notes R. Bot. Gard. Edinb. 42, 529-557 (1985). - CHANDEL, K.P.S., LESTER, R.N., STARLING, R.J.: Bot. J. Linn. Soc. 89, 85-96 (1984). - CHANG, T.T.: Iowa State J. Res. 59, 325-496 (1985). - CHAUHAN, K.P.S., BRANDHAM, P.E.: Kew Bull. 40, 567-571 (1985a); - ibid. 40, 745-758 (1985b). - CHEN, C.-J.: Acta Phytotax. Sinica 23, 444-456 (1985). - CHEN, Z.-Y., SCAGEL, R.K., MAZE, J.: Can. J. Bot. 64, 1654-1663 (1986). - CHRISTENSEN, P.B.: Grana 25, 95-117 (1986). - CHRISTMANN, M.: Bot. Jahrb. Syst. 106, 379-390 (1986). - CHOU, C.-H., HWANG, Y.-H.: Bot. Bull. Acad. Sinica 26, 155-170 (1985). - CHOU, C.-H., SHEEN, S.-S., HWANG, Y.-H.: Bot. Bull. Acad. Sinica 25, 177-189 (1984). - CHOUDRY, A.S.: J. Sci. Hiroshima Univ., Ser. B, Div. 2, 19, 57-109 (1984). - CLAY, K., ANTONOVICS, J.: Evolution 39, 335-348 (1985). - CLAY, K., LEVIN, D.A.: Ecology 67, 37-45 (1986). - CLASSEN, R.: Beitr. Biol. Pflanz. 60, 383-402 (1985). - CLAWSON, D.L.: Econ. Bot. 39, 56-67 (1985). - COLEY, P.D., BRYANT, J.P., CHAPIN, F.S.: Science 230, 895-899 (1985). - CONRAN, J.G., CLIFFORD, H.T.: Nord. J. Bot. 5, 215-219

(1985). - COOK, C.D.K.: Aquat. Bot. 21, 157-164 (1985). - COOK, C.D.K., NICHOLS, M.S.: Bot. Helv. 96, 213-267 (1986). - COOK, C.D.K., URMI-KÖNIG, K.: Aquat. Bot. 21, 111-156 (1985). - COODE, M.J.E.: Kew Bull. 40, 479-507 (1985). - COOPER-DRIVER, G.A., SWAIN, T., CONN, E.E. (eds.): Chemically Mediated Interactions Between Plants and Other Organisms. (Recent Advances in Phytochemistry, Vol. 19), 246 pp. New York, London: Plenum Press 1985. - COPE, T.A., STACE, C.A.: Watsonia 15, 309-320 (1985). - CORADIN, L., GIANNASI, D.E., PRANCE, G.T.: Brittonia 37, 169-178 (1985). - CORNER, E.J.H.: Allertonia 4, 125-168 (1985a); - Biol. J. Linn. Soc. 25, 187-195 (1985b). - COUDERC, H., GORENFLOOT, R., MORET, J., SIAMI, A.: Bull. Soc. Bot. France, Actual. Bot. 132, 63-71 (1985). - COUHENOR, M.B.: Ann. Missouri Bot. Gard. 72, 852-863 (1985). - COUVET, D., GOUYON, P.-H., KJELLBERG, F., VALDEYRON, G.: Compt. Prend. Acad. Sci. Paris 300, Sér. III, 665-668 (1985a). - COUVET, D., HENRY, J.-P., GOUYON, P.-H.: Am. Nat. 126, 294-299 (1985b). - CRANE, P.R.: Ann. Missouri Bot. Gard. 72, 716-793 (1985); - Linn. Soc. Symp. Ser. 12, 179-202 (1986). - CRANE, P.R., STOCKEY, R.: Ann. Missouri Bot. Gard. 73, 382-393 (1986). - CRANE, E., WALKER, P., DAY, R.: Directory of Important Honey Sources, 384 pp. Gerrards Cross, Bucks.: International Bee Research Assoc. 1984. - CRAWFORD, D.J., ORNDUFF, R., VASEY, M.C.: Am. J. Bot. 72, 1177-1184 (1985). - CRAWFORD, D.J., STUESSY, T.F., SILVA, O.M.: Plant Syst. Evol. 153, 133-139 (1986). - CREMERS, G., SELL, Y.: Bull. Mus. Natl. Hist. Nat., Sect. B, Adansonia 8, 63-76 (1986). - CREPET, W.L.: Ann. Missouri Bot. Gard. 71, 607-630 (1984). - CREPET, W.L., TAYLOR, D.W.: Science 228, 1087-1089 (1985). - CRIBB, P.J.: Kew Bull. 41, 615-692 (1986). - CRIBB, P.J., HAWKES, J.G.: Experimental evidence for the origin of *Solanum tuberosum* subspecies *andigena*, 383-404. In: see ref. D'ARCY 1986. - CROAT, T.B.: Monogr. Syst. Bot. 14, 1-204 (1985). - CRONQUIST, A.: Ann. Missouri Bot. Gard. 71, 780-782 (1984). - CRUDEN, R.W., LYON, D.L.: Bot. Gaz. 146, 143-149 (1985). - CUATRECASAS, J.: Fl. Neotr. 2, Suppl., 1-103 (1985). - CUNHA, M.C. DA SILVA: Sellowia 37, 5-41 (1985). - CZECZUGA, B.: Biochem. Syst. Evol. 14, 13-15 (1986).

DAHLGREN, R., BREMER, K.: Cladistics 1, 349-368 (1985). - DAHLGREN, R., LU, A.-M.: Nord. J. Bot. 5, 321-330 (1985). - DAHLGREN, R., THORNE, R.F.: Ann. Missouri Bot. Gard. 71, 633-699 (1984). - D'ARCY, W.G. (ed.): Solanaceae. Biology and Systematics, 603 pp. New York: Columbia University Press 1986; - Ann. Missouri Bot. Gard. 73, 117-127 (1986b). - DAVIDSE, G., ELLIS, R.P.: Ann. Missouri Bot. Gard. 71, 994-1012 (1984). - DAVIS, IV, T.: Jaltomata in the Tarahumara indian region of Northern Mexico, 405-411. In: see ref. D'ARCY 1986. - DELENDICK, T.J.: Taxon 34, 96-101 (1985). - DEL PERO MARTINEZ, M.A.: Biochem. Syst. Ecol. 13, 253-256 (1985). - DEL PERO MARTINEZ, M.A., SWAIN, T.: Biochem. Syst. Ecol. 13, 391-402 (1985). - DEMYANAOVA, E.I.: Bot. Zh. 70, 1289-1301 (1985). - DERRICK, L.N., HEYWOOD, V.H., SCHEEPEN, J. van: Syst. Assoc. Spec. Vol. 29, 53-69 (1986). - DETIENNE, P., LOUREIRO, A.A., JACQUET, P.: Acta Amazonica 13, 831-867 (1983). - DEWEY, D.R.: The genomic system of classification as a guide to intergeneric hybridization within the perennial Triticeae, 209-272. In: Gene Manipulation and Plant Improvement, ed. J.P. GUSTAFSON. New York: Plenum Press 1984. - DICKINSON, T.A.: Can. J. Bot. 64, 2738-2747 (1986). - DICKINSON, T.A., PHIPPS, J.B.: Syst. Bot. 10, 322-337 (1985); - Am. J. Bot. 73, 116-130 (1986). - DICKISON, W.C.: Syst. Bot. 10, 214-221 (1986a); - Brittonia 38, 249-259 (1986b); - Bot. Gaz. 147, 312-326 (1986c). - DI FULVIO, T.E., COCCUCCI, A.E.: Kurtziana 18, 13-21 (1986). - DIRZO, R., SARUKHÁN, J. (ed.): Perspectives on Plant Population Ecology, 478 pp. Sunderland: Sinauer 1984. - DITTRICH, M.: Bot. Jahrb. Syst. 107, 591-609 (1985). - DOEBLEY, J.F.: Ann. Missouri Bot. Gard. 71, 1100-1113 (1984). - DOEBLEY, J.F., GOODMAN, M.M., STUBER, C.W.: Am. J. Bot. 72, 629-639 (1985); - ibid. 73, 64-69 (1986). - DONOGHUE, M.J.: J. Arnold Arb. 66, 421-469 (1985). - DOWNIE, S.R., DENFORD, K.E.: Can. J. Bot. 64, 1355-

1372 (1986). - DOYLE, J.A., DONOGHUE, M.J.: Syst. Assoc. Spec. Vol. 31, 177-198 (1986a); - Bot. Rev. 52, 421-431 (1986b). - DOYLE, J.J., LADEN, B.F., BEACHY, R.N.: Biochem. Syst. Ecol. 13, 123-132 (1985). - DRANSFIELD, J., COMBER, J.B., SMITH, G.: Kew Bull. 41, 575-613 (1986). - DRUMMOND, B.A.: Coevolution of Ithomiine butterflies and solanaceous plants, 307-327. In: see ref. D'ARCY 1986. - DUKE, J.A., AYENSU, E.S.: Medicinal Plants of China, 705 pp. Algonac, MI: Reference Publications 1985. - DU PUY, D.J., FORD-LLOYD, B.V., CRIBB, P.J.: Kew Bull. 40, 421-434 (1985). - DUVICK, D.N.: Econ. Bot. 40, 289-297 (1986).

ECKENWALDER, J.E., BARRETT, S.C.H.: Syst. Bot. 11, 373-391 (1986). - EDGAR, E.: N.Z. J. Bot. 24, 425-503 (1986). - EDMONDS, J.M.: Bot. J. Linn. Soc. 92, 1-38 (1986). - EDWARDS, S.R., DAVIS, G.M., NEVLING, L.I.: The Systematics Community. Educational background for curators, sources of funding, etc., 275 pp. Lawrence, KS: Association of Systematics Collections c/o Museum of Natural History, University of Kansas 1985. - ELDREDGE, N.: Annu. Rev. Ecol. Syst. 17, 351-369 (1986). - ELISENS, W.J.: Syst. Bot. Monogr. 5, 1-97 (1985). - ELLENBERG, H.: Flora 176, 169-188 (1985). - ELLSTRAND, N.C., ANTONOVICS, J.: Evolution 39, 657-666 (1985). - EMERENCIANO, V. de P., KAPLAN, M.A.C., GOTTLIEB, O.R.: Biochem. Syst. Ecol. 13, 145-166 (1985). - EMERENCIANO, V. de P., KAPLAN, M.A.C., GOTTLIEB, O.R., BONFANTI, M.R. de M., FERREIRA, Z.S., COMEGNO, L.M.A.: Biochem. Syst. Ecol. 14, 585-589 (1986). - ENDRESS, P.K.: Plant Syst. Evol. 152, 1-28 (1986a); - Ann. Missouri Bot. Gard. 73, 297-324 (1986b); - Veröff. Geobot. Inst. ETH, Stiftung Rübel, Zürich 87, 100-111 (1986c). - ERBAR, C., LEINS, P.: Bot. Jahrb. Syst. 105, 379-400 (1985). - ERBEN, M.: Mitt. Bot. Staatssamml. München 21, 339-740 (1985). - EVANS, S.V., FELLOWS, L.E., BELL, E.A.: Biochem. Syst. Ecol. 13, 271-302 (1985). - EZCURRA, C.: Darwiniana 26, 219-284 (1985).

FALLEN, M.E.; Am. J. Bot. 72, 572-579 (1985); - Bot. Jahrb. Syst. 106, 245-286 (1986). - FARNSWORTH, N.R., SOEJARTO, D.D.: Econ. Bot. 39, 231-240 (1985). - FARRIS, M.A., MITTON, J.B.: Am. J. Bot. 72, 1719-1725 (1985). - FARRON, C.: Bot. Helv. 95, 59-72 (1985). - FAVRE-DUCHARTRE, M.: Homologies and phylogeny, 698-734. In: see ref. JOHRI 1984. - FELSENSTEIN, J.: Syst. Zool. 34, 300-311 (1985b); - Am. Nat. 125, 1-15 (1985b); - Evolution 39, 783-791 (1985c); - Syst. Zool. 34, 300-311 (1985d). - FERGUSON, I.K.: Kew Bull. 40, 25-34 (1985). - FERGUSON, I.K., PEARCE, K.J.: Linn. Soc. Symp. Ser. 12, 283-296 (1986). - FERNÂNDEZ CASAS, J., SERNA, A.S. de la: Treb. Inst. Bot. Barcelona 10, 1-174 (1985). - FERNÂNDEZ-PERALTA, A.Mª., GONZALEZ-AGUILERA, J.J.: Plant Syst. Evol. 152, 167-183 (1986). - FERREIRA, V.G., PARENTE, J.P., PINTO, M.C.F.R., PINTO, A.V., PORTER, B., SILVA, M.M., MOUTINHO, J.L.: Biochem. Syst. Ecol. 14, 199-202 (1986). - FEUER, S.: Pollen Spores 28, 123-156 (1986). - FEUER, S.M., KUIJT, J.: Ann. Missouri Bot. Gard. 72, 187-212 (1985). - FEUER, S., NIEZGODA, C.J., NEVLING, L.I.: Am. J. Bot. 72, 1871-1890 (1985). - FIALA, K.L., SOKAL, R.R.: Evolution 39, 609-622 (1985). - FORCELLA, F., WOOD, H.: Aust. J. Bot. 34, 455-461 (1986). - FORD, H.: Biol. J. Linn. Soc. 25, 169-186 (1985). - FORD-LLOYD, B., JACKSON, M.: Plant Genetic Resources: An Introduction to Their Conservation and Use, 146 pp. London: Arnold 1986. - FORGET, P.M., LEBBE, J., PUIG, H., VIGNES, R., HIDEUX, M.: Bot. J. Linn. Soc. 93, 205-223 (1986). - FOSTER, S.A.: Bot. Rev. 52, 260-299 (1986). - FOWLER, S.V., LAWTON, J.H.: Am. Nat. 126, 181-195 (1985). - FREEMAN, B., TURNER, D.W.: Aust. J. Bot. 33, 393-408 (1985). - FREITAG, H.: Not. R. Bot. Gard. Edinb. 42, 355-489 (1985). - FRENCH, J.C.: Bot. Gaz. 146, 521-533 (1985); - Am. J. Bot. 73, 434-449 (1986). - FRENCH, J.C., TOMLINSON, P.B.: Kew Bull. 41, 561-574 (1986). - FRISCHKNECHT, P.M., ULMER-DUFEK, J., BAUMANN, T.W.: Phytochemistry 25, 613-616 (1986). - FROMM-TRINTA, E.: Bradea 4(29), 188-210 (1985). - FROST-OLSEN, P., HOLM-NIELSEN, L.B.: Rep. Bot. Inst. Univ. Aarhus 14, 1-39 (1986). - FRYXELL, P.A.: Sida 11,

62-91 (1985). - FUNK, V.A.: Ann. Missouri Bot. Gard. 72, 681-715 (1985); - Taxon 35, 311-313 (1986). - FUNK, V.A., CRACRAFT, J.: Ann. Missouri Bot. Gard. 72, 591-595 (1985). - FURNESS, C.A.: Pollen Spores 27, 307-320 (1985).

GADEK, P.A., QUINN, C.J.: Phytomorphology 24, 267-272 (1985). - GAMAL, M., HEMLEBEN, V.: Plant Syst. Evol. 154, 63-77 (1986). - GANDERS, F.R., DENNY, S.K., TSAI, D.: Can. J. Bot. 63, 533-538 (1985). - GARBUTT, K., BAZZAZ, F.A., LEVIN, D.A.: Am. J. Bot. 72, 640-648 (1985). - GARDNER, C.S.: Selbyana 9, 130-146 (1986). - GIANNASI, D.E.: Ann. Missouri Bot. Gard. 73, 417-437 (1986). - GIBBS, P.E.: Plant Syst. Evol. 154, 285-323 (1986). - GIBBY, M., WESTFOLD, J.: Plant Syst. Evol. 153, 205-222 (1986). - GILES, B.E., LEFKOVITCH, L.P.: Plant Syst. Evol. 153, 181-197 (1986). - GILMARTIN, A.J., BROWN, G.K.: Am. J. Bot. 73, 387-397 (1986). - GIVEN, D.R.: What the conservationist require of ex-situ collections. Abstract, IUCN conference "Botanic Gardens and the World Conservation Strategy" (1985). - GLEN, H.: Bothalia 16, 203-226 (1986). - GLOVER, D.E., BARRETT, S.C.H.: Evolution 40, 1122-1131 (1986). - GODLEY, E.J.: N.Z. J. Bot. 23, 687-706 (1985). - GOEBEL, K.: Die Entfaltungsbewegungen der Pflanzen und deren teleologische Deutung, 483 pp. Jena: Fischer 1920. - GOETGHEBEUR, P.: Taxon 34, 617-632 (1985). - GOETGHEBEUR, P., COUDIJZER, J.: Bull. Jard. Nat. Belg. 55, 207-259 (1985). - GOLDBERG, A.: Smithsonian Contrib. Bot. 58, 1-314 (1986). - GOLDBLATT, P.: Ann. Missouri Bot. Gard. 72, 277-447 (1985); - Bull. Mus. Natl. Hist. Nat., Sect. B, Adansonia 8, 117-132 (1986a); - Ann. Missouri Bot. Gard. 73, 102-116 (1986b). - GOLDBLATT, P., HENRICH, J.E., RUDALL, P.: Ann. Missouri Bot. Gard. 71, 1013-1020 (1984). - GOLDBLATT, P., TOBE, H., CARLQUIST, S., PATEL, V.C.: Ann. Missouri Bot. Gard. 72, 167-183 (1985). - GOLDMAN, D.A., WILSON, M.F.: Bot. Rev. 52, 157-194 (1986). - GONNET, J.-F.: Biochem. Syst. Ecol. 14, 409-415 (1986). - GOODNIGHT, C.J.: Evolution 39, 545-558 (1985). - GORNALL, R.J.: Nord. J. Bot. 6, 257-275 (1986). - GORNALL, R.J., BOOM, B.A.: Bot. J. Linn. Soc. 90, 1-71 (1985). - GOTTLIEB, L.D., WARWICK, S.I., FORD, V.S.: Syst. Bot. 10, 484-495 (1985). - GOTTLIEB, O.R., KUBITZKI, K.: Planta Medica 50, 380-385 (1984). - GOTTSBERGER, G., AMARAL, A.: Ber. Dtsch. Bot. Ges. 97, 391-410 (1984). - GOTTSBERGER, G., SILBERBAUER-GOTTSBERGER, I.: Beitr. Biol. Pflanz. 59, 465-473 (1984). - GOULD, S.J., VRBA, E.S.: Paleobiology 8, 4-15 (1982). - GRAHAM, S.A.: Ann. Missouri Bot. Gard. 71, 757-779 (1984); - J. Arnold Arb. 66, 395-420 (1985). - GRAHAM, A., NOWICKE, J., SKVARLA, J.J., GRAHAM, S.A., PATEL, V., LEE, S.: Am. J. Bot. 72, 1012-1031 (1985). - GRANT, W.F. (ed.): Plant Biosystematics, 674 pp. Toronto: Academic Press 1984. - GRAU, J., HOPF, H.: Bot. Jahrb. Syst. 107, 251-268 (1985). - GRAYUM, M.H.: Am. J. Bot. 72, 1565-1577 (1985); - Plant Syst. Evol. 151, 145-161 (1986a); - Linn. Soc. Symp. Ser. 12, 313-327 (1986b). - GRIME, J.P., SCHACKLOCK, J.M.L., BAND, S.R.: New Phytol. 100, 435-445 (1985). - GRIMES, J.W.: Am. J. Bot. 73, 1222-1229 (1986). - GROENENDAHL, J. v.: Acta Bot. Neerl. 34, 357-358 (1985). - GROLL, U., MAHN, E.-G.: Flora 178, 93-110 (1986). - GUINET, P.: Linn. Soc. Symp. Ser. 12, 297-311 (1986). - GUNN, C.R., Fruits and seeds of genera in the subfamily Mimosoideae (Fabaceae). USDA Techn. Bull. 1681, 194 pp. (1984).

HAINES, R.W., LYE, K.A.: Lidia 1, 149-179 (1986). - HAINSWORTH, F.R., WOLF, L.L., MERCIER, T.: J. Ecol. 73, 263-270 (1985). - HAKKI, M.I.: Bot. Jahrb. Syst. 107, 379-394 (1985). - HALL, G.D., LANGENHEIM, J.H.: Biochem. Syst. Ecol. 14, 61-69 (1986). - HALVORSEN, T., BORGEN, L.: Sommerfeltia 3, 1-103 (1986). - HAMILTON, C.W.: Am. J. Bot. 72, 1081-1088 (1985). - HAMMER, K.: Kulturpflanze 32, 11-34 (1984). - HAMMER, K., LEHMANN, C.O., PERRINO, P.: Biol. Zentralbl. 104, 511-517 (1985). - HANELT, P.: Flora 176, 99-116 (1985); - Syst. Assoc. Spec. Vol. 29, 140-156 (1986). - HANSEN, B.: Bot. Jahrb. Syst. 106, 359-377 (1986). - HANSEN, C.: Bot. Jahrb. Syst. 106, 15-22 (1985). - HAQUE, M.Z., GODWARD,

M.B.E.: Bot. J. Linn. Soc. 89, 321-340 (1984). - HARBORNE, J.B.: Systematic significance of variations in defense compounds in the Solanaceae, 328-344. In: see ref. D'ARCY 1986; - Phytochemistry 25, 1641-1643 (1986b). - HARBORNE, J.B., TOMES-BARBERAN, F.A., WILLIAMS, C.A., GIL, M.I.: Phytochemistry 25, 2811-2816 (1986). - HARE, J.D., KENNEDY, G.K.: Evolution 40, 1031-1043 (1986). - HARI GOPAL, B., MOHAN RAM, H.Y.: Plant Syst. Evol. 148, 239-246 (1985). - HARLAN, J.R., DE WET, J.M.J.: Syst. Assoc. Spec. Vol. 29, 71-76 (1986). - HARPER, J.L., ROSEN, B.R., WHITE, J. (eds.): Phil. Trans. R. Soc. London B 313, 1-250 (1986). - HARTMANN, H., LIEDE, S.: Bot. Jahrb. Syst. 106, 433-485 (1986). - HARTMANN, T.: Plant Syst. Evol. 150, 15-34 (1985). - HARVEY, W.: Bot. J. Linn. Soc. 90, 201-208 (1985). - HEEL, W.A. van: Blumea 31, 273-276 (1986). - HEGNAUER, R.: Chemotaxonomie der Pflanzen, Band 7, Nachträge zu Band 1 und 2, 804 pp. Basel: Birkhäuser 1986. - HEISER, C.B.: Of Plants and People, 237 pp. Norman: University of Oklahoma Press 1985. - HELENURM, K., GANDERS, F.R.: Evolution 39, 753-765 (1985). - HEMSLEY, A.J., FERGUSON, I.K.: Ann. Missouri Bot. Gard. 72, 570-590 (1985). - HENDERSON, A.: Bot. Rev. 52, 221-259 (1986). - HENNESSY, E.F.F.: Bothalia 15, 505-530 (1985). - HELLWIG, F.: Mitt. Bot. Staatssamml. München 21, 1-47 (1985). - HENRICKSON, J., FLYR, L.D.: Sida 11, 107-172 (1985). - HESSE, M.: Plant Syst. Evol. 153, 37-48 (1986). - HESSE, M., MORAWETZ, W., EHRENDORFER, F.: Plant Syst. Evol. 148, 253-285 (1985). - HEYWOOD, J.S.: J. Heredity 77, 146-150 (1986). - HEYWOOD, J.S., LEVIN, D.A.: Evolution 39, 1076-1086 (1985); - Oecologia 68, 285-290 (1986). - HEYWOOD, V.H.: Botanic Gardens Conservation Strategy, 82 pp. Gland (Switzerland): WWF and IUCN World Conservation Centre 1985; - Syst. Assoc. Spec. Vol. 29, 420-424 (1986). - HIDEUX, M., ABADIE, M.: Linn. Soc. Symp. Ser. 12, 35-48 (1986). - HILGER, H.H.: Bot. Jahrb. Syst. 105, 323-378 (1985). - HILGER, H.H., BALZER, M., FREY, W., PODLECH, D.: Plant Syst. Evol. 148, 291-312 (1985). - HILLIARD, O.M., BURTT, B.L.: Notes R. Bot. Gard. Edinb. 42, 171-225 (1985); - ibid. 43, 407-438 (1986). - HILLIS, D.M., DAVIS, S.K.: Evolution 40, 1275-1288 (1986). - HILU, K.W.: Taxon 34, 102-114 (1985). - HODGSON, J.G., MACKEY, J.M.L.: New Phytol. 104, 497-515 (1986). - HOFMANN, U., LUDEWIG, J.: Bot. Jahrb. Syst. 105 , 401-431 (1985). - HOFMANN, U., SPECHT, A.K.: Beitr. Biol. Pflanz. 61, 79-85 (1986). - HOLDEN, J.H.W., WILLIAMS, J.T.: Crop Genetic Reserves: Conservation and Evolution, 296 pp. London: Allen and Unwin 1984. - HOLWERDA, B.C., JANA, S., CROSBY, W.L.: Genetics 114, 1271-1291 (1986). - HOPPE, J.R.: Bot. Jahrb. Syst. 105, 497-581 (1985). - HSIAO, T.H.: Specificity of certain chrysomelid beetles for Solanaceae, 345-363. In: see ref. D'ARCY 1986. - HU CHI-MING; Acta Phytotax. Sinica 23, 355-368 (1985). - HUBER, H.: Bot. Jahrb. Syst. 107, 269-276 (1985). - HUDSON, W.D.: Relationships of domesticated and wild *Physalis philadelphica*, 416-432. In: see ref. D'ARCY 1986. - HUENNEKE, L.F.: Am. J. Bot. 72, 152-158 (1985). - HUISKES, A.H.L., SCHAT, H., ELENBAAS, P.F.M.: Acta Bot. Neerl. 34, 271-282 (1985). - HUNT, D.R.: Kew Bull. 41, 107-129 (1986).

IETSWAART, J.H., KLIPHUIS, E.: Acta Bot. Neerl. 34, 157-170 (1985). - IHLENFELDT, H.-D.: Ber. Dtsch. Bot. Ges. 98, 409-423 (1985). - IHLENFELDT, H.-D., BITTRICH, V.: Bot. Jahrb. Syst. 105, 289-322 (1985). - ILTIS, H.H., DOEBLEY, J.F. Zea - a biosystematical odyssey, 587-616. In: see ref. GRANT 1984. - INAMDAR, J.A., MOHAN, J.S.S., BAGAVATHI SUBRAMANIAN, R.: Feddes Rep. 97, 147-160 (1986). - INGROUILLE, M.J., STACE, C.A.: Nord. J. Bot. 5, 113-125 (1985). - IOS WORKING PARTY: Bradleya 4, 65-78 (1986).

JACKSON, J.B.C., BUSS, L.W., COOK, R.C. (eds.): The Population Biology and Evolution of Clonal Organisms. New Haven: Yale University Press 1985. - JACOBSEN, N.: Nord. J. Bot. 5, 31-50 (1985). - JACQUARD, P., HEIM, G., ANTONOVICS, J. (eds.): Genetic Differentiation and Dispersal in Plants, 452 pp. Berlin, Heidelberg, New York: Springer 1986. - JACQUES-FELIX, H.: Bull. Mus. Natl. Hist. Nat., Sect. B, Adansonia 6,

383-451 (1984); - ibid. 7, 3-58 (1985); - 7, 377-383 (1985). - JAHNKE, C.: Trop. Subtrop. Pflanzenw. 57, 1-146 (1986). - JAKUPOVIC, J., KAHNKE, J., SCHUSTER, A., METWALLY, M.A., BOHLMANN, F.: Phytochemistry 25, 1133-1142 (1986). - JANSEN, R.K.: Syst. Bot. Monogr. 8, 1-115 (1985). - JANZEN, D.H.: Oikos 46, 281-283 (1986a); - Annu. Rev. Ecol. Syst. 17, 305-324 (1986b); - ibid. 17, 595-636 (1986c). - JARRET, R.L., LITZ, R.E.: J. Heredity 77, 183-188 (1986). - JEANMONOD, D.: Candollea 40, 35-56 (1985). - JEFFREY, C.: Kew Bull. 41, 873-943 (1986). - JÉRÉMIE, J., JEUNE, B.: Bull. Mus. Natl. Hist. Nat., Sect. B, Adansonia 7, 213-238 (1985). - JIN XIOBAI: Kew Bull. 41, 379-391 (1986). - JOHNS, T., OSMAN, S.F.: Biochem. Syst. Ecol. 14, 651-655 (1986). - JOHNSON, D.E.: Huntia 6, 3-121 (1985). - JOHNSON, L.A.S., BRIGGS, B.G.: Ann. Missouri Bot. Gard. 71, 700-756 (1984). - JOHNSON, N.D., BRAIN, S.A.: Biochem. Syst. Ecol. 13, 5-9 (1985). - JOHRI, B.M. (ed.): Embryology of Angiosperms, 830 pp. Berlin, Heidelberg, New York: Springer 1984. - JOLIVET, P.: Les fourmis et le plants. Un example de coévolution, 232 pp. Paris: Société Nouvelle de Editions Boubée 1986. - JONES, J.H.: Ann. Missouri Bot. Gard. 73, 228-275 (1986). - JØRGENSEN, R.B.: Hereditas 104, 273-291 (1986). - JÜRGENS, N.: Ber. Dtsch. Bot. Ges. 98, 425-446 (1985); - Mitt. Inst. Allg. Bot. Hamburg 21, 139-365 (1986).

KADEREIT, J.W.: Bot. Jahrb. Syst. 108, 1-16 (1986a); - Notes R. Bot. Gard. Edinb. 44, 25-43 (1986b); - Bot. Jahrb. Syst. 106, 221-244 (1986c). - KADEREIT, J.W., BRIGGS, D.: New Phytol. 99, 155-169 (1985). - KÄSTNER, A.: Flora 176, 73-93 (1985); - ibid. 178, 111-138 (1985a). - KAMELINA, O.P., PROSKURINA, O.B.: Bot. Zh. 70, 721-730 (1985). - KARASAWA, K., SAITO, K.: Bull. Hiroshima Bot. Gard. 5, 1-69 (1982). - KAUL, R.B.: Am. J. Bot. 72, 1962-1977 (1985); - Ann. Missouri Bot. Gard. 73, 284-296 (1986). - KAUR, A., JONG, K., SANDS, V.E., SOEPADMO, E.: Bot. J. Linn. Soc. 92, 75-88 (1986). - KEATING, R.C.: Ann. Missouri Bot. Gard. 71, 801-823 (1984). - KEDDAM-MALPLANCHE; M.: Mem. Mus. Natl. Hist. Nat., N.S., Sér B, Bot. 29, 1-109 (1985). - KEELER, K.H.: Oikos 44, 407-414 (1985). - KEIGHERY, G.J.: Bot. Jahrb. Syst. 106, 107-113 (1985). - KELLOGG, E.A.: J. Arnold Arb. 66, 201-242 (1985). - KENTON, A., JONES, K.: Chromosoma 92, 176-184 (1985). - KENTON, A.Y., RUDALL, P.J., JOHNSON, A.R.: Bot. Gaz. 147, 342-354 (1986). - KESSELI, R.V., JAIN, S.K.: Theor. Appl. Genet. 71, 292-299 (1985). - KHANH TRAN CONG: Feddes Rep. 96, 99-120 (1985); - ibid. 97, 713-751 (1986). - KING, G.J.: Plant Syst. Evol. 151, 163-174 (1986). - KINLOCH, B.B., WESTFALL, R.D., FORREST, G.I.: New Phytol. 104, 703-729 (1986). - KIRCHER, P.: Diss. Bot. 94, 1-218 (1986). - KIRKPATRICK, K.J., DECKER, D.S., WILSON, H.D.: Econ. Bot. 39, 289-299 (1985). - KLUSKA, T.: Acta Biol. Cracov. 28, 1-29 (1986). - KNABEN, G.S.: Bot. Helv. 95, 177-191 (1985). - KNOWLES, P., GRANT, M.C.: Can. J. Bot. 63, 722-727 (1985). - KNOX, R.B.: The pollen grain, 197-272. In: see ref. JOHRI 1984. - KÖNIG, G., RIMPLER, H.: Phytochemistry 24, 1245-1248 (1985). - KÖRNER, C., MEUSEL, H.: Flora 178, 209-232 (1986). - KONDO, K., LAVARACK, P.S.: Bot. J. Linn. Soc. 88, 317-333 (1984). - KONING, R. de, SOSEF, M.S.M.: Blumea 30, 279-318 (1985). - KOLAKOWSKI, A.A.: Bot. Zh. 70, 2-11 (1985); - ibid. 71, 1155-1166 (1986). - KOSTERMANS, A.J.G.H.: Taxon 34, 426-435 (1985); - Ginkgoana 6, 1-171 (1986). - KRÄHENBÜHL, M.: Bull. Soc. Neuchâteloise Sci. Nat. 107, 39-54 (1984). - KRESS, W.J.: Selbyana 9, 2-22 (1986a); - Linn. Soc. Symp. Ser. 12, 329-345 (1986). - KRIEBER, M., ROSE, M.R.: Annu. Rev. Ecol. Syst. 17, 465-485 (1986). - KUBITZKI, K.: Taxon 36, 21-28 (1987). - KUKKONEN, J.: Ann. Bot. Fenn. 23, 107-120 (1986). - KUNZE, H.: Beitr. Biol. Pflanz. 60, 93-140 (1985); - ibid. 61, 221-234 (1986). - KVIST, L.P., PEDERSEN, J.A.: Biochem. Syst. Ecol. 14, 385-405 (1986).

LADIZINSKY, G.: Econ. Bot. 39, 191-199 (1985). - LADIZINSKY, G., VAN OSS, H.: Bot. J. Linn. Soc. 89, 97-100 (1984). - LA FRANKIE, J.V.: Am. J. Bot. 73, 1258-1260 (1986a); - J. Arnold Arb. 67, 371-439 (1986b). -

LAMMERS, T.G., STUESSY, T.F., SILVA, O.M.: Plant Syst. Evol. 152, 243-266 (1986). - LANDRUM, L., STEVENSON, D.: Syst. Bot. 10, 155-162 (1986). - LANDSMAN, J., DENNO, E.S., HIGGINS, T.J.V., APPLEBY, C.A., KORFF, A.A., PEACOCK, W.J.: Nature 324, 166-168 (1986). - LANE, M.A.: Syst. Bot. 10, 7-28 (1985). - LANGENHEIM, J.H., CONVIS, C.L., MACEDO, C.A., STUBBLEBINE, W.H.: Biochem. Syst. Ecol. 14, 41-49 (1986a). - LANGENHEIM, J.H., MACEDO, C.A., ROSS, M.K., STUBBLEBINE, W.H.: Biochem. Syst. Ecol. 14, 51-59 (1986b). - LANGTON, F.A.: Syst. Assoc. Spec. Vol. 29, 263-276 (1986). - LARSEN, B.B.: Nord. J. Bot. 6, 427-446 (1986). - LAUBENFELS, D.J. de: Blumea 30, 251-278 (1985). - LA VALVA, V., SABATO, S., GIGLIANO, G.S.: Taxon 34, 191-196 (1985). - LAVIN, M.: Biochem. Syst. Ecol. 14, 71-73 (1986). - LAWRENCE, M.E.: Aust. J. Bot. 33, 197-232 (1985). - LEACH, L.C.: A Revision of *Stapelia* L. (Asclepiadaceae), 157 pp. Johannesburg: Aloe Books 1985. - LEE, G.J.: Am. J. Bot. 73, 230-235 (1986). - LEENHOUTS, P.W.: Blumea 31, 219-234 (1985); - ibid. 31, 373-436 (1986). - LEEUWENBURG, A.J.M. (ed.): Meded. Landbouwhoogesch. Wageningen 83-7 (1983); - Agric. Univ. Wageningen Papers 85-2, 1-83 (1985a); - ibid. 85-3, 1-122 (1985b). - LEINS, P., ERBAR, C.: Bot. Jahrb. Syst. 106, 53-60 (1985a); - ibid. 107, 343-368 (1985b). - LEJOLY, J., LISOWSKI, S.: Bull. Jard. Bot. Nat. Belg. 55, 27-60 (1985). - LES, D.H.: Syst. Bot. 11, 549-558 (1986); - ibid. 10, 338-346 (1985). - LESTER, R.N., NIAKAN, L.: Origin and domestication of the Scarlet Eggplant, *Solanum aethiopicum* from *S. anguivi* in Africa, 433-456. In: see ref. D'ARCY 1986. - LESTER, R.N., ROBERTS, P.A.: Ann. Missouri Bot. Gard. 73, 128-133 (1986). - LESTER, R.N., HAKIZA, J.J.H., STAVROPOULOS, N., TEIXEIRA, M.M.: Syst. Assoc. Spec. Vol. 29, 283-307 (1986). - LE THOMAS, A., MORAWETZ, W., WAHA, M.: Linn. Soc. Symp. Ser. 12, 375-388 (1986). - LEVIN, D.A.: Evolution 39, 1275-1281 (1985). - LEVIN, D.A., SCHMIDT, K.P.: Am. J. Bot. 72, 1404-1409 (1985). - LEVIN, G.A.: Syst. Bot. 11, 515-530 (1986a); - Ann. Missouri Bot. Gard. 73, 29-85 (1986b). - LEVINE, D.A., ANDERSON, G.J.: Evolution of diocey in an American *Solanum*, 264-273. In: see ref. D'ARCY 1986. - LEWIS, J.: Syst. Assoc. Spec. Vol. 29, 115-137 (1986). - LIDEN, M.: Opera Bot. 88, 1-133 (1986). - LIGHTBODY, J.P.: Biotropica 17, 339-342 (1985). - LIMA, M.P.M. de: Rodriguesia 37(62), 53-78 (1985). - LINDER, H.P.: Bothalia 15, 387-503 (1985). - LINDER, H.P., FERGUSON, I.K.: Grana 24, 65-76 (1985). - LINDROTH, R.L., BATZLI, G.O., SEIGLER, D.S.: Biochem. Syst. Ecol. 14, 597-602 (1986). - LINS, A.P., YOSHIDA, M., GOTTLIEB, O.R., GOTTLIEB, H.E., KUBITZKI, K.: Bull. Soc. Chim. Belg. 95, 737-748 (1986). - LLOYD, D.G., WEBER, C.J.: N.Z. J. Bot. 24, 135-162 (1986). - LOBREAU-CALLEN, D., JEREMIE, J.: Grana 25, 155-158 (1986). - LÖRCHER, H., WEBERLING, F.: Trop. Subtrop. Pflanzenw. 47, 1-31 (1984). - LORENCE, D.H.: Ann. Missouri Bot. Gard. 72, 1-165 (1985). - LOUP, C.: Premières données sur l'architecture des Cactaceae, 42 pp. Montpellier: Thèse 1983. - LOWREY, T.K., CRAWFORD, D.J.: Syst. Bot. 10, 64-72 (1985). - LU, A.-M.: Nord. J. Bot. 5, 229-240 (1985). - LÜTTGE, U.; Naturwissenschaften 72, 557-566 (1985). - LUNDGREN, L., NORELIUS, G., STENHAGEN, G.: Nord. J. Bot. 5, 315-320 (1985).

MAAS, P.J.M., RÜBSAMEN, T.: Flora Neotrop. 40, 1-55 (1986). - MAAS, P.J.M., RUYTERS, P.: Flora Neotrop. 41, 1-93 (1986). - MAAS, P.J.M., MAAS-VAN DE KRAMER, H., BENTHEM, J. van, SNELDERS, H.C.M., RÜBSAMEN, T.: Flora Neotrop. 42, 1-189 (1986). - MADDOX, G.D., CAPPUCINO, N.: Evolution 40, 862-863 (1986). - MAESEN, L.J.G. van der: Agric. Univ. Wageningen Papers 85-4, 1-225 (1985a); - ibid. 85-1, 1-132 (1985b). - MAI, D.H.: Feddes Rep. 97, 1-28 (1986). - MARKHAM, K.R., VILAIN, C., MOLLOY, B.P.J.: Phytochemistry 24, 2607-2609 (1985). - MARTIN, P.G., DOWD, J.M.: Ann. Missouri Bot. Gard. 73, 442-448 (1986a), - Taxon 35, 469-475 (1986b). - MARTIN, P.G., BOULTER, D., PENNY, D.: Taxon 34, 393-400 (1985). - MARTIN, P.G., DOWD, J.M., MORRIS, C., SYMON, D.E.: Aust. J. Bot. 34, 187-195 (1986). - MARTINI, F., PAIERO, P.: Atti. Ist. Ecol. Selvic. Univ. Stud. Padova 3, 107-242 (1984). - MASTENBROEK, O.,

BREDERODE, J. van: Biochem. Syst. Ecol. 14, 165-181 (1986). - MAUTHE, S., BACHMANN, K., CHAMBERS, K.L., PRICE, H.J.: Biol. Zentralbl. 104, 1-20 (1985). - MAYO, S.J., GILBERT, M.G.: Kew Bull. 41, 261-278 (1986). - MAZE, J., SCAGEL, R.K., BOHM, L.R.: Can. J. Bot. 64, 1733-1738 (1986). - McCLENAGHAN, L.R., BEAUCHAMP, A.C.: Evolution 40, 315-322 (1986). - McLACHLAN, D., ARNASON, T., LAM, J.: Biochem. Syst. Ecol. 14, 17-23 (1986). - MEDAN, D.: Bot. Jahrb. Syst. 105, 205-262 (1985). - MEEUSE, A.D.J.: Anatomy of Morphology, 277 pp. Leiden: Brill and Backhuys 1986. - MERLIN, C.M., GRANT, W.F.: Can. J. Bot. 64, 1069-1074 (1986). - MICHAELS, H.J., BAZZAZ, F.A.: Ecology 67, 27-36 (1986). - MILLER, E.H. (ed.): Museum Collections: Their Roles and Future in Biological Research, 222 pp. Occ. Papers Brit. Columb. Prov. Mus. 25 (1985). - MILLER, H.A., LAMB, S.H.: Oaks of North America, 327 pp. Happy Camp: Naturegraph Publishers 1985. - MICHENER, D.C.: Brittonia 38, 360-361 (1986). - MITCHELL-OLDS, T., WALLER, D.M.: Evolution 39, 533-544 (1985). - MOHANA RAO, P.R.: Flora 178, 243-249 (1986). - MØLGAARD, P.: Nord. J. Bot. 5, 203-213 (1985). - MØLLER, J.D., RASMUSSEN, H.: Bot. J. Linn. Soc. 89, 53-76 (1984). - MONOD, T.: Bull. Mus. Natl. Hist. Nat., Sect. B, Adansonia 8, 103-115 (1986). - MOORE, H.E., UHL, N.W.: Allertonia 3, 313-402 (1984). - MORALDO, B.: Webbia 40, 203-278 (1986). - MORAN, N.A.: Evolution 40, 1044-1050 (1986). - MORAWETZ, W.: Bot. Jahrb. Syst. 107, 329-342 (1985); - Plant Syst. Evol. 154, 147-173 (1986a); - ibid. 152, 49-100 (1986b). - MORAWETZ, W., WAHA, M.: Plant Syst. Evol. 150, 119-141 (1985). - MOTHES, K., SCHÜTTE, H.R., LUCKNER, M. (eds.): Biochemistry of Alkaloids, 406 pp. Weinheim: VCH 1985. - MULCAHY, D.L.: The relationships between self-incompatibility, pseudo-compatibilty, and self-compatibility, and self-compatibility, 229-235. In: see ref. GRANT 1984; - Preface, vii-xi. In: see ref. MULCAHY et al. 1986. - MULCAHY, D.L., MULCAHY, G.B., OTTAVIANO, E.: Biotechnology and Ecology of Pollen, 528 pp. New York, Berlin, Heidelberg: Springer 1986. - MÜLLER-DOBLIES, D., MÜLLER-DOBLIES, U.: Bot. Jahrb. Syst. 107, 17-47 (1985). - MÜLLER, J.: Blumea 31, 161-218 (1985). - MUNIYAMA, M., PHIPPS, J.B.: Can. J. Bot. 63, 1319-1324 (1985). - MURRAY, M.G.: Biol. J. Linn. Soc. 26, 69-81 (1985). - MUSSELMAN, L.J., AYENSU, E.S.: Taxonomy and biosystematics of *Striga*, 37-45. In: see ref. AYENSU 1984.

NAHRSTEDT, A.: Plant Syst. Evol. 150, 35-47 (1985). - NARAYANA, P.S., NARAYANA, L.L.: Plant Syst. Evol. 154, 137-145 (1986). - NASCIMENTO, J.C., LANGENHEIM, J.H.: Bioch. Syst. Evol. 14, 615-624 (1986). - NEFF, N.A.: Syst. Zool. 35, 110-123 (1986). - NEGASSA, M.: Hereditas 104, 41-48 (1986). - NEGBI, M., SARGENT, J.A.: Bot. J. Linn. Soc. 93, 247-258 (1986). - NEUFFER, B., HURKA, H.: Plant Syst. Evol. 152, 277-298 (1986).

NEVO, E., GERECHTER-AMITAI, Z., BEILES, A., GOLDENBERG, E.M.: Plant Syst. Evol. 153, 13-30 (1985a); - NEVO, E., ATSMUN, D., BEILES, A.: Plant Syst. Evol. 150, 205-222 (1985b). - NEVO, E., BEILES, A., KAPLAN, D., STORCH, N., ZOHARY, D.: Plant Syst. Evol. 153, 141-164 (1986a). - NEVO, E., BEILES, A., ZOHARY, D.: Biol. J. Linn. Soc. 27, 355-380 (1986b). - NEVO, E., BEILES, A., KAPLAN, D., GOLDENBERG, E.M., OLSVIG-WHITTAKER, L., NAVEH, Z.: Evolution 40, 13-20 (1986). - NICHOLLS, K.W., BOHM, B.A., ORNDUFF, R.: Biochem. Syst. Ecol. 13, 261-263 (1985). - NICHOLLS, M.S.: Plant Syst. Evol. 150, 291-301 (1985); - ibid. 153, 243-258 (1986). - NIELSEN, I., BARETTA-KUIPERS, T., GUINET, P.: Opera Bot. 76, 1-120 (1984). - NIKLAS, K.J.: Bot. Rev. 51, 328-386 (1985). - NILSSON, L.A.: Nord. J. Bot. 5, 407-419 (1985). - NILSSON, L.A., JONSSON, L., RASON, L., RANDRIANJOHANI, E.: Biol. J. Linn. Soc. 26, 1-19 (1985). - NILSSON, S.: Linn. Soc. Symp. Ser. 12, 359-374 (1986). - NOGLER, G.A.: Gametophytic apomixis, 474-518. In: see ref. JOHRI 1984a; - Bot. Helv. 94, 411-422 (1984b). - NOGUCHI, J.: J. Sci. Hiroshima Univ., Ser. B, Div. 2, 20, 29-193 (1986). - NORDAL, I., LAANE, M.M., HOLT, E., STAUBO, I.: Nord. J. Bot. 5, 15-30 (1985). - NORDENSTAM,

B.: Opera Bot. 87, 1-41 (1986). - NOWICKE, J.W., PATEL, V., SKVARLA, J.J.: Am. J. Bot. 72, 1106-1113 (1985). - NOZZOLILLO, C., McNEILL, J.: Can. J. Bot. 63, 1066-1071 (1985). - NYANANYO, B.L.: Feddes Rep. 97, 767-769, 771-773 (1986). - NYBOM, H.: Feddes Rep. 5, 521-525 (1986).

OHRI, D., KHOSHOO, T.N.: Plant Syst. Evol. 153, 119-132 (1986). - OJALA, A., ROUSI, A.: Ann. Bot. Fenn. 23, 289-303 (1986). - OKADA, H.: Bot. Mag. Tokyo 99, 289-299 (1986). - OLMEDILLA, A., DELCASSO, D., DELSENY, M., CAUWET-MARC, A.-M.: Plant Syst. Evol. 150, 263-274 (1985). - OLOWOKUDEJO, J.D.: Feddes Rep. 97, 837-845 (1986). - OLSEN, J.: Plant Syst. Evol. 149, 47-63 (1985). - ORCHARD, A.E.: Brunonia 8, 173-291 (1985). - ØSTERBYE, U.: Hereditas 104, 61-73, 131-144, 305 (1986). - OUSTED, S.: Bull. Jard. Bot. Nat. Belg. 55, 421-459 (1985).

PACHECO, P., CRAWFORD, D.J., STUESSY, T.F., SILVA, O.M.: Am. J. Bot. 72, 989-998 (1985). - PACINI, E., FRANCHI, G.G., HESSE, M.: Plant Syst. Evol. 149, 155-185 (1985). - PALMER, J.D., ZAMIR, D.: Proc. Natl. Acad. Sci. USA 79, 5006-5010 (1982). - PALMER, J.D., JORGENSEN, R.A., THOMPSON, W.F.: Genetics 109, 195-213 (1985). - PARKER, M.A.: Evolution 39, 713-723 (1985); - Syst. Assoc. Spec. Vol. 29, 99-114 (1986). - PARKES, D.M., HALLAM, N.D.: Aust. J. Bot. 32, 595-604 (1984). - PATEL, V.C., SKVARLA, J.J., RAVEN, P.H.: Ann. Missouri Bot. Gard. 71, 858-969 (1984). - PATEL, V., SKVARLA, J.J., FERGUSON, I.K., GRAHAM, A., RAVEN, P.H.: Am. J. Bot. 72, 407-413 (1985). - PEDLEY, L.: Bot. J. Linn. Soc. 92, 219-254 (1986). - PELLMYR, O.: Oecologia (Berlin) 68, 304-307 (1986). - PELLMYR, O., THIEN, L.B.: Taxon 35, 76-86 (1986). - PEREIRA, E., LEME, E.M.C.: Bradea 4(32), 219-254 (1986). - PERL-TREVES, R., GALUN, E.: Theor. Appl. Genet. 71, 417-429 (1985). - PERL-TREVES, R., ZAMIR, D., NAVOT, N., GALUN, E.: Theor. Appl. Genet. 71, 430-436 (1985). - PETERSEN, F.P., FAIRBROTHERS, D.E.: Bull. Torrey Bot. Club 112, 43-52 (1985). - PFISTERER, R., PODLECH, D.: Mitt. Bot. Staatssamml. München 22, 571-604 (1986). - PHILIPSON, W.R.: Am. J. Bot. 72, 1954-1961 (1985); - Blumea 30, 389-415 (1985). - PHILIPSON, W.R., PHILIPSON, M.N.: Notes R. Bot. Gard. Edinb. 44, 1-23 (1986). - PHILLIPSON, J.D., ROBERTS, M.F., ZENK, M.H.: The Chemistry and Biology of Isoquinoline Alkaloids, 304 pp. Berlin, Heidelberg, New York: Springer 1985. - PHIPPS, J.B.: Problems of hybridity in the cladistics of *Crataegus* (Rosaceae), 417-438. In: see ref. GRANT 1984. - PICKERSGILL, B.: Syst. Assoc. Spec. Vol. 29, 191-209 (1986). - PICMAN, A.K.: Biochem. Syst. Ecol. 14, 255-281 (1986a); - ibid. 14, 361-364 (1986b). - PILS, G.: Plant Syst. Evol. 149, 19-45 (1985). - PINKAS, R., ZAMIR, D., LADIZINSKY, G.: Plant Syst. Evol. 151, 131-140 (1985). - PLOWMAN, T.: The ethnobotany of *Coca* (Erythroxylum spp. Erythroxylaceae), 62-111. In: see ref. PRANCE and KALLUNKI 1984. - PLUCKNETT, D.L., SMITH, N.J.H.: Econ. Bot. 40, 298-309 (1986). - POGGIO, L., HUNZIKER, J.H.: J. Heredity 77, 43-48 (1986). - POIRIER-HAMON, S., PERNES, J.: C.R. Acad. Sci. 302, Sér. III, 319-324 (1986). - POLECHKO, M.A., CLARKSON, R.B.: Biochem. Syst. Ecol. 14, 33-39 (1986). - POLLACK, J.C., DANCIK, B.P.: Can. J. Bot. 63, 201-210 (1985). - PONCY, O.: Mém. Mus. Natl. Hist. Nat., N.S., Sér B, Bot. 31, 1-124 (1985). - POSLUZNY, U., CHARLTON, W.A., JAIN, D.K.: Bot. J. Linn. Soc. 92, 323-342 (1986). - POTTS, B.M.: Aust. J. Bot. 34, 305-329 (1986). - PRAGLOWSKI, J., GRAFSTRÖM, E.: Grana 24, 11-21 (1985). - PRAKASH, N., FOREMAN, D.B., GRIFFITH, S.J.: Aust. J. Bot. 32, 605-612 (1984). - PRANCE, G.T., KALLUNKI, J.A. (eds.): Ethnobotany in the Neotropics, 156 pp. New York: New York Botanical Garden 1984. - PRENDERGAST, H.D.V., HATTERSLEY, P.W.: Aust. J. Bot. 33, 317-336 (1985). - PRENTICE, H.C.: Bot. J. Linn. Soc. 27, 179-189 (1986a); - Syst. Assoc. Sep. Vol. 29, 21-32 (1986b). - PRICE, H.J., CHAMBERS, K.L., BACHMANN, K., RIGGS, J.: Biol. Zentralbl. 104, 269-276 (1985). - PRICE, S.C., SWARD, W.L., WEDBERG, H.L.: Bot. Gaz. 146, 150-156 (1985). - PRIMACK, R.B.: Annu. Rev. Ecol. Syst. 16, 15-37 (1985). - PROKSCH, P.: Plant Syst. Evol. 150, 89-100 (1985). - PRUS-GLOWACKI, W.,

SZWEYKOWSKI, J., NOWAK, R.: Silvae Genet. 34, 162-170 (1985). - PUFF, C.: A Biosystematic Study of the African and Madagascan Rubiaceae-Anthospermeae, 535 pp. Berlin, Heidelberg, New York: Springer 1986. - PUTZ, F.E., HOLBROOK, N.M.: Selbyana 9, 61-69 (1986).

RAAMSDONK, L.W.D. van: Plant Syst. Evol. 149, 287-297 (1985a); - ibid. 150, 179-190 (1985b); - Nord. J. Bot. 6, 525-544 (1986). - RATHCKE, B., LACEY, E.P.: Annu. Rev. Ecol. Syst. 16, 179-214 (1985). - RAVEN, P.H.: Ann. Missouri Bot. Gard. 71, 631-632 (1984). - RAYBURN, A.L., PRICE, H.J., SMITH, J.D., GOLD, J.R.: Am. J. Bot. 72, 1610-1617 (1985). - RAYNAL-ROQUES, A.: Bull. Mus. Natl. Hist. Nat., Sect. B, Adansonia 7, 123-133 (1985). - REES, H.: Nuclear DNA variation and the homology of chromosomes, 87-96. In: see ref. GRANT 1984. - REES, S.B., HARBORNE, J.B.: Phytochemistry 24, 2225-2231 (1985). - REID, C., DYER, R.A.: A Review of the South African Species of *Cyrtanthus*, 68 pp. La Jolla: American Plant Life Society 1984. - REITSMA, J.M.: Meded. Landbouwhoogesch. Wageningen 83-9, 21-53 (1983). - RENNER, S.C.: Selbyana 9, 104-111 (1986). - RENUKA, C., SWARUPANANDAN, K.: Blumea 31, 313-318 (1986). - RENVOIZE, S.A.: The grasses of Bahia, 301 pp. Kew: Royal Botanic Gardens 1984. - REYNAUD, C.: Bull. Soc. Bot. France, Lett. Bot. 133, 167-177 (1986). - REZNICEK, A.A., CATLING, P.M.: Taxon 35, 495-501 (1986). - RHOADES, D.F.: Am. Nat. 125, 205-238 (1985). - RICHARDS, A.J.: Plant Breeding Systems, 529 pp. Winchester: Allen and Unwin 1986. - RIDDERNUMANN, J.W.A., WIRIADINATA, H.: Reinwardtia 10, 139-205 (1985). ROBBRECHT, E., PUFF, C.: Bot. Jahrb. Syst. 108, 63-137 (1986). - ROBINSON, H.: Taxon 35, 309-311 (1986). - ROBINSON, H., MARTICORENA, C.: Smithsonian Contrib. Bot. 64, 1-50 (1986). - ROCKWOOD, L.L.: Biotropica 17, 32-39 (1985). - RODRIGUEZ RIOS, R.: Mitt. Bot. Staatssamml. München 22, 97-201 (1986). - ROHWER, J.G.: Plant Syst. Evol. 152, 47-48 (1986a); - Mitt. Inst. Allg. Bot. Hamburg 20, 3-278 (1986b). - ROHWER, J.G., KUBITZKI, K.: Bot. Jahrb. Syst. 107, 129-135 (1985). - ROMERO, E.J.: Ann. Missouri Bot. Gard. 73, 276-283 (1986). - ROOD, S.B., CAMPBELL, J.S., DESPINS, T.: Can. J. Bot. 64, 1382-1388 (1986). - ROSENBLUM, I.R., BASILE, D.V.: Am. J. Bot. 71, 52-64 (1984). - ROSS, M.D., JONES, W.T.: Theor. Appl. Genet. 71, 284-288 (1985). - ROSSIGNOL, L., ROSSIGNOL, M.: Bull. Mus. Natl. Hist. Nat., Sect. B, Adansonia 7, 67-80 (1985). - ROSTANSKI, K.: Feddes Rep. 96, 3-14 (1985). - ROTHSCHILD, M., NASH, R.J., BELL, E.A.: Phytochemistry 25, 1853-1854 (1986). - ROUSI, A., HUTTUNEN, H., HYRKÄS-LYYTIKÄINEN, K.: Nord. J. Bot. 5, 127-141 (1985). - RUDALL, P.: Bot. Gaz. 145, 524-534 (1984); - Nord. J. Bot. 6, 277-289 (1986). - RÜBSAMEN, T.: Diss. Bot. 92, 1-310 (1986). - RÜFFELE, L.: Gleditschia 13, 95-100 (1985). - RURY, P.M., DICKISON, W.C.: Structural correlations amon wood, leaves, and plant habit, 495-540. In: Contemporary Problems in Plant Anatomy, eds. R.A. WHITE, W.C. DICKISON, 598 pp. Orlando: Academic Press 1984. - RUTISHAUSER, R.: Beitr. Biol. Pflanz. 59, 375-424 (1984). - RUTISHAUSER, R., SATTLER, R.: Bot. Jahrb. Syst. 107, 415-455 (1985); - Can. J. Bot. 64, 1987-2019 (1986). - RYDING, O.: Symp. Bot. Uppsalienses 25, vi, 1-152 (1986).

SABATO, S., DE LUCA, P.: Am. J. Bot. 72, 1353-1363 (1985). - SAINT-MARTIN, M.: Bull. Soc. Bot. France, Lett. Bot. 133, 137-153 (1986). - SARKAR, A.K.: Biol. Zentralbl. 105, 621-639 (1986). - SARMIENTO, G., GOLDSTEIN, G., MEINZER, F.: Biol. Rev. 60, 315-355 (1985). - SCHAARSCHMIDT, H.: Feddes Rep. 96, 345-361 (1985). - SCHILL, R., BAUMM, A., WOLTER, M.: Plant Syst. Evol. 148, 185-214 (1985). - SCHILL, R., WOLTER, M.: Nord. J. Bot. 6, 321-324 (1986). - SCHLARBAUM, S.E., TSUCHIYA, T.: Bot. Gaz. 146, 264-267 (1985). - SCHLEE, D.: Ökologische Biochemie. Berlin, Heidelberg, New York: Springer 1986.

SCHLICHTING, C.D.: Annu. Rev. Ecol. Syst. 17, 667-693 (1986). - SCHLICHTING, C.D., LEVIN, D.A.: Biol. J. Linn. Soc. 29, 37-47 (1986). - SCHMID, R.: Taxon 35, 476-491 (1986). - SCHMID, R., CARLQUIST, S.,

HUFFORD, L.D., WEBSTER, G.: Bot. J. Linn. Soc. 89, 119-152 (1984). - SCHMIDT, K.P., LEVIN, D.A.: Evolution 39, 396-404 (1985). - SCHMITT, J., ANTONOVICS, J.: Evolution 40, 817-829 (1986a); - ibid. 40, 830-836 (1986b). - SCHMITZ, J., FROEBE, H.A.: Bot. Jahrb. Syst. 106, 337-357 (1986). - SCHNEE, B.K., WALKER, D.M.: Am. J. Bot. 73, 376-386 (1986). - SCHOLZ, H.: Bot. Jahrb. Syst. 106, 419-426 (1986). - SCHULTZE-MOTEL, J. (ed.): Rudolf Mansfeld. Verzeichnis landwirtschaftlicher und gärtnerischer Kulturpflanzen (ohne Zierpflanzen), 4 vols., 1998 pp. Berlin, Heidelberg, New York: Springer 1986. - SCHWARZ-SOMMER, Z., GIERL, A., CUYPERS, H., PETERSON, P.A., SAEDLER, H.: EMBO J. 4, 591-597 (1985). - SCHWARZACHER-ROBINSON, T.: Plant Syst. Evol. 154, 259-274 (1986). - SEALY, J.R.: Bot. J. Linn. Soc. 92, 117-159 (1986). - SEARCY, K.B., MULCAHY, D.L.: Am. J. Bot. 72, 1700-1706 (1985). - SEAVEY, S.R., BAWA, K.S.: Bot. Rev. 52, 195-219 (1986). - SEBSEBE, S.: Symb. Bot. Upsalienses 25, 101 pp. (1985). - SEETHARAM, Y.N.: Trav. Sect. Sci. Techn. Franc. Pondichéry 21, 1-81 (1985). - SEMPLE, J.C.: Observations on IOPB 1983 and notes on the discussions among participants, 655-658. In: see ref. GRAN 1984. - SEREBRYAKOVA, T.I., PAVLOVA, N.R.: Bot. Zh. 71, 154-167 (1986). - SETTEN, A.K. van, KOEK-NOORMAN, J.: Bot. Jahrb. Syst. 108, 17-50 (1986). - SHARMA, A.K., CHATTOPADHYAY, S.: Bot. Helv. 94, 355-370 (1984). - SHAW, R.G.: Evolution 40, 492-505 (1986). - SHMIDA, A.: Ann. Missouri Bot. Gard. 72, 184-186 (1985). - SHORE, J.S., BARRETT, S.H.: Syst. Bot. 10, 308-321 (1985). - SIEGERIST, E.S.: Bot. Mus. Leafl. 30, 199-222 (1986). - SILANDER, J.A.: Evolution 39, 1034-1052 (1985). - SIMPSON, M.G.: Grana 24, 77-92 (1985a); - ibid. 24, 23-31 (1985b). - SINNOTT, Q.P.: Rhodora 87, 189-286 (1985). - SIQUEIRA, J.C. de: Pesq. Bot. 37, 1-111 (1985). - SIVARAJAN, V.V.: Introduction to the Principles of Plant Taxonomy, 253 pp. New Delhi, Bombay, Calcutta: Oxford and IGH 1984. - SLEUMER, H.: Bot. Jahrb. Syst. 105, 449-480 (1985); - Blumea 31, 245-269 (1986). - SLUIS, W.G. van der: Plant Syst. Evol. 149, 253-286 (1985). - SMALL, E.: Hybridization in the domesticated-weed-wild complex, 195-210. In: see ref. GRANT 1984. - SMILEY, J.T., WISDOM, C.S.: Biochem. Syst. Ecol. 13, 305-312 (1985). - SMITH, A.P.: Oecologia (Berlin) 69, 283-287 (1986). - SMITH, D.L., SCOTT, R.C.: Bot. J. Linn. Soc. 90, 73-103 (1985). - SMITH, J.S.C., GOODMAN, M.M., STUBER, C.W.: Econ. Bot. 39, 12-24 (1985). - SMITH, L.B., WASSHAUSEN, D.C., GOLDING, J., KARGEANNES, C.E.: Smithsonian Contrib. Bot. 60, 1-548 (1986). - SMITH, R.M.. Notes R. Bot. Gard. Edinb. 42, 261-314 (1985); - ibid. 43, 439-466 (1986). - SOHMER, S.H. (ed.): Bernice Pauahi Bishop Museum, Spec. Publ. 74, v + 79 pp. (1985). - SOJÁK, J.: Bot. Jahrb. Syst. 106, 145-210 (1986). - SOKAL, R.R.: Am. Nat. 126, 729-749 (1985); - Annu. Rev. Ecol. Syst. 17, 423-442 (1986). - SOKAL, R.R., FIALA, K.L., HART, G.: Syst. Zool. 33, 387-407 (1984). - SOLADOYE, M.O.: Kew Bull. 40, 291-386 (1985). - SOLADOYE, M.O., CRANE, P.R.: Grana 24, 145-160 (1985). - SOLO'EVA, N.M., PIMENOV, M.G., VASIL'EVA, M.G., ZIGAREVA, N.N., TURKOV, V.D.: Plant Syst. Evol. 151, 89-101 (1985). - SOLTIS, D.E., RIESEBERG, L.H.: Am. J. Bot. 73, 310-318 (1985). - SOULE, J.: Glossary for Horticultural Crops, 898 pp. New York: Wiley 1985. - SPENCER, K.C., SEIGLER, D.S.: Biochem. Syst. Ecol. 13, 23-24 (1985a); - ibid. 13, 421-431 (1985b). - SPENCER, K.C., SEIGLER, D.S., FRALEY, S.W.: Biochem. Syst. Ecol. 13, 433-435 (1985). - SPIES, J.J., DU PLESSIS, H.: Bothalia 15, 591-596 (1985). - SPJUT, R.W.: Econ. Bot. 39, 266-288 (1985). - STEAD, J.W., STYLES, B.T.: Bot. J. Linn. Soc. 89, 249-275 (1984). - STEBBINS, G.L.: Bot. J. Linn. Soc. 88, 149-164 (1984); - Ann. Missouri Bot. Gard. 72, 824-832 (1985). - STEBBINS, G.L., BASILE, D.V.: Evolution 40, 422-425 (1986). - STEINBRÜCK, G., SCHLEGEL, M., DAHLSTRÖM, I., RÖTTGER, B.: Plant Syst. Evol. 153, 229-241 (1986). - STEINHARTER, T.P., COOPER-DRIVER, G.A., ANDERSON, G.J.: Biochem. Syst. Ecol. 14, 299-305 (1986). - STEINMEYER, B., WÖHRMANN, K., HURKA, H.: Flora 177, 323-334 (1985). - STERN, W.L., PRIDGEON, A.M., LUER, C.A.: Bot. J. Linn. Soc. 91, 457-471 (1985). - STEVENS, P.F.: Kew Bull. 40, 457-465 (1985); - J. Arnold Arb. 67, 313-339 (1986). - STRONG, D.R.,

LAWTON, J.H., SOUTHWOOD, R.: Insects on plants: Community Patterns and Mechanisms, 313 pp. Cambridge, MA: Harvard University Press 1984. - STUETZEL, T.: Diss. Bot. 71, 1-138 (1984); - Flora 177, 335-344 (1985). - STUHLFAUTH, T., FOCK, H., HUBER, H., KLUG, K.: Biochem. Syst. Ecol. 13, 447-453 (1985). - STYLES, B.T. (ed.): Syst. Assoc. Spec. Vol. 29, 1-435 (1986). - SUMERE, C.F. van, LEA, P.J. (eds.): The Biochemistry of Plant Phenolics, 483 pp. Oxford: Clarendon Press 1985. - SUNDBERG, M.D.: Pollen Spores 27, 15-28 (1985). - SUNDBERG, M.D., ORR, A.R.: Am. J. Bot. 73, 1699-1712 (1986). - SYMON, D.E.: J. Adelaide Bot. Gard. 8, 1-171 (1985a); - Blumea 31, 319-328 (1986b). - SYTSMA, K.J., GOTTLIEB, L.D.: Evolution 40, 1248-1261 (1986a); - Proc. Natl. Acad. Sci. USA 83, 5554-5557 (1986b). - SYTSMA, K.J., SCHAAL, B.A.: Evolution 39, 582-608 (1985).

TAKAHASHI, H.: Bot. Mag. Tokyo 99, 137-154 (1986). - TAKHTAJAN, A.L., MEYER, N.R., KOSENKO, V.N.: Bot. Zh. 70, 153-162 (1985). - TAKHTAJAN, A.L., SHILKINA, I.A., YATSENKO-KHMELEVSKY, A.A.: Bot. Zh. 71, 1203-1206 (1986). - TARODA, N., GIBBS, P.E.: Notes R. Bot. Gard. Edinb. 44, 105-140 (1986). - TAYLOR, T.N., ZAVADA, M.S.: Linn. Soc. Symp. Ser. 12, 165-178 (1986). - TERABAYASHI, S.: Mem. Fac. Sci. Kyoto Univ., Ser. Biol. 10, 73-90 (1985a); - Acta Phytotax. Geobot. 36, 1-13 (1985b). - TERPO, A.: Feddes Rep. 96, 73-87 (1985). - TERRELL, E.E., LEWIS, W.H., ROBINSON, H., NOWICKE, J.W.: Am. J. Bot. 73, 103-115 (1986). - THANIKAIMONI, G.: Linn. Soc. Symp. Ser. 12, 119-136 (1986). - THIEN, L.B., BERNHARDT, P., GIBBS, G.W., PELLMYR, O., BERGSTRÖM, G., GROTH, I., McPHERSON, G.: Science 227, 540-543 (1985). - THOMAS, W.W.: Mem. New York Bot. Gard. 37, 1-116 (1984). - THOMASSON, J.R.: Ann. Missouri Bot. Gard. 72, 843-851 (1985). - THOMPSON, K.: New Phytol. 103, 597-601 (1986). - THULIN, M.: Symb. Bot. Upsalienses 25, 51 pp. (1985). - TIEMANN, A.: Diss. Bot. 82, 1-202 (1985). - TIFFNEY, B.H.: Ann. Missouri Bot. Gard. 71, 551-576 (1984); - ibid. 73, 394-416 (1986). - TILLICH, H.-J.: Gleditschia 13, 63-73 (1985). - TOBE, H., KEATING, R.C.: Bot. Mag. Tokyo 98, 291-316 (1985). - TOBE, H., RAVEN, P.H.: Ann. Missouri Bot. Gard. 71, 836-843 (1984a); - ibid. 71, 844-852 (1984b); - ibid. 72, 451-468 (1985); - Am. J. Bot. 73, 475-488 (1986). - TOBE, H., RAVEN, P.H., GRAHAM, S.A.: Taxon 35, 13-20 (1986). - TOMLINSON, P.B., WILDER, G.J.: Bot. Gaz. 145, 535-549 (1984). - TRÂN DÎNH LŶ: FEDDES Rep. 97, 235-273, 405-468, 607-689 (1986). - TREPLIAKOVA, T.E.: Bot. Zh. 70, 1051-1057 (1985). - TUCKER, S.C.: Origin of symmetry in flowers, 351-395. In: Contemporary Problems in Plant Anatomy, eds. R.A. WHITE, W.C. DICKISON, 598 pp. Orlando: Academic Press 1984. - TURNBULL, J.W., GRIFFIN, A.R.: Syst. Assoc. Spec. Vol. 29, 157-189 (1986). - TURNER, B.L.: Plant Syst. Evol. 150, 237-262 (1985a); - Taxon 34, 85-88 (1985). - TURNER, G.W.: Am. J. Bot. 73, 1178-1192 (1986).

UEDA, K.: Bot. Mag. Tokyo 99, 333-349 (1986). - UHLARZ, H.: Beitr. Biol. Pflanz. 60, 277-291 (1985). - URBANSKA, K.M.: Plant reproductive strategies, 211-228. In: see ref. GRANT 1984; - Bot. Helv. 95, 81-97 (1985).

VALANT-VETSCHERA, K.M.: Bot. Rev. 51, 1-51 (1985). - VANDER KLOET, S.P.: Taxon 34, 440-447 (1985). - VAN HULST, R., THERIAULT, A., SHIPLEY, B.: Can. J. Bot. 64, 1443-1449 (1986). - VASCONCELLOS-NETO, J.: Interactions between Ithomiinae (Lepidoptera:Nymphalidae) and Solanaceae, 364-377. In: see ref. D'ARCY 1986. - VASIL'EVA, M.G., KLUJUJKOV, E.V., PIMENOV, M.G.: Plant Syst. Evol. 149, 71-88 (1985). - VELDKAMP, J.F., KONING, R. de, SOSEF, M.S.M.: Blumea 31, 281-307 (1986). - VERDCOURT, B.: Kew Bull. 40, 1-23 (1985). - VERKERKE, W.: J. Arnold Arb. 66, 353-394 (1985a); - Beitr. Biol. Pflanz. 60, 341-351 (1985). - VERLAQUE, R.: Rev. Cytol. Biol. Végét.-Bot. 8, 117-243 (1985); - ibid. 9, 5-72, 97-176 (1986). - VICKERY, J.W., JACOBS, S.W.L.,

EVERETT, J.: Telopea 3, 1-132 (1986). - VICKERY, R.K.: Biosystematics 1983, 1-24. In: see ref. GRANT 1984. - VINCENT, L.P.D.: S. Afr. J. Bot. 51, 209-252 (1985). - VINK, W.: Blumea 31, 39-55 (1985). - VITEK, E.: Plant Syst. Evol. 151, 241-249 (1986). - VLIET, G.J.C.M., BAAS, P.: Ann. Missouri Bot. Gard. 71, 783-800 (1984). - VOGEL, S.: Trop. Subtrop. Pflanzenw. 54, 1-168 (1986). - VOYTENKO, V.F., OPARINA, S.N.: Bot. Zh. 70, 865-875 (1985). - VUILLE, C., KÜPFER, P.: Bull. Soc. Neuchâteloise Sci. Nat. 108, 123-134 (1985).

WAGNER, W.L., STOCKHOUSE, R.E., KLEIN, W.M.: Nonogr. Syst. Bot. 12, 1-103 (1985). - WAHA, M.: Plant Syst. Evol. 150, 165-177 (1985). - WALKER, J.W., WALKER, A.G.: Ann. Missouri Bot. Gard. 71, 444-463 (1984); - Linn. Soc. Symp. Ser. 12, 203-217 (1986). - WALLACE, G.D.: Aliso 11, 393-415 (1986). - WALT, J.J.A. van der: Bothalia 15, 345-385 (1985). - WALTERS, S.M.: New Phytol. 104, 527-546 (1986). - WALTERS, S.M. et al. (ed.): The European Garden Flora. Vol. I: Pteridophyta, Gymnospermae, Angiospermae-Monocotyledons (Part I), 430 pp. Cambridge: Cambridge University Press 1986. - WANNAN, B.S., WATERHOUSE, J.T., GADEK, P.A., QUINN, C.J.: Biochem. Syst. Ecol. 13, 105-108 (1985). - WARWICK, S.I., BLACK, L.D.: Can. J. Bot. 64, 1632-1643 (1986). - WARWICK, S.I., GOTTLIEB, L.D.: Evolution 39, 1236-1241 (1985). - WASER, N.M., RPICE, M.V.: Am. J. Bot. 72, 1726-1732 (1985). - WASMUND, O., STUBBE, W.: Plant Syst. Evol. 154, 79-88 (1986). - WATANABE, K.: Plant Species Biol. 1, 99-116 (1986). - WATANABE, K., YAHARA, T.: Bot. Mag. Tokyo 97, 87-105 (1984). - WATANABE, K., CARTER, C.R., SMITH-WHITE, S.: Can. J. Bot. 27, 410-420 (1985). - WATERMAN, P.G.: Phytochemistry 25, 3-17 (1986). - WATKINSON, A.R., WHITE, J.: Some life history consequences of modular construction in plants, 31-51. In: see ref. HARPER et al. 1986. - WATSON, E.M.: Aust. J. Bot. 34, 481-504 (1986a); - ibid. 34, 505-522 (1986b). - WATSON, L., CLIFFORD, H.T., DALLWITZ, M.J.: Aust. J. Bot. 33, 433-484 (1985). - WATSON, L., DALLWITZ, M.J., JOHNSTON, C.R.: Aust. J. Bot. 34, 223-230 (1986a). - WATSON, L., AIKEN, S.G., DALLWITZ, M.J., LEFKOVITCH, L.P., DUBE, M.: Can. J. Bot. 64, 53-70 (1986b). - WEBB, C.J.: Constraint on the evolution of plant breeding systems and their relevance to systematics, 249-270. In: see ref. GRANT 1984. - WEBB, C.J., LLOYD, D.G.: N.Z. J. Bot. 24, 163-178 (1986). - WEBER, A., VOGEL, S.: Plant Syst. Evol. 154, 237-250 (1986). - WEBERLING, F.: Bot. Jahrb. Syst. 107, 395-414 (1985). - WEBERLING, F., HILDENBRAND, M.: Beitr. Biol. Pflanz. 61, 3-20 (1986). - WEI ZHI: Acta Phytotax. Sinica 23, 185-208 (1985). - WEIR, J.S., KIEW, R.: Biol. J. Linn. Soc. 27, 113-132 (1986). - WENDEL, J.F., PARKS, C.R.: Am. J. Bot. 72, 52-65 (1985). - WERFF, H. van der, RICHTER, H.G.: Syst. Bot. 10, 166-173 (1985). - WERK, K.S., EHLEREINGER, J.: Evolution 40, 1334-1337 (1986). - WESTFALL, R.H., GLEN, H.F., PANAGOS, M.D.: Bot. J. Linn. Soc. 92, 65-73 (1986). - WHALEN, M.D.: Gentes Herbarum 12, 179-282 (1984). - WHALEN, M.D., COSTICH, D.E.: Andromonoecy in Solanum, 284-302. In: see ref. D'ARCY 1986. - WHEELER, Q.D.: Syst. Zool. 35, 102-109 (1986). - WHITCOMBE, J.R., MURPHY, P.J.: Theor. Appl. Genet. 71, 736-741 (1986). - WHITE, F.: Bothalia 16, 143-168 (1986). - WICKENS, G.E., GOODIN, J.R., FIELD, D.V. (eds.): Plants for Arid Lands, 452 pp. London: Allen and Unwin 1985. - WIEGLEB, G.: Tuexenia 6, 365-377 (1986). - WILDE, W.J.J.O. de: Gard Bull. Sing. 37, 115-179 (1984); - ibid. 38, 55-144, 185-225 (1985). - WILLERT, D.J. von: Adv. Bot. Res. 11, 157-191 (1985). - WILLIAMS, C.A., HARBORNE, J.B., GOLDBLATT, P.: Phytochemistry 25, 2135-2154 (1986). - WILMOT-DEAR, M.: Kew Bull. 40, 243-271 (1985). - WILSON, E.O.: Science 230, 1227 (1985). - WILSON, J.B., PARTRIDGE, T.R.: Taxon 35, 1-12 (1986). - WINK, M.: Plant Syst. Evol. 150, 65-81 (1985). - WINKLER, S.: Beitr. Biol. Pflanz. 61, 283-314 (1986). - WOLTER, M., SCHILL, R.: Grana 24, 139-143 (1985); - Trop. Subtrop. Pflanzenw. 56, 1-93 (1986). - WOLTER FILHO, W., ROCHA, A.I. da; YOSHIDA, M., GOTTLIEB, O.R.: Phytochemistry 24, 1991-1993 (1985). - WOOD, H., DEGABRIELE, R.: Aust. J. Bot. 33, 677-685 (1985).

YATAGAI, M., SATO, T.: Biochem. Syst. Ecol. 14, 469-478 (1986). - YATSKIEVYCH, G., MASON, C.T.: Syst. Bot. 11, 531-548 (1986). - YEO, P.F.: Bot. J. Linn. Soc. 89, 1-36 (1984); - Hardy Geraniums, 192 pp. Beckenham (Kent): Croom Helm 1985. - YOSHIE, F., SAKAI, A.: Can. J. Bot. 63, 2150-2158 (1985).

ZAVADA, M.S., DILCHER, D.L.: Ann. Missouri Bot. Gard. 73, 348-381 (1986). - ZAVADA, M.S., TAYLOR, T.N.: Plant Syst. Evol. 154, 31-39 (1986). - ZAVARIN, E., SNAJBERG, K.: Biochem. Syst. Ecol. 14, 1-11 (1986). - ZELTNER, L.: Bot. Helv. 95, 47-57 (1985). - ZENTGRAF, J., BACHMANN, K., CHAMBERS, K.L., PRICE, H.J.: Plant Syst. Evol. 151, 103-119 (1985). - ZIBURSKI, A., KADEREIT, J., LEINS, P.: Flora 178, 233-242 (1986). - ZULVAGA, F.O., SODERSTROM, T.R.: Smithsonian Contrib. Bot. 59, 1-63 (1985).

Dr. HANS-HELMUT POPPENDIECK
Institut für Allgemeine Botanik und
Botanischer Garten
Ohnhorststraße 18
D-2000 Hamburg 52, FRG

E. Geobotany

I. Areal- und Florenkunde (Floristische Geobotanik)

Von ECKEHART J. JÄGER

1. Forschungstrends

In allen Ländern wird an neuen Floren gearbeitet, die auch kritische Taxa, genaue Gesamt- und Lokalverbreitungsangaben enthalten und meistens auch die Höhenverbreitung beschreiben.

Die Gefäßpflanzenflora Europas, der meisten Gebiete Nordamerikas, der arktischen Zone, Mittelsibiriens, des Fernen Ostens, Australiens, Südafrikas und einiger Teilgebiete Südamerikas werden vollständig kartiert. Oft werden dazu Datenspeicher verwendet. Auch in Lokalfloren, Monographien und Roten Büchern werden jährlich Tausende von neuen Verbreitungskarten publiziert. Auf ihrer Grundlage ist u.a. eine objektive pflanzengeographische Gebietsgliederung oder eine Beurteilung der Gefährdung der Arten möglich.

Bei der Untersuchung der ökologischen Ursachen der Arealgrenzen verwendet man zunehmend Experimente zu den Toleranzbereichen der Produktion und einzelner Lebensvorgänge, zur Verbreitungsbiologie und zur Konkurrenz.

Das Verhältnis der Chorogenese und der Artbildungsvorgänge (evolutionäre Migration, regionale Isolation, periphere Divergenz, Fernverbreitungs-Migration, adaptive Radiation, hybridogene und cytologische Artbildung) wird viel diskutiert. Die Vikarianz-Biogeographie, die nur die Bildung geographischer Barrieren, allenfalls seltene Fernverbreitungsvorgänge als Artbildungsmechanismen ansieht und aufgrund der Übertragung der Sippen-Cladogramme in den Raum selbständige Argumente zur Erdgeschichte gewinnen will, wird hier nicht als Fortschritt gewertet, da sie die Vielfalt der Artbildungsmechanismen, die unterschiedliche Evolutionsgeschwindigkeit und die Arealdynamik vernachlässigt.

Das Interesse an anthropogenen Florenveränderungen nimmt zu. Die Invasion fremdländischer und der Rückgang heimischer Arten, die Auswirkung dieser Vorgänge in den Ökosystemen, die Pflanzenverbreitung in urbanen Ökosystemen, die Veränderungen der Unkrautflora, die Prognose der weiteren Florenduchmischung und der anthropogenen Arealerweiterung sind hier zentrale Themen.

Diese fünf Arbeitsrichtungen werden im vorliegenden Bericht behandelt. Das umfangreichere Kapitel über neue Floren kann als Ergänzung der Standardfloren-Bibliographie von FRODIN (1984) dienen.

2. Zusammenfassende Darstellungen

Die reich illustrierte Pflanzengeographie von KORNAŚ und MEDWECKA-KORNAŚ (1986) ist die gründlichste neue Darstellung der Phytochorologie. Die

Progress in Botany, Vol. 49

ökologischen, besonders aber die historischen Verbreitungsursachen, das Einwirken des Menschen bis hin zur Geographie der Kulturpflanzen werden ausführlich besprochen und mit zahlreichen gut ausgewählten Beispielen aus der Literatur der ganzen Welt illustriert. BROWN und GIBSON (1983) konzentrieren sich in ihrem Versuch einer Synthese der Zoo- und Phytogeographie stärker auf theoretische Themen (kritische Einschätzung der Vikarianz-Biogeographie und der Insel-Äquilibriums-Theorie, sympatrische und allopatrische Artbildung, Ursachen der Diversität).

3. Neue Floren

Für monographische Floren ist der Zeitaufwand wegen der Menge des Materials sehr gestiegen: Im vorigen Jahrhundert bewältigte ein Taxonom jährlich die Bearbeitung von durchschnittlich 250 Arten, jetzt nur noch von 15 - 50 Arten. Bei der Planung neuer Floren sollte daher die Wahl unter den verschiedenen Möglichkeiten (Monographische Flora, Florenhandbuch, Exkursionsflora und Enumeration) so getroffen werden, daß die Bearbeitung in absehbarer Zeit abgeschlossen werden kann (HEYWOOD 1984).

Auch außerhalb Europas gibt es nur wenige Länder, in denen seit 1960 nicht eine Flora verfaßt oder begonnen worden wäre. Die Vorbildwirkung besonders guter Floren und die bessere internationale Zusammenarbeit schlägt sich in der Qualität nieder.

a) Europa und Orient (außer Sowjetunion) (vgl. Prog. Bot. 47, 308)

Die erfreulichsten Neuerscheinungen sind hier zweifellos die Floren der Iberischen Halbinsel und der griechischen Gebirge. Die vorzüglich illustrierte Flora Iberica (1986, 1. Bd. von ca. 15, Lycopodiaceae - Papaveraceae) enthält detaillierte Verbreitungsangaben nach dem engen Netz der politischen Provinzen und Höhenverbreitungsgrenzen, die Endemiten sind besonders gekennzeichnet, zwischen belegten und nicht belegten Angaben wird unterschieden. In der Flora der griechischen Gebirge (STRID 1986, 1. Bd. von 2, choripetale Dicotylen) sind alle über (1500)-1800 m vorkommenden Arten berücksichtigt und die Fundorte nach den Zahlen in einer Karte leicht zu finden, auch die Höhenverbreitung wird angegeben. Hier ist eine große Zahl von Neufunden zusammengefaßt.

Im Mediterrangebiet sind nun noch Albanien, Jugoslawien und das Tiefland Griechenlands "relativ weiße" Flecken. Die mustergültige Flora der Türkei wurde mit dem 9. Band abgeschlossen (DAVIS 1985, Juncaceae - Gramineae, 114 Karten von ca. 500 Taxa), ebenso mit dem 4. Band die Flora Palaestina, die auch Jordanien umfaßt (FEINBRUN-DOTHAN 1986, 2470 Gefäßpflanzen-Arten, in der Türkei 3 x so viele) und mit dem 3. Band die Flora von Syrien und Libanon (MOUTERDE 1983, 1984). Die Flora Libyens ist mit der Bearbeitung von 133 Familien zu etwa 80% fertig (JAFRI und EL-GADI 1985-1986). Die Liste der marokkanischen Flora (HANSEN und SUNDING 1985) umfaßt in der 3. Auflage 3125 Arten, davon ist allderdings ein Drittel nicht heimisch. Der 3. Band der Med-Checklist verzeichnet die exakte Nomenklatur und die Ländervorkommen für die Arten der Familien von den Convolvulaceae bis zu den Lamiaceae (alphabetische Folge), die Asteraceae stehen noch aus (GREUTER et al. 1986). Vor der monographischen, mit vielen Herbarfotos illustrierten Flora Iranica erschienen 1986 die Mimosaceae, Caesalpiniaceae und Anthemideae (*Anthemis* mit 47, *Tanacetum* mit 54, *Artemisia* mit 64 Arten besonders artenreich), nur 20% des Gesamtumfangs dieser Flora fehlen noch.

Für die noch vor wenigen Jahrzehnten nur randlich erforschte Arabische Halbinsel wird in England (Kew und British Museum) eine große Flora erarbeitet. Erschienen ist der 1. Band (von 2, Dicotylen excl. Asteraceae, DAOUD 1985) einer Flora von Kuwait und eine einbändige Flora von Quatar (BATANOUNY 1981, 312 Spermatophyten).

In den mittel- und nordeuropäischen Ländern wurden neue Florenbände für Polen (Fl. Polski 1985, starke Berücksichtigung der Synanthropen, Diskussion kritischer Taxa, genaue Höhenverbreitung), die Slowakei (Fl. Slovenska, 4/1-2, 1984-1985, 114 Punktkarten), Slowenien (MARTINČIČ und SUŠNIK 1984, Exkursionsflora), Finnland (HÄMET-AHTI et al. 1984, kleine Karten für alle Arten, einbändig) und Skandinavien publiziert (LID 1985, einbändig). HEGIs Fl. von Mitteleuropa enthält in den neuen Lieferungen (seit 1985) für jede Art zonale und regionale Arealdiagnosen nach dem Verfahren von MEUSEL et al., die eine vergleichbare Synthese der chorologischen Merkmale von Floren und Vegetationseinheiten gestatten. Die Fortsetzung der Flora der Seealpen von BRIQUET durch CHARPIN und SALANON (1985) verbessert die Möglichkeiten einer Florenanalyse auch für die bisher noch relativ wenig floristisch erforschten Westalpen.

Unter den zahlreichen neuen europäischen Lokalfloren (z.B. Wales: ELLIS 1983, mit Gitternetzkarten; Insel Elba: FOSSI INNAMORATI 1983; unteres Loire-Tal: CORILLON 1981; Elsaß: Flore d'Alsace 1982) sind jetzt auch zunehmend Stadtfloren, die spezifische Verbreitungsbilder auch für sonst allgemein verbreitete Arten zeigen (z.B. Stockholm: LINDBERG 1983; Dublin: JACKSON und SKEFFINGTON 1984; London: BURTON 1983, über 1200 anthropochore Arten berücksichtigt, 750 Arten im 2 x 2 km-Raster kartiert).

b) Sowjetunion (vgl. Prog. Bot. 47, 309)

Die wichtigste neue Gebietsflora ist die auf 10 Bände veranschlagte, voll illustrierte Flora des gesamten Fernen Ostens von der Wrangel-Insel bis Wladiwostok, einem Gebiet, das mit 3,5 Millionen km^2 größer als Indien ist. Der vorliegende 1. Band (Sosud. rast. 1985, Lycopodiaceae, Juncaceae, Poaceae, 488 Arten) enthält Punktkarten für alle Arten, insgesamt sind 4000 Arten zu erwarten. Auch in diesem Gebiet nimmt der Einfluß der Neophyten rasch zu, wie ein Florenkonspekt von VOROŠILOV (1986) erkennen läßt (3215 Arten insgesamt, 52 heimische und 63 Neophyten mehr als 1982). Die einbändige Exkursionsflora des Magadaner Gebietes (CHOCHRJAKOV 1985) gestattet es, für 1/3 der Fläche des Fernen Ostens die Verbreitungsbilder aller Gefäßpflanzen zu präzisieren.

Die sowjetischen Republikfloren für Grusinien und Turkmenien erscheinen schon in 2. Auflage (Fl. Gruzii Bd. 9, 1984: Violaceae-Cornaceae, Bd. 10, 1985: Pyrolaceae-Verbenaceae, leider grusinisch, einige Punktkarten, und Opr. rast. Turkm. Bd. 3, 1985: Choripetalen), unvollendet ist nur die auch chorologisch besonders ausführliche tadshikische Flora, von der jetzt im 7. Band die Apiaceae bis Verbenaceae, im 8. Bd. die Plumbaginaceae bis Plantaginaceae bearbeitet wurden (Fl. Tadžik. 1984, 1986). Der zusammenfassende Mittelasienkonspekt ist wichtig u.a. zur Korrektur der recht unzuverlässigen Verbreitungsangaben der älteren Flora von Kasachstan (Opred. rast. sredn. Azii Bd. 7, 1983: Geraniaceae - Cornaceae). Die 52 Punktkarten im Band 9/2 (Fabaceae) der Arktičeskaja Flora (1986) zeigen die Zunahme der Funddichte in Nordsibirien.

Floren kleinerer Gebiete der Sowjetunion enthalten häufig auch eine Analyse der systematischen Struktur und der geographischen Florenelemente sowie Schlußfolgerungen über die Florengenese. So führt BAJTENOV (1985) die geringe Zahl der Endemiten (2,9% von 839 Arten) in der von ihm bearbeiteten Hochgebirgsflora des Nordtienschan (incl. Dsung. Alatau) auf die junge Hebung dieser Gebirge und das vorwiegend postpleistozäne Alter ihrer Flora zurück. Im 4464 m hohen, an der Grenze von Mittel- und Zentralasien gelegenen Dsungarischen Alatau sind nach GOLOSKOKOV (1984) 62 = 3% der 2168 Arten endemisch. Höher ist der Endemitenanteil im kolchischen Laubwaldrefugialgebiet: von 1978 Gefäßpflanzen der Flora Abchasiens sind 319 = 16,1% kolchische Endemiten (KOLAKOVSKIJ Bd. 3, 1985; Bd. 4, 1986; vollständig).

Vielen solcher Florenanalysen liegt die von TOLMAČEV ausgearbeitete Methodik des Vergleichs "konkreter Floren" zugrunde, die in einem Sammelband seiner chorologischen Arbeiten (TOLMAČEV 1986) enthalten ist. Für die Halbinsel Kanin in der europäischen Arktis verwendet SERGIENKO (1986) den mathematischen Vergleich von 11 konkreten Floren für die objektive pflanzengeographische Gliederung und die Abgrenzung der Arktis. Weitere Floren kleinerer Gebiete: Mittelrussische Meschtschera zwischen Oka und Kljasma östlich Moskau: ALEKSEEV et al., 1. Bd. von 2; südöstliche Ukraine: KONDRATJUK und BUDRA 1985, 42 von 1817 Arten neu, beide Floren mit Punktkarten; Podolien: ZAVERUCHA 1985, pflanzengeographische Analyse, Florengenese; Kurilen-Insel Kunaschir: ALEKSEEVA 1983; 45 von 883 Arten neu für die Sowjetunion).

c) Süd- und Ostasien

Die Flora of Pakistan, die östlich an die Fl. Iranica anschließt und dem Vorbild dieser Flora folgt, ist mit der Bearbeitung von 20 Familien in den letzten beiden Jahren rasch der Vollendung nähergekommen. Neu sind z.B. die Euphorbiaceae (Fasc. 172, 1986), Primulaceae, Solanaceae und Orchidaceae. Nach dem Abschluß des Konspektes der nepalischen Flora, in dem Schlüssel, Synonymik, Gesamtverbreitung und lokale Verbreitung der 6530 Samenpflanzen enthalten sind und der daher schon eher als Florenhandbuch gelten kann (HARA et al. 1982), liegen nun zwei von zehn Lieferungen einer ähnlichen, aber noch ausführlicheren und chorologisch genaueren Flora von Bhutan vor, der ersten dieses Landes (GRIERSON und LONG 1983, 1984).

In der neuen monographischen Fl. of India wurden in 18 Faszikeln bis 1984 kleinere Gruppen bearbeitet (z.B. *Prunus*), so daß der Abschluß vorläufig nicht abzusehen ist. Die Florenhandbücher für einzelne der 22 Unionsstaaten (Fl. of India Ser. II) sind dagegen in begrenzter Zeit zu bewältigen. Für einige dieser Staaten gibt es schon Floren vom Beginn dieses Jahrhunderts, in den letzten Jahren kamen Floren für Jammu, Himahal Pradesh, Tripura, Karnataka (Südwesten der Dekkan-Halbinsel, 1. Bd. von 3, Dicotylen, SALDANHA 1985) und den südöstlich angrenzenden Staat Tamil Nadu dazu (bis auf einen Supplementband vollständig, MATTHEW 1981-1983). Diese neuen Floren zeigen wieder, daß die indischen (wie die afrikanischen) Tropen artenärmer sind als gemeinhin angenommen wird. Selbst der besonders artenreiche Staat Karnataka hat mit 3924 Arten auf 192.000 km^2 keine höhere Artendichte als z.B. Italien.

Für Meghalaya liegt der erste Band einer Forest Flora vor (HARIDASAN und RAO 1985), für Madhya Pradesh in der Monographien-Reihe Fl. of India Ser. IV die Poaceae (ROY 1984) und Papaveraceae (DEBNATH und NAYAR 1986). In der Serie III der Fl. of India schließlich erscheinen Distriktfloren, in den letzten Jahren für Distrikte in den Staaten Rajasthan und West Bengal. In der neuen Flora von Bangladesh (KHAN 1972-1985) wurden in den vergangenen 14 Jahren 39 kleinere Familien in unsystematischer Reihenfolge publiziert. Auch die Flora von Ceylon, von der schon fünf gewichtige Bände vorliegen (DASSANAYAKE und FOSBERG Bd. 4, 1983; Bd. 5, 1985) enthält die Familien in der Reihenfolge der Bearbeitung.

Von den zahlreichen chinesischen Floren sind dem Sprachunkundigen leider nur die wisschenschaftlichen Namen, die Diagnosen neuer Arten und die klaren Abbildungen zugänglich, ein einfacher Zahlenschlüssel für die Verbreitungsangaben wäre wünschenswert. In der monographischen Fl. Reipubl. Popularis Sinicae sind mit 39 von 80 Bänden (z.T. mit mehreren Teilbänden) schon fast die Hälfte der Gefäßpflanzen erschienen (1985 z.B. Apiaceae, Rosoideae, Astereae). Die Verbreitungsangaben nach den politischen Provinzen sind recht großzügig, da z.B. Xinjiang größer als die Mongolei ist.

Neue Teilgebiets- und Provinzfloren erschienen in den letzten 6 Jahren für die Innere Mongolei (MA 1978-1985, vollständig, illustriert), die chinesischen Wüstengebiete (Fl. in desertis, Bd. 1 Monocotylen, Dicotylen z.T., 1985), Nordostchina (Fl. Plantarum Herbacearum 1958-1980, Monocotylen, Dicotylen bis Apiaceae), Peking (Fl. of Beijing Bd. 1, 1984: Farne, Gymnospermen, Dicotylen p.p.), das Tsinglingschan-Gebiet (Fl. Tsinglingensis 1974-1983, Farne, Monocotylen und Dicotylen fast vollständig), die Provinzen Hebei (= Hopeh, Fl. Hebeiensis Bd. 1, 1986: Farne, Gymnospermen, Dicotylen bis Leguminosen), Hubei (= Hupeh, Fl. Hupehensis Bd. 1 und 2 von 4, 1978-1980, Gymnospermen und Dicotylen p.p.), Sichuan (= Szetschwan, Fl. Sichuanica Bd. 1, 1981: Dicotylen p.p.; Bd. 2, 1983: Gymnospermen), Yunnan (Fl. Yunnanica Bd. 1-3, 1977-1983, mono- und dicotyle Familien in unsystematischer Folge, 2 Bände Index 1984) und Xizang (= Tibet, Fl. Xizangica Bd. 1, 1983; Bd. 2, 1985; 5 Bände mit ca. 6000 Arten geplant). Noch liegt die Funddichte relativ locker; wie rasch aber die Erkundung vorangeht, zeigt die Tatsache, daß die große Zahl der lokalen Listen in Tibet bereits eine Berechnung des Gefälles der Florenelemente als Grundlage für eine Florengebietsgliederung gestattet (FRANKENBERG und DU 1983).

Die Flore du Cambodge, du Laos et du Viêt-Nam ist ein Beispiel für den recht langsamen Fortschritt der großen monographischen Floren. In den letzten 8 Jahren wurden ca. 300 Arten bearbeitet, zuletzt erschienen Bd. 21: Scrophulariaceae und Bd. 22: Bignoniaceae, beide 1985. Mit Bd. 4/1+2 (Caesalpinioideae, Mimosoideae, 1984, 1985) wurde die Fl. of Thailand fortgesetzt, die ziemlich detaillierte Verbreitungsangaben enhält. In insgesamt 8 voluminösen Bänden werden die Farne Japans abgebildet, ihre Fundorte aufgelistet (bis 20 Seiten für 1 Art!) und kartiert (KURATA and NAKAIKE, Bd. 4, 1985).

d) Australien, Malesien, Ozeanien

Neben der vielbändigen modernen Flora des australischen Kontinents (Prog. Bot. 47, 311) sind in den letzten Jahren für alle Bundesstaaten und Territorien Australiens außer dem Nordterritorium neue Floren erschienen oder begonnen worden. Die Fl. von Südaustralien liegt vollständig in der 4. Auflage vor (JESSOP und TOELKEN 1986), für Westaustralien gibt es eine verbesserte Fassung der Liste von GREEN (1985) und eine Exkursionsflora von GRIEVE und BLACKALL (1982). Die umfassende Florenserie von New South Wales (JOHNSON 1984) endet mit der 106. Lieferung (Fabaceae 2) und wird künftig in der Zeitschrift Tuatara fortgesetzt. In der Flora des westlichen Teils dieses Staates sind die meisten Arten abgebildet und viele ökologische Daten aufgenommen (CUNNINGHAM et al. 1982). Dieses aride Gebiet ist mit 2000 Gefäßpflanzen auf 450000 km^2 ziemlich artenarm. Aus dem ganzen Bundesstaat, für den 1916 nur 3500 Arten angegeben wurden, sind jetzt aber 6000 Arten bekannt (Liste von JACOBS und PICKARD 1981, Verbreitung nach 13 ökogeographischen Teilgebieten angegeben). Für den Nordosten des Staates liegt eine gute mehrbändige Flora vor (BEADLE und BEADLE 1971-1984, Bd. 1-5: Farne, Gymnospermen, Dicotylen fast vollständig). Eines der floristisch reichsten Gebiete, auch an synanthropen Arten, ist die Umgebung von Sydney (einbändige illustrierte Flora von BEADLE et al. 1982). Für Queensland, den zweitgrößten Bundesstaat, sind vier Regionalfloren vorgesehen. Die des Südostens enthält auf einem Gebiet von der Größe beider deutscher Staaten 3600 Arten (STANLEY und ROSS, 1. und 2. Bd. von drei Bänden, Dicotylen vollständig, 1983, 1986). Die Flora Zentralaustraliens (JESSOP 1981) beherbergt auf der zehnfachen Fläche nur 1992 Arten. Das von 75 Mitarbeitern in nur 3 Jahren verfaßte illustrierte Florenhandbuch dieses Gebietes erfaßt die ariden Teile von fünf Staaten und trägt dadurch zur Vereinheitlichung der Artabgrenzung bei.

Der erste Teil einer Florenliste der mikronesischen Inseln zeigt, wieviel leichter sich die Farne (207 Arten) ausbreiten als die Gymnospermen (19 Arten, aber nur *Cycas circinalis* heimisch, FOSBERG et al. 1982). In der monographischen, mit Verbreitungskarten ausgestatteten Flore de la Nouvelle Calédonie wurden in der 10.-13. Lieferung seit 1982 neun Familien bearbeitet. Die umfangreiche Gebirgsflora von Neuguinea liegt mit dem 4. Band vollständig vor (VAN ROYEN 1970-1983).

e) Afrika südlich der Sahara

Neu ist als erste vollständige Flora von Togo das einbändige Florenhandbuch von BRUNEL et al. (1984, Belegzitate nach ökogeographischen Regionen geordnet), eine einbändige Flora von Côte d'Ivoire (3660 Gefäßpflanzen, d.h. wenig mehr als in mitteleuropäischen Gebieten gleicher Größe; chorologische Analyse, AKE ASSI 1984) und eine Lokalflora der Loma-Berge in Sierra Leone (JAEGER und ADAM 1980 ff., nur 9 Endemiten, 53 Arten sonst nur in den ostafrikanischen Gebirgen, Disjunktionen u.a. durch Klimaschwankungen erklärt). Für die Kap-Flora gibt es einen einbändigen Katalog der 8505 Arten. Von diesen sind 5832 = 68,2% endemisch, mehr als in allen anderen Teilen Afrikas (BOND und GOLDBLATT 1984).

Die übrigen Standardfloren wurden fortgesetzt:
Niger: 2. Supplement zum Katalog, 74 Arten neu, jetzt 1178 (LEBRUN et al. 1983).
Fl. da Guine-Bissau (bis Fasc. 6: Fl. da Guine Portuguesa): Fasc. 1-8 (1971-1984) u.a. Leguminosae, Malvaceae.
Flore du Cameroun: bisher 1600 von 8000 Arten, Bd. 21-28 (1980-1985), Brassic., Balsamin., Loranth., Melastomat., Combret., Bignoniaceae.
Flore du Gabon: bisher ca. 25%, Bd. 25-27 (1983-1985), Melastomat. Mor., Gesner., Bignoniaceae.
Flore d'Afrique Centrale: Orchidaceae (1984, 1987).
Äthiopien: Leguminosen, 613 Arten, 35 neu, 64 endemisch (THULIN 1983). Neues Florenprojekt 1980 begonnen.
Fl. of Tropical East Africa: Poaceae (1982), Orchidaceae (1984, 1986), Amaranth., Zingiber., Araceae und 6 kleine Familien (1986).
Flore du Rwanda: Bd. 1-3 Dicotylen vollständig (TROUPIN 1978-1985).
Fl. Zambesiaca: bisher ca. 50%, Bd. 7 (1983, 1984), Crassul., Valerian., Dipsac., Campanul., Apocyn., Loganiaceae.
Fl. de Moçambique: noch unter 10%, 1969-1983 meist kleinere Familien, zuletzt z.B. Melastomat., Lythr., Cycad., Apiaceae.
Fl. of Southern Africa: Bd. 7/2 Ixioideae; 33/6 Gnaphaliinae (1983); 21/1 Tiliaceae (1984); 4/2 Xyridaceae-Juncaceae; 14 Crassulaceae; 28 Lamiaceae (1985); 31/1 Rubioideae (1986); Pteridophyta (1986).
Flore de Madagascar: 1982-1985 7 Familien, u.a. Brassic., Elaeocarp., Loganiaceae.
Flore des Mascareignes: Arecaceae (1984), Urtic., Moraceae und 7 kleine Familien (1985).

f) Nordamerika (vgl. Prog. Bot. 45, 322)

Seit der 1983 gegebenen Zusammenstellung und dem Standardflorenführer von FRODIN (1984) erschienen von der Nord- bis zur Südspitze der beiden amerikanischen Kontinente sehr viele neue Floren. Im Norden der arktischen Ellesmere-Insel kommen noch 125 Arten Höherer Pflanzen vor, einzelne steigen dort noch bis 1220 m auf; die sechs größten Familien stellen 62% aller Arten (SOPER und POWELL 1985). Die Aktualisierung der Flora von Prince Edward Island (ERSKINE 1985) ergab 60 neue Taxa.

Intensiv durchforscht sind schon seit längerer Zeit die nordöstlichen USA (neu: Fl. von New Jersey, HOUGH 1984; 2. Bd. von drei der illustrierten Fl. von Michigan, County-Raster-Karten für alle Arten, VOSS 1985) und der Westen (Bibliographie von über 100 Lokalfloren: BOWERS 1982, neu z.B. Fl. von San Diego Cy./Calif., BEAUCHAMP 1986, und von West-Washington, CRESO 1984). Umfangreiche neue Floren und Verbreitungskartenwerke gibt es jetzt aber auch aus dem Mittleren Westen (Fl. der Great Plains von Ost-Montana bis Nordost-Texas, West-Missouri und West-Minnesota, einbändig, BARKLEY 1986; South Dakota mit vielen Neufunden aus den Black Hills, BRUGGEN 1985) und aus den großen Beckenlandschaften (CRONQUIST et al. 1984, Bd. 4: Sympetalen außer Asteraceae, Verbreitungsangaben grob; ebenso in der Fl. von Montana von DORN 1984, genau dagegen in der Fl. von Utah von WELSH 1982-1986, erschienen die Rosaceae, Asteraceae, Salicaceae, Cactaceae, Chenopodiaceae, Saxifragaceae und Apiaceae). Aus dem artenreichen, aber relativ schwach erforschten Südosten der USA liegen keine neuen Staaten-Floren vor, aber eine Liste der 3049 Gefäßpflanzen von Georgia (COILE und JONES 1985) und eine für Louisiana (MAC ROBERTS 1984).

In Mexico hatte nur Lower California eine moderne Flora, nun kommen die rasch fortgesetzte Fl. de Veracruz (1982-1983 30 Familien, automatisch hergestellte Punktkarten für alle Arten), die Flora des Beckens von Mexico (2. von drei Bänden, Dicotylen vollständig, RZEDOWSKI und RZEDOWSKI 1985) und die Fl. Novogaliciana dazu (Staaten Jalisco, Colima und angrenzende Gebiete; Fagales, Poaceae, Asteraceae - 750 Arten!, Orchidaceae, MC VAUGH 1983-1985).

g) Mittel- und Südamerika (vgl. Prog. Bot. 45, 322)

Seit dem letzten Bericht sind hier drei monographische, illustrierte, in Lieferungen ohne systematische Reihenfolge erscheinende Länderfloren begonnen worden: die Fl. Colombia (Fasc. 1-4, 1983-1986, Magnoliaceae, *Crotalaria* u.a., mit Verbreitungskarten), die Fl. of the Guianas, die alle drei ehemaligen europäischen Kolonien umfaßt, von denen nur die zentrale (Surinam) eine fast vollständige Flora hatte, und die Fl. von Paraguay (SPICHINGER und BOCQUET 1985-1987, 8 Familien), außerdem eine Flora des venezolanischen Guyana (STEYERMARK 1984, ca. 50% endemische Arten).

In der einbändigen Synopsis der Flora von Puerto Rico (LIOGIER und MARTORELL 1982) sind Status, Verbreitung und Höhenstufen angegeben. Zu der bereits vorliegenden Blütenpflanzen-Flora von Jamaika wurden die Farne ergänzt (PROCTOR 1985), die hier mit 579 Arten besonders reich vertreten sind. Vom gleichen Autor (1984) stammt eine Flora der Cayman-Inseln mit 601 heimischen und 80 synanthropen Arten. Erfreulich rasch fortgesetzt wurden die folgenden, in Prog. Bot. 45, 321 ff. zitierten Floren: Fl. Neotropica Monographs (seit 1983 13 Bände, zuletzt 1986 Bd. 42 Burmanniaceae), Niederländische Antillen (Bd. 3, 1984), Hispaniola (Bd. 2, 1983, 1985), Costa Rica (u.a. Caryophyllales 1983), Surinam (Bd. 5/1,4, 1984, geht nun in der obengenannten Fl. of the Guianas auf), Venezuela (Bd. 11/2 und 4/2, 1982), Hochland von Guyana (MAGUIRE et al. 1984, Hubschrauber-Expeditionen auf die unzugänglichen Tepuis), Ecuador (seit 1983 10 Lieferungen mit 25 Familien), Santa Catarina/Brasilien (3 Bände Poaceae 1981-1983), Jujuy/Argentinien (Bd. 8, 1983, Sympetalae bis Solanaceae) und Patagonien (Bd. 4, 1984, Rosales, Fabales). Für Chile legen MARTICORENA und QUEZADA (1985) einen Katalog der 5215 Arten mit Status-Angaben vor, eine neue Landesflora ist nicht in Aussicht.

Vier ausführliche Lokalfloren aus dem tropischen Amerika ermöglichen einen Vergleich des Artenreichtums, der systematischen Struktur und biologischer Spektren: Rio Palenque/Ecuador, Barro Colorado Island/Panama, Jauneche/Ecuador (DODSON et al. 1986) und die für die karibischen Tiefländer repräsentative von La Selva/Costa Rica (HAMMEL 1986, 1900 Arten auf 1500 ha).

4. Floristische Kartierung

Immer häufiger ist die Verbreitung bereits in den Floren kartiert. Wenn die Längen- und Breitengrade aus den Herbaretiketten in Datenspeicher aufgenommen werden, lassen sich die Karten automatisch herstellen (MAGILL et al. 1983: Vergleich handgezeichneter Karten mit Computerkarten aus Südafrika; automatisch hergestellte Karten in der Flora de Veracruz; KRASNOBOROV et al. 1986: Datenspeicher für Sibirien). Bei lockerer Fundpunktdichte dienen Karten in den Floren der anschaulichen Dokumentation, bei höherer Dichte der Nachweise (z.B. Sowjetischer Ferner Osten, *Sosud. rast.* 1985) tragen sie zur ökogeographischen Sippencharakterisierung bei. In den im ersten Abschnitt genannten Florenbänden sind über 10.000 Verbreitungskarten enthalten.

In fast allen Ländern Nord-, Mittel- und Südwesteuropas sind Kartierungen der gesamten Gefäßpflanzenfloren entweder abgeschlossen oder in Arbeit. Besonders erfreulich ist der raschere Fortgang der Flora Europaea-Kartierung (JALAS und SUOMINEN 1986: Bd. 7, Silenoideae, bis Karte 1508), der nun einen Abschluß in der Mitte des nächsten Jahrhunderts zu extrapolieren gestattet. Parallel zur Fl. der ČSR wird ein Florenatlas produziert (Raster 12 × 11 km). Die erste Lieferung enthält 315 Karten der Farne, Magnoliidae und Caryophyllidae, darunter auch mehrere "kritische" Taxa (SLAVÍK 1986). Die Kartierung der letzteren wird allerdings weiterhin meistens von Spezialisten durchgeführt, z.B. für *Ranunculus auricomus* aggr. in Bayern von BORCHERS-KOLB (1985) oder für *Rubus* in Westfalen von WEBER (1985, 127 Karten). Der Atlas der polnischen Flora soll erst nach dem Jahr 2000 erscheinen (Übersicht über die weiteren floristischen Arbeiten: BROWICZ 1985). Schon 1987 kann dagegen die Kartierung der 1500 Taxa für den polnischen Karpaten-Atlas abgeschlossen werden (MEDWECKA-KORNAŚ und KORNAŚ 1985). In der DDR ist mit der Publikation eines Florenatlas im 5,5 × 5,5 km-Raster 1990 zu rechnen.

Für das nördliche Teilgebiet zeigen FUKAREK und HUSE (1985) anhand von datierten Punktkarten von 80 dort ausgestorbenen Arten, daß der Rückgang bereits in der 2. Hälfte des 19. Jahrhunderts erfolgte, während die jetzt vom Aussterben bedrohten Arten durch die Intensivierung der Landwirtschaft seit 1960 verschwinden. Auf der Grundlage des engen Gitternetzes von 4 × 4 km wurden in Belgien und Nachbargebieten auch schon die Hepaticae mit Computer-Hilfe kartiert (SCHUMACKER 1984, 169 kurz kommentierte Karten, 75 Mitarbeiter). Mit 26 Karten im 10 × 10 km UTM-Raster, dazu 19 Deck-Karten zur Interpretation und einem Einführungsband wurde eine Kartierung der katalanischen Länder begonnen (ORCA 1985). Die Flora von Lettland wird im 8 × 8 km-Raster kartiert (4. Serie: Chorologija 1986, 69 Punktkarten seltener Arten). Mit 593 kommentierten Punktkarten im 3. Band wurde die Kartierung der Bäume und Sträucher der Sowjetunion abgeschlossen (SOKOLOV et al. 1986, ausführlicher Begleittext). Die erste lokale Kartierung in einem Tropengebiet (624 Gehölzkarten aus Ghana) benutzen HALL und SWAINE (1981) als Grundlage für eine floristische Vegetationsgliederung. Auch aus dem australischen Nordterritorium gibt es einen Atlas der Gefäßpflanzen (DUNLOP und BOWMAN 1986, 979 Punktkarten).

Auch zur Raumbewertung in Stadtgebieten läßt sich die Rasterkartierung verwenden (MAAS 1985). KUNICK (1987) schlägt dazu für Zentraleuropa eine Auswahl von 100 Zeigerfplanzen vor. Besonders aber durch das Interesse an der Kontrolle der Umweltzerstörung erhält die Kartierung starken Auftrieb. Viele der Roten Bücher enthalten Karten der gefährdeten Pflanzenarten (z.B. Krasnaja kniga 1985). Sehr aussagekräftig für das Ausmaß des Artenrückganges ist der Atlas van de Nederlandse Flora, in dessen 2. Band (1985) die heutige Verbreitung von 544 seltenen Arten mit den bis 1950 festgestellten Vorkommen verglichen wird. Die Arealdynamik wird in jedem Einzelfall sachkundig interpretiert.

Bei lokalen Schutzbestrebungen muß auch die Lage der Vorkommen im Gesamtareal beachtet werden. KNAPP (1986) schildert deshalb die Grenzsituation der Trockenrasen-Orchideen in der DDR.

An die Genauigkeit von Gesamtverbreitungskarten können heute immer höhere Anforderungen gestellt werden. HULTÉN hatte schon 1976 Zirkumpolarkarten aller nordeuropäischen Gefäßpflanzenarten im Manuskript fertiggestellt. Damit wurde die Arealtypisierung und die geographische Florenanalyse auch für dieses Florengebiet (wie vorher nur für Alaska) auf eine überprüfbare Grundlage gestellt. Die Karten wurden nach HULTÉNs Tod z.T. ergänzt und jetzt in drei repräsentativen Bänden publiziert (HULTÉN und FRIES 1986). Allerdings konnten die neuen Floren der Türkei, Sowjetunion, Mongolei, Chinas, z.T. auch die Flora Europaea nicht mehr berücksichtigt werden. In der Genauigkeit und ökogeographischen Aussage bleiben die Karten daher hinter denen bei MEUSEL et al. (1978) zurück.

5. Höhenverbreitung

Die Höhenverbreitung wird jetzt in den meisten Floren besser berücksichtigt. Ihre Änderung in zonaler und peripher-zentraler Richtung untersucht KUVAEV (1982) für das Gebiet der Sowjetunion. Daraus geht das starke Ansteigen mancher Höhengrenzen im extrem kontinentalen Klima, aber auch die Absenkung der Untergrenzen von arktisch-alpinen Arten hervor. Nach LANG (1985) reichen im Bergland von Szechuan die Orchideen am kontinentalen Westhang bis 4400 m aufwärts, am Osthang nur bis 4000 m, die epiphytischen nur bis 1800 m; dabei ist der Osthang aber artenreicher. In einer Nachuntersuchung des "Elevationseffektes" (vgl. Prog. Bot. 32, 294) fordert HENGEVELD (1985) eine dynamischere Betrachtung der Höhenverbreitung und die Berücksichtigung der Häufigkeit. Ähnlich wie an der Nordgrenze auf den Britischen Inseln hat *Tilia platyphyllos* an ihrer Höhengrenze in den Westkarpaten wärmezeitliche, natürliche Reliktvorkommen bei 1380 m auf relativ warmen, nährstoffreichen Standorten (JENÍK 1986).

6. Klimaökologische Arealinterpretation

a) Wuchsformen und Vegetationsformationen

Seit HUMBOLDT werden die Arealgrenzen der Wuchsformen und Vegetationsformationen in erster Linie auf Klimaunterschiede zurückgeführt. Die hierzu durchgeführten Vergleiche faßt WOODWARD (1987) in einem Buch über Klima und Pflanzenverbreitung zusammen. Die Korrelation erweist sich in immer neuen Untersuchungen als sehr eng (z.B. SOWELL 1985 für Nordamerika).

Einer der Fälle, in denen die aktuelle Vegetation dem Klima nicht entspricht, sind die Pampas: die Humidität dieses waldfreien Gebietes ist für Baumwuchs hoch genug. BOX (1986) vergleicht den Wuchsformenbestand von 150 Punkten dieses Gebietes mit den Klimaformeln von THORNTWAITE und HOLDRIDGE und findet u.a., daß die Nischen für einige Lebensformen nicht besetzt sind. Eine weitere überraschende Diskrepanz zwischen Klima und Wuchsform ist das Vorkommen sommergrüner *Nothofagus*-Wälder im südamerikanischen Winterregengebiet. Nach GOLTE (1983) ist es durch das Wasserspeichervermögen der tiefgründigen Böden zu erklären, auf die die sommergrünen Arten hier beschränkt sind. Wie auf der Nordhemisphäre sind sie an Winterschnee und ein sommerliches Tagesmittel von >10°C gebunden. Wenn die Vorhersage morphologischer Merkmale, etwa des

Blattflächenindex (WOODWARD 1987), mit der tatsächlichen Verbreitung nur wenig übereinstimmt, so liegt das oft daran, daß die verwendeten Klimaformeln die Klimaabläufe nicht differenziert genug wiedergeben.

Die Kontinentalität als summarischer Ausdruck verschiedener Klimafaktoren ist für einen ersten Vergleich der Pflanzenareale gut geeignet. Die Berechnung der Kontinentalität nach dem GAMSschen Index für die Pyrenäen ergibt nach IZARD et al. (1985), daß dort im Unterschied zu den Alpen eine geschlossene kontinentale Innenzone fehlt. Nur einige Täler an beiden Enden des zentralen Rückens sind so kontinental, daß sie nicht *Fagus*, sondern eher *Pinus* und *Abies* beherbergen.

b) Arealgrenzen einzelner Taxa

Für die Verbreitung einzelner Taxa kann der Vergleich des Verlaufs der Arealgrenzabschnitte mit Klimalinien recht differenzierte Hinweise auf die begrenzenden Faktoren geben. Gute Punktkarten zeigen, daß die meisten *Amaranthus*-Arten im Oberrheingebiet an Sommerwärme gebunden sind (HÜGIN 1986). Auch LANDOLT (1986) findet für die Lemnaceae aufgrund des Vergleiches von Punktkarten und Klimalinien Korrelationen der Nordgrenzen mit der Temperatur des wärmsten Monats; deutlich gestaffelt sind bei den verschiedenen Arten auch die höchsten ertragenen Sommermaxima. Nach HOEK (1984) ist die weltweite Verbreitung der Gattungen der Rhodophyta mit der Wassertemperatur korreliert. Die Nordostgrenze von *Carpinus betulus* ist am ehesten mit Temperatursummen über 0°C, absoluten Minima und Apriltemperaturen als Ausdruck der Länge der Vegetationsperiode korreliert (KOŽARINOV 1985).

Unterstützt werden solche Aussagen durch langjährige Beobachtungen am Standort und durch Kulturexperimente, die in England z.B. bei *Tilia*, *Sedum*, *Eupatorium*, *Verbena* und *Geum* angestellt wurden (WOODWARD 1987). Die Phasen des Lebenszyklus reagieren dabei unterschiedlich, und diese Unterschiede sind artabhängig. Bei *Larix* sollen die Erwachsenen, bei *Betula* die Jungpflanzen unter dem Kühlerwerden des Klimas der letzten Jahrzehnte an der Nordgrenze mehr gelitten haben (WOODWARD 1987). Bei *Poa bulbosa* findet OFIR (1986), daß hohe Temperaturen für die Auflösung der Sommerruhe wichtig sind. Bei dem Laubmoos *Macromitrium* können jenseits der 6°-Januar-Isotherme die ♂ Protonemata keine Knospen bilden. Die Pflanze kann in kälteren Gebieten nur dadurch Sporogone hervorbringen, daß auf den ♀ Gametophyten Zwergmännchen gebildet werden (UNE 1985). Die Arealdifferenzierung des Schwesternartenpaares *Carlina vulgaris*/*C. stricta* in Gebieten mit unterschiedlicher Länge der Vegetationsperiode und unterschiedlichen Jahres-Niederschlagskurven spiegelt sich in der lokalen Verbreitung in den Alpen, aber auch in zahlreichen morphologischen und physiologischen Parametern (KÖRNER und MEUSEL 1986).

c) Ausfüllung der klimatischen Areale

Eine entscheidende Voraussetzung für klimaökologische Schlußfolgerungen aus Arealgrenzen ist die Ausfüllung der potentiellen Areale. Für *Nothofagus cunninghamii* reichen Temperatur- und Feuchte-Werte nach BUSBY (1986) allgemein zur Erklärung der Arealgrenzen aus, einzig in Nordost-Victoria scheint eine unvollständige Ausbreitung seit der letzten Eiszeit vorzuliegen. Bäume reagieren mit Arealerweiterung oder -schrumpfung auf Klimaschwankungen langsamer als Annuelle. Errechnet wurde, daß bei Bäumen eine Klimaschwankung mindestens 15 Jahre dauern muß, um mit Ausbreitung beantwortet zu werden, aber mindestens 164 Jahre, um eine Arealreduktion zu bewirken. Bei Annuellen soll schon eine Klimaabweichung von 16 Tagen einen Effekt in der Verbreitung haben können (WOODWARD 1987). Nach DAVIS et al. (1986) geht aus pollenanalytischen Daten hervor, daß eine Barriere von 100 km Breite (Große Seen, USA) die Ausbreitung von *Fagus* um 1000 Jahre verzögerte, während sie für die windverbreitete *Tsuga canadensis* keine wesentliche Rolle spielte. Nach WEBB (1986) reagiert die Waldvegetation auf Klimaschwankungen von 500 - 1000 Jahren mit deutlichen Arealveränderungen. Dabei sind Aussagen wie "Erwärmung" nicht differenziert genug: Bei gleichbleibendem Jahresmittel, aber abnehmenden Juni- und ansteigenden Januar-Temperaturen konnte sich *Picea* nach Süden, *Fagus* nach Norden ausbreiten.

7. Chorogenese und Artbildung

Aus der Verbindung der HENNIGschen Cladistik mit CROIZATs Panbiogeographie ist vor etwa 10 Jahren die Vikarianz-Biogeographie entstanden. Ihre Vertreter wollen die historische Biogeographie durch eine exakte Methodik und die Falsifizierbarkeit der Aussagen vom erzählenden Stil auf die Ebene der Wissenschaftlichkeit heben und sie aus der Stagnation befreien. Ein Gegensatz wird konstruiert zwischen den Vikariisten, die ihre geistigen Vorfahren in HUMBOLDT, DE CANDOLLE und CROIZAT sehen, und den Dispersalisten, zu denen Biogeographen wie DARWIN, WALLACE oder AXELROD und RAVEN (z.B. in der Untersuchung zur Genese der Flora der Cordilleren 1985) gezählt werden. Die ersteren sehen die Ursache von Disjunktionen in der Entstehung von Barrieren, die die Areale trennen, die letzteren angeblich nur in der Ausbreitung über bereits existierende Barrieren hinweg (CRAW und GIBBS 1984).

Die Methode der Vikarianz-Biogeographie besteht in der Feststellung der abgestuften Verwandtschaft von Gebieten aufgrund übereinstimmender Cladogramme ihrer Endemitentaxa und aufgrund der Hypothese, gleiche Disjunktionen müßten auf die gleiche Geschichte hindeuten. In den Florengebiets-Cladogrammen werden die endemischen Taxa wie Apomorphien, Schwesterngruppenbeziehungen wie Synapomorphien, Vikarianz wie Cladogenese und Fernverbreitung wie Hybridisierung im Taxon-Cladogramm bewertet. Aus der Verwandtschaft der Gebiete soll sich eine generelle, von geologischen Hypothesen unabhängige Theorie der Erdgeschichte ergeben. Die bisherigen Ergebnisse sind zweifelhaft, sie gipfeln z.B. bei HUMPHRIES und PARENTI (1986) in der Konstruktion einer "Präpangäa", in der die australen Gebiete der Erde im Norden, die borealen in der Mitte und die tropischen im Süden lagen, weil anders die beidhemisphaerisch temperaten Disjunktionen nicht zu erklären seien!

Dieses Denkschema nimmt wenig Rücksicht auf die Fossilien, die Arealdynamik, die Klimadifferenzierung der geologischen Epochen, die an Zonalität und Ozeanität gebundenen Ausbreitungsmöglichkeiten und die unterschiedlichen Artbildungsprozesse. Es sollte daher in diesen "Fortschritten" vielleicht nicht erwähnt werden, es wird aber gegenwärtig heiß diskutiert (JEFFREY 1983; NORDAL 1985a über mehrere Symposien; HEADS 1985; WEBSTER 1987; HEADS and CRAW 1987) und von manchen Taxonomen für die Rekonstruktion der Chorogenese erprobt (z.B. ANDERBERG 1985: westmediterrane Inulinae; WESTON et al. 1984: Cladogramme von *Boronia* und *Boronella* als Beleg für das größere Alter der Abtrennung von Neukaledonien gegenüber der West-Ost-Gliederung von Australien). Andere Autoren verwenden vikarianzbiogeographische Analysen zum Nachweis ehemaliger Fernverbreitungsvorgänge (z.B. ROBSON 1987 bei *Hypericum* sect. *Brathys* in Südamerika) oder beurteilen die Ergebnisse der Areal-Cladogramme als uninformativ (HILL und CAMUS 1986 für die Marattiales, Hinweis auf Arealexpansion und -retraktion; LINDER 1983 für die Orchidaceae-Disinae in Afrika).

Fossilien können bei Rekonstruktionen der Chorogenese nicht nur zur Feststellung des Mindesalters der Sippe dienen, wie HUMPHRIES und PARENTI (1986) angeben. Bei der Analyse der europäischen Wasserpflanzen seit dem Tertiär zeigt MAI (1985) am Beispiel von *Trapa* und anderen Gattungen, daß Fossilien Evolutionswege erkennen lassen.

Mit welch großen Arealverschiebungen zu rechnen ist, geht aus dem tertiären Vorkommen von *Casuarina* in Israel und von Winteraceen oder Cupanieen in Nordamerika hervor (COETZEE und MULLER 1984). Den neotropischen *Buxus*-Sippen ähnliche Formen fanden sich im Tertiär Europas (KÖHLER 1985). *Trigonobalanus* ist jetzt auf Südostasien und Columbien beschränkt, war aber im Tertiär in der Holarktis verbreitet und kam wohl von Norden nach Südamerika (VAN DER HAMMEN und CLEEF 1983). Das tertiäre Areal der Cycadeen reichte von Alaska bis nach Argentinien (BALDUZZI et al. 1985). Selten läßt sich belegen, daß ein Relikt wirklich in seinem Areal überdauert hat wie im Fall von *Betula nana*, die als Postglazialrelikt im französischen Zentralmassiv seit dem Präboreal konstant nachzuweisen ist (VASARI und VASARI 1985). Die isolierten Vorkommen westarktischer Elemente in Skandinavien sind nach NORDAL (1985b) nicht auf Eiszeit-Überdauerung zurückzuführen. Auch die sudanischen Elemente in der Exklave am Toten Meer, die gewöhnlich als Miozänrelikte angesehen werden, sind wegen ihrer geringen taxonomischen Differenzierung und der Eignung zur Fernverbreitung zum großen Teil eher auf postglaziale Einwanderungen zurückzuführen (SHMIDA und ARONSON 1986). Daß ähnliche Disjunktionen auf ein ähnliches Alter hindeuten, lehnt TIFFNEY (1985) in einer Untersuchung der ostasiatisch-ostamerikanischen Disjunktionen ausdrücklich ab.

Geographische Ursprungszentren sind bei alten Sippen aus der rezenten Verbreitung kaum zu ermitteln. So ist die Hypothese, daß die Mangrove in Südostasien entstanden sei, kaum zu belegen (MEPHAM 1983). VAN DER WALT und VORSTER (1983) können den Ursprung von *Pelargonium* in Südafrika durch das Vorkommen der ursprünglichsten sect. *Eumorpha* im tropischen Ostafrika zwar in Frage stellen, aber nicht sicher in die Tropen verlegen. Ebenso wird durch die Zuordnung der nordaustralischen *Regelia punicea* zu der bisher ausschließlich südwestaustralischen Gattung *Regelia* nur die Unsicherheit von Hypothesen über die Herkunft demonstriert (BARLOW 1986).

Bei jungen Differenzierungen aber sind die paläogeographischen und paläoklimatologischen Verschiebungen gering. Hier können z.B. karyologische Daten durchaus Hinweise auf Herkunft und Ausbreitungsrichtungen geben. Neue Beispiele dafür sind *Centaurium pulchellum* (Diploide in Israel, ZELTNER 1985) und *Linum tenuifolium* (NICHOLLS 1986, eigenes iberisches Entfaltungszentrum durch Diploide belegt) oder einige Orophyten in Nordwestafrika (GALLAND und KÜPFER 1984). Auch bei den Kulturpflanzen ist die Ausbreitungsgeschichte anhand der abgestuften genetischen Vielfalt zu verfolgen. *Vicia faba* erreichte aus einem ostmediterranen Primärzentrum in einer ersten Ausbreitungswelle Abessinien, Mitteleuropa und Westasien, in einer 2. Phase (Mittelalter) China und in der Neuzeit Kalifornien, Südamerika und Südafrika (HAMMER et al. 1986).

Aus der Genomanalyse von *Ornithogalum* geht die Rolle der Hybridisierung hervor, die in Kontaktzonen nach dem Eisrückgang möglich war (RAAMSDONCK 1986). Auch bei den Pteridophyten Nordamerikas ist hybridogene Artbildung in einem Ausmaß verbreitet (LELLINGER 1985), das vikarianzbiogeographische Rekonstruktionen kaum zuläßt. Die unterschiedlichen Vorgänge bei der Ausbreitung und Speciation faßt TRYON (1986) in einer umfassenden Analyse der Biogeographie der Farne zusammen: evolutionäre Migration, regionale Isolation, periphere Divergenz, Langstrecken-Migration, Hybrid-Speciation und cytologische Speciation. Die Rolle dieser Speciationsmechanismen (und anderer, z.B. der adaptiven Radiation) in den verschiedenen Verwandtschaftskreisen wird in Zukunft aufgrund taxonomisch und chorologisch gut erforschter Beispiele noch viel zu diskutieren sein (Zusammenfassung des gegenwärtigen Standes: MOORE 1984). Bei den Gehölzen im tropischen Afrika ist nach WHITE (1984) die geographische Speciation bei weitem vorherrschend.

Als ökologisch-geographische Transgressoren bezeichnete WHITE (1983) solche Arten, die über das Areal des Verwandtschaftskreises weit hinausreichen. Es handelt sich auch in größeren Gattungen dabei stets nur um Einzelfälle. Bei *Chrysobalanus*, *Hirtella*, *Gyrocarpus* und anderen Gattungen ist ihre transatlantische Disjunktion wohl auf Fernverbreitung zurückzuführen. Argumente für die Fernverbreitung liefern aber vor allem die Floren landferner Inseln, die auch bei geringer Ausdehnung und geringem Alter von zahlreichen Diasporen erreicht werden. Die Entfernung von St. Helena betrug schon zur Zeit der Entstehung der Insel vor 14,5 Millionen Jahren etwa 2000 km von Afrika und 3000 km von Südamerika. Unter den 36 endemischen (von 45 heimischen) Arten haben die stärker isolierten Sippen heute eher Beziehungen zum humiden Südamerika, Indomalesien und Madagaskar. Sie konnten von Südwestafrika aus mit Südostwinden und dem Benguelastrom hierher gelangen, bevor dort der Regenwald verschwand. Auch die isolierten holzigen Asteraceen-Gattungen zeigen solche geographischen Beziehungen, sie werden daher von CRONK (1987) nicht als Ergebnis spezieller Evolutionsprozesse auf den Inseln, sondern als Relikt mit Beziehungen zum Regenwald aufgefaßt. Die weniger isolierten, offenbar jüngeren Endemiten St. Helenas stehen dagegen meist Elementen der ariden Küstengebiete Westafrikas nahe. Schließlich sind auch ökologische Lizenzen für die Wanderungsmöglichkeiten ausschlaggebend. Trotz der geologischen Geschichte ist Neuguinea zwar reich an australischen Gattungen, aber das malesische Element ist doch stärker (STEENIS 1985).

8. Synanthrope Areal- und Florenveränderungen

Eingeschleppte Pflanzen stellen heute in den Festlandfloren 2 - 25% des Artenbestandes, auf Inseln und in lokalen urbanen Floren bis über 50%. Die Zahl der ausgestorbenen Arten nimmt von Jahr zu Jahr geometrisch zu. Kein Wunder also, daß das Interesse an synanthropen Arealveränderungen in allen Kontinenten wächst.

a) Beurteilung des Status

Für die Gruppierung der Anthropophyten gibt es zwar schon viele Begriffssysteme (Vergleich bei ZIZKA in ZIZKA und FLECKENSTEIN 1985, der die SCHROEDERsche Terminologie als besonders logisch empfiehlt), aber in vielen Einzelfällen ist es schwierig, den Status der Arten (heimisch, alteingebürgert, nach 1500 neueingebürgert, ephemer usw.) zu bestimmen. WEBB (1985) zählt zu den Kriterien für die Beurteilung des Status Fossilfunde, Nachweisgeschichte, Standort, Areal, Neigung zur synanthropen Ausbreitung in anderen Gebieten, genetische Abweichungen, Reproduktionsbiologie und vermutliche Einschleppungsmittel. PARNELL (1986) nennt außerdem das Vorkommen oligophager Insekten (auch spezifische Bestäuber) und belegt damit, daß *Impatiens noli-tangere* in Südengland nicht heimisch ist. In England muß nach WEBB (1985) für eine große Zahl von angeblich heimischen Arten das Indigenat angezweifelt werden. Für Korsika geben GAMISANS et al. (1985) nur 4,4% eingeschleppte Arten an, unter den 2546 "heimischen" sind aber sicher sehr viele alteingebürgerte, in Kreta soll ihr Anteil 30% ausmachen! Außerhalb Europas gibt es über archäophytische Ausbreitungen höchstens vage Vermutungen. In Australien wurde bis 1840 kaum auf anthropochore Pflanzen geachtet (KLOOT 1985).

Für den Nachweis der frühen *Juglans*-Ausbreitung in Mitteleuropa kann ISENBERG (1986) Pollenfunde verwenden (schon in der vorrömischen Eisenzeit kultiviert, aber nicht synanthrop ausgebreitet), während KALIS (1984) für *Picea* in den Hochvogesen nur Alter und Einführungsdaten; POZGAJ (1986) für *Quercus frainetto* in der Slowakei Vergesellschaftung und Standortverhältnisse als Hinweis auf das Indigenat anführen können.

b) Ausbreitungsgeschichte von Neophyten

Für die Ausbreitungsgeschichte von Neophyten gibt es inzwischen schon Hunderte von gut datierten Einzeluntersuchungen, zuletzt z.B. von STUCKEY (1985) für *Najas marina* und *Epipactis helleborine* und von WIRDLECHNER (1983) für *Chaenorrhinum minus* in Nordamerika, von KÖCK (1986) für *Corispermum leptoterum* und von SCHULZ (1984) für *Galinsoga* in Europa. Wie schon vor 100 Jahren KERNER bemerkt hat, ist bei der Ausbreitung oft (aber nicht immer) eine Lag-Phase von einigen Jahrzehnten zu beobachten, in der die eingeschleppte Art konstant anwesend ist, ohne sich stark auszubreiten. Bei *Epilobium ciliatum* in Europa dauerte diese Phase etwa 40 Jahre (JÄGER 1986). Die Beschreibung der Ausbreitung in einem stochastischen Modell (MOLLISON in KORNBERG und WILLIAMSON 1986) kann solche Besonderheiten nicht erfassen.

Die Grenzen des kohärenten potentiellen Areals werden von den Anthropophyten in der Phase der explosiven Verbreitung gewöhnlich in 100 - 200 Jahren ausgefüllt (JÄGER 1986), isolierte Vorposten z.T. erst wesentlich später (z.B. *Juncus tenuis* erst 1978 im klimatisch isolierten, relativ ozeanischen Baikalsee-Gebiet, KOVTONJUK 1985). Interessant für die Beurteilung der Konkurrenz sind Fälle von "Neophyten-Zwillingen" wie *Corispermum marschallii* - *C. leptopterum*, *Azolla caroliniana* - *A. filiculoides*, *Galinsoga parviflora* - *G. ciliata*, *Bidens connata* - *B. frondosa*, *Cuscuta gronovii* - *C. lupuliformis*, *Elodea canadensis* - *E. nuttallii*, *Eragrostis minor* - *E. pilosa*, in denen eine eingeschleppte Art durch eine nahe verwandte, später eingeschleppte mehr oder weniger verdrängt wird (gute Analysen im Band 2 des Atlas van de Nederlandse Flora). Dabei sind die Erstnachweisdaten der ersten Art zuverlässiger, weil diese im Gegensatz zu der ähnlichen zweiten sogleich als fremd erkannt und gemeldet wird. Bei *Galinsoga ciliata* sind deshalb nachbestimmte Herbarbelege aus Europa z.T. mehrere Jahrzehnte älter als die ersten Fundmeldungen in der Literatur, bei der zuerst eingeschleppten *G. parviflora* stimmen beiden Datengruppen überein (SCHULZ 1984).

c) Prognose der Ausbreitung

Trotz der genauen Beobachtung der Arealexpansions- und -regressionsvorgänge gestaltet sich die Prognose künftiger anthropogener Arealveränderungen schwierig. Man sucht die Eignung der Pflanzen zur synanthropen Ausbreitung auf bestimmte Eigenschaften zurückzuführen (ökologische, phänotypische und genetische Plastizität, Anpassung an gestörte Standorte als r-Strategen usw., KORNBERG und WILLIAMSON 1986), aber die Nützlichkeit dieser Eigenschaften ist sehr von den wechselnden anthropogenen Einflüssen abhängig.

Einige Unkräuter sind heute schon großflächig vom Aussterben bedroht (für Skandinavien SVENSON und WIGREN 1986, Segetalunkräuter stärker als Dorfunkräuter gefährdet), andere Arten nehmen kompensatorisch ihren Platz ein. Gegenwärtig sind Pflanzen eutropher Standorte favorisiert (HILBIG und JAGE 1984, Auswertung wiederholter Kartierung der Häufigkeit) oder auch Halophyten, die sich an tausalzbehandelten Straßen ausbreiten (SCOTT 1985: 12 Arten in England, darunter der Pilz *Agaricus bernardii*). In welchem Maße die Grenzen des potentiellen Areals durch anthropogene Standortveränderung oder durch die anthropogen geförderte Sippendifferenzierung erweitert werden, läßt sich kaum vorhersehen. *Echinochloa crus-galli* konnte durch Verkürzung ihres Lebenszyklus ihr Areal in Amerika weit nach Norden ausdehnen (POTVIN 1986), ebenso *Raphanus raphanistrum* in Europa (PISTRICK 1986).

In den extrem veränderten Stadtbiotopen differenziert sich auch die Verbreitung der Ubiquisten stark (KUNICK 1987, vgl. S. 350). Sehr urbanophil sind in Zentraleuropa thermophile Arten wie *Buddleia davidii* und *Ailanthus glandulosa*, urbanophob dagegen z.B. die Gentianaceae und Cyperaceae (WITTIG et al. 1985).

d) Auswirkung der Ausbreitung von Neophyten

Nur wenige anthropochore Pflanzen dringen in die ungestörte natürliche Vegetation ein. Der Grad der Immunität der Vegetation ist von den anthropogenen Störungen, aber auch von der natürlichen Dynamik der Pflanzengesellschaften abhängig (SUKOPP und TREPL 1987). Daher sind Flußtäler bevorzugte Wege der Neophytenausbreitung, in Natal z.B. für *Chromolaena (Eupatorium) odorata* aus Amerika und *Melia azedarach* aus Asien (MACDONALD 1983). Bei dem heute erreichten Grad der anthropogenen Beeinflussung der Vegetation gibt es viele unerwartete und unerwünschte Auswirkungen der Ausbreitung einzelner Arten (KOWARIK und SUKOPP 1986), z.B. Veränderung naturnaher Wälder durch *Robinia pseudacacia*. Das kaukasische *Rhododendron ponticum* ist in den Wäldern Englands Ursache starker Artenverarmung (KORNBERG und WILLIAMSON 1986). *Impatiens parviflora* aus dem Tienschan war in Mitteleuropa lange auf anthropogene Standorte beschränkt, breitete sich aber in den letzten Jahren auch in kaum gestörten Laubmischwäldern aus (Statistik der Fundorts- und Standortsdaten: TREPL 1984). Durch anthropogene Standortveränderungen werden auch räumliche Kreuzungsbarrieren durchbrochen, und die Sippenbildung kann durch erneute Durchmischung gleichsam rückgängig gemacht werden (BROCHMANN 1987 für Kanarenpflanzen).

e) Artenverlust

Das schlimmste Ergebnis der anthropogenen Florenveränderung ist der Artenverlust. In einer Analyse von Lokalfloren in der Umgebung von Basel seit dem 17. Jahrhundert zeigt MEIER-KÜPFER (1985) das starke Anwachsen der Verluste in den letzten Jahrzehnten. Welche Arten gefährdet und schützenswert sind, ist nicht nur nach der Seltenheit zu beurteilen. *Puccinellia distans*, die sich nach 1974 stark ausbreitete, stand vorher im Rheinland schon als stark gefährdet im Roten Buch (REICHERT 1984).

Für den Artenschutz sind die "Roten Bücher" nur ein erster Schritt, auch wenn darin aufgrund der Verbreitung und Biologie der Arten Maßnahmen zur Erhaltung vorgeschlagen werden. Am effektivsten ist der Schutz ausreichend großer Populationen an den natürlichen Standorten. Von DIERSSEN und DIERSSEN (1984) wird deshalb für die Moorstandorte im Schwarzwald auch der Erhaltungszustand kartiert. Die Erhaltung eines repräsentativen Restes der Mannigfaltigkeit setzt die Einsicht voraus, daß die Erde nicht nur Produktionsmittel ist.

Literatur

AKE ASSI, L.: Flore de la Côte d'Ivoire, 1206 S. Abidjan: Universität 1984. - ALEKSEEV, E.B., KISELEVA, K.V., NOVIKOV, V.C., OKTJABREVA, N.B., TICHOMIROV, V.N., CICIEV, A.V.M.: Opredelitel' rastenij Meščery, Čast' 1, 240 S. Moskva: Universitet 1986. - ALEKSEEVA, A.M.: Flora ostrova Kunašir, 129 S. Vladivostok: Akad. Nauk SSSR 1983. - ANDERBERG, A.A.: Studies in the Inuleae-Inulinae (Compositae), 105 S. Stockholm: Universitet 1985. - Arktičeskaja Flora SSSR, IX/2, 188 S. Leningrad: Nauka 1986. - Atlas van de Nederlandse Flora. 2. Zeldzame en vrij zeldzame planten, 349 S. Utrecht: Bohn, Scheltema, Holkema 1985. - AXELROD, D.I., RAVEN, P.H.: J. Biogeogr. 12, 21-47 (1985).

BAJTENOV, M.S.: Vysokogornaja flora Severnogo Tjan-Šana, 230 S. Alma-Ata: Nauka 1985. - BALDUZZI, A., DE LUCA, P., SABATO, S.: Delpinoa N.S. 23-24, 185-202, "1981-1982" (1985). - BARKLEY, T.M. (ed.): Flora of the Great Plains, 1392 S. Lawrence: Univ. of Kansas Press 1986. - BARLOW, B.A.: Brunonia 9, 89-97 (1986). - BATANOUNY, K.H.: Ecology and Flora of Quatar, 245 S. Doha: Univ. of Quatar 1981. - BEADLE, N.C.W., BEADLE, L.D.: Students Flora of North Eastern New South Wales, Parts 1-5, 69, 198, 100, 142, 207 S. Armidale: Dept. of Botany, Univ. of New England 1971, 1972, 1986, 1980, 1984. - BEADLE, N.C.W., EVANS, O.D., CAROLIN, R.C., TINDALE, M.D.: Flora of the Sydney Region, 720 S. Wellington: Reed 1982. - BEAUCHAMP, E.M.: A Flora of San Diego County, California, 241 S. National City: Sweetwater River Press 1986. - BOND, P., GOLDBLATT, P.: Plants of the Cape Flora, 455 S. J. South Afr. Botany, Suppl. 13 (1984). - BORCHERS-KOLB, E.: Mitt. Bot. München 21, 49-300 (1985). - BOWERS, J.E.: Great Basin Naturalist 42, 105-112 (1982). - BOX, E.O.: Veröff. Geobot. Inst. Rübel Zürich 91, 181-216 (1986). - BROCHMANN, C.: Blyttia 45, 2-11 (1987). - BROWICZ, K.: Kosmos (PRL) 34, 427-439 (1985). - BROWN, J.H., GIBSON, A.C.: Biogeography, 643 S. St. Louis: Mosby 1983. - BRUGGEN, T. VAN: The Vascular Plants of South Dakota, 476 S. Iowa State Univ. Press 1985. - BRUNEL, J.F., HIEPKO, P., SCHOLZ, H.: Englera 4, 1-751 (1984). - BURTON, R.M.: Flora of the London Area, 225 S. London: Natur. Hist. Soc. 1983. - BUSBY, J.R.: Aust. J. Ecol. 11, 1-7 (1986).

CHARPIN, A., SALANON, R.: Boissiera 36, 1-258 (1985). - CHOCHRJAKOV, A.P.: Flora Magadanskoj oblasti, 396 S. Moskva: Nauka 1985. - Chorologija flory Latvijskoj SSR, 110 S. Riga: Zinatne 1986. - COETZEE, J.A., MULLER, J.: Ann. Missouri Bot. Garden 71, 1088-1099 (1984). - COILE, N.C., JONES, S.B.: The Georgia Plant List. Athens/Georgia: Univ. of Georgia, Dept. of Botany 1985. - CORILLON, R.: Flore et Vegetation de la Vallee de la Loire, 1 (Texte) 717 S. Paris: Jouve 1981. - CRAW, R.C., GIBBS, G.W.: Tuatara 27, 1-75 (1984). - CRESO, I.: Vascular Plants of Western Washington, 532 S. Tacoma/Washington: I. Creso 1984. - CRONK, Q.C.B.: New Phytologist 105, 509-520 (1987). - CRONQUIST, A., HOLMGREN, A.H., HOLMGREN, N.H., REVEAL, J.L., HOLMGREN, P.K.: Intermountain Flora, Vol. 4, 573 S. Bronx: N.Y. Bot. Garden 1984. - CUNNINGHAM, G.M., MULHAM, W.E., MILTHORPE, P.L., LEIGH, J.H.: Plants of Western New South Wales, 776 S. Sydney: Gvt. Printer "1981", 1982.

DAOUD, H.S.: Flora of Kuwait, Vol. 1, XI+224 S. London u.a.: KPI; Univ. of Kuwait 1985. - DASSANAYAKE, M.D., FOSBERG, F.R. (eds.): A Revised Handbook to the Flora of Ceylon, Vol. 4, 532 S.; Vol. 5, 476 S. Peradeniya: Univ. of Sri Lanka 1983, 1985. - DAVIS, M.B., WOODS, K.D., WEBB, S.L., FUTYMA, R.P.: Vegetatio 67, 93-103 (1986). - DAVIS, P.H. (ed.): Flora of Turkey and the East Aegean Islands, Vol. 9, 724 S. Edinburgh: Univ. Press 1985. - DEBNATH, H.S., NAYAR, M.P.: The Poppies of Indian Region (Papaveraceae), 192 S. Fl. of India Ser. 4. Howrah: Bot. Surv. India 1986. - DIERSSEN, B., DIERSSEN, K.: Beih. Veröff. Naturschutz Landschaftspflege Baden-Württ. 39, 1-512 (1984). - Distributiones plantarum africanarum 28 (Karte 925-956), 29 (Karte 957-986). Meise: Jard. Bot. Nation. Belgique 1986. - DODSON, C.H., GENTRY, A.H., VALVERDE, F.M.: Selbyana 8, XXIX+1-512 (1986). - DORN, R.D.: Vascular Plants of Montana. Cheyenne/Wyoming: Mountain West Publ. 1984. - DUNLOP, C.R., BOWMAN, D.M.J.S.: Atlas of the Vascular Plant Genera of the Northern Territory, 116 S., 979 Karten. Austr. Flora and Fauna Series 6, 1986.

ELLIS, R.G.: Flowering Plants of Wales, 338 S. Cardiff: Nation. Mus. Wales 1983. - ERSKINE, D.S.: The Plants of Prince Edward Island. Ottawa: Agricult. Canada Res. Branch 1985.

FEINBRUN-DOTHAN, N.: Flora Palaestina IV (Text und Tafeln) 462 S., 525 Tafeln. Jerusalem: Israel Acad. Sci. Human. 1986. - Flora Colombia, Vol. 1, 123 S.; 2, 83 S.; 3, 178 S.; 4, 118 S. Bogota: Univ. Nacional 1983, 1982, 1984, 1986. - Flora da Guine-Bissau, ed. Jardim e Museu Agricola do Ultramar. Fasc. 1-8, 328 S. Lisboa: Ministerio do Ultramar 1971-1984. - Flora de Moçambique. Fasc. 1-61. Lisboa: Junta de Invest. Cient. do Ulramar 1969-1981. - Flora de Veracruz. Fasc. 21-49. Xalapa: Instituto Nacion. Invest. Recurs. Biot. 1982-1986. - Flora Gruzii, Bd. 9, 343 S.; Bd. 10, 389 S. Tbilisi: Mecniereba 1984, 1985. - Flora Hebeiensis, Tomus 1, 831 S. Beijing: Academia Sinica 1986. - Flora Hupehensis, Tomus 1 und 2. Wuhan: Inst. of Botany 1978, 1979. - Flora Iberica, ed. S. CASTROVIEJO, Vol. 1, 575 S. Madrid. Real Jard. Bot. 1986. - Flora in desertis Reipublicae Populorum Sinicorum, Tomus 1, 546 S. Beijing: Academia Sinica 1985. - Flora Iranica, ed. K.H. RECHINGER, Lfg. 158, 234 S.; 159, 2 S.; 160, 11 S.; 161, 15 S. Graz: Akad. Druck- und Verlagsanstalt 1986. - Flora of Beijing, Vol. 1, 710 S. Beijing: Academia Sinica 1984. - Flora of India, Fasc. 13-17. Howrah: Botan. Garden 1983, 1984. - Flora of Pakistan, Fasc. 153-172. Islamabad: Pakistan Agricult. Res. Council 1984-1986. - Flora of Southern Africa, Vol. 33/7 (1983) 325 S.; 7/2 (1983) 76 S.; 21/1 (1984) 44 S.; 4/2 (1985) 96 S.; 14 (1985) 244 S.; 28 (1985) 247 S.; 31/1,2 (1986) 79 S.; Pteridophyta (1986) 292 S. Pretoria: Botan. Res. Institute. - Flora of Thailand, Vol. 4/1+2, 129+92 S. Bangkok: Applied Sci. Res. Corporation of Thailand 1984, 1985. - Flora of the Guianas. A. Phanerograms. Fam. 192-195, 73 S.; Fam. 85, 107 S. Königstein: Koeltz 1985, 1986. - Flora of Tropical East Africa, ed. R.M. POLHILL. Orchidaceae, Teil 2 (1984) 175 S.; Teil 3 (1986) 200 S.; 9 kleine Familien (1985) 301 S. Rotterdam: Balkema. - Flora plantarum herbacearum Chinae boreali-orientalis, Vol. 1 (1985) 75 S.; 2 (1959) 120 S.; 3 (1975) 242 S.; 4 (1980) 239 S. Beijing: Academia Sinica. - Flora Polski. Rośliny naczyniowe, Tom 4, 306 S. Warszawa, Krakow: Państ. Wyd. Nauk. 1985. - Flora Reipublicae Popularis Sinicae, Vol. 16/1 (1985) 213 S.; 37 (1985) 516 S.; 38 (1986) 167 S.; 47/1 (1985) 140 S.; 55/2 (1985) 282 S.; 74 (1985) 392 S. Beijing: Academia Sinica. - Flora Sichuanica, Vol. 1 (1981) 509 S.; 2 (1983) 250 S. Beijing: Academia Sinica. - Flóra Slovenska, Vol. 4/1+2, 763 S. Bratislava: Vyd. Slovensk. Akad. Vied. 1984, 1985. - Flora Tadžikskoj SSR, Vol. 7 (1984) 563 S.; 8 (1986) 519 S. Leningrad: Nauka. - Flora Tsinglingensis, Vol. 1/1-4, ca. 2200 S. (1974, 1984, 1981, 1983); 2 (1974) 246 S.; 3 (1978) 231 S. Beijing: Acadmia Sinica. - Flora Xizangica, Vol. 1 (1983) 791 S.; 2 (1985) 956 S. Beijing: Academia Sinica. - Flora Yunnanica, ed. Kunming Bot. Inst., Vol. 1 (1977) 870 S.; 2 (1979) 889 S.; 3 (1983) 795 S.; Index Florae Yunnanensis, 2 Vols. (1984) 2259 S. Beijing: Academia Sinica. - Flora Zambesiaca, Vol. 7/1+2 (1983, 1984) 502 S. London: British Mus. (Natur. Hist.). - Flore d'Afrique Centrale (Zaire- Rwanda-Burundi). Orchidaceae 1, 246 S. Bruxelles: Jard. Bot. Nation. Belg. 1984. - Flore d'Alsace, Plaine Rhénane, Vosges, Sundgau. 621 S. Strasbourg: Inst. de Botanique 1982. - Flore de la Nouvelle Calédonie et Dépendances, Vol. 10 (1982) 302 S., 11 (1982) 182 S.; 12 (1983) 123 S.; 13 (1984) 91 S. Paris. Mus. Nation. Hist. Natur. - Flore de Madagascar et des Comores. Fam. 84-87 (1982) 64 S.; Fam. 61 (1984) 10 S.; Fam. 167 (1984) 107 S.; Fam. 125 (1985) 53 S. Paris: Mus. Nation. Hist. Natur. - Flore des Mascareignes: La Réunion, Maurice, Rodrigues. Fam. 189 (Arecaceae) (1984) 34 S.; Fam. 161-169 (1985) 81 S. Mauritius: Sugar Ind. Res. Inst.; Paris: Off. Rech. Sci. et Techn. Outre-Mer. - Flore du Cambodge, du Laos et du Viêt-Nam, Vol. 21 (1985) 217 S.; Vol. 22 (1985) 72 S. Paris: Mus. Nation. Hist. Natur. - Flore du Cameroun, Vol. 21-28. Paris: Mus. Nation. Hist. Natur. 1980-1985. - Flore du Gabon, Vol. 25 (1983) 167 S.; 26 (1984) 276 S.; 27 (1985) 61 S. Paris: Mus. Nation. Hist. Natur. - FOSBERG, F.R., SACHET, M.-H., OLIVER, R.: Micronesia 18, 23-82 (1982). - FOSSI INNAMORATI, T.: Webbia 36, 273-411 (1983). - FRANKENBERG, P.,

DU, Z.: Abh. Akad. Wiss. Lit. Mainz, Mathem.-naturwiss. Kl. 1983, 72-92 (1983). - FRODIN, D.G.: Guide to the Standard Floras of the World, 619 S. Cambridge u.a.: Univ. Press 1984. - FUKAREK, F., HUSE, M.: Natur Naturschutz Mecklenburg 21, 3-59 (1985).

GALLAND, N., KÜPFER, P.: Webbia 38, 473-490 (1984). - GAMISANS, J., ABOUCAYA, A., ANTOINE, C., OLIVIER, L.: Candollea 40, 571-582 (1985). - GOLOSKOKOV, V.P.: Flora Džungarskogo Alatau, 221 S. Alma-Ata: Nauka 1984. - GOLTE, W.: Abh. Akad. Wiss. Lit. Mainz, Mathem.-naturwiss. Kl. 1983, 9-51 (1983). - GREEN, J.W.: Census of Vascular Plants of Western Australia, 312 S. South Perth: Western Austr. Dept. Agriculture 1985. - GREUTER, W., BURDET, H.M., LONG, G. (eds.): Med-Checklist, Vol. 3, Convolvulaceae-Labiatae, 395 S. Genève: Conserv. Jard. Bot. 1986. - GRIERSON, A.J.C., LONG, D.G.: Flora of Bhutan, Vol. 1/1+2, 462 S. Edinburgh: Royal Bot. Garden 1983, 1984. - GRIEVE, B.J., BLACKALL, W.E.: How to Know Western Australian Wildflowers, 525 S. Nedlands: Univ. of Western Austr. Press 1982.

HALL, J.B., SWAINE, M.D.: Distribution and Ecology of Vascular Plants in a Tropical Rain Forest - Forest Vegetation in Ghana, 383 S. The Hague: Junk 1981. - HÄMET-AHTI, L., SUOMINEN, J., ULVINEN, T., UOTILA, P., VUOKKO, S.: Retkeilykasvio, 544 S. Helsinki: Suomen Luonnonsuolejun Tuki Oy 1984. - HAMMEL, B.E.: Selbyanna 9, 149-155 (1986). - HAMMER, K., HANELT, P., LEHMANN, C.O.: Biol. Zentralbl. 105, 199-205 (1986). - HANSEN, A., SUNDING, P.: Sommerfeltia 1, 1-167 (1985). - HARA, H., CHATER, A.O., WILLIAMS, L.H.J.: An Enumeration of Flowering Plants of Nepal, Vol. 3, 226 S. London: Brit. Mus. Natur. Hist. 1982. - HARIDASAN, K., RAO, R.R.: Forest Flora of Meghalaya, Vol. 1 (1985) 450 S.; 2 (1987) 485 S. - HEADS, M.: Taxon 34, 474-492 (1985). - HEADS, M., CRAW, R.: Taxon 36, 305 (1987). - HEGI, G.: Illustrierte Flora von Mitteleuropa, Bd. I/3/3, 80 S. Berlin, Hamburg: Parey 1985. - HENGEVELD, R.: Blumea 30, 353-361 (1985). - HEYWOOD, V.H.: Designing Floras of the Future, S. 397-410. In: Current Concepts in Plant Taxonomy, eds. V.H. HEYWOOD, D.M. MOORE, 432 S. London: Academic Press 1984. - HILBIG, W., JAGE, H.: Acta Bot. Slov. 1984 A Suppl. 1, 61-73 (1984). - HILL, C.R., CAMUS, J.M.: Bull. Brit. Mus. (Natur. Hist.) Bot. Ser. 14, 219-300 (1986). - HOEK, C. VAN DEN: Helgoländ. Meeresuntersuch. 38, 227-257 (1984). - HOUGH, M.Y.: New Jersey Wild Plants, 414 S. Harmony, NJ: Harmony Press 1984. - HÜGIN, G.: Phytocoenologia 14, 289-379 (1986). - HULTÉN, E., FRIES, M.: Atlas of North European Vascular Plants North of the Tropic of Cancer, Vol. 1-3, zus. 1172 S. Königstein: Koeltz 1986. - HUMPHRIES, C.J., PARENTI, L.R.: Cladistic Biogeography, 98 S. Oxford: Clarendon Press 1986.

ISENBERG, E.: Abh. Westfäl. Mus. Naturkunde 48, 457-469 (1986). - IZARD, M., CASANOVA, H., DEVAU, B., PAUTOU, G.: Acta Oecol. Oecol. Gen. 6, 317-344 (1985).

JACKSON, P.W., SKEFFINGTON, M.S.: Flora of Inner Dublin, 174 S. Dublin: Thomas Prior House, ohne Jahr (1984?). - JACOBS, S.W.L., PICHARD, J.: Plants of New South Wales, 226 S. Sydney: N.S.W. Gvt. Printer 1981. - JAEGER, P., ADAM, J.-G.: Boissiera 32, 5-301 (1980), 33, 1-397 (1981). - JAFRI, S.M.H., EL-GADI, A.: Flora of Libya. Fasc. 117-133. Tripoli: Al Faateh University 1985, 1986. - JÄGER, E.J.: Wiss. Z. Univ. Halle, Mathem.-Naturwiss. 35, 122-134 (1986). - JALAS, J., SUOMINEN, J.: Atlas Florae Europeae, Vol. 7 (Silenoideae), 229 S. Helsinki: Committee Mapping Flora of Europe 1986. - JEFFREY, C.: Kew Bull. 38, 1-10 (1983). - JENÍK, J.: Preslia 58, 199-210 (1986). - JESSOP, J. (ed.): Flora of Central Australia, 537 S. Sydney: Reed 1981. - JESSOP, J.P., TOELKEN, H. (eds.): Flora of South Australia, Vol. 1-4, 2248 S. Plympton, S.A.: Govt. Printing Division 1986. -

JOHNSON, L.A.S. (ed.): Flora of New South Wales. 101: Fabaceae, Part 2, S. 93-178. Sydney: Royal Bot. Gardens 1984.

KALIS, A.J.: Revue Paleobiologique 1984, Vol. spec., 103-115 (1984). - KHAN, M.S.: Flora of Bangladesh, Fasc. 1-27. Dacca: Agricult. Res. Council 1972-1985. - KLOOT, P.M.: J. Adelaide Bot. Gard. 7, 145-157 (1985). - KNAPP, H.D.: Archiv Naturschutz Landschaftsforsch. Berlin 26, 65-89 (1986). - KÖCK, U.-V.: Gleditschia 14, 305-325 (1986). - KÖHLER, E.: Feddes Repert. 96, 663-675 (1985). - KOLAKOVSKIJ, A.A.: Flora Abchazii, Vol. 3 (1985) 292 S.; Vol. 4 (1986) 402 S. Tbilisi: Mecniereba. - KONDRATJUK, E.N., BUDRA, R.I.: Konspekt flory jugo-vostoka Ukrainy, 272 S. Kiev: Naukova Dumka 1985. - KORNAŚ, J., MEDWECKA-KORNAŚ, A.: Geografia Roślin, 528 S. Warszawa: Państw. Wyd. Naukowe 1986. - KORNBERG, H., WILLIAMSON, M.H.: Phil. Transact. R. Soc. London B 314, 501-742 (1986). - KÖRNER, C., MEUSEL, H.: Flora 178, 209-232 (1986). - KOVTONJUK, N.K.: Bot. Zh. 70, 1416-1417 (1985). - KOWARIK, J., SUKOPP, H.: Universitas 41, 828-845 (1986). - KOŽARINOV, A.V.: Klimatičeskaja determinacija severo-vostočnoj granicy graba obyknovennogo v Belorussii. In: Geografičeskie aspekty racional'nogo prirodopol'zovanija. Minsk: Nauka 1985. - Krasnaja kniga SSSR, 2. Bd., 487 S. Moskva: Lesnaja promyšlennost' 1985. - KRASNOBOROV, I.M., BARANOV, V.S., NIGMATULIN, R.S.: Bot. Zh. 71, 821-825 (1986). - KUNICK, W.: Wiss. Beitr. Martin-Luther-Univ. Halle-Wittenberg 1987/4 (P 26), 101-119 (1987). - KURATA, S., NAKAIKE, T.: Illustrations of Pteridophyta of Japan, Vol. 4, 850 S. Tokyo: Univ. of Tokyo Press 1985. - KUVAEV, V.B.: Osobennosti vysotnogo raspredelenija rastenij v uslovijach resko kontinental'nogo klimata, 83-95. In: Prostranstvennaja struktura ekosistem. Leningrad: Nauka 1982.

LANDOLT, E.: Veröff. Geobot. Inst. ETH Zürich 71, 1-566 (1986). - LANG, KAI-YUNG: Acta Phytotaxon. Sin. 23, 418-428 (1985). - LEBRUN, J.P., BOUDOURESQUE, E., DULIEU, D., GARBA, M., SAADOU, M., ROUSSEL, B.: Bull. Soc. Bot. France, Lett. Bot. 130, 249-256 (1983). - LELLINGER, D.B.: A Field Manual of the Ferns and Fern Allies of the United States and Canada, 389 S. Washington: Smithsonian Inst. Press 1985. - LID, J.: Norsk, svensk og finsk flora. (Ny utgave ved O. GJAEREVOLL), 837 S. Oslo: Norske samlaget 1985. - LINDBERG, P.S.: Stockholms Floran, 280 S. Stockholm: Norstedt 1983. - LINDER, H.P.: Bothalia 14, 565-570 (1983). - LIOGIER, H.A., MARTORELL, L.F.: Flora of Puerto Rico and Adjacent Islands: A Systematic Synopsis, 342 S. Rio Piedras/Puerto Rico: Universidad 1982.

MA, Y.C. (ed.): Flora Intramongolica, Vol. 1-8, 2667 S. Huhhot: Dept. of Biology, Univ. of Inner Mongolia 1978-1985. - MAAS, S.: Floristische Rasterkartierungen als Informationssystem für die ökologische Bewertung von Städten, 197 S. Saarbrücken: Universität 1984. - MACDONALD, I.A.W.: Bothalia 14, 949-959 (1983). - MAC ROBERTS, D.T.: The Vascular Plants of Louisiana, 165 S. Shreveport: Louisiana State Univ. 1984. - MAGILL, R.E., GIBBS RUSSELL, G.E., MORRIS, J.W., GONSALVES, P.: Bothalia 14, 481-495 (1983). - MAGUIRE, B., STEYERMARK, J.A., FRODIN, D.G.: Mem. N.Y. Bot. Garden 38, 1-82 (1984). - MAI, D.H.: Flora 176, 449-511 (1985). - MARTICORENA, C., QUEZADA, M.: Gayana 42, 1-157 (1985). - MARTINČIČ, A., SUŠNIK, F.: Mal Flora Slovenije, 793 S., Ljubljana: Državna Založba Slovenije 1984. - MATTHEW, K.M.: Materials for a Flora of the Tamilnadu Carnatic, Vol. 1, 2, 3/1, 2154 S. Tiruchirapalli: Rapinat Herbarium 1981, 1983. - MC VAUGH, R.: Flora Novo-Galiciana, Vol. 12 (Compositae, 1984) 1157 S.; Vol. 14 (Gramineae, 1983) 448 S.; Vol. 16 (Orchidaceae, 1985) 368 S. Ann Arbor: Univ. of Michigan Press. - MEDWECKA-KORNAŚ, A., KORNAŚ, J.: Zesz. Naukowe Uniw. Jagiellońsk., Prace Bot. 13, 7-15 (1985). - MEIER-KÜPFER, H.: Beitr. Geobot. Landesaufnahme Schweiz 62/1+2, 1-448 (1985). - MEPHAM, R.H.: South Afr. J. Bot. 2, 1-8 (1983). - MEUSEL, H., JÄGER, E., RAUSCHERT,

S., WEINERT, E.: Vergleichende Chorologie der zentraleuropäischen Flora, Bd. 2. Jena: Fischer 1978. (Bd. 3 im Druck). - MOORE, D.M.: Taxonomy and Geography, 219-234. In: Current Concepts in Plant Taxonomy, eds. V.H. HEYWOOD, D.M. MOORE. London: Academic Press 1984. - MOUTERDE, P.: Nouvelle Flore du Liban et de la Syrie, Vol. 3/1-4 und Atlas, 578 S., 352 Tafeln. Berlin-Dahlem: OPTIMA Secretariat; Beirut: Dar el-Machreq 1983, 1984.

NICHOLLS, M.S.: Plant Syst. Evol. 153, 243-258 (1986). - NORDAL, I.: Nordic J. Bot. 5, 14 (1985a). - NORDAL, I.: Blyttia 43, 185-193 (1985b).

OFIR, M.: Ann. Bot. 58, 81-89 (1986). - Opredelitel' rastenij Srednej Azii, Vol. 7, 413 S. Taškent: FAN 1983. - Opredelitel' rastenij Turkmenistana, Vol. 3, 202 S. Aŝchabad: Ylym 1985. - O.R.C.A. (Organització per a la Cartografia de les Plantes Vasculares dels Països Catalans): Corologia de la Flora dels Països Catalans, Vol. Introduct., 79 S., 26 Karten, 19 Folien-Karten. Barcelona: Inst. Estud. Catalans 1985.

PARNELL, J.A.N.: Watsonia 16, 83 (1986). - PISTRICK, K.: Systematische Untersuchungen in der Gattung *Raphanus*, 97 S. Dissert. Halle: Martin-Luther-Universität 1986. - POTVIN, C.: J. Ecol. 74, 915-923 (1986). - POZGAJ, J.: Preslia (Praha) 58, 211-221 (1986). - PROCTOR, G.R.: Flora of the Cayman Islands, 834 S. London: Majestys Station. Office 1984; - Ferns of Jamaica, 610 S. London: Brit. Mus. (Natur. Hist.) 1985.

RAAMSDONK, L.W.D. VAN: Nordic J. Bot. 6, 525-544 (1986). - REICHERT, H.: Mitt. Pollichia 71, 117-122 ("1983", 1984). - ROBSON, N.K.B.: Bull. Brit. Mus. Natur. Hist. (Bot.) 16, 1-106 (1987). - ROY, G.P.: Grasses of Madhya Pradesh, 180 S. (Flora of India Ser. 4). Allahabad: Bot. Survey of India 1984. - RZEDOWSKI, J., RZEDOWSKI, G.C.: Flora Fanerogamica del Valle de México, Vol. 2, 674 S. México: Compania Edit. Continental, S.A. 1985.

SALDANHA, C.J.: Flora of Karnataka, Vol. 1, 535 S. New Delhi: Pauls Press, Rotterdam: Balkema, Oxford: IBN Publ. 1984. - SCHULTE, W.: Florenanalyse und Raumbewertung im Bochumer Stadtbereich, 394 S. Bochum: Ruhruniversität, Geogr. Institut 1985. - SCHULZ, D.: Acta Bot. Slov. 1984 A Suppl. 1, 285-296 (1984). - SCHUMACKER, R. (ed.): Atlas de distribution des bryophytes de Belgique, du Grand-Duché de Luxembourg et des régions limitrophes, Vol. 1, Anthocerotae, Hepaticae, 61 S., 169 Karten. Meise: Jard. Bot. Nation. Belg. 1984. - SCOTT, N.E.: Watsonia 15, 381-386 (1985). - SERGIENKO, V.G.: Flora poluostrova Kanin, 147 S. Leningrad: Nauka 1986. - SHMIDA, A., ARONSON, J.A.: Ann. Missouri Bot. Garden 73, 1-28 (1986). - SLAVÍK, B.: Fytogeograficke syntezy ČSR, Vol. 1, 199 S. Pruhonice: Bot. Ustav ČSAV 1986. - SOKOLOV, S.JA., SVJAZEVA, O.A., KUBLI, V.A.: Arealy derev'ev i kustarnikov SSSR, Vol. 3, 181 S., 92 Kartenblätter. Leningrad: Nauka 1986. - SOPER, J.H., POWELL, J.M.: Botanical Studies in the Lake Hazen Region, Northern Ellesmere Island, Northwest Territories, Canada, 67 S. Ottawa: National Mus. Canada 1985. - Sosudistye rastenija Sovetskogo Dal'nego Vostoka, Vol. 1 (Lycopodiaceae, Juncaceae, Poaceae), 398 S. Leningrad: Nauka 1985. - SOWELL, J.B.: Vegetatio 60, 103-111 (1985). - SPICHINGER, R., BOQUET, G. (eds.): Flora del Paraguay. Fasc. 1-4, 36, 45, 137 S. Genève: Conservat. et Jard. Bot. 1983-1987. - STANLEY, T.D., ROSS, E.M.: Flora of South-Eastern Queensland, Vol. 1 (1983) 545 S.; Vol. 2 (1986) 623 S. Queensland Dept. of Primary Industries. - STEENIS, C.G.G.J. VAN: Brunonia 8, 349-372 (1985). - STEYERMARK, J.A.: Ann. Missouri Bot. Garden 71, 297-340 (1984). - STRID, A.: Mountain Flora of Greece, Vol. 1, 822 S. Cambridge: Cambridge Univ. Press 1986. - STUCKEY, R.L.: Bartonia 51, 2-16 (1985). - SUKOPP, H., TREPL, L.: Ecol. Stud. 61, 245-276 (1987). - SUOMINEN, J.: Memoranda Soc. Fauna Flora Fenn. 62,

79-82 (1986). - SVENSSON, R., WIGREN, M.: Symbolae Bot. Upsalienses 25/43-74 (1986).

THULIN, M.: Leguminosae of Ethiopia, 223 S. (Opera Botanica 68). Copenhagen 1983. - TIFFNEY, B.H.: J. Arnold Arboretum 66, 73-94 (1985). - TOLMAČEV, A.I.: Metody sravnitel'noj floristiki i problemy florogeneza, 195 S. Novosibirsk: Nauka 1986. - TREPL, L.: Über *Impatiens parviflora* DC. als Agriophyt in Mitteleuropa, 400 S. Vaduz: Cramer 1984. - TROUPIN, G.: Flore du Rwanda, Vol. 1 (1978) 413 S.; Vol. 2 (1983) 603 S.; Vol. 3 (1985) 729 S. Butare/Rwanda: Inst. Nation. Rech. Sci. - TRYON, R.: Bot. Rev. 52, 118-156 (1986).

UNE, K.: J. Hattori Bot. Lab. 59, 523-529 (1985).

VAN DER HAMMEN, T., CLEEF, A.M.: J. Biogeogr. 10, 437-440 (1983). - VAN DER WALT, J.A.A., VORSTER, P.J.: Bothalia 14, 517-523 (1983). - VAN ROYEN, P.: The Alpine Flora of New Guinea, Vol. 1-4, 3516 S. Vaduz: Cramer 1979, 1980, 1982, 1983. - VASARI, A., VASARI, V.: Ecol. Mediterr. 11, 65-68 (1985). - VOROŠILOV, V.N.: Spisok sosudistych rastenij Sovetskogo Dal'nego Vostoka, S. 139-200. In: Florističeskie issledovanija v raznych rajonach SSSR. Moskva: Nauka 1985. - VOSS, E.G.: Michigan Flora, Vol. 2, 744 S. Ann Arbor: Univ. of Michigan, Herbarium 1985.

WEBB, D.A., Watsonia 15, 231-236 (1985). - WEBB, T.: Vegetatio 67, 75-91 (1986). - WEBER, H.E.: Abh. Westfäl. Mus. Naturkunde Münster 47/3, 1-452 (1985). - WEBSTER, G.L.: Taxon 36, 306-307 (1987). - WELSH, S.L.: Great Basin Naturalist 42, 1-44 (1982); - ibid. 43, 179-357, 531-550 (1983); - ibid. 44, 52-69, 183-209 (1984); - ibid. 45, 155-172 (1985); - ibid. 46, 66-106 (1986). - WESTON, P.H., CAROLIN, R.C., ARMSTRONG, J.A.: Austr. J. Bot. 32, 187-203 (1984). - WHITE, F.: Bothalia 14, 395-403 (1983). - WIRDLECHNER, M.P.: Canad. J. Bot. 61, 179-187 (1983). - WITTIG, R., DIESING, D., GÖDDE, M.: Flora 177, 265-282 (1985). - WOODWARD, F.I.: Climate and Plant Distribution, 174 S. Cambridge: Cambridge Univ. Press 1987.

ZAVERUCHA, B.V.: Flora Volyno-Podolii i ee genezis, 190 S. Kiev: Naukova Dumka 1985. - ZELTNER, L.: Botanica Helvetica 95, 47-57 (1985). - ZIZKA, G., FLECKENSTEIN, S.: Botanische Untersuchungen in Nordnorwegen, 179 S. Vaduz: Cramer 1985.

Dr. ECKEHART J. JÄGER
Sektion Biowissenschaften der
Martin-Luther-Universität
Neuwerk 21
DDR-4020 Halle/Saale

II. The History of Flora and Vegetation During the Quaternary

By BURKHARD FRENZEL

1. Reconstruction of Past Types of Vegetation and of Climate by Means of Mathematical Procedures

The CLIMAP research group was the first to reconstruct past climates quantitatively by means of transfer functions from present-day distribution patterns of faunal communities, as far as these are primarily controlled by climate. This has also stimulated paleobotanical research. The basic concept is that the huge botanical formations, like tundra, the boreal forest, the nemoral forests or even smaller plant communities, are controlled by climate only in so far as a direct dependence of vegetation on climate exists. It is implicitly suggested that plant or animal communities may have changed their geographical ranges during the past, but not their composition. On the other hand, it is well known that migrations of plant and animal taxa during the Holocene have not terminated even up till now. The first prerequisite is thus not fulfilled, because evidently not all distribution areas are governed by climate only, but by a wealth of historical factors as well. Moreover, it becomes increasingly clear that the present-day plant and animal communities did not migrate as such during the Quaternary, without having undergone strong changes in composition. So it is in general not permissible to give fossil plant communities modern names, since these very often (and animal communities as well) have had no modern equivalents. From this it would follow that all attempts to reconstruct past climates quantitatively from fossil plant and animal assemblages in general meet with serious difficulties. To overcome these, much research work has been done. Although it must be admitted that the difficulties mentioned do exist, it may be questioned how long into the past present-day plant formations may be followed as a help in reconstructing paleoecological conditions. Here three possibilities are at hand: the interrelations between modern vegetation types, their recent pollenflora in surface samples, and the prevailing climatical conditions may be investigated, or it may be asked how the present-day pollen or macrofossil floras are formed, and finally special climate indicators, e.g., certain highly susceptible taxa or certain geological processes, may be used. Research has progressed along all of these three lines.

If annually layered sediments are used to reconstruct the development of past climates, a phenological calendar is necessary. I. BORTENSCHLAGER et al. (1984) have contributed much to solving this problem for the surroundings of Innsbruck, Austria. On the other hand, JANSSEN (1984) has shown for the Myrtle Lake Peatland, Minnesota, that the local vegetation finds its equivalents very well in the corresponding present-day pollen flora, the regional pollen influx being of only minor importance there. The opposite was shown to be true for the arid and semi-arid regions of Syria and Lebanon (BOTTEMA and BARKOUDAH 1979); and whereas

Progress in Botany, Vol. 49

the recent pollen flora seems to reflect the regional vegetation in the coastal mountains of British Columbia appreciably well, even here the pollen transport seems to tend upwards (McLENNAN and MATHEWES 1984). Above the upper tree limit of the Alps this situation changes drastically, and evidently pollen of only a few herbaceous plant taxa offer reliable means for reconstructing the former position of the alpine tree limit (JOCHIMSEN 1986). Here winds, time of flowering, rate of pollen production etc. seem to be of prime importance. An accumulation of sporomorphs in laminated lake sediments by melting glacier ice on which the sporomorphs had sedimented previously is discussed by LARSEN et al. (1984), and the contribution of wind activity and weathering to changes in fossil leaf assemblages is shown by FERGUSON (1985). Of course, not all sediment types are equally suitable for the preservation of pollen grains. This holds true most of all for some minerogenic sediments which may lead to misinterpretations of the interrelations between pollen rain, pollen assemblages in surface samples and the contributing vegetation (PONS 1984). GRONER (1985) points to serious difficulties, when sediments of partially water-filled karst caves are pollenanalytically investigated, if these caves are huge and if their caption area comprises various altitudinal vegetation belts. Under these conditions, it seems nearly impossible to reconstruct the prevailing vegetation or the stratigraphical position of the sediments under consideration. This is a warning against a too rapid interpretation of pollen floras in cave sediments. On the other hand, BASTIN and SCHNEIDER (1984) have shown that stalagmites can serve as good pollen traps of the former vegetation. However, the limitations of this approach are not well understood.

The other method mentioned is to investigate the interrelations between present-day zonal vegetation types and the concomitant pollen assemblages in surface samples. This was done from a general point of view by WEBB et al. (1978) and by OVERPECK et al. (1985) as to the sources of error and the most appropriate method of research. ANDREWS and NICHOLS (1981) have done the same, concentrating themselves on Central Northern Canada. Yet according to ANDREWS et al. (1980), various attempts to arrive at the same aim may cause serious discrepancies, as could be shown for the Eastern Canadian Arctic. On the other hand, ADAM and WEST (1983) report only the ratio between the pollen sum of *Quercus* to *Quercus* plus *Pinus* in evaluating past climates in California. It may be questioned whether this method is applicable under certain climatic conditions only, or whether it is an oversimplification. The values obtained by the authors seem to favor the second view.

As has been pointed out earlier, the evaluation of past climates and past vegetation types from modern pollen floras suffers from the fact that the plant communities have changed repeatedly during the past. So RITCHIE (1983) tries, as far as past climates are concerned, to compare the paleoclimate data obtained by these methods with those originating in the former distribution pattern of climate indicators. This is done very critically, and the differences between the results of both these methods are clearly shown.

To enable the investigations mentioned, a zonation of the pollen diagrams is necessary, if not enough ^{14}C-datations are ready at hand. D'ANTONI (1983) favors the r-mode cluster analysis, whereas JANSSEN (1980) indicates the necessity of choosing for each scientific aim the most appropriate zonation method, avoiding a standardization. I feel that he is right. Using statistical methods, the principal component analysis is often preferred. In general, it is only the first-order component which is taken into consideration. Yet RITCHIE and YARRANTON (1978) convincingly demonstrate that the second- or third-order component, from an ecological point of view, are repeatedly much more decisive.

Much work has been done toward better understanding of the interrelations between present-day vegetation types, the corresponding pollen floras, and climate. This has been done intensively by LAMB (1984) for Labrador, by P.A. DELCOURT et al. (1983) for the southeastern United States, by P.A. DELCOURT et al. (1984) for eastern North America, by BARTLEIN et al. (1984) for the northern Midwest of the USA, and by ANDREWS and DIAZ (1981) for Northern Canada. The investigations are based on very thorough investigations of the recent spontaneous vegetation, repeatedly using the forest descriptions at the very beginning of the European impact on the North American forests.

2. Dating Methods

A cornerstone for all the investigations mentioned is the reliable dating method of pollen samples. It is well known that the radiocarbon content of the atmosphere has changed repeatedly, so that the starting point for ^{14}C-datations, i.e., the assumption that the radiocarbon content of the atmosphere has always been constant, is not valid. The best means to test the magnitude of deviations from true ages by the ^{14}C-method is by analyzing the ^{14}C-ages of tree rings, the real ages of which are exactly known. According to LEUSCHNER and DELORME (1984), the Göttingen tree-ring sequence dates back now to the year 4008 B.C., and BITVINSKAS et al. (1978) report on the Lithuanian tree-ring calendar, up to about 5800 B.P.. Here pine and oak trees from peat bogs and river terraces were used, contributing a wealth of material for the periods 1950 to 2300 B.P., 4200 to 4800 B.P., and for 5300 to 5800 B.P. The intervening periods here seem to lack tree trunks nearly completely. It would be most promising to obtain more comparable material like this from various regions of Northern Eurasia, to learn something about the regional differences in river activity, and also in climate. TEGNER et al. (1977) report on the Swedish dendrochronology time scale which is much shorter than the others. Yet the most interesting results have been obtained by the ^{14}C-datation of south Central European oak and pine chronologies and of those of the famous bristlecone pine (*Pinus aristata* and *P. longaeva*) (VOGEL et al. 1986; STUIVER and BECKER 1986; DE JONG et al. 1986; KROMER et al. 1986; BECKER and KROMER 1986; STUIVER et al. 1986). It is very stimulating to learn that the ^{14}C-ages of South-German oaks do not follow the sinuoidal curve of ^{14}C-ages against real ages before 6200 B.C. as they have done up to this date (KROMER et al. 1986). This implies that the Younger Dryas to Preboreal transition did not happen as is generally accepted, at about 10,300 B.P., but at about 11,280 calendar years B.P. (BECKER and KROMER 1986; STUIVER et al. 1986). Much more time is thus available for biological processes than was thought hitherto. On the other hand, this means that the ^{14}C-content in the atmosphere must have been much higher at that time than today, and it might be that even strong fluctuations in ^{14}C-production had occurred. This seems to be corroborated by ANDREE et al. (1986), comparing the ^{14}C-ages of extremely small particles of macrofossils from terrestrial plants (0.5 mg to 3.0 mg per sample; accelerator mass spectrometry) with ^{14}C-ages of gyttja samples, in which the remains of the terrestrial plants mentioned were found. The landplants give in general ages approximately 800 years too young compared with the lake sediments, during late glacial times, and in Younger Dryas time much stronger deviations seem to have occurred, as has already been deduced by FRENZEL (1983). This points to the urgent necessity of extending the ^{14}C-calibration by the help of tree rings to late glacial times. To me it seems that this will not be easy for longer parts of the European late glacial, but the huge wood masses occurring in late glacial river terraces of Western and Central Siberia may become most interesting. Moreover, the ^{14}C-fluctuations occurred extremely abruptly, within 20 years or so (VOGEL et al. 1986). This must have been caused by abrupt changes in ^{14}C-production, not in the exchange with ^{14}C-sinks (cf. FRENZEL 1975), and it can be shown that a strong deviation of ^{14}C-ages against real ages occurred at about 9200 to 9800 B.P., at the very time when most important changes in the ecological conditions in terrestrial habitats took place (BECKER and KROMER 1986); so they cannot be dated exactly by the radiocarbon method.

HARBOTTLE et al. (1979) report on ^{14}C-dating facilities of extremely small samples, the real ages of which are lower than 2000 years; according to WIGLEY and MULLER (1981), the ^{14}C-/^{13}C-fractionation, caused by reservoir effects, is of only minor importance for the reliability of ^{14}C-datations, whereas STUCKENRATH et al. (1979) deal with some difficulties in choosing the most appropriate types of samples from the area of perennially frozen ground in northernmost Canada. In view of strong discrepancies in the ^{14}C-ages of various fractions of the same sample, it is recommended to use only the NaOH-insoluble material.

As to younger parts of the Quaternary, other methods than the ^{14}C-datation are repeatedly used. ANDERSON and ANDERSON (1981) discuss the importance of measuring the thickness of weathering rinds on quartzite clasts, and PORTER (1981) as well as CALKIN and ELLIS (1980) favor lichenometry. Yet HOLZHAUSER (1984) warns against this method in view of the fact that the growth rate of even the best-suited lichens is controlled by a wealth of ecological factors, climate being only one of them.

SZABO (1980) discusses problems encountered in the U/Th-method, applied to the datation of fossil bones, and LOBOV (1976), as well as HÜTT and RAUKAS (1977), warn against a too uncritically used thermoluminescence datation (TL, see also DĄBROWSKI et al. 1980). Electron spin resonance (ESR) was introduced as a dating method into Quaternary geology some years ago. IKEYA and MIKI (1980) discuss some of the principal problems of this datation method, whereas KATZENBERGER and GRÜN (1985), as well as KATZENBERGER and HAUSMANN (1986), report on some results obtained up till now. Yet here, as with other datation methods, the starting point for all attempts must be the exact knowledge of the geological setting of the sample to be examined. It may be questioned whether this prerequisite has been always fulfilled when interglacial samples of the Eurasian coasts of the Arctic Ocean have been dated by this method (LINKE et al. 1985).

3. The Problem of the Last but One Interglacial

It is in general held that the last two interglacials in Europe were the Eemian (the younger) and the Holsteinian (the older), respectively. It is further suggested that the Holsteinian (i.e., in Russia the Likhvinian) has an age of approximately 325,000 to 340,000 years B.P. (KAPLIN 1976), or of 370,000 to 400,000 years B.P. (GLAZEK et al. 1980; STREMME 1985). Yet the question is whether this sequence is correct or whether there was still another interglacial period intercalated (see Fortschritte der Botanik 44, 406, 1982).

The famous interglacial site of Hamarnia in the foothill region of the northeastern Carpathian Mts. is held to be characterized by a strong share of *Fagus*. Yet according to new investigations by ŚRODOŃ (1984), this is not true. On the contrary, it can be shown that the former determination of certain pollen grains as being those of *Fagus* was wrong, and that this interglacial of until now unknown stratigraphical position should at best be looked upon as of Holsteinian age (= Mazovian age). Very rich macrofossil floras of approximately the same age are described by KRUTOUS (1979) for the Pripjat region in Byelorussia (as to human remains in France, see BILLY and VALLOIS 1977). On the other hand, DE BEAULIEU and MONJUVENT (1985) describe the evolution of an interglacial vegetation at Pompillon, Val de Lans en Vercors, which strongly resembles that of other sites of allegedly Holsteinian age, yet differs clearly in its vegetation sequence from that of an interglacial in the vicinity of Bern, Switzerland, which was held by WELTEN (1982) to be of Holsteinian age, by counting the interglacials within this geological profile from top to bottom. So the authors think that the interglacial at Pompillon and that from the neighborhood of Bern (Meikirch), described by WELTEN, differ in their ages. Those interglacials which were here claimed to be of Holsteinian age, seem to be characterized on the northern and western flanks of the Alps by a strong share of *Fagus* and of *Pterocarya*. Yet this does not hold true for the interglacial flora from Stuttgart-Bad Cannstatt (ADAM 1986ab; REIFF 1986; WAGNER 1984, 1986) which is in general also held to be of Holsteinian age. Its botanical macrofossil flora is appreciably modern (ADAM 1986a), lacking *Fagus*, and presumably also lacking *Pterocarya*. This site has given U/Th-ages of approximately 200,000 or 250,000 years (WAGNER 1984, 1986), and it is correlated, due to the rich assemblages of various artifacts with the site of Bilzingsleben, Thuringia, which has yielded U/Th-ages of 228,000 (+ 17,000/- 12,000) years (GLAZEK et al. 1980). Comparable data have been reported for an interglacial period in the southern part of Western Siberia (CEJTLIN 1982). Provided these dates are correct, the time of formation of this interglacial sequence should correspond to deep-sea stage 7, so it should be situated between the Holsteinian proper and the Eemian, provided the Holsteinian is older than deep-sea stage 7. At Bilzingsleben, also the rich plant and animal remains have no exotic character, and taxa typical for the Holsteinian interglacial are lacking. As to the geological, paleontological, and paleobotanical situation at Bilzingsleben, see the monograph by MANIA et al. (1980). In Marsworth, UK, also a sequence of two interglacial periods has been described, following each other, and which resemble each other in their evolution of vegetation and in their fossil faunas, yet which must be of different ages, the older one having given U/Th-ages of approximately 150,000 to 170,000 years (GREEN et al. 1984). Here, too, the flora of both these interglacials is modern, lacking taxa characteristic for the Holsteinian, and it is thought that the older interglacial period is situated between the classical Holsteinian and the Eemian proper. KOSMOWSKA-SUFFCZYNSKA and SZCZEPANEK (1981) describe from the Sandomierz-area, Central Poland, an interglacial from Karsy, the vegetation history of which had an intermediate position between that of the classical Holsteinian and of the Eemian. It is connected with the Lublin interglacial which is thought to have an intermediate stratigraphical position between the Holsteinian and the Eemian interglacials. The fossil-bearing sediment at Karsy is situated on top of material, the TL-dates of which point to ages of 188,000 to 190,000 B.P. RADTKE et al. (1982) and HENNIG et al. (1983) describe marine and river terraces, respectively, which in the Mediterranean area or in the Hungarian Plain gave the same U/Th- and ESR-ages, and FEYLING-HANSEN (1982) describes foraminifera faunas from the North Atlantic (61°12,4' N, 1°50' E) which should date from an interglacial period which was clearly older than the Eemian, but which was not as temperate as that of the Holsteinian proper, and which is believed to have had an intermediate stratigraphical position. On the other hand, LINKE et al. (1985) state hat the typical Holsteinian sequence from Cuxhaven and Hamburg (Northern Germany) have given ESR-ages of about 195,000 to 223,000 years B.P. This would fit into the span of time just mentioned for the interglacial period between the classical Holsteinian and the Eemian. So the discrepancies between so-called absolute datings and the geological situations are readily visible. Compare in this respect TUFFREAU et al. (1982), who show for Northwestern France, Somme valley, that here, too, possibly more interglacial periods occurred than was hitherto thought. On the other hand, STEPHAN (1983) doubts the reliability of older descriptions of the type section of the so-called Odintsovo interglacial in Western Russia which is believed to have occurred between the Holsteinian and the Eemian interglacials, respectively (see Fortschritte der Botanik 44, 406, 1982), since it seems to be possible, from a geotechnical point of view, that the sediments there have been distorted. For contemporaneous tundra fauna and forest tundra vegetation on the river Pechora, Komi ASSR, USSR, see AGADZHANJAN and ISAJCEV (1976).

The age problems discussed here are not only worth mentioning from a stratigraphical point of view, all the more since this, of course, influences our understanding of climate history. On the other hand, it might prove to become a crucial problem to differentiate exactly between two succeeding interglacial periods, the ages of which might differ from one another by about 80,000 to 100,000 years, yet whose vegetation histories strongly resembled each other. If this should be correct, it would mean that the geographical setting of the glacial refugia and the immigration facilities during the preceding glacial periods must also have resembled each other. This contradicts the view generally held in Europe that to the north of the Alps the interglacial periods of different ages can be reliably differentiated from one another by their modes of development of the climax vegetation; yet this principle has already been broken by the group of the Cromer interglacials.

4. The Eemian Interglacial

The general opinion is that deep-sea stage 5e is the equivalent of the Eemian interglacial, yet the validity of this correlation could up till now not be proven. So it is greatly appreciated that TURON (1984) has isotopically and palynologically shown that both these phases correspond to each other exactly (deep sea boring, 42° 06′ N, 09° 41′ W). Evidently the accumulation of cold abyssal waters had begun there already during the *Carpinus* phase of the Western Europe vegetation history. This would mean that the beginning of an accumulation of inland ice in Greenland and in Antarctica had happened at the same time as the sea level reached its maximal position (ZAGWIJN 1983) and when warm and moist conditions prevailed in Central Europe (MENKE 1981; KONECKA-BETLEY and MARUSZCZAK 1976). Synchroneously the decalcification of the interglacial soils had finished in Northern Germany and the formation of oligotrophic soils begun (MENKE 1981). The formation of inland ice must have occurred appreciably rapidly, since the sea level had dropped already during the ensuing pine-spruce-phase of the Central European Eemian sequence by about 7 m (ZAGWIJN 1983). It is suggested that at that time no icecaps existed in Scandinavia (TURON 1984). According to TL-dates, the Eemian transgression in Fjøsanger, near Bergen, Norway, had an age of about 105,100 to 119,000 years (HÜTT et al. 1983). This also fits very well into the picture of the deep sea stage 5e being the equivalent of the Eemian interglacial. Marine molluscan faunas on Jameson Land, East Greenland, enable the suggestion that during full-interglacial times the water temperature there was appreciably higher than at present, roughly resembling that of recent southernmost Greenland (PETERSEN 1982). Comparable observations were made in the Sarsbukta, Spitzbergen (FEYLING-HANSSEN and ULLEBERG 1984). These warm climates are also proven by the vast distribution pattern of *Emys orbicularis* at that time, having then been a common inhabitant of Great Britain

(STUART 1979). It is very important that former reports on the occurrence of Eemian marine sediments in the lower course of the Weichsel river, Northern Poland, can now be corroborated (MAKOWSKA 1982; DROZDOWSKI 1986). Evidently the Eemian Transgression which had reached south as far as Graudenz (Grudziądz) was preceded here at the end of the Saalian glaciation by the "Stuhm-Transgression" and was followed at the very beginning of the last glaciation by the "Mniszek-Transgression" (MAKOWSKA 1982). The timing and distribution pattern of marine Eemian sediments and ingressions into Schleswig-Holstein are thoroughly investigated by MENKE and TYNNI (1984).

The synchroneous vegetation history of North America seems to be a matter of debate. WATTS (1971) has dealt with these problems intensively, as far as Southern Georgia and Central Florida are concerned. It is true that in the southeastern USA last-interglacial sediments are undoubtedly rare. On the other hand, P.A. DELCOURT (1980) and H.R. DELCOURT et al. (1983) have described some sites from Alabama which seem to prove that during full glacial times of the last glaciation an exacting broadleaved and mixed forest vegetation had thrived, rich in thermophilous and hygrophilous plant taxa. It is thought that there must have existed at that time a very steep climatological gradient, leading from nearly subarctic conditions immediately to the south of the inland ice to warm-temperate conditions in Alabama. It is strange to note that the pollen flora mentioned seems to occur in minerogenic sediments only, its pollen content being very low. This changes drastically at the sharp transition to the overlying organogenic sediments. I suggest that the pollen flora mentioned in reality dates from interglacial times which fits much better into the picture of extremely cold climate conditions of full glacial times in the southern Appalachian Mts., described by P.A. DELCOURT (1985) and P.A. and H.R. DELCOURT (1985). The same view is held by WATTS (1983). The problem seems to be a crucial one, because, if the view held by the DELCOURTs should be right, this would mean an extremely strange climatological situation during full glacial times and, on the other hand, the existence of large refuge areas for the warmth- and moisture-loving taxa. Yet, if it should prove to be right that the exacting vegetation mentioned was of interglacial age, it should at first be suggested that it dated from the last interglacial and it would, moreover, indicate climates which are well known for this very interglacial from various parts of the northern hemisphere, yet, on the other hand, any indications as to full glacial refugia of thermophilous and hygrophilous taxa would be lacking. Here, much more research work must be done.

Several years ago, WOILLARD (1978) described in the southwestern Vosges Mts., France, a remarkable sequence of events of approximately Eemian age. Here it could be shown that the Eemian proper was followed by two warm-climate periods of interglacial rank, St. Germain I and St. Germain II. This has caused some discussion, because during St. Germain I and II climate must have been very warm and appreciably moist and mild. This strongly contrasts with other sites in which this interglacial seems to have been followed by two or three interstadial phases with only subalpine coniferous forests or even with forest steppes in Central Europe (for this discussion, see MENKE and TYNNI 1984). So observations by DE BEAULIEU and REILLE (1984a,b) and DE BEAULIEU et al. (1980) from the site of Les Echets near Lyon, France, are welcome. They are exactly comparable with those made by WOILLARD in the Vosges Mts. just mentioned, and comparable observations were also made by BASTIN and SCHNEIDER (1984) on stalagmites of the Belgian karstic area. Moreover, it is interesting to note that CEJTLIN (1982) reported on the fossil soils of the last interglacial in northern Siberia, the Kazancev interglacial. They seem to consist repeatedly of two fossil soils which are developed one upon the other. They are held to date from about 110,000 to 140,000 years B.P., so it might be that in reality St. Germain I and II do not correspond to the interstadial periods just mentioned which had caused the spread of subalpine forests only. Yet here, too, much more research work is needed. For the Eemian history of the northern North Sea basin: see KNUDSEN (1984, 1985). The herbivore faunas of some parts of the interglacial and the beginning of the ensuing last glaciation are dealt with for a former volcano in the Mainz-Koblenz area, Western Germany, by BOSINSKI et al. (1983).

According to MARTINEZ-TUDELA et al. (1986), at about 111,000 B.P. there had thrived in Northeastern Spain (Rio Matarraña, Beceite) a thermophilous and hygrophilous vegetation in which *Adiantum capillus-veneris*, *Quercus faginea*, *Acer opalus*, *Buxus sempervirens*, *Fraxinus excelsior*, *Fr. angustifolia*, *Vitis vinifera* ssp. *silvestris* and others were characteristic elements. It is held that the climate must have been cooler and moister than at present, yet it may be questioned whether this interpretation is correct. At any rate, the macrofossil flora seems to be the first one for northeastern Spain to date from the last interglacial. In the same paper, observations on the floras of travertines dating from 267,000 ± 32,000 B.P. are reported. The Eemian interglacial from Nakło (Nakel) in northern Poland: NORYŚKIEWICZ (1979). Presumably last-interglacial plant remains from Mühlhausen, Thuringia, DDR, are described by PATZELT (1985). The geological situation of the last interglacial from Zeifen, southern Bavaria: MÜLLER and ZIEGLER (1978). Synchroneous human settlements in the Levant, Tabun Cave: JELINEK (1982).

5. Paleoecological Problems of the Last Glaciation

It has already been stressed that two young interglacials may have existed in Europe, the vegetation history of which had strongly resembled each other. So methods to differentiate safely between these interglacials rather than relying only on vegetation history are needed. According to DĄBROWSKI et al. (1980), paleomagnetism may help, since both the Eemian interglacial and that at about 260,000 to 290,000 B.P. were succeeded by paleomagnetic reversals, yet the directions of magnetization were opposite to one another. So a reliable starting point for paleoecological investigations of the ensuing last glaciation can be obtained. This is important, because the question as to the number and timing of inland ice advances during the last glaciation of Europe is being discussed anew. From several regions of northern Europe, besides the well-known glaciation after about 22,000 B.P., one or even two much older Weichselian glaciations have been reported, dating already from its beginning (Denmark: PETERSEN 1984, 1985; Southern Norway: VORREN 1977; VORREN and ROALDSET 1977; lower course of the river Weichsel: DROZDOWSKI and FEDORWICZ 1987; French Alps: HANNSS 1985; BRUN and HANNSS 1987; North Sea: SISSONS 1981). This corroborates earlier observations on anaglacial ice advances of the last glaciation in North America and in Northern Siberia, yet causes some problems as to the glacial and vegetation history of Europe during the last glaciation, provided the observed ice advances are dated correctly. The general paleogeographic problems of these ice advances are discussed for the Northern Hemisphere by KIND (1982). The glaciation history of the Late Weichselian ice advances is discussed for Western Scotland by BOULTON et al. (1981), for Finland and Scotland by HARKNESS (1981), for the former Rhine glacier area by KELLER and KRAYSS (1987), and for Austria by VAN HUSEN (1981/1982). VORREN (1982) reports on most interesting observations on the full glacial vegetation (about 18,000 to 10,000 B.P.) on Andöya in the Troms area, northern Norway. Here, a vegetation mostly dominated by grasses and *Draba*, with only a small admixture of *Ranunculus* and *Artemisia*, had developed. *Koenigia* seems to have been lacking here, only immigrating during the Alleröd interstadial of the late glacial.

It is well known that for vast areas of the Northern Hemisphere loess steppes were most characteristic during full glacial times. So investigations on the paleoecological situation of loess formation in Poland are welcome, dealing with problems of the prevailing wind directions, soil formation, mineralogical and trophic situation of the loesses etc. (general problems: MARUSZCZAK and MORAWSKI 1976; RÓŻYCKI 1976; Upper Silesia: RACZKOWSKI 1976; Holy Cross Mts.: LINDNER 1976; CHLEBOWSKI and LINDNER 1976; Lublin area: MARUSZCZAK and RACINOWSKI 1976; UZIAK and POMIAN 1976; RACINOWSKI 1976; HARASIMIUK and HENKIEL 1976; the north German lowlands: MERKT 1968).

The synchroneous paleoecological situation in the northeastern part of the Great Hungarian Plain is intensively discussed by BORSY et al. (1982). These extreme ecological conditions must have threatened man and the herbivore faunas considerably, so a summary by SACHSE-KOZŁOWSKA (1978) of relevant Polish findings is welcome. The evolution of man during the last glaciation in Europe: PROTSCH and SEMMEL (1978). Interstadial and full glacial faunas of Europe are dealt with intensively by BOCHEŃSKI (1981: bird fauna of Poland), CZYZEWSKA and USNARSKA (1980: reindeer finds near Kielce, Poland, seasonal distribution pattern and age classes), KUBIAK (1980a,b: mammoth in Poland). Evidently, the mammoth had survived in France, Switzerland and Poland up to the beginning of the late glacial. It is thought that its disappearance from there was caused by a complex of adverse events, like changes of climate, vegetation, hunting and perhaps also by degeneration (KUBIAK 1980a).

An interesting interstadial vegetation of unknown age is reported by SOBOLEWSKA (1976) from Southern Poland. Here during this period only some trees seem to have survived within the prevailing herb steppe, like pine, a few examples of spruce, *Betula* sect. *Albae*, *Pinus cembra*, *Larix*, *Abies* and some thermophilous broadleaved species. Comparable observations were made by WEGMÜLLER (1984a) for the Swiss Jura Mts., where during an interstadial period or at the very end of the last interglacial, a very poor coniferous vegetation seems to have thrived, even this being completely lacking during the Oldest Dryas of the late glacial. This contrasts strongly with results obtained by LEROI-GOURHAN (1984a,b) in Western Europe and in France. Here it is held that thermophilous trees had survived in sheltered sites along the river valleys during full glacial times. The same is discussed very cautiously by BASTIN and SCHNEIDER (1984) for pollen floras of full glacial stalagmites in France and Belgium. Yet in general these cave sediments suffer from redeposition of older pollen floras (see GRONER 1985), so it might be questioned whether the problem of full glacial refuge areas can be reliably investigated by pollen analysis of cave sediments.

The history of distribution areas of certain plant taxa is dealt with repeatedly. According to FOLLIERI et al. (1986), *Zelkova* was one of the most characteristic elements of the interglacial and interstadial forest floras of Central Italy up to the end of the early Weichselian period. *Buxus sempervirens* has evidently survived the adverse climatic conditions of the last glaciation in northeastern Spain, Corsica, in the vicinity of the Southern Alps and in northwestern Greece, and it has spread since then rapidly and discontinuously during the first part of the Holocene, even reaching Central England (WEGMÜLLER 1984b). According to ŚRODOŃ (1985), *Fagus silvatica* has never existed in Poland since the end of the Tertiary. This is corroborated by reinvestigations of Pleistocene interglacial sites from which *Fagus* pollen has previously been reported. It is thought that beech was strongly favored by the neolithic forest clearances, although it had already spontaneously immigrated into Poland. One wonders how the strange distribution pattern of beech during the Holsteinian and the Eemian interglacials in Central Europe might be explained, if it should be correct that all previous reports of *Fagus* pollen grains in Pleistocene sediments of Eastern Central Europe must be explained by redeposition, since *Fagus* occurred during these interglacials in certain phases of the interglacial vegetation history only, i.e., during the dominance of shade-loving plants. *Cryptomeria japonica* had survived full glacial conditions of the last glaciation in various sheltered sites of Japan, as was shown by TSUKADA (1982), constructing for this most interesting tree species an isochrone map of distribution since the beginning of the late glacial. WEBB (1987) constructed pollen maps since 18,000 B.P. (at intervals of 2000 years) for *Artemisia*, Cyperaceae, *Picea*, *Betula*, *Alnus*, *Abies* and *Pinus* for the eastern part of North America, and according to HEBDA and MATHEWES (1984), the immigration of *Thuja plicata* can be traced now from its glacial refuge areas, to the south of the Vancouver-Seattle area, to its modern sites, here favoring the evolution of native cultures. The competition between the immigrating *Pinus strobus* and oak trees in Minnesota during the Holocene is analyzed by JACOBSON (1979). The history of the Central European distribution and immigration of *Juglans regia*: ISENBERG (1986). Here the question of introduction by the Romans versus spontaneous, much older immigration to Central Europe is discussed intensively.

The full glacial vegetation history of northwestern Alaska is dealt with by ANDERSON (1985). It is held that pollen influx and sedimentation of minerogenic material into the lake basin studied resembled that of present-day shrub tundra communities and it is thought that the former and the recent vegetation there resembled each other, which may be doubted. As in Eastern Siberia, the amelioration of climate had begun weakly at about 14,000 B.P., enabling the immigration of birch, willow and poplar. The climatic optimum seems to have happened there at about 11,000 to 8,000 B.P., so that spruce could immigrate slowly from east to west. JANSSENS and BARNOSKY (1985) and BARNOSKY (1985) investigate the vegetation history, beginning with full glacial times of the last glaciation in southwestern Washington. Here not only the development of the moss communities is dealt with (JANSSENS and BARNOSKY 1985), but the history of the altitudinal belts and of vegetation zones as well (BARNOSKY 1985). It is stressed that some last glacial vegetation types cannot be compared with living ones.

The Sierra Nevada region, California, and its surroundings are in general held to be important last glacial refuge areas of exacting plant and animal species. According to COLE (1983), at that time the chaparral communities were lacking there, and *Sequoia* seems also to have migrated. LANNER and DEVENDER (1981) describe the Holocene formation of the present-day desert scrub communities of the Chihuahua Desert, replacing there the full glacial *Pinus remota* and *P. edulis* stands. *P. cembroides* seems to have immigrated only after the last glaciation. On the other hand, it can be demonstrated by DEVENDER (1985) that the Bolson de Mapimi, Chihuahua, was characterized in its lowlands by the present-day xeric scrub communities even during full glacial times. Both these papers cited are most interesting, thanks to the rich macrofossil floras from *Neotoma* middens. Higher up in the mountains (Boulder County, Colorado), the full glacial herb vegetation had found no modern equivalents, consisting, e.g., of *Artemisia*, *Juniperus*, Chenopodiaceae/Amaranthaceae and *Koenigia islandia*. In this paper the vertical migrations and the changes in composition of the vegetation belts, beginning with full glacial times, are discussed (LEGG and BAKER 1980). As in the Chihuahua Desert, the full glacial vegetation of the Cuatro Ciénagas Basin, Coahuila, Mexico, to some extent resembled that of today, so that the increase of moisture cannot have been strong there during full glacial times.

Whereas the vegetation of the full glacial loess steppes is appreciably well known in Northern Eurasia, this does not hold true for North America, so reports by BAKER et al. (1986a,b) on the geobotanical conditions in Western Illinois and in Iowa during full glacial times are welcome. It can be shown that the loess areas of Western Illinois were nearly devoid of vegetation at that time. The problems of the full glacial vegetation in the southeastern and eastern parts of North America have already been mentioned. Here it is a question of whether thermophilous and hygrophilous forest communities might have existed or not at that time in the southeastern USA. More information is found in H.R. DELCOURT (1985), P.A. DELCOURT (1985), P.A. DELCOURT and H.R. DELCOURT (1984, 1985), WATTS (1970, 1980, 1983), together with its paleoclimatological implications.

The vegetation history of presumably some parts of the last glaciation in the eastern Pamir Mts., Central Asia, is discussed by ZHEZHEL (1980). During the stadials, an astonishingly species-rich herb-steppe vegetation seems to have occurred there. In this investigation, various species of *Ephedra* and of the Chenopodiaceae are differentiated. I feel that this might be overemphasized as to the Chenopodiaceae. During interstadial times, some stands of pine, spruce and birch seem to have spread within the otherwise dominating *Artemisia*-Chenopodiaceae steppe. The observations of ZHAO et al. (1984) report on the full glacial to Holocene vegetation history of the Beijing Plain, Northern China. Here from 23,000 to 15,000 B.P. a steppe or desert-steppe vegetation had spread, with only rare stands of coniferous trees. This vegetation had followed the previously dominating pine and fir forests, and it was replaced from about 15,000 B.P. by forests composed of *Picea*, *Larix*, together with some *Tilia*, *Betula* and *Quercus*. The synchroneous vegetation of Taiwan was described by TSUKADA (1967). Simultaneously the southwest monsoon in the northern part of the Indian Ocean seems to have been appreciably weaker than at present, so that less humidity was transported onto the continent from the ocean. On the other hand, the northeast monsoon may have been stronger than today (DUPLESSY 1982). At the same time, Tasmania was clad by a tundra-like vegetation which was replaced from about 15,000 B.P. by forests. Here man had immigrated at about 23,000 B.P., when the island was connected with the Australian continent by lowering of the sea level (LEWIN 1982). During long periods of the last glaciation, the Galápagos Islands suffered from extreme drought which had already begun before 40,000 B.P., and it was only after 10,000 to 8,600 B.P. that the modern fern vegetation reached its present-day position in the higher mountains there (COLINVAUX and SCHOFIELD 1976). A wealth of information on the history of climate, vegetation and paleoecology in the northern part of Mali, Western Africa, is reported by PETIT-MAIRE (1986). Here the paleoecological evolution is followed from full glacial and even early last glacial times up to the Holocene. This is corroborated to some extent by DEVINEAU (1984), focusing on modern distribution patterns of various vegetation types on the Côte d'Ivoire, West Africa, which cannot be explained by modern climates or by the activity of man. This paper contains very interesting results as to the influence of grazing and browsing of the spontaneous herbivores on the vegetation.

6. The Late Glacial of the Last Glaciation

The Polish late glacial has been investigated intensively. It can be shown that the then existing vegetation of northern Poland was much richer in present-day tundra plants (LATAŁOWA 1982; ZACHOWICZ et al. 1982; PAWLIKOWSKI et al. 1982; NORYŚKIEWICZ 1982) than that in the southeastern or southern parts of Poland (LATAŁOWA 1976; BAŁAGA 1982). On the other hand, in several of these papers, which deal with the Holocene vegetation history as well, including the impact of man, the role of *Populus* cf. *tremula* and even of *Larix* in wide areas of Poland becomes evident. At that time there must have existed a strange mixture of northern oceanic and of continental plant communities in the eastern part of Central Europe. According to BOCHEŃSKI (1981), this vegetation seems to have been very appropriate for an extremely rich bird fauna which was much richer in various bird species than the Holocene vegetation. It is well known that the volcanic ash layer of the Laacher See eruption during the second part of the Alleröd interstadial is a good marker for studies in the late glacial vegetation history of the western part of Europe. According to MANGERUD et al. (1984), this was paralleled by a volcanic ash layer, dating from Younger Dryas time in Western Norway and in the adjacent Norwegian Sea. The vegetation belts, at that time moving rapidly northward, nourished even in the northern part of Central Europe a very rich and diverse herbivorous and carnivorous fauna which is dealt with by AARIS-SØRMØEN and PETERSEN (1984) for Denmark. The development of the late glacial glaciation on the Isle of Skye and the synchroneous vegetation history there are discussed by WALTHER (1984). For paleoecological problems of the late glacial in Schleswig-Holstein, Northern Germany, including the "Meiendorf-Intervall" before the Bölling-Alleröd complex, see BOKELMANN et al. (1983) and BOCK et al. (1985). LEROI-GOURHAN (1984c) discusses plant geographical problems of the Alleröd interstadial in Western Germany. It is thought that the prehistoric sites of Gönnersdorf and Andernach (just to the north of Koblenz) date from exactly the same time, although the vegetation of Gönnersdorf was characterized by *Pinus* stands, whereas that of Andernach was governed by birch, nearly completely lacking in *Pinus*. It is held that this was caused by local climatic conditions and it is emphasized that the zonation of pollen diagrams of the late glacial, without taking the local situations duly into consideration, might be misleading. This holds true most of all for the Alps, as can be shown by SEIWALD (1980) for the surroundings of Brixen, Northern Italy, BORTENSCHLAGER (1984) for the Lanser See near Innsbruck, BURGA (1983) for Graubünden, Switzerland, LANG (1984) for the Simplon area, Central Switzerland, and FURRER et al. (1984) for Graubünden. The regions studied nearly always became free of ice some time before 14,000. The same seems to hold true for the surroundings of St.-Laurent-du-Pont, Isère, French Alps (BIJU-DUVAL and HANNSS 1985), whereas it is held that the Swiss Plateau, some 15 km to the northwest of Bern, may have become free of ice already at about 16,000 B.P. (AMMANN and TOBOLSKI 1983). Comparable dates were reported by FELBER (1981) for Carinthia, Austria (in this paper a wealth of dates about the Holocene immigration of some forest tree species can be found). It seems to become clear now that at that time larch was of significance only in the southern and eastern Alps (e.g., for the Brixen area: SEIWALD 1980), yet that this tree species was lacking at that time in the Central Swiss area (cf. BURGA 1983; FURRER et al. 1984). Here, deciduous trees like birch, alder, maple (?) and poplar seem to have been much more important. As was observed earlier, the so-called Older Dryas deterioration of climate can in general no longer be detected in modern pollen diagrams, e.g., FURRER et al. (1984), BORTENSCHLAGER (1984), BURGA (1983), SEIWALD (1980). The same holds true for the Swiss Plateau near Bern (Lobsigensee, AMMANN and TOBOLSKI 1983; AMMANN et al. 1983, 1984). It is therefore very unlikely that certain late glacial readvances of glaciers should have happened during the Older Dryas time. This was discussed by FURRER et al. (1984). The strong occurrence of *Betula nana* at certain sites of the Swiss Plateau during the Ia to Ib transition, i.e., at the very beginning of the Bölling-Alleröd complex, is discussed by GAILLARD (1983), indicating its ability to act as a pioneer species, yet it seems that the dwarf birch, at least today, cannot be classified as a pioneer plant. The amelioration of climate seems to have been considerable (AMMANN et al. 1983). Simultaneously, the melting of ground ice in Siberia had begun, forming the sediments of the so-called Edom series, with its strong paludification (KAPLINA and LOZHKIN 1979). This amelioration of climate could also be felt in northwestern-most Canada (RITCHIE 1985), yet here it seems to have begun much more slowly than was deduced from the formation of a mire vegetation on permafrost in northeastern-most Siberia (KAPLINA and LOZHKIN 1979). A thorough investigation of the influx rates of pollen grains into the former lake basin investigated in northwestern-most Canada reveals that the strongest amelioration of climate had occurred there at about 10,000 B.P. (see also ANDERSON 1985). This was strongly camouflaged by immigration history and by competition between the various plant species studied (RITCHIE 1985). The same difficulties in interpreting the reasons for the immigration history of various plant species and in the formation of certain vegetation types are met with by MATHEWES (1985) in Southern British Columbia. In this paper most interesting observations on the Holocene differentiation of various vegetation types at the forest to prairie transition can be found. This holds true for the late glacial and Holocene history of vegetation and of climate in the mountainous regions of the Western United States, too (BAKER 1983), where strong emphasis is laid on the regional evolution patterns of vegetation in this ecologically very complicated area. The late glacial and Holocene vegetation and climate history of the Lake Valencia Basin, Venezuela, is thoroughly analyzed by LEYDEN (1985). Here the development from the late glacial arid conditions to the Holocene subtropical forest vegetation is worth mentioning. It can be clearly shown that this area cannot have served as a glacial refuge area for thermophilous and hygrophilous species during the last glaciation. The obvervations are interpreted on the background of the late glacial to Holocene climate history of the northern part of South America.

References

AARIS-SØRMOEN, K., PETERSEN, K.S.: Boreas 13, 29-33 (1984). - ADAM, D.P., WEST, G.J.: Science 219, 168-170 (1983). - ADAM, K.D.: Fundber. Baden-Württ. 11, 25-61 (1986a). - ADAM, K.D.: Fundber. Baden-Württ. 11, 92-100 (1986b). - AGADZHANJAN, A.K., ISAJCEV, K.J.: 278-299. In: Problemy obshchej fizicheskoj geografii i paleogeografii, eds. A.K. AGADZHANJAN, O.P. DOBRODEEV. Moskva: Univ. 1976. - AMMANN, B., TOBOLSKI, K.: Rev. de Paléobiol. 2, 163-180 (1983). - AMMANN, B., CHAIX, L., EICHER, U., ELIAS, S.A., GAILLARD, M.J., HOFMANN, W., SIEGENTHALER, U., TOBOLSKI, K., WILKINSON, B.: Rev. de Paléobiol. 2, 221-227 (1983). - AMMANN, B., CHAIX, L., EICHER, U., ELIAS, S.A., GAILLARD, M.J., HOFMANN, W., SIEGENTHALER, U., TOBOLSKI, K., WILKINSON, B.: 69-73. In: Climatic Changes on a Yearly to Millenial Basis, eds. N.A. MÖRNER, W. KARLÉN. Amsterdam: Reidel 1984. - ANDERSON, L.W., ANDERSON, D.S.: Arctic Alp. Res. 13, 25-31 (1981). - ANDERSON, P.M.: Quat. Res. 24, 307-321 (1985). - ANDREE, M., OESCHGER, H., SIEGENTHALER, U., RIESEN, T., MOELL, M., AMMANN, B., TOBOLSKI, K.: Radiocarbon 28, 2A, 411-416 (1986). - ANDREWS, J.T., DIAZ, H.F.: Quat. Res. 16, 373-389 (1981). - ANDREWS, J.T., NICHOLS, H.: Arctic Alp. Res. 13, 387-408 (1981). - ANDREWS, J.T., MODE, W.N., DAVIS, P.T.: Arctic Alp. Res. 12, 41-64 (1980). - ANTONI, H.L. DE: Revista Acad. Colomb. de Ciencias Exactas, Fis., Nat. 15, Nr. 58, 83-86 (1983).

BAKER, R.G.: 109-127. In: Late-quaternary environments of the United States, 2, The Holocene, ed. H.E. WRIGHT. Univ. of Minnesota 1983. - BAKER, R.G., BETTIS, E.A., RHODOS, R.S., HALLBERG, G.R.: Amer. Quat. Assoc. Ninth Bienn. Meet., 1 (1986a). - BAKER, R.G., SULLIVAN, A.E., HALLBERG, G.R., HORTON, D.G., CARTER, K.Z., SCHWERT, D.P.: Amer. Quat. Assoc. Ninth Bienn. Meet., 2 (1986b). - BAŁAGA, K.: Acta Palaeobot.

XXII, 7-22 (1982). - BARNOSKY, C.W.: Quat. Res. 23, 109-122 (1985). - BARTLEIN, P.J., WEBB, T., FLERI, E.: Quatern. Res. 22, 361-374 (1984). - BASTIN, B., SCHNEIDER, A.-M.: Kölner Geogr. Arb. 45, 87-93 (1984). - BEAULIEU, J.-L. DE, MONJUVENT, G.: Bull. Assoc. Franç. pour l'étude du Quatern., 75-83 (1985). - BEAULIEU, J.-L. DE, REILLE, M.: Boreas 13, 111-132 (1984a). - BEAULIEU, J.-L. DE, REILLE, M.: Géogr. phys. et quatern. 38, 3-9 (1984b). - BEAULIEU, J.-L. DE, EVIN, J., MANDIER, P., MONTJUVENT, G., REILLE, M.: Mém. du Mus. Nat. d'Hist. Nat., sér. B, Bot. 27, 123-136 (1980). - BECKER, B., KROMER, B.: Radiocarbon 28, 2B, 961-967 (1986). - BIJU-DUVAL, J., HANNSS, CH.: Rev. Géogr. Alp. LXXIII, 313-320 (1985). - BILLY, G., VALLOIS, H.V.: L'Anthropol. 81, 411-458 (1977). - BITVINSKAS, T., DERGACHEV, V., DAUKANTAS, A., LIJVA, A., SUURMAN, S., SHULIJA, K.: 51-55. In: Uslovija sredy i radialnyj prirost derevev, ed. T.T. BITVINSKAS. Kaunas: Inst. Bot. Akad. Nauk Litovskoj SSR 1978. - BOCHEŃSKI, Z.: Folia Quatern. 54, 1-24 (1981). - BOCK, W., MENKE, B., STREHL, E., ZIEMUS, H.: Eiszeitalter Gegenwart 35, 161-180 (1985). - BOKELMANN, K., HEINRICH, D., MENKE, B.: Offa 40, 199-239 (1983). - BORSY, Z., CSONGOR, E., SZABO, J.: Quatern. Stud. Hungary, 193-208 (1982). - BORTENSCHLAGER, I., BORTENSCHLAGER, S., FRANK, A.: Ber. nat.-med. Ver. Innsbruck 71, 213-243 (1984). - BORTENSCHLAGER, S.: Diss. Bot. 72, 71-79, Festschr. WELTEN (1984). - BOSINSKI, G., KULEMEYER, J., TURNER, E.: Archäol. Korrespondenzbl. 13, 415-428 (1983). - BOTTEMA, S., BARKOUDAH, Y.: Pollen et Spores 21, 427-480 (1979). - BOULTON, G.S., CHROSTON, P.N., JARVIS, J.: Boreas 10, 39-51 (1981). - BRUN, A., HANNSS, CH.: N. Jb. Geol. Paläontol. Mh 1987, 129-145 (1987). - BURGA, C.A.: Jber. Natf. Ges. Graubünden 100, 135-149 (1983).

CALKIN, P.E., ELLIS, J.M.: Arctic Alp. Res. 12, 245-264 (1980). - CEJTLIN, S.M.: 150-158. In: Problemy geologii i istorii chetvertichnogo perioda (antropogena), eds. M.N. ALEKSEEV, S.M. CEJTLIN. Moskva: Nauka 1982. - CHLEBOWSKI, R., LINDNER, L.: Z badań czwartorzędu w Polsce 18, 293-305 (1976). - COLE, K.: Quat. Res. 19, 117-129 (1983). - COLINVAUX, P., SCHOFIELD, E.K.: J. Ecol. 64, 1013-1028 (1976). - CZYZEWSKA, T., USNARSKA, K.: Folia Quatern. 51, 17-30 (1980).

DĄBROWSKI, A., TKACZ, M., TUCHOŁKA, P.: Quatern. Stud. in Poland 2, 7-12 (1980). - DELCOURT, H.R.: Ecol. Medit. 11, 9-16 (1985). - DELCOURT, H.R., DELCOURT, P.A., SPIKER, E.C.: Ecology 64, 874-887 (1983). - DELCOURT, P.A.: Ecology 61, 371-386 (1980). - DELCOURT, P.A.: Ecol. Medit. 11, 17-26 (1985). - DELCOURT, P.A., DELCOURT, H.R.: Palaeogeogr., Palaeoclim., Palaeoecol. 48, 263-284 (1984). - DELCOURT, P.A., DELCOURT, H.R.: 191-220. In: Field Trips in the Southern Appalachians, Field Trips 1-5, 7, ed. N.B. WOODWARD. Knoxville: Southeaster Sect. Geol. Soc. Amer. 1985. - DELCOURT, P.A., DELCOURT, H.R., DAVIDSON, J.L.: Rev. Palaeobot. Palynol. 39, 1-45 (1983). - DELCOURT, P.A., DELCOURT, H.R., WEBB, TH.: AASP Contrib. Series Nr. 14, Amer. Assoc. Stratigr. Palynol. Foundation, 131 (1984). - DEVENDER, T.L.: Quat. Res. 24, 346-353 (1985). - DEVINEAU, J.L.: Structure et Dynamique de quelques forêts tropophiles de l'ouest africain (Côte d'Ivoire). Trav. des Chercheurs de la Stat. de Lamto (Côte d'Ivoire) Nr. 5, p. 295. Univ. d'Abidjan 1984. - DROZDOWSKI, E.: Acta Universitatis Nicolai Copernici, Geografia XXI, 135-141 (1986). - DROZDOWSKI, E., FEDORWICZ, S.: Boreas 16, 139-153 (1987). - DUPLESSY, J.C.: Nature 295, 494-498 (1982).

FELBER, H.: Radiocarbon 23, 322-326 (1981). - FERGUSON, D.K.: Rev. Palaeobot. Palynol. 46, 117-188 (1985). - FEYLING-HANSSEN, R.W.: Bull. geol. Soc. Denmark 31, 29-47 (1982). - FEYLING-HANSSEN, R.W., ULLEBERG, K.: Polar Res. 2, n.s. 77-106 (1984). - FOLLIERI, M., MAGRI, D., SADORI, L.: New Phytol. 103, 269-273 (1986). - FRENZEL, B.: Proc. WMO/IAMAP Sympos. Long-Term Climat. Fluct. WMO No. 421 (104-118 (1975). - FRENZEL, B.: 91-166. In: Urgeschichte in Baden-Württemberg, ed. H.J. MÜLLER-BECK.

Stuttgart: Theiss 1983. - FURRER, G., MAISCH, M., BURGA, C.: Diss. Bot. 72, 87-115, Festschr. WELTEN (1984).

GAILLARD, M.J.: Rev. de Paléobiol. 2, 181-188 (1983). - GŁAZEK, J., HARMON, R.S., NOWAK, K.: Acta Geol. Polon. 30, Nr. 1, 1-14 (1980). - GREEN, C.P., COOPE, G.R., CURRANT, A.P., HOLYOAK, D.T., IVANOVICH, M., JONES, R.L., KEEN, D.H., McGREGOR, D.F.M., ROBINSON, J.E.: Nature 309, 778-781 (1984). - GRONER, U.K.: Palynologie der Karsthöhlensedimente im Hölloch, Zentralschweiz, p. 172. Diss. Phil. Fak. II, Zürich (1985).

HANNSS, CH.: 157-163. In: ESR-Dating and Dosimetry, eds. M. IKEYA, T. MIKI. Tokyo: Ionics 1985. - HARASIMIUK, M., HENKIEL, A.: Z badań czwartorzędu w Polsce 18, 177-184 (1976). - HARBOTTLE, G., SAYRE, E.V., STOENNER, R.W.: Science 206, 683-685 (1979). - HARKNESS, D.D.: Radiocarbon 23, 252-304 (1981). - HEBDA, J.R., MATHEWES, R.W.: Science 225, 711-713 (1984). - HENNIG, G.J., GRÜN, R., BRUNNACKER, K., PÉCSI, M.: Eiszeitalter Gegenwart 33, 9-19 (1983). - HOLZHAUSER, H.: Zur Geschichte der Aletschgletscher und des Fieschergletschers. p. 449. Diss. Phil. Fak. II, Zürich (1984). - HÜTT, G.I., RAUKAS, A.V.: Bjull. Kom. po izuch. chetvert. per. 47, 77-83 (1977). - HÜTT, G., PUNNING, J.M., MANGERUD, J.: Boreas 12, 227-231 (1983). - HUSEN, D. VAN: Mitt. österr. geol. Ges. 74/75, 197-230 (1981/82).

IKEYA, M., MIKI, T.: Science 207, 977-979 (1980). - ISENBERG, E.: Abhandl. Landesmus. Naturkde. Münster 48, H. 2/3, 457-469 (1986).

JACOBSON, G.L.: J. Ecol. 67, 697-726 (1979). - JANSSEN, C.R.: Bull. Assoc. franç. pour l'étude du Quatern. 1980, 171-176 (1980). - JANSSEN, C.R.: Ecological Monographs 54, 2, 213-252 (1984). - JANSSENS, I.A., BARNOSKY, C.W.: Rev. Palaeobot. Palynol. 46, 97-116 (1985). - JELINEK, A.J.: Science 216, 1369-1375 (1982). - JOCHIMSEN, M.: Zum Problem des Pollenfluges in den Hochalpen. Diss. Bot. 90, p. 241 (1986). - JONG, A.F.M. DE, BECKER, B., MOOK, W.G.: Radiocarbon 28, 2B, 939-941 (1986).

KAPLIN, P.A.: 150-162. In: Problemy obshchej fizicheskoj geografii i paleogeografii, eds. A.K. AGADZHANJAN, O.P. DOBRODEEV. Moskva: Univ. 1976. - KAPLINA, T.N., LOZHKIN, A.V.: Izv. Akad. Nauk SSSR, ser. geol. 1979, Nr. 2, 69-76 (1979). - KATZENBERGER, E., HAUSMANN, R.: Abstr. INQUA Subcomm. Europ. Quatern. Stratigr. Sympos. Hamburg, p. 13, 1986. - KATZENBERGER, O., GRÜN, R.: Nucl. Tracks 10, 885-890 (1985). - KELLER, O., KRAYSS, E.: Geogr. Helvet. 1987, Nr. 2, 169-178 (1987). - KIND, N.V.: 213-226. In: Problemy geologii i istorii chetvertichnogo perioda (antropogena), eds. M.N. ALEKSEEV, S.M. CEJTLIN. Moskva: Nauka 1982. - KNUDSEN, K.L.: Bull. geol. soc. Denmark 32, 169-180 (1984). - KNUDSEN, K.L.: Boreas 14, 311-324 (1985). - KONECKA-BETLEY, K., MARUSZCZAK, H.: Z badań czwartorzędu w Polsce 18, 185-209 (1976). - KOSMOWSKA-SUFFCZYNSKA, D., SZCZEPANEK, K.: Folia Quatern. 54, 25-41 (1981). - KROMER, B., RHEIN, M., BRUNS, M., SCHOCH-FISCHER, H., MÜNNICH, K.O., STUIVER, M., BECKER, B.: Radiocarbon 28, 2B, 954-960 (1986). - KRUTOUS, E.A.: 132-144. In: Sovetskaja paleokarpologija (itogi i perspektivy), eds. G.J. GORECKIJ, V.P. GRICHUK. Moskva: Nauka 1979. - KUBIAK, H.: Folia Quatern. 51, 31-45 (1980a); - ibid. 47-56 (1980b).

LAMB, H.F.: J. Ecol. 72, 37-59 (1984). - LANG, G.: Verhandlg. Ges. f. Ökol. 12, 383-390 (1984). - LANNER, R.M., DEVENDER, TH.R. VAN: Quat. Res. 15, 278-290 (1981). - LARSEN, E., EIDE, F., LONGVA, O., MANGERUD, J.: Arctic Alp. Res. 16, 137-160 (1984). - LATAŁOWA, M.: Acta Palaeobot. XVII, 55-80 (1976); - ibid. XXII, 47-63 (1982). - LEGG, TH.E., BAKER, R.G.: Arctic Alp. Res. 12, 319-333 (1980). - LEROI-GOURHAN, A.: Histoire et Archéologie - Les Dossiers, Nr. 87, 20-24 (1984a). - LEROI-GOURHAN, A.: Gallia Préhistoire XIXe suppl., 111-116 (1984b). - LEROI-

GOURHAN, A.: Rev. de Paléobiol., vol. spéc., 117-118 (1984c). - LEUSCHNER, H.-H., DELORME, A.: Forstarchiv 55, 1-4 (1984). - LEWIN, R.: Science 218, 988 (1982). - LEYDEN, B.W.: Ecology 66, 1279-1295 (1985). - LINDNER, L.: Z badań czwartorzędu w Polsce 18, 307-316 (1976). - LINKE, G., KATZENBERGER, O., GRÜN, R.: Quatern. Science Rev. 4, 319-331 (1985). - LOBOV, E.S.: 258-263. In: Opredelenie absoljutnogo vozrasta rudnykh mestorozhdenij i molodykh magmaticheskikh porod, eds. G.D. AFANASJEV, A.P. VINOGRADOV, L.N. OVCHINNIKOV, E.K. GERLING, T.B. PEKARSKAJA. Moskva: Nauka 1976.

MAKOWSKA, A.: Biul. Inst. Geologiczn. 343, Geology of Poland 5, 31-49 (1982). - MANGERUD, J., LIE, S.E., FURNES, H., KRISTIANSEN, I.L., LØMO, L.: Quat. Res. 21, 85-104 (1984). - MANIA, D., TOEPFER, V., VLČEK, E.: Bilzingsleben I, Homo erectus - seine Kultur und seine Umwelt. Veröff. Landesmus. f. Vorgesch. Halle 32, p. 176. Berlin 1980. - MARTINEZ-TUDELA, A., CUENCA, R.F., SANTISTEBAN, B.C., GRÜN, R., HENTZSCH, B.: Quaternary Climate in Western Mediterranean; Proc. Symp. Climatic Fluct. during the Quaternary in the West. Mediterranean Regions, 307-324 (1986). - MARUSZCZAK, H., MORAWSKI, J.: Z badań czwartorzędu w Polsce 18, 63-76 (1976). - MARUSZCZAK, H., RACINOWSKI, R.: Z badań czwartorzędu w Polsce 18, 211-225 (1976). - MATHEWES, R.W.: 397-422. In: Climatic change in Canada 5: Critical Periods in the Quaternary Climatic History of Northern North America. Syllogeus Nr. 55, ed. C.R. HARRINGTON (1985). - McLENNAN, D.S., MATHEWES, R.W.: Canad. J. Bot. 62, 2154-2164 (1984). - MENKE, B.: Verh. naturwiss. Ver. Hamburg, N.F. 24, 123-132 (1981). - MENKE, B., TYNNI, R.: Geol. Jb. A76, 3-120 (1984). - MERKT, J.: Geol. Jb. 86, 107-112 (1968). - MEYER, E.R.: Ecology 54, 982-995 (1973). - MÜLLER, M., ZIEGLER, H.: Geol. Karte von Bayern 1:25 000. Erläuter. Blatt Nr. 8042 Waging am See, p. 97. München: Bayer. Geol. L.A. 1978.

NORYŚKIEWICZ, B.: Quatern. Studies in Poland 1, 53-64 (1979); - Acta Palaeobot. XXII, 65-83 (1982).

OVERPECK, J.T., WEBB, T., PRENTICE, I.C.: Quat. Res. 23, 87-108 (1985).

PATZELT, G.: Mühlhäuser Beiträge, H. 8, 79-85 (1985). - PAWLIKOWSKI, M., RALSKA-JASIEWICZOWA, M., SCHÖNBORN, W., STUPNICKA, E., SZEROCZYŃSKA, K.: Acta Palaeobot. XXII, 85-116 (1982). - PETERSEN, K.S.: Malacologia 22, 721-726 (1982); - Striae 20, 75-78 (1984); - J. Danish Archaeol. 4, 7-22 (1985). - PETIT-MAIRE, N.: Episodes 9, 7-16 (1986). - PILCHER, J.R., BAILLIE, M.G.L., SCHMIDT, B., BECKER, B.: Nature 312, 150-152 (1984). - PONS, A.: Bull. Soc. bot. France 131, lettres bot. 49-53 (1984). - PORTER, S.C.: Arctic Alp. Res. 13, 11-23 (1981). - PROTSCH, R., SEMMEL, A.: Eiszeitalter Gegenwart 28, 200-210 (1978).

RACINOWSKI, R.: Z badań czwartorzędu w Polsce 18, 227-247 (1976). - RACZKOWSKI, W.: Z badań czwartorzędu w Polsce 18, 329-343 (1976). - RADTKE, U., HENNIG, G.J., MANGINI, A.: Eiszeitalter Gegenwart 32, 49-55 (1982). - REIFF, W.: Fundber. Baden-Württemb. 11, 1-24 (1986). - RITCHIE, J.C.: 157-170. In: Glacial Lake Agassiz, eds. J.T. TELLER, L. CLAYTON. Geol. Assoc. Canada Special Paper 26 (1983); - Ecology 66, 612-621 (1985). - RITCHIE, J.C., YARRANTON, G.A.: J. Ecol. 66, 199-212 (1978). - RÓZYCKI, S.Z.: Z badań czwartorzędu w Polsce 18, 77-92 (1976).

SACHSE-KOZŁOWSKA, E.: Folia Quatern. 50, p. 37 (1978). - SEIWALD, A.: Ber. nat.-med. Ver. Innsbruck 67, 31-72 (1980). - SISSONS, J.B.: Boreas 10, 1-17 (1981). - SOBOLEWSKA, M.: Acta Palaeobot. XVII, 39-44 (1976). - ŚRODOŃ, A.: Acta Palaeobot. XXIV, 53-66 (1984); - Acta Palaeobot. XXV, 119-137 (1985). - STEPHAN, H.-J.: Eiszeitalter Gegenwart 33, 177-180 (1983). - STREMME, H.E.: Bull. Assoc. franç. pour l'étude du Quatern. 159-166 (1985). STUART, A.J.: Boreas 8, 359-371 (1979). - STUCKENRATH,

R., MILLER, G.H., ANDREWS, J.T.: Arctic Alp. Res. 11, 109-120 (1979). - STUIVER, M., BECKER, B.: Radiocarbon 28, 2B, 863-910 (1986). - STUIVER, M., KROMER, B., BECKER, B., FERGUSON, C.W.: Radiocarbon 28, 2B, 969-979 (1986). - SZABO, B.J.: Arctic Alp. Res. 12, 95-100 (1980).

TEGNÉR, G., BARTHOLIN, T.S., BRÅTHEN, A., LÖFSTRAND, L.: Fornvännen 72, 25-32 (1977). - TSUKADA, M.: Palaeogeogr., Palaeoclimat., Palaeoecol. 3, 49-64 (1967); - Ecology 63, 1091-1105 (1982). - TUFFREAU, A., MUNAUT, A.V., PUISSEGUR, J.J., SOMMÉ, J.: Bull. Soc. Préhist. Franç. 78, 291-305 (1982). - TURON, J.-L.: Nature 309, 673-676 (1984).

UZIAK, S., POMIAN, J.: Z badań czwartorzędu w Polsce 18, 265-274 (1976).

VOGEL, J.C., FULS, A., VISSER, E., BECKER, B.: Radiocarbon 28, 2B, 935-938 (1986). - VORREN, T.O.: Boreas 6, 247-257 (1977); - Quatern. Stud. Poland 3, 119-128 (1982). - VORREN, T.O., ROALDSET, E.: Boreas 6, 53-69 (1977).

WAGNER, E.: Germania 62, 229-267 (1984); - Fundber. Baden-Württemb. 11, 62-91 (1986). - WALTHER, M.: Geomorphologische Untersuchungen zum Spätglazial und Frühholozän in den Cuillin Hills (Insel Skye, Schottland), p. 178, Diss. FU Berlin (1984). - WATTS, W.A.: Ecology 51, 17-33 (1970); - Ecology 52, 676-690 (1971); - Ann. Rev. Ecol. Syst. 11, 387-409 (1980); - 294-310. In: Late-Quaternary Environments of the United States, 1. The Late Pleistocene, ed. S.C. PORTER. Minneapolis: Univ. Minnesota Press 1983. - WEBB, TH.: Vegetatio 69, 177-187 (1987). - WEBB, TH., LASESKI, R.A., BERNABO, J.C.: Ecology 59, 1151-1163 (1978). - WEGMÜLLER, S.: Rev. de Paléobiol., vol. spéc. 243-247 (1984a); - Diss. Bot. 72, 333-344, Festschrift WELTEN (1984b). - WELTEN, M.: Pollenanalytische Untersuchungen im jüngeren Quartär des nördlichen Alpenvorlandes der Schweiz. Beitr. zur Geol. Karte der Schweiz, N.F. 156 (1982). - WIGLEY, T.M.L., MULLER, A.B.: Radiocarbon 23, 173-190 (1981). - WOILLARD, G.: Quat. Res. 9, 1-21 (1978).

ZACHOWICZ, J., PRZYBYŁOWSKA-LANGE, W., NAGLER, J.: Acta Palaeobot. XXII, 141-161 (1982). - ZAGWIJN, W.H.: Geol. en Mijnb. 1983, 437-450 (1983). - ZHAO, X., Sun, X., ZHANG, Y., HUANG, X.: Scientia Sinica, ser. B. 27, Nr. 11, 1183-1196 (1984). - ZHEZHEL, O.N.: 90-93. In: Paleomikrofitologicheskie issledovanija dlja tseleij stratigrafii, ed. L.A. PANOVA. Trudy Minist. Geol. SSSR, nov. ser. 305. Leningrad: Nauka 1980.

Prof. Dr. Dr. h.c. BURKHARD FRENZEL
Botanisches Institut der
Universität Hohenheim
D-7000 Stuttgart 70 (Hohenheim), FRG

III. Vegetation Science (Sociobiological Geobotany)

By KLAUS DIERSSEN

1. Introduction

In recent years, studies of vegetation science and a more autecological approach have been increasingly combined by different working groups. This is due, for instance, to the combination of population and production ecology and investigations on community succession.

Furthermore, the influence of the application of numerical methods is still increasing all over the world. This tendency to a more generalizing description by mathematical methods often appears to be combined with neglect of a detailed presentation of the floristical data basis of the investigations, which, however, is useful from a synsystematical point of view. Of some importance for synsystematics is the second, mostly unaltered version of the generally accepted code of phytosociological nomenclature (BARKMAN et al. 1986).

2. Monographs and Bibliographies

The well-presented textbook by ELLENBERG (1986) has been prepared in its 4th edition, but without essential new information since the preceding one. An outline of the development of the flora and vegetation of the Alps and neighbouring mountains has been published by OZENDA (1985), a textbook on temperate to arctic zones from Europe to Northern Asia by WALTER and BRECKLE (1986). The latter combines descriptions of zonal vegetation types with more detailed ecophysiological aspects, but without phytosociological comparisons.

CHABOT and MOONEY (1985) gave a survey of the terrestrial plant communities of North America including deserts, alpine and arctic areas. This informative book gives biome characteristics, a floristic comparison, information on the physical environment, structural features, life cycle adaptations and ecophysiological aspects.

For those working with coastal vegetation, the publication edited by BEEFTINK et al. (1985) is essential. Some aspects will be presented in Section 4b. Concerning numerical methods, WILDI (1986) gives a short and informative introduction to theory and application (see also WILDI 1985). PEET (1985) has combined a synopsis on theory and methods of community analysis, gradient analysis, aspects on syndynamics and species diversity from 20 papers previously published (1982-1984) in the journal Vegetatio.

Progress in Botany, Vol. 49

Recent bibliographies concern phytosociological literature from Portugal (CORREIRA and PINTO DA SILVA 1986), the ČSSR (NEUHÄUSLOVA-NOVOTNA and NEUHÄUSL 1987) and France (GÉHU and BIGOT 1987). Of special interest for conservation problems is the bibliography on human impact on the groundwater of plant communities (SCHIPPER and GROOTJANS 1986).

3. Methods and General Results

a) Succession Studies

Detailed studies on vegetation dynamics are of central interest for an understanding of the vegetation composition in specific sites and for applied purposes, i.e. nature protection. SCHMIDT (1986) studies vegetation development under different ploughing regimes. With spring ploughing, Therophyta aestivalia, geophytes and elements of the class *Artemisietea* increase, whereas with summer ploughing, *Chenopodietalia* and *Molinietalia* species predominate (near Göttingen, northern BRD). FISCHER (1986) compared population dynamics and spontaneous settlement of representative species on loessial slopes in SW Germany. Sown and unsown areas show different reactions; from a conservational point of view, the establishment of specific edaphic species in unsown areas is useful.

Long-term development, structure and distribution of shrub forests combined with grazing and farming have been examined in Siegerland, BRD (POTT 1985), including palynological methods and studies on recent vegetation composition. KULLMANN (1986) investigated the demography of *Betula pubescens* ssp. *tortuosa* in permanent plots of contrasting habitats close to the birch tree limit in Central Sweden.

b) Population Structure of Vegetation

The combination of population biology and phytosociological aspects is traditional in the Soviet approach to vegetation science. An excellent survey is presented in the textbook of WHITE (1985). The distribution pattern of species in a given site may be characterized by its life strategy and changing habitat conditions in different years. The reaction of *Pulegium vulgare* to different flooding regimes of four consecutive years has been tested in the Netherlands, the effects of flood on the vegetation composition depending on the date of inundation. A short, later summer flood has a stronger impact than a longer inundation in spring and early summer (SYKORA 1986). KASHIMURA (1985) studied the distribution of *Pyrola* sp., *Vaccinium vitis-idaea*, *Arctostaphylos uva-ursi*, *Calluna vulgaris* and *Erica tetralix* in comparison to microtopographic gradients. Potential seed production, another key factor, varied between 17 and 39 kg ha^{-1} in *Polygalo-Nardetum* grasslands (JAKRLOVA 1985). Ecological demands, population structure and phytosociology are summarized, i.e. for *Anemone nemorosa* (SHIRREFFS 1985), *Melampyrum arvense* (MATTHIES 1986), *Jasione montana* (PARNELL 1985), *Orchis militaris* (FARRELL 1985) or the *Carex muricata* aggregate (DAVID and KELCEY 1985).

c) Numerical Methods

According to the increased influence of the application of numerical methods and modelling on vegetation science, a new periodical, *Coenoses*, started in spring 1986. Its objective is to promote study of the theoretical basis of community and population studies. From a rich literature, three publications on the subject of this theoretical basis are presented: GIBSON and GREIG-SMITH (1986) introduce a method of quantifying community spatial patterns by ordination analysis. FISCHER (1986) presents possibilities of predicting site factors from ordination models by means of multiple-regression analysis. A short sketch of different approaches in vegetation science using microcomputers is given by STÖHR and BÖCKER (1986).

4. Vegetation of the Temperate Zone and Adjacent Regions

a) Temperate European Forests and Shrublands

The distribution pattern of herbs is controlled by different ecological parameters. WITTIG and NEITE (1985) studied the presence of acid indicators around the trunk base of *Fagus sylvatica* in limestone and loess beechwoods, discussing their microdistribution and the resulting phytosociological problems.

Gradients with the floristic composition of semi-natural deciduous woods in NE Ireland were compared in respect to soil acidity, soil mineral nutrient status and the degree of soil waterlogging as the principal factors influencing the ground flora (COOPER 1985). Some publications on forest and shrubland vegetation types on a supraregional level are presented for oak-hornbeam forests with special application to NW Germany (DIERSCHKE 1986), for *Corylus avellana* coppices and the soil properties on the West Norwegian coast (RØSBERG and ØVSTEDAL 1987), for the *Salicetum triandro-viminalis* from Czech and Moravian rivers (NEUHÄUSLOVA 1985) and *Prunus mahaleb* shrubland in Central Europe (MÜLLER 1986).

b) Coastal Vegetation

The above-mentioned symposium volume on coastal vegetation (BEEFTINK et al. 1985) is a mine of European geobotanical research on coastal ecosystems. Vegetation descriptions include studies of halophilous vegetation from Italy to SE Norway and the Baltic Sea. Population biology, life history and germination behaviour are described for different European saltmarsh plants, as well as for those from coastal dunes. Ecophysiological studies include, for instance, salt resistance, osmotic adaptations and interactions between salinity and nitrogen limitation. Of special interest in solving conservation problems are experiences in mowing and moderate grazing, as well as studies on vegetation development on newly embanked sites, the latter with special reference to the Netherlands.

Relationships between major vegetation zones and tidal cycles and soil aeration were studied by measuring redox potentials (ARMSTRONG et al. 1985). The increasing occurrence of *Agropyron repens* communities in NW German saltmarshes is pointed out by GLAHN (1986).

c) Weed, Ruderal and Urban Vegetation

Changes in the floristic composition of weed communities are reported from different parts of Europe, i.e. BERNHARDT (1986), KULP and CORDES (1986). A drastic change both in vegetation composition and structure is obvious, caused by a decrease of "stenök" species due to increasing farming intensity. Generally, a stabilization of the floristic delimitation of weed communities can be noticed (HÜPPE 1986).

Ruderal vegetation types were used to characterize the intensity of human influence in Western Bohemia, CSSR (PYSEK 1985). The vegetation development of urban rural sites shows a decrease of "village" weed communities (GUTTE 1986). Their function for recreation and nature conservation is discussed for Berlin (KOWARIK 1986).

Some synsystematic surveys concern ruderal vegetation in general (ELIAS 1986), the alliance *Malvion neglectae* (MUCINA 1986) and observations about immigration, habitat conditions and phytosociological position of Neophytes (FISCHER 1986).

5. Vegetation of Tropical and Subtropical Areas

TOMLINSON (1986) wrote a textbook on mangroves, dealing with their ecology, floristics and biogeography, as well as with shoot systems, water relations, salt balance, growth, reproduction and morphological adaptations of characteristic species. GARTLAN et al. (1986) investigated by correspondence analysis ordination the floristic gradients correlated with altitude, slope, soil phosphorus and potassium of the Korup Forest Reserve, Cameroun. Swamp and dry forests with a different amount of available soil phosphorus in the Douala-Edea Forest Reserve, Cameroun, were compared (NEWBERRY et al. 1986).

A phytogeographic characteristic and considerations of the evolution of the flora of Cuba are given by BORHIDI (1985). BRACK EGG (1986) presents an outline of the phytogeographical regions of Peruvia, whereas GUTTE (1985, 1986b) gives more detailed information about the subalpine vegetation (3600-3900 m NN) of Central Peruvia, including dry meadows, wet grassland and rock vegetation, as well as alpine grassland types.

References

ARMSTRONG, W., WRIGHT, E.J., LYTHE, S., GAYNARD, T.J.: J. Ecol. 73, 323-339 (1985).

BARKMAN, J.J., MORAVEC, J., RAUSCHERT, S.: Vegetatio 67 (3), 145-195 (1986). - BEEFTINK, W.G., ROZEMA, J., HUISKES, A.H.L.: Ecology of Coastal Vegetation, 598 pp. (Vegetatio 61, 62) Dordrecht: Junk 1985. - BERNHARDT, K.-G.: Tuexenia 6, 37-52 (1986). - BORHIDI, A.: Acta Bot. Hung. 31, 3-34 (1985). - BRACK EGG, A.: Bol. Lima 8 (44), 57-70 (1986).

CHABOT, B.F., MOONEY, H.A. (eds.): Physiological Ecology of North American Plant Communities, 351 pp. New York 1985. - COOPER, A.: Vegetatio 64, 67-74 (1985). - CORREIRA, A.J., PINTO DA SILVA, A.R.: Excerpta Bot. B 24 (3), 161-189 (1986).

DAVID, R.W., KELCEY, J.G.: J. Ecol. 73, 1021-1039 (1985). - DIERSCHKE, H.: Tuexenia 6, 299-323 (1986).

ELIAS, P.: Feddes Repert. 97 (3-4), 197-221 (1986). - ELLENBERG, H.: Vegetation Mitteleuropas mit den Alpen, 4. ed., 988 pp. Stuttgart: Ulmer 1986.

FARRELL, L.: J. Ecol. 73 (1985). - FISCHER, A.: Veröff. Natursch. Landschaftspfl. Baden-Württ. 61, 349-390 (1986). - FISCHER, H.: Tuexenia 6, 405-414 (1986). - FISCHER, W.: Gleditschia 14 (2), 291-304 (1986).

GARTLAN, J.S., NEWBERRY, D.Mc C., THOMAS, D.W., WATERMAN, P.G.: Vegetatio 65, 131-148 (1986). - GÉHU, J.-M., BIGOT, M.: Excerpta Bot. B 25 (1), 55-80 (1987). - GIBSON, D.J., GREIG-SMITH, P.: Vegetatio 66, 41-47 (1986). - GLAHN, H. VON: Drosera 86 (2), 119-131 (1986). - GUTTE, P.: Wiss. Z. Karl-Marx-Univ. Leipzig, math.-naturwiss. Reihe 34 (4), 357-401 (1985); - Arch. Naturschutz Landschaftsforsch. 26 (2), 99-104 (1986a); - Feddes Report. 97, 319-371 (1986b).

HÜPPE, J.: Abh. Westfäl. Mus. Naturkd. 48 (2-3), 209-221 (1986).

JAKRLOVA, J.: Ekologie (CSSR) 4 (2), 177-183 (1985).

KASHIMURA, T.: Vegetatio 60, 57-65 (1985). - KOWARIK, I.: Tuexenia 6, 75-98 (1986). - KULLMANN, L.: Vegetatio 65, 13-20 (1986). - KULP, H.-G., CORDES, H.: Tuexenia 6, 25-36 (1986).

MATTHIES, D.: Tuexenia 6, 3-20 (1986). - MUCINA, L.: Fol. Geobot. Phytotax. 21 (4), 337-448 (1986). - MÜLLER, T.: Abh. Westfäl. Mus. Naturkd. 48 (2-3), 143-155 (1986).

NEUHÄUSLOVA, Z.: Preslia 57, 313-333 (1985). - NEUHÄUSLOVA-NOVOTNA, Z., NEUHÄUSL, R.: Excerpta Bot. B 25 (1), 1-54 (1987). - NEWBERRY, D.Mc C., GARTLAN, J.S., McKEY, D.B., WATERMAN, P.G.: Vegetatio 65, 149-162 (1986).

OZENDA, P.: La végétation de la chaine alpine dans l'espace montagnard Européen, 344 pp., Paris (1985).

PARNELL, J.A.N.: J. Ecol. 73, 341-358 (1985). - PEET, R.K. (ed.): Plant Community Ecology, 332 pp. Dordrecht: Junk 1985. - POTT, R.: Abh. Westfäl. Mus. Naturkd. 47 (4), 75 pp. (1985). - PYSEK, P.: Fol. Mus. Rer. Nat. Bohem. Occ. Bot. 22, 35 pp. (1985).

RØSBERG, J., ØVSTEDAL, D.O.: Nord. J. Bot. 7 (2), 169-185 (1987).

SCHIPPER, P.C., GROOTJANS, A.P.: Excerpta Bot. B 24 (3), 217-230 (1986). - SCHMIDT, W.: Tuexenia 6, 53-74 (1986). - SHIRREFFS, D.A.: J. Ecol. 73, 1005-1020 (1985). - STÖHR, M., BÖCKER, R.: Gött. Flor. Rundbr. 20 (1), 24-52 (1986). - SYKORA, K.V.: Tuexenia 6, 249-260 (1986).

TOMLINSON, P.B.: The Botany of Mangroves, 413 pp. Cambridge 1986.

WALTER, H., BRECKLE, S.: Spezielle Ökologie der Gemäßigten und Arktischen Zonen Euro-Nordasiens, 587 pp. Stuttgart: Ulmer 1986. - WHITE, J.: The Population Structure of Vegetation, 669 pp. Dordrecht: Junk 1985. - WILDI, O.: Tuexenia 6, 379-390 (1985); - Veröff. Geobot. Inst. ETH Zürich 90, 226 pp. (1986). - WITTIG, R., NEITE, H.: Vegetatio 64, 113-119 (1985).

KLAUS DIERSSEN
Botanisches Institut
der Universität
Olshausenstraße 40
D-4301 Kiel, FRG

IV. Ecosystem Research (Ecological Geobotany)

By Wolfgang Schmidt

1. General

The increasing number of ecologically orientated publication makes it impossible to give a complete review of ecosystem research during the last 2 to 3 years. This report concentrates mainly on publications from ecological journals. Studies from biological, botanical, or applied journals are rarely taken into consideration.

The volumes edited by COOLEY and GOLLEY (1984) and PRICE et al. (1984) are two of several recent attempts to define future directions in ecology. A major goal is to address the widening gap between theoretical and empirical ecology, representing approaches from the microbial to the ecosystem level. Concepts and theories of modern ecology, as well as ecosystem research, were presented in the textbooks of PUTMAN and WRATTEN (1984), McINTOSH (1985), BEGON et al. (1986). Main aspects of ecosystem research are contained in the books on community ecology edited by GILLER (1984), STRONG et al. (1984) and DIAMOND and CASE (1986).

The study of ecological principles and problems on a worldwide basis is the aim of a book edited by SOUTHWICK (1985). The *Physiological Ecology of North America plant communities* edited by CHABOT and MOONEY (1985) clearly demonstrates that physiological ecology is a powerful tool for revealing how plants are mechanistically integrated with their natural environments, and how environmental and genetic constraints can regulate the energy and nutrient gateways in ecosystems.

WARING and SCHLESINGER's (1985) book provides an excellent introduction to forest ecosystems for beginning graduate students in both ecology and forestry. The increasing need for detailed and accurate measurements of physical processes operating over and within forested areas of the earth is emphasized by the book edited by HUTCHINSON and HICKS (1985). In a second edition, WHITMORE (1984) reviewed the knowledge of the tropical rainforest environment, the ecology of rainforest plants and animals and of rainforest ecosystems within the context of the forests of the Far East. Ecological effects of stream water regulation and treatments of wastewater on floodplain and swamp forest ecosystems were discussed in the books edited by PENKA et al. (1985) and EWEL and ODUM (1985). JAKUCS (1985) records the results of a comprehensive programme of ecological research in the *Quercetum petraea-cerris* forest community in the hills of northern Hungary. ELLENBERG et al. (1986) summarized the results of the Solling project, the well-known German contribution to the IBP.

The large amount of ecological work and the vast literature on deserts is concentrated in two volumes edited by EVENARI et al. (1985, 1986), while KRUCKEBERG (1985) described the unique vegetation on serpentine

Progress in Botany, Vol. 49

soils and its ecology. Unifying concepts of agricultural ecosystms were presented in a book edited by LOWRANCE et al. (1984).

The aim of a book edited by FASHAM (1984) is to bring together theoretical ecologists and biological oceanographers to assess the present and future capability for measuring ecological fluxes and incorporating these data into models. The book is an invaluable reference source for the state-of-the-art in understanding the processes underlying energy and material flows in the pelagic realm of the oceans. The ecology of different aquatic ecosystems was summarized by WETZEL (1983), MEYERS and STRICKLER (1984), SIOLI (1984), WHITTON (1984) and MOORE and SEED (1986).

Compartmental models are a useful representation of ecological systems and indeed are used widely to describe the storage and flow of conservative substances (energy or matter) in ecosystems. In several papers, PATTEN and HIGASHI (1984), HIGASHI (1986a,b), HIGASHI and PATTEN (1986) extended the compartmental model of ecosystems open to their environments by including self-loop flows at compartments. Furthermore, a general method for evaluating the effects of storages on the cycling efficiency of ecosystems and the relationship between direct and indirect influences is developed. A new approach, called pansystems analysis, has been proposed by BAILIAN (1986) to aid in constructing models, pattern recognition, stability analysis and formal simplification for ecosystem studies. Different models of aquatic ecosystems were discussed by ALEKSEEV and KORNILOVSKY (1985), ALEKSEEV et al. (1984) and GODA and MATSUOKA (1986). Aspects of self-regulation of ecosystems are the theme of a short essay by TROJAN (1984). The experimental approach for the study of ecosystems is emphasized by LIKENS (1985).

2. Productivity

a) Biomass and Net Primary Production

α) General, Modeling, and Methods. The book of BEADLE et al. (1985) is a useful introduction to the literature for researchers embarking on a study of photosynthesis in relation to plant production in terrestrial environments. It integrates agronomic, ecological and physiological considerations. Computer analysis of plant primary production and water relations was applied by TENHUNEN et al. (1985). LIETH (1985) gives an introduction into the procedure of surface-process modelling in ecology. The procedure is explained using global productivity models.

DICKERMAN et al. (1986) used frequently repeated measurements of growth characteristics in undisturbed study plots, and replicated harvests of shoots from nearby plots, to measure biomass accumulation. The data were used to calculate net annual aboveground production using seven computational procedures.

β) Algae and Aquatic Ecosystems. A synopsis of production and energy flow of Lake Nakuru (East Africa) is presented by VARESCHI and JACOBS (1985). The trophic structure of this equatorial alkaline-saline lake shows no predictable long-term continuity. During 5 years it had bloom of the cyanophyte *Spirulina platensis* that persisted at least 2 years; it had periods with low algal densities, and in addition it had various transitional phases with dramatic fluctuations of species composition and density.

The phytoplankton and phosphorus content of two parts of the Lake of Constance have been analyzed monthly throughout 30 years and have been statistically evaluated by LEHN (1985). The equations for the primary production, that have been calculated especially for these parts of the lake, have been extended to a general equation for lakes. This equation allows calculation of a possible primary production of a lake by its average depth and the concentration of phosphorus before the beginning of the productive season. In order to develop a method of predicting and assessing water reservoir eutrophication, which is one of the significant environmental problems of hydroelectric power plants, and to propose effective measures for the conservation of water quality, simulation methods, taking into account stratified flow and plankton dynamics in reservoirs, have been presented by MIYANAGA (1986), VOINOV and SVIREZHEV (1984), SUMMERS (1985) and WLOSINSKI and COLLINS (1985). Results from 5 years research, the *Ecological basis for the management of rivers and reservoirs* in Poland were presented by KOWALCZEWSKI et al. (1985a,b,c), SIMM (1985), DUSOGE et al. (1985), WORONIECKA (1986) and WOJCIECHOWSKA (1986). Models of the eutrophic Lake Müggelsee and the oligotrophic Lake Stechlin (G.D.R.) have been developed by KOZERSKI et al. (1984) and KOSCHEL and KASPRZAK (1984), respectively. A mathematical simulation model was used by LUDDEN et al. (1985) to ascertain the relation between the diffusion of oxygen and inorganic carbon into layers of marine microphytes and the carbon metabolism of these microphytes. The model predicted upper limits of primary production and biomass observed in well-developed natural populations. A system of simulation models on the outbreak of a large-scale phytoplankton bloom in eutrophicated coastal waters is developed by KISHI and IKEDA (1986). The models consist of two parts: one is to determine hydraulic conditions of water movements in the sea, and the other to calculate biological interactions among phytoplankton organisms and nutrients under the diffusive and advective transportation associated with calculated water movement.

Intrinsic rates of growth of the freshwater macrophyte *Salvinia molesta* were measured by ROOM and THOMAS (1986) every week over periods of 16 weeks to 2 years at seven sites located near the equator to 33°S. Among the sites, intrinsic rates of growth and net assimilation rates were generally significantly correlated with air temperature and NPK content of the plant. Maximum photosynthesis and growth rates in *Salvinia molesta* and *Eichhornia crassipes* occurred in well-developed communities in mid-summer, when irradiance and temperatures were highest. The rates reported by SALE et al. (1985) are low to medium compared with those of crop plants under similar conditions. These two emergent macrophytes are often regarded as noxious weeds because of their ability to cover large water surfaces very rapidly. It is concluded that this ability depends not on any intrinsic photosynthetic advantage compared to other plants, but on their free-floating habit and many-branched growth pattern, which enable them to remain in an active vegetative form until the water surface is covered. A model of *Eichhornia crassipes* growth and yield was developed, calibrated, and validated by LORBER et al. (1984) for Florida conditions. Net annual productivity and annual carbon budgets were determined by BOSTON and ADAMS (1986) for four populations of *Littorella uniflora* and *Isoetes macrospora*. CAM contributed 45 to 55% of the annual carbon gain for the study populations.

Soil algal productivity and biomass were monitored by SHIMMEL and DARLEY (1985) in a conventional tillage agricultural system. Total annual productivity (as carbon) was estimated to be 39 g m^{-2}.

γ) Herbs and Grasses in Different Ecosystems, Mainly Grasslands. *Cyperus papyrus* swamps at Lake Naivasha, Kenya and at Busoro, Rwanda, were investigated by JONES and MUTHURI (1985). Aerial biomass was 3245 g m^{-2} at Naivasha, and 1384 g m^{-2} at Busoro while the culm densities were 12.7 m^{-2} and 17.9 m^{-2}, the area of photosynthetic tissue per unit ground area (umbel area index) was 7.0 and 8.1, respectively. Lower night temperatures, reducing respiratory losses, may account for the higher biomass at Naivasha. Differences in the root environment with lower conductivity, redox potential and nitrogen concentration at Busaro suggest that growing conditions at Busoro were poorer than at Naivasha.

An annual net primary shoot production for *Spartina anglica* of 220 and 280 g C m^{-2} was recorded by JACKSON et al. (1986a,b) on a saltmarsh in England. Root production was estimated at 180 and 230 g C m^{-2} yr^{-1}. Live shoot biomass reached a peak in late summer, exceeding 150 g C m^{-2}. The relationship between above-ground biomass of *Spartina alterniflora* and sediment oxidation status, available ammonium concentration and salinity was investigated by HOWES et al. (1986) at 21 sites in a New England saltmarsh. Sediment oxidation gave the best prediction of above-ground biomass. Measurements in nitrogen-fertilized areas indicate that increasing plant production leads to more oxidized sediments. The effects of fire and herbivory on the standing crops and primary production of the marsh species *Spartina anglica, S. alterniflora, Typha latiflora Scirpus lacustries*, and *Distichlis spicata* were investigated by JACKSON et al. (1986b), BERTNESS (1985) and SMITH and KADLEC (1985). Average net aerial primary production in a low-salinity *Hibiscus moscheutos* marsh was estimated by CAHOON and STEVENSON (1986) to be slightly over 1200 g m^{-2} yr^{-1}.

Growth and flowering of *Eriophorum vaginatum* were observed by SHAVER et al. (1986a) over 4 years at 34 sites in Alaska. They found a strong correlation between the average number of thawing-degree days during the growing season and the peak-season leaf mass per tiller. In another paper, SHAVER et al. (1986b) conclude that the seasonal peak in whole-plant *Eriophorum vaginatum* biomass is closely regulated by whole-plant nutrient content. The amount of growth achieved per unit area of ground within a growing season (i.e. net production) is regulated mainly by nutrient reserves at the start of the year and by uptake after mid-July. The contribution of woody stems to the aboveground net primary production of several species of tundra dwarf shrubs was documented by SHAVER (1986) at four sites in Alaska. Results showed that stem secondary growth was an important component in all species and sites, ranging from 13 to 49% of total aboveground net primary production. The aboveground biomass of *Andromeda polifolia, Empetrum hermaphroditum*, and *Rubus chamaemorus* on hummocks on a subarctic peat bog was estimated by WALLEN (1986) to be 23 g m^{-2}, 64 g m^{-2} and 12 g m^{-2} respectively; the belowground coarse root and rhizome biomass was 150 g m^{-2}, 213 g m^{-2} and 220 g m^{-2} respectively, and the total fine root (< 0.5 mm) biomass was 1100 g m^{-2}, 600 g m^{-2} and 350 g m^{-2} respectively. *Andromeda* and *Rubus* had thus 98%, and *Empetrum* 90% of their living biomass belowground. Biomass distribution and shoot growth of the rhizomatous dwarf shrub *Salix herbacea* were studied by WIJK (1986a,b) in alpine snow-bed sites in northern Sweden. Within a transect which spanned a sharp gradient in snow cover, a 20% increase in growing season length was associated with a fivefold increase in *S. herbacea* shoot growth. However, *S. herbacea* showed its maximum biomass were snow persisted for the longest period, probably due to competition and herbivore damage.

A plant biomass simulation model developed by PARTON and SINGH (1984) for shortgrass prairie was adapted to a contrasting, more productive tropical grassland site. Both sites were dominated by C_4 grasses. No structural change in the model was found to be necessary. However, parameters related to photosynthetic allocation, effects of nitrogen and soil water potential on growth, and root death had to be modified. Plant growth models for grassland ecosystems were examined by HANSON et al. (1985) and MacNEIL et al. (1985), while WHITE (1984a,b) presented a validation of a multispecies sward and multispecies herbivore model across nine treatments of a sheep grazing trial. Sheep grazing with different foraging efficiencies in a Dutch mixed grassland was investigated by DE LEEUW and BAKKER (1986). Low species diversity was related to very low efficiency of foraging, whereas species diversity was relatively high for foraging efficiencies ranging from moderate to high. Results by SMITH and ALLCOCK (1985) indicate that the presence of indigenous grasses does not inevitably reduce productivity below that of swards consisting entirely of improved species.

Primary productivity and herbivory were studied in detail by McNAUGHTON (1985), COUGHENOUR (1983) and COUGHENOUR et al. (1984a,b,1985) in the dry-subhumid African savannas of the Serengeti National Park, Tanzania. Basic state variables measured were aboveground plant biomass inside permanent and temporary fences, and outside fences. Control productivity (cPn) was calculated from biomass dynamics inside permanent fences. Temporary fences were moved in concert with grazing by the region's abundant ungulates to estimate actual aboveground primary producitivity (aPn). Short-term productivities could be extremely high, exceeding 30 g m^{-2} d^{-1}. Annual cPn was linearly related to rainfall and averaged 357 g m^{-2} yr^{-1} over the year and 1.89 g m^{-2} d^{-1} during the growing season. Actual aPn was substantially greater than cPn at most sites, averaging 664 g m^{-2} yr^{-1}. Growing season aPn averaged 3.78 g m^{-2} d^{-1}. Rain had an increasing effect on primary productivity as the wet season progressed and plant biomass accumulated. Net aerial primary production of a grassland in Nairobi National Park was estimated by DESHMUKH (1986). As a result of the long rains (March-April) net primary production was 651 g m^{-2} and as a result of the short rains (October-December) 420 g m^{-2}. These estimates were more than twice the peak biomass observed in each season.

The productivity of three *Agave* species was calculated by NOBEL (1984, 1985) and NOBEL and QUERO (1986) using physiological responses to environmental parameters. The net productivity of *A. deserti*, a common succulent perennial of the Sonoran Desert, over a 5-month period was 0.57 kg m^{-2} ground area, a large value for a desert CAM plant. The productivity of *A. lechuguilla*, a commercially harvested plant that occurs over vast areas of the Chihuahuan Desert, was 0,38 kg m^{-2} yr^{-1}, which, although much less than for agricultural crops, is still much greater than the average productivity for desert ecosystems. The productivity of 6-yr-old plants of *A. fourcroydes*, a plant extensively cultivated in the Yucatan peninsula of Mexico, was 1.60 kg m^{-2} yr^{-1}. This is apparently the highest annual productivity reported for a CAM plant and is only slightly lower than for most C_3 or C_4 agricultural crops.

The estimation of root biomass and production is one of the most significant and challenging problems facing grassland ecologists. Estimates of net root production are usually based on a series of root biomass data collected over a growing season by the harvest method. Since the "true" value of net root production is unknown, it is not possible to evaluate the accuracy of the estimate. In two papers VOGT et al. (1986) and LAUENROTH et al. (1986) discussed the real or imaginary problem of overestimation of net root production. Belowground biomass in a second-year *Festuca pratensis* meadow was estimated by HANSSON and ANDREN (1986) using two different field methods: soil coring and ingrowth mesh bags. The lowest production (0.19 kg m^{-2}) was considered to be an underestimate, since root death and decomposition during the growing season were not considered. The highest estimate (0.46 kg m^{-2}) was suggested to be the most accurate estimate of the belowground production. Spatial and temporal patterns of root activity in a species-rich alluvial grassland were measured by FITTER (1986) using a technique involving the simultaneous injection to different depths in soil of three chemical tracers (Li, Rb, Sr). Using both root activity and productivity data, species were grouped into two main guilds, one active in spring and one in summer. Correlations of aboveground biomass with root activity at different depths revealed that species of the spring guild were more active in the 5 - 15 cm horizons and those of the summer guild at 15 - 25 cm. These patterns suggest that rooting depth and time of activity are strongly linked: early-active species tend to be less productive and shallower-rooted, and this combination of characters allows them to escape from competition with more productive species, by being active at a time when deeper soil layers are less hospitable. Root growth responses to defoliation in two *Agropyron* bunchgrasses were observed by RICHARDS (1984) in the field with an improved root periscope technique. Reduction of root growth following defoliation appears to be an effective mechanism to aid reestablishment of the photosynthetic canopy and the root:shoot balance. As such it contributes to both herbivory tolerance and maintenance of competitive ability.

δ) Woody Plants and Forest Ecosystems. The environment, structure and biomass of four evergreen Dipterocarpaceae rainforest sites in India are described by RAI and PROCTOR (1986a). Data on growth rates and population structure of Moraceae trees have been extracted by PRIMACK et al. (1985) from long-term forestry studies in Sarawak, an area at the centre of species richness of Old World tropical forests. Mortality patterns and stand turnover rates in a wet tropical forest in Costa Rica were determined by LIEBERMAN et al. (1985) over a 13-yr period. This primary lowland forest appears to be among the most dynamic of tropical forests studied to data, with an annual loss of stems of 2% and a consequent stand half-life of 34 years. Vegetation biomass of a subtropical floodplain forest in Puerto Rico was partitioned as follows: leaves 9.8% (75% are palm leaves), wood 68.4%, and roots 21.8%. Net aboveground primary carbon productivity was 876 g m^{-2} yr^{-1} (FRANGI and LUGO 1985).

Dry matter production was determined by KADEBA and ADUAYI (1986) in a 10-yr-old *Pinus caribaea* stand planted at a subhumid tropical savanna site in Nigeria. A simulation model of a *Pinus* flatwoods ecosystem in Florida was constructed by GOLKIN and EWEL (1984) to determine rough estimates of annual ecosystem primary production, respiration and transpiration while SHARPE et al. (1985) described a plant growth model for semi-arid woodlands. Annual pollen production rates in a 35-yr-old *Pinus densiflora* forest in Japan were found by SAITO and TAKEOKA (1985) to range from 55 to 96 kg ha^{-1} yr^{-1}.

RUNNING et al. (1986) reported the first attempt to measure LAI (leaf area index) of coniferous forests using optical scanners of satellite resolution. LAI of a 450-yr-old *Pseudotsuga menziesii* stand was estimated by MARSHALL and WARING (1986) from litterfall, light interception, sapwood cross-sectional area, and tree diameter. Estimates made by the first three techniques were similar, but the estimate based on tree diameter appears to be inaccurate; therefore, the exceedingly high LAI previously reported for old *Pseudotsuga* forests are unreliable. Sapwood cross-sectional area varies in correspondence with the canopy area and therefore is a better estimator of leaf area of large trees. Maximum LAI estimates based on sapwood area are similar to those for other temperate coniferous forests. Data on leaf area and leaf area production in different plant forms over an altitudinal range in New Zealand were given by HAASE (1986) and KÖRNER et al. (1986).

Biomass formation in three shrub layer species of a *Quercus-Carpinus* forest was estimated by VOOKOVA (1986) under Central European conditions. The biomass of the herbaceous layer in three flood-plain forests of Slovakia was estimated by KUBICEK and SOMSAK (1985).

The relationships between fine roots, net primary production, and soil nitrogen availability were studied by NADELHOFFER et al. (1985) and ABER et al. (1985) at 13 temperate forest sites. Provided that fine-roots biomass and N turnover rates were similar within individual sites, allocation of production to belowground biomass also increased relative to increases in soil N availability. Furthermore, the proportion of total net primary production allocated to belowground biomass did not decrease with increased N availability. According to SHARMA and AMBASHT (1986) the annual nodule production of 244 - 492 kg ha^{-1} contributed significantly to the belowground production of *Alnus nepalensis* plantation stands in the Eastern Himalayas.

A comparison of tree growth models was given by DALE et al. (1985) while DOWNING and WEBER (1984) analyzed published data to produce equations that predict rates of net production and net harvestable production of forests. These equations can be applied between the latitudes of 31° and 65° N and S, and use common biotic and abiotic site descriptors as independent variables.

b) Respiration (see also Sect. 4a)

Based on the differential discrimination between $^{13}CO_2$ and $^{12}CO_2$ during photosynthetic CO_2 fixation by C_3 and C_4 plants, the relative contribution of organic soil material of varying age can be checked by measuring the $\delta^{13}C$ value of CO_2 evolved during soil respiration. The CO_2 from a soil cropped only with C_3 plants for at least 7 years showed a $\delta^{13}C$ value very close to that of C_3 plant tissue. Correspondingly, a soil bearing the C_4 plant *Zea mays* for at least 6 years had a $\delta^{13}C$ value for the respiratory CO_2 quite near to that of a typical C_4 plant. Therefore, SCHÖNWITZ et al. (1986) concluded that with soil standardized for cultivation and cropping, a quantification of the part that fresh organic matter and especially the last-year plant residues play in soil respiration, can be achieved. Stable carbon isotope ratios of roots and soil organic matter were measured by DZUREK et al. (1985) to determine if changes in the relative dominance of two shrub species had occurred in salt-desert community. In a forest investigated by SCHLESER and JAYASEKERA (1985) a $\delta^{13}C$ difference of 5‰ has been determined from bottom to top of a *Fagus sylvatica* tree, which results in a CO_2 contribution from the soil of about 22% for the lower forest strata, while the total contribution of soil-respired CO_2 accounts for about 5% of the overall assimilation.

The annual rates of soil respiration in three *Chamaecyparis obtusa* stands ranged from 4.54 to 6.40 t C ha^{-1} yr^{-1} which were small compared with the rates reported in southern Japan before. The rate of soil respiration increased exponentially with the increase of temperature (TSUTSUMI et al. 1985). The eddy-correlation technique was used by BALDOCCHI et al. (1986) to measure CO_2 efflux from the floor of a deciduous forest. Midday values of CO_2 efflux typically ranged between 0.30 and 0.45 mg m^{-2} s^{-1}. These values were somewhat larger than those from previous chamber studies, which typically ranged from 0.2 to 0.3 mg m^{-2} s^{-1}. Daytime CO_2 efflux from the canopy floor was correlated with air temperature. A suppression in CO_2 efflux was observed late in the afternoon and a burst in CO_2 efflux was observed near dusk, which was independent of temperature.

The influence of rain, tidal wetting and relative humidity on release of CO_2 by standing-dead salt marsh plants was investigated by NEWELL et al. (1985). The CO_2-release rates are within the range of maximal rates reported for decaying terrestrial litter, and are as high as most rates reported for *Spartina alterniflora* decaying under continuously wetted or submerged conditions.

Atmospheric ethylene (C_2H_4) levels were examined by SAWADA (1985a) and SAWADA et al. (1985) in three forest types and above paddy fields during the burning of agricultural wastes in Japan.

c) Carbon Cycle

The topical subject of increasing CO_2 content is the theme of two books published by WOODWELL (1984) and SMIL (1985). Increases of 10 - 20% since 1900 and a predicted further doubling before the end of the next century have serious implications for warming of the earth, melting of polar ice, concomitant sea-level rises and shifts in climatic zones. This can lead to shifts in agricultural zones, changes in world food supply and so to world economics and trade. The causes of the increase in atmospheric CO_2 can be traced to two sources: the combustion of fossil fuels and the destruction of vegetation. The amount of CO_2 released by fossil fuel combustion is known with some precision, whereas the amount released from biota and soils is uncertain. The temporal response of forests to CO_2-induced climate changes was examined by SOLOMON (1986) for eastern North America. A forest stand simulation model was used with the assumption that climate will change at a constant rate as atmospheric CO_2 doubles, and then as CO_2 doubles again. The simulated effects of changing monthly temperature and precipitation included a distinctive dieback of extant trees at most locations, with only partial recovery of biomass in areas of today's temperate deciduous forest. Boreal conifers also underwent diebacks and were replaced by deciduous hardwoods more slowly in the north than in the south. Transient responses in species composition and carbon storage continued as much as 300 years after simulated climate changes ceased. The interaction of increasing atmospheric CO_2 and soil nitrogen on the carbon balance of tundra microcosms was investigated by BILLINGS et al. (1984). They conclude that atmospheric CO_2 is not now limiting net ecosystem production in the tundra and that its direct effects will be slight even at double the present concentration. The most probable effects of CO_2 in the coastal tundra will be through its indirect effects on temperature, water table, peat decomposition, and the availability of soil nutrients. A model ecosystem composed of garden soil and a mixture of *Trifolium repens* and *Lolium perenne* was used by OVERDIECK and BOSSEMEYER (1985) to study long-term effects of CO_2 enrichment on CO_2 gas exchange. Needs of modelling plant and ecosystem responses are discussed by DAHLMAN (1985) and REYNOLDS and ACOCK (1985) with regard to their potential for predicting and evaluating possible scenarios of vegetation response to elevated CO_2 levels.

The importance of studying the cycling of carbon in order to obtain an analytical insight into forest ecosystem mechanisms is emphasized by NAKANE (1986). Carbon balance of a tropical *Dalbergia sissoo* forest stand in India based on carbon input in litter production (1.97 t C ha^{-1} yr^{-1}) and carbon output in soil respiration (6.83 t C ha^{-1} yr^{-1}) revealed that annual CO_2-C output was 3.47 times higher than the total input by litter (RAJVANSHI and GUPTA 1986). During regeneration of a deciduous braodleaf forest (KAWAGUCHI and YODA 1986) and a *Pinus densiflora* stand (NAKANE et al. 1986) in Japan the forest ecosystems changed from a CO_2 source to CO_2 sink, relative to the atmosphere, fairly early in the regeneration.

Organic soil-wetlands, particularly those in the temperate zone, under natural conditions, are net carbon sinks and hence are important links in the global cycling of CO_2 and other athmospheric gases. Widespread drainage for agriculture has significantly altered the carbon balance. A computer model was used by ARMENTANO and MENGES (1986) to track the consequent changes in the carbon balance of nine wetland regions. The net carbon sink in Finland and the U.S.S.R. has been reduced by 21 - 33%, in Western European wetlands by nearly 50%, and in Central Europe the sink has been completely lost.

Stable isotopes of sulfur and carbon were used by PETERSON et al. (1986) to trace the dominant flows of organic matter from producers to macroconsumers in a *Spartina alterniflora* saltmarsh. *Spartina* detritus and plankton were both shown to be much more important as organic matter sources for marsh macroconsumers than either sulfur-oxidizing bacteria or organic matter derived from terrestrial inputs. The losses of organic matter as methane (0.1 - 0.3 mol m^{-2} yr^{-1}) and dissolved organic carbon (0 - 3 mol m^{-2} yr^{-1}) and burial (7.4 mol m^{-2} yr^{-1}) were measured by HOWES et al. (1985) in order to construct a carbon budget for the sediments of a short *Spartina alterniflora* salt marsh. These data, when combined with the estimates of carbon mineralization, gave an estimate for the organic carbon loading to the sediments of 68 - 78 mol m^{-2} yr^{-1}.

3. Energy and Water Balance

a) Energy Flow and Efficiency

In a book entitled *Energy and Ecology*, GATES (1985) carefully explores the environmental and ecological consequences of the extraction, conversion, use and disposal of existing and potential sources of commercial energy. The volume edited by FASHAM (1984) on *Flows of Energy and Materials in Marine Ecosystems: Theory and Practice* resulted from an international meeting whose purpose was "to bring together theoretical ecologists and biological oceanographers to assess the present and future capability for measuring ecological fluxes and incorporating these data into models". The principle that energy does not cycle in ecosystems is examined by PATTEN (1985a,b) by path and flow analyses of a small, steady state, energy model of an oyster reef community. A simple random path method for the analysis of flow networks is presented by BRANER (1985). By analyzing food webs formed by flower heads of Asteraceae and associated insects it was shown by ZWÖLFER (1985) that "informational processes" are essential for the control of energy flow. A system model of energy flows in arable lowland grassland in Japan is considered by SHIYOMI et al. (1986). Energy flows under optimal stocking rate are as follows, in kJ m^{-2} yr^{-1}: global solar radiation (ca. 473×10^4) → photosynthetically active radiation (191×10^4) → energy fixed in above-ground plant portion and grazed by cattle (17460) → net secondary production (761).

Two canopy radiative transfer models were tested by BALDOCCHI et al. (1985) against solar radiation measurements made in an oak-hickory forest. Comparable measurements of some transformation processes of solar radiation in a spruce monoculture and on a clear cutting area covered by an unclosed 10-yr-old stand have been carried out by PIVEC and KUCERA (1986). The 1980 ash fall from Mount St. Helens was experimentally reproduced by BLACK and MACK (1986) in 1982 by applying silt-sized ash to an *Artemisia-Agropyron* stand in Washington. Compared to the adjacent control site, ash caused an immediate increase in albedo from 13% to 28%, while other parameters of the energy budget were simultaneously lowered: net radiation by 20%, soil surface temperatures by as much as 10°C, and soil heat flux by as much as 50%. The effects of population density, row spacing and intercropping on the interception and utilization of solar radiation by *Sorghum bicolor* and *Vigna unguiculata* were measured by REES (1986) in semi-arid conditions in Botswana.

Caloric values of 18 woody plant species as well as the energy balance of the aboveground shrub layer biomass in different forest ecosystems of the Karpaty Mountains were measured by VOOKOVA (1985a,b).

b) Transpiration and Water Cycle

Periodic flooding, poor soil aeration, intensive year-round rainfall, and low atmospheric saturation vapour pressure deficits are believed by FRANGI and LUGO (1985) to be the main driving forces of a subtropical floodplain forest in Puerto Rico, which exhibits many characteristics typical of lowland rain forests and floodplain wetlands. Throughfall, stemflow, interception, runoff, transpiration, and evapotranspiration accounted for 81.6, 9.8, 8.6, 77.7, 13.7, and 22.3% of annual rainfall (3725 mm) respectively. Factors affecting water outflow beyond the rooting zone were studied by KNIGHT et al. (1985), using data from eight contrasting stands of *Pinus contorta* forest. The results indicate that stands differing in site or habitat type experience different rates of water losses at different times during the snow melt season, and contribute differentially to streamwater quality and hydrograph shape. The diurnal course of the xylem water flow in a solitary *Salix fragilis* tree in a wet grassland was measured by CERMAK et al. (1984) using the tissue heat-balance method. During the summer period, about 33 mature (polycormic) trees per ha may drain 100% of water consumed by the present-day sedge-grass marsh.

A new continuous simulation model for transpiration from *Salix* energy forest is formulated and solved numerically by KOWALIK and ECKERSTEN (1984). The model has wider application than simply for *Salix*; a change in the computer simulation from one plant species to another may be achieved by altering the root, xylem and leaf parameters of the model. A simulation model of a *Pinus* flatwoods ecosystem in Florida was constructed by GOLKIN and EWEL (1984) to determine rough estimates of annual ecosystem transpiration (990 mm yr^{-1}) while HUILLET and LAUGA (1985) presented a soil-plant-water model with a case study in a forested catchment. In rain-fed agriculture, most of the efforts in modelling have been concerned with surface runoff. Significant strides could be made by a concerted effort to combine the knowledge gained in irrigated and rain-fed agricultural modelling to develop more comprehensive models (SKOGERBOE 1986). A description is presented by BAJKIEWICZ-GRABOWSKA (1985a) of the water cycle in four Masurian lakes (Poland).

Subterranean termites affect the hydrology of Chihuahuan desert systems by enhancing water infiltration and retention of top soil. The presence of a *Larrea tridentata* shrub canopy and litter layer cancels any effect of subterranean termites on hydrological parameters. Since approximately 2/3 of the area is not under shrub canopies, subterranean termites are considered by ELKINS et al. (1986) to be essential for the maintenance of the soil water characteristics that support the present vegetation.

4. Mineral Cycle

a) Decomposition

The decomposition process of wood litter under field conditions is discussed by YONEDA (1985, 1986). A comparison of litter dynamics and nitrogen recycling among riverine, fringe, and basin mangroves in Florida investigated by TWILLEY et al. (1986a) suggests that the relative influence of export and decomposition on litter dynamics depends on the magnitude of tidal inundation.

Inland basin mangroves have low nitrogen recycling efficiency in litterfall, but higher internal recycling of litter on the forest floor, due in part to a longer litter residence time, nitrogen retranslocation prior to leaf senescence, and nitrogen immobilization during litter decomposition and peat deposition. Decomposition and changes in nitrogen and organic-chemical content of six types of forest litter were studied by McCLAUGHERTY et al. (1985) in five adjacent forests. Decomposition rates of transplanted *Acer saccharum* leaves and *Acer rubrum* wood were not correlated with mineralization rates in all five stands, indicating that nitrogen mineralization rates do not affect initial decomposition rates. However, mineralization rates were correlated with decomposition rates of the native dominant foliage litter. Chemical composition of litter affected decomposition rates and patterns. Soluble substances and litters relatively rich in solubles disappeared rapidly during early stages of decomposition. Eventually, slowly disappearing acid-soluble and acid-insoluble substances dominated the pattern of mass loss in all litters. The dynamics of organic matter and nutrient elements in the forest floor of *Pinus contorta* forests were investigated by

YAVITT and FAHEY (1986). About 35% of the original mass of leaf litter remained after 84 - 96 months; about 30% of annual release from the forest floor resulted from leaching of organic compounds.

Decomposition of leaf litter from two chaparral shrubs, *Ceanothus megacarpus* (an evergreen sclerophyll) and *Salvia mellifera* (a drought-deciduous species) was studied by SCHLESINGER (1985) using litter bag methods for a 3-yr period. Turnover times of dry mass were calculated by MITCHELL et al. (1986) to be 14.5 and 11.4 yr for leaf litter of the South African fynbos species *Leucospermum parile* and sterile culms of *Thamnocortus punctatus*, respectively, but male flowerheads of *T. punctatus* showed no significant decay.

SCHAEFER et al. (1985) measured mass losses of both buried and surface litter of six litter types in a North American desert. There was no correlation between rates of mass loss and percent lignin, carbon-nitrogen ratio, or lignin-nitrogen ration. WHITFORD et al. (1986) tested the hypotheses that rates of decomposition in a desert should be higher following single large rain events of 25 mm than evenly spaced 6 mm events and that supplemental rainfall should result in higher populations of soil biota. There were no significant differences in mass losses of *Larrea tridentata* leaf litter between plots receiving water supplementation and plots without added water. A single event pulse compared with weekly pulses of rainfall during the normal "dry" period resulted in no differences in mass losses. The litter-bag technique was used by DESHMUKH (1985) to estimate decomposition rates of standing-dead grass and grass litter in Nairobi National Park. Decomposition was extremely slow.

An annual loss of standing dead material equivalent to 29 - 32% of annual net primary shoot production of *Spartina anglica* in a salt-marsh was estimated by JACKSON et al. (1986b) due to microbes and microfauna, with an additional loss of 3% yr^{-1} due to meio-fauna. Approximately 15 - 20% yr^{-1} of annual net primary shoot production was exported from the site of production as particulate matter. The decomposition rate of *Hibiscus moscheutos* stem material in a low-salinity marsh was estimated by CAHOON and STEVENSON (1986) to be on the order of 7 - 8 yrs, or up to five times slower than stems and leaves of other typical brackish marsh species. The rates of decomposition were compared by TWILLEY et al. (1986b) among six aquatic plants occurring in an estuarine environment. Short-time (24 h) and long-time (4 - 6 d) decomposition of phytoplankton cells were investigated by HANSEN et al. (1986) under in situ conditions in four Danish lakes.

b) Mineral Cycle

α) General. HUTZINGER (1984, 1985) edited another two parts of *The Handbook of Environmental Chemistry* dealing with the behaviour of natural and anthropogenic chemicals in the environment and the processes which control their cycling and distribution. A book edited by TINSLEY and DARBYSHIRE (1984) reports the proceedings of a conference on biological processes and soil fertility, while SMIL (1985) describes human "interference" in the global cycles of carbon, nitrogen, and sulfur. The particular importance of gaseous nitrogen losses from warm areas, especially those where soils are poorly buffered, compared with temperate regions, where losses of nitrogen by leaching may be more important, is discussed by various specialists in a book edited by FRENEY and SIMPSON (1983). ROBERTSON and ROSSWALL (1986) have calculated a nitrogen cycle budget for West Africa. Deposition of phosphorus from the atmosphere was measured by LEWIS et al. (1985) in the montane region of Colorado. Deposition of phosphorus consistently reached a sharp peak in early summer, near the beginning of maximum biological activity. The study indicates transport of filterable, water-soluble phosphorus substance, probably of biogenic origin, in large amounts over a short period of time. This transport mechanism, heretofore unsuspected, may account for significant long-distance phosphorus transport.

β) Aquatic Ecosystems. The book of SALOMONS and FÖRSTNER (1984) aims to provide an account of presently available knowledge of trace metal transport through the hydrological cycle with emphasis on the processes involved. The topic is covered in four chapters dealing, respectively, with sources, distributions and reactivity of trace metals in (1) the atmosphere, (2) soils, lakes and rivers, (3) estuarine and coastal regions and (4) the oceans. Methods to study nitrogen transformations in the oceans and new concepts and understanding of the marine N cycle are summarized and presented in a book edited by CARPENTER and CAPONE (1983). The eutrophication of marine ecosystems, rivers, lakes and reservoirs is recognized worldwide as a serious problem of water quality management. Therefore the results of many field studies have been published during the last 3 years (DITTMER et al. 1985; BAJKIEWICZ-GRABOWSKA 1985b; GOSZCZYNSKA 1985; LAWACZ et al. 1985; PENCZAK et al. 1985; PLANTER and WISNIEWSKI 1985; LAWACZ 1985; FAIRCHILD et al. 1985; GROBLER 1986; HART et al. 1986). Mathematical models of nutrient dynamics in aquatic ecosystems including the aspect of eutrophication are presented by JORGENSEN et al. (1986a,b), MATSUOKA et al. (1986), VIRTANEN et al. (1986), NAKANISHI et al. (1986), KAWASAKI et al. (1986), OHUCHI et al. (1986) and BOLLA and KUTAS (1984).

Rates of nutrient regeneration were compared by TWILLEY et al. (1986b) among six aquatic plants occurring in an estuarine environment. Nitrogen content of all plant tissues increased during decomposition, yet reductions of C:N ratios were only observed for plants with initial C:N > 20. N:P ratios generally increased due to a much higher leaching for P (10 - 40% of initial P) compared with N (1 to 10% of original N). According to BOWDEN (1986) both internal recycling and translocation make the tidal marsh plant community less dependent on nitrogen imported from the river. Submerged plant biomass and aquatic plant species distribution were related to sediment nutrient (N, P, K) concentrations (ANDERSON and KALFF 1986). Nitrogen metabolism of the shallow and deep-water phytoplankton in a subalpine lake was investigated by PRISCU et al. (1985) while MULHOLLAND et al. (1985) analyzed seasonal variations of phosphorus spiralling in a woodland stream.

γ) Grasslands and Herb Communities. Nitrogen effects of faeces of snow geese and saltmarsh vegetation were studied by BAZELY and JEFFERIES (1985) in order to determine how goose grazing increased the net aboveground primary production of forage species, *Puccinellia phryganodes*, and *Carex subspathacea*. A study was made by DELAUNE et al. (1986) of nitrogen cycling and the effect of added inorganic nitrogen on biomass production in a freshwater marsh of *Panicum hemitomon*. Nitrogen is apparently a limiting nutrient since supplemental nitrogen increased aboveground biomass by 40%. Nutrient turnover in *Molinia coerulea* and *Carex acutiformis* at fens and in a litter meadow of Central Europe were investigated by EGLOFF (1985) and PFADENHAUER and TWENHÖVEN (1986). From August through October *M. coerulea* exhibits highly efficient retranslocation of phosphorus into coarse roots and internodes, followed by nitrogen and potassium.

The growth rate of *Eriophorum vaginatum* in Alaskan tundra is probably controlled mainly by the availability of internal nutrient reserves rather than by an effect of temperature upon synthesis or transport of organic-carbon compounds required for growth. In spring, internal nutrient reserves from stems support a near-maximal growth rate. In late summer, growth slows because nutrient accumulation in stems takes precedence over its incorporation in growth. If nutrient availability increases in late summer, continued growth is possible, because the nutrient demands of both storage and growth can be met (CHAPIN III et al. 1986; SHAVER et al. 1986a). Accumulation of nitrogen and phosphorus in woody stems of several species of tundra shrubs occurred at slower rates, presumably due to internal reuse of these elements, but still accounted for 5 - 19% of the annual net aboveground N accumulation and 6 - 24% of the P accumulation. One implication of these results presented by SHAVER (1986) is a faster rate of N and P turnover in the vegetation than previously believed, and a greater annual nutrient demand. Processes controlling movement, storage, and export of phosphorus were studied by RICHARDSON and MARSHALL (1986) in a fen peatland. An annual P budget completed for the fen ecosystem revealed that plant uptake requirements were 7 - 9 kg ha^{-1} yr^{-1}, but 35% of aboveground P uptake by plants was returned to the peatland surface via litterfall. Permanent storage of organic P in peat ranged between 2 and 5 kg ha^{-1} yr^{-1} under natural levels of P input. Both microbial uptake and soil exchange capacity controlled the amount of P made available for plant growth. BUTTLEMAN and GRIGAL (1985) use the Rb/K ratio to evaluate potassium nutrition of forested peatlands. McKNIGHT et al. (1985) described the biogeochemistry of aquatic humic substances in an ombrotrophic floating-mat *Sphagnum* bog. Inorganic elements concentrations are given by MALMER and WALLEN (1986) for above- and belowground fractions of two dwarf shrubs, *Andromeda polifolia* and *Empetrum hermaphroditum*, growing on a peat bog in subarctic Skandinavia. When combined with biomass estimates from the same site, it was shown that 80% of the total plant content of the elements is in the fine roots. The distributions of carbon, nitrogen and phosphorus in a moss community-soil system developed on a cold desert in Antarctica were investigated by INO and NAKATSUBO (1986). Nitrogen fixed by cyanobacteria flowed from these organisms to the moss but little N was transported within the moss body. It seems that absorption of P from soil is difficult for the moss, because of the paucity of water necessary for the movement of P and the suppressed growth of the moss due to the arid conditions.

Nitrogen content of bulk precipitation and throughfall (canopy leachates) was measured by SEASTEDT (1985) on annually burned and unburned tallgrass prairie dominated by *Andropogon gerardii*. Measurements of carbon, nitrogen, and phosphorus were carried out by SCHIMEL et al. (1985) in the soils of a hillslope of shortgrass steppe. Trophic interactions and nitrogen cycling in a semi-arid grassland soil were investigated by INGHAM et al. (1985, 1986a,b) under field site conditions and in microcosms. Plant responses to different phosphorus nutritional levels and defoliation on C_4 graminoids from the Serengeti Plains (Tanzania) were measured by McNAUGHTON and CHAPIN III (1985) to evaluate potential limiting interactions between energy and nutrient flows in this grazing ecosystems. The possible ecological significance of the rapid phosphorus transfer from dying roots to living plants of *Lolium perenne* and *Plantago lanceolata* is discussed by RITZ and NEWMAN (1985).

Patterns of nitrogen cycling were studied by GROFFMAN et al. (1986) in N-fertilized and unfertilized agroecosystems with heavy weed infestations. Nitrification, denitrification, and leaching losses of N were greater in fertilized treatments than in unfertilized treatments, especially after residue input following harvest of the summer crop. Watershed-level agroecosystem studies are essential to relate land management to the external environmental effects produced by agricultural nutrients and to enhance the understanding of agricultural nutrient cycles. LOWRANCE et al. (1985) determined nutrient budgets for agricultural watersheds in the Georgia Coastal Plain while JAKUBOWSKA and SUCHECKA (1985), STASIAK and TATUR (1985) and TATUR and STASIAK (1985) studied the pollution of surface waters and lakes in agricultural watersheds in Poland. Nitrogen washing out form the soil of an intensively farmed area in Czechoslovakia was highest in the spring period. A second less significant maximum was found by GABRIS and

SUJANSKA (1985) in the autumn period. Results by TOTH and SVETLOSANOV (1985) showed a strong suppression of nitrification process in water environment due to continual increase of agrochemicals input to water reservoirs.

Successional changes in the available soil nitrogen may be a key to understand the dynamics of secondary succession. Factors affecting the accumulation of nitrogen were determined by PALMER and CHADWICK (1985) in *Festuca rubra* and *Lolium perenne* swards established on colliery spoil. The 7-yr-old swards had been sown with *Trifolium repens* or supplied with N fertilizer since establishment. *T. repens* plots and fertilizer N plots had accumulated 376 and 273 kg N ha^{-1} respectively and unameliorated spoil 58 kg N ha^{-1} since establishment. During this time fertilizer plots had received 500 kg N ha^{-1} and *T. repens* plots no N fertilizer. The level of N accumulation was very low compared to soils, but the rate of accumulation was similar to that found on other derelict land. The use of a legume is considered by PALMER and CHADWICK (1985) to be the most efficient way of accumulating N in colliery spoil. Nitrogen supply was measured by SCHULZ et al. (1985) at two different sites on a covered dump. Recultivation with *Lolium perenne* leads to bioretention of most of the N, whereas under the annual vegetation of undisturbed succession high amounts of N are leached during the first year of succession. Nitrogen-limited growth in plants from different successional stages was studied by TILMAN (1986). There was a highly significant tendency for early successional species to grow more rapidly at low nitrogen levels and to acquire more nitrogen per plant from nitrogen-poor soils than late successional species. However, late successional species did not grow more rapidly at high nitrogen levels than early successional species. These results are consistent with the hypothesis that early successional species are dominant following old field abandonment because of their ability to compete for soil nitrogen. Nitrification potentials of forest floors from an old-field chronosequence were measured by THORNE and HAMBURG (1985). It appears that decreases in nitrification potential during this secondary succession could have resulted from changing forest floor pH and its effects on microorganisms.

δ) Forest Ecosystems. In a book, JORDAN (1985) aims "to review the studies pertaining to nutrient cycling in the humid tropics, to build a case for the critical nature of nutrients in this region, and to discuss the implications of nutrient scarcity for sustained productivity in both disturbed and undisturbed ecosystems". A comparative analysis on the rate of nutrient fluxes was carried out by CUEVAS and MEDINA (1986) in three Amazonian forest ecosystems. N/P ratios indicate that the mixed forest on Tierra Firme is limited by P availability, while low N availability predominates in the tall Amazon Caatinga. A Bana forest on podsolized sands appears limited by both N and P. These differences probably relate to variations in degree of sclerophylly and leaf duration which determine leaf nutrient concentrations in the Amazonian ecosystems studied. Net nitrogen mineralization was extremely high in a lowland rainforest on volcanic soil studied by VITOUSEK and DENSLOW (1986) in Costa Rica (822 kg N ha^{-1} yr^{-1}), while available phosphorus concentrations were relatively low. In these forests, the root-throw zone of treefall gaps had significantly lower nitrogen mineralization and phosphorus concentrations than the crown-fall zone or the zone along the fallen trunk. Low nutrient availability in the root-throw zone could select for colonizing species distinct from those establishing in other zones of treefall gaps. Stemflow inputs of magnesium and potassium were measured by HERWITZ (1986) under heavy rainfall conditions in two tropical forest sites in Australia.

The chemical composition of the litterfall in four Indian rainforests showed large differences between sites. Magnesium concentration was generally high and samples from two sites had the highest concentration of this element recorded from tropical forest litterfall (RAI and PROCTOR 1986b). Tropical montane rainforests in general may be short of N, P, K and Ca compared to the most nutrient-rich tropical lowland rainforests (TANNER 1985). Runoff losses of nitrogen (9.98 kg N ha^{-1} yr^{-1}) and of phosphorus (0.46 kg P ha^{-1} yr^{-1}) from a nutrient-poor undisturbed tropical moist rain forest in Venezuela were unexpectedly high. The data presented by LEWIS (1986) are consistent with the hypothesis that mature forested ecosystems have high yields of critical nutrients because they are not accumulating biomass. Compared to a rainfall phosphorus input to a watershed of a subtropical floodplain forest in Puerto Rico of 0.63 kg P ha^{-1} yr^{-1}, leaching from the canopy was high (1.67 kg P ha^{-1} yr^{-1}), as was the loss of P from the watershed (6.11 kg P ha^{-1} yr^{-1}). Phosphorus-conserving mechanisms included a high rate of translocation in palms (5.04 kg ha^{-1} yr^{-1}). In spite of these mechanisms, there was a net P loss from the watershed that ranged from 1.36 to 5.44 kg P ha^{-1} yr^{-1} (FRANGI and LUGO 1985).

Mean annual nitrogen fixation by *Inga jinicuil*, a tropical leguminous shade tree in Mexican coffee plantations, equalled 35 kg N ha^{-1} yr^{-1}, which constitutes a significant nitrogen input to the coffee ecosystem (ROSKOSKI and VAN KESSEL 1985). Foliar nutrient relations of non-N_2-fixing and potentially N_2-fixing deciduous trees at five savanna woodland sites in Tanzania were compared by HÖGBERG (1986) with data on soil nutrients. Nodulation and nitrogenase activity were confirmed in mature specimens of 12 of the 14 potentially N_2-fixing legume species studied. The concentrations of nutrients in leaves and the relations between them indicate that N limits growth in non N_2-fixing species at most sites, while P is limiting in N_2-fixing species. Nutrient retention by different types of savanna ecosystems was investigated by KELLMAN and SANMUGADAS (1985), KELLMAN et al. (1985), KELLMAN (1985) and KELLMAN and CARTY (1986) in Belize, Central America, during wet- and dry-season conditions, in the absence of and after fire. They conclude that these oligotrophic soil-vegetation systems are highly retentive of nutrients and are not subject to large short-term losses in the absence of fire. Nutrient distribution in a 10-yr-old *Pinus caribaea* stand planted in a subhumid tropical savanna site in Nigeria was determined by KADEBA and ADUAYI (1986). Soil nutrient capital and nutrient removals in native eucalypt forest harvesting were estimated by TURNER and LAMBERT (1986) in Australia together with potential impacts of these removals on the nutrient capital balance. It is expected that at least four forest rotations (320 yrs) would be required before any detectable change of phosphorus and calcium would occur within the forest communities. The other nutrients had considerably greater soil nutrient reserves.

Changes in distribution and cycling of N, P, K, Mg, and Ca were investigated by GHOLZ et al. (1985) in plantations of *Pinus elliotti* using an age sequence. N cycling of intensively managed *Pinus taeda* plantations of South Carolina was examined by VITOUSEK and MATSON (1985), BIRK and VITOUSEK (1986) and VITOUSEK and ANDARIESE (1986). Nutrient retention in the snow-dominated *Pinus contorta* ecosystems appears to be primarily dependent on evergreen leaf area, duration of the vernal transpiration period, and high carbon/nutrient ratios of the forest floors. Net losses of limiting nutrients probably occur primarily in pulses after abiotic perturbations such as fire (KNIGHT et al. 1985; YAVITT and FAHEY 1986). Annual N and P requirements and internal redistributions were determined by MEIER et al. (1985) for above- and belowground vegetation component of young and mature subalpine *Abies amabilis* stands. Nitrogen mineralization rates in five adjacent Wisconsin forest stands ranged from 29 to 125 kg ha^{-1} yr^{-1} (McCLAUGHERTY et al. 1985). Autumnal resorption and accretion of copper, iron, zinc, and manganese were measured by KILLINGBECK (1985) in the foliage of five gallery forest tree species.

The effect of herbivory on the cycling of nutrients was investigated by WARING and PITMAN (1985), DZIADOWIEC and PLICHTA (1985), HOLLINGER (1986) and SCHROEDER (1986). The amount of nitrogen reaching soil during the mass occurrence of *Lymantria monacha* in a *Pinus sylvestris* forest increased three times, and of phosphorus more than twice. During the outbreak of *Phryganidia californica*, total N and P flow to the ground more than doubled from isolated *Quercus agrifolia* trees.

These results suggest that these herbivore moth species have significant effects on the nutrient cycling beneath these trees. The existence of green islands around wood-ant nests in otherwise damaged *Betula* forests has been explained by two alternative hypotheses: (1) predation by ants protects the trees against defoliators, and (2) the ants ameliorate tree vigor by concentrating soil nutrients. According to NIEMELÄ and LAINE (1986) the size of green islands and the nitrogen content of the soil and foliage do not support the nutrition hypothesis.

5. Ecosystem Development: Succession, Stability and Man's Influence on Ecosystem Structure and Processes

a) General

Although there has long been an awareness amongst ecologists of the importance of the dynamics of ecosystems, the emphasis on the processes and effects of disturbance is fairly recent. Disturbances can be external to the ecosystem or generated within it and their study now involves a wide variety of aquatic and terrestrial ecosystem types and concerns a large variety of temporal and spatial scales. The phrase "patch dynamics" has been adopted as a convenient description of the focus of the research which includes not only the study of disturbances but also phenomena such as the filling and changes within patches, the relation of patches to one another and the matrix, and the flows of organisms, materials, and energy among patches. PICKETT and WHITE (1985) have recognized the need of a unifying framework for the diverse research in disturbance and patch dynamics and have produced an important synthesis which has a broad ecological relevance. Systems theory formulation of ecological disturbance is given by GERRITSEN and PATTEN (1985) and DeANGELIS et al. (1985). The notions of the indices and criteria for a measurement of ecosystem stability were discussed by SVETLOSANOV (1985a,b). Topics such as succession and resilience of ecosystems and ecotoxicology are covered in a book on *Ecology, Impact Assessment, and Environmental Planning* by WESTMANN (1985) while HALL et al. (1985) edited a volume of conference papers on *Economics of Ecosystem Management*.

Forests, vital to maintain environmental stability and the fuelwood and other wood needs of society, are disappearing at an alarming rate throughout the world. Agricultural expansion into about 10 million hectares of new land accounts for most of the 11.6 million hectares of forest land being lost each year. According to PIMENTEL et al. (1986) the most promising policy to reduce deforestation and increase food supplies is to double crop yields over the next two to three decades on the most productive lands. This would require successful control of soil erosion plus the increased use of fertilizers and other agricultural inputs that help increase crop yield. The results presented by MAHN and KÄSTNER (1985) contribute to a better understanding of the complexity of human caused disturbances in agroecosystems, elucidating some relationships between the herbicide stress and its effects on different structural levels.

The term "ecotoxicology" is being used with increasing frequency, but the interpretation of what this developing area of research is really about varies widely. Twenty-six papers selected from those presented at a conference entitled *Ecotoxicology* were edited by RASMUSSEN (1984) and gave a good record of the current state of the knowledge in this field. Methods for estimating risk of chemical injury on ecosystems were part of the book edited by VOUK et al. (1985). Methods for scaling biological response and fluxes of potentially toxic substances in the soil-vegetation complex were described by SHIRAZI et al. (1984) and FRÄNZLE et al. (1985).

For ecologists requiring an introduction to the subject of nuclear power and environment the books of EHRLICH et al. (1984) and EICHHOLZ (1985) give some useful information. A mathematical description of the dispersion of radioactive pollution in surface water is presented by CHYTIL (1984).

b) Aquatic Ecosystems

A dynamic spatial simulation model of coastal wetland habitat succession was designed by SKLAR et al. (1985) to project habitat changes as a function of marsh type, while JOHNSON et al. (1985) described the succession of wetland vegetation in an evolving river delta. Internal factors controlling peatland-lake ecosystem development were examined by KRATZ and DEWITT (1986). Their results underscore the importance of spatial dynamics in peatland development. BARKO and SMART (1986) tested sediment-related mechanisms of growth limitation in submersed macrophytes during lake succession.

Aerial photographs show freshwater vegetation changes since the 1950's in a lowland tropical swamp in Papua New Guinea. Records of diatom assemblages and plant seeds in sediment cores confirm that the lake was at one time dominated by a mixed emergent swamp, that this was replaced by floating-leaved plants and that the latter were replaced by the present open water. Water-level change is an obvious cause, but cannot explain all features of the vegetational changes observed by OSBORNE and POLUNIN (1986). Evidence is presented to show that nutrient enrichment could have been influential. These two influences cannot be separated easily in the present record because the same cause – increase in sewage and urban run-off with human development in the catchment – has led to both. The restoration of two lowland lakes in England by isolation from nutrient-rich water sources with and without removal of its recently deposited phosphorus-rich sediments was investigated by MOSS et al. (1986). A reduction in phytoplankton growth followed soon after isolation and sediment removal, and the water became clear and has remained so up today. Long-term chemical and phytoplankton analysis were made by MORLING et al. (1985) in softwater lakes in Sweden that have experienced severe acidification during a 16-yr period. Results indicated that acidification effects are readily reversible once sulfate loading is reduced. Model aquatic ecosystems were established and analyzed by FEIND et al. (1985), SHIKANO and KURIHARA (1985) and RECKNAGEL (1985) to investigate the long term effects of several water pollutants.

The impact of civil engineering works on the successions of communities in a fluvial system was studied by BRAVARD et al. (1986) in a methodological and predictive approach applied to a section of the Rhone River, France. To study the impact of an altered hydrologic regime on the growth and succession of a coastal forested floodplain in South Carolina, a model was developed by PEARLSTINE et al. (1985). The model predicted a loss of up to 97% of the existing bottomland forest.

c) Grasslands and Herb Communties, Old-Field Succession

GOLDBERG and TURNER (1986) report on vegetation changes and population dynamics during a 72-yr period within permanent plots in Sonoran Desert vegetation. Coverage of most species responded strongly to regimes of extremely wet or extremely dry years; the response of density to climatic extremes was somewhat less strong. Therefore, the relative cover of the dominants was generally similar within a given plot over the entire time sequence. Several investigators have suggested that portions of the perennial grassland of the Serengeti Plains, Tanzania, have recently been converted to annual-grass and shrub-dominated communities due to overgrazing and disturbances by wildebeest, which have increased fivefold in population since 1961. Study sites that were described and mapped in the 1960's and 1970's were revisited by BELSKY (1985, 1986a) in 1982. She found no changes in species composition in any of the communities. Furthermore, BELSKY (1986b,c) investigated the effects of small, physical disturbances on the structure and species composition and of grazing on colonization and regrowth in four grassland communities of the Serengeti Plains. The depth of disturbance significantly affected 28% of the species; seedlings were more abundant in the deeply disturbed and sprouting species were more abundant in shallowly disturbed plots. Grazing significantly affected 23% of the species. Most species escaped the effects of grazing by being prostrate, resprouting from ground-level or underground organs, or by growing before or after the periods of heaviest grazing. After the elimination of grazing, short and sexually-reproducing species disappeared from the short- and mid-grass communities, which became dominated by tall, vegetatively-reproducing species. The mosaic and tall-grass communities were unchanged.

The ecological development of reclaimed coal mine waste near the timberline the Canadian Rockies was examined by comparing six sites revegetated with a seeded grass-legume sward, ranging in age from 0 to 6 years, a naturally revegetated disturbed area, and a native grassland (FYLES et al. 1985). Studies were made by DAVIS et al. (1985) and RICHARDSON and EVANS (1986) of the factors thought to be responsible for slow development of vegetation on a limestone quarry floor, and on the interdependence of cover and species richness. Species establishment after 3 yrs on artificially revegetated ski slopes and the nature of species associations were investigated by THOMPSON and HUTCHINSON (1986) in relation to altitude, slope angle and soil properties. Changes in species richness and abundance following experimental disturbance were studied by ARMESTO and PICKETT (1985) in a 7-yr-old field dominated by *Solidago canadensis* and a 2-yr-old field dominated by *Ambrosia artemisiifolia*. The influence of cadmium on a secondary succession during the first years of vegetation development was investigated by CORNELIUS (1985). The soil pollution had no significant influence on species composition, but on the abundance of some dominant plants.

Changes in phytocoenose structure and plant biomass in early phases of old-field succession in Poland were studied by SYMONIDES (1985a,b) and SYMONIDES and BOROWIECKA (1985). Regardless of the succession phase, the biomass is characterized by an aggregated type of spatial structure; it is primarily the structure intensity that varies. The study of CID-BENEVENTO and WERNER (1986) combined the use of demographic variables and physiological criteria to explain species distribution patterns of summer-flowering old-field and woodland annuals. The responses of *Polygonum pensylvanicum*, an early successional annual, and of *P. virginianum*, a late successional perennial, were examined by LEE et al. (1986) along discontinuous gradients of soil moisture, light and nutrient availability. Leaf dynamics and carbon gain as well as differential costs of sexual and vegetative reproduction were calculated by JURIK (1985) and JURIK and CHABOT (1986) for five populations of *Fragaria virginiana* and *F. vesca* on sites representing a gradient of successional regrowth. The effects of two stem gallmakers, fertilization, and ramet isolation on nutrient and biomass allocation in *Solidago altissima*, a dominant species in many North American old-fields, was investigated by McCREA et al. (1985) and ABRAHAMSON and McCREA (1986). Leaf-cutting ant herbivory in successional and agricultural tropical ecosystems was examined by BLANTON and EWEL (1985) while BROWN and HYMAN (1986) described successional communities of plants and phytophagous Coleoptera from the colonization of bare ground, through permanent pastureland to birch woodland of known successional age.

d) Forest Succession

A stochastic-deterministic simulation model of shrub succession was developed by STEINHORST et al. (1985), while KACHI et al. (1986) described revegetation following clear-cutting of a *Pinus densiflora* forest by a stochastic model. HOBBS and MOONEY (1986) studied community changes following shrub invasion of a grassland in California. Exclosures suggested that herbivory by small mammals in the closed shrub stand may be important in reducing the abundance of herbaceous species following shrub invasion of grassland. The effects of insect and mammal herbivores on the regrowth of shrub seedlings and crown sprouts were studied by MILLS (1986) during the early postfire succession period in Californian chaparral. Herbivore pressures were sufficient to eliminate a competitive advantage of *Ceanothus greggii* in the absence of herbivory. In a Danish heathland, invasion of *Quercus* shrub has taken place, the succession rate being approximately 300 m during the last 100 years. Colonization experiments with radioactive acorns showed that rodent acorn dispersal can explain the observed succession rate of *Quercus* into the heathland (JENSEN and NIELSEN 1986). The nutrient contents of throughfall under a *Betula pendula* and a *Pinus sylvestris* canopy were measured and compared by ALCOCK and MORTON (1985) with rainfall collected at an adjacent heathland site as part of a study of the effects of woodland development on the nutrient status of heathland. The annual nutrient exchange budgets of the tree canopies represented at net leaching of K and Mg and a net absorption of Ca and possibly also of N and P. If the rates of nutrient uptake and loss from the three ecosystems are similar then, on the basis of these results alone, it might be expected that the K and Mg content of the soils under the woodland sites would increase, whilst the Ca, and possibly the P and N contents, may decrease. The role of fire on vegetation structure and nutrient cycling in South African fynbos (sclerophyllous shrubland) succession was investigated by VAN WILGEN et al. (1985), VAN WILGEN and RICHARDSON (1985) and STOCK and LEWIS (1986).

Growth of woody species has two components, one pertaining to increment of biomass, the other to the architectonic arrangement of dry matter in space. Only the combination of both components explained the competitive strength of species that dominate in different stages of a hedgerow succession in Central Europe as well as of tree species of a tropical forest in India. Plant architecture, biomass increment and assimilate partitioning of woody species in a hedgerow were determined by KÜPPERS (1985). Those species with lower dry matter, nitrogen, and phosphorus costs for occupying and shading space during canopy development were the stronger competitors and replaced pioneers, even though their rate of CO_2 uptake was low. According to SHUKLA and RAMAKRISHNAN (1986), the extension and radial growth rates of early-successional tropical tree species were very rapid compared with late-successional species. Faster growth over a longer annual growth period results in the sparse branch arrangement in early-successional species which facilitates greater leaf exposure. The slower growth rate over a shorter period results in densely packed canopies with mutual shading of leaves of late-successional species. Gap-phase regeneration of trees was described by BROKAW (1985) in a tropical forest in Panama. Gap formation produces patches that differ markedly in tree population dynamics, species composition, and growth rate. Some biological and chemical changes in a soil following replacement of rainforest with *Araucaria cunnighamii* were investigated by HOLT and SPAIN (1986) in Australia. They suggested that the decline of organic matter and associated nutrients in the *Araucaria* forest soil results from the loss of a large proportion of the plant nutrients from the ecosystem when the original rainforest was cleared. The subsequent establishment and growth of an *Araucaria* plantation then lowered the existing soil nutrient supplies by transfer of nutrients into the vegetation. Different aspects of forest succession in North American forest ecosystems were discussed by GLITZENSTEIN et al. (1986), CLARK (1986), STEWART (1986) and FOSTER

(1985). The pattern of primary succession on an Alaskan floodplain dominated by *Salix alaxensis*, *Alnus tenuifolia*, *Populus balsamifera*, and *Picea glauca* resulted largely from a complex interaction of life history, facilitative, competitive, and stochastic processes (WALKER et al. 1986; WALKER and CHAPIN III 1986). The growth strategies of typical pioneer and climax species of forest ground cover were studied by HOLM and KELLOMÄKI (1984), based on models for ground cover succession.

e) Air Pollution, Effects of Acid Precipitation and Forest Dieback

The past decade has seen an increase in public and political awareness of the problems of air pollution, as well as major advances in knowledge in many aspects of the subject. The books edited by STERN et al. (1984), STRAUSS and MAINWARING (1984), SHEEHAN et al. (1984) and NURNBERG (1985) are dealing with general problems and recent developments in the field of air pollution while the books of BUBENICK (1984), FARMER (1984) and THE ACID RAIN FOUNDATION (1985) provide a more complete picture of the acid rain situation and its effects on ecosystems. The book edited by NRIAGU and DAVIDSON (1986) deals with different aspects of the subject of toxic or trace metals in the atmosphere, spanning a very wide range from chemistry through engineering to atmospheric modelling. As a result of high emissions, increased density of emittors and general use of tall stocks, the present ambient air pollution situation in Central Europe is characterized by the large-scale occurrence of persistent exposures to pollutant mixtures. Therefore, special attention should be given to latent injury as a result of pollutant accumulation and long-term exposures and to reactions of ecosystems (GUDERIAN 1986). Models of forest ecosystems, based on data of *Picea abies* and of a mixed-conifer forest, were developed by METZLER et al. (1986) and KERCHER and AXELROD (1984) to demonstrate the possible dynamics of forest dieback caused by air pollution and fire.

The influence of acid rain on the quantitative distribution and chemical composition of forest ecosystem compartments was investigated by many authors. The highest deposition rates of important elements in rain throughfall and stemflow were found by BLOCK and BARTELS (1986) in the highly polluted area of Nordrhein-Westfalen (West Germany) under *Picea abies* followed by *Pinus sylvestris*. Strong polluted are also the stem flows of *Fagus sylvatica* and *Quercus rubra*. The stemflow water contains acid depositions and modifies the soil properties in root space. At the high-stemflow side of root area the brown forest soil beneath old *Fagus sylvatica* trees changed during a few decades from pH 5.0 - 6.2 to pH 3.2 - 4.2 (GLAVAC and KOENIES 1986). For many years not only acid components but also some toxic heavy metals belong to the long-range transported air pollutants and are deposited even in forest ecosystems far away from industrial areas. Measurements of ecologically relevant heavy metals were made by KAHLE and BRECKLE (1986), FANGMEIER et al. (1986), SCHMIDT and MAYER (1986) and SCHULTZ (1986) in different West German forest ecosystems, while MANKOVSKA (1986) monitored the effect of sulfur dioxide and heavy metals emitted from a smelter on a *Pinus sylvestris* forest in Poland. The effects of acid deposition and heavy metal accumulation in forest floor were studied by LETTL (1985, 1986), TONNER et al. (1985), BIENKOWSKI et al. (1986) and SCHÄFER (1986) in relation to soil microflora, decomposition and nitrogen mineralization.

Soil acidification and vegetation changes in deciduous forest in Sweden were investigated by FALKENGREN-GRERUP (1986). The average pH change in the humus layer was -0.78 and -0.23 in soils originally studied 30 - 35 and 15 - 20 years ago, respectively. The number of species had increased in spite of pH decreases, reaching a maximum at pH 4.0 - 5.0, while the total cover of the field layer was unchanged. Without excluding other factors, FALKENGREN-GRERUP (1986) suggests that the increased nitrogen deposition and the increased acidity in the humus layer might cause some of the cover changes in the presented species. A comparison of the floral composition of the herb layer of a *Fagus sylvatica* forest community of West Germany in 1976 with the composition in 1983 indicated that, in 1983, the sites were more favourable to acidophilic plant species than in 1976 (WITTIG et al. 1985).

References

ABER, J.D., MELILLO, J.M.; NADELHOFFER, K.J., McCLAUGHERTY, C.A., PASTOR, J.: Oecologia 66, 317-321 (1985). - ABRAHAMSON, W.G., McCREA, K.D.: Oecologia 68, 174-180 (1986). - ALCOCK, M.R., MORTON, A.J.: J. Ecol. 73, 625-632 (1985). - ALEKSEEV, V.V., KORNILOVSKY, A.N.: Ecol. Model. 28, 217-229 (1985). - ALEKSEEV, V.V., KOKORIN, A.O., LYAMIN, M.Y.: Ecol. Model. 25, 31-46 (1984). - ANDERSON, M.R., KALFF, J.: J. Ecol. 74, 953-961 (1986). - ARMENTANO, T.V., MENGES, E.S.: J. Ecol. 74, 755-774 (1986). - ARMESTO, J.J., PICKETT, S.T.A.: Ecology 66, 230 240 (1985).

BAILIAN, L.: Ecol. Model. 32, 227-236 (1986). - BAJKIEWICZ-GRABOWSKA, E.: Ekol. Polska 33, 201-223 (1985a); - ibid. 33, 173-200 (1985b). - BALDOCCHI, D.D., HUTCHISON, B.A., MATT, D.R., McMILLAN, R.T.: J. Appl. Ecol. 22, 539-555 (1985). - BALDOCCHI, D.D., VERMA, S.B., MATT, D.R., ANDERSON, D.E.: J. Appl. Ecol. 23, 967-975 (1986). - BARKO, J.W., SMART, R.M.: Ecology 67, 1328-1340 (1986). - BAZELY, D.R., JEFFERIES, R.L.: J. Appl. Ecol. 22, 693-703 (1985). - BEADLE, C.L., LONG, S.P., IMBAMBA, S.K., HALL, D.O., OLEMBO, R.J.: Photosynthesis in Relation to Plant Production in Terrestrial Environments. Natural Resources and Environment Series, Vol. 18, 156 pp. Oxford: United Nations Environmental Programme and Tycooly Publ. 1985. - BEGON, M., HARPER, J.L., TOWNSEND, C.R.: Ecology: Individuals, Populations and Communities, 876 pp. Sunderland, MA: Sinauer 1986. - BELSKY, A.J.: J. Appl. Ecol. 22, 449-460 (1985); - J. Ecol. 74, 841-856 (1986a); - ibid. 74, 419-437 (1986b); - ibid. 74, 937-951 (1986c). - BERTNESS, M.D.: Ecology 66, 1042-1055 (1985). - BIENKOWSKI, P., FISCHER, Z., GOZDZIEWICZ, Z.: Ekol. Polska 34, 75-86 (1986). - BILLINGS, W.D., PETERSON, K.M., LUKEN, J.O., MORTENSEN, D.A.: Oecologia 65, 26-29 (1984). - BIRK, E.M., VITOUSEK, P.M.: Ecology 67, 69-79 (1986). - BLACK, R.A., MACK, R.N.: Ecology 67, 1289-1302 (1986). - BLANTON, C.M., EWEL, J.J.: Ecology 66, 861-869 (1985). - BLOCK, J., BARTELS, U.: Verh. Ges. Ökol. 14, 257-264 (1986). - BOLLA, M., KUTAS, T.: Ecol. Model. 26, 115-143 (1984). - BOSTON, H.L., ADAMS, M.S.: Oecologia 68, 615-622 (1986). - BOWDEN, W.B.: Ecology 67, 88-99 (1986). - BRANER, M.: Ecol. Model 28, 165-180 (1985). - BRAVARD, J.-P., AMOROS, C., PAUTOU, G.: Oikos 47, 92-111 (1986). - BROKAW, N.V.L.: Ecology 66, 682-687 (1985). - BROWN, V.K., HYMAN, P.S.: J. Ecol. 74, 963-975 (1986). - BUBENICK, D.V. (ed.): Acid Rain Information Book, 396 pp. New Jersey: Noyes 1984. - BUTTLEMAN, C.G., GRIGAL, D.F.: Oikos 44, 253-256 (1985).

CAHOON, D.R., STEVENSON, J.C.: Ecology 67, 1341-1350 (1986). - CARPENTER, E.J., CAPONE, D.G. (eds.): Nitrogen in the Marine Environment, 900 pp. New York: Academic Press 1983. - CERMAK, J., JENIK, J., KUCERA, J., ZIDEK, V.: Oecologia 64, 145-151 (1984). - CHABOT, B.F., MOONEY, H.A. (eds.): Physiological Ecology of North American Plant Communities, 351 pp. New York: Chapman and Hall 1985. - CHAPIN III, F.S., SHAVER, G.R., KEDROWSKI, R.A.: J. Ecol. 74, 167-195 (1986). - CHYTIL, I.: Ecol. Model. 26, 145-153 (1984). - CID-BENEVENTO, C.R., WERNER, P.A.: J. Ecol. 74, 857-880 (1986). - CLARK, J.S.: Ecol. Monographs 56, 259-277

(1986). - COOLEY, J.H., GOLLEY, F.B. (eds.): Trends in Ecological Research for the 1980's. NATO Conference Series: Ecology, Vol. 7, 344 pp. New York: Plenum Press 1984. - CORNELIUS, R.: Verh. Ges. Ökol. 13, 627-630 (1985). - COUGHENOUR, M.B.: Ecol. Model. 26, 203-230 (1984). - COUGHENOUR, M.B., McNAUGHTON, S.J., WALLACE, L.L.: Ecol. Model. 23, 101-134 (1984a); - ibid. 26, 177-201 (1984b); - Oecologia 68, 80-86 (1985). - CUEVAS, E., MEDINA, E.: Oecologia 68, 466-472 (1986).

DAHLMAN, R.C.: Ecol. Model. 29, 77-106 (1985). - DALE, V.H., DOYLE, T.W., SHUGART, H.H.: Ecol. Model. 29, 145-169 (1985). - DAVIS, B.N.K., LAKHANI, K.H., BROWN, M.C., PARK, D.G.: J. Appl. Ecol. 22, 473-490 (1985). - DeANGELIS, D.L., WATERHOUSE, J.C., POST, W.M., O'NEILL, R.V.: Ecol. Model. 29, 399-419 (1985). - DeLAUNE, R.D., SMITH, C.J., SARAFYAN, M.N.: J. Ecol. 74, 249-256 (1986). - DE LEEUW, J., BAKKER, J.P.: J. Appl. Ecol. 23, 781-793 (1986). - DESHMUKH, I.: Oecologia 67, 147-149 (1985); - J. Appl. Ecol. 23, 115-123 (1986). - DIAMOND, J., CASE, T.J. (eds.): Community Ecology, 665 pp. New York: Harper and Row 1986. - DICKERMAN, J.A., STEWART, A.J., WETZEL, R.G.: Ecology 67, 650-659 (1986). - DITTMER, J., JUODAITIS, U., RACHOR, E.: Verh. Ges. Ökol. 13, 37-40 (1985). - DOWNING, J.A., WEBER, L.A.: Ecol. Model. 23, 227-241 (1984). - DUSOGE, K., BOWNIK-DYLINSKA, L., EJSMONT-KARABIN, J., SPODNIEWSKA, I., WEGLENSKA, T.: Ekol. Polska 33, 455-479 (1985). - DZIADOWIEC, H., PLICHTA, W.: Ekol. Polska 33, 715-728 (1985). - DZUREC, R.S., BOUTTON, T.W., CALDWELL, M.M.; SMITH, B.N.: Oecologia 66, 17-24 (1985).

EGLOFF, T.: Verh. Ges. Ökol. 13, 127-137 (1985). - EHRLICH, P.R., SAGAN, C., KENNEDY, D., ROBERTS, W.O.: The Nuclear Winter: the Cold and the Dark, 227 pp. London: Sidgwick & Jackson 1984. - EICHHOLZ, G.G.: Environmental Aspects of Nuclear Power, 683 pp. Michigan: Lewis 1985. - ELKINS, N.Z., SABOL, G.V., WARD, T.J., WHITFORD, W.G.: Oecologia 68, 521-528 (1986). - ELLENBERG, H., MAYER, R., SCHAUERMANN, J. (eds.). Ökosystemforschung - Ergebnisse des Sollingprojekts, 507 pp. Stuttgart: Ulmer 1986. - EVENARI, M., NOY-MEIR, I., GOODALL, D.W. (eds.). Hot Deserts and Arid Shrublands A. Ecosystems of the World, 12A, 365 pp. Amsterdam: Elsevier 1985; - Hot Deserts and Arid Shrublands B., Ecosystems of the World, 12B, 451 pp. Amsterdam: Elsevier 1986. - EWEL, K.C., ODUM, H.T. (eds.): Cypress Swamps, 472 pp. Gainsville: University of Florida Press 1985.

FAIRCHILD, G.W., LOWE, R.L., RICHARDSON, W.B.: Ecology 66, 465-472 (1985). - FALKENGREN-GRERUP, U.: Oecologia 70, 339-347 (1986). - FANGMEIER, A., STEUBING, L., GNITTKE, J.: Verh. Ges. Ökol. 14, 269-278 (1986). - FARMER, P.: Acid Rain and the Environment 1980-1984. A Select Bibliography, 108 pp. Letchworth: Technical Communications 1984. - FASHAM, M.J.R. (ed.): Flows of Energy and Materials in Marine Ecosystems: Theory and Practice, 533 pp. New York: Plenum Press 1984. - FEIND, D., ZIERIS, F.-J., HUBER, W.: Verh. Ges. Ökol. 13, 359-367 (1985). - FITTER, A.H.: Oecologia 69, 594-599 (1986). - FOSTER, D.R.: J. Ecol. 73, 517-534 (1985). - FRÄNZLE, O., KUHNT, D., KUHNT, G.: Verh. Ges. Ökol. 13, 323-340 (1985). - FRANGI, J.L., LUGO, A.E.: Ecol. Monographs 55, 351-369 (1985). - FRENEY, J.R., SIMPSON, J.R. (eds.): Gaseous Loss of Nitrogen from Plant-Soil Systems, 317 pp. The Hague: Nijhoff/ Junk 1983. - FYLES, J.W., FYLES, I.H., BELL, M.A.M.: J. Appl. Ecol. 22, 239-248 (1985).

GABRIS, L., SUJANSKA, L.: Ecology (CSSR) 4, 61-73 (1985). - GATES, D.M.: Energy and Ecology, 377 pp. Sunderland, MA: Sinauer 1985. - GERRITSEN, J., PATTEN, B.C.: Ecol. Model. 29, 383-397 (1985). - GHOLZ, H.L., FISHER, R.F., PRITCHETT, W.L.: Ecology 66, 647-659 (1985). - GILLER, P.S.: Community Structure and the Niche. Outline Studies in

Ecology, 176 pp. New York: Chapman and Hall 1984. - GLAVAC, V., KOENIES, H.: Verh. Ges. Ökol. 14, 293-298 (1986). - GLITZENSTEIN, J.S., HARCOMBE, P.A., STRENG, D.R.: Ecol. Monographs 56, 243-258 (1986). - GODA, T., MATSUOKA, Y.: Ecol. Model. 31, 11-32 (1986). - GOLDBERG, D.E., TURNER, R.M.: Ecology 67, 695-712 (1986). - GOLKIN, K.R., EWEL, K.C.: Ecol. Model. 24, 113-136 (1984). - GOSZCZYNSKA, W.: Ekol. Polska 33, 241-269 (1985). - GROBLER, D.C.: Ecol. Model. 31, 237-247 (1986). - GROFFMAN, P.M., HOUSE, G.J., HENDRIX, P.F., SCOTT, D.E., CROSSLEY Jr., D.A.: Ecology 67, 80-87 (1986). - GUDERIAN, R.: Verh. Ges. Ökol. 14, 245-256 (1986).

HAASE, P.: Oecologia 68, 456-458 (1986). - HALL, D.O., MYERS, N., MARGARIS, N.S. (eds.): Economics of Ecosystem Management. Tasks for Vegetation Science, 14, 244 pp. Dordrecht: Junk 1985. - HANSEN, L., KROG, G.F., SONDERGAARD, M.: Oikos 46, 37-44 (1986). - HANSON, J.D., PARTON, W.J., INNIS, G.S.: Ecol. Model. 29, 131-144 (1985). - HANSSON, A.-C., ANDREN, O.: J. Appl. Ecol. 23, 657-666 (1986). - HART, B.T., OTTAWAY, E.M., NOLLER, B.N.: Ecol. Model. 31, 249-265 (1986). - HERWITZ, S.R.: Oecologia 70, 423-425 (1986). - HIGASHI, M.: Ecol. Model. 32, 137-147 (1986a); - ibid. 32, 243-250 (1986b). - HIGASHI, M., PATTEN, B.C.: Ecol. Model. 31, 69-77 (1986). - HOBBS, R.J., MOONEY, H.A.: Oecologia 70, 508-513 (1986). - HÖGBERG, P.: J. Appl. Ecol. 23, 675-688 (1986). - HOLLINGER, D.Y.: Oecologia 70, 291-297 (1986). - HOLM, J., KELLOMÄKI, S.: Ecol. Model. 23, 135-150 (1984). - HOLT, J.A., SPAIN, A.V.: J. Appl. Ecol. 23, 227-237 (1986). - HOWES, B.L., DACEY, J.W.H., TEAL, J.M.: Ecology 66, 595-605 (1985). - HOWES, B.L., DACEY, J.W.H., GOEHRINGER, D.D.: J. Ecol. 74, 881-898 (1986). - HUILLET, T., LAUGA, J.: Ecol. Model. 27, 235-250 (1985). - HUTCHINSON, B.A., HICKS, B.B. (eds.): The Forest-Atmosphere Interaction. Proceedings of the Forest Environmental Measurements Conference held and Oak Ridge, TN, October 1983, 684 pp. Dordrecht: Reidel 1985. - HUTZINGER, O. (ed.): The Natural Environment and the Biogeochemical Cycles. The Handbook of Environmental Chemistry IC, 220 pp. Berlin, Heidelberg, New York: Springer 1984; - (ed.): The Handbook of Environmental Chemistry, Vol. 1. The Natural Environment and the Biogeochemical Cycles, Part D, 246 pp. Berlin, Heidelberg, New York: Springer 1985.

INGHAM, E.R., TROFYMOW, J.A., INGHAM, E.R., COLEMAN, D.C.: Ecol. Monographs 55, 119-140 (1985). - INGHAM, E.R., TROFYMOW, J.A., AMES, R.N., HUNT, H.W., MORLEY, C.R., MOORE, J.C., COLEMAN, D.C.: J. Appl. Ecol. 23, 597-614 (1986a); - ibid. 23, 615-630 (1986b). - INO, I., NAKATSUBO, T.: Ecol. Res. 1, 59-69 (1986).

JACKSON, D., HARKNESS, D.D., MASON, C.F., LONG, S.P.: Oikos 46, 163-170 (1986a). - JACKSON, D., LONG, S.P., MASON, C.F.: J. Ecol. 74, 647-662 (1986b). - JAKUBOWSKA, L., SUCHECKA, T.: Ekol. Polska 33, 547-561 (1985). - JAKUCS, P. (ed.): Ecology of an Oak Forest in Hungary, 546 pp. Budapest: Akademai Kiado 1985. - JENSEN, T.S., NIELSEN, O.F.: Oecologia 70, 214-221 (1986). - JOHNSON, W.B., SASSER, C.E., GOSSELINK, J.G.: J. Ecol. 73, 973-986 (1985). - JONES, M.B., MUTHURI, F.M.: J. Ecol. 73, 481-491 (1985). - JORDAN, C.F.: Nutrient Cycling in Tropical Forest Ecosystems, 190 pp. Chichester: Wiley 1985. - JORGENSEN, S.E., KAMP-NIELSEN, L., CHRISTENSEN, T., WINDOLF-NIELSEN, J., WESTERGAARD, B.: Ecol. Model. 32, 165-182 (1986a). - JORGENSEN, S.E., KAMP-NIELSEN, L., JORGENSEN, L.A.: Ecol. Model. 32, 251-266 (1986b). - JURIK, T.W.: Oecologia 66, 394-403 (1985). - JURIK, T.W., CHABOT, B.F.: Oecologia 69, 296-304 (1986).

KACHI, N., YASUOKA, Y., TOTSUKA, T., SUZUKI, K.: Ecol. Model. 32, 105-117 (1986). - KADEBA, O., ADUAYI, E.A.: OIKOS 46, 237-242 (1986). - KAHLE, H., BRECKLE, S.-W.: Verh. Ges. Ökol. 14, 265-267 (1986). - KAWAGUCHI, H., YODA, K.: Jap. J. Ecol. 35, 551-563 (1986). - KAWASAKI,

S., GROMIEC, M.J., HANAKI, K., MATSUMOTO, J.: Ecol. Model. 32, 183-190 (1986). - KELLMAN, M.: J. Ecol. 73, 963-972 (1985). - KELLMAN, M., CARTY, A.: J. Appl. Ecol. 23, 211-226 (1986). - KELLMAN, M., SANMUGADAS, K.: J. Ecol. 73, 935-951 (1985). - KELLMAN, M., MIYANISHI, K., HIEBERT, P.: J. Ecol. 73, 953-962 (1985). - KERCHER, J.R., AXELROD, M.C.: Ecol. Model. 23, 165-184 (1984). - KILLINGBECK, K.T.: Ecology 66, 283-286 (1985). - KISHI, M., IKEDA, S.: Ecol. Model. 31, 145-174 (1986). - KNIGHT, D.H., FAHEY, T.J., RUNNING, S.W.: Ecol. Monographs 55, 29-48 (1985). - KÖRNER, CH., BANNISTER, P., MARK, A.F.: Oecologia 69, 577-588 (1986). - KOSCHEL, R., KASPRZAK, P.: Ecol. Model. 26, 97-101 (1984). - KOWALCZEWSKI, A., PERLOWSKA, M., PRZYLUSKA, J.: Ekol. Polska 33, 389-405 (1985a); - ibid. 33, 407-422 (1985b); - ibid. 33, 423-438 (1985c). - KOWALIK, P.J., ECKERSTEN, H.: Ecol. Model. 26, 251-284 (1984). - KOZERSKI, H.-P., SCHELLENBERGER, G., BEHRENDT, H., MOHAUPT, V.: Ecol. Model. 26, 103-113 (1984). - KRATZ, T.K., DeWITT, C.B.: Ecology 67, 100-107 (1986). - KRUCKEBERG, A.R.: California Serpentines: Flora, Vegetation, Geology, Soils and Management Problems, 180 pp. Berkeley: University of California Press 1985. - KUBICEK, F., SOMSAK, L.: Ecology (CSSR) 4, 347-365 (1985). - KÜPPERS, M.: Oecologia 66, 343-352 (1985).

LAUENROTH, W.K., HUNT, H.W., SWIFT, D.M., SINGH, J.S.: Ecology 67, 580-582 (1986). - LAWACZ, W.: Ekol. Polska 33, 357-381 (1985). - LAWACZ, W., GOSZCZYNSKA, W., KUROWSKI, C., TOMASZEWSKI, K.: Ekol. Polska 33, 271-300 (1985). - LEE, S.H., ZANGERL, A.R., GARBUTT, K., BAZZAZ, F.A.: Oecologia 68, 606-610 (1986). - LEHN, H.: Verh. Ges. Ökol. 13, 347-352 (1985). - LETTL, A.: Ecology (CSSR) 4, 19-31 (1985); - ibid. 5, 313-329 (1986). - LEWIS, W.M. Jr.: Ecology 67, 1275-1282 (1986). - LEWIS, W.M., Jr., GRANT, M.C., HAMILTON, S.K.: Oikos 45, 428-432 (1985). - LIEBERMAN, D., LIEBERMAN, M., PERALTA, R., HARTSHORN, G.S.: J. Ecol. 73, 915-924 (1985). - LIETH, H.: Verh. Ges. Ökol. 13, 403-414 (1985). - LIKENS, G.E.: J. Ecol. 73, 381-396 (1985). - LORBER, M.N., MISHOE, J.W., REDDY, P.R.: Ecol. Model. 24, 61-77 (1984). - LOWRANCE, R., STINNER, B.R., HOUSE, G.J. (eds).: Agricultural Ecosystems: Unifying Concepts, 233 pp. Chichester: Wiley 1984. - LOWRANCE, R.R., LEONARD, R.A., ASMUSSEN, L.E., TODD, R.L.: Ecology 66, 287-296 (1985). - LUDDEN, E., ADMIRAAL, W., COLIJN, F.: Oecologia 66, 50-59 (1985).

MAHN, E.-G., KÄSTNER, A.: Oikos 44, 185-190 (1985). - MALMER, N., WALLEN, B.: Oikos 46, 200-206 (1986). - MANKOVSKA, B.: Ecology (CSSR) 5, 71-79 (1986). - MARSHALL, J.D., WARING, R.H.: Ecology 67, 975-979 (1986). - MATSUOKA, Y., GODA, T., NAITO, M.: Ecol. Model. 31, 201-219 (1986). - MacNEIL, M.D., SKILES, J.W., HANSON, J.D.: Ecol. Model. 29, 57-76 (1985). - McLAUGHERTY, C.A., PASTOR, J., ABER, J.D., MELILLO, J.M.: Ecology 66, 266-275 (1985). - McCREA, K.D., ABRAHAMSON, W.G., WEIS, A.E.: Ecology 66, 1902-1907 (1985). - McINTOSH, R.P.: The Background of Ecology: Concept and Theory, 383 pp. New York: Cambridge University Press 1985. - McKNIGHT, D., THURMAN, E.M., WERSHAW, R.L., HEMOND, H.: Ecology 66, 1339-1352 (1985). - McNAUGHTON, S.J.: Ecol. Monographs 55, 259-294 (1985). - McNAUGHTON, S.J., CHAPIN III, F.S.: Ecology 66, 1617-1629 (1985). - MEIER, C.E., GRIER, C.C., COLE, D.W.: Ecology 66, 1928-1942 (1985). - METZLER, W., GOCKERT, D., KRIEGER, H., SCHÄFER, H.: Verh. Ges. Ökol. 14, 319-334 (1986). - MEYERS, D.G., STRICKLER, J.R. (eds.): Trohpic Interactions within Auqatic Ecosystems. AAAS Selected Symposium 85, 472 pp. Boulder, CO: Westview Press 1984. - MILLS, J.N.: Ecology 67, 1637-1649 (1986). - MITCHELL, D.T., COLEY, P.G.F., WEBB, S., ALLSOPP, N.: J. Ecol. 74, 977-993 (1986). - MIYANAGA, Y.: Ecol. Model. 31, 133-144 (1986). - MOORE, P.G., SEED, R. (eds.): The Ecology of Rocky Coasts: Essays Presented to J.R. LEWIS, 467 pp. New York: Columbia University Press 1986. - MORLING, G., FORSBERG, C., WETZEL, R.G.: Oikos 44, 324-330 (1985). - MOSS, B., BALLS, H., IRVINE, K., STANSFIELD, J.: J. Appl. Ecol. 23, 391-414 (1986). - MULHOLLAND, P.J., NEWBOLD, J.D., ELWOOD, J.W., WEBSTER, J.R.: Ecology 66, 1012-1023 (1985).

NADELHOFFER, K.J., ABER, J.D., MELILLO, J.M.: Ecology 66, 1377-1390 (1985). - NAKANE, K.: Jap. J. Ecol. 36, 29-39 (1986). - NAKANE, K., TSUBOTA, H., YAMAMOTO, M.: Ecol. Res. 1, 47-58 (1986). - NAKANISHI, H., UKITA, M., KAWAI, Y.: Ecol. Model. 31, 105-123 (1986). - NEWELL, S.Y., FALLON, R.D., RODRIGUEZ, R.M.C., GROENE, L.C.: Oecologia 68, 73-79 (1985). - NIEMELÄ, P., LAINE, K.J.: Oecologia 68, 476-478 (1986). - NOBEL, P.S.: Oecologia 64, 1-7 (1984); - J. Appl. Ecol. 22, 157-173 (1985). - NOBEL, P.S., QUERO, E.: Ecology 67, 1-11 (1986). - NRIAGU, J.O., DAVIDSON, C.I. (eds.): Toxic Metals in the Atmosphere. Wiley Series in Advances in Environmental Science and Technology, Vol. 17, 635 pp. Chichester: Wiley 1986. - NURNBERG, H.W. (ed.): Pollutants and their Ecotoxicological Significance, 515 pp. Chichester: Wiley 1985.

OHUCHI, A., FUKUOKA, J., MIYAKOSHI, A., SUZUKI, M.: Ecol. Model. 32, 149-163 (1986). - OSBORNE, P.L., POLUNIN, N.V.C.: J. Ecol. 74, 197-210 (1986). - OVERDIECK, D., BOSSEMEYER, D.: Angew. Bot. 59, 179-198 (1985).

PALMER, J.P., CHADWICK, M.J.: J. Appl. Ecol. 22, 249-257 (1985). - PARTON, W.J., SINGH, J.S.: Ecol. Model. 23, 151-163 (1984). - PATTEN, B.C.: Verh. Ges. Ökol. 13, 271-284 (1985a); - Ecol. Model. 28, 1-71 (1985b). - PATTEN, B.C., HIGASHI, M.: Ecol. Model. 25, 69-83 (1984). - PEARLSTINE, L., McKELLAR, H., KITCHENS, W.: Ecol. Model. 29, 283-302 (1985). - PENCZAK, T., MOLINSKI, M., GALICKA, W., PREJS, A.: Ekol. Polska 33, 301-309 (1985). - PENKA, M., VYSKOT, M., KLIMO, E., VASICEK, F.: Floodplain Forest Ecosystem. Part 1. Before Water Management Measures. Developments in Agriculture and Managed-Forest Ecology, Vol. 15A, 466 pp. New York: Elsevier 1985. - PETERSON, B.J., HOWARTH, R.W., GARRITT, R.H.: Ecology 67, 865-874 (1986). - PFADENHAUER, J., TWENHÖVEN, F.L.: Flora 178, 157-166 (1986). - PICKETT, S.T.A., WHITE, P.S. (eds.): The Ecology of Natural Disturbance and Patch Dynamics, 472 pp. Orlando, FL: Academic Press 1985. - PIMENTEL, D., DAZHONG, W., EIGENBRODE, S., LANG, H., EMERSON, D., KARASIK, M.: Oikos 46, 404-412 (1986). - PIVEC, J., KUCERA, J.: Ecology (CSSR) 5, 261-269 (1986). - PLANTER, M., WISNIEWSKI, J.: Ekol. Polska 33, 329-344 (1985). - PRICE, P.W., SLOBODCHIKOFF, C.N., GAUD, W.S. (eds.): A New Ecology: Novel Approaches to Interactive Systems, 515 pp. New York: Wiley 1984. - PRIMACK, R.B., ASHTON, P.S., CHAI, P., LEE, H.S.: Ecology 66, 577-588 (1985). - PRISCU, J.C., AXLER, R.P., GOLDMAN, C.R.: Oikos 45, 137-147 (1985). - PUTMAN, R.J., WRATTEN, S.D.: Principles of Ecology, 338 pp. Berkeley, CA: University of California Press 1984.

RAI, S.N., PROCTOR, J.: J. Ecol. 74, 439-453 (1986a); - ibid. 74, 455-463 (1986b). - RAJVANSHI, R., GUPTA, S.R.: Flora 178, 251-260 (1986). - RASMUSSEN, L. (ed.): Ecotoxicology. Ecological Bulletins, 36, 170 pp. Sweden: Swedish Research Council 1984. - RECKNAGEL, F.: Ecol. Model. 27, 221-234 (1985). - REES, D.J.: J. Appl. Ecol. 23, 917-928 (1986). - REYNOLDS, J.F., ACOCK, B.: Ecol. Model. 29, 107-129 (1985). - RICHARDS, J.H.: Oecologia 64, 21-25 (1984). - RICHARDSON, C.J., MARSHALL, P.E.: Ecol. Model. 56, 279-302 (1986). - RICHARDSON, J.A., EVANS, M.E.: J. Appl. Ecol. 23, 317-332 (1986). - RITZ, K., NEWMAN, E.I.: Oikos 45, 174-180 (1985). - ROBERTSON, G.P., ROSSWALL, T.: Ecol. Monographs 56, 43-72 (1986). - ROOM, P.M., THOMAS, P.A.: J. Appl. Ecol. 23, 1013-1028 (1986). - ROSKOSKI, J.P., VAN KESSEL, C.: Oikos 44, 306-312 (1985). - RUNNING, S.W., PETERSON, D.L., SPANNER, M.A., TEUBER, K.B.: Ecology 67, 273-276 (1986).

SAITO, H., TAKEOKA, M.: Jap. J. Ecol. 35, 67-76 (1985). - SALE, P.J.M., ORR, P.T., SHELL, G.S., ERSKINE, D.J.C.: J. Appl. Ecol. 22, 125-137 (1985). - SALOMONS, W., FÖRSTNER, U.: Metals in the Hydrocycle, 349 pp. Berlin, Heidelberg, New York: Springer 1984. - SAWADA, S.: Jap. J. Ecol. 35, 215-223 (1985). - SAWADA, S., NAKAHATA, K., TOTSUKA, T.: Jap. J. Ecol. 35, 453-459 (1985). - SCHAEFER, D., STEINBERGER, Y., WHITFORD,

W.G.: Oecologia 65, 382-386 (1985). - SCHÄFER, H.: Verh. Ges. Ökol. 14, 309-318 (1986). - SCHIMEL, D., STILLWELL, M.A., WOODMANSEE, R.G.: Ecology 66, 276-282 (1985). - SCHLESER, G.H., JAYASEKERA, R.: Oecologia 65, 536-542 (1985). - SCHLESINGER, W.H.: Ecology 66, 1353-1359 (1985). - SCHMIDT, M., MAYER, R.: Verh. Ges. Ökol. 14, 279-285 (1986). - SCHÖNWITZ, R., STICHLER, W., ZIEGLER, H.: Oecologia 69, 305-308 (1986). - SCHROEDER, L.A.: Ecology 67, 1628-1636 (1986). - SCHULTZ, R.: Verh. Ges. Ökol. 14, 287-292 (1986). - SCHULZ, V.M., BÖLLING, C., MATHES, K.: Verh. Ges. Ökol. 13, 809-815 (1985). - SEASTEDT, T.R.: Oecologia 66, 88-92 (1985). - SHARMA, E., AMBASHT, R.S.: J. Appl. Ecol. 23, 689-701 (1986). - SHARPE, P.J.H., WALKER, J., PENRIDGE, L.K., WU, H.: Ecol. Model. 29, 189-213 (1985). - SHAVER, G.R.: Ecology 67, 660-669 (1986). SHAVER, G.R., FETCHER, N., CHAPIN III, F.S.: Ecology 67, 1524-1535 (1986a). - SHAVER, G.R., CHAPIN III, F., GARTNER, B.L.: J. Ecol. 74, 257-278 (1986b). - SHEEHAN, P., MILLER, D., BUTLER, G., BOURDEAU, P. (eds.): Effects of Pollutants at the Ecosystem Level. SCOPE 22, 443 pp. New York: Wiley 1984. - SHIKANO, S., KURIHARA, Y.: Jap. J. Ecol. 35, 297-305 (1985). - SHIMMEL, S.M., DARLEY, W.M.: Ecology 66, 1439-1447 (1985). - SHIRAZI, M.A., LIGHTHART, B., GILLETT, J.: Ecol. Model. 23, 203-226 (1984). - SHIYOMI, M., AKIYAMA, T., TAKAHASHI, S.: Ecol. Model. 32, 119-135 (1986). - SHUKLA, R.P., RAMAKRISHNAN, P.S.: J. Ecol. 74, 33-46 (1986). - SIMM, A.T.: Ekol. Polska 33, 439-453 (1985). - SIOLI, H. (ed.): The Amazon: Limnology and Landscape Ecology of a Mighty Tropical River and its Basin. Monographiae Biologicae, Vol. 56, 763 pp. Boston, MA: Junk 1984. - SKLAR, F.H., COSTANZA, R., DAY, J.W., Jr.: Ecol. Model. 29, 261-281 (1985). - SKOGERBOE, G.V.: Ecol. Model. 31, 45-59 (1986). - SMIL, V.: Carbon-Nitrogen-Sulfur: Human Interference in Grand Biospheric Cycles. Modern Perspectives in Energy, 459 pp. New York: Plenum Press 1985. - SMITH, A., ALLCOCK, P.J.: J. Appl. Ecol. 22, 185-198 (1985). - SMITH, L.M., KADLEC, J.A.: Ecology 66, 259-265 (1985). - SOLOMON, A.M.: Oecologia 68, 567-579 (1986). - SOUTHWICK, C.H. (ed.): Global Ecology, 325 pp. Sunderland, MA: Sinauer 1985. - STASIAK, K., TATUR, A.: Ekol. Polska 33, 225-232 (1985). - STEINHORST, R.K., MORGAN, P., NEUENSCHWANDER, L.F.: Ecol. Model. 29, 35-55 (1985). - STERN, A.C., BOUBEL, R.W., TURNER, D.B., FOX, D.L.: Fundamentals of Air Pollution, 2nd ed., 530 pp. New York: Academic Press 1984. - STEWART, G.H.: Ecology 67, 534-544 (1986). - STOCK, W.D., LEWIS, O.A.M.: J. Ecol. 74, 317-328 (1986). - STRAUSS, W.J., MAINWARING, S.J.: Air Pollution, 154 pp. London: Arnold 1984. - STRONG, D.R. Jr., SIMBERLOFF, D., ABELE, L.G., THISTLE, A.B. (eds.): Ecological Communties: Conceptual Issues and the Evidence, 613 pp. Princeton, NJ: Princeton University Press 1984. - SUMMERS, J.K.: Ecol. Model. 28, 279-309 (1985). - SVETLOSANOV, V.A.: Ecology (CSSR) 4, 427-433 (1985a); - Ecol. Model. 28, 311-322 (1985b). - SYMONIDES, E.: Tuexenia 5, 259-271 (1985a); - Ekol. Polska 33, 61-79 (1985b). - SYMONIDES, E., BOROWIECKA, M.: Ekol. Polska 33, 81-102 (1985).

TANNER, E.V.J.: J. Ecol. 73, 553-568 (1985). - TATUR, A., STASIAK, K.: Ekol. Polska 33, 233-240 (1985). - TENHUNEN, J.D., LANGE, O.L., RADERMACHER, G.: Verh. Ges. Ökol. 13, 415-427 (1985). - THE ACID RAIN FOUNDATION (eds.): Air Pullutants: Effects on Forest Ecosystems. Proceedings of Symposium Held at St. Paul, MN, May 1985, 439 pp. Blackhawk Hills, St. Paul, MN: The Acid Rain Foundation 1985. - THOMPSON, J.D., HUTCHINSON, I.: J. Appl. Ecol. 23, 239-250 (1986). - THORNE, J.F., HAMBURG, S.P.: Ecology 66, 1333-1338 (1985). - TILMAN, D.: Ecology 67, 555-563 (1986). - TINSLEY, J., DARBYSHIRE, J.F. (eds.): Biological Process and Soil Fertility. Developments in Plant and Soil Sciences, 403 pp. The Hague: Nijhoff/Junk 1984. - TONNER, G., NECKER, U., KUNZE, C.: Verh. Ges. Ökol. 13, 511-517 (1985). - TOTH, D., SVETLOSANOV, V.: Ecology (CSSR) 4, 75-96 (1985). - TROJAN, P.: Ecosystem Homeostasis, 132 pp. Dordrecht: Junk 1984. - TSUTSUMI, T., NISHITANI, Y., SAKAI, M.: Jap. J. Ecol. 35, 207-214 (1985). - TURNER, J., LAMBERT, M.J.: Oecologia

70, 140-148 (1986). - TWILLEY, R.R., LUGO, A.E., PATTERSON-ZUCCA, C.: Ecology 67, 670-683 (1986a). - TWILLEY, R.R., EJDUNG, G., ROMARE, P., KEMP, W.M.: Oikos 47, 190-198 (1986b).

VAN WILGEN, B.W., RICHARDSON, D.M.: J. Appl. Ecol. 22, 955-966 (1985). - VAN WILGEN, B.W., LE MAITRE, D.C., KRUGER, F.J.: J. Appl. Ecol. 22, 207-216 (1985). - VARESCHI, E., JACOBS, J.: Oecologia 65, 412-424 (1985). - VIRTANEN, M., KOPONEN, J., DAHLBO, K., SARKKULA, J.: Ecol. Model. 31, 185-199 (1986). - VITOUSEK, P.M., ANDARIESE, S.W.: Oecologia 68, 601-605 (1986). - VITOUSEK, P.M., DENSLOW, J.S.: J. Ecol. 74, 1167-1178 (1986). - VITOUSEK, P.M., MATSON, P.A.: Ecology 66, 1360-1376 (1985). - VOGT, K.A., GRIER, C.C., GOWER, S.T., SPRUGEL, D.G., VOGT, D.J.: Ecology 67, 577-579 (1986). - VOINOV, A.A., SVIREZHEV, YU.M.: Ecol. Model. 23, 277-292 (1984). - VOOKOVA, B.: Ecology (CSSR) 4, 33-42 (1985a); - ibid. 4, 367-376 (1985b); - ibid. 5, 135-146 (1986). - VOUK, V.B., BUTLER, G.C., HOEL, D.G., PEAKALL, D.B. (eds.): SCOPE 26. Methods for Estimating Risk of Chemical Injury: Human and Non-Human Biota and Ecosystems, 680 pp. Chichester: Wiley 1985.

WALKER, L.R., CHAPIN III, F.S.: Ecology 67, 1508-1523 (1986). - WALKER, L.R., ZASADA, J.C., CHAPIN III, F.S.: Ecology 67, 1243-1253 (1986). - WALLEN, B.: Oikos 46, 51-56 (1986). - WARING, R.H., PITMAN, G.B.: Ecology 66, 889-897 (1985). - WARING, R.H., SCHLESINGER, W.H.: Forest Ecosystems - Concepts and Management, 340 pp. Orlando, FL: Academic Press 1985. - WESTMAN, W.E.: Ecology, Impact Assessment, and Environmental Planning, 532 pp. New York: Wiley 1985. - WETZEL, R.G. (ed.): Periphyton of Freshwater Ecosystems. Proceedings of the First International Workshop on Periphyton of Freshwater Ecosystems, Växjö, Sweden, 1982, 346 pp. The Hague: Junk 1983. - WHITE, E.G.: Ecol. Model. 24, 137-157 (1984a); - ibid. 24, 241-262 (1984b). - WHITFORD, W.G., STEINBERGER, Y., MacKAY, W., PARKER, L.W., FRECKMAN, D., WALLWORK, J.A., WEEMS, D.: Oecologia 68, 512-515 (1986). - WHITMORE, T.C.: Tropical Rain Forests of the Far East, 352 pp. Oxford: Clarendon Press 1984. - WHITTON, B.A. (ed.): Ecology of European Rivers, 644 pp. Oxford: Blackwell 1984. - WIJK, S.: J. Ecol. 74, 675-684 (1986a); - ibid. 74, 685-692 (1986b). - WITTIG, R., BALLACH, H.-J., BRANDT, C.J.: Angew. Bot. 59, 219-232 (1985). - WLOSINSKI, J.H., COLLINS, C.D.: Ecol. Model. 29, 303-313 (1985). - WOODWELL, G.M. (ed.). The Role of Terrestrial Vegetation in the Global Carbon Cycle: Measurement by Remote Sensing. SCOPE 23, 247 pp. Chichester: Wiley 1984. - WOJCIECHOWSKA, W.: Ekol. Polska 34, 133-143 (1986). - WORONIECKA, U.: Ekol. Polska 34, 111-124 (1986).

YAVITT, J.B., FAHEY, T.J.: J. Ecol. 74, 525-545 (1986). - YONEDA, T.: Jap. J. Ecol. 35, 57-66 (1985); - ibid. 36, 117-129 (1986).

ZWÖLFER, H.: Verh. Ges. Ökol. 13, 285-294 (1985).

Prof. Dr. WOLFGANG SCHMIDT
Systematisch-Geobotanisches Institut
Untere Karspüle 2
D-3400 Göttingen, FRG

F. Special Topics

I. Seed Dispersal

By SUSANNE S. RENNER

1. Introduction

In the last few years, hundreds of publications dealing with seed dispersal have appeared. The years 1982-83 seem a good starting point for the present review because several important summaries of the theoretical and empirical aspects of seed dispersal appeared about that time: VAN DER PIJL (1982, 3rd revised and enlarged edition), HOWE and SMALLWOOD (1982), JANZEN (1983c), KUBITZKI (ed., 1983), and WILLSON (1983b). Books by LEIGH et al. (1982) and SUTTON et al. (1983) also contain good chapters on dispersal.

While systematic study of the dispersal of seeds started with biogeographers impressed with instances of long-distance dispersal and morphologists interested in dispersal mechanisms, the focus is now on experimental analyses (often in the field) of the average dispersal distance of a seed and on its fate immediately after dispersal. The broad implications of the dispersal of seeds for such diverse fields as population biology, succession dynamics, and plant and animal coevolution are just beginning to be understood.

Diaspores may be dispersed by several different agents, but the morphology of a diaspore will often indicate the major means of dispersal: hence, the predictive value of syndromes (syndromes are morphological, physiological, phenological, and life-history traits associated with a particular type of seed dispersal). Morphological classifications of fruits like those of ROOSMALEN (1985) and SABATIER (1983) therefore permit one to assess the potential for plant-frugivore relationships in a particular flora.

The great degree of overlap among different dispersers and the many two-phase systems (e.g., explosive release, followed by ant dispersal) make a classification difficult. In my mind, the main dichotomy in dispersal from a plant's point of view is between those systems that involve a reward for the dispersal agent and those that do not. Obviously, only reward-dependent dispersal systems can be mutualistic and lead to coevolution. Three types of rewards may be distinguished, fruit pulp, elaiosomes, and endosperm (i.e., offspring) and the section of this review dealing with reward-dependent systems is subdivided according to these main categories. Dispersal without a reward is not always the same as abiotic dispersal; for example, many fruits and seeds are dispersed by adhesion to animals.

Due to space considerations, I have mostly excluded autecological studies of frugivorous birds and mammals, which often contain data on fruits eaten. Observations on the dispersal of particular taxa are also found

Progress in Botany, Vol. 49

in taxonomic papers, monographs, and floras; see, e.g., the "Flora Neotropica" series, Parkia (HOPKINS 1983), the Meliaceae (WHITE 1986), or the "Flora of Barro Colorado Island" (CROAT 1978). However, these are not studies of dispersal biology as defined above.

2. General and Theoretical Aspects

The central question in the theory of dispersal is: Why should dispersal of seeds away from the parent be of positive selective value? This question arises because the dispersal of propagules away from "known" appropriate parental sites leads to the waste of those seeds which end up in unfavorable localities. Thus, dispersing does not help an individual to get a larger average progeny in an unpredictable and heterogeneous environment: average progeny is exactly equal for (partially) dispersing and nondispersing populations (METZ et al. 1983). Desert plants have in fact evolved characteristics that severely restrict dispersal (see below under wind dispersal).

In general, spatial dispersal in plants is limited (ref. in HARPER 1977). Typically, the density of propagules is highest near the parent and declines with distance. Models of dispersal curves, which show the frequency distribution of the distances traveled by the seeds (GREEN 1983; GERITZ et al. 1984; K.G. MURRAY 1986; HORVITZ and SCHEMSKE 1986b), imply that dispersal curves are molded by natural selection: the shapes of these curves should differ for species with different distributions and densities of favorable sites for the germination of their seeds.

Ecologists have proposed two main selective advantages of dispersal: the avoidance of conspecific competition and of those seed predators whose activity is density-dependent. Mortality of seedlings due to competition, predation, and pathogens is indeed disproportionately high near parents, where the densities of seedlings are highest (AUGSPURGER 1983a,b; HOWE et al. 1985; BECKER and WONG 1985; further ref. in CLARK and CLARK 1984). The pattern of seedling recruitment affects adult distributions and thus the overall pattern of survival in a given area. In particular, it has been argued that density-dependent predation lowers the local density of individual species, opens habitats for additional species, and thereby maintains diversity (JANZEN 1970; CONNELL 1971). The relative contributions to the maintenance of diversity of this and other factors (such as natural disturbances on various scales) are still a matter of debate.

Dispersal in time via soil seed banks may allow plants to withstand short-term elimination. A computer simulation using data on the frequency and size of favorable germination sites (i.e., tree fall gaps) in cloud forest in Costa Rica (K.G. MURRAY 1986) shows that the capacity for a two-year dormancy may increase a species' reproductive success by as much as 565%. In contrast, the model developed by HORVITZ and SCHEMSKE (1986b), combining data from their intensive study of *Calathea ovandensis* in Veracruz (Mexico) with data on tree fall dynamics, shows that for species with the particular life-history traits of *Calathea ovandensis*, the selective advantage of seed dormancy is negligible compared to that conveyed by local dispersal to favorable sites.

3. Dispersal and Gene Flow

Very little is known about the extent of gene flow via seed dispersal. In *Mimulus guttatus* (Scrophulariaceae), gene exchange among the populations along canyons in Utah is via the mostly water-dispersed seeds (WASER et al. 1982); from these authors' study, it seems that gene flow via seeds is not enough to explain the similarities and differences among the populations. WASER et al. may be underestimating the possibilities for gene flow via seeds in *M. guttatus* since this species is dispersed by wind and deer as well as water (VICKERY et al. 1986). But an analysis of the relationships between local genetic differentiation (assessed electrophoretically) and modes of seed dispersal for nine completely unrelated Panamian tree species, dispersed by birds, bats, or explosive mechanisms, also seems to indicate that compared with the effect of pollination mechanisms, seed dispersal does not strongly influence the genetic structure of populations (HAMRICK and LOVELESS 1986). Clearly, comparative studies of the genetic structure among closely related species are necessary in order to understand the relative importance of pollination and seed dispersal in organizing gene flow patterns.

In attempting to correlate the pollination system of *Carduus* species with the differential dispersal of the heteromorphic fruits (well known in composites; ref. in HARPER 1977), OLIVIERI et al. (1983) contend that in these species paternal parentage and fruit dispersal are correlated. Because incoming pollinators, mainly bees, land (and presumably deposit pollen) first in the flowers' centers, the dispersable inner fruits should mostly result from outcrossing. The bees then work their way to the flowers' margins and the nondispersing outer fruits should mostly result from inbreeding. Breeding system and fruit dispersal are clearly correlated in *Impatiens capensis*; in this species, the largely outcrossed seeds from chasmogamous flowers are thrown farther than are obligately self-fertilized ones from cleistogamous flowers (SCHMITT et al. 1985). SCHMITT et al. found that approximately 15% of the chasmogamous seeds were thrown 100 cm or more from the pods which dehisce explosively, whereas only 2% of the cleistogamous seeds attained this distance. The difference did not result from differences in seed mass but was due to pod morphology (particularly pod length was positively correlated with dispersal distance). The above two examples of outcrossed seeds being dispersed further than selfed ones fit in well with expectations that heterozygosity is of selective value for individuals potentially encountering new and different environments.

4. Dispersal by Wind

Dispersal by wind has sometimes been considered to be inefficient (e.g., SALISBURY 1976) relative to dispersal by animals because anemochory, which may involve expensive morphological structures, is indiscriminate, placing propagules in every type of habitat. The problem here is how to properly measure dispersal capacities (GREENE and JOHNSON 1986).

In the Leguminosae (AUGSPURGER 1987) wind dispersal occurs in approximately 22 of the caesalpinioid genera, 42 of the mimosoid genera, and 20% of the genera in the Faboideae. The temperate herbaceous genera have specialized into cylindrical inflated fruits or have developed use of accessory structures; the woody tropical members are characterized largely by having indehiscent samaras or dehiscent fruits with winged seeds. Wind pollination and wind dispersal almost never co-occur. A survey of morphological dispersal adaptations in relation to growth form in 5900 species of Compositae shows that annuals have a higher percentage of species with either adhesive structures or else without dispersal structures than perennials (VENABLE and LEVIN 1983). Habit notwithstanding, weeds have a greater proportion of species with no dispersal structures than nonweeds. For annuals and weeds, long-lived seeds and a large local seed pool may be a better strategy than dispersal because of the extremely low probability of reaching a suitable habitat.

Extrapolation from laboratory experiments (AUGSPURGER 1986b) on the aerodynamic traits of the fruits of 34 anemochorous tree species from different families leads to mean dispersal distances between 22 and 149 m. For a subset of these species, rates of descent at a wind speed of 1.75 ms^{-1} correlate weakly with the shade tolerance of their seedlings (investigated earlier, AUGSPURGER 1984): species that depend on light gaps as seedlings have lighter seed mass and/or larger surface area of their diaspores and thus they potentially disperse over broader areas increasing their chance of encountering a gap. Seedling mortality in *Platypodium elegans* and 17 other wind-dispersed tree species (AUGSPURGER 1983a,b; AUGSPURGER and KELLY 1984) is mostly due to fungal pathogens. After one year, the seedlings require the microclimatic conditions of light gaps for continued survival and growth.

To study the spatial distribution (seed shadow) of wind-dispersed diaspores, AUGSPURGER and FRANSON (1987) performed a series of field experiments using artificial fruits mimicking those of *Tachigalia versicolor*, *Aspidosperma cruentum*, and *Platypodium elegans*. Hundreds of artificial fruits of 15 models and painted in different colors were employed, controlling for mass, area, and wing-loading of the "fruits", height of release, variability in wind speed and direction, and surrounding topography. Multiple trials tested dispersal under different wind speeds. It was found that wind speed overrode the effects of variation in mass or fruit surface area. Thus, variations in seed number per fruit, e.g., in *Platypodium elegans* (AUGSPURGER 1986a), *Lonchocarpus pentaphyllus* (AUGSPURGER and HOGAN 1983), and *L. costaricensis* (JANZEN 1982j), influence fruit weights but may have only negligible influence on dispersal distances. Plant height and wind speed are also more important than fruit weight or morphology in the dispersal of *Asclepias syriaca* (MORSE and SCHMITT 1985) and of Dutch grassland herbs (VERKAAR 1983).

Adaptations which increase the lodging ability of fruits or seeds are important for desert plants to ensure that a diaspore stays at a site long enough to allow germination and rooting. Such negative anemochory is apparently common in Californian desert plants (MADDOX and CARLQUIST 1985) and in true deserts in Israel (ELLNER and SHMIDA 1981).

Winds, especially hurricanes, are responsible for the long-distance dispersal of *Andropogon bicornis* from the Caribbean to southern Florida where it has recently established itself (CAMPBELL 1983).

5. Dispersal by Water

A guide to tropical drift seeds and fruits is provided by GUNN and DENNIS (1976). The role of water dispersal in Amazonia is stressed by KUBITZKI (1986). However, there appears to be no recent study of the implications of water dispersal.

6. Adhesive Dispersal

SORENSEN (1986) gives a concise review of fruit dispersal by adhesion; the mechanisms of dispersal and establishment in the grasses are reviewed by DAVIDSE (1987).

Epizoochorous dispersal systems probably evolved from anemochorous ones: both involve no reward for the dispersal agent, and a shift between these two systems is often achieved by minor morphological changes (this is exemplified by the genus *Geum*; RIDLEY 1930; VAN DER PIJL 1982). Propagules stuck to animals travel farther than seeds carried internally or by the wind (SORENSEN 1986). This is because they can be carried by the dispersal agent indefinitely. In contrast, the distance traveled by fruits carried internally is limited by gut passage rates. The hooks, spines, and barbs of epizoochorous fruits often have additional functions as nicely demonstrated by PEART (1984, and earlier studies), who observed the behavior of 2193 germinating grass diaspores of 11 species over a five-year period. Three gross structural forms (actively-awned, passively-awned, and unawned) represent distinct strategies for the establishment of seedlings. Of 28 Brazilian savanna grass species, 11 are principally epizoochorous or facultatively anemochorous (SILBERBAUER-GOTTSBERGER 1984).

Burrs of *Xanthium occidentale* attach to the hair of cattle and horses and the wool of sheep; they are also dispersed by water currents, by trampling animals, and by being catapulted away when a bird bends the plants and causes them to flip (LIDDLE and ELGAR 1984). The proportion of burrs following these various pathways at different sites depends mainly on occasional events, such as flooding or the presence of stock. In an investigation of epizoochorous diaspores differing in adherence capability, the nature of the surface proved more important in promoting adherence than the weight or size of the diaspores (CARLQUIST and QUINN 1985). Also, the adherence of the fruits and seeds to natural fur and feathers was comparable to their adherence to three types of fabrics, which may facilitate future experimental work.

7. Dispersal by Frugivores

a) General

A short review of the conceptual history of fruit-frugivore studies, chapters on plant and frugivore strategies, on the consequences of seed dispersal, and on community aspects of frugivory and seed dispersal are contained in a book edited by ESTRADA and FLEMING (1986).

The concept of coevolution, describing evolutionary changes in two or more species which are acting as selective forces on each other, is used with increasing caution in the context of seed dispersal. The differences in the lengths of the generation times and in the evolutionary life spans of the partners involved as well as the lack of specificity of the relationships likely result in great delays and occasional directional changes in the mutual selection. Because of the usually broad arrays of plant and animal species involved, JANZEN's (1980) hedged version, "diffuse coevolution", referring to coevolutionary relationships between groups of species, is more precise in the case of dispersal systems. The majority of studies, however, are appropriately described as dealing with mutualistic interactions.

Mutualisms can occur between species that have or have not lived together for a long time. Angiosperm species last on average 27-38 million years, bird and mammal species 0.5-4 million years (HERRERA 1985b). Even if this estimate exaggerates the evolutionary lifetime of angiosperms, it is reasonably certain that a flowering plant is dispersed by various animal species during its evolution. In general, the time necessary for coevolution to occur may be long: "The major benefits to plants of dispersal will take considerable time in feeding back to the bats, at best several bat generations" (HEITHAUS 1982). How many more generations of bats are needed for their behavior to influence the plants?

Plants need not direct the dispersal of their seeds to particular locations with a degree of exactness comparable to pollen dispersal. Therefore, in seed dispersal, constancy and specificity in the disperser are not necessary for the interaction to work and be of mutual benefit. This fundamental difference between pollination systems and seed dispersal partly explains the virtual absence of species-specific interactions in seed dispersal (WHEELWRIGHT and ORIANS 1982; JANZEN 1983b; HOWE and WESTLEY 1986). Another reason, applicable in the case of widespread plant species, is that they may be dispersed by different disperser classes in a variety of ways. This is shown for *Opuntia* (GONZALEZ-ESPINOSA and QUINTANA-ASCENCIO 1986; JANZEN 1986), *Muntingia calabura* (WEBB 1984; FLEMMING et al. 1985), and *Cecropia* (ESTRADA and COATES-ESTRADA 1984; ESTRADA et al. 1984; CHARLES-DOMINIQUE 1986). The fruits of one species of *Cecropia* are eaten by as many as 48 animal species (ESTRADA et al. 1984).

Another reason for the general lack of specificity in fruit-frugivore relations are the strong seasonality and frequent scarcity of fruits. Even in the tropics, evidence for this is now overwhelming (see book by LEIGH et al. 1982; HOWE and SMALLWOOD 1982; TERBORGH 1983, 1986; GAUTIER-HION et al. 1985b; SABATIER 1985, and ref. in these works). The results of a two-year survey of understory plants (excluding palms, the fruits of which are not usually taken by small frugivores) in Costa Rica (DENSLOW et al. 1986) provide good evidence of the food scarcity: 60% of the plants bore an average of ten or fewer fruits per plant.

Still, dispersers have undoubtedly selected fruit traits. A study of a tropical rain forest in Gabon, Africa (GAUTIER-HION et al. 1985a) shows this clearly. In this study, botanists and zoologists combined forces and analyzed the diets of seven species of large canopy birds, 19 species of rodents, seven species of ruminants, and six species of monkeys all of which feed on plants (belonging to 122 species in 41 families) over one year. The animals showed an outstanding lack of specificity in their fruit choice. However, birds and monkeys definitively prefer brightly colored berries or drupes with succulent pulp or arillate seeds with no protective cover whilst ruminants, rodents, and elephants disperse large, dull-colored, dry fibrous, well-protected fruits. Unfortunately, no survey of frugivorous bats was included in the study.

Exceptions from the general lack of tight relationships in zoochory are some species of mistletoes and their bird dispersers, certain pines and their birds, and some ant-dispersed species (see below). Interestingly, these three cases of close coevolution involve the three different types of reward offered to dispersers by plants, pulp, seeds, and elaiosomes.

b) Birds

α) General. For a review of fruit dispersal by eastern North American birds, see WILLSON (1986); for a review of the behavior, morphology, and nutrition of neotropical frugivorous birds, MOERMOND and DENSLOW (1985); for a discussion of the adaptations to frugivory in Mediterranean seed-dispersing birds, HERRERA (1984a); for a review of habitat-consumer influences in frugivorous birds, HERRERA (1985a).

Ornithologists initially distinguished between general and specialized frugivorous birds and then went on to suggest that this basic dichotomy coincides not only with the nutritional quality of the fruits these birds eat (specialists eating high-quality fruits), but also with the quality of the dispersal they provide (SNOW 1971; McKEY 1975). However, a rigid classification of frugivorous birds into specialists and generalists, with few exceptions, is not supported by comparison of diets, behaviors, morphologies, or disperser functions (M.S. FOSTER and McDIARMID 1983; MOERMOND and DENSLOW 1985; WHEELWRIGHT et al. 1984). Frugivores are not automatically more efficient at feeding on fruit than birds that eat fruits and insects; also, adaptations of specialists may exist only at a particular level. For example, an obligate frugivore may be efficient in digesting and assimilating nutrients in fruits but inefficient in handling them.

Botanists may be surprised by assertions that, in any one area, fruit attributes (such as color, size, composition, etc.; see below) are molded by the seasonally changing demands of the bird population (e.g., SNOW 1981; HERRERA 1982c). Bird ecologists, however, have great faith in their objects of study as agents of selection. For example, in Israel, "spring fruiting has been selected against because spring migratory birds are eager to get as quickly as possible to their breeding territories and are not interested in any type of food" (IZHAKI and SAFRIEL 1985).

In contrast to this is a growing realization (cf. HERRERA 1982c and 1986) that other factors, such as risks while feeding (SNOW and SNOW 1986) or the presence of other fruiting trees, are often the overriding factors governing avian fruit choice. A model by MARTIN (1985) integrates the behavioral, ecological, and morphological constraints influencing fruit choice by birds. It assumes that the potential yield of a fruit to a bird is a function of pulp mass and quality, discounted by the bird's body size, the time spent locating and eating the fruit, and the risks while picking the fruit.

β) Fruit Color. Many studies confirm the notion (DARWIN 1859; RIDLEY 1930) that colorful, most often black or red, displays of ripe fruits are attractive to birds (JANSON 1983; KNIGHT and SIEGFRIED 1983; MORDEN-MOORE and WILLSON 1982; WILLSON 1983a; WILLSON and THOMPSON 1982; WILLSON and MELAMPY 1983; WILLSON and HOPPES 1986; for a critique of the sampling methods of WILLSON and collaborators, GREIG-SMITH 1986). WHEELWRIGHT and JANSON (1985) note that there are also many bird fruits that are not brightly colored. Assessing STILES' (1982) hypothesis that "early autumnal color changes have evolved through their effect on increased fruit dispersal", WILLSON and HOPPES (1986) find that while there is evidence that migrant birds may use foliar color as a signal for finding fruit sources, there is no strong evidence that fruit color was selected because of its effects on dispersal agents.

γ) Fruit Accessibility. DENSLOW and MOERMOND (1982) find that fruit removal by bird dispersers and rodent seed predators is influenced differently by the placement of the infructescences. Birds prefer fruits near horizontal petioles to those along the main stem and out of reach from convenient perches (DENSLOW and MOERMOND 1985). The terminal placement of infructescences may maximize the probability of removal by birds and minimize the risk of seed destruction by rodents. Obviously, birds that take fruits primarily on the wing are not influenced by the same aspects of fruit accessibility as those that take fruits from perches (MOERMOND et al. 1986; DENSLOW et al. 1986).

δ) Fruit Size. Though field observations have shown the broad dietary overlap among birds, one fruit attribute that constrains their choice is size. Plants producing small fruits attract more species of birds than large-fruited plants because most fruit-eating birds typically swallow fruits whole (SNOW 1971; WHEELWRIGHT 1985a; PRATT and STILES 1983, 1985). Seed size (and pulp texture) also influence the physical treatment of seeds in bird guts (LEVEY 1986). While large-gaped birds eat even very small fruits, large fruits exclude narrow-gaped birds. In this respect, plants producing large fruits may be at a disadvantage. Also, large fruits are more likely to have their seeds dropped under the parent tree because large, entirely frugivorous birds frequently make long visits to fruiting plants during which they defecate seeds (PRATT and STILES 1983, 1985). Frugivorous birds of paradise remain in feeding trees for an average of 4.5 min; fruit pigeons stay 10 min on average (BEEHLER 1987).

The seed dispersal capabilities of the cassowaries (*Casuarius casuarius*, *C. bennetti*), large, flightless, frugivorous birds of the rain forests in Australia and New Guinea, are remarkable (STOCKER and IRVINE 1983; PRATT 1983). Viable diaspores of dozens of plant species, some diaspores weighing up to 350 g, were found in the dung of the two species. Birds of paradise are apparently the dispersers of the large *Amorphophallus* seeds, which they regurgitate within 30 min (PECKOVER 1985).

ε) Fruit Composition. The least known factor that might affect avian fruit choice and therefore seed dispersal is the chemical composition of fruits (HERRERA 1982b). FOSTER (1978) and FOSTER and McDIARMID (1983) provide data on the composition of many species of tropical fruits. One of the most lipid-rich fruits as yet known is that of *Trichilia cuneata* (Meliaceae), which is attractive to a broad array of specialized and opportunistic frugivorous birds (FOSTER and McDIARMID 1983). Similarly large potential disperser coteries have also been observed on other arillate fruit species (McDIARMID et al. 1977; SKUTCH 1980; TRAINER and WILL 1984).

SORENSEN (1983), using artificial fruits of flour dough with added natural fruit flavors, shows that thrushes (*Turdus* spp.) avoid several species of fruits containing identified toxins. SORENSEN, like HERRERA (1982a,c), speculates that some plants have evolved fruit toxins to prevent particular bird taxa from consuming their fruits and dispersing their seeds to unfavorable habitats. Toxic secondary compounds are frequently present in the pulp of ripe fruit: as many as one-third of European fruit-producing plant genera have fruits toxic to man (HERRERA 1982a). Also, birds differ in the strength of their preferences for certain fruits when presented with a choice (LEVEY et al. 1984). In some situations, birds prefer fruits that yield the highest energy gain, i.e., that have high pulp/seed ratios. Because gut volume limits fruit uptake, seed regurgitation, which results in rapid elimination of nonnutritional ballast, allows birds to obtain particularly high rates of energy gain (SORENSEN 1984). In *Berberis hispanica*, a bird-dispersed shrub of southern Spain, ovipositing flies influence the number of seeds, which in turn influences the choice of fruits by birds because birds prefer healthy, large fruits (HERRERA 1984b). HERRERA (1984c) also calls attention to the competition between birds and insect predators on ripe fruits.

The heavy reliance of the principal overwintering bird dispersers on fall and winter fruits in the northern hemisphere is thought by some workers (SNOW 1971; HERRERA 1982c,d, 1984d; STILES and WHITE 1982) to have selected for the higher lipid content sometimes observed in winter fruits, when compared to fruits that ripen and are dispersed in the summer. Other studies found no higher lipid content in winter fruits (JOHNSON et al. 1985; PIPER 1986a). Also, BOROWICZ and STEPHENSON (1985) noted that in two simultaneously fruiting *Cornus* species, fat-rich fruits were not preferred over less rich ones. PIPER (1986a) has shown that in aviary experiments with 11 species of birds, lipid content per fruit was not correlated with fruit consumption.

STAPANIAN's (1982a,b) study of seasonal variation in fruit composition at a community scale, strongly supported his prediction that plant species fruiting during seasons of high caterpillar abundances, the birds' favorite food, would have higher ratios of pulp energy per seed than plants fruiting during seasons of low caterpillar abundance (STAPANIAN 1982a,b).

A fact, sometimes overlooked in the studies reviewed in the above section, is that a considerable proportion of modern floras consists of species which evolved elsewhere and did not coevolve with their current coinhabitants.

ς) Fruiting Phenology. Generally, the idea (SNOW 1965) that plants adjust fruiting phenologies to match the needs of bird dispersal agents, is discredited (e.g., GAUTIER-HION et al. 1985b). Two lines of reasoning discount this idea: first, a species' seeds are usually dispersed by many species of birds and, secondly, the partners usually do not overlap for most of their respective ranges. It is probably impossible to demonstrate whether character displacement of fruiting times due to competition for dispersers has occurred; many tropical birds (bats and monkeys, see below) depend on a constant supply of fruits and probably adapted their movements and reproductive schedules to the seasonal abundance of this supply.

It has long been known that Lauraceae are particularly important for large numbers of fruit-eating birds (SNOW 1971; WHEELWRIGHT 1983, 1985b, 1986; SANTANA and MILLIGAN 1984). The seeds of *Ocotea tonduzii*, for example, are dispersed by at least 18 bird species, including generalists such as flycatchers and migrating thrushes (WHEELWRIGHT et al. 1984). In Costa Rica, quetzals (*Pharomachrus mocimo*, Trogonidae), specialized frugivores, depend mostly on the drupes of the 22 sympatric species of laurels and the phenologies and habitat distributions of the plants appear to dictate the birds' seasonal movements. Lauraceous fruits make up 60-80% of all fruits eaten by these birds. Over a six- to seven-year period, overall fruit production was high during three years and low during four (WHEELWRIGHT 1986). Within populations, individual laurel trees sometimes fruited out of phase or failed to fruit altogether. In any month of the year, at least one species could be found with ripe fruits. WHEELWRIGHT (1986) infers "mutual dependence and, possibly, general coevolution between quetzals and lauraceous trees". The fossil record for Lauraceae fruits and for oilbirds (Steatornithidae), which are another group of specialized frugivores known to presently feed on Lauraceae, extends back approximately 50 million years (WING and TIFFNEY 1987). It is thus reasonable to assume that frugivorous birds, and not only quetzals, have been dispersing the morphologically and chemically (HERRERA 1986) conservative laurels from the early Tertiary on.

η) Crop Size and the Fruiting Environment. There is a strong correlation between the occurrence of dioecism and fleshy fruits. According to sexual selection theory, in dioecious plants male and female functions do not compete for the limited resources; thus, an increase in female reproductive effort is possible. BAWA (1980) and GIVNISH (1980), searching for explanations (other than the promotion of out-crossing) for the evolution of dioecism, hypothesized that, among other ecological factors, seed dispersal by animals, mainly birds, has selected for dioecism. This hypothesis rested on the assumption that fruit crops are larger in dioecious species and that the proportion of seeds dispersed increases disproportionately with the size of the fruit crop. The latter assumption is not sustained by the (now) available data (HOWE 1983; JORDANO 1982; DAVIDAR and MORTON 1986; McDONNELL et al. 1984; MOORE and WILLSON 1982; LEVEY et al. 1984; PIPER 1986b; K.G. MURRAY 1987).

Undoubtedly, large fruit crops attract the attention of many individuals and species of seed dispersers; for example, *Ficus* is one of the staples for many frugivores. This is confirmed in Cameroon (BREITWISCH 1983), Mexico (SCOTT and MARTIN 1984; COATES-ESTRADA and ESTRADA 1986), and Costa Rica (JORDANO 1983; ROBERTS and HEITHAUS 1986). Two specialist birds of paradise species feed to 80% on figs (BEEHLER 1983). In a single tree, birds, mostly parrots (who carefully remove and eat the seeds and discard the pulp), may eat an estimated average of 20,800 figs per tree per day; an estimated 4.4 million fig seeds may be lost each day mostly to agaonid wasps. Only 6.3% of the seeds leaving the tree per day are actually dispersed undamaged by birds (JORDANO 1983). Figs are also very important for bats and monkeys (see below).

θ) Habitat Influence on Bird Dispersal and Vice Versa. The fruit-bearing plants influence the kinds and numbers of birds attracted to a certain habitat. Simultaneously fruiting species sometimes compete for dispersers and thus the "fruiting environment" affects the probability that a plant will have its own seeds dispersed (HOWE 1983; MANASSE and HOWE 1983; HERRERA 1984e; DENSLOW et al. 1986; PIPER 1986b).

An example of how bird behavior influences the seed shadow of bird-dispersed species is that the seedlings of these plants tend to appear along fence rows, hedges, and other perches (HOLTHUIJZEN and SHARIK 1985; CARTER JOHNSON and ADKISSON 1985; McDONNELL and STILES 1983; McDONNELL 1986). This is clearly of importance in plant succession. Another example of the influence of bird behavior on seedling recruitment is provided by the flowerpeckers (Diacaeidae), who after feeding on a certain number of mistletoe fruits (Loranthaceae and Viscaceae), squat on a branch and within minutes defecate the sticky seeds. The seeds are stuck to the branch by lateral rubbing of the bird's bottom (DOCTERS VAN LEEUWEN 1954). Further details of the coevolution between some of the mistletoes and their dispersers are provided by DAVIDAR (1983a,b, 1987).

c) Bats

Authoritative accounts of the foraging strategies of plant-visiting bats and of the coevolution between bats and plants are given by FLEMING (1982) and HEITHAUS (1982).

Muntingia, *Chlorophora*, *Cecropia*, *Ficus*, and *Piper* species produce roughly half of the wet season fruits eaten by the bat *Carollia* in Santa Rosa National Park, Costa Rica, where many studies have been done. Analyses of the fruits taken by bats (HERBST 1986; DINERSTEIN 1986) bear on the long-standing question of how frugivorous bats meet their nitrogen requirements, particularly during breeding seasons. It appears that several of the preferred fruits, with the exception of figs, are adequate nitrogen sources for all but lactating bats. Nursing mothers may have to switch to a special protein-rich diet of *Piper* fruits, which are available in most months of the year (FLEMING 1985). Bats apparently do not eat fruits of Lauraceae, Burseraceae, and Rubiaceae (FLEMING 1986).

Bats also take large, dry seeds, for example, the 3.6-cm-long seeds of *Gonystylus macrophyllus*, an Indonesian member of the Thymelaeaceae. Because of their large size, the bats accidentally lose some of these seeds during flight and they thus escape being eaten (KEVAN and GASKELL 1986). Removal of the fruit pulp by bats may increase seed survival: in *Andira inermis*, the cleaned seeds do not become infested with weevils that normally kill between 70-100% of the seeds (JANZEN 1982h).

d) Primates

For a recent list of frugivorous primate genera, see HOWE (1987). The role of primates in fruit dispersal is still little known. It seems that the fruit diet of Old World monkeys overlaps to a large extent with that of birds (GAUTIER-HION et al. 1985a), whereas New World monkeys differ substantially from birds in their fruit preferences (JANSON et al. 1986).

Howler monkeys (*Alouatta palliata*), among the most ubiquitous arboreal mammals in the neotropics, in Veracruz (Mexico), spend 80% of their time eating fruits. In some cases they act as seed dispersers; however, the seeds of one of their preferred fruits, the laurel *Nectandra ambigens*, are wasted under the parent tree since they only eat the pericarp and immediately drop the rest (ESTRADA and COATES-ESTRADA 1984, 1986). The monkeys, covering daily distances between 10 and 839 m, respond to the temporal and spatial dispersion of fruiting trees and search out individual ones, which are consistently used year after year.

The use of fruit resources in pristine tropical lowland forest in southeastern Peru by five species of monkeys is closely analyzed by TERBORGH (1983). In southeastern Peru, the brown capuchin, *Cebus apella*, consumes the fruits of over 130 species of plants (JANSON et al. 1986). Seeds of 98% of these species are defecated intact.

Chimpanzees in Gombe National Park, Tanzania, spend 60% of their total feeding time eating fruit; 38 fruit species account for 80.5% of this time (WRANGHAM and WATERMAN 1983). The ripe fruits of these species are low in tannin, while a high content of secondary compounds is present in unripe ones. The diet of *Cercopithecus cephus*, studied in Gabon (SOURD and GAUTIER-HION 1986), consists of up to 80% fruits. The fruits of 23 species accounted for over 75% of this. Interestingly, these were not the most abundant species; apparently, the monkeys' favorite fruits are acidic.

As indicated above (Sect. 7b), *Cecropia* fruits are taken by many animals; the longer transit time of *Cecropia* fruits in the gut of monkeys, as opposed to the faster passage through bats, affects the seed coat more strongly and eliminates dormancy (VAZQUEZ-YANES and OROZCO-SEGOVIA 1986).

e) Other Non-flying Mammals

The selective influences of large, herbivorous Pleistocene and pre-Pleistocene mammals, such as giant ground sloths, giant armadillos, elephant-like gomphotheres, and native horses, on the neotropical flora is emphasized by JANZEN (JANZEN and MARTIN 1982; JANZEN 1986, and papers cited below; for a critique, HOWE 1985). The interaction between this substantial fauna of large mammals and the plants lasted until 10,000 years ago (JANZEN 1986) and was responsible for the evolution of a "megafauna dispersal syndrome". A list of 38 fruit species as candidates for a reconstruction of this dispersal syndrome (JANZEN and MARTIN 1982) includes morphologically very diverse fruits, e.g., the spiny fruits of the Tiliaceae *Apeiba*, Annonaceae, and figs. Many of the species listed by JANZEN and MARTIN as having been shaped by the Pleistocene megafauna have living dispersal agents. For example, *Manilkaria zapota*, *Brosimum alicastrum*, and *Spondias mombin* are monkey-dispersed (ESTRADA and COATES-ESTRADA 1986); *B. alicastrum* is a favorite with *Artibeus* bats (C. HANDLEY pers. comm.); *Andira inermis* is bat-dispersed (R.B. FOSTER et al. 1986); *Apeiba* is eaten by monkeys and birds; *Byrsonima crassifolia* is bird-dispersed (EMMONS and M.S. FOSTER pers. comm.). However, it seems that the large fruits of *Guazuma ulmifolia* (Sterculiaceae) (JANZEN 1982b), *Crescentia alata* (Bignoniaceae) (JANZEN 1982c,e), *Spondias mombin* (Anacardiaceae) (JANZEN 1985b), *Opuntia* (JANZEN 1986), and *Hymenaea courbaril* (Leguminosae) (HALLWACHS 1986) are poorly adapted to their present dispersers.

A long-term monitoring program of the fruiting of about 400 *Enterolobium cyclocarpum* (Leguminosae) trees in Santa Rosa National Park in Costa Rica (JANZEN 1982a,d,f,g,i,k; 1985a) demonstrates the impact of variations in timing, fruit crop sizes, seed sizes, and seed predation on dispersal which, today, is by horses, cows, tapirs, and peccaries.

Facultative consumption and dissemination of small seeds by large herbivores is an important method of long-distance dispersal (RIDLEY 1930; JANZEN 1983c; SILBERBAUER-GOTTSBERGER 1984). JANZEN (1984) proposes that some plants may offer foliage as a "fruit" to attract grazing or browsing mammals, which then inadvertently disperse their seeds (for a critique, see COLLINS and UNO 1985). Black bears, which move up to 32 km per day, like to eat fruits and swallow them unchewed (ROGERS and APPLEGATE 1983). For large-seeded fruits, for example, plums, the bears may be important long-distance dispersal agents. In contrast, kangaroos are apparently seasonal and mostly destructive frugivores (CLIFFORD and DRAKE 1985).

f) Fishes

The only comprehensive treatment to date on the role of fish in seed dispersal is that of GOULDING (1980).

Few plant adaptations for regular dispersal of their seeds by fruit-eating fish are known. A possible case is presented by KUBITZKI (1985) from Brazilian Amazonia. The fleshy envelope of the seeds of *Gnetum* species is usually red and the "fruits" are bird-dispersed. In at least one species, *G. venosum*, ripe fruit are dark gray, 6-8 cm long, and have a great amount of needlelike sclereids in all but the outermost layer of fruit pulp. They are eaten by catfish who strip the outer fleshy layer and regurgitate the seed.

8. Dispersal by Ants

Two recent books summarize current knowledge and theory of ant-plant mutualisms: BEATTIE (1985) and BUCKLEY (ed., 1982). The latter includes articles on plants' use of ants for dispersal and on the concentration of myrmecochory in Australia and South Africa.

The selective advantages that have been suggested for myrmecochory, reflect those envisioned for seed dispersal in general (see Sect. 2): seeds are removed from the vicinity of the parent where predation is highest, may be abandoned at nutrient-enriched nest sites (BEATTIE and CULVER 1983, and their earlier work), or, when buried in ant nests, may remain unharmed by the high temperatures of wildfire (BOND et al. 1984).

The majority of diaspores of the myrmecochorous *Carex pilulifera* drop within 40 cm from the center of the parent plant where predation by beetles and mice is high. Subsequent ant dispersal usually takes the seeds up to 2 m away (KJELLSSON 1985a,b); it is unclear whether this is enough to reduce predation. HORVITZ and SCHEMSKE (1986c) show the great differences in foraging behavior among five ant species; only two regularly moved seeds beyond the parent plant shadow, one of these moved the seeds about 75 cm, another moved them 12 times further; another ant species interfered with dispersal.

Although nest soil is significantly enriched in nitrate-nitrogen, magnesium, iron, manganese, and cadmium, the growth of seedlings of an ant-dispersed plant is not enhanced by this soil (HORVITZ and SCHEMSKE 1986a). RICE and WESTOBY (1986), who tested the nutrient-rich nest site hypothesis for 15 species in Australian sclerophyllous vegetation, find that soil close to the radicles of ant-dispersed seeds is not higher in total N or available P than that near to non-myrmecochorous seeds.

In South Africa as many as 1300 species of plants or 20% of the Cape flora may be myrmecochorous (BOND and SLINGSBY 1983). There, the introduced Argentine ant, *Iridomyrmex humilis*, which is replacing the native ants, does not store seeds below the soil but leaves them on the soil surface, exposed to predators, and thus endangers some species of ant-dispersed Proteaceae (BOND and SLINGSBY 1984).

It is well known that rodents prey heavily on most elaiosome-bearing seeds. In different *Asarum canadense* popuations, seed predation by rodents is the most important factor explaining variations in seedling recruitment (HEITHAUS 1986), and very high predation by rodents has also been found in the ant-dispersed *Jeffersonia diphylla* (B.H. SMITH et al. 1986). Fresh elaiosomes of *Corydalis aurea* (Fumariaceae) repel a common seed-eating mouse (in cage experiments), and attract at least nine species of ants (HANZAWA et al. 1985). The fact that many Fumariaceae contain alkaloids may be relevant here. In *Viola nuttallii*, the seeds are ballistically ejected between 9 AM and 1 PM. TURNBULL and CULVER (1983) contend that this is adaptive, since most ants and few rodents are active in the morning.

A selective advantage for the diplochorous dispersal system in *Viola* is suggested by BUELOW-OLSEN (1984). Seeds that land in an area less searched by ants will contribute to a long-lasting seed bank in the soil because relatively few of them germinate during the first year; seeds found by ants and carried into the nests usually germinate in the spring. Ballistochory and myrmecochory are a common combination, and if enhanced germination in the seeds handled by the ants is widespread, ballistochory might ensure that some of them escape the ants and enter the soil seed bank.

A number of newly reported cases of ant dispersal (GADDY 1986), including trailing arbutus (*Epigaea*, Ericaceae; CLAY 1983), support the notion (BEATTIE and CULVER 1981) that myrmecochory is widespread in the eastern deciduous forest of North America. In western American redwood forests, on the other hand, ant-dispersed *Trillium ovatum* often remains undispersed because of low ant activity (MESLER and LU 1983). Myrmecochory in this and in a few other species in the Rocky Mountains (TURNBULL et al. 1983) may be a remnant of more favorable early Miocene conditions.

Adaptations for myrmecochory and ornithochory in 20 species of *Acacia* (DAVIDSON and MORTON 1984) suggest to the authors that the two dispersal systems are evolutionarily close (an idea also expressed by BERG 1958). In one species there is geographic variation: at one locality, the seeds resemble those of exclusively myrmecochorous species, which have white arils, and at another they resemble those of exclusively ornithochorous species with colored arils. Some *Acacia* species are initially dispersed by birds and subsequently by ants. The similarity in the nitrogen-enriched microhabitats encountered in the early germination stage by ant-dispersed and bird-dispersed seeds is supposed to facilitate a shift from one system to the other (evidently starting from a potentially diplochorous one).

ROBERTS and HEITHAUS (1986) report on the rearrangement of fig seeds on the ground by seed-harvesting ants. Though many seeds are likely to perish inside nests (figs require light for germination), some may benefit from being left isolated along trails where they experience less sibling competition.

9. Dispersal by Seed-Hoarding Animals

Seed-hoarding animals often "plant" seeds in ideal germination locations; however, since they are seed predators, this may be a most precarious dispersal system (and wasteful?). A list of animals known to cache food is included in a review of the evolution of food caching by birds and mammals (C.C. SMITH and REICHMAN 1984).

The closed cones and wingless seeds of pines of the *Pinus cembra* group are adaptations for seed dispersal by seed-hoarding Nutcrackers (*Nucifraga* spp., Corvidae). Continuing earlier studies (VANDER WALL and BALDA 1977; TOMBACK 1982, 1986; LINHART and TOMBACK 1985), TOMBACK (1983) concludes that the associations between the Eurasian Nutcracker and several species of Eurasian pines are the result of mutual selective interaction during their evolutionary history (i.e., coevolved). The development of cone characteristics facilitating efficient seed harvest by the birds is counteracted by the necessity of protection against rodent predators (BENKMAN et al. 1984).

During a three-year study in Michigan, squirrels and chipmunks (*Eutamias*) removed and eventually destroyed *Carya glabra* seeds thoroughly; SORK (1983a,b) believes that under present-day conditions hickory relies on unusually large mast years combined with unusually low mammal densities for its reproduction. To monitor seed dispersal by mammals, large fruit can be foil-wrapped or metal-tagged and subsequently retrieved with a metal detector (STAPANIAN and SMITH 1978; SORK 1984).

Food-caching squirrels influence the invasion of prairies in Kansas by *Juglans nigra* (STAPANIAN and SMITH 1986), and food-caching Blue jays (*Cyanocitta cristata*) that of man-made habitat islands in Wisconsin by beeches and oaks (CARTER JOHNSON and ADKISSON 1985). In North America, the rate at which oaks advanced north 10,000 years ago, about 350 m per year, matches the dispersal distances achieved via nut-carrying Blue jays.

Scatter-hoarding agoutis (SMYTHE et al. 1982) and squirrels (EMMONS 1984) are the dispersers of the hard fruits of *Dipteryx*, *Scheelea*, *Astrocaryum*, and of *Hymenaea courbaril* (HALLWACHS 1986). Burial by agoutis enhances germination in *Gustavia* (Lecythidaceae; SORK 1985).

10. Modes of Dispersal in Plant Communities

The importance of long-distance dispersal is usually assessed by a quantification of the dispersal modes found in the flora of an island or in a remote continental habitat. For the cloud forest on a mountain range rising from arid lowlands on the coast of Colombia, SUGDEN (1982) presents such an analysis; hardly any of the 126 species of vascular plants in the cloud forest are found in the surrounding lowlands. Forty-one percent of the flora is ornithochorous, 39% anemochorous, and 5% epizoochorous. The plants most likely come from the nearest extensive moist area, 250 km away. The dispersal mechanisms found in a deciduous forest in the southern part of the Venezuelan coastal cordillera (WIKANDER 1984), are anemochory (42%), zoochory (30%), barochory (19%), and autochory (9%). Anemochory dominates at the higher elevations, whereas zoochory is more common in lower levels of the forest.

The results of an interesting study done in the natural sere on a riverbank in Peruvian Amazonia (R.B. Foster et al. 1986) suggest that plants with wind- and bat-dispersed seeds are more frequent in the early successional stages on the beaches than are bird- or terrestrial mammal-dispersed ones. This agrees well with findings in French Guiana (FORESTA et al. 1984) that only bats but not birds generate seed shadows in large open areas up to 50 m from forest edges. Of 90 pioneer species in French Guiana, 51% produce berries, mostly with seeds smaller than 5 mm, and an individual may produce 2.3 million seeds during a life span of four years (PRÉVOST 1983).

Dispersal modes in plant communities are obviously also influenced by the plants' habits. For example, epizoochory is most frequent in herbs (SORENSEN 1986). Many opportunistic epiphytes rely on birds for dispersal (SCHIMPER 1888; RIDLEY 1930). While 40% of all the neotropical Melastomataceae have capsules (the rest have berries), 85% of the 227 epiphytic ones have bird-dispersed berries (RENNER 1986).

The biogeographic distribution of five fruit types (drupe, berry, pod, capsule, and nut) and dispersal mechanisms (extrapolated from the fruit types) in the southern African tree flora (1340 spp.) is analyzed and statistically related to environmental variables by KNIGHT (1984, 1986). Some of the results seem self-evident; e.g., "those species dispersed by birds showed a pronounced similarity to the biogeography of the berry-bearing species". The average size of berries and drupes is larger under tropical conditions; however, berries are larger still in hot xeric climates, water perhaps being an important disperser attractant (KNIGHT 1986).

11. Conservation of Dispersal Systems

From much of the above it is clear that most zoochorous plants are dispersed by arrays of animal species, so that when one animal becomes extinct, the remainder continue to disperse the plant's seeds. However, in any one area, species of *Ficus*, *Piper*, or Lauraceae are limiting resources for the animals. Loss of such keystone plant species (GILBERT 1980; HOWE 1984; TERBORGH 1986) can cause waves of extinctions (LOVEJOY et al. 1986). Also, a reserve's flora is constantly changed by the seed rain from the surrounding habitat (JANZEN 1983a; LOVEJOY et al. 1986). Establishment of small reserves will not prevent the extinction of, e.g., quetzals, since these birds, at any particular time and place, depend on a relatively low number of species of lauraceous trees, the fruiting of which determines their seasonal migrations (WHEELWRIGHT 1983).

Studies of reward-based dispersal systems thus provide strong arguments for the conservation of communities. However, the great role of chance and the scarcity of tight relationships between animals and plants make such studies particularly difficult.

Acknowledgments. H. Howe, B. Simpson, C. Augspurger, M. Foster, G. Davidse, B. Beehler, P. Davidar, C.H. Janson, F. White, S. Wing and B. Tiffney kindly permitted me to read papers they had in press. This review has profited from criticism by C. Sobrevila, L. Emmons, M. Foster, S. Smith, and S. Mazer.

References

AUGSPURGER, C.K.: Oikos 40, 189-196 (1983a); - J. Ecol. 71, 759-771 (1983b); - Ecology 65, 1705-1712 (1984); - Biotropica 18, 45-50 (1986a); - Am. J. Bot. 73, 353-363 (1986b); - in: Advances in Legumes Biology, eds. J.L. ZARUCCI, C.M. STIRTON. The Missouri Botanical Garden 1987 (in press). - AUGSPURGER, C.K., FRANSON, S.E.: Ecology 68, 27-42 (1987). - AUGSPURGER, C.K., HOGAN, K.P.: Am. J. Bot. 70, 1031-1037 (1983). - AUGSPURGER, C.K., KELLY, C.K.: Oecologia 61, 211-217 (1984).

BAWA, K.S.: Annu. Rev. Ecol. Syst. 11, 15-39 (1980). - BEATTIE, A.J.: The Evolutionary Ecology of Ant-Plant Mutualism. Cambridge, NY: Cambridge Univ. Press 1985. - BEATTIE, A.J., CULVER, D.C.: Ecology 62, 107-115 (1981); - Oecologia 56, 99-103 (1983). - BECKER, P., WONG, M.: Biotropica 17, 230-237 (1985). - BEEHLER, B.: Auk 100, 1-12 (1983); - Patterns of frugivory and the evolution of birds of paradise. Symp. Tropical Frugivory, Int. Ornith. Congr. Ottawa, June 1986 (in press). - BENKMAN, C.W., BALDA, R.P., SMITH, C.C.: Ecology 65, 632-642 (1984). - BERG, R.Y.: Skr. Norske Vid. Akad. Oslo (Math.-Nat.) 1, 5-35 (1958). - BOND, W., SLINGSBY, P.: S. Afr. J. Sci. 79, 231-233 (1983); - Ecology 65, 1031-1037 (1984). - BOND, W.J., VLOK, J., VIVIERS, M.: J. Ecol. 72, 209-221 (1984). - BOROWICZ, V.A., STEPHENSON, A.G.: Oecologia 67, 435-441 (1985). - BREITWISCH, R.: Biotropica 15, 125-128 (1983). - BUCKLEY, R.C. (ed.): Ant-Plant Interactions in Australia. The Hague: Junk 1982. - BUELOW-OLSEN, A.: Am. Midl. Nat. 112, 251-260 (1984).

CAMPBELL, C.S.: Rhodora 85, 65-72 (1983). - CARLQUIST, S., QUINN, P.: Aliso 11, 167-177 (1985). - CARTER JOHNSON, W., ADKISSON, C.S.: Am. Midl. Nat. 113, 319-324 (1985). - CHARLES-DOMINIQUE, P.: pp. 119-135 in: see ref. ESTRADA and FLEMING 1986. - CLARK, D.A., CLARK, D.B.: Am. Nat. 124, 769-788 (1984). - CLAY, K.: Bull. Torrey Bot. Club 110, 166-169 (1983). - CLIFFORD, H.T., DRAKE, W.E.: J. Trop. Ecol. 1, 373-374 (1985). - COATES-ESTRADA, R., ESTRADA, A.: J. Trop. Ecol. 2, 349-357 (1986). - COLLINS, S.L., UNO, G.E.: Am. Nat. 125, 866-872 (1985). -

CONNELL, J.H.: pp. 298-312 in: Dynamics of Populations, eds. P.J. DEN BOER, G.R. GRADWELL. Wageningen: PUDOC 1971. - CROAT, T.B.: Flora of Barro Colorado Island. Stanford, CA: Stanford Univ. Press 1978.

DARWIN, C.: On the Origin of Species. London: Murray 1859. - DAVIDAR, P.: Biotropica 15, 32-37 (1983a); - Oecologia 60, 271-273 (1983b); - Fruit structure in mistletoes and its consequence for seed dispersal. Biotropica 19, 137-139 (1987). - DAVIDAR, P., MORTON, E.S.: Ecology 67, 262-265 (1986). - DAVIDSE, G.: pp. 143-155 in: Grass. Systematics and Evolution, ed. T. SODERSTROM. Smithsonian Institution Press 1987. - DAVIDSON, D.W., MORTON, S.R.: Ecology 65, 1038-1051 (1984). - DENSLOW, J.S., MOERMOND, T.C.: Oecologia 54, 170-176 (1982). - DENSLOW, J.S., MOERMOND, T.C.: pp. 245-254 in: The Botany and Natural History of Panama, eds. W. D'ARCY, M.D. CORREA A. St. Louis, MO: Missouri Botanical Garden 1985. - DENSLOW, J.S., MOERMOND, T.C., LEVEY, D.J.: pp. 37-44 in: Frugivores and Seed Dispersal, eds. A. ESTRADA, T.H. FLEMING. Dordrecht: Junk 1986. - DINERSTEIN, E.: Biotropica 18, 307-318 (1986). - DOCTORS VAN LEEUWEN, W.M.: Beaufortia 4, 103-207 (1954).

ELLNER, S., SHMIDA, A.: Oecologia 51, 133-144 (1981). - EMMONS, L.H.: Biotropica 16, 210-222 (1984). - ESTRADA, A., COATES-ESTRADA, R.: Am. J. Primat. 6, 77-91 (1984); - pp. 93-104 in: see ref. ESTRADA and FLEMING 1986. - ESTRADA, A., FLEMING, T.H. (eds.): Frugivores and Seed Dispersal. Dordrecht: Junk 1986. - ESTRADA, A., COATES-ESTRADA, R., VAZQUEZ-YANES, C.: Biotropica 16, 315-318 (1984).

FLEMING, T.H.: pp. 287-325 in: Ecology of Bats, ed. T.H. KUNZ. New York, London: Plenum Press 1982; - Ecology 66, 688-700 (1985); - pp. 105-118 in: see ref. ESTRADA and FLEMING 1986. - FLEMING, T.H., WILLIAMS, C.F., BONACCORSO, F.J., HERBST, L.H.: Am. J. Bot. 72, 383-391 (1985). - FORESTA, H. DE, CHARLES-DOMINIQUE, P., ERARD, C., PREVOST, M.F.: Rev. Ecol. (Terre Vie) 39, 369-400 (1984). - FOSTER, M.S.: Trop. Ecol. 19, 131-154 (1978). - FOSTER, M.S., McDIARMID, R.W.: Biotropica 15, 26-31 (1983). - FOSTER, R.B., ARCE, B.J., WACHTER, T.S.: pp. 357-370 in: see ref. ESTRADA and FLEMING 1986.

GADDY, L.L.: Bull. Torrey Bot. Club 113, 247-251 (1986). - GAUTIER-HION, A., DUPLANTIER, J.-M., QURIS, R., FEER, F., SOURD, C., DECOUX, J.-P., DUBOST, G., EMMONS, L., ERARD, C., HECKETSWEILER, P., MOUNGAZI, A., ROUSSILHON, C., THIOLLAY, J.-M.: Oecologia 65, 324-337 (1985a). - GAUTIER-HION, A., DUPLANTIER, J.-M., EMMONS, L., FEER, F., HECKETSWEILER, P., MOUNGAZI, A., QURIS, R., SOURD, C.: Rev. Evol. (Terre Vie) 40, 405-434 (1985b). - GERITZ, S.A.H., DE JONG, T.J., KLINKHAMER, P.G.L.: Oecologia 62, 219-221 (1984). - GILBERT, L.E.: pp. 11-34 in: Conservation Biology, eds. M.E. SOULE, B.A. WILCOX. Sunderland, MA: Sinauer 1980. - GIVNISH, T.J.: Evolution 34, 959-972 (1980). - GONZALEZ-ESPINOSA, M., QUINTANA-ASCENCIO, P.F.: pp. 237-284 in: see ref. ESTRADA and FLEMING 1986. - GOULDING, M.: The Fishes and the Forest. Berkeley: Univ. California Press 1980. - GREEN, D.S.: Oecologia 56, 356-358 (1983). - GREENE, D.F., JOHNSON, E.A.: Oecologia 68, 629-631 (1986). - GREIG-SMITH, P.W.: Am. Nat. 127, 246-251 (1986). - GUNN, C.R., DENNIS, J.V.: World Guide to Tropical Drift Seeds and Fruits. New York: Quadrangle/ The New York Times Bk. Co. 1976.

HALLWACHS, W.: pp. 285-304 in: see ref. ESTRADA and FLEMING 1986. - HAMRICK, J.L., LOVELESS, M.D.: pp. 211-223 in: see ref. ESTRADA and FLEMING 1986. - HANZAWA, F.M., BEATTIE, A.J., HOLMES, A.: Am. J. Bot. 72, 1707-1711 (1985). - HARPER, J.L.: Population Biology of Plants. London: Academic Press 1977. - HEITHAUS, E.R.: pp. 327-367 in: Ecology of Bats, ed. T.H. KUNZ. New York, London: Plenum Press 1982; - pp. 199-210 in: see ref. ESTRADA and FLEMING 1986. - HERBST, L.H.: Biotropica

18, 39-44 (1986). - HERRERA, C.M.: Am. Nat. 120, 218-241 (1982a); - ibid. 120, 819-822 (1982b); - Ecology 63, 773-785 (1982c); - Evolution 36, 1299-1314 (1982d); - Ecology 65, 609-617 (1984a); - Oikos 42, 166-170 (1984b); - ibid. 42, 203-210 (1984c); - Ecol. Monogr. 54, 1-23 (1984d); - Oecologia 63, 386-393 (1984e); - pp. 341-365 in: Habitat Selection in Birds, ed. M.L. CODY. New York: Academic Press 1985a; - Oikos 44, 132-141 (1985b); - pp. 5-18 in: see ref. ESTRADA and FLEMING 1986. - HOLTHUIJZEN, A.M.A., SHARIK, T.L.: Am. Midl. Nat. 113, 200-202 (1985). - HOPKINS, H.C.: Bot. J. Linn. Soc. (London) 87, 135-167 (1983). - HORVITZ, C.C., SCHEMSKE, D.W.: Oecologia 70, 318-320 (1986a); - pp. 169-186 in: see ref. ESTRADA and FLEMING 1986 (1986b); - Biotropica 18, 319-323 (1986c). - HOWE, H.F.: pp. 211-227 in: Tropical Rain Forest: Ecology and Management, eds. S.L. SUTTON, T.C. WHITMORE, A.C. CHADWICK. Oxford: Blackwell 1983; - Biol. Conserv. 30, 261-281 (1984); - Am. Nat. 125, 853-865 (1985); - pp. 123-189 in: Seed Dispersal, ed. D. MURRAY. Sidney: Academic Press 1986. - HOWE, H.F., SMALLWOOD, J.: Annu. Rev. Ecol. Syst. 13, 201-218 (1982). - HOWE, H.F., WESTLEY, L.C.: pp. 185-215 in: Plant Ecology, ed. M.J. CRAWLEY. London: Blackwell 1986. - HOWE, H.F., SCHUPP, E.W., WESTLEY, L.C.: Ecology 66, 781-791 (1985).

IZHAKI, I., SAFRIEL, U.N.: Oecologia 67, 40-43 (1985).

JANSON, C.H.: Science 219, 187-189 (1983). - JANSON, C.H., STILES, E.W., WHITE, D.W.: pp. 83-92 in: see ref. ESTRADA and FLEMING 1986. - JANZEN, D.H.: Am. Nat. 104, 501-528 (1970); - Evolution 34, 611-612 (1980); - Am. J. Bot. 69, 1169-1178 (1982a); - ibid. 69, 1240-1250 (1982b); - ibid. 69, 1258-1268 (1982c); - Oikos 38, 150-156 (1982d); - Biotropica 14, 149-152 (1982e); - Ecology 63, 1887-1900 (1982f); - Anim. Behav. 30, 483-489 (1982g); - Brenesia 19/20, 165-170 (1982h); - ibid. 19/20, 209-219 (1982i); - ibid. 19/20, 363-368 (1982j); - ibid. 19/20, 425-429 (1982k); - Oikos 41, 402-410 (1983a); - Biol. J. Linn. Soc. 20, 103-113 (1983b); - pp. 232-262 in: Coevolution, eds. D.J. FUTUYMA, M. SLATKIN. Sunderland, MA: Sinauer 1983c; - Am. Nat. 123, 338-353 (1984); - Biotropica 17, 322-325 (1985a); - J. Trop. Ecol. 1, 131-155 (1985b); - Annu. Rev. Ecol. Syst. 17, 595-636 (1986). - JANZEN, D.H., MARTIN, P.S.: Science 215, 19-27 (1982). - JOHNSON, R.A., WILLSON, M.F., THOMPSON, J.N., BERTIN, R.I.: Ecology 66, 819-827 (1985). - JORDANO, P.: Oikos 38, 183-193 (1982); - Biotropica 15, 38-41 (1983).

KEVAN, P.G., GASKELL, B.H.: Biotropica 18, 76-78 (1986). - KJELLSSON, G.: Oecologia 67, 416-423 (1985a); - ibid. 67, 424-429 (1985b). - KNIGHT, R.S.: J. Biogeogr. 11, 501-513 (1984); - ibid. 13, 99-108 (1986). - KNIGHT, R.S., SIEGFRIED, W.R.: Oecologia 56, 405-412 (1983). - KUBITZKI, K. (ed.): Dispersal and Distribution. Sonderbd. Naturwiss. Verein Hamburg, Bd. 7, 1983; - An. Acad. Brasil. Ciênc. 57, 513-516 (1985); - pp. 192-206 in: Amazonia, eds. G.T. PRANCE, T.E. LOVEJOY. Oxford, New York: Pergamon Press 1986.

LEIGH, E.G., RAND, A.S., WINDSOR, D.M. (eds.): The Ecology of a Tropical Forest: Seasonal rhythms and long-term changes. Washington, DC: Smithsonian Inst. Press 1982. - LEVEY, D.J.: pp. 147-158 in: see ref. ESTRADA and FLEMING 1986. - LEVEY, D.J., MOERMOND, T.C., DENSLOW, J.S.: Ecology 65, 844-850 (1984). - LIDDLE, M.J., ELGAR, M.A.: Bot. J. Linn. Soc. (London) 88, 303-315 (1984). - LINHART, Y.B., TOMBACK, D.F.: Oecologia 67, 107-110 (1985). - LOVEJOY, T.E., BIERREGAARD, R.O. Jr., RYLANDS, A.B., MALCOLM, J.R., QUINTELA, C.E., HARPER, L.H., BROWN, K.S. Jr., POWELL, A.H., POWELL, G.V.N., SCHUBART, H.O.R., HAYS, M.B.: pp. 257-285 in: Conservation Biology, ed. M.E. SOULE. Sunderland, MA: Sinauer 1986.

MADDOX, J.C., CARLQUIST, S.: Aliso 11, 77-96 (1985). - MANASSE, R.S., HOWE, H.F.: Oecologia 59, 185-190 (1983). - MARTIN, T.E.: Oecologia 66, 563-573 (1985). - McDIARMID, R.W., RICKLEFS, R.E., FOSTER, M.S.: Biotropica 9, 9-25 (1977). - McDONNELL, M.J.: Bull. Torrey Bot. Club 113, 6-11 (1986). - McDONNELL, M.J., STILES, E.W.: Oecologia 56, 109-116 (1983). - McDONNELL, M.J., STILES, E.W., CHEPLICK, G.P., ARMESTO, J.J.: Am. J. Bot. 71, 895-901 (1984). - McKEY, D.: pp. 155-191 in: Coevolution of Animals and Plants, eds. L.E. GILBERT, P.H. RAVEN. Austin: University of Texas Press 1975. - MESLER, M.R., LU, K.L.: Am. J. Bot. 70, 1460-1467 (1983). - METZ, J.A.J., DE JONG, T.J., KLINKHAMER, P.G.L.: Oecologia 57, 166-169 (1983). - MOERMOND, T.C., DENSLOW, J.S.: pp. 856-897 in: Neotropical Ornithology, eds. P.A. BUCKLEY, M.S. FOSTER, E.S. MORTON, R.S. RIDGELY, F.G. SMITH. Ornithol. Monogr. Vol. 36, 1985. - MOERMOND, T.C., DENSLOW, J.S., LEVEY, D.J., SANTANA, C.: pp. 137-146 in: see ref. ESTRADA and FLEMING 1986. - MOORE, L.A., WILLSON, M.F.: Can. J. Bot. 60, 557-560 (1982). - MORDEN-MOORE, A.L., WILLSON, M.F.: Can. J. Bot. 60, 1554-1560 (1982). - MORSE, D.H., SCHMITT, J.: Oecologia 67, 372-379 (1985). - MURRAY, D.R. (ed.): Seed Dispersal. Sidney: Academic Press 1986. - MURRAY, K.G.: pp. 187-198 in: see ref. ESTRADA and FLEMING 1986; - Am. Nat. 129, 18-31 (1987).

OLIVIERI, I., SWAN, M., GOUYON, P.-H.: Oecologia 60, 114-117 (1983).

PEART, M.H.: J. Ecol. 72, 437-453 (1984). - PECKOVER, W.S.: Guinea. Aroideana 8, 70-71 (1985). - PIJL, L. VAN DER: Principles of Dispersal in Higher Plants, 3. edn. Berlin, Heidelberg, New York: Springer 1982. - PIPER, J.K.: Oikos 46, 303-310 (1986a); - Can. J. Bot. 64, 1050-1054 (1986b). - PRATT, T.K.: Emu 82, Suppl., 283-285 (1983). - PRATT, T.K., STILES, E.W.: Am. Nat. 122, 797-805 (1983); - Biotropica 17, 314-321 (1985). - PRÉVOST, M.-F.: Rev. Ecol. (Terre Vie) 38, 121-145 (1983).

RENNER, S.S.: Selbyana 9, 104-111 (1986). - RICE, B., WESTOBY, M.: Ecology 67, 1270-1274 (1986). - RIDLEY, H.N.: The Dispersal of Plants Throughout the World. London: Reeve 1930. - ROBERTS, J.T., HEITHAUS, E.R.: Ecology 67, 1046-1051 (1986). - ROGERS, L.L., APPLEGATE, R.D.: J. Mammal. 64, 310-311 (1983). - ROOSMALEN, M.G.M.: Fruits of the Guianan Flora. Utrecht: Institute of Systematic Botany, Utrecht University 1985.

SABATIER, D.: Fructification et dissemination en foret guyanaise. Thèse de Doctorat 3e cycle, University of Montpellier, 338 pp. 1983; - Rev. Ecol. (Terre Vie) 40, 289-320 (1985). - SALISBURY, E.J.: Proc. R. Soc. London B 192, 323-329 (1976). - SANTANA, C.E., MILLIGAN, B.G.: Biotropica 16, 152-154 (1984). - SCHIMPER, A.F.W.: Die epiphytische Vegetation Amerikas. Bot. Mitt. Tropen. II. Jena: Fischer 1888. - SCHMITT, J., EHRHARDT, D., SCHWARTZ, D.: Am. Nat. 126, 570-575 (1985). - SCOTT, P.E., MARTIN, R.F.: Biotropica 16, 319-323 (1984). - SILBERBAUER-GOTTSBERGER, I.: Plant Syst. Evol. 147, 1-27 (1984). - SKUTCH, A.F.: Condor 82, 31-42 (1980). - SMITH, B.H., RONSHEIM, M.L., SWARTZ, K.R.: Am. J. Bot. 73, 1416-1426 (1986). - SMITH, C.C., REICHMAN, O.J.: Annu. Rev. Ecol. Syst. 15, 329-351 (1984). - SMYTHE, N., GLANZ, W.E., LEIGH, E.G. Jr.: pp. 227-238 in: see ref. LEIGH et al. 1982. - SNOW, D.W.: Oikos 15, 274-281 (1965); - Ibis 113, 194-202 (1971). - SNOW, D.W., SNOW, B.K.: pp. 159-164 in: see ref. ESTRADA and FLEMING 1986. - SORENSEN, A.E.: Oecologia 56, 117-120 (1983); - J. Anim. Ecol. 53, 545-557 (1984); - Annu. Rev. Ecol. Syst. 17, 443-463 (1986). - SORK, V.L.: Am. Midl. Nat. 109, 81-88 (1983a); - Ecology 64, 1049-1056 (1983b); - ibid. 65, 1020-1022 (1984); - Biotropica 17, 130-136 (1985). - SOURD, C., GAUTIER-HION, A.: J. Anim. Ecol. 55, 235-244 (1986). - STAPANIAN, M.A.: Ecology 63, 1422-1431 (1982a); - ibid. 63, 1432-1443 (1982b). - STAPANIAN, M.A., SMITH, C.C.: Ecology 59, 884-896 (1978); - J. Mammal. 67, 326-332 (1986). - STILES,

E.W.: Am. Nat. 120, 500-509 (1982). - STILES, E.W., WHITE, D.W.: Am. Nat. 120, 823-827 (1982). - STOCKER, G.C., IRVINE, A.K.: Biotropica 15, 170-176 (1983). - SUDGEN, A.M.: Biotropica 14, 208-219 (1982). - SUTTON, S.L., WHITMORE, T.C., CHADWICK, A.C. (eds.): Tropical Rain Forest: Ecology and Management. Oxford: Blackwell 1983.

TERBORGH, J.: Five New World Primates: A Study in Comparative Ecology. Princeton, NJ: Princeton Univ. Press 1983; - pp. 330-344 in: Conservation Biology, ed. M.E. SOULE. Sunderland, MA: Sinauer 1986. - TOMBACK, D.F.: J. Anim. Ecol. 51, 451-467 (1982); - pp. 179-223 in: Coevolution, ed. M.H. NITECKI. Chicago, London: The Univ. Chicago Press 1983; - Madroño 33, 100-110 (1986). - TRAINER, J.M., WILL, T.C.: Auk 101, 193-195 (1984). - TURNBULL, C.L., CULVER, D.C.: Oecologia 59, 360-365 (1983). - TURNBULL, C.L., BEATTIE, A.J., HANZAWA, F.M.: Southwest. Nat. 28, 289-293 (1983).

VANDER WALL, S.B., BALDA, R.P.: Ecol. Monogr. 47, 89-109 (1977). - VAZQUEZ-YANES, C., OROZCO-SEGOVIA, A.: pp. 71-77 in: see ref. ESTRADA and FLEMING 1986. - VENABLE, D.L., LEVIN, D.A.: Plant Syst. Evol. 143, 1-16 (1983). - VERKAAR, H.J., SCHENKEVELD, A.J., KLASHORST, M.P. VAN DE: New Phytol. 95, 335-344 (1983). - VICKERY, R.K., PHILLIPS, D.R., WONSAVAGE, P.R.: Am. Midl. Nat. 116, 206-208 (1986).

WASER, N.M., VICKERY, R.K., PRICE, M.V.: Evolution 36, 753-761 (1982). - WEBB, C.J.: Biotropica 16, 37-42 (1984). - WHEELWRIGHT, N.T.: Auk 100, 286-301 (1983); - Ecology 66, 808-818 (1985a); - Oikos 44, 465-477 (1985b); - pp. 21-35 in: see ref. ESTRADA and FLEMING 1986. - WHEELWRIGHT, N.T., JANSON, C.H.: Am. Nat. 126, 777-799 (1985). - WHEELWRIGHT, N.T., ORIANS, G.H.: Am. Nat. 119, 402-413 (1982). - WHEELWRIGHT, N.T., HABER, W.A., MURRAY, K.G., GUINDON, C.: Biotropica 16, 173-192 (1984). - WHITE, F.: Bothalia 16, 143-168 (1986). - WIKANDER, T.: Biotropica 16, 276-283 (1984). - WILLSON, M.F.: Bull. Torrey Bot. Club 110, 298-303 (1983a); - Plant Reproductive Ecology. New York: Wiley 1983b; - Curr. Ornithol. 3, 223-279 (1986). - WILLSON, M.F., HOPPES, G.H.: pp. 55-69 in: see ref. ESTRADA and FLEMING 1986. - WILLSON, M.F., MELAMPY, M.N.: Oikos 41, 27-31 (1983). - WILLSON, M.F., THOMPSON, N.J.: Can. J. Bot. 60, 701-713 (1982). - WING, S.L., TIFFNEY, B.H.: The reciprocal interaction of vertebrate herbivory and angiosperm evolution. Review of Palaeobotany and Palynology 50, 179-210 (1987). - WRANGHAM, R.W., WATERMAN, P.G.: Biotropica 15, 217-222 (1983).

Dr. SUSANNE S. RENNER
Botanical Institute
University of Aarhus
Nordlandsvej 68
DK-8240 Risskov
Denmark

Plant Name Index

Subject Index

MIX
Papier aus verantwortungsvollen Quellen
Paper from responsible sources
FSC® C105338

If you have any concerns about our products,
you can contact us on
ProductSafety@springernature.com

In case Publisher is established outside the EU,
the EU authorized representative is:
Springer Nature Customer Service Center GmbH
Europaplatz 3, 69115 Heidelberg, Germany

Printed by Libri Plureos GmbH
in Hamburg, Germany